Intermediate Alge
A Discovery Approach

Preliminary Edition

Intermediate Algebra: A Discovery Approach

Preliminary Edition

David Arnold
College of the Redwoods

Addison-Wesley Publishing Company

Reading, Massachusetts • Menlo Park, California • New York
Don Mills, Ontario • Wokingham, England • Amsterdam • Bonn
Sydney • Singapore • Tokyo • Madrid • San Juan • Milan • Paris

Sponsoring Editor: Jason A. Jordan
Production Supervisor: Mona Zeftel
Cover Designer: Paul Montie of Farenheit Design
Copy Editor: Stephanie Argeros-Magean
Proofreader: Cynthia Insolio Benn
Compositor: Galley Graphics
Manufacturing Coordinator: Evelyn M. Beaton

Library of Congress Cataloging-in-Publication Data

Arnold, David, 1945–
    Intermediate algebra : a discovery approach / David Arnold. —
  Prelim. ed.
     p. cm.
  Includes index.
  ISBN 0-201-62873-2
  1. Algebra.   I. Title.
QA154.2.A76 1994
512-9—dc20   94-14420
                                    CIP

1 2 3 4 5 6 7 8 9 10 BAH 9897969594

Preface

This book offers a unique approach to the study of intermediate algebra.

Distinctive Features

The Discovery Method

Although many of the fundamental concepts of intermediate algebra are developed in this text in a traditional manner, the examples, problems and activities in this text are designed to help students achieve a level of success that may once have been beyond their grasp. There are four fundamental ways that this text will help students succeed in the study of intermediate algebra.

The Graphing Method. Graphing is introduced in the first section of Chapter One and is used as a fundamental problem-solving tool throughout the text.

The Analytical Method. There are traditional analytical techniques that students must master to be successful in intermediate algebra. This text emphasizes that the analytical solution *must match* the graphical solution.

The Numerical Method. Much can be gained from analyzing a problem numerically. Students will be asked to complete and analyze tables of numerical values throughout the text. Numerical solutions *must match* the graphical and analytical solutions.

The Writing Method. This text encourages students to use writing to justify conclusions, summarize results, and explain the significance of their solutions.

The Graphing Calculator

The graphing calculator is an important component of the discovery process. I have chosen to use the TI-82 Graphing Calculator from Texas Instruments. However, there are a number of comparable graphing calculators that can also be used with this text.

To use the graphing calculator effectively, adequate support must be provided. With many intermediate algebra texts, this support comes in the form of supplementary materials. This is *not* the case with this text. In the fourth section of Chapter one, students are shown how to execute the basic commands required to draw the graph of a function. Later sections will introduce and explain other capabilities of the calculator as needed. I feel that it is an advantage that students have a text that *explicitly* shows them how to use their calculator as an effective tool for problem solving.

Intermediate algebra texts often claim that activities with the graphing calculator are provided. However, on closer examination, these activities have been appended at the end of a problem set, then labeled with an appropriate icon deeming them as graphing calculator exercises. I feel that this approach does not work. Almost all sections of this text have examples of *effective* graphing calculator use. These examples provide students with *specific* directions and often exhibit resulting windows that students can use for comparison with results on their own calculators. Problem sets are interspersed with graphing calculator activities. However, because of the unique approach taken with the graphing calculator in this text, students will find that they are using their calculator as a problem solving tool *all of the time,* even when the problem is not marked specifically as a graphing calculator exercise.

Unique Problem Sets

Making Connections. I believe that the problem sets in this text set it apart from other intermediate algebra texts currently on the market. These unique problem sets are one of the major strengths of the text. Perhaps an example at this point would be informative.

A problem from a traditional text. Solve the following equation for x: $x^2 - 4x = 0$.

The same problem in this text. Consider the function f defined by the equation $f(x) = x^2 - 4x$. Complete the following table of points that satisfy the equation of the function f.

x	-1	0	1	2	3	4	5
$f(x)$							

a) Set up a coordinate system on a sheet of paper. Scale and label each axis. Plot each point from the table on your coordinate system and use these points to help draw a complete graph of the function f.

b) Draw a number line below your graph and shade and label the solutions of $f(x) = 0$ on your number line.

c) Use an analytical method to find the solutions of $f(x) = 0$. If these solutions do not compare favorably with those found in part (b), check your work for error.

Note that this problem employs the graphing method, the analytical method, the numerical method, and asks the student to organize and compare the results from each method.

Applications. It is important that students make connections with the techniques of intermediate algebra and the world around them. In addition to the unusual mix of applications found in most intermediate algebra texts, attempts have been made to create more meaningful applications. Here is an example.

(TI-82) When the chase plane *Liberty One* spots the shuttle *Columbia* on its search radar, a distance of 150 miles separates the two aircraft. *Liberty One* is closing the distance between the two craft at a constant rate of 10.3 miles per minute. Let D represent the distance separating the two craft and let t represent the amount of time that has passed since *Columbia* showed up on *Liberty One*'s search radar.

a) Express D as a function of t. Load this equation in T1 in the Y= menu of the TI-82.

b) Find the amount of time required by *Liberty One* to close the gap between the two aircraft to 20 miles. Begin by loading the constant function $y = 20$ in Y2 in the Y= menu. Adjust the WINDOW parameters so that the point of intersection of the two graphs is showing in your viewing window. Set up a coordinate system on your homework paper and duplicate the image from your viewing window onto your coordinate system. Label the axes with D and t and indicate the scale on each axis by including the WINDOW parameters on each axis.

c) Use the intersection utility in the CALC menu to find the time it takes *Liberty One* to close the gap to 20 miles. Draw a number line below your coordinate system and shade and label this solution on your number line.

d) Use an analytical method to find the amount of time required by *Liberty One* to close the gap between the aircraft to 20 miles. Include units with your answer and write a short sentence explaining the meaning of your answer.

This problem incorporates the graphing method, the analytical method, graphing calculator skills, and asks the student to explain the meaning of the solution in writing.

Drill for Skill. I have also included a sufficient number of traditional drill and skill type problems.

The Power of Visualization

When students make connections between the image on their calculator screen and the results of analysis, their chances for success are significantly increased. Consequently, this text is packed with graphs, diagrams, and artwork that enable students to make connections through visualization. There are over 900 pieces of artwork in the text. Also, the exercises consistently ask students to provide graphs, diagrams, and artwork with their problem solutions. Visualization is not a sometimes thing.

Function Notation

Function notation causes many a student to struggle. Therefore, I introduce function notation in section 3 of Chapter one and practice its use throughout the remaining chapters of the text. This gives students the entire semester to practice and absorb the power of function notation.

Supplementary Materials

Sequences of Series Chapters

Available upon request from Addison-Wesley.

Answers to Selected Exercises

Answers to selected exercises are provided in the back of the text.

Student Solutions Manual

There is a more comprehensive solutions manual available from Addison-Wesley.

TI-81 Programs

There are a number of programs provided in an appendix in the back of the text that enable the TI-81 graphing calculator to behave much like a TI-82 calculator.

Acknowledgments

I am grateful to a number of people who have lent their support over the years that I have worked on this project.

First, I am grateful for the support of my wife Mary and our children Jason, Emily, Tim, Liz, and Jane. Without your love and support I could not do what I do.

Next, I wish to acknowledge the appreciation of my colleagues at the College of the Redwoods, one of the finest, caring, and creative math departments with which I have ever been associated. Thanks to Michelle Olsen, Sandra Taylor, Mike Butler, Kevin Yokoyama, Don Hickethier, and Sandy Vrem. You are all a source of daily inspiration. I would also like to thank my good buddy Allen Martin at Loyola High School, one of the finest teachers I have ever known, who was the backboard off which I bounced many of the ideas for this book.

I would like to thank all of the associate faculty at College of the Redwoods who have used my materials in the classroom. Your comments and criticisms have shaped this work. Thanks to Don Bellairs, Tammy Leslie, Renee Ward, Steve Jackson, Rex Sinclair, and Cheryl Coppin.

I am intended to all of my wonderful students who exhibited so much patience as this project progressed. Your enthusiasm for learning makes each day a joy to experience. It's been a thrill to work with all of you.

I would also like to thank faculty from around the country for their reviews of these materials. Their comments and criticisms helped shape the project.

Jo Cannon	Baylor University
Rob Farinelli	Community College of Allegheny-South Campus
Catherine Gardner	Grand Valley State University
Scott Guth	Mount San Antonio College
Nancy Hyde	Broward Community College-Central Campus
Jackie Lefebvre	Illinois Central College
Beverly Michaels	University of Pittsburgh-Main Campus
Francis Ventola	Brookdale Community College
Elizabeth Wade	Hiwasee College
Charles M. Wheeler	Montgomery College-Rockville Campus
Brenda Wood	Florida Community College-South Campus

I would especially like to thank two copyeditors, Paula Dempsey and Stephanie Magean, who taught me how to write.

Finally, I would like to thank the folks at Addison-Wesley. Thanks to Jason Jordan, my editor, who was there whenever I called and to Mona Zeftel, my production supervisor.

David Arnold

Contents

Intermediate Algebra: A Discovery Approach

Preliminary Edition

Chapter 1

Introduction to Functions—
Notation and Graphs

1.1 Linear Applications for Data Analysis

In many algebra texts, students are asked to graph points in the real plane—a task that has little connection with the real world. I can remember middle school teachers who attempted to make graphing points more attractive by giving us sets of points that when plotted and connected with line segments produced pictures of cats and dogs. This was a nice attempt to put some life into the plotting of points.

In today's information-gathering society, people collect huge sets of data and there are real tasks that require the plotting of points. If you are going to learn how to plot points on a coordinate system, why not use some of these sets of data as your data points? This is precisely how I will introduce the plotting of points in this book. Real-world sets of data points will be used and you will learn how to scale axes appropriately, plot data, fit some sort of line or curve to the data, and use models to make predictions. Let's begin with an example from the physics laboratory.

The Relationship Between Mass and Distance

Jim is hanging blocks of various mass on a spring in the physics lab. He notices that the spring will stretch further if he adds more mass to the end of the spring. He is soon convinced that the distance the spring will stretch depends on the amount of mass attached to it. He decides to take some measurements. He records the amount of mass attached to the end of the spring and then measures the distance that the spring stretched. Here is Jim's data:

Mass (grams)	50	100	150	200	250	300
Distance stretched (centimeters)	1.2	1.9	3.1	4.0	4.8	6.2

> *If you are free to change the value of a variable as you wish, then that variable is referred to as the independent variable. If you have a variable whose value somehow depends on the value of the independent variable, then that variable is called the dependent variable.*

Jim believes that the distance the spring is stretched somehow depends on the amount of mass attached to the end of the spring. This is a somewhat general idea, and he would like to model precisely how the distance depends on the mass. One way that Jim can accomplish this is to graph the data and then examine the graph for a possible relationship in the distance-mass data.

Visualizing the Data. Jim must first decide which data to put on the horizontal axis and which to put on the vertical axis. Should he put the mass on the horizontal axis, or should it be the distance stretched? A similar decision needs to be made for the vertical axis. There are some useful guidelines that can help Jim.

> *It is traditional to put the independent variable on the horizontal axis and the dependent variable on the vertical axis.*

With this in mind, Jim thinks about how to conduct his experiment. He decides that he is completely free to choose how much mass to attach to the spring, and the distance the spring stretched depends on that decision. He makes the mass the *independent* variable and the distance stretched the *dependent* variable. The mass will be placed on the horizontal axis, and the distance stretched on the vertical axis. Jim must now decide how to scale each axis.

Tip	When scaling axes, you should have two thoughts in mind: (1) Try to fit all your data on the coordinate system; and (2), choose a scale that makes your graph easy to read and interpret. We plan to use an entire sheet of graph paper for each graph in this section.

Jim glances at his data set and makes the following decision: tick marks on the horizontal axis will represent 25 grams and on the vertical axis they will represent 0.5 centimeters, as shown below.

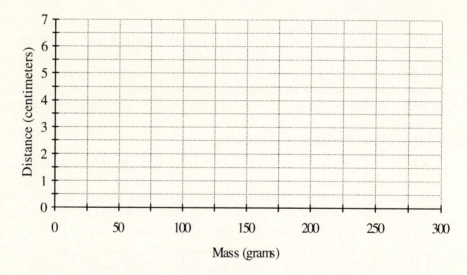

Next, Jim plots his data points on this set of coordinate axes:

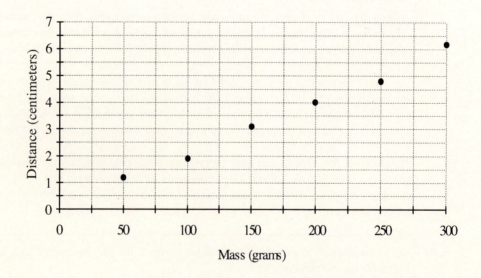

The relationship between the distance stretched and the amount of mass added to the end of the spring appears to be *linear*. A line can be drawn fairly close to all of the data points, although it is not necessary that the line go through each of the points. Rather, it is only important that the line is close to all the data points. Draw a line on Jim's graph that fits the data points as nearly as possible.

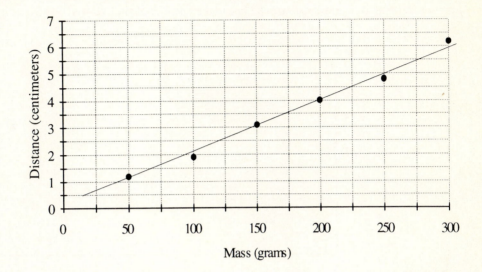

Making Predictions from a Graph.   Jim is pleased to realize that he can now use his completed graph to make some predictions. He wonders how far the spring will stretch if he attaches a mass of 175 grams to it. Since mass is located on the horizontal axis, he locates a mass of 175 grams on the horizontal axis, then draws a vertical line to his newly drawn *line of best fit*. Then, from the point where his vertical line intersects the line of best fit, Jim draws a horizontal line over to the vertical axis, as shown below.

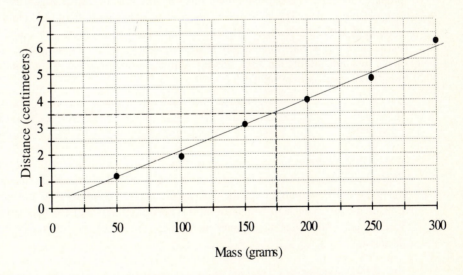

The vertical axis represents distance stretched, so Jim reads his answer on the vertical axis, 3.5 centimeters.

Testing Predictions.   Now, Jim wonders how well will his model will fit reality. He attaches a mass of 175 grams to the spring, then measures the distance the spring stretches. Even though the spring actually stretches 3.7 centimeters, which is different from his model's prediction of 3.6

centimeters, Jim is satisfied because his prediction closely models his real-world measurement. He then goes on to make more predictions from his model and compares them to reality by making appropriate measurements with his mass-spring apparatus.

Jim would now like to predict the amount of mass necessary to stretch the spring 2.1 centimeters. Since the distance stretched is located on the vertical axis, he locates 2.1 cm on the vertical axis, draws a horizontal line to the line of best fit, followed by a vertical line to the horizontal axis.

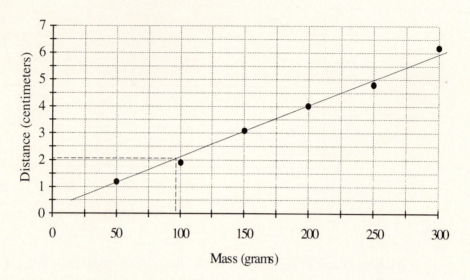

It would appear that the mass is a little shy of 100 grams, possibly 93 grams.

The Relationship Between Celsius and Fahrenheit Temperatures

Dave and Melody are lab partners in Milo Johnson's afternoon chemistry lab. Professor Johnson has prepared an experiment to help them discover the relationship between the Celsius and Fahrenheit temperature scales. The experiment consists of a beaker full of ice and two thermometers, one calibrated in the Fahrenheit scale, the other in the Celsius scale. Dave and Melody use a Bunsen burner to heat ice water, eventually bringing the water to the boiling point. Every few minutes they make two temperature readings, one in Fahrenheit, one in Celsius. The data that they record during their laboratory session follows:

Celsius	4.0	18	30	51	70	85	100
Fahrenheit	39	65	86	122	159	186	210

Although they could just as easily make the Celsius measurement the dependent variable and the Fahrenheit measurement the independent variable, they decide that the Celsius measurement will be the *independent* variable, and the Fahrenheit measurement will be the *dependent* variable. The Celsius measurement will be placed on the horizontal axis and the Fahrenheit measurement on the vertical axis. Note that sometimes there is a certain amount of subjectivity when making this decision.

Visualizing the Data. Melody and Dave must now arrive at a plan for scaling each of the axes. For the Celsius scale, on the horizontal axis, they decide that each tick mark will represent 10 units. For the Fahrenheit scale each tick mark will represent 20 units. At this point in their planning, their coordinate system has the following appearance.

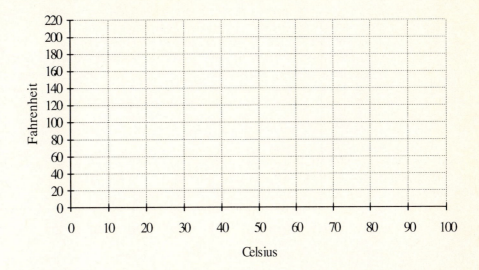

The lab partners now plot each of their data points on the coordinate system. Here are the results:

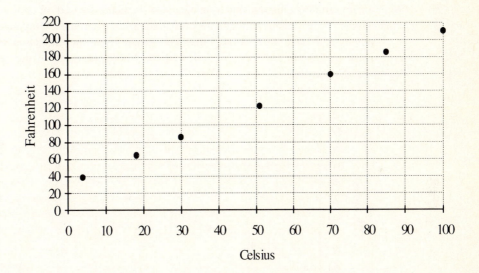

The relationship between the Fahrenheit and Celsius temperatures appears to be *linear*, and a line can be drawn to fit the data points quite closely. Once again, it is not necessary that this line of best fit pass through each of the data points, but it is important that Dave and Melody try to minimize the distance between this line and each of the data points. The line of best fit that the partners agree on is shown below.

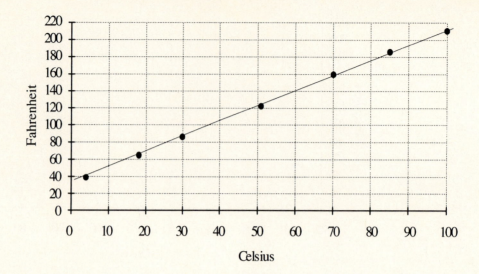

Making Predictions from the Graph.

When this phase of the experiment has been completed, Professor Johnson asks his students to make some predictions based on their work. His first question to the class is "If the Celsius temperature is 40, what is the Fahrenheit temperature?"

Melody and Dave set to work. First, they locate 40 on the horizontal axis. Then they draw a vertical line to their line of best fit, followed by a horizontal line to the vertical axis. Here is a diagram of their work.

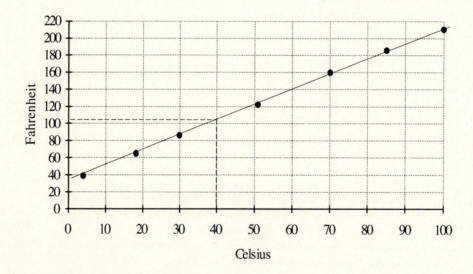

Dave and Melody report to Professor Johnson that they estimate the Fahrenheit reading to be about 105. Professor Johnson congratulates them on a job well done then asks another question: "When the Fahrenheit temperature is 100, what is the Celsius temperature?"

The students return to their lab table to work on this new question. First, they locate 100 on the vertical axis. Then they draw a horizontal line to their line of best fit, followed by a vertical line to the horizontal axis:

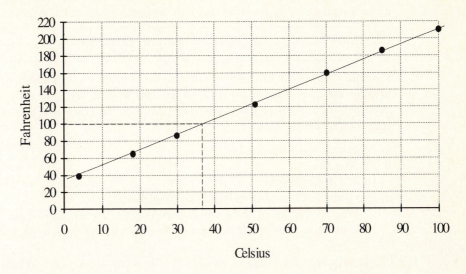

The pair quickly obtains their answer, 37, which they present to Professor Johnson for analysis. "This is outstanding work," comments Professor Johnson.

Dave and Melody head out the lab door to their next class, brimming with confidence. They have both learned that taking measurements and then drawing a graph can be a useful tool for prediction.

Exercises for Section 1.1

1. Gloria and Allen are working in the chemistry lab on an experiment involving temperature and pressure. They have enclosed a container and attached a device that records the pressure inside the container. Initially, the temperature of the container is 75°F, and the pressure inside the container is 15 pounds per square inch. They use a Bunsen burner to begin heating the container. As the temperature inside the container rises, they record both temperature and pressure at regular intervals. Here is their data:

Temperature (°F)	75	87	102	122	130	147	163	195
Pressure (psi)	15	17	21	25	26	29	33	39

Although they are free to adjust the temperature on the burner, the pressure inside the container depends on the temperature. They decide to make the temperature the *independent* variable and place it on the horizontal axis. They place the *dependent* variable, the pressure inside the container, on the vertical axis.

a) Scale each axis appropriately.
b) Plot the data.
c) Draw a line of best fit.
d) Use your model to predict the following:
   i)   What will be the pressure inside the container when the temperature reading is 110°F ?
   ii) What will be the temperature inside the container when the pressure reading is 35 pounds per square inch?

2. Edna runs a copying service where she uses a very sophisticated and expensive copying machine. Once a machine such as this is purchased and placed in service in a business, it immediately begins to *depreciate*—to lose value due to normal wear and tear. Edna is allowed to make a deduction on her taxes for this depreciation. When Edna first bought the machine, its initial value

was $20,000. Using a table from the Internal Revenue Service, she is able to determine the value of the machine after each year of use. Here are some values of the machine:

Years in use	0	3	4	5	7	9
Value	20,000	14,000	12,000	10,000	6,000	2,000

Since the value of the machine for tax purposes depends on the number of years the machine has been in use, the years in use will be the *independent* variable, while the value will be the *dependent* variable. Years in use will be placed on the horizontal axis, while the value will be placed on the vertical axis.

a) Scale each axis appropriately.
b) Plot the data.
c) Draw a line of best fit.
d) Use your model to make the following predictions:
    i) What will the value of the copying machine be after six years of use?
    ii) How many years of use must pass before the value of the copying machine is zero?

3. People have been fascinated by the mile run for many years. Since records for the mile run were first kept, there has been a steady decline in the world record. Listed below are some of the runners who broke the world record and the year in which the record was set.

Year	Record Time	Runner	Country
1937	4:06.4	Sidney Wooderson	Great Britain
1942	4:04.6	Gunder Haegg	Sweden
1943	4:02.6	Arne Andersson	Sweden
1945	4:01.4	Gunder Haegg	Sweden
1957	3:57.2	Derek Ibbotson	Great Britain
1958	3:54.5	Herb Elliott	Australia
1962	3:54.4	Peter Snell	New Zealand
1966	3:51.3	Jim Ryun	United States

Because the record time seems to depend on the number of years that have passed, place the year on the horizontal axis, and the record time on the vertical axis. There are a number of ways to accomplish this. For example, let $t = 0$ represent the year 1937, and let each tick mark thereafter represent the passage of one year. For the record time, we suggest that you convert each time into seconds. For example, 4:06.4 would equal 246.4 seconds. Fill in the remaining entries in the chart below.

Year	t	Record Time	R	Runner	Country
1937	0	4:06.4	246.4	Sidney Wooderson	Great Britain
1942		4:04.6		Gunder Haegg	Sweden
1943		4:02.6		Arne Andersson	Sweden
1945		4:01.4		Gunder Haegg	Sweden
1957		3:57.2		Derek Ibbotson	Great Britain
1958		3:54.5		Herb Elliott	Australia
1962		3:54.4		Peter Snell	New Zealand
1966		3:51.3		Jim Ryun	United States

a) Place the t-values on the horizontal axis and scale this axis appropriately.

b) Place the R-values on the vertical axis and scale this axis appropriately. It is not necessary to begin with zero. In fact, it is probably adequate to begin with 230 and scale upward to 250 by ones.

c) Plot the data and draw a line of best fit.

d) Use your line to predict in which year the mile was first run in less than four minutes (240.0 seconds). The actual year was 1954, when Roger Bannister from Great Britain ran the mile in a record time of 3:59.4.

4. Willie is walking his dog one evening under a streetlight. His eye strays to his shadow, which lengthens as he walks away from the source of light. Willie has been studying geometry all week in school, and the way his shadow lengthens catches his interest. He decides to go home to get his tape measure so that he can take some measurements. When he returns to the streetlight, he measures his distance from it, then measures the length of his shadow. He then walks a little farther away and repeats his measurements. After a while, he has the following data:

Distance from streetlight (meters)	0	5	10	15	20	25	30
Length of shadow (meters)	0	2.9	5.9	9.2	12.1	14.9	18.0

Since the length of the shadow depends upon the distance from the streetlight, Willie decides to let the distance from the streetlight be the *independent* variable and places it on the horizontal axis. The length of the shadow thus becomes the *dependent* variable.

a) Scale each axis appropriately.
b) Plot the data and draw a line of best fit.
c) Use your model to make the following predictions.
   i) What will be the length of Willie's shadow if he stands a distance of 18 meters from the streetlight?
   ii) How far is Willie standing from the streetlight if the length of his shadow is 5 meters?

5. Professor Butler is running a geometry lab experiment in his pre-algebra classes. He has divided his students into teams of three members each. The previous day Professor Butler had asked students to bring in cylindrical cans of various diameters, and the classroom is full of different sized containers. Alissa, Dawn, and George select seven cans from the collection. Professor Butler then asks them to make two measurements of each can: the diameter and the circumference. Here are the results:

Diameter (centimeters)	5.0	5.6	6.1	6.5	6.9	7.3	8.0
Circumference (centimeters)	15.7	17.4	19.3	20.4	21.8	22.9	24.9

Although an argument can be made that the diameter of each can depends on its circumference, the team decides to make the diameter the *independent* variable and the circumference the *dependent* variable.

a) Scale each axis appropriately.
b) Plot the data and draw the line of best fit.
c) Use your model to make the following predictions.
   i) What will be the circumference of the can if the diameter is 5.8 centimeters?
   ii) What will be the diameter of the can if the circumference is 18.0 centimeters?

6. Ellie Sue is sitting with her great-grandmother on the front porch of her home in SmallTown U.S.A. "Do you know what the temperature is tonight?" asks Ellie's great-grandmother. "It's 70 degrees Fahrenheit, my dear." Ellie Sue glances around the front porch, but she knows there is no thermometer anywhere. "How do you know it's 70 degrees, Grandma?" asks Ellie Sue. "I counted the number of cricket chirps I heard in one minute's time," Grandma replied. "Then I performed some mental arithmetic and came up with an answer of 70 degrees." Ellie Sue, needless to say, is

amazed and over the remaining evenings of the summer, she keeps a record of temperature and cricket chirps. Here are some of her measurements:

Chirps per minute	55	83	99	119	134	149	178
Temperature (°F)	50	58	60	67	69	74	79

Although you can make the case that the number of cricket chirps will depend on the temperature, Grandma implied that her calculation of temperature depends upon the number of chirps per minute. Ellie thus made the number of chirps the *independent* variable and made the temperature the *dependent* variable.

a) Scale each axis appropriately.
b) Plot the data and draw a line of best fit.
c) Use your model to predict the temperature if the number of cricket chirps in one minute is 140.

7. Loggers in a certain part of Oregon can determine the height of a particular kind of tree by Community College thought this implausible, so they took their instruments and headed into the forest. When they found the variety of tree they were searching for, they began to make their measurements. With their sophisticated surveying instruments, it was easy to measure the height of the tree, and a simple tape measure was enough to find the circumference of the tree. Here are their measurements on several trees.

Circumference (feet)	19	23	30	35	38	43	50
Height (feet)	56	68	92	108	110	129	152

Because the loggers computed the height of the tree based on its circumference, the students made the circumference the *independent* variable and the height of the tree the *dependent* variable.

a) Scale each axis appropriately.
b) Plot the data and draw a line of best fit.
c) Use your model to make the following predictions.
   i)   If the circumference of a tree is 27 feet, what is its height?
   ii) If the height of a tree is 80 feet, what is its circumference?

8. Professor Holt is presenting a demonstration of the resistance in a copper wire to the students in his basic electricity class. Separated into teams of three, students are given meters to measure the resistance and several lengths of copper wire. Louis, John, and Sonja's team measures the resistance in seven different lengths of copper wire. Here are their results.

Length (feet)	5	10	15	20	25	30	35
Resistance (10^{-2} ohms)	1.08	3.6	5.4	7.1	8.9	10.7	12.5

Since the resistance seems to depend on the length of the copper wire, the team decides that the length will be the independent variable and the resistance will be the dependent variable.

a) Scale each axis appropriately.
b) Plot the data and draw the line of best fit.
c) Use your model to predict each of the following.
   i)   What will be the resistance in a piece of copper wire that is 23 feet long?
   ii) If the resistance in a length of copper wire is 6.0×10^{-2} ohms, what is the length of the copper wire?

1.2 Graphing Functions in the Real Plane

It's time to leave the real world of applications for the abstract world of pure mathematics. In the last section you were asked to plot real-world data on a coordinate system. In this section you will do the opposite. You will be given an equation, and you will be asked to sketch its graph in the real plane.

Graphing Ordered Pairs in the Real Plane

The notation (x, y) is called an *ordered pair*. A set of ordered pairs is called a *relation*.

A relation is a set of ordered pairs.

Example 1   Graph the following relation, or set of ordered pairs, in the real plane:

$$R = \left\{(4,3),(-3,2),(-2,-4),(5,-2)\right\}$$

Solution   The real plane is divided into four quadrants by the vertical and horizontal axes. Right and up mark the positive direction on the axes, and left and down mark the negative direction, as shown:

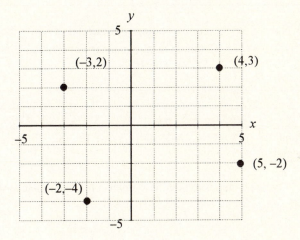

Important Note: In this text, if no scale is indicated on the x- and y-axes, it is assumed that each tick mark represents one unit. In all other cases the scale on each axis will be clearly indicated..

Using a Rule to Define a Relation

There are several ways of representing a relation. You can list all of the elements of the relation in a set such as $R = \left\{(4,3),(-3,2),(-2,-4),(5,-2)\right\}$ or you can represent the relation with a graph. There is yet another way to represent a relation: with a rule or an equation that relates x and y. For example, we can define a relation to be the set of all ordered pairs that satisfy the equation $y = 2x + 3$.

Graphing Relations That Are Defined by Equations or Rules

What exactly are you being asked when you are asked to graph an equation? You are asked to *graph all the points that satisfy the equation*. When you are finished, you will have plotted an infinite set of points that satisfy an equation; so many, in fact, that the finished product will look like a solid curve or line. This leads us to a fundamental concept.

> *If a point satisfies the equation, it will be on the graph of the equation. Also, if a point is on the graph of the equation, that point will satisfy the equation.*

Example 2   Sketch the graph of $y = 2x + 3$.

Solution   You are being asked to graph every point that satisfies the equation $y = 2x + 3$. This is an awesome task, as there are an infinite number of points that satisfy this equation. So set your sights a little lower at first and just try to graph a few points that satisfy the equation; you can worry about graphing the rest of the points later. Start by selecting some arbitrary x-values.

x	−3	−2	−1	0	1	2	3
y							

Now, find the corresponding y-values. If $x = -3$, then $y = 2(-3) + 3 = -3$. If $x = -2$, then $y = 2(-2) + 3 = -1$. Continue in this manner until the table is completely filled in. Here is the final result:

x	−3	−2	−1	0	1	2	3
y	−3	−1	1	3	5	7	9
Points to plot	(−3,−3)	(−2,−1)	(−1,1)	(0,3)	(1,5)	(2,7)	(3,9)

Your next step is to plot these points on a coordinate system, as follows:

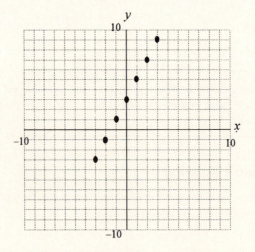

The graph shows *exactly seven* points that satisfy the equation—your instruction was to plot *all* the points that satisfy the equation. You have a long way to go. Or do you?

Try plotting a few more points: If $x = \frac{1}{2}$, then $y = 2\left(\frac{1}{2}\right) + 3 = 4$. Similarly, when $x = \frac{3}{2}$ and $x = \frac{5}{2}$, then $y = 6$ and $y = 8$, respectively. Plot these new points, as follows:

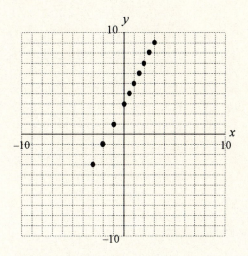

A pattern is emerging; the points all appear to lie on a line. You now have two choices: you can continue plotting points one at a time until you have plotted every point that satisfies the equation, or you can devise a faster technique to accomplish this task. Take your pencil and draw a smooth line through the points that have been plotted. When you draw the smooth line, *you are not connecting the dots, you are filling in the rest of the points that satisfy the equation.*

When using this technique, you must exercise caution. Are you sure that you know where the rest of the plotted points will fall? Have you established enough of a pattern with the few points you have plotted? If your answer is yes, then you can feel confident that you have an accurate graph of the equation. But what if your answer is no? What do you do then? What you must do is to plot more points until you feel that you know what the graph will look like, and you should use whatever *x*-values will accomplish this task.

Tip	Choose whatever x-values you believe are appropriate. If you are still not sure what the graph will do, select more x-values and plot more points until you are confident of the graph's shape.

Following these rules, and scaling down oversized points to their true infinitesimal size, here is the final graph of $y = 2x + 3$:

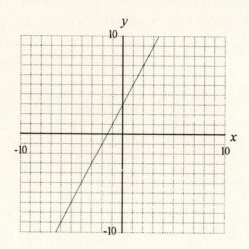

Example 3   Sketch the graph of $y = x^2$.

Solution   Choose x-values that you think are appropriate for the problem. (You can use the x-values from the previous example.) If $x = -3$, then $y = (-3)^2 = 9$. If $x = -2$, then $y = (-2)^2 = 4$. Continue in this manner to fill in the rest of the table.

x	−3	−2	−1	0	1	2	3
y	9	4	1	0	1	4	9
Points to plot	(−3,9)	(−2,4)	(−1,1)	(0,0)	(1,1)	(2,4)	(3,9)

Plotting these points results in the following graph:

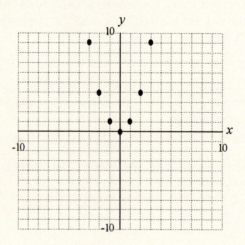

Now, you must make a decision. Have you plotted enough points? Are you sure that you know what the shape of the graph will be? If the answer is no, then you need to plot more points before you continue. However, if you are confident that the graph will not suddenly shift its pattern, then you can go ahead and draw a smooth curve through your points, keeping in mind that what you are really doing is filling in the rest of the points that satisfy the equation. The final graph follows:

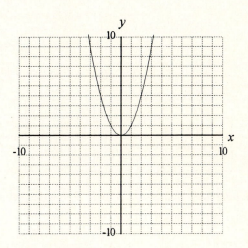

Complete Graphs

A complete graph shows all the important features of a graph, such as its highs and lows.

It is difficult to determine whether you have captured all the important features of a graph, and you cannot expect to become an expert overnight. It is important to thoroughly investigate each graph before declaring that you have all the essentials. Also, it is probably better to graph too many points than too few.

Example 4   Sketch the graph of $y = -x^2 + 4x + 5$.

Solution   First, select some x-values.

x	-3	-2	-1	0	1	2	3
y							

Now, compute the corresponding y-value for each value of x.

$$y = -(-3)^2 + 4(-3) + 5 = -9 - 12 + 5 = -16$$

$$y = -(-2)^2 + 4(-2) + 5 = -4 - 8 + 5 = -7$$

$$y = -(-1)^2 + 4(-1) + 5 = -1 - 4 + 5 = 0$$

$$y = -(0)^2 + 4(0) + 5 = 5$$

$$y = -(1)^2 + 4(1) + 5 = -1 + 4 + 5 = 8$$

$$y = -(2)^2 + 4(2) + 5 = -4 + 8 + 5 = 9$$

$$y = -(3)^2 + 4(3) + 5 = -9 + 12 + 5 = 8$$

x	−3	−2	−1	0	1	2	3
y	−16	−7	0	5	8	9	8
Points to plot	(−3,−16)	(−2,−7)	(−1,0)	(0,5)	(1,8)	(2,9)	(3,8)

Plotting these points results in the following picture:

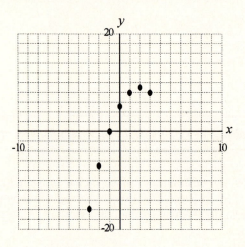

It appears that the graph is beginning to turn downward. What will happen if you travel farther to the right? Try adding a few more x-values to your table.

x	4	5	6	7
y	5	0	−7	−16
Points to plot	(4,5)	(5,0)	(6,−7)	(7,−16)

Now add these points to your graph.

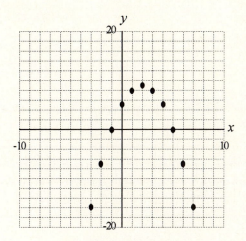

Do you feel that you have a *complete* graph, one that exhibits all the essential character of the graph? You can now assume that if you were to fill in the rest of the points that satisfy the equation your graph would look as follows:

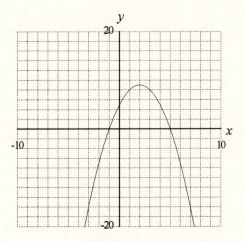

Example 5   Sketch the graph of $y = \sqrt{x-2}$.

Solution   You cannot take the square root of a negative number. An expression such as $\sqrt{-4}$ is undefined. There is no real number, that, when squared, will equal -4. Therefore, the first acceptable x-value to use in this equation is $x = 2$. Any x-value lower than 2 is unacceptable; you can use any x-values that are greater than or equal to 2. Assume you choose the following x-values: 2, 3, 6, 11, and 18. Compute the corresponding y-values to see why these particular x-values were chosen.

$$y = \sqrt{2-2} = 0$$

$$y = \sqrt{3-2} = 1$$

$$y = \sqrt{6-2} = 2$$

$$y = \sqrt{11-2} = 3$$

$$y = \sqrt{18-2} = 4$$

Notice that the *x*-values chosen made the computation of *y*-values easy. Now place these values in a table.

x	2	3	6	11	18
y	0	1	2	3	4
Points to plot	(2,0)	(3,1)	(6,2)	(11,3)	(18,4)

These points can be plotted as follows.

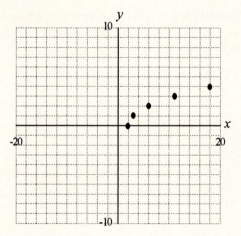

The graph appears to begin at (2,0) and to slowly increase as you move to the right. Assuming that this behavior continues, fill in the rest of the points that satisfy the equation. Remember, you are not drawing a smooth curve through your points; you are filling in the rest of the points that satisfy the equation.

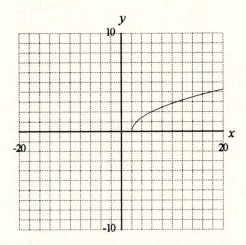

Absolute value is defined as follows:

> **The expression $|x|$ is read "the absolute value of x."**
>
> $$|x| = \begin{cases} x & \text{if } x \geq 0 \\ -x & \text{if } x < 0 \end{cases}$$

For example, $|7| = 7$, $|0| = 0$, and $|-4| = -(-4) = 4$.

Example 6   Sketch the graph of $y = |x|$.

Solution   Create the following table of points.

x	-3	-2	-1	0	1	2	3
y	3	2	1	0	1	2	3
Points to plot	$(-3,3)$	$(-2,2)$	$(-1,1)$	$(0,0)$	$(1,1)$	$(2,2)$	$(3,3)$

Plotting these points results in the following:

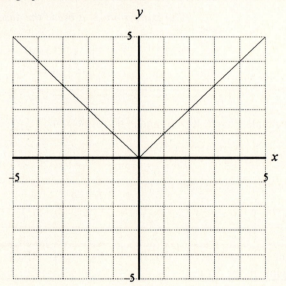

It appears that the graph consists of two lines emanating from the origin (0,0) at 45° angles, one to the left and one to the right. Assuming that this pattern continues, fill in the rest of the points that satisfy the equation. Your graph will look as follows:

Exercises for Section 1.2

1. Consider the equation $y = -2x + 4$. The following table shows several points that satisfy this equation:

x	−4	−3	−2	−1	0	1	2	3	4
y	12	10	8	6	4	2	0	−2	−4
Points to plot	(−4,12)	(−3,10)	(−2,8)	(−1,6)	(0,4)	(1,2)	(2,0)	(3,−2)	(4,−4)

a) Double-check the calculations in the table.

b) Set up a coordinate system on a sheet of graph paper and clearly indicate the scale on each coordinate axis. Plot each point from the table on your coordinate system.

c) Plot the rest of the points that satisfy the equation.

2. Consider the equation $y = x^2 - 4$. The following table shows several points that satisfy this equation:

x	−4	−3	−2	−1	0	1	2	3	4
y	12	5	0	−3	−4	−3	0	5	12
Points to plot	(−4,12)	(−3,5)	(−2,0)	(−1,−3)	(0,−4)	(1,−3)	(2,0)	(3,5)	(4,12)

a) Double-check the calculations in the table.

b) Set up a coordinate system on a sheet of graph paper and clearly indicate the scale on each coordinate axis. Plot each point from the table on your coordinate system.

c) Plot the rest of the points that satisfy the equation.

3. Consider the equation $y = |x - 3|$. The following table shows several points that satisfy this equation:

x	−1	0	1	2	3	4	5	6
y	4	3	2					
Points to plot	(−1,4)	(0,3)	(1,2)					

a) Double-check the calculations.

b) Copy and complete the table of points on a sheet of graph paper.

c) Set up a coordinate system on a sheet of graph paper and clearly indicate the scale on each coordinate axis. Plot each point from the table on your coordinate system.

d) Plot the rest of the points that satisfy the equation.

4. Consider the equation $y = \sqrt{x - 3}$. The following table shows several points that satisfy this equation:

x	3	4	7	12	19	28
y			2	3		
Points to plot			(7,2)	(12,5)		

a) Double-check the calculations.

b) Copy and complete the table of points on a sheet of graph paper.

c) Set up a coordinate system on a sheet of graph paper and clearly indicate the scale on each coordinate axis. Plot each point from the table on your coordinate system.

d) Plot the rest of the points that satisfy the equation.

5. Consider the equation $y = x^3 - 12x$. The following table shows several points that satisfy this equation:

x	−4	−3	−2	-1	0	1	2	3	4
y	−16	9	16						
Points to plot	(−4,−16)	(−3,9)	(−2,16)						

a) Double-check the calculations.
b) Copy and complete the table of points on a sheet of graph paper.
c) Set up a coordinate system on a sheet of graph paper and clearly indicate the scale on each coordinate axis. Plot each point from the table on your coordinate system.
d) Plot the rest of the points that satisfy the equation.

6. Consider the equation $y = (x + 3)^2$. The following table shows several points that satisfy this equation:

x	−7	−6	−5	−4	−3	−2	−1	0	1
y							4	9	16
Points to plot							(−1,4)	(0,9)	(1,16)

a) Double-check the calculations.
b) Copy and complete the table of points on a sheet of graph paper.
c) Set up a coordinate system on a sheet of graph paper and clearly indicate the scale on each coordinate axis. Plot each point from the table on your coordinate system.
d) Plot the rest of the points that satisfy the equation.

7. If $y = 3x - 2$, copy and complete the following table of points on graph paper. Set up a coordinate system near your table of points and plot the points from the table on your coordinate system. Use your plotted points to help draw a complete graph of the equation.

x	−4	−3	−2	−1	0	1	2	3	4
y									
Points to plot									

8. If $y = 4 - x^2$, copy and complete the following table of points on graph paper. Set up a coordinate system near your table of points and plot the points from the table on your coordinate system. Use your plotted points to help draw a complete graph of the equation.

x	−4	−3	−2	−1	0	1	2	3	4
y									
Points to plot									

9. If $y = 5 - |x|$, copy and complete the following table of points on graph paper. Set up a coordinate system near your table of points and plot the points from the table on your coordinate system. Use your plotted points to help draw a complete graph of the equation.

x	−4	−3	−2	−1	0	1	2	3	4
y									
Points to plot									

10. If $y = x^2 - 6x - 7$, copy and complete the following table of points on graph paper. Set up a coordinate system near your table of points and plot the points from the table on your coordinate system. Use your plotted points to help draw a complete graph of the equation.

x	−2	−1	0	1	2	3	4	5	6	7	8
y											
Points to plot											

11. If $y = 5 + 4x - x^2$, copy and complete the following table of points on graph paper. Set up a coordinate system near your table of points and plot the points from the table on your coordinate system. Use your plotted points to help draw a complete graph of the equation.

x	−3	−2	−1	0	1	2	3	4	5	6	7
y											
Points to plot											

12. If $y = \sqrt{x + 4}$, copy and complete the following table of points on graph paper. Set up a coordinate system near your table of points and plot the points from the table on your coordinate system. Use your plotted points to help draw a complete graph of the equation.

x	−4	−3	0	5	12	21
y						
Points to plot						

13. If $y = \frac{3}{5}x + 2$, copy and complete the following table of points on graph paper. Set up a coordinate system near your table of points and plot the points from the table on your coordinate system. Use your plotted points to help draw a complete graph of the equation.

x	−20	−15	−10	−5	0	5	10	15	20
y									
Points to plot									

14. For each of the following equations, create a table of points that satisfy the equation on a sheet of graph paper. Set up a coordinate system near your table of points and plot the points from your table on your coordinate system. Plot the rest of the points that satisfy the equation. Be sure that you have a *complete* graph, one that exhibits all the essential features of the graph. Keep adding additional points to your table and to your graph until you feel that you have captured all the important characteristics of the graph.

a) $y = |x + 5|$ b) $y = (x - 5)^2$ c) $y = x^2 + 6x - 8$ d) $y = 9 + 8x - x^2$

e) $y = 5 - 2|x|$ f) $y = 9 - x^2$ g) $y = -\sqrt{x - 3}$ h) $y = \frac{3}{2}x + 4$

15. Make a duplicate of the following graph on a sheet of graph paper.

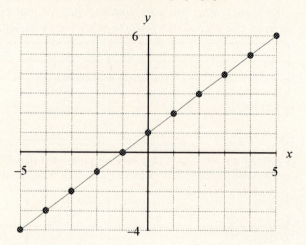

Read the graph and complete the following table on graph paper. Be sure to double-check the calculations in the completed entries.

x	−5	−4	−3	−2	−1	0	1	2	3	4	5
y	−4	−3	−2								
Points to plot	(−5,−4)	(−4,−3)	(−3,−2)								

16. Make a duplicate of the following graph on a sheet of graph paper.

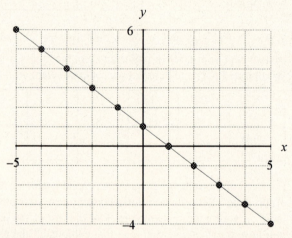

Read the graph and complete the following table on graph paper. Be sure to double-check the calculations in the completed entries.

x	−5	−4	−3	−2	−1	0	1	2	3	4	5
y				3			0			−3	
Points to plot				(−2,3)			(1,0)			(4,−3)	

17. Make a duplicate of the following graph on a sheet of graph paper.

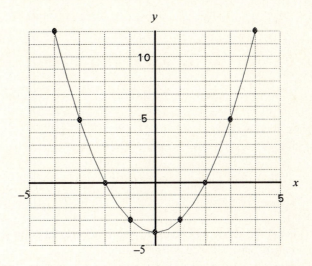

Read the graph and complete the following table on graph paper. Be sure to double-check the calculations in the completed entries.

x	−4	−3	−2	−1	0	1	2	3	4
y			0			−3			
Points to plot			(−2,0)			(1,−3)			

18. Make a duplicate of the following graph on a sheet of graph paper.

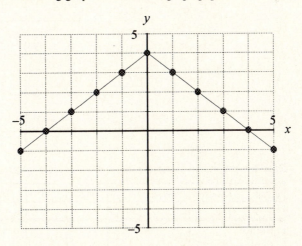

Read the graph and complete the following table on graph paper.

x	−5	−4	−3	−2	−1	0	1	2	3	4	5
y											
Points to plot											

19. Make a duplicate of the following graph on a sheet of graph paper.

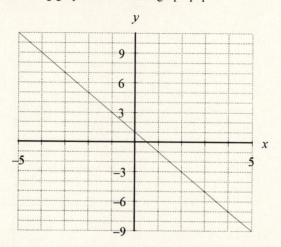

Read the graph and complete the following table on graph paper. Be sure to double-check the calculations in the completed entries.

x	−5	−4	−3	−2	−1	0	1	2	3	4	5
y					3			−3			
Points to plot					(−1,3)			(2,−3)			

20. Make a duplicate of the following graph on a sheet of graph paper.

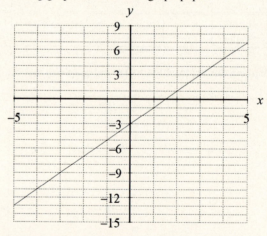

Read the graph and complete the following table on graph paper.

x	−5	−4	−3	−2	−1	0	1	2	3	4	5
y											
Points to plot											

21. Consider the equation $y = |x|$. On a sheet of graph paper, copy and complete the following table of points that satisfy this equation.

x	-7	-6	-5	-4	-3	-2	-1	0	1	2	3	4	5	6	7
y															
Points to plot															

a) Set up a coordinate system on graph paper where each tick mark on the x-axis and the y-axis represents one unit. Plot the points from the table on your coordinate system. Fill in the rest of the points that satisfy the equation.

b) Set up a second coordinate system on your graph paper. Let each box on the x-axis represent one unit and each box on the y-axis represent two units. Plot the points from the table on your coordinate system. Fill in the rest of the points that satisfy the equation.

c) Set up a third coordinate system on your graph paper. Let each box on the x-axis represent one unit and each box on the y-axis represent one-half of a unit. In other words, two boxes on the y-axis represent one unit. Plot the points from the table on your coordinate system. Fill in the rest of the points that satisfy the equation.

d) Explain in your own words how the different scales on the y-axis affect your graph of the equation $y = |x|$.

22. Consider the equation $y = |x|$. On a sheet of graph paper, copy and complete the following table of points that satisfy this equation.

x	-7	-6	-5	-4	-3	-2	-1	0	1	2	3	4	5	6	7
y															
Points to plot															

a) Set up a coordinate system on graph paper where each tick mark on the x-axis and the y-axis represents one unit. Plot the points from the table on your coordinate system. Fill in the rest of the points that satisfy the equation.

b) Set up a second coordinate system on your graph paper. Let each box on the x-axis represent two units and each box on the y-axis represent one unit. Plot the points from the table on your coordinate system. Fill in the rest of the points that satisfy the equation.

c) Set up a third coordinate system on your graph paper. Let each box on the x-axis represent one-half of a unit. In other words, two boxes on the x-axis represent one unit. Let each box on the y-axis represent one unit. Plot the points from the table on your coordinate system. Fill in the rest of the points that satisfy the equation.

d) Explain in your own words how the different scales on the y-axis affect your graph of the equation $y = |x|$.

1.3 Function Notation

This section introduces the function—a key concept in the study of mathematics and a unifying theme of this book. We will also introduce a powerful new symbolism called function notation. Finally, we will examine some practical examples where we will express one quantity as a function of another quantity.

The Domain and Range of a Relation

Recall that a relation is a set of ordered pairs. If we denote the ordered pairs of our relation with the symbolism (x, y), then the *domain* of the relation is the set of x-values of the ordered pairs. The *range* of a relation is the set of y-values of the ordered pairs.

Example 1   What is the domain and range of the relation $R = \{(1,2),(1,3)(5,7)\}$?

Solution   The domain of the relation R is the set of x-values of the ordered pairs of R. Therefore, the domain is $\{1,5\}$. The range of the relation R is the set of y-values of the ordered pairs of R. Therefore, the range is $\{2,3,7\}$.

The Definition of a Function

A *function* is a special type of relation.

> *A relation is called a function if and only if each domain element is paired with exactly one range element.*

Example 2   Is the relation in Example 1 a function?

Solution   Note that 1 is an element of the domain of the relation $R = \{(1,2),(1,3)(5,7)\}$. Since 1 is paired with the range values 2 and 3, the relation R is not a function. To be a function, a relation must pair each domain object with exactly one range object.

Example 3   Is the relation $T = \{(1,2),(4,5),(7,9)\}$ a function?

Solution   The domain of the relation T is $\{1, 4, 7\}$. The range of the relation T is $\{2, 5, 9\}$. Each domain object is used exactly once. Each domain object is paired with exactly one range object. Therefore, the relation T is a function.

We will use this preliminary definition of a function for the time being, but we will discuss the function concept in greater depth in Chapter 6. At this time let's look at an extremely important and useful bit of notation called *function notation*.

Consider the equation $y = 4x^2 + 3$. Suppose that you are asked to find the value of y when x is 5. You would substitute 5 for x, then calculate the corresponding value of y. Your work would look something like this:

$$y = 4x^2 + 3$$
$$y = 4(5)^2 + 3$$
$$y = 4(25) + 3$$
$$y = 100 + 3$$
$$y = 103$$

To summarize, when x equals 5, the corresponding value of y is 103.

This is a simple computation, but the instruction that produced this computational response was rather wordy. Here is what you were asked to do, expressed as simply as possible: "Consider the equation $y = 4x^2 + 3$. If x equals 5, what is the corresponding value of y?" Is there some way to rephrase this question in a manner that makes it less wordy and less cumbersome? The answer is yes, if we use *function notation*.

Consider again the equation $y = 4x^2 + 3$. Replace the y in this equation with the notation $f(x)$, as follows:

$$f(x) = 4x^2 + 3$$

Note that y and f(x) are interchangeable.

When reading aloud the notation $f(x)$, we say "f of x." It is important to note that $f(x)$ *does not mean "f times x."*

What is the advantage here? Why is it that mathematicians prefer $f(x) = 4x^2 + 3$ to $y = 4x^2 + 3$? Consider the question asked earlier: "If x equals 5, what is the corresponding value of y?" With function notation, this question becomes "What is $f(5)$?" Simply put, $f(5)$ means to *substitute* 5 for x, as follows:

$$f(x) = 4x^2 + 3$$
$$f(5) = 4(5)^2 + 3$$
$$f(5) = 4(25) + 3$$
$$f(5) = 100 + 3$$
$$f(5) = 103$$

The last line of our computation is read aloud as "f of 5 is 103." Thus, $f(5)$ represents the y-value that is paired with $x = 5$.

x - value

y - value

Thus, $f(5)$ is a y-value. In addition, it is a *particular* y-value. It is the y-value that is paired with the x-value of 5.

Example 4   If $f(x) = 2x^2 + 3x + 7$, find $f(2)$.

Solution   To calculate $f(2)$, you simply substitute 2 for x in the equation, as follows:

$$f(x) = 2x^2 + 3x + 7$$
$$f(2) = 2(2)^2 + 3(2) + 7$$
$$f(2) = 2(4) + 6 + 7$$
$$f(2) = 8 + 6 + 7$$
$$f(2) = 21$$

Reading aloud the last line of your work, you would say "f of 2 is 21." The x-value is 2 and the y-value is $f(2)$ or 21.

Example 5   If $g(x) = \dfrac{x}{x+5}$, find $g(3)$.

Solution   Note that the letter g is being used to denote this function (g is a function because each x-value substituted into this equation will produce exactly one y-value). You may select a letter of your choice when naming a function. To find $g(3)$, you simply substitute 3 for x in the equation, as follows:

$$g(x) = \frac{x}{x+5}$$

$$g(3) = \frac{3}{3+5}$$

$$g(3) = \frac{3}{8}$$

Reading aloud the last line of your work, you would say "g of 3 is 3/8." The x-value is 3 and the y-value (or function value) is $g(3)$ or 3/8.

Example 6   Sketch a complete graph of the equation $g(x) = |x + 2|$.

Solution   Since $g(x)$ and y are interchangeable, you can label your points x and $g(x)$ instead of x and y when you set up your table of points.

x	-3	-2	-1	0	1	2	3
$g(x)$							

Now, substitute each value of x into the function. For example, compute $g(-3)$. Remember, this notation simply means substitute -3 for x in the equation.

$$g(x) = |x+2|$$

$$g(-3) = |-3+2|$$

$$g(-3) = |-1|$$

$$g(-3) = 1$$

In a similar manner, $g(-2)$ is computed as follows:

$$g(x) = |x+2|$$

$$g(-2) = |-2+2|$$

$$g(-2) = 0$$

Here is the rest of the work required to fill in the table of points.

$$g(-1) = |-1+2| = 1$$

$$g(0) = |0+2| = 2$$

$$g(1) = |1+2| = 3$$

$$g(2) = |2+2| = 4$$

$$g(3) = |3+2| = 5$$

Inserting the results of these computations into a table results in the following:

x	-3	-2	-1	0	1	2	3
$g(x)$	1	0	1	2	3	4	5
Points to plot	$(-3,1)$	$(-2,0)$	$(-1,1)$	$(0,2)$	$(1,3)$	$(2,4)$	$(3,5$

Plotting these points results in the following graph:

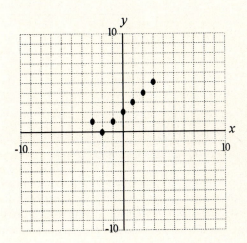

The graph will probably continue upward and to the right as indicated here. However, what happens to the left of $x = -3$? You need to plot a few more points in order to determine the behavior of the graph to the left of $x = -3$. Try adding the following x-values to the table: $-7, -6, -5,$ and -4.

x	-7	-6	-5	-4	-3	-2	-1	0	1	2	3
$g(x)$					1	0	1	2	3	4	5
Points to plot					$(-3,1)$	$(-2,0)$	$(-1,1)$	$(0,2)$	$(1,3)$	$(2,4)$	$(3,5)$

Now evaluate the function g at each of these new x-values.

$$g(-7) = |-7+2| = 5$$
$$g(-6) = |-6+2| = 4$$
$$g(-5) = |-5+2| = 3$$
$$g(-4) = |-4+2| = 2$$

Add these points to your table.

x	-7	-6	-5	-4	-3	-2	-1	0	1	2	3
$g(x)$	5	4	3	2	1	0	1	2	3	4	5
Points to plot	$(-7,5)$	$(-6,4)$	$(-5,3)$	$(-4,2)$	$(-3,1)$	$(-2,0)$	$(-1,1)$	$(0,2)$	$(1,3)$	$(2,4)$	$(3,5)$

Plotting these additional points results in the following graph.

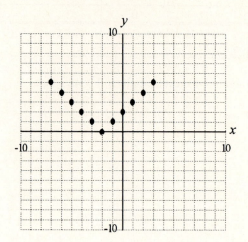

This graph is more *complete* than the previous graph. One branch of our graph will probably continue upward and to the left, while the other branch will continue upward and to the right. There is an important lesson to learn here. If your sketch does not show all of the important features of the graph, then you should compute and plot additional points until you feel that your sketch exhibits all the important behavior of the function. Finally, plotting the rest of the points that satisfy the equation $g(x) = |x + 2|$ results in the following graph.

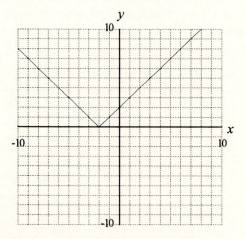

Expressing One Quantity as a Function of Another

Example 7   Express the circumference, C, of a circle as a function of its radius, r.

Solution   You must come up with a rule for the circumference of a circle in terms of its radius. Start by drawing a picture.

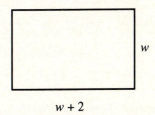

You can find the circumference of a circle by multiplying the number π by the diameter of the circle. The diameter of the circle is twice its radius; therefore, you can write the following:

$$C = \pi(2r) \text{ or } C = 2\pi r$$

The circumference of the circle is expressed as a function of its radius.

Example 8   Express the area, A, of a circle as a function of its radius, r.

Solution   You can find the area of a circle by multiplying the number π by the square of the radius. Hence,

$$A = \pi r^2.$$

The area of the circle is expressed as a function of its radius.

Example 9   Melissa is digging a rectangular garden in her backyard. The length of her garden is two more feet than its width. Express the area, A, of Melissa's garden as a function of its width, w.

Solution   First draw a rectangle and represent its width (in feet) by the variable w.

w

$w + 2$

Since the length of the rectangle is two more feet than its width, the length is $w + 2$. Finally, since the area of a rectangle is computed by multiplying its width and length, the area A of the plot, as a function of w, is given by the following equation:

$$A = w(w + 2).$$

Example 10   Jack makes ceramic pots. His initial costs for tools and supplies total \$100. In addition, for each pot that Jack makes, his costs increase by \$8. Express Jack's total costs, C, as a function of x, the number of pots that he makes.

Solution   If Jack does not make a single pot, his costs will be \$100, his initial outlay for tools and supplies. If Jack makes one pot, his costs will be $C = 100 + 8$ dollars. If Jack makes two pots, his costs will be $C = 100 + 8(2)$ dollars. If Jack makes three pots, his costs will be $C = 100 + 8(3)$ dollars. A pattern is emerging. If Jack makes x pots, his costs will be given by the following equation:

$$C = 100 + 8x$$

Jack's costs, C, are expressed as a function of x, the number of pots made.

Example 11   If Jack sells his pots for \$50 each at the county fair, express Jack's revenue, R, as a function of x, the number of pots sold.

Solution   To compute Jack's revenue, multiply $50 by the number of pots, x, that are sold. Therefore, Jack's revenue will be given by the following equation.

$$R = 50x$$

Jack's revenue, R, is expressed as a function of x, the number of pots sold.

Example 12   Suppose that Jack manages to sell all of his pots. Express Jack's profit, P, as a function of x, the number of pots made and sold.

Solution   Profit is computed by subtracting Jack's costs from his revenue.

$$\text{Profit} = \text{Revenue} - \text{Cost}$$
$$P = 50x - (100 + 8x)$$

You can simplify this result by distributing the minus sign and combining like terms, as follows:

$$P = 50x - 100 - 8x$$
$$P = 42x - 100$$

Jack's profit, P, is expressed as a function of x, the number of pots made and sold.

Exercises for Section 1.3

1. In each of the following problems, fill the blank spaces with the number 10 and simplify:

   a) $f(\ ) = 3(\ ) + 7$

   b) $g(\ ) = (\ )^2 + 1$

   c) $h(\ ) = \sqrt{(\ ) - 6}$

   d) $f(\ ) = 2(\ )^2 - 3(\ ) + 5$

   e) $g(\ ) = |3 - (\ )|$

   f) $h(\ ) = (\ )^3 - 2(\ )^2 - 11$

2. (TI-82) Check each of the results from Exercise 1 by entering the following keystrokes on your TI-82 and pressing the ENTER key:

   a) $3 \times 10 + 7$

   b) $10^2 + 1$

   c) $\sqrt{}\ (10 - 6)$

   d) $2 \times 10^2 - 3 \times 10 + 5$

   e) ABS$(3 - 10)$

   f) $10^3 - 2 \times 10^2 - 11$

   Note: There are two strikingly similar keys on your calculator, the negative sign $\boxed{(-)}$ key and the subtraction sign $\boxed{-}$ key. The negative sign appears on the button to the immediate left of the ENTER key and is used to negate numbers. It is not needed in this exercise. The subtraction sign appears on the button immediately above the addition $\boxed{+}$ button and is used to subtract numbers. In this exercise, use the subtraction sign exclusively. If you reverse the roles of these two keys, then the TI-82 will respond with a syntax error message.

3. (TI-82) Enter each of the following sets of keystrokes on your calculator and write a sentence explaining the difference between the two computations.

   a) -2^2

   b) $(-2)^2$

   Note: The negative sign used in this exercise is on the $\boxed{(-)}$ button to the immediate left of the ENTER key. Note the use of parentheses in (b).

4. (TI-82) Enter each of the following sets of keystrokes on your calculator and write a sentence explaining the difference between the two computations.

   a) $2 \times -3 \wedge 2$ b) $2 \times (-3) \wedge 2$

   Note: The negative sign used in this exercise is on the $\boxed{(\text{-})}$ button to the immediate left of the ENTER key. Note the use of parentheses in (b).

5. In each of the following problems, fill the blank spaces with the number -2 and simplify:

   a) $f(\ ) = 2(\ ) - 5$ b) $g(\ ) = -3(\ ) - 8$ c) $h(\ ) = 2(\ )^2 - 4(\ ) - 8$

   d) $f(\ ) = 5 - 3(\ ) - (\ )^2$ e) $g(\ ) = \sqrt{7 - (\ )}$ f) $h(\ ) = 4(\ )^3 - 9$

6. (TI-82) Check each of the results from Exercise 5 by entering the following keystrokes on your TI-82 and pressing the ENTER key:

   a) $2 \times (-2) - 5$ b) $-3 \times (-2) - 8$ c) $2 \times (-2) \wedge 2 - 4 \times (-2) - 8$

   d) $5 - 3 \times (-2) - (-2) \wedge 2$ e) $\sqrt{\ } (7 - (-2))$ f) $4 \times (-2) \wedge 3 - 9$

   Hint: Do not interchange the roles of the $\boxed{(\text{-})}$ button and the $\boxed{-}$ button.

7. Given that $f(x) = 2x + 3$ and $g(x) = x^2 + 4$, evaluate each of the following mentally. Check your result on your calculator.

   a) $f(3)$ b) $g(3)$ c) $f(-5)$ d) $g(-5)$

8. Consider the function $h(x) = |x - 2|$. Evaluate each of the following mentally. Check your result on your calculator.

   a) $h(5)$ b) $h(1)$

9. Consider the function $f(x) = x^2 - 4x - 8$. Evaluate each of the following mentally. Check your result on your calculator.

   a) $f(3)$ b) $f(-2)$

10. Consider the function $f(x) = \dfrac{x+6}{2}$ and $g(x) = \dfrac{x+6}{x+2}$. Evaluate each of the following mentally. Check your result on your calculator.

    a) $f(4)$ b) $g(2)$

11. (TI-82) Check each of the results from Exercise 10 by entering the following keystrokes on your TI-82 and pressing the ENTER key:

    a) $(4 + 6) \div 2$ b) $(2 + 6) \div (2 + 2)$

    Note the use of parentheses to ensure computation is carried out in the desired order.

12. Given $g(x) = |3x - 5|$ and $h(x) = \left| \dfrac{x-4}{2} \right|$, evaluate each of the following mentally. Check your result on your calculator.

    a) $g(2)$ b) $h(-2)$

13. (TI-82) Check each of the results from Exercise 12 by entering the following keystrokes on your TI-82 and pressing the ENTER key:

    a) $\text{ABS}(3\times2-5)$ b) $\text{ABS}((\text{-}2-4)\div2)$

14. (TI-82) Use your calculator to find $f(1.23)$ for each of the following functions:

    a) $f(x)=1.23x-3.125$ b) $f(x)=-3.23x-1.45$
    c) $f(x)=|4.1-5.3x|$ d) $f(x)=3.2x^2-1.4x-2.8$
    e) $f(x)=1.8x^3-2.5x^2-7.4x+1.1$

15. We have completed a few entries in the following table of points that satisfy the equation $f(x) = 3x + 2$. Be sure to double-check these results. Copy and complete the table on a sheet of graph paper. Set up a coordinate system on your graph paper near your table and carefully plot each point in the table on your coordinate system. Based on the points plotted thus far, graph the rest of the points that satisfy the equation of the function.

x	−5	−4	−3	−2	−1	0	1	2	3	4	5
$f(x)$	−13			−4					11		
Points to plot	(−5,−13)			(−2,−4)					(3,11)		

16. A few entries have been completed in the following table of points that satisfy the equation $f(x) = |x+3|$. Be sure to double-check these results. Copy and complete the table on a sheet of graph paper. Set up a coordinate system on your graph paper near your table and carefully plot each point in the table on your coordinate system. Based on the points plotted thus far, graph the rest of the points that satisfy the equation of the function.

x	−6	−5	−4	−3	−2	−1	0
y		2					3
Points to plot		(−5,2)					(0,3)

17. A few entries have been completed in the following table of points that satisfy the equation $g(x) = (x + 1)^2$. Be sure to double-check these results. Copy and complete the table on a sheet of graph paper. Set up a coordinate system on your graph paper near your table and carefully plot each point in the table on your coordinate system. Based on the points plotted thus far, graph the rest of the points that satisfy the equation of the function.

x	−6	−5	−4	−3	−2	−1	0	1	2	3	4
$g(x)$	25				1			4			
Points to plot	(−6,25)				(−2,1)			(1,4)			

18. An entry has been completed in the following table of points that satisfy the equation $h(x)=\sqrt{x-3}$. Be sure to double-check this result. Copy and complete the table on a sheet of graph paper. Set up a coordinate system on your graph paper near your table and carefully plot each point in the table on your coordinate system. Based on the points plotted thus far, graph the rest of the points that satisfy the equation of the function.

x	3	4	7	12	19	28
$h(x)$			2			
Points to plot			(7,2)			

19. A few entries have been completed in the following table of points that satisfy the equation $f(x) = 5 + 4x - x^2$. Be sure to double-check these results. Copy and complete the table on a sheet of graph paper. Set up a coordinate system on your graph paper near your table and carefully plot each point in the table on your coordinate system. Based on the points plotted thus far, graph the rest of the points that satisfy the equation of the function.

x	−3	−2	−1	0	1	2	3	4	5	6	7
$f(x)$	−16	−7									−16
Points to plot	(−3,−16)	(−2,−7)									(7,−16)

20. For each of the following functions, do the following:

   i)   On graph paper, set up a table of points that satisfy the equation of the function.

   ii) Set up a coordinate system near your table and plot each of the points from your table on your coordinate system.

   iii) If you feel that your graph is not complete, add more points to your table of points and plot them.

   iv) Based on the points plotted thus far, graph the rest of the points that satisfy the equation of the function.

   a) $f(x) = 5 - 2x$ b) $g(x) = 2x + 3$ c) $f(x) = (x + 3)^2$

   d) $g(x) = |x - 5|$ e) $h(x) = x^2 - 8x - 10$ f) $f(x) = 9 - 4x - x^2$

21. Copy the following graph of the function f onto a sheet of graph paper.

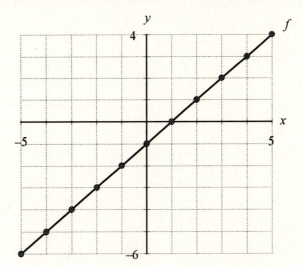

   Put a copy of the following table on your graph paper near the graph of f. Read the graph of the function f and fill in the remaining entries of the table. Be sure to double-check the given computations.

x	−5	−4	−3	−2	−1	0	1	2	3	4	5
f(x)	−6	−5									4
Points to plot	(−5,−6)	(−4,−5)									(5,4)

22. Copy the following graph of the function g onto a sheet of graph paper.

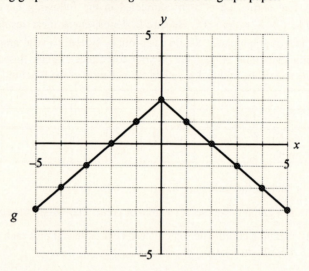

Put a copy of the following table on your graph paper near the graph of g. Read the graph of the function g and fill in the remaining entries of the table. Be sure to double-check the given computations.

x	−5	−4	−3	−2	−1	0	1	2	3	4	5
g(x)	−3	−2									−3
Points to plot	(−5,−3)	(−4,−2)									(5,−3)

23. Copy the following graph of the function g onto a sheet of graph paper.

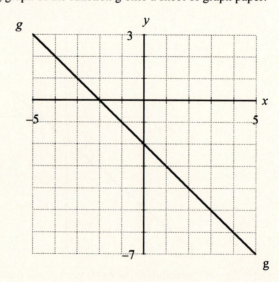

Put a copy of the following table on your graph paper near the graph of g. Read the graph of the function g and fill in the remaining entries of the table. Be sure to double-check the given computations.

x	−5	−4	−3	−2	−1	0	1	2	3	4	5
g(x)	3	2									−7
Points to plot	(−5,3)	(−4,2)									(5,−7)

24. Copy the following graph of the function h onto a sheet of graph paper.

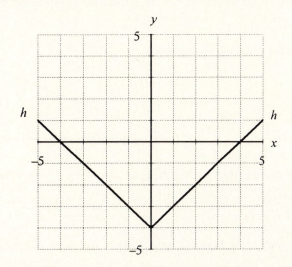

Put a copy of the following table on your graph paper near the graph of h. Read the graph of the function h and fill in the remaining entries of the table. Be sure to double-check the given computations.

x	−5	−4	−3	−2	−1	0	1	2	3	4	5
h(x)	1	0									1
Points to plot	(−5,1)	(−4,0)									(5,1)

25. The length of a rectangle is three feet longer than twice its width. If w represents the width of the rectangle, express the length of the rectangle as a function of w. Include a clearly labeled diagram with your work.

26. The area of a rectangle is 100 square feet. If w represents the width of the rectangle, express the length of the rectangle as a function of w. Include a clearly labeled diagram with your work.

27. The perimeter of a rectangle is 200 feet. If w represents the width of the rectangle, express the length of the rectangle as a function of w. Include a clearly labeled diagram with your work.

28. Alana rents out roller skates for use on the boardwalk on the beach. She charges a flat fee of \$2 plus \$1.50 for each hour that the skates are used. If t represents the number of hours that the skates are in use, express the rental charge, R, as a function of t.

29. The following circles are concentric (they have the same center).

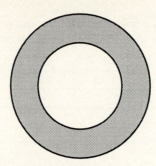

If the radius of the inner circle is r and the radius of the outer circle is 4, express the area of the shaded region, A, as a function of r.

30. Consider the following triangle.

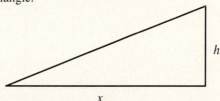

Suppose that the length of the base, x, is 8 inches longer than the altitude of the triangle, h.

a) Express the altitude of the triangle, h, as a function of x.
b) Express the area of the triangle as a function of x.

Note: Recall that the area of a triangle is one-half of the base times the altitude.

31. Henry is a master craftsman with leather. His specialty is leather driving gloves used by drivers at the Indianapolis 500 speedway. His fixed costs are $200 for tools. Thereafter, for each pair of gloves he makes, his costs increase by $8. Let x represent the number of pairs of gloves that Henry makes, and express his costs, C, as a function of x.

32. Suppose that Henry sells his gloves for $50 a pair. Let x represent the number of pairs of gloves that Henry sells. Express Henry's revenue, R, as a function of x.

33. Express Henry's profit, P, as a function of x, the number of pairs of gloves he makes and sells. What is Henry's profit if he makes and sells 100 pairs of gloves?

1.4 Introducing the Graphing Calculator

Today there are many kinds of calculators on the market—calculators for insurance agents and real estate agents, calculators for statistics and for electronics, even calculators to organize messages and phone numbers. More importantly, these calculators are affordable, about the price of a pair of athletic shoes or a Nintendo Game Boy.

A remarkable innovation of the last few years has been the introduction of the graphing calculator. Companies such as Casio, Hewlett-Packard, Sharp, and Texas Instruments have all developed various types of programmable graphing calculators.

In this book we will use the TI-82 graphing calculator from Texas Instruments. If you have a TI-81 model, much of the keystroking is identical to that of the TI-82. Programs have been included in the appendix that will make your TI-81 duplicate some of the features of the TI-82.

Much of what you will do here can be duplicated on other brands of graphing calculators. If your calculator can draw a graph, adjust the window, and trace along a graph, then it will probably be adequate for use with this text. Note that there are supplementary materials available from the publisher of this book that illustrate the use of many other brands of graphing calculators.

Drawing the Graph of an Equation

Example 1 shows you how to draw a graph of an equation on the TI-82. In no time at all you will be able to draw the complete graphs of some very difficult equations.

Example 1   Draw the graph of $y = x^2$ on your TI-82 graphing calculator.

Solution   Push the Y= menu button on the top row of your calculator. The following menu will appear on your screen.

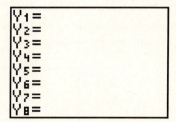

You have the ability to enter eight separate equations on the TI-82. If you have any equations currently entered in your TI-82, use the arrow keys and the CLEAR button to delete them. Use the arrow keys to place the cursor after $Y_1 =$ and enter the equation $y = x^2$ in the following manner:

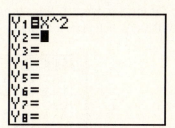

You can capture the X by pushing the $\boxed{X,T,\theta}$ button on your keyboard (the $\boxed{X|T}$ on the TI-81). The $\boxed{\wedge}$ key is located just above the division button and is the exponentiation key. If you wish, you could also type the following, using the $\boxed{x^2}$ button on your calculator.

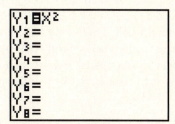

Assuming you have successfully entered your equation, how do you get the calculator to draw the graph of your equation? If you push the ZOOM button on the top row of the TI-82, the following menu appears.

The downarrow by the number 7 indicates that there are more entries in this menu. For now, you will not need them. The next step is to select 6:ZStandard (6:Standard on the TI-81). You can accomplish this in one of two ways:

Either hit the down-arrow key several times until the 6 is highlighted, then press the ENTER key.

Or push the number 6 on the numeric keypad of your calculator.

Either of these methods will work on all of the menus on the TI-82. You should now have the following image in your viewing window:

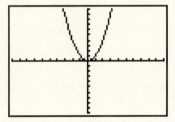

What Does 6:ZStandard Mean?   What happens when we select 6:ZStandard from the ZOOM menu? Push the WINDOW button on the top row of the TI-82 and the following appears:

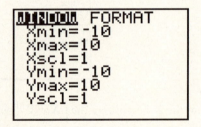

Xmin is the left side of the viewing window and is equal to −10; Xmax is the right side and is equal to 10. Xscl is the distance between tick marks on the *x*-axis and is equal to 1. If you count the number of tick marks from the origin to the right side of the window, you will see that there are precisely 10, each one representing 1 unit. Ymin is the bottom of the viewing window and is equal to −10; Ymax is the top and is equal to 10. Yscl is the distance between tick marks on the *y*-axis and is equal to 1.

What happens if you change one of these window parameters? For example, try changing Ymin to zero. Use the cursor keys to move the cursor down to Ymin. Enter a 0 from the numeric keypad and press ENTER.

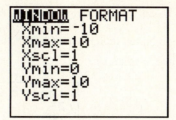

Push the GRAPH button on the top row of buttons on your calculator. The graph is refreshed, using the changes in the WINDOW parameters.

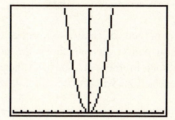

Note that the bottom of the viewing window is now zero, which is precisely the change that was made in the WINDOW parameters. Remember when you set Ymin equal to zero? If you push the ZOOM button and select 6:ZStandard once again, the following graph appears in the viewing window.

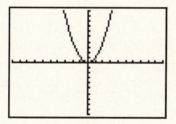

Note that this is your original graph. Push the WINDOW button again.

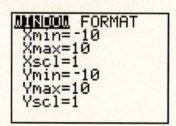

Notice that Ymin equals −10 again! You are now looking at the so-called "standard" viewing window. Any time you wish to set your WINDOW parameters to these standard default settings, push the ZOOM button and select 6:ZStandard.

Tip	If you set the WINDOW parameters to values other than the standard values, you must push the GRAPH button in order to keep your settings. If you push the ZOOM button, followed by 6:ZStandard, your WINDOW settings will be replaced with the "standard" WINDOW parameters.

Example 2   Use the TI-82 to sketch a complete graph of $y = x^2 - 8x - 9$.

Solution   Press the Y= menu button and use the CLEAR button to delete the last problem. Enter the equation $y = x^2 - 8x - 9$ in Y1, as shown:

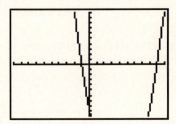

Use the $\boxed{x^2}$ button on your calculator to produce the exponent of 2 on the variable X. You must also use the subtraction button located above the addition button on your keyboard since those are subtraction operators in the equation, not negative signs. Push the ZOOM button and select 6:ZStandard. These actions will produce the following graph in your viewing window:

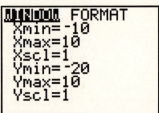

As you move from left to right across the viewing window, it appears that the graph of $y = x^2 - 8x - 9$ falls from the top of the viewing window, plummets out of sight at the bottom, then rises again, only to disappear again at the top. There must be some sort of turning point beyond the bottom of the viewing window. Try to experiment with the WINDOW parameters in order to locate and display the turning point of the graph. Press the WINDOW button and change Ymin to −20, as shown:

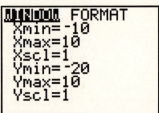

Tip	You must use the negative sign, which is located to the immediate left of the ENTER key, when entering -20 for Ymin. Do not use the subtraction button.

Press the GRAPH button to see the result of this change.

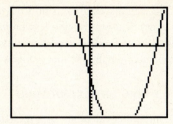

Not quite enough! Change Ymin to −30.

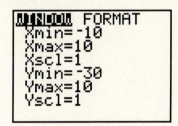

Push the GRAPH button to see the effect of changing Ymin to −30.

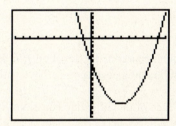

That's much better! You now have a window that shows the turning point of your graph. After some further experimentation, you'll find that the following adjustments to the WINDOW parameters present an even nicer final image in your viewing window.

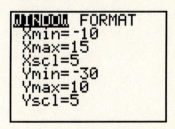

Push the GRAPH button to capture the following image in your viewing window.

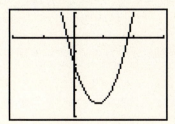

The right side of the viewing window is now 15, giving a bit of balance. Also, the tick marks on the *x*- and *y*-axes are every 5 units, giving a less crowded appearance and making the divisions easier to read on both axes. It is this sort of experimenting, tweaking the WINDOW parameters, that enables you to capture a *complete* graph, one that exhibits all the important features of the graph.

The Graphing Calculator as a Tool for Exploration

Think of the TI-82 graphing calculator as a tool for exploration. With its ability to quickly sketch the graphs of functions, you are free ask the question "What if?" without the fear that this question will produce large amounts of extra work. Such visualization and exploration will lead to a more thorough understanding of the concepts of intermediate algebra, and you will find yourself thinking about this material at a higher level than ever before.

Exercises for Section 1.4

1. (TI-82) Enter each of the following equations in Y1 of the Y= menu, push the ZOOM button, then select 6:ZStandard. Set up a coordinate system on a sheet of graph paper and clearly indicate the scale on each coordinate axis. Copy the image in your viewing window onto your coordinate system.

   a) $f(x) = 2x - 1$ b) $h(x) = 5 - 3x$

   Hint: Enter the equation of the function *f* in the following manner: Y1 = 2X − 1. Note that you must use the subtraction operator, not the negative sign.

2. (TI-82) Enter each of the following equations in Y1 of the Y= menu, push the ZOOM button, then select 6:ZStandard. Set up a coordinate system on a sheet of graph paper and clearly indicate the scale on each coordinate axis. Copy the image in your viewing window onto your coordinate system.

   a) $f(x) = -3x + 4$ b) $g(x) = -2X - 1$

   Hint: Enter the the equation of the function *g* in the following manner: Y1 = -2X − 1. You must use the negative sign in front of the 2 and the subtraction symbol between the -2X and 1. Note the difference.

3. (TI-82) Enter each of the following equations in Y1 of the Y= menu, push the ZOOM button, then select 6:ZStandard. Set up a coordinate system on a sheet of graph paper and clearly indicate the scale on each coordinate axis. Copy the image in your viewing window onto your coordinate system.

   a) $g(x) = x^2 - 4$ b) $f(x) = 9 - x^2$ c) $h(x) = 5$

   d) $f(x) = -3$ e) $f(x) = x^2 - 2x - 3$ f) $g(x) = 5 - 4x - x^2$

4. (TI-82) Enter the equation $f(x) = |x| - 3$ in Y1 in the Y= menu in the following manner.

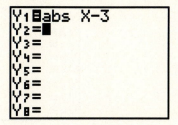

Push the ZOOM button, then select 6:ZStandard. Set up a coordinate system on a sheet of graph paper and clearly indicate the scale on each coordinate axis. Copy the image in your viewing window onto your coordinate system.

Hint: The ABS is located above the x^{-1} key on your TI-82. You must push the 2nd key to access it.

5. (TI-82) Enter the equation $f(x) = |x - 3|$ in Y1 in the Y= menu in the following manner.

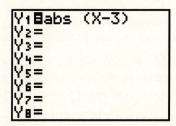

Push the ZOOM button, then select 6:ZStandard. Set up a coordinate system on a sheet of graph paper and clearly indicate the scale on each coordinate axis. Copy the image in your viewing window onto your coordinate system.

Hint: The ABS is located above the x^{-1} key on your TI-82. You must push the 2nd key to access it. Note the use of parentheses.

6. (TI-82) Enter the equation $f(x) = \sqrt{x + 2}$ in Y1 in the Y= menu in the following manner.

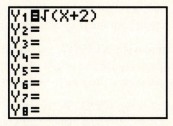

Push the ZOOM button, then select 6:ZStandard. Set up a coordinate system on a sheet of graph paper and clearly indicate the scale on each coordinate axis. Copy the image in your viewing window onto your coordinate system.

Hint: The $\sqrt{}$ symbol is located above the x^2 key. You will have to use the 2nd key to access it. Note the use of parentheses.

7. (TI-82) Enter the equation $f(x) = \sqrt{x} + 2$ in Y1 in the Y= menu in the following manner.

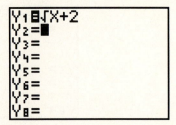

Push the ZOOM button, then select 6:ZStandard. Set up a coordinate system on a sheet of graph paper and clearly indicate the scale on each coordinate axis. Copy the image in your viewing window onto your coordinate system.

Hint: The $\sqrt{}$ symbol is located above the x^2 key. You will have to use the 2nd key to access it.

8. Enter the equation $f(x) = x^2 - 10x - 12$ in Y1 in the Y= menu. Push the ZOOM key and select 6:ZStandard.

a) Set up a coordinate system on a sheet of graph paper and copy the image in your viewing window onto your coordinate system. Be sure to indicate the WINDOW settings on the axes.

b) Change the following WINDOW parameter: Xmax = 20. Press the GRAPH button. Set up a second coordinate system on graph paper and copy the image in your viewing window onto your coordinate system. Be sure to indicate the WINDOW settings on the axes.

c) Change the following WINDOW parameter: Ymin = −50. Press the GRAPH button. Set up a third coordinate system on graph paper and copy the image in your viewing window onto your coordinate system. Be sure to indicate the WINDOW settings on the axes.

d) Change the following WINDOW parameters: Xscl = 5 and Yscl = 5. Press the GRAPH button. Set up a fourth coordinate system on graph paper and copy the image in your viewing window onto your coordinate system. Be sure to indicate the WINDOW settings on the axes.

9. (TI-82) Enter the equation $h(x) = x^3 - 9x^2 + 12x - 14$ in Y1 in the Y= menu as follows: Y1 = X^3 − 9X^2 + 12X − 14. Push the ZOOM button and select 6:ZStandard.

a) Set up a coordinate system on a sheet of graph paper and copy the image in your viewing window onto your coordinate system. Be sure to indicate the WINDOW settings on the axes.

b) Change the following WINDOW parameter: Ymin = −70. Press the GRAPH button. Set up a second coordinate system on graph paper and copy the image in your viewing window onto your coordinate system. Be sure to indicate the WINDOW settings on the axes.

c) Change the following WINDOW parameter: Yscl = 10. Press the GRAPH button. Set up a third coordinate system on graph paper and copy the image in your viewing window onto your coordinate system. Be sure to indicate the WINDOW settings on the axes.

10. (TI-82) Enter the function $f(x) = x^4 - 15x^2 + 12$ in Y1 in the Y= menu as follows: Y1 = X^4 - 15X^2 + 12. Push the ZOOM button and select 6:ZStandard.

a) Set up a coordinate system on a sheet of graph paper and copy the image in your viewing window onto your coordinate system. Be sure to indicate the WINDOW settings on the axes.

b) Change the following WINDOW parameters: Ymin = −50 and Ymax = 20. Press the GRAPH button. Set up a second coordinate system on graph paper and copy the image in your viewing window onto your coordinate system. Be sure to indicate the WINDOW settings on the axes.

c) Change the following WINDOW parameter: Yscl = 10. Press the GRAPH button. Set up a third coordinate system on graph paper and copy the image in your viewing window onto your coordinate system. Be sure to indicate the WINDOW settings on the axes.

11. (TI-82) Enter the function $f(x) = (x+15)(x-12)$ in Y₁ in the Y= menu as follows: Y₁ = $(X+15)(X-12)$. Push the ZOOM button and select 6:ZStandard.

   a) Set up a coordinate system on a sheet of graph paper and copy the image in your viewing window onto your coordinate system. Be sure to indicate the WINDOW settings on the axes.

   b) Change the following WINDOW parameters: Xmin = −23, Xmax = 20, and Xscl = 5. Press the GRAPH button. Set up a second coordinate system on graph paper and copy the image in your viewing window onto your coordinate system. Be sure to indicate the WINDOW settings on the axes.

   c) Change the following WINDOW parameters: Ymin = −300, Ymax = 300, and Yscl = 50. Press the GRAPH button. Set up a third coordinate system on graph paper and copy the image in your viewing window onto your coordinate system. Be sure to indicate the WINDOW settings on the axes.

12. (TI-82) Enter the function $f(x) = (14-x)(x+15)$ in Y₁ in the Y= menu as follows: Y₁ = $(14-X)(X+15)$. Push the ZOOM button and select 6:ZStandard.

   a) Set up a coordinate system on a sheet of graph paper and copy the image in your viewing window onto your coordinate system. Be sure to indicate the WINDOW settings on the axes.

   b) Change the following WINDOW parameters: Xmin = −22, Xmax = 21, and Xscl = 5. Press the GRAPH button. Set up a second coordinate system on graph paper and copy the image in your viewing window onto your coordinate system. Be sure to indicate the WINDOW settings on the axes.

   c) Change the following WINDOW parameters: Ymin = −300, Ymax = 300, and Yscl = 50. Press the GRAPH button. Set up a third coordinate system on graph paper and copy the image in your viewing window onto your coordinate system. Be sure to indicate the WINDOW settings on the axes.

13. (TI-82) Consider the function $f(x) = x^3 - 64x$. Enter this function in the Y= menu in the following manner: Y₁ = X^3 − 64X. After some experimentation with the WINDOW parameters, we produced the following image in our viewing window.

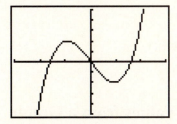

Set up a coordinate system on a sheet of graph paper and copy this image onto your coordinate system. Experiment with the WINDOW parameters on your calculator until you reproduce this image in your viewing window. Place your final WINDOW parameters on your coordinate axes.

14. (TI-82) Consider the function $f(x) = (x+10)(x-1)(x-15)$. Enter this function in the Y= menu in the following manner: Y₁ = $(X+10)(X-1)(X-15)$. After some experimentation with the WINDOW parameters, we produced the following image in our viewing window.

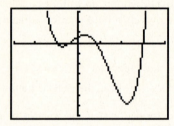

Set up a coordinate system on a sheet of graph paper and copy this image onto your coordinate system. Experiment with the WINDOW parameters on your calculator until you reproduce this image in your viewing window. Place your final WINDOW parameters on your coordinate axes.

15. Consider the function $f(x)=(x+5)(x+2)(x-4)(x-15)$. Enter this function in the Y= menu in the following manner: $Y_1 = (X+5)(X+2)(X-4)(X-15)$. After some experimentation with the WINDOW parameters, we produced the following image in our viewing window.

Set up a coordinate system on a sheet of graph paper and copy this image onto your coordinate system. Experiment with the WINDOW parameters on your calculator until you reproduce this image in your viewing window. Place your final WINDOW parameters on your coordinate axes.

16. (TI-82) It's time to experiment on your own. For each of the following equations, perform the following tasks:

   i)   Set up a coordinate system on a sheet of graph paper.

   ii) Enter the equation into the Y= menu on your calculator. Experiment with the WINDOW parameters until you feel the image in your viewing window is a *complete* graph, one that shows all the important features of the graph.

   iii) Copy this final image onto your coordinate system. Place your WINDOW parameters on the axes of your coordinate system.

   a) $f(x)=15-2x^2$ b) $g(x)=2x^2-20$ c) $h(x)=2|x|-18$

   d) $f(x)=(x+5)(x-7)$ e) $g(x)=(3-x)(x+5)$ f) $h(x)=\left(x^2-4\right)\left(x^2-16\right)$

1.5 Reading Function Values from a Graph

If you have an equation representing a function, function values for various values of x can be computed simply by substituting the x-values into the equation. For example, if $f(x)=2x+3$, then $f(10)=2(10)+3=23$.

In this section, we will discuss how to *estimate* function values by reading a graph. Consider a point (x,y) on the graph of a function f, as shown below.

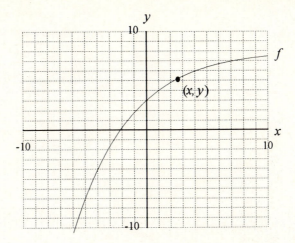

f (x) and y are interchangeable, so f (x) is y and y is f (x).

The point (x,y) will be renamed $(x, f(x))$.

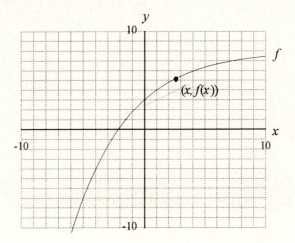

The idea that f (x) represents the y-value that is paired with x will play a fundamental role in all our ensuing work.

If you draw a vertical line from point $(x, f(x))$ to the x-axis, then you will be able to read the x-value of this point on the x-axis. If you draw a horizontal line from the point $(x, f(x))$ to the y-axis, then you will be able to read $f(x)$, or the y-value of this point, on the y-axis, as shown:

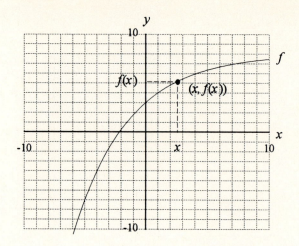

f(x) **is the y-value that is paired with x.**

Example 1 Consider the following graph of the function *f*. Use the graph to estimate $f(2)$.

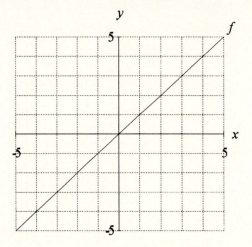

Solution Remember, $f(2)$ is a *y*-value. It is the *y*-value that is paired with an *x*-value of 2. Locate the number 2 on the *x*-axis, draw a vertical line to the graph of the function, then a horizontal line to the *y*-axis, as shown:

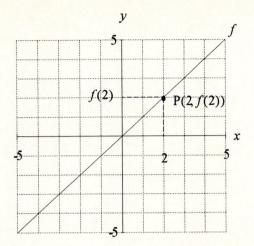

The coordinates of point P are $(2, f(2))$. The y-value of point P is $f(2)$. Read the y-value of this point on the y-axis. It appears that $f(2)$, the y-value of the point P, is approximately 2. Therefore, $f(2) = 2$.

Example 2   Use the following graph to find $g(-1)$.

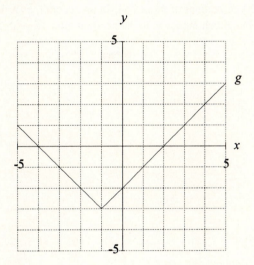

Solution   Remember, $g(-1)$ is a y-value. It is the y-value that is paired with the x-value -1. Therefore, locate -1 on the x-axis, draw a vertical line to the graph of the function, followed by a horizontal line to the y-axis, where you read the answer for $g(-1)$, as follows:

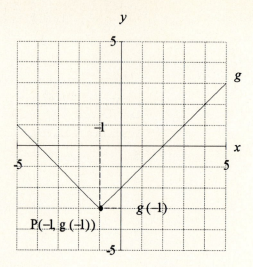

The coordinates of point P are $(-1, g(-1))$. The y-value of point P is $g(-1)$. It appears that $g(-1)$, the y-value of point P, is approximately -3. Therefore, $g(-1) = -3$.

Example 3   The amount of money in a savings account increases with time, because of the interest on the account. The amount of money, *A,* in an account, as a function of time, *t,* is shown in the following graph. How much money is in the account after 15 years?

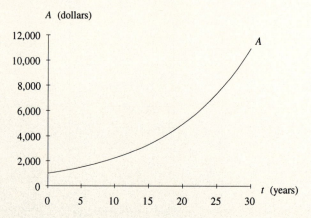

Solution   Note that the x-axis and y-axis have been labeled the t-axis and A-axis, respectively. Remember, $A(15)$, the amount of money in the account after 15 years, is an A-value. It is the A-value that is paired with a t-value of 15. We locate 15 on the t-axis, and draw a vertical line to the graph of the function, followed by a horizontal line to the A-axis, as shown:

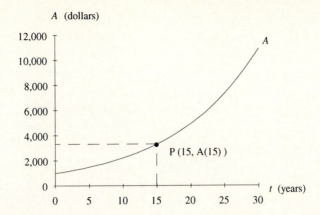

The coordinates of point P are $(15, A(15))$. The A-value of point P is $A(15)$. It appears that $A(15)$, the A-value of point P, lies somewhere between \$2,000 and \$4,000. It can be estimated that $A(15)$ is \$3,300. Therefore, $A(15) = \$3,300$.

Remember, this is only an estimation. If someone else were to estimate that the A-value of point P, $A(15)$, were \$3,400, that would be fine. This rough graph of A leaves itself open to such interpretation. If there were a scale on the A-axis finer than the present one, you could make a more precise estimate.

Reversing the Problem Situation.   Up to this point, you have been given the x-value and have been asked to find the corresponding function value or y-value. You will now perform the opposite operation. You will be given the function value and will then be asked to find the corresponding x-value(s). Let's start with an example.

Example 4   Use the following graph of the function f to find all values of x such that $f(x) = 3$.

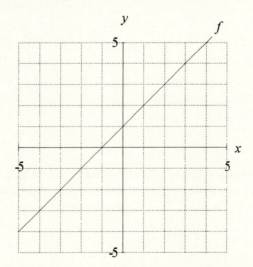

Solution   This time 3 is a function value, or y-value. Locate 3 on the y-axis, and draw a horizontal line to the graph of the function, followed by a vertical line to the x-axis, as shown:

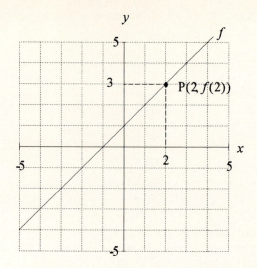

Because the $f(2) = 3$, the solution of $f(x) = 3$ is $x = 2$.

Example 5   Use the following graph of *f* to find all values of *x* such that $f(x) = 2$.

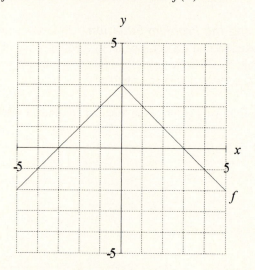

Solution   2 is a function value, or *y*-value. Locate 2 on the *y*-axis, and draw a horizontal line to the graph of the function, followed by vertical lines to the *x*-axis, as shown:

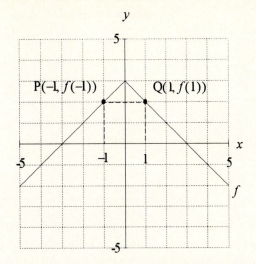

Because the horizontal line intersects the graph of f twice., there are two x-values that are paired with the function value 2. Because $f(-1) = 2$ and $f(1) = 2$, there are two solutions of $f(x) = 2$, $x = -1$ and $x = 1$.

Example 6   Julie is counting amoebas in a petri dish in the biology lab. The number of amoebas present, N, at time t, is given in the graph below. How much time will pass until Julie will see 64 amoebas in her petri dish?

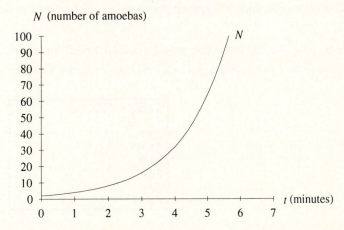

Solution   Note that the x-axis and y-axis have been labeled as the t-axis and N-axis, respectively. Locate 64, the number of amoebas, on the N-axis. Draw a horizontal line to the function, followed by a vertical line to the t-axis, as shown:

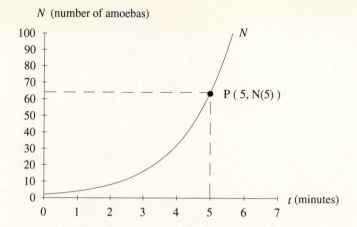

It appears that the vertical line hits the *t*-axis at *approximately* 5 minutes. Therefore, $N(5) = 64$, and it will be *approximately* 5 minutes before Julie will see 64 amoebas in her petri dish.

Using the TRACE Button on the TI-82 Graphing Calculator

Once the graph of a function has been displayed on the viewing screen of a TI-82 graphing calculator, it is possible to identify the coordinates of points on the graph by using the TRACE button.

Example 7   Sketch the graph of $y = x^2 - 2x - 3$ on the TI-82. Use the TRACE button and the cursor keys to identify the coordinates of several points on the graph of the function.

Solution   First, push the Y= menu button and use the CLEAR button to remove the last problem from the menu. Enter the equation $y = x^2 - 2x - 3$ into Y1 of the Y= menu in the following manner.

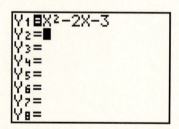

Push the ZOOM button and select 6:ZStandard to obtain the following image in your viewing window.

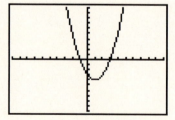

Now, push the TRACE button on the top row of buttons. The trace cursor comes on, and the coordinates of the trace cursor are displayed at the bottom of the screen.

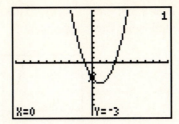

The number 1 in the top right corner of the viewing window indicates that you are tracing along the function that was entered in Y1. If you entered a function in Y2, and you were tracing along the function entered in Y2, then a 2 would appear in the top right corner of your viewing window. You can use the right and left arrow keys to move the trace cursor along the graph. The coordinates of the trace cursor are displayed on the bottom of the screen. For example, if the trace cursor is moved a bit more to the right, the following image appears.

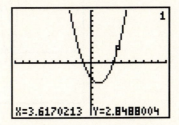

Your coordinates may differ slightly, particularly if you have WINDOW settings other than those used here. This trace capability can be extremely useful. You may wish to experiment by moving the trace cursor around and noting the updated coordinates.

Zoom Integer—A Great Method for Completing Tables

Push the ZOOM button again. You will get the following menu.

Remember that down arrow next to menu item 7? This arrow indicates that there are hidden menu items below. Use the down arrow on your keyboard to scroll down below menu item 7.

Highlight menu item 8 and press ENTER. Alternately, simply press 8 on your keyboard to select menu item 8. The following screen appears:

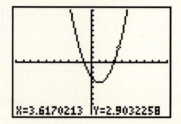

It seems that very little has happened. There is a secret here. *YOU MUST NOW PRESS THE ENTER KEY!* You will be rewarded with the following screen:

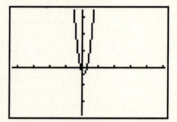

Where is the reward? This is not even a very good viewing window for this graph. Press the TRACE button again. The following screen will appear.

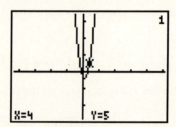

It doesn't matter if your cursor is in a slightly different position. Try moving the cursor around with the right and left arrow keys. Note that the *x*-value of the cursor is incremented or decremented by 1 each time you move the trace cursor.

You can use this feature (8:ZInteger) to complete many of the tables in this book. This method will also work on the TI-81.

A Method for Completing Tables

    1. Enter your function in Y1 in the Y= menu.
    2. Push ZOOM and select 6:ZStandard.
    3. Push ZOOM and select 8:ZInteger.
    4. Press the ENTER key.
    5. Press the TRACE button.

Exercises for Section 1.5

1. On the following graph of *f*, locate the point $(2, f(2))$.

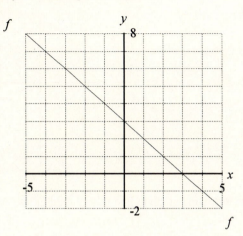

2. On the following graph of *g*, locate the point $(1, g(1))$.

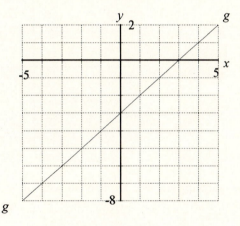

3. On the following graph of *h*, locate the point $(0, h(0))$.

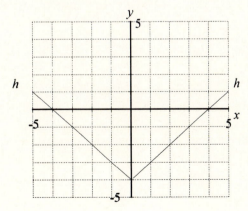

4. On the following graph of f, locate the point $(-2, f(-2))$.

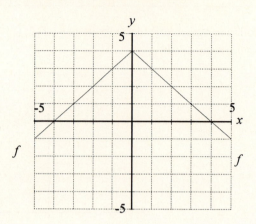

5. Read the following graph to find the value of $g(2)$.

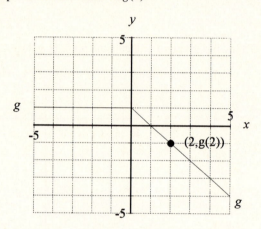

6. Read the following graph of h to find the value of $h(-3)$.

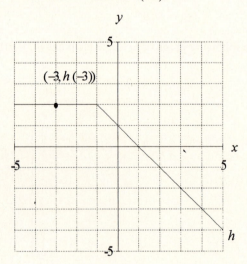

7.   Read the following graph of *f* to find the value of *f*(1).

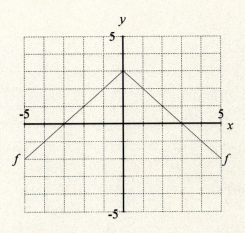

8. Read the following graph of *h* to find the value of *h*(−2).

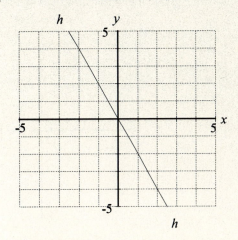

9. Read the following graph of *g* to find the value of *g*(0).

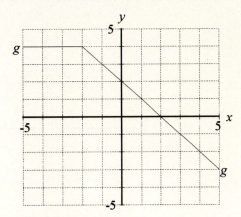

10. Use the following graph of f to find all values of x so that $f(x) = 2$.

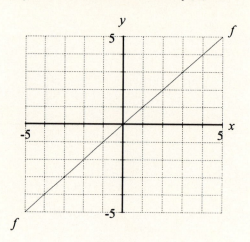

11. Read the following graph of g to find all values of x so that $g(x) = -3$.

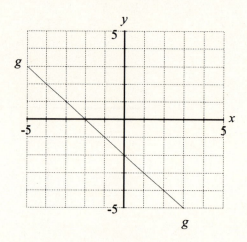

12. Read the following graph of h to find all values of x so that $h(x) = 2$.

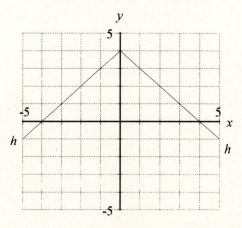

13. Mitchell and Debra are working in their nuclear physics laboratory on an experiment that will show how radioactive materials decay. Initially, they have 250 milligrams of a radioactive substance. Over time, it begins to decay. The amount present is a function of the amount of time that has passed. In the following graph, the behavior depicted is called *radioactive decay*.

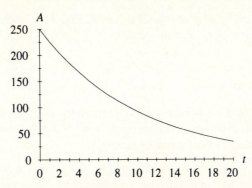

The amount A of radioactive material present at any time t is measured in milligrams, while the time is measured in minutes. Answer the following questions.

a) How much material will be present after 13 minutes?
b) How much material will be present after 18 minutes?
c) After how many minutes will the amount of material slip below 100 milligrams?

14. Samuel is an apprentice typesetter, new to the job. He is not very good at first, but he is highly motivated and determined to improve. As the days go by, Samuel is setting more and more lines of type. In the following graph, L represents the number of lines of type that Samuel can set in one day, and t represents the number of days that Samuel has been on the job.

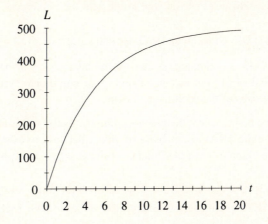

This type of graph is often called a *learning curve*. The number of lines of type that Samuel can set in a day is steadily increasing as he gains experience. Now, answer the following questions.

a) How many lines of type was Samuel setting after 5 days on the job?
b) How many lines of type was he setting after 8 days on the job?
c) What appears to be Samuel's maximum output?
d) How many days had Samuel spent on the job before he set more than 400 lines of type per day?

15. (TI-82) Use your calculator to draw the graph of $f(x) = 4 - 3x - x^2$.

a) Set up a coordinate system on a sheet of graph paper and copy the image in the viewing window of your calculator onto your coordinate system. Indicate your WINDOW settings on the axes of the coordinate system.

b) The graph of f is called a *parabola*. The turning point of the parabola, in this case the point with the maximum y-value on the graph, is called the *vertex* of the parabola. Use the TRACE capability of your calculator to estimate the coordinates of this vertex and place this estimate on your graph near the vertex.

16. (TI-82) Use your calculator to draw the graph of $f(x) = x^2 - 5x - 6$. Change the following WINDOW parameters: Ymin = −15, Yscl = 5.

    a) Set up a coordinate system on a sheet of graph paper and copy the image in the viewing window of your calculator onto your coordinate system. Indicate your WINDOW settings on the axes of the coordinate system.

    b) The graph of f is called a parabola. The turning point of the parabola, in this case the point with the minimum y-value on your graph, is called the vertex of the parabola. Use the TRACE capability of your calculator to estimate the coordinates of this vertex and place this estimate on your graph near the vertex.

17. (TI-82) Use your calculator to draw the graph of $f(x) = 0.5(x + 7)$.

    a) Set up a coordinate system on a sheet of graph paper and copy the image in the viewing window of your calculator onto your coordinate system. Indicate your WINDOW settings on the axes of the coordinate system.

    b) The point where the graph of f intercepts the y-axis is called the y-intercept of the graph of f. Use the TRACE capability of your calculator to estimate the coordinates of the y-intercept and place this estimate on your graph near the y-intercept.

18. (TI-82) Use your calculator to draw the graph of $f(x) = x^2 - 4x - 4$.

    a) Set up a coordinate system on a sheet of graph paper and copy the image in the viewing window of your calculator onto your coordinate system. Indicate your WINDOW settings on the axes of the coordinate system.

    b) The points where the graph of f intercepts the x-axis are called the x-intercepts of the graph of f. Use the TRACE capability of your calculator to estimate the coordinates of the x-intercepts of the graph of f and place these estimates on your graph near the x-intercepts.

19. (TI-82) Use your calculator to draw the graph of $f(x) = x^3 - 50x$. Change the following WINDOW parameters: Ymin = −300, Ymax = 300, Yscl = 50.

    a) Set up a coordinate system on a sheet of graph paper and copy the image in the viewing window of your calculator onto your coordinate system. Indicate your WINDOW settings on the axes of the coordinate system.

    b) If you completed part (a) properly, you should have the following image in your viewing window.

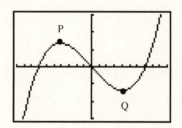

Mathematicians call point P a local maximum and point Q a local minimum. Use the TRACE capability of your calculator to estimate the coordinates of points P and Q and place these estimates on your graph near the points P and Q.

20. Consider the function represented by the equation $f(x) = x^2 + 4x - 6$. Enter this equation into the Y= menu in the following manner: $Y_1 = X^2 + 4X - 6$.

    a) Press the ZOOM button and select 6:ZStandard. Press the ZOOM button again and select 8:ZInteger. Press the ENTER ENTER key. Press the TRACE TRACE button and use the arrow keys to trace along the curve, completing the following table of points as you trace.

x	−7	−6	−5	−4	−3	−2	−1	0	1	2	3
$f(x)$											
Points to plot											

    b) Draw a coordinate system on a sheet of graph paper and plot each of the points from the table on your coordinate system. Plot the rest of the points that satisfy the equation of the function.

21. Consider the function represented by the equation $f(x) = 9 + 8x - x^2$. Enter this equation into the Y= menu in the following manner: $Y_1 = 9 + 8X - X^2$.

    a) Press the ZOOM button and select 6:ZStandard. Press the ZOOM button again and select 8:ZInteger. Press the ENTER key. Press the TRACE button and use the arrow keys to trace along the curve, completing the following table of points as you trace.

x	−2	−1	0	1	2	3	4	5	6	7	8	9	10
$f(x)$													
Points to plot													

    b) Draw a coordinate system on a sheet of graph paper and plot each of the points from the table on your coordinate system. Plot the rest of the points that satisfy the equation of the function.

22. The graph of the equation $f(x) = 5 - x$ is shown below. Point P is on the graph of f.

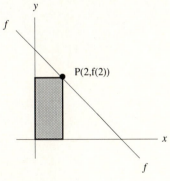

    What is the area of the shaded rectangle?

23. The graph of the equation $g(x) = 9 - x^2$ is shown below. Point P is on the graph of g.

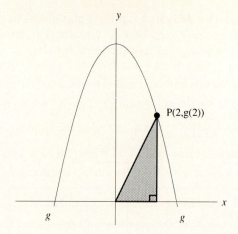

What is the area of the shaded triangle?

24. Consider the graph of $f(x) = 10 - 2x$ shown below. P is a point on the graph of f.

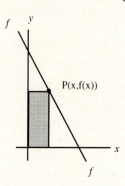

Express the area of the shaded rectangle as a function of x.

25. Consider the graph of $g(x) = 9 - x^2$ shown below. Point P is on the graph of g.

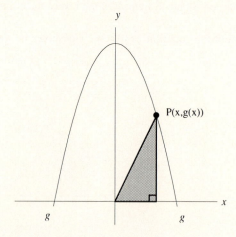

Express the area of the shaded triangle as a function of x.

1.6 Comparing Functions

In this section we'll be comparing functions, asking questions such as the following: For which values of x is $f(x) < g(x)$? For which values of x is $f(x) = g(x)$? For which values of x is $f(x) > g(x)$? Let's begin with an example.

Example 1   Consider the functions f and g, defined by the following equations: $f(x) = x^2 + 3$ and $g(x) = 1 - 4x$. Compare $f(-1)$ and $g(-1)$.

Solution   First, we compute $f(-1)$ and $g(-1)$.

$$f(x) = x^2 + 3 \qquad\qquad g(-1) = 1 - 4x$$
$$f(-1) = (-1)^2 + 3 \qquad g(-1) = 1 - 4(-1)$$
$$f(-1) = 4 \qquad\qquad g(-1) = 5$$

Thus, $f(-1) = 4$ and $g(-1) = 5$. Because 4 is less than 5, $f(-1) < g(-1)$. Note that $x = -1$ is a solution of $f(x) < g(x)$.

Using a Table to Compare Functions

Compare the functions represented by the equations $f(x) = 2x + 3$ and $g(x) = 7 - 2x$ for several values of x. Select several values of x and compute the corresponding function values for both f and g. Arrange your work in the table, as shown:

x	$f(x)$	$g(x)$
−3	−3	13
−2	−1	11
−1	1	9
0	3	7
1	5	5
2	7	3
3	9	1
4	11	−1
5	13	−3

For which values of x are $f(x)$ and $g(x)$ equal? By examining the table, you can see that $f(x)$ and $g(x)$ are equal when $x = 1$. In fact, 1 seems to be a critical value. When x is less than 1, the function values of the function f are less than the function values of the function g. When x is greater than 1, the function values of f are greater than the function values of g.

Example 2   Consider the graphs of the functions *f* and *g*. Compare $f(1)$ and $g(1)$.

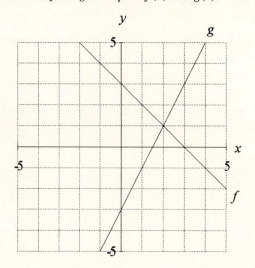

Solution   First, read the graph and find $f(1)$ and $g(1)$. Locate 1 on the *x*-axis and draw vertical lines to the graphs of *f* and *g*, followed by horizontal lines to the *y*-axis, as shown:

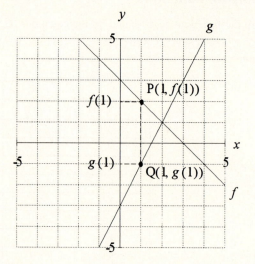

The *y*-value of point P is $f(1)$, which is 2. Therefore, $f(1) = 2$. The *y*-value of point Q is $g(1)$, which is −1. Therefore, $g(1) = -1$. Thus, $f(1) = 2$ and $g(1) = -1$. Because 2 is greater than −1, $f(1) > g(1)$.

It is not important to know the exact values of $f(1)$ and $g(1)$ in this problem. Point P is *above* point Q so the *y*-value of point P must be greater than the *y*-value of point Q. Therefore, $f(1) > g(1)$.

> *Saying $f(1) > g(1)$ is another way of saying that the graph of f is above the graph of g when x = 1.*

Example 3   Use the graphs of *f* and *g* to compare $f(2)$ and $g(2)$.

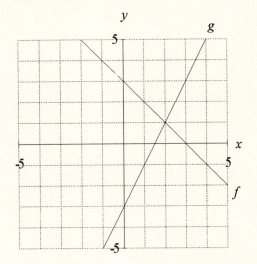

Solution   Locate 2 on the *x*-axis and draw vertical lines to the graphs of *f* and *g*, followed by horizontal lines to the *y*-axis, as shown:

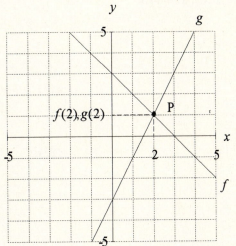

The *y*-value of point P is $f(2)$, which is 1. Therefore, $f(2) = 1$. The *y*-value of point P is also $g(2)$; therefore, $g(2) = 1$ and $f(2) = g(2)$.

Once again, it is not important to know the exact values of $f(2)$ and $g(2)$. Since the functions *f* and *g* intersect at point P, the function values must be equal at point P.

Saying $f(2) = g(2)$ is another way of saying that the graphs of f and g intersect when x = 2.

Example 4 Use the graphs of f and g to compare $f(3)$ and $g(3)$.

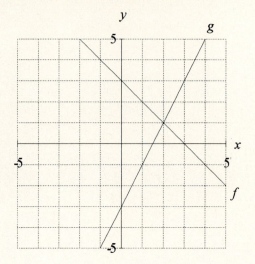

Solution First locate 3 on the x-axis. Then draw vertical lines to the graphs of f and g, followed by horizontal lines to the y-axis, as shown:

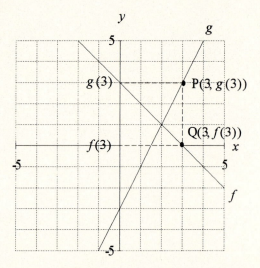

The y-value of point P is $g(3)$, which is 3. Therefore, $g(3) = 3$. The y-value of point Q is $f(3)$, which is 0. Therefore, $f(3) = 0$. Thus, $f(3) = 0$ and $g(3) = 3$, which means that $f(3) < g(3)$.

Once again, it is not important to know the exact values of $f(3)$ and $g(3)$. Since point Q is *below* point P, the y-value of point Q must be less than the y-value of point P.

Saying $f(3) < g(3)$ is another way of saying that the graph of f is below the graph of g when $x = 3$.

A Summary of Comparing Functions

> *To find the solution of the inequality f(x) > g(x), read the graph of f and g and note where the graph of f is above the graph of g.*

> *To find the solution of the equation f(x) = g(x), read the graph of f and g and note where the graphs of f and g intersect.*

> *To find the solution of the inequality f(x) < g(x), read the graph of f and g and note where the graph of f is below the graph of g.*

Example 5   Use the graphs of f and g to solve the equation $f(x) = g(x)$ for x.

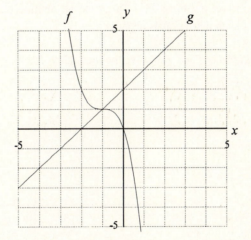

Solution   To find the value of x such that $f(x) = g(x)$, examine the graphs to find where they intersect. It appears that the x-value of the point of intersection of the two graphs is -1. Because $f(-1) = g(-1)$, the solution of $f(x) = g(x)$ is $x = -1$. Now, there is a nice way of summarizing your results. Draw a number line below the graph, drop a dotted line from the point of intersection of the graphs of f and g to the number line, and shade and label the solution on the number line, as follows:

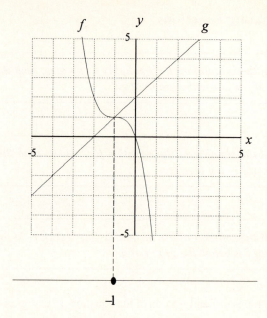

The solution set is $\{x:x=-1\}$. This style of notation is called *set-builder notation* and is read "The set of all x such that x equals negative one."

Example 6 Use the graphs of f and g to solve the inequality $f(x)<g(x)$ for x.

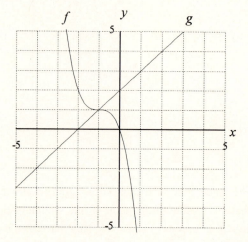

Solution To find where $f(x)<g(x)$, examine your sketch and interpret where the graph of f is below the graph of g. Whenever you are to the right of the point of intersection of the two graphs, the graph of f is below the graph of g. Use a number line to summarize your answer. Draw a number line under the graph, then drop a vertical, dotted line from the point of intersection to the number line. Because the graph of f is below the graph of g for all values of x to the right of the point of intersection, shade all points on the number line to the right of -1, as follows:

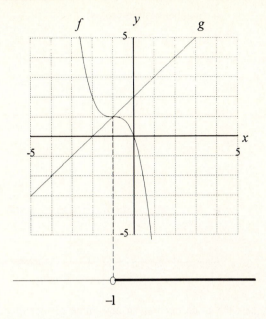

The solution set is $\{x : x > -1\}$. This notation is called set-builder notation and is read "The set of all x such that x is greater than negative one."

Example 7   Sally owns a huge apartment complex, with a total of 50 units available as rentals. The following graph shows the number of units rented, N, as a function of the price per rental unit, p. For what values of p will the number of rental units fall below 30?

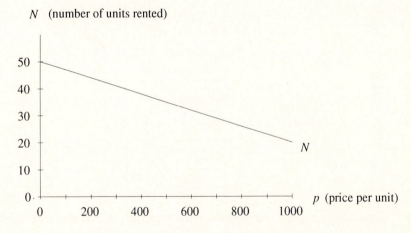

Solution   Begin by drawing the horizontal line $N = 30$. You need to find where $N < 30$ and, therefore, where the graph of N is below the horizontal line $N = 30$. It appears that the graph of N is below the horizontal line $N = 30$ to the right of the point of intersection.

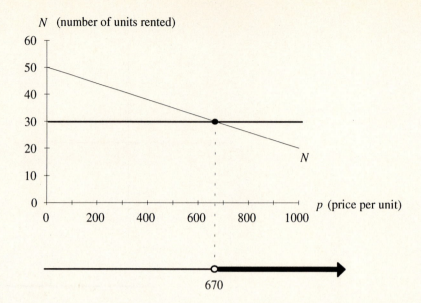

If Sally sets the price per unit at any value greater than $670, she will rent fewer than 30 of her apartment units. The solution set is $\{p: p > 670\}$. This notation is read "The set of all p such that p is greater than 670."

Example 8   Use the graphs of f and g to solve the inequality $f(x) > g(x)$ for x.

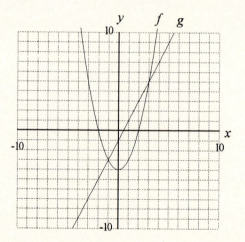

Solution   Examine your graph and interpret where the graph of f is above the graph of g. If x is to the left of -1 or if x is to the right of 3, the graph of f is above the graph of g.

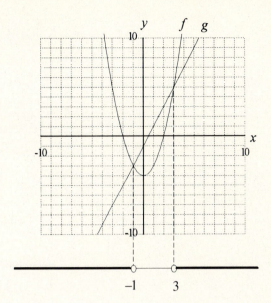

Note the empty circles on the number line. This is a way of emphasizing that $x = -1$ and $x = 3$ are *not* included in the solution. In set-builder notation, the solution set is $\{x: x < -1 \text{ or } x > 3\}$. This is read "The set of all x such that x is less than -1 or x is greater than 3." In later sections we will have a lot to say about the logic of the word *or* and its appropriateness in this situation. Suffice it to say, for the moment, that x cannot be less than -1 *and* greater than 3 at the same time. Therefore, the word *or* is used in this situation.

Example 9   Use the graphs of f and g to solve the inequality $f(x) \le g(x)$ for x.

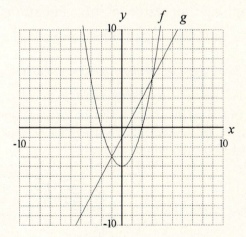

Solution   Note the $\le$ symbol. Find all values of x such that $f(x)$ is less than or equal to $g(x)$. Find where the graph of f is below the graph of g and also include where the graph of f intersects the graph of g. Between -1 and 3 on the number line, the graph of f is below the graph of g. In addition, the graphs of f and g intersect when $x = -1$ or $x = 3$.

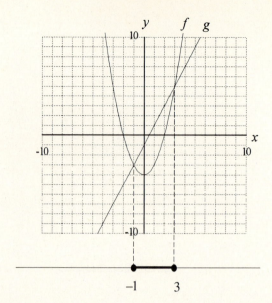

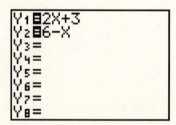

Note how $x = -1$ and $x = 3$ have been shaded on the number line. This is a way of emphasizing that $x = -1$ and $x = 3$ are *included* in the solution. In set-builder notation, the solution set can be written as $\{x : x \geq -1 \text{ and } x \leq 3\}$. This is read "The set of all x such that x is greater than or equal to -1 and x is less than or equal to 3."

A Note on Set-builder Notation.   In Section 2.7 the logic of *and* and *or* will be rigorously discussed. Don't worry at present whether you completely understand the set-builder notation. If you know how to shade a solution on a number line below a graph, that will suffice for now. Think of this as your first, not your last, attempt at understanding set-builder notation.

Using the TI-82 Graphing Calculator to Solve Equations

Let's now see how to use the TI-82 graphing calculator to solve equations.

Example 10   Use the TI-82 graphing calculator to solve the following equation for x: $2x + 3 = 6 - x$.

Solution   First, note that you are being asked to solve an equation of the form $f(x) = g(x)$, where $f(x) = 2x + 3$ and $g(x) = 6 - x$. Use the following strategy to solve this equation: Graph each side of the equation, then estimate the x-value of the point of intersection. Begin by loading your functions in Y1 and Y2 in the following manner:

```
Y1∎2X+3
Y2∎6-X
Y3=
Y4=
Y5=
Y6=
Y7=
Y8=
```

Push the ZOOM button and select 6:ZStandard, capturing the following image in your viewing window:

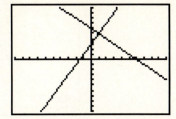

At this point, you could push the TRACE button and move the trace cursor over the point of intersection to obtain a rough estimate of the *x*-value of the point of intersection. However, the TI-82 has a utility that finds points of intersection with great accuracy. This utility is located in the CALC menu which is located on the calculator case above the TRACE button. You must use the 2nd key to access this menu. Press the 2nd key, followed by the TRACE button. The following menu opens on your screen:

Select 5:Intersect. Your screen will look as follows:

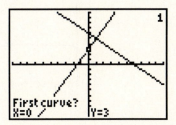

The calculator has placed the cursor on the graph of Y1. What does that message in the lower left corner of the screen mean? Since you can draw up to eight curves on the calculator, the calculator needs to know which two curves your point of intersection lies on. The calculator is asking if the presently selected curve is one you wish to use. Press the ENTER key to signal the calculator that you wish to use the selected curve. The calculator responds with the following screen:

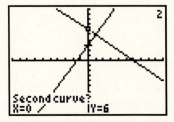

Note that the trace cursor is now on the graph of Y2. The calculator wishes to know if this is the second curve that you wish to use. Press the ENTER key to signal that you wish to use this selected curve. The calculator responds with the following screen:

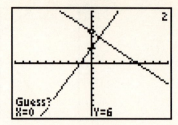

The calculator is waiting for you to make a reasonable guess at the point of intersection. Use the arrow keys to move the trace cursor reasonably close to the point of intersection, as shown in the following screen:

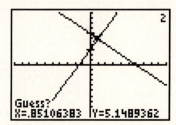

Now that you have made your guess, press the $\boxed{\text{ENTER}}$ key. The calculator will work with the curves that you selected, begin with your guess, and use an algorithm to find the coordinates of the point of intersection with a great deal of accuracy.

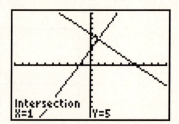

The x-value of the point of intersection is 1. Therefore, the solution of $2x + 3 = 6 - x$ is $x = 1$.

Check. Substitute $x = 1$ in the original equation:

$$2x + 3 = 6 - x$$
$$2(1) + 3 = 6 - 1$$
$$5 = 5$$

The solution checks.

Using the TI-82 to Solve Inequalities

Inequalities can also be solved with the TI-82 graphing calculator. Suppose, for example, you are asked to solve the inequality $x + 2 > 3 - x$. Graph the equation $f(x) = x + 2$, graph the equation $g(x) = 3 - x$, then use the intersection utility in the CALC menu to find the x-value of the point of intersection. You would then interpret where the graph of f is *above* the graph of g, and shade your solution on a number line in the appropriate manner.

Exercises for Section 1.6

1. Consider the functions $f(x) = 2x + 3$ and $g(x) = 12 - x$. Complete the following table.

x	$f(x)$	$g(x)$
−1		
0		
1		
2		
3		
4		
5		
6		
7		

   a) For which value of x is $f(x) = g(x)$?
   b) For which values of x is $f(x) > g(x)$?
   c) For which values of x is $f(x) < g(x)$?

2. Consider the functions $f(x) = 4 - x$ and $g(x) = 2x + 7$. Complete the following table.

x	$f(x)$	$g(x)$
−5		
−4		
−3		
−2		
−1		
0		
1		
2		
3		

   a) For which value of x is $f(x) = g(x)$?
   b) For which values of x is $f(x) > g(x)$?
   c) For which values of x is $f(x) < g(x)$?

3. Consider the functions $f(x) = x^2 - x - 2$ and $g(x) = x + 1$. Complete the following table.

x	$f(x)$	$g(x)$
−4		
−3		
−2		
−1		
0		
1		
2		
3		
4		

a) For which value of x is $f(x) = g(x)$?

b) For which values of x is $f(x) > g(x)$?

c) For which values of x is $f(x) < g(x)$?

Hint: Enter each function in the Y= menu, press ZOOM , and select 6:ZStandard. Press ZOOM and select 8:ZInteger. Press TRACE and use the trace cursor to help fill in the table. Try pushing the up-arrow on your keyboard to trace along your second curve.

4. Consider the functions $f(x) = 16 - x^2$ and $g(x) = -2x + 13$. Complete the following table.

x	$f(x)$	$g(x)$
−5		
−4		
−3		
−2		
−1		
0		
1		
2		
3		
4		
5		

a) For which value of x is $f(x) = g(x)$?

b) For which values of x is $f(x) > g(x)$?

c) For which values of x is $f(x) < g(x)$?

Hint: Enter each function in the Y= menu, press ZOOM , and select 6:ZStandard. Press ZOOM and select 8:ZInteger. Press TRACE and use the trace cursor to help fill in the table. Try pushing the up-arrow on your keyboard to trace along your second curve.

5. Consider the graphs of the functions f and g.

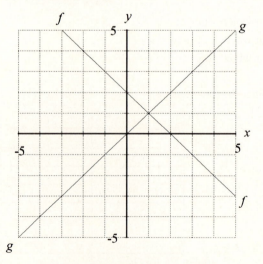

a) Use the graphs of f and g to help complete the following table.

x	$f(x)$	$g(x)$
−3		
−2		
−1		
0		
1		
2		
3		

b) For which value of x is $f(x) = g(x)$?

c) For which values of x is $f(x) > g(x)$?

d) For which values of x is $f(x) < g(x)$?

6. Consider the graphs of the functions g and h.

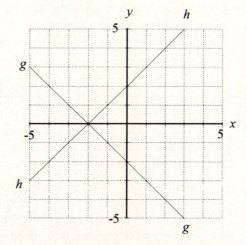

a) Use the graphs of g and h to help complete the following table.

x	$g(x)$	$h(x)$
−5		
−4		
−3		
−2		
−1		
0		
1		
2		
3		

b) For which value of x is $g(x) = h(x)$?

c) For which values of x is $g(x) > h(x)$?

d) For which values of x is $g(x) < h(x)$?

7. Consider the graphs of the functions f and g.

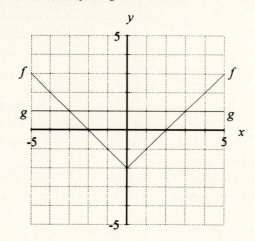

a) Use the graphs of f and g to help complete the following table.

	$f(x)$	$g(x)$
–5		
–4		
–3		
–2		
–1		
0		
1		
2		
3		
4		
5		

b) For which value of x is $f(x) = g(x)$?

c) For which values of x is $f(x) > g(x)$?

d) For which values of x is $f(x) < g(x)$?

e) For which values of x is $f(x) \le g(x)$?

8. Consider the graphs of the functions f and h.

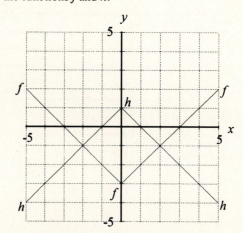

a) Use the graphs of f and h to help complete the following table.

x	$f(x)$	$h(x)$
−5		
−4		
−3		
−2		
−1		
0		
1		
2		
3		
4		
5		

b) For which value of x is $f(x) = h(x)$?

c) For which values of x is $f(x) > h(x)$?

d) For which values of x is $f(x) < h(x)$?

e) For which values of x is $f(x) \leq h(x)$?

9. Consider the following graph of the functions f and g.

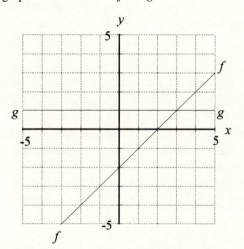

a) Set up a coordinate system on graph paper and copy the graphs of f and g onto your coordinate system. Draw a number line below your graph of f and g and drop a dotted line from the point of intersection of the graphs of f and g to the number line. Shade and label the solution of $f(x) = g(x)$ on your number line. Use set-builder notation to describe your solution set.

b) Draw a second number line below the first and extend the dotted line from the point of intersection to the second number line. Shade and label the solution of $f(x) < g(x)$ on the second number line. Use set-builder notation to describe your solution set.

c) Draw a third number line below the first and second number lines and extend the dotted line from the point of intersection to the third number line. Shade and label the solution of $f(x) > g(x)$ on the third number line. Use set-builder notation to describe your solution set.

10. Consider the graphs of the functions *f* and *g*.

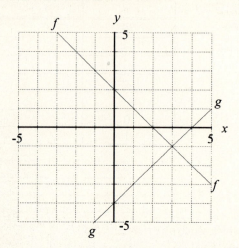

a) Set up a coordinate system on graph paper and copy the graphs of *f* and *g* onto your coordinate system. Draw a number line below your graph of *f* and *g* and drop a dotted line from the point of intersection of the graphs of *f* and *g* to the number line. Shade and label the solution of $f(x) = g(x)$ on your number line. Use set-builder notation to describe your solution set.

b) Draw a second number line below the first and extend the dotted line from the point of intersection to the second number line. Shade and label the solution of $f(x) < g(x)$ on the second number line. Use set-builder notation to describe your solution set.

c) Draw a third number line below the first and second number lines and extend the dotted line from the point of intersection to the third number line. Shade and label the solution of $f(x) > g(x)$ on the third number line. Use set-builder notation to describe your solution set.

11. The *demand* is the number of units the public is willing to purchase. The demand, *D*, for a certain item is often a function of its price. As the price goes up, the public is less willing to purchase the item. The *supply* is the number of units that the manufacturers produce. The supply, *S*, is also a function of the unit price. As the price of the unit goes up, the manufacturers are willing to produce more of the item. The graphs of the demand and supply functions are shown below:

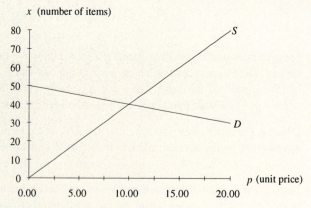

a) Copy this graph onto a sheet of graph paper.

b) The point where these graphs intersect is called the *equilibrium point*. Draw a number line below your graph and drop a dotted line from the point of intersection of the supply and demand functions to the number line. Shade and label the *p*-value of the equilibrium point on the number line. Explain, in your own words, the meaning of your answer.

c) For what unit prices will the demand outweigh the supply? Draw a second number line below your first and shade the solution to this question. Explain, in your own words, how this situation might come about.

12. Consider the following graphs of *f* and *g*.

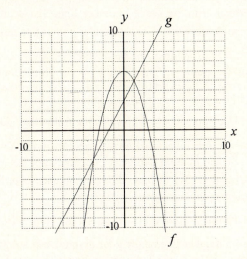

a) Set up a coordinate system on graph paper and copy the graphs of *f* and *g* onto your coordinate system. Draw a number line below your graph of *f* and *g* and drop dotted lines from the points of intersection of the graphs of *f* and *g* to the number line. Shade and label the solutions of $f(x) = g(x)$ on the number line. Use set-builder notation to describe your solution set.

b) Draw a second number line below the first and extend the dotted lines from the points of intersection to the second number line. Shade and label the solution of $f(x) < g(x)$ on the second number line. Use set-builder notation to describe your solution set.

c) Draw a third number line below the first and second number lines and extend the dotted lines from the points of intersection to the third number line. Shade and label the solution of $f(x) > g(x)$ on the third number line. Use set-builder notation to describe your solution set.

13. Consider the following graphs of *f* and *g*.

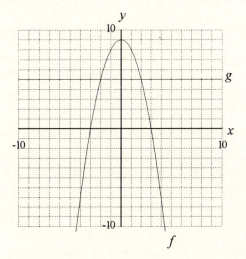

a) Set up a coordinate system on graph paper and copy the graphs of f and g onto your coordinate system. Draw a number line below your graph of f and g and drop dotted lines from the points of intersection of the graphs of f and g to the number line. Shade and label the solutions of $f(x) = g(x)$ on the number line. Use set-builder notation to describe your solution set.

b) Draw a second number line below the first and extend the dotted lines from the points of intersection to the second number line. Shade and label the solution of $f(x) < g(x)$ on the second number line. Use set-builder notation to describe your solution set.

c) Draw a third number line below the first and second number lines and extend the dotted lines from the points of intersection to the third number line. Shade and label the solution of $f(x) > g(x)$ on the third number line. Use set-builder notation to describe your solution set.

14. (TI-82) Use your calculator to solve the following equation for x: $2x + 3 = 7$.

a) Enter the functions f and g in Y1 and Y2 in the Y= menu, like this: Y1 $= 2X + 3$, Y2 $= 7$. Push the $\boxed{\text{ZOOM}}$ button and select 6:ZStandard. Set up a coordinate system on a sheet of graph paper and clearly indicate the scale on each coordinate axis. Copy the image in your viewing window onto your coordinate system.

b) Use the intersection utility on your calculator to find the point of intersection.

c) Draw a number line below your graph. Drop a dotted line from the point of intersection of the graphs of f and g to the number line. Shade and label your solution of $2x + 3 = 7$ on the number line.

d) Check your solution in the equation $2x + 3 = 7$.

15. (TI-82) Use your calculator to help solve the following inequality for x: $3 - x > 1$.

a) Enter the functions $f(x) = 3 - x$ and $g(x) = 1$ into Y1 and Y2 in the Y= menu, like this: Y1 $= 3 - X$, Y2 $= 1$. Push the $\boxed{\text{ZOOM}}$ button and select 6:ZStandard. Set up a coordinate system on a sheet of graph paper and clearly indicate the scale on each coordinate axis. Copy the image in your viewing window onto your coordinate system.

b) Use the intersection utility on your calculator to find the point of intersection.

c) Draw a number line below your graph. Drop a dotted line from the point of intersection of the graphs of f and g to the number line. Shade and label your solution of $3 - x > 1$ on the number line. Use set-builder notation to describe your solution set.

16. (TI-82) Use your calculator to help solve the following inequality for x: $4 - x < 2x - 2$.

a) Enter the functions $f(x) = 4 - x$ and $g(x) = 2x - 2$ into Y1 and Y2 in the Y= menu, like this: Y1 $= 4 - X$, Y2 $= 2X - 2$. Push the $\boxed{\text{ZOOM}}$ button and select 6:ZStandard. Set up a coordinate system on a sheet of graph paper and clearly indicate the scale on each coordinate axis. Copy the image in your viewing window onto your coordinate system.

b) Use the intersection utility on your calculator to find the point of intersection.

c) Draw a number line below your graph. Drop a dotted line from the point of intersection of the graphs of f and g to the number line. Shade and label your solution of $4 - x < 2x - 2$ on the number line. Use set-builder notation to describe your solution set.

17. Mr. Luigi is the owner of a shoe manufacturing business. His *costs* are defined to be the amount of money he must spend to manufacture shoes. His costs are a function of the number of pairs of shoes he makes. The more shoes he makes, the higher his costs will be. His *revenues* are defined to be the amount of money he makes from selling shoes. His revenues are a function of the

number of pairs of shoes he makes and sells. The more shoes he makes and sells, the higher his revenues will be. Shown below are the graphs of Mr. Luigi's costs function, C, and his revenues function, R.

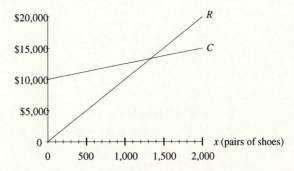

a) The point where the graphs intersect is called the *break-even* point. Copy the graphs of Mr. Luigi's costs and revenues functions onto your homework paper. Draw a number line below your graph. Drop a dotted line from the point of intersection of the cost graph and revenue graph to the number line and shade and label the x-value of the break-even point on the number line. Explain, in your own words, the meaning of your answer.

b) When the revenues are more than the costs, Mr. Luigi is making a *profit*. For what values of x does Mr. Luigi make a profit ? Shade the solution of this question on a second number line below the graph on your homework paper.

1.7 Comparing Functions with Zero

In this section we'll compare functions with zero. For example, we may try to find where a function is greater than *or* equal to zero. Or, perhaps we'll try to find where a function is less than zero. We'll also discuss the relationship between the zeros of a function and the x-intercepts of the graph of the function.

Example 1   Use the following graph of f to find all values of x such that $f(x) = 0$.

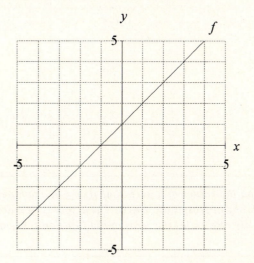

Solution   Remember, y and $f(x)$ are interchangeable. If you wish to find a solution of $f(x) = 0$, then you must find a point on the graph of f whose y-value is zero.

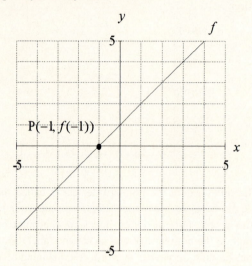

The y-value of point P equals zero. Therefore $f(-1) = 0$, and $x = -1$ is a solution of $f(x) = 0$. Place your solution on a number line in a manner similar to that used in the previous section.

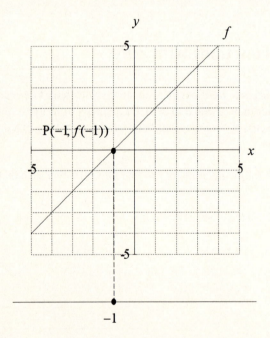

To find the values of x such that f(x) = 0, examine the graph of f and determine where the graph of f intercepts the x-axis. These values of x are called the x-intercepts of the graph of f. They are also called the zeros of the function f, because they make the function value equal to zero.

Example 2   Use the following graph to help find the zeros of the function f.

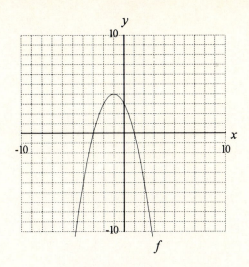

Solution   Find all values of x such that $f(x) = 0$. The function value $f(x)$ is equal to zero at the places where its graph crosses the x-axis.

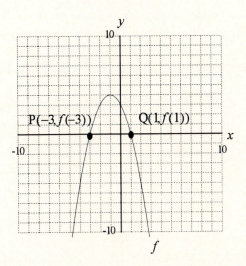

Points P and Q are called the *x-intercepts* of the graph of f. The function value (y-value) of point P is zero, so $f(-3) = 0$. Therefore, $x = -3$ is a solution of $f(x) = 0$ and is called a *zero* of the function f. The function value (y-value) of point Q is zero, so $f(1) = 0$. Therefore, $x = 1$ is a solution of $f(x) = 0$ and is called a *zero* of the function f. Place your solution on a number line in a manner similar to that used in the previous section.

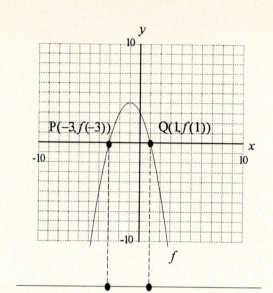

In set-builder notation, the solution set of $f(x) = 0$ is $\{x : x = -3 \text{ or } x = 1\}$. This notation is read as follows: "The set of all x such that x equals -3 or x equals 1." Note the use of the word *or*.

Example 3   Use the graph of f to find all values of x such that $f(x) > 0$.

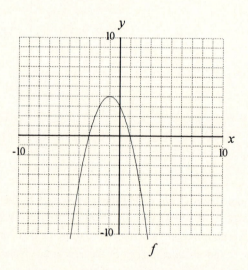

Solution   A point on the graph of f will have a function value (y-value) greater than zero if that point is above the x-axis. First, determine the zeros of the function f. These can be obtained by noting where the graph of f intercepts the x-axis, as in the last example. Finally, the graph of f is above the x-axis for x-values between -3 and 1.

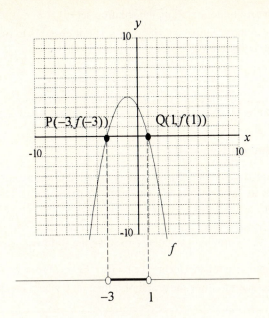

In set-builder notation, the solution set of $f(x) > 0$ is $\{x : x > -3 \text{ and } x < 1\}$. This is read, "The set of all x such that x is greater than -3 and x is less than 1."

> ***To find solutions of $f(x) > 0$, examine where the graph of the function f is above the x-axis.***

Example 4   In the last example, find all values of x such that $f(x) < 0$.

Solution   A point on the graph of f will have a function value (y-value) less than zero if that point is below the x-axis. It would appear that the graph of f is below the x-axis for x-values to the left of -3 or to the right of 1.

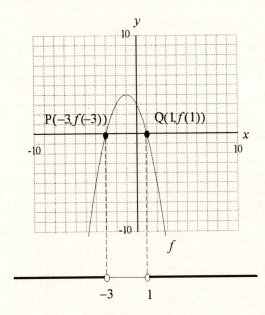

In set-builder notation, the solution set of $f(x) < 0$ is $\{x : x < -3 \text{ or } x > 1\}$. Note the use of the word *or* in the solution set. Because it is not possible to be to the left of -3 *and* the right of 1 at the same time, the word *or* is used.

> *To find solutions of $f(x) < 0$, examine where the graph of the function f is below the x-axis.*

Using the Root Finder on the TI-82

There are some subtle differences between the terms, but essentially the words root, zero, and x-intercept are equivalent. We can use the root-finding utility of the TI-82 to find the zeros of a function.

Example 5   Use the TI-82 graphing calculator to solve the following equation for x: $x^2 - 4x - 6 = 0$

Solution   This equation is of the form $f(x) = 0$, where $f(x) = x^2 - 4x - 6$. Try the following strategy: Use your calculator to draw the graph of f, then note where the graph crosses the x-axis.

Load the equation $f(x) = x^2 - 4x - 6$ into the Y= menu.

Push the ZOOM button and select 6:ZStandard to capture the following image.

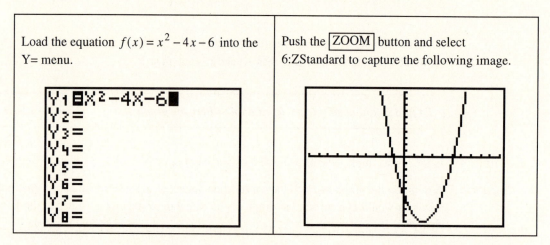

You could push the TRACE button and use the arrow keys to place the trace cursor over each x-intercept to obtain rough approximations for the zeros of the function f. However, the TI-82 has a utility that will find the zeros of a function with a great deal of accuracy. You can find this utility in the CALC menu which is located on the calculator case above the TRACE button.

Press 2nd CALC to open the following menu.

Select 2:root to start the root finding utility.

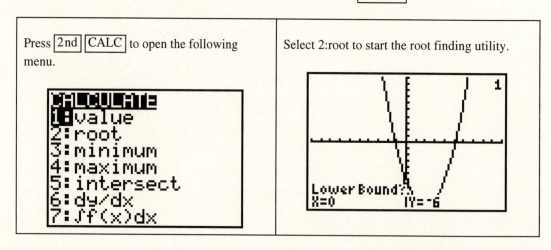

Since a function can have more than one zero, the calculator is asking you for an interval within which you wish it to search for the zero. It wants the lower bound of the search interval.

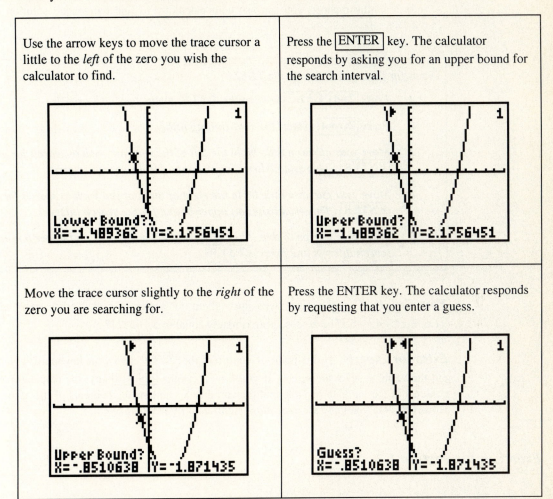

| Use the arrow keys to move the trace cursor a little to the *left* of the zero you wish the calculator to find. | Press the ENTER key. The calculator responds by asking you for an upper bound for the search interval. |
| Move the trace cursor slightly to the *right* of the zero you are searching for. | Press the ENTER key. The calculator responds by requesting that you enter a guess. |

Note the reminder marks at the top of the screen. These are the *x*-values you entered for the lower and upper bound of the search interval. When entering guess, you must select an *x*-value in this search interval. Since our cursor is already in the search interval, you may accept the present *x*-value as your guess by pressing the ENTER key. The calculator responds by finding the zero, as follows:

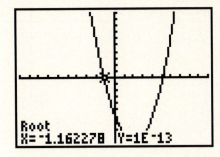

One of the solutions of $x^2 - 4x - 6 = 0$ is approximately $x = -1.162278$. Note the *y*-value. What does $Y = 1E-13$ mean? This is scientific notation and it means $Y = 1.0 \times 10^{-13}$. Because the exponent is -13, you must move the decimal point 13 places to the left. Therefore, Y =

0.0000000000001, which is very close to zero. This is as it should be because you were searching for a *zero* of the function.

In a similar manner, you can use your calculator to find the second solution of the equation $x^2 - 4x - 6 = 0$.

Steps for finding a zero with the TI-82.

1.   *Press* 2nd CALC *to open the* **CALC** *menu.*

2.   *Press* 2:root *to start the root-finding utility.*

3.   *Move your cursor a little bit to the left of the root you wish to search for and press* ENTER *. This establishes a lower bound for the search interval.*

4.   *Move your cursor a little bit to the right of the root you wish to search for and press* ENTER *. This establishes an upper bound for the search interval.*

5.   *Move your cursor so that it lies between the lower bound and upper bound of the search interval and press* ENTER *.*

You Try It. Use the five steps just given to search for the other zero of the function $f(x) = x^2 - 4x - 6$. The zero is approximately equal to 5.1622777.

Extra for Experts. At this point, if you press the X,T,θ on your keyboard, your calculator will put the variable X on the screen. If you now press the ENTER key, the calculator will display the contents of the variable X. Note that the variable X contains the contents of the zero that was found. However, note that there are even *more* significant figures displayed.

Exercises for Section 1.7

1. What is the *x*-intercept of the graph of the function *f*?

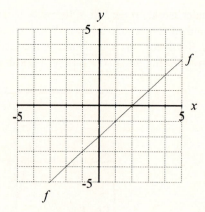

2. What is the x-intercept of the graph of the function g?

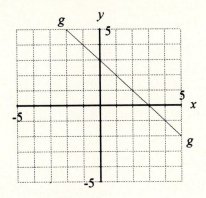

3. Consider the following graph of the function f.

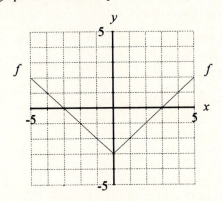

   a) Set up a coordinate system on a sheet of graph paper and copy the graph of the function f onto your coordinate system.

   b) Draw a number line below your graph. Shade and label the solutions of $f(x) = 0$ on your number line as done in Example 2 in this section. Use set-builder notation to describe your solution set.

4. Consider the following graph of the function f.

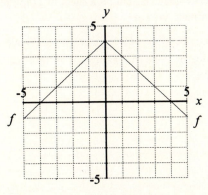

   a) Set up a coordinate system on a sheet of graph paper and copy the graph of the function f onto your coordinate system.

   b) Draw a number line below your graph. Shade and label the solutions of $f(x) = 0$ on your number line as done in Example 2 in this section. Use set-builder notation to describe your solution set.

5. Consider the following graph of the function *f*.

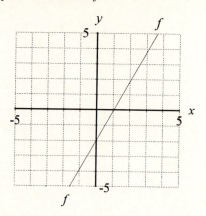

a) Set up a coordinate system on a sheet of graph paper and copy the graph of the function *f* onto your coordinate system.

b) Draw a number line below your graph. Shade and label the solutions of $f(x) > 0$ on your number line as done in Example 3 in this section. Use set-builder notation to describe your solution set.

6. Consider the following graph of the function *g*.

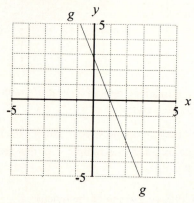

a) Set up a coordinate system on a sheet of graph paper and copy the graph of the function *g* onto your coordinate system.

b) Draw a number line below your graph. Shade and label the solutions of $g(x) < 0$ on your number line as done in Example 3 in this section. Use set-builder notation to describe your solution set.

7. Consider the following graph of the function h.

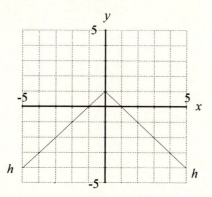

a) Set up a coordinate system on a sheet of graph paper and copy the graph of the function h onto your coordinate system.

b) Draw a number line below your graph. Shade and label the solutions of $h(x) > 0$ on your number line as done in Example 3 in this section. Use set-builder notation to describe your solution set.

8. Consider the following graph of the function g.

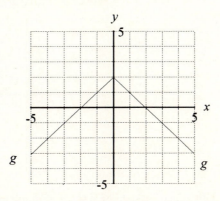

a) Set up a coordinate system on a sheet of graph paper and copy the graph of the function g onto your coordinate system.

b) Draw a number line below your graph. Shade and label the solutions of $g(x) < 0$ on your number line as done in Example 3 in this section. Use set-builder notation to describe your solution set.

9. Consider the following graph of the function h .

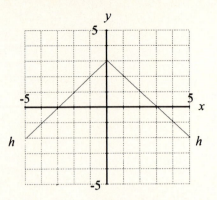

a) Set up a coordinate system on a sheet of graph paper and copy the graph of the function h onto your coordinate system.

b) Draw a number line below your graph. Shade and label the solutions of $h(x) > 0$ on your number line as done in Example 3 in this section. Use set-builder notation to describe your solution set.

10. Consider the following graph of the function f .

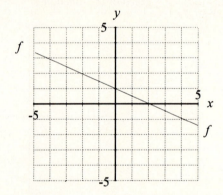

a) Set up a coordinate system on a sheet of graph paper and copy the graph of the function f onto your coordinate system.

b) Draw a number line below your graph. Shade and label the solutions of $f(x) \leq 0$ on your number line as done in Example 3 in this section. Notice the $\leq$ symbol. You are being asked to find where the graph of f is *below* or *on* the x-axis. Use set-builder notation to describe your solution set.

11. Consider the following graph of the function f.

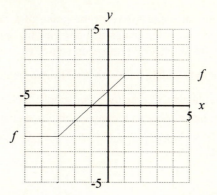

a) Set up a coordinate system on a sheet of graph paper and copy the graph of the function f onto your coordinate system.

b) Draw a number line below your graph. Shade and label the solutions of $f(x) \geq 0$ on your number line as done in Example 3 in this section. Notice the $\geq$ symbol. You are being asked to find where the graph of f is *above* or *on* the x-axis. Use set-builder notation to describe your solution set.

12. (TI-82) Consider the function $f(x) = 1.23x + 4.56$.

a) Use your calculator to sketch the graph of f. Set up a coordinate system on a sheet of graph paper and copy the image in your viewing window onto the coordinate system. Be sure to include the WINDOW settings on your axes.

b) Use the root-finding utility on the TI-82 to find the x-intercept of the function f.

c) Draw a number line below your graph. Shade and label the solutions of $f(x) > 0$ on your number line as done in Example 3 in this section. Use set-builder notation to describe your solution.

13. (TI-82) Consider the function $f(x) = -1.32x - 4.55$.

a) Use your calculator to sketch the graph of f. Set up a coordinate system on a sheet of graph paper and copy the image in your viewing window onto the coordinate system. Be sure to include the WINDOW settings on your axes.

b) Use the root-finding utility on the TI-82 to find the x-intercept of the function f.

c) Draw a number line below your graph. Shade and label the solutions of $f(x) \geq 0$ on your number line as done in Example 3 in this section. Notice the $\geq$ symbol. You are being asked to find where the graph of f is *above* or *on* the x-axis. Use set-builder notation to describe your solution.

14. (TI-82) Consider the function $f(x) = 1.2x^2 - 2.3x - 4.6$.

a) Use your calculator to sketch the graph of f. Set up a coordinate system on a sheet of graph paper and copy the image in your viewing window onto the coordinate system. Be sure to include the WINDOW settings on your axes.

b) Use the root-finding utility on the TI-82 to find the x-intercept of the function f.

c) Draw a number line below your graph. Shade and label the solutions of $f(x) < 0$ on your number line as done in Example 3 in this section. Use set-builder notation to describe your solution.

15. (TI-82) Consider the function $f(x) = 5.62 + 2.44x - 1.15x^2$.

   a) Use your calculator to sketch the graph of f. Set up a coordinate system on a sheet of graph paper and copy the image in your viewing window onto the coordinate system. Be sure to include the WINDOW settings on your axes.

   b) Use the root-finding utility on the TI-82 to find the x-intercept of the function f.

   c) Draw a number line below your graph. Shade and label the solutions of $f(x) \le 0$ on your number line as done in Example 3 in this section. Notice the $\le$ symbol. You are being asked to find where the graph of f is *below* or *on* the x-axis. Use set-builder notation to describe your solution.

1.8 Summary and Review

Important Visualization Tools

Remember, y and $f(x)$ are interchangeable; $f(x)$ is the y-value that is paired with x.

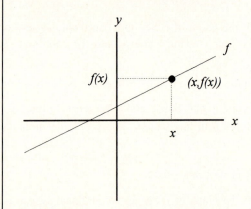

To use a graph to find the value of $f(x)$, first locate x on the x-axis. Draw a vertical line to the graph of the function f, followed by a horizontal line to the y-axis. Read the value of $f(x)$ on the y-axis.

The solution of $f(x) = g(x)$ is the x-value of the point of intersection of the graphs of f and g.

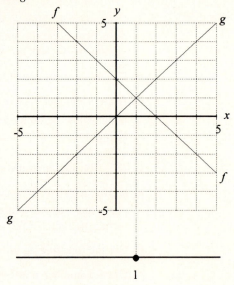

The solution set is $\{x : x = 1\}$.

The solution of $f(x) > g(x)$ is the set of all x-values for which the graph of f is *above* the graph of g.

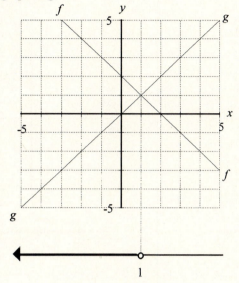

The solution set is $\{x : x < 1\}$.

The solution of $f(x) < g(x)$ is the set of all x-values for which the graph of f is *below* the graph of g.

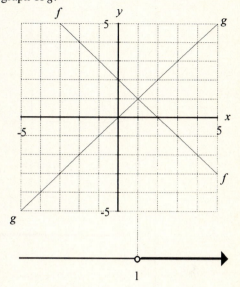

The solution set is $\{x : x > 1\}$.

The solutions of $f(x) = 0$ are called *zeros* of the function. The *zeros* of a function are found by noting where the graph of f *intercepts* the x-axis.

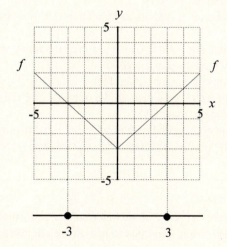

The solution set is $\{x : x = -3 \ \text{or} \ x = 3\}$.

The solutions of $f(x) > 0$ are found by noting where the graph of f is *above* the x-axis.

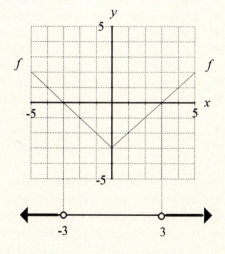

The solution set is $\{x : x < -3 \ \text{or} \ x > 3\}$.

The solutions of $f(x) < 0$ are found by noting where the graph of f is *below* the x-axis.

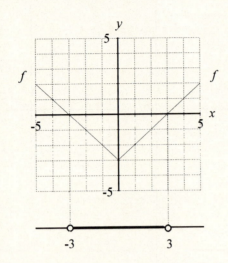

The solution set is $\{x : x > -3 \text{ and } x < 3\}$.

The solutions of $f(x) \leq 0$ are found by noting where the graph of f is *below* or *on* the x-axis.

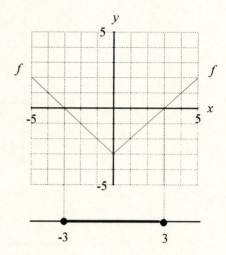

The solution set is $\{x : x \geq -3 \text{ and } x \leq 3\}$.

A Review of Key Words and Phrases

☛ The variable to be chosen is called the *independent* variable. The variable that depends on the choice of an independent variable is called the *dependent* variable. (page 1)

☛ When drawing the *line of best fit*, the line should pass as close to as many points as possible. (page 3)

☛ A *relation* is a set of ordered pairs. (page 11)

☛ If a point satisfies the equation of a function, then that point is on the graph of the function. Therefore, a graph of an equation is the set of all points that satisfy the equation. (page 12)

☛ A *complete* graph shows all the important features of a graph. (page 15)

☛ The *domain* of a relation is the set of x-values of the ordered pairs of the relation. The *range* of the relation is the set of y-values of the ordered pairs of the relation. (page 28)

☛ A relation is a *function* if each x-value in its domain is paired with exactly one y-value in its range. (page 28)

☛ If $f(x) = 3x + 2$ and you are asked to find $f(5)$, simply substitute 5 for x as follows: $f(5) = 3(5) + 2 = 17$. (page 29)

☛ $f(x)$ and y are interchangeable. (page 51)

☛ $f(x)$ is the y-value that is paired with x. (page 51)

☛ The *x-intercept* of the graph of a function is where the graph of the function intercepts the x-axis. (page 90)

☛ If $f(x) = 0$, then x is called a *zero* of the function f. It is important to note the relationship between the zeros of a function and the x-intercepts of the graph of the function. (page 90)

TI-82 Keywords and Menus

☛ Entering functions in the Y= menu. (page 41)

☛ The ZOOM menu and 6:ZStandard. (page 41)

☛ Changing the WINDOW parameters. (page 42)

☛ Using the TRACE utility. (page 58)

☛ Using ZOOM and 8:ZInteger to complete tables with ease. (page 59)
☛ Using the CALC menu and 5:intersect to solve equations. (page 79)
☛ Using the CALC menu and 2:root to find zeros of a function. (page 96)

Chapter Review Exercises

1. Alphonso and Maria are rolling ball bearings down a long inclined plane in their high school physics lab. They have stopwatches to record the amount of time that the ball has been rolling and a special instrument that records the speed of the ball at any instant desired. They start the ball bearing from rest, let it go, and take their measurements at various intervals of time as the ball begins to accelerate down the inclined plane. Here are their data:

Time (seconds)	0	2	4	6	8	10
Speed (meters per second)	0	3.4	6.6	10.4	13.6	17.2

a) Since the speed of the ball bearing *depends* on the amount of time that it has been rolling down the inclined plane, make the time the independent variable and place it on the horizontal axis. Make the speed the dependent variable and place it on the vertical axis. Use an entire sheet of graph paper, scale each axis appropriately (you might want to consider the predictions you will be asked to make), and plot each of the data points from the table.

b) Draw a line of best fit, then use your graph to make the following predictions:

   i) What is the speed of the ball bearing after 5 seconds? Indicate how you found your answer on your graph by using a set of horizontal and vertical dotted lines in the manner presented in the examples.

   ii) Approximately how many minutes must pass before the ball bearing reaches a speed of 12 meters per second? Indicate how you found your answer on your graph by using a set of horizontal and vertical dotted lines in the manner presented in the examples.

2. For each of the following equations, set up a separate coordinate system on a sheet of graph paper. Create a table of points that satisfy the equation and plot each of the points from your table on your coordinate system. Finally, graph the rest of the points that satisfy the equation. If you do not feel that your graph is *complete*, continue adding points to your table and plotting them on your coordinate system until your are confident that you have shown all the important features of the graph in your sketch.

   a) $y = 3 - 2x$ b) $y = |x| - 4$ c) $y = |x - 4|$

   d) $y = (x + 4)^2$ e) $y = x^2 + 4$ f) $y = \sqrt{x - 2}$

   g) $y = \sqrt{x} - 2$ h) $y = x^2 - 6x - 8$ i) $y = 9 - 6x - x^2$

   j) $y = |x| + |x - 4|$

   Hint: Check each solution on your calculator when you are finished. Also, you may find 8:ZInteger in the ZOOM menu of your calculator useful for completing your tables of points.

3. For each of the following functions, find $f(9)$.

   a) $f(x) = 3x + 5$ b) $f(x) = |x| - 6$ c) $f(x) = |x - 6|$

   d) $f(x) = \sqrt{x + 16}$ e) $f(x) = \sqrt{x} + 4$ f) $f(x) = 9 - x^2$

4. In the two concentric squares shown below, the outer square has a side of length x and the inner square has a side of length 5. Express the area of the shaded region as a function of x.

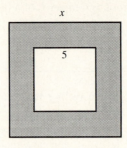

x

5

5. The sum of two numbers is 10. Let x represent one of the numbers. Express the product P of the two numbers as a function of x.

6. The length of a rectangle is 4 more feet than twice the width of the rectangle. Let w represent the width of the rectangle.

   a) Express the length, L, of the rectangle as a function of its width, w.

   b) Express the perimeter, P, of the rectangle as a function of w.

   c) Express the area, A, of the rectangle as a function of w.

7. In the two concentric circles shown below, the radius of the outer circle is 8 and the radius of the inner circle is x. Express the area of the shaded region as a function of x.

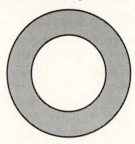

8. Use your calculator to find $f(-1.23)$ for each of the following functions.

   a) $f(x) = 2.23x^2$ b) $f(x) = -3.14x^2$

   Note: You might look at Exercise 4 in Section 1.3 again.

9. Use your calculator to find $f(-1.25)$ for each of the following functions.

   a) $f(x) = 2x^2 - 3x - 9$ b) $f(x) = 3 - 2x - x^2$

10. Use your calculator to find $f(2.55)$ for each of the following functions.

    a) $f(x) = \sqrt{9.15 - x}$ b) $f(x) = |x - 8.1234|$ c) $f(x) = \dfrac{x+2}{x}$ d) $f(x) = \dfrac{x+4}{2-3x}$

11. Consider the following graph of the function f.

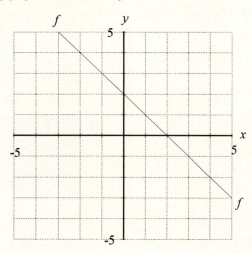

Use the graph of f to evaluate each of the following:

a) $f(-3)$ b) $f(-2)$ c) $f(0)$ d) $f(3)$

12. Shown below is the graph of the function $f(x) = 5 - x$. The point P is on the graph of f.

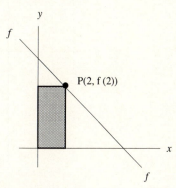

What is the area of the shaded rectangle?

13. Shown below is the graph of the function $g(x) = 9 - x^2$. The point P is on the graph of g. What is the area of the shaded triangle?

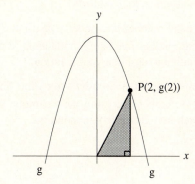

14. Shown below is the graph of $f(x) = 16 - x^2$, with a rectangle inscribed inside. Express the area of the shaded rectangle as a function of x.

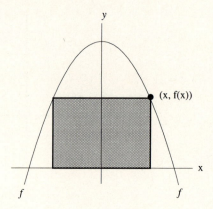

15. Consider the following graphs of the functions f and g.

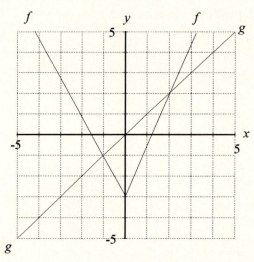

a) Set up a coordinate system on a sheet of graph paper. Copy the graphs of f and g onto your coordinate system.

b) Draw a number line below your graph. Shade and label the solutions of $f(x) = g(x)$ on your number line using the technique presented in Example 8 in Section 1.6. Use set-builder notation to describe your solution.

c) Draw a second number line under your first. Shade and label the solutions of $f(x) > g(x)$ on your second number line using the technique presented in Example 8 in Section 1.6. Use set-builder notation to describe your solution.

d) Draw a third number line under your first. Shade and label the solutions of $f(x) \le g(x)$ on your second number line using the technique presented in Example 9 in Section 1.6. Use set-builder notation to describe your solution.

16.   Consider the following graph of the function f.

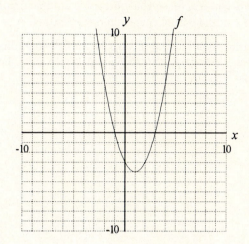

a) Set up a coordinate system on a sheet of graph paper. Copy the graph of f onto your coordinate system.

b) Draw a number line below your graph. Shade and label the solutions of $f(x) = 0$ on your number line using the technique presented in Example 3 in Section 1.7. Use set-builder notation to describe your solution.

c) Draw a second number line under your first. Shade and label the solutions of $f(x) > 0$ on your second number line using the technique presented in Example 3 in Section 1.7. Use set-builder notation to describe your solution.

d) Draw a third number line under your second. Shade and label the solutions of $f(x) \le 0$ on your second number line using the technique presented in Example 3 in Section 1.7. Use set-builder notation to describe your solution.

17. Alicia throws a ball into the air. The graph below shows the height of the ball above ground level as a function of the amount of time that has passed since the ball left Alicia's hand.

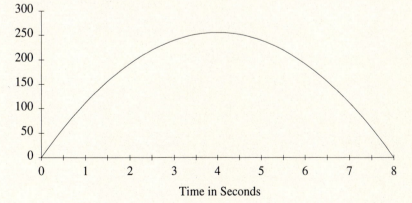

a) Set up a coordinate system on a sheet of graph paper and copy this graph onto your coordinate system.

b) How long does it take the ball to return to the ground? *Note*: You are being asked to find when the height is zero, so you must solve the equation $h = 0$. Draw a number line below your graph and shade and label your solution to $h = 0$ on your number line in the usual manner.

c) During what interval of time is the ball more than 150 feet in the air? Draw a number line below your graph and shade and label your solution to $h > 150$ on your number line in the usual manner.

Chapter 2
Linear Functions

2.1 Introduction to Lines—The Plumber's Bill

Emily needs some plumbing done on her house, so she calls Honest Abe, the local plumber, and asks him what his charges will be. Abe replies that he charges a flat fee of $20 to come out to a customer's home and $10 for each hour he works on a repair.

Modeling with a Graph

The amount of Abe's bill depends on the number of hours he works. Therefore, the number of hours worked is the independent variable and will be placed on the horizontal axis. The amount of the bill is the dependent variable and will be placed on the vertical axis.

Tip	It is permissible to choose a different scale on each axis. Try to choose a scale that makes the plotting of a graph as easy as possible.

Abe charges $10 for each hour he works on a repair. It seems natural to let each tick mark on the horizontal axis represent 1 hour of labor and each tick mark on the vertical axis represent $10.

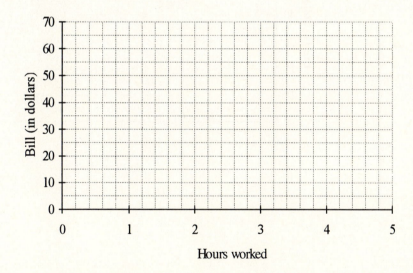

At time zero, Abe has done no work at all, so Emily's bill will be $20, which is represented by the point (0,20) on the graph. Abe charges $10 for each hour of labor. Therefore, whenever Abe performs an additional hour of labor, Emily's bill increases by $10. Starting at point (0,20), draw a horizontal line 1 unit to the right, representing 1 hour of labor, then a vertical line upward 10 units, representing $10, to arrive at point (1,30) as shown:

111

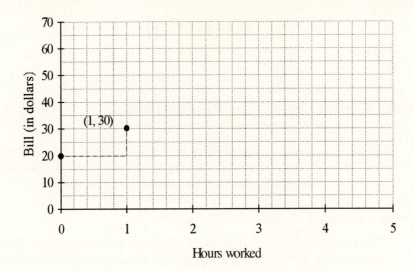

This process can be repeated for several points. Every time you move 1 unit to the right, which represents an additional hour of labor, you will move upward 10 units, representing a $10 increase in Emily's bill.

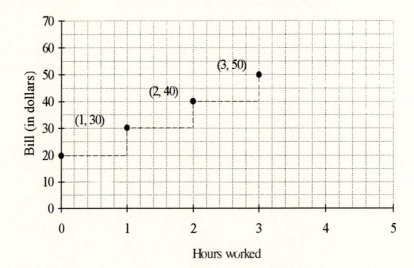

If you continue plotting points in this manner, the graph will be a line.

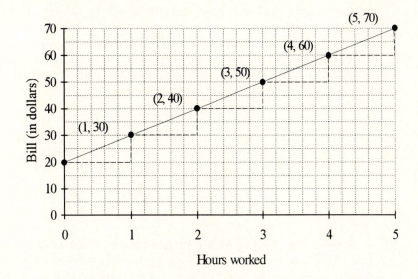

The rate that Honest Abe charges for repairs, $10 per hour, controls the angle of ascent of the line and is called the slope of the line.

Using This Model to Make Predictions. Once mathematicians have created a graph that models a problem situation, they like to use their graph to make predictions. For example, if Abe works for 4 hours, what will be the amount of Emily's bill? To answer this question, you first locate 4 hours on the horizontal axis, then draw a vertical line to the graph, followed by a horizontal line to the vertical axis. The answer to the question lies on the vertical axis.

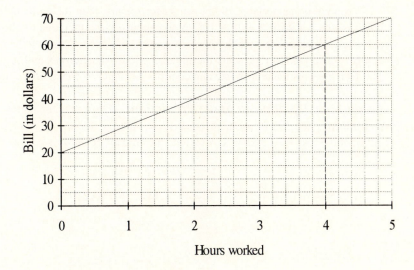

Emily's bill will be approximately $60.

Modeling with an Equation

Let t represent the number of hours that Abe puts in on the job and let A represent the amount of Emily's bill in dollars. How would you express A as a function of t? Your first step might be to search for a *pattern* in the method in which Emily's bill is computed.

If Abe does no work at all, Emily's bill will be $20. Thereafter, Abe charges $10 for each hour he works on repairs. If Abe works for an hour, Emily's bill will be $30. If he works for 2 hours, her bill will be $40, and so on. These results can be arranged in a table:

t	A
0	20
1	30
2	40
3	50
4	60

When you are looking for a pattern, it can be a disadvantage to compute your results in this manner. This table reveals nothing of the relationship between the amount of Emily's bill and the number of hours that Abe works on repairs. Try to arrange your results in a different manner. If Abe makes no repairs, Emily's bill will be 20 + 10(0) dollars. If he works for an hour, Emily's bill will be 20 + 10(1) dollars. If he works for 2 hours, Emily's bill will be 20 + 10(2) dollars, and so on, as shown in the following table.

t	A
0	20 + 10(0)
1	20 + 10(1)
2	20 + 10(2)
3	20 + 10(3)
4	20 + 10(4)

A pattern is emerging. If Abe works for t hours, then the amount of Emily's bill, in dollars, is given by the equation $A = 20 + 10t$. You can also write the equation as follows: $A = 10t + 20$. It is extremely important to note the *position and significance* of the numbers 10 and 20 in the equation $A = 10t + 20$. The 20 represents the initial fee that Abe charges and is plotted on the vertical axis. The 10 represents the hourly rate and controls the angle of ascent of the line.

Using Your Equation to Make Predictions.   Once mathematicians determine an equation for their model, they like to use their equation to make predictions. For example, what will be Emily's bill if Abe works for 4 hours? To compute the amount of Emily's bill, we let $t = 4$ in your equation.

$$A = 20 + 10t$$

$$A = 20 + 10(4)$$

$$A = 60$$

The amount of Emily's bill is $60.

It is important to note how our graphical solution and numerical solution agree.

Exercises for Section 2.1

1. Nayra is a mechanic. Each month she puts aside part of her income in a savings account in the hopes that one day she will have enough to open her own shop. For the present, she rents space in a shop owned by Leo Loncar. Leo charges Nayra $10 per day to rent one of the lifts in his shop. Nayra specializes in tune-ups. For each tune-up she performs, her costs are $20. Sketch a graph showing Nayra's costs for one day as a function of the number of tune-ups she performs in a day.

    Since Nayra's costs depend on the number of tune-ups, make the number of tune-ups the independent variable and place it on the horizontal axis. Let x represent the number of tune-ups that Nayra performs in a day. Make Nayra's costs the dependent variable and place it on the vertical axis. Let C represent Nayra's costs for a day's work.

    Let each tick mark on the x-axis represent one tune-up and each tick mark on the C-axis represent $10. Copy the following coordinate system onto a sheet of graph paper.

    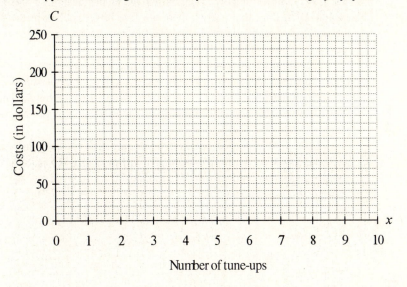

    a) If Nayra does not perform any tune-ups, her costs are still $10. Plot point (0,10) on your coordinate system.

    b) For each tune-up performed, Nayra's costs go up an additional $20. Beginning at point (0,10), move one tune-up to the right and $20 up and plot a new data point. What are the coordinates of this new point?

    c) Each time you move one tune-up to the right, you must move $20 up and plot a new data point. Repeat this process until you reach a point with an x-value of 10 tune-ups.

    d) Draw a line through your data points.

    e) Use your graph to make the following predictions: What will Nayra's costs be if she performs 8 tune-ups?

    f) The table below lists Nayra's costs and the number of tune-ups that she performs. Some entries have been completed for you. Fill in the remaining entries.

Number of tune-ups (x)	Costs (C)
0	$10 + 20(0)$
1	$10 + 20(1)$
2	$10 + 20(2)$
3	
4	
5	

g) Express Nayra's costs, C, as a function of the number of tune-ups, x, that she performs in one day. Use your equation to predict Nayra's costs if she performs 8 tune-ups.

2. Jenny builds a rabbit hutch behind her barn. She places 50 rabbits in the hutch, then locks the door and leaves. Unfortunately, there is a flaw in the design of the hutch and the rabbits begin to escape. Each hour that passes, there are five fewer rabbits in the hutch. Sketch a graph showing the number of rabbits in the hutch as a function of the amount of time that has passed since Jenny locked the door of the hutch.

   The number of rabbits in Jenny's hutch depends on the amount of time that has passed. Make time the independent variable and place it on the horizontal axis. Let t represent the amount of time that has passed since Jenny locked the door of the hutch. The number of rabbits will be the dependent variable and will be placed on the vertical axis. Let N represent the number of rabbits in Jenny's hutch.

   Since five rabbits escape each hour, each tick mark on the N-axis will represent five rabbits. Each tick mark on the horizontal t-axis will represent the passing of 1 hour. Copy the following coordinate system onto a sheet of graph paper.

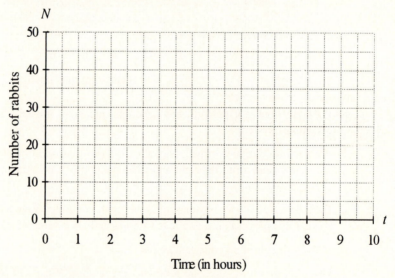

a) Initially, at time zero, there were 50 rabbits in the hutch. Plot point (0,50) on your coordinate system.

b) As each hour passes, there will be five fewer rabbits in the hutch. Beginning at point (0,50), move 1 hour to the right, then five rabbits down and plot a new data point. What are the coordinates of this new point?

c) Each time you move 1 hour to the right, you must move 5 units (rabbits) down and plot a new data point. Repeat this process until you reach a point with a t-value of 10 hours.

d) Draw a line through your data points.

e) Use your graph to make the following prediction: If 6 hours have elapsed since Jenny locked the door of the hutch, how many rabbits remain in the hutch?

f) The following table lists the number of rabbits present and the amount of time that has passed since Jenny locked the door of the hutch. Some of the entries have been completed for you. Fill in the remaining entries.

Time (t)	Number of Rabbits (N)
0	$50 - 5(0)$
1	$50 - 5(1)$
2	$50 - 5(2)$
3	
4	
5	

g) Express the number of rabbits present, N, as a function of the amount of time, t, that has passed since Jenny locked the door of the hutch. Use your equation to predict the number of rabbits remaining in the hutch after 6 hours.

3. It has been a dry season in Fortuna, and only 200 gallons of water remain in the Farnhams' water tank. The weather changes and the constant rain begins to raise the level of the water table. With their pump running at full speed, the Farnhams manage to pump an additional 100 gallons of water into their water tank each day. Of course, this rate will not last forever, but for 10 days the Farnhams manage to pump an additional 100 gallons of water into their water tank each day. Draw a graph showing the volume of water in the Farnhams' tank as a function of the number of days that their pump has been running.

   The volume of water depends on the number of days that the pump has been running. Make the number of days the independent variable and place it on the horizontal axis. Let t represent the number of days that the pump has been running. Make the volume of water in the tank the dependent variable and place it on the vertical axis. Let V represent the volume of water in the tank.

   Let each tick mark on the t axis represent one day. Because the volume of water in the Farnhams' water tank increases by 100 gallons with each day, make each tick mark on the V axis represent 100 gallons. Copy the following coordinate system onto a sheet of graph paper.

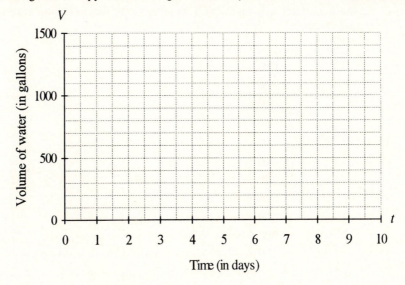

a) When the Farnhams started their pump, there were only 200 gallons of water in the tank. Plot the point (0,200) on your coordinate system.

b) The Farnhams' pump places an additional 100 gallons into their water tank every day. Beginning at point (0,200), move one day to the right and 100 gallons upward and plot a new data point. What are the coordinates of this new point?

c) Each time you move to the right one day, you must move 100 gallons upward and plot a new data point. Repeat this process until you reach a point whose *t*-value is 10 days.

d) Draw a line through your data points.

e) Use your graph to make the following prediction: How much water will be in the Farnhams' water tank after 8 days?

f) The following table lists the volume of water in the Farnhams' water tank and the number of days that the pump has been working. Some of the entries have been completed for you. Double-check our work, then fill in the remaining entries.

Time (t)	Volume (V)
0	200 + 100(0)
1	200 + 100(1)
2	200 + 100(2)
3	
4	
5	

g) Express the volume of water in the Farnhams' water tank, V, as a function of the number of days their pump has been working, t. Use your equation to predict the volume of water in the tank after 8 days.

4. Mr. Fisher is riding in a hot-air balloon at an altitude of 1000 feet. The pilot begins a slow descent, adjusting the controls so that the balloon loses 100 feet in altitude every minute. Sketch a graph showing Mr. Fisher's altitude as a function of time.

   Because altitude depends on the amount of time that has passed, make the altitude the dependent variable and place it on the vertical axis. Let the variable h represent the altitude of the balloon. Make the time the independent variable and place it on the horizontal axis. Let t represent the amount of time that has passed since the balloon began its descent.

   Let each tick mark on the vertical axis represent 100 feet and we will let each tick mark on the horizontal axis represent 1 minute. Copy the following coordinate system onto a sheet of graph paper.

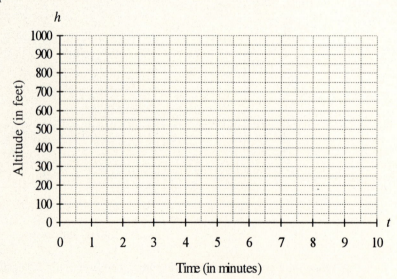

a) When the pilot first begins the descent, the altitude of the balloon is 1000 feet. Plot point (0,1000) on your coordinate system.

b) For each minute that passes, the altitude of the balloon decreases by 100 feet. Beginning at point (0,1000), move 1 minute to the right and 100 feet down and plot a new data point. What are the coordinates of this new point?

c) Each time you move 1 minute to the right, you must move 100 feet downward and plot a new data point. Repeat this process until you reach a point with a t-value of 8 minutes.

d) Draw a line through your data points.

e) Use your graph to make the following prediction: What will be Mr. Fisher's altitude after 6 minutes of descent?

f) The table below lists Mr. Fisher's altitude at various times during the descent. Some of the entries have been completed for you. Fill in the remaining entries.

Time (t)	Altitude (h)
0	$1000 - 100(0)$
1	$1000 - 100(1)$
2	
3	
4	
5	

g) Express Mr. Fisher's altitude, h, as a function of the number of minutes that the balloon has been descending, t. Use your equation to predict Mr. Fisher's altitude after 6 minutes of descent.

5. Alex and Alicia are spending a day hiking in northern California. When they begin their hike, their elevation is 100 feet above sea level. Each hour of their hike, they manage to gain an additional 200 feet in altitude.

a) Set up an appropriately scaled coordinate system on a sheet of graph paper and draw the graph of Alex and Alicia's altitude as a function of the amount of time that has passed since they started their walk.

b) Use your graph to make the following predictions: What altitude will the hikers reach after 6 hours?

c) Express the altitude, A, as a function of the time, t, Alex and Alicia have been hiking. Use your equation to predict the altitude after 6 hours of hiking.

6. Hank works in a sheet-metal factory where he cuts parts for elevators. He gets paid a base salary of $20 per day plus $4 for every piece he cuts.

a) Set up an appropriately scaled coordinate system on a sheet of graph paper and draw the graph of Hank's daily pay as a function of the number of pieces he cuts.

b) Use your graph to make the following prediction: What will Hank's daily pay be if he cuts 10 pieces?

c) Express Hank's daily pay, P, as a function of the number of pieces, x, that Hank cuts in one day. Use your equation to predict Hank's daily pay if he cuts 10 pieces.

7. Shanti is nearing the end of a beautiful day of sailing on a lake in northern Minnesota. It is 5 P.M. and time to go home. Shanti turns the boat toward a dock 2000 yards away. She catches a good

wind and begins to decrease the distance between her boat and the dock at a rate of 150 yards per minute.

a) Set up an appropriately scaled coordinate system on a sheet of graph paper and draw a graph of the distance between Shanti's boat and the dock as a function of the amount of time that has passed since 5 P.M.

b) Use your graph to make the following prediction: After 11 minutes of sailing toward the dock, how far is Shanti's boat from the dock?

c) Express Shanti's distance from the dock, d, as a function of the amount of time that has passed since 5 P.M., t. Use your equation to predict Shanti's distance from the dock after 11 minutes of sailing.

8. Kay has built a fabulous go-cart, the envy of all the kids in town. Spurred by a promise that they will also get a ride, several friends agree to give Kay a running start before her go-cart hits a small downgrade. They huff and puff and push and manage to get the speed of the go-cart up to 2 miles per hour before it reaches the crest of the downgrade. At that point, everyone lets go, and Kay in her go-cart begins to pick up speed at the rate of 0.4 miles per hour every second.

a) Set up an appropriately scaled coordinate system on a sheet of graph paper and draw the graph of Kay's speed as a function of the amount of time that has passed since everyone stopped pushing. *Hint:* Let each tick mark on the vertical axis represent 0.4 miles per hour.

b) Use your graph to make the following prediction: What will Kay's speed be 20 seconds after everyone stops pushing?

c) Express Kay's speed, S, as a function of the time that elapses after the pushing stops, t. Use your equation to predict Kay's speed 20 seconds after the pushing stops.

9. When the Wellington Bank closes at 6 P.M., there are 100 customers waiting in line. Abel the security guard locks the doors of the bank. Customers already in the bank are allowed to leave the bank when their business is finished, but no new customers are allowed to enter the bank. The available tellers begin to serve the customers at a rate of 12 customers per minute.

a) Set up an appropriately scaled coordinate system on a sheet of graph paper and draw the graph of the number of customers remaining in the bank as a function of the amount of time since the bank has closed.

b) Use your graph to make the following prediction: How many customers remain in the bank 8 minutes after closing time?

c) Express the number of customers in the bank, N, as a function of the number of minutes that have passed since closing time, t.

10. Sebastian waves good-bye to his brother, who is talking to a group of his friends approximately 20 feet away. Sebastian then begins to walk away from his brother at a rate of 4 feet per second.

a) Set up an appropriately scaled coordinate system on a sheet of graph paper and draw the graph of Sebastian's distance from his brother as a function of the amount of time that has passed since Sebastian waved good-bye to his brother.

Hint: Let each tick mark on the vertical axis represent 4 feet and each tick mark on the horizontal axis represent 1 second.

b) Use your graph to make the following prediction: What will be Sebastian's distance from his brother 10 seconds after waving good-bye?

c) Express Sebastian's distance from his brother, d, as a function of the amount of time that has passed since waving to his brother, t. Use this equation to predict Sebastian's distance from his brother 10 seconds after waving good-bye.

11. Leonardo runs a small business making wicker baskets that he sells to flower shops. The following graph shows Leonardo's costs as a function of the number of baskets he makes in a day.

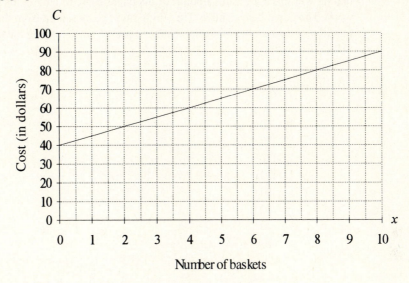

a) If Leonardo makes no baskets at all, what will be his costs ?

b) At what rate are Leonardo's costs increasing?

c) Express Leonardo's costs, C, as a function of the number of baskets, x, that Leonardo makes in a day.

12. Giuseppe has created a beautiful display of apples in his grocery store in Seattle. The apples are some of Washington's finest and are a popular item with shoppers that day. The following graph shows the number of apples in the display as a function of the amount of time that has passed since the store opened for business.

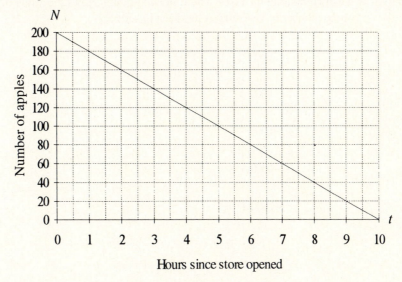

a) What was the initial number of apples in the display when the store opened?

b) At what rate are the apples leaving the display?

c) Express the number of apples in the display, *N*, as a function of the amount of time that has passed since the store opened, *t*.

13. Timothy has had his eyes on a Super Nintendo game for quite a spell. He makes a solemn pledge that he will place a certain amount of his paper route money in his piggy bank at the end of each week. The graph below shows Timothy's savings, measured from the day he made his pledge, as a function of the number of weeks that have passed since making the pledge.

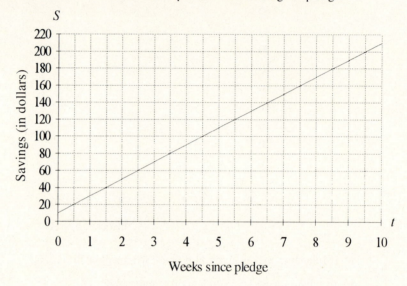

Weeks since pledge

a) How much money was in Timothy's piggy bank on the day he made his pledge?

b) At what rate is Timothy's money increasing?

c) Express the amount of Timothy's savings, *S*, as a function of the number of weeks, *t*, that have passed since his pledge.

14. William runs a car rental agency, and he is allowed to depreciate the value of his automobiles when computing his taxes. The following graph shows the value of a particular car as a function of the number of years that is has been in service.

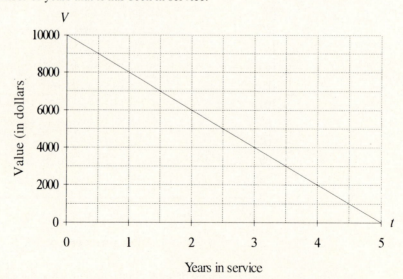

Years in service

a) What was the original value of the car?

b) At what rate is the value of the car depreciating?

c) Express the value of the car, *V*, as a function of the number of years that the car has been in service, *t*.

2.2 The Slope of a Line

In the previous section, we established a relation between the rate of increase or decrease and the *slope* of a line. For example, we saw that Honest Abe's rate of $10 per hour controlled the angle of ascent of the line. In this section we will formally define the *slope* of a line, provide an easy computational formula for computing it, and consider more applications involving rates.

Example 1   On January 1, Melissa deposits $10 in the bank. Thereafter, the amount of money in her account increases at the rate of $5 every 2 months. Sketch a graph of the amount of money in Melissa's account as a function of the number of months that have passed since January 1.

Solution   The amount in Melissa's account depends upon the amount of time that has passed. Let *A* represent the amount in her account and place it on the vertical axis. Let *t* represent the number of months that have passed since January 1 and place it on the horizontal axis. Begin by plotting the point (0,10) on the coordinate system. Since the rate is $5 every 2 months, each time you move 2 months to the right, you must move upward $5, as follows:

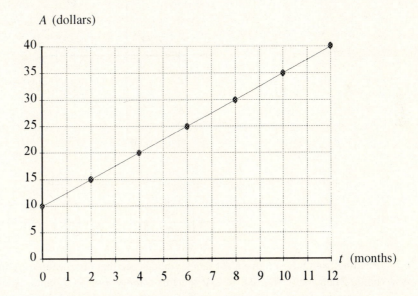

The rate, $5 every 2 months, is called the *slope* of the line and controls the rate of ascent (or descent in some cases) of the line. If the rate were higher, say $10 every 2 months, then the line would rise at a faster rate. If the rate were lower, say $1 every 2 months, the line would rise at a slower rate.

Example 2   Consider the following coordinate system:

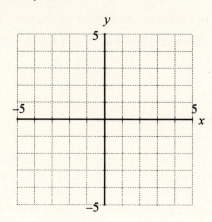

Plot the point $(-5,-1)$. Suppose that the value of y increases 1 unit in the y-direction for every 2 units in the x-direction. Sketch the graph of y as a function of x and note any patterns you see.

Solution   After plotting the point $(-5,-1)$, each time the x-value increases by 2 units, we must increment the y-value by 1 unit, as follows:

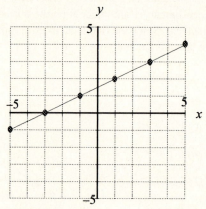

Some of the points from the graph can be arranged in a table.

x	-5	-3	-1	1	3	5
y	-1	0	1	2	3	4

Reading from left to right in the table, you can see that the x-values are increasing in increments of 2 units. Mathematicians like to say that the *change in x* is 2 units. The y-values are increasing in increments of 1 unit, which means that the *change in y* is 1 unit. The rate at which the y-values are changing with respect to the x-values is $\frac{1}{2}$, 1 y-unit for every 2 x-units.

The Definition of the Slope of a Line.

The most natural definition that can be made for the slope of a line is as follows:

The slope of a line is defined to be the change in y divided by the change in x. It is the rate at which the y-values are changing with respect to the x-values.

Example 3   Use the definition above to find the slope of the following line:

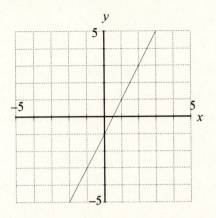

Solution   It is important to note the scale on each axis. On both the *x*- and *y*-axis, each tick mark represents 1 unit. Select two points P and Q on the line.

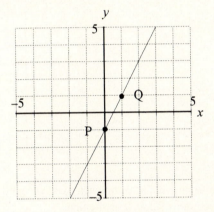

As you move from point P to point Q, the change in *x* is 1 unit and the change in *y* is 2 units.

$$\text{Slope} = \frac{\text{Change in } y}{\text{Change in } x} = \frac{2}{1} = 2$$

Note that the slope is *positive*. This is consistent with the fact that the *y*-values are *increasing* as you move from left to right along the line. A different choice can be made for the points P and Q.

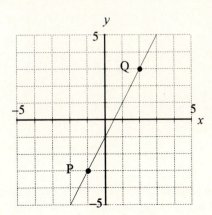

As you move from point P to point Q, the change in x is 3 units and the change in y is 6 units.

$$\text{Slope} = \frac{\text{Change in } y}{\text{Change in } x} = \frac{6}{3} = 2$$

The slope remains unchanged regardless of the points selected.

Tip	When choosing the points P and Q on the line, try to select points whose coordinates are easy to interpret.

Example 4   Find the slope of the following line.

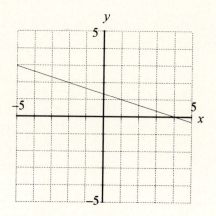

Solution   It is important to note the scale on each axis. On both the x- and y-axis, each tick mark represents 1 unit. Select two points P and Q on the line.

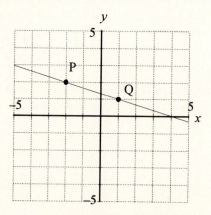

As you move from point P to point Q, the change in x is 3 units and the change in y is −1 units.

$$\text{Slope} = \frac{\text{Change in } y}{\text{Change in } x} = \frac{-1}{3} = -\frac{1}{3}$$

Note that the slope is *negative*. This is consistent with the fact that the y-values are decreasing as you move from left to right along the line.

A Visual Image. The following images can be useful when computing the slope of a line:

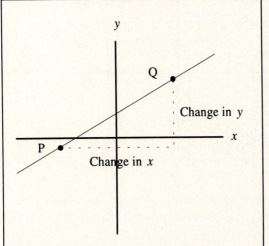 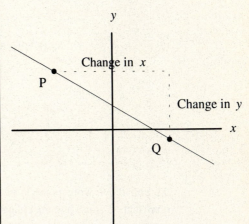

In this case the slope is *positive*. As you move from point P to point Q, the change in *x* is positive and the change in *y* is positive. As you move from left to right along the line, the *y*-values are *increasing*.

In this case the slope is *negative*. As you move from point P to point Q, the change in *x* is positive and the change in *y* is negative. As you move from left to right along the line, the *y*-values are *decreasing*.

Example 5   Find the slope of the following line:

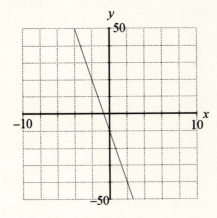

Solution   It is important to note the scale on each axis. Each tick mark on the *x*-axis represents 2 units. Each tick mark on the *y*-axis represents 10 units. Select two points P and Q on the line.

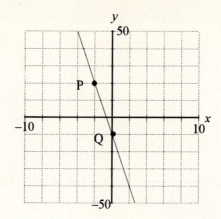

As you move from point P to point Q, the change in x is 2 units (each tick mark on the x-axis represents 2 units) and the change in y is -30 units (each tick mark on the y-axis represents 10 units).

$$\text{Slope} = \frac{\text{Change in } y}{\text{Change in } x} = \frac{-30}{2} = -15$$

The Slope Formula. In the last example, you could easily overlook the scale on the axes and calculate an incorrect slope. There is a slope formula that will greatly simplify your work.

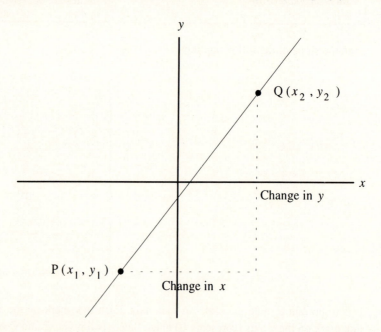

As you move from point P to point Q, you can calculate the change in x by subtracting the x-coordinate of the point P from the x-coordinate of the point Q.

$$\text{Change in } x = x_2 - x_1$$

As you move from point P to point Q, you can calculate the change in y by subtracting the y-coordinate of the point P from the y-coordinate of the point Q.

$$\text{Change in } y = y_2 - y_1$$

The slope of the line is the change in y divided by the change in x.

$$\text{Slope} = \frac{\textbf{Change in } y}{\textbf{Change in } x} = \frac{y_2 - y_1}{x_2 - x_1}$$

Example 6   Calculate the slope of the line that goes through the points P($-3,-2$) and Q($4,4$).

Solution   Draw the graph and use it to help find the slope of the line.

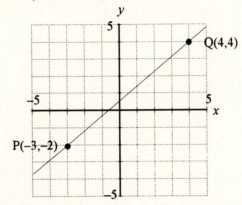

As you move from point P to point Q on the line, the change in x is 7 units and the change in y is 6 units. Therefore, the slope is $\frac{6}{7}$.

Now, use your new slope formula to compute the slope.

$$\text{Slope} = \frac{y_2 - y_1}{x_2 - x_1} = \frac{4 - (-2)}{4 - (-3)} = \frac{6}{7}$$

This result agrees exactly with the result captured from the graph.

Example 7   Use the slope formula to compute the slope of the line through points P($-5,6$) and Q($7,-1$).

Solution   Simply use the slope formula:

$$\text{Slope} = \frac{y_2 - y_1}{x_2 - x_1} = \frac{-1 - 6}{7 - (-5)} = -\frac{7}{12}$$

The following computation shows that it does not matter which points you select as (x_1, y_1) and (x_2, y_2).

$$\text{Slope} = \frac{y_2 - y_1}{x_2 - x_1} = \frac{6 - (-1)}{-5 - 7} = -\frac{7}{12}$$

As long as you are consistent, it does not matter which way you subtract. Simply pick a direction to subtract and stay with it.

Example 8   It costs Alfredo $375 to make 50 wallets and $650 dollars to make 100 wallets. Assuming that Alfredo's costs are a linear function of the number of wallets that he makes, at what rate are his costs increasing?

Solution   Since Alfredo's costs depend on the number of wallets that he makes, let C represent his costs and place this variable on the vertical axis. Let x represent the number of wallets that he makes and place this variable on the horizontal axis. Scale the axes, plot the data points, and draw a line through the data points, as follows:

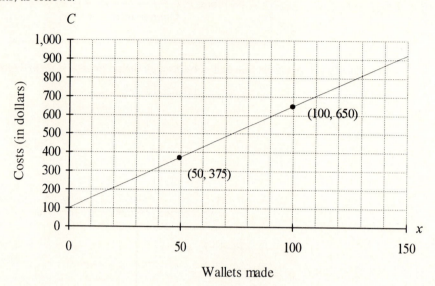

You can find the rate at which Alfredo's costs are increasing by using the slope formula. Remember, slope is a rate.

$$\text{Rate} = \frac{650 \text{ dollars} - 375 \text{ dollars}}{100 \text{ wallets} - 50 \text{ wallets}}$$
$$= \frac{275 \text{ dollars}}{50 \text{ wallets}}$$
$$= 5.50 \text{ dollars} / \text{wallet}$$

When you attach units to the value of the slope, the slope becomes more meaningful. In this case, the slope of the line is 5.50 dollars/wallet. For each additional wallet that Alfredo makes, his costs increase by $5.50.

Exercises for Section 2.2

1. Set up a coordinate system on a sheet of graph paper and copy the following image onto your coordinate system.

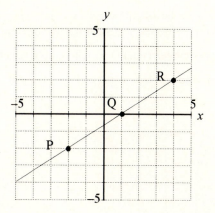

   a) As you move from point P to point Q along the line, what is the change in x and the change in y? Clearly state these changes on your homework paper. Use the change in x and the change in y to compute the slope of the line.

   b) As you move from point P to point R along the line, what is the change in x and the change in y? Clearly state these changes on your homework paper. Use the change in x and the change in y to compute the slope of the line.

   Note: The phrase "clearly state the change in x" asks that you write a complete sentence on your paper that begins as follows: "As I move from point P to point Q, the change in x is . . ." You should create a similar statement stating the slope of the line. Always use complete sentences to express your results.

2. Set up a coordinate system on a sheet of graph paper and copy the following image onto your coordinate system.

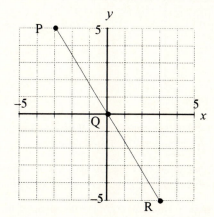

   a) As you move from point P to point Q along the line, what is the change in x and the change in y? Clearly state these changes on your homework paper. Use the change in x and the change in y to compute the slope of the line.

   b) As you move from point P to point R along the line, what is the change in x and the change in y? Clearly state these changes on your homework paper. Use the change in x and the change in y to compute the slope of the line.

Note: The phrase "clearly state the change in *x*" asks that you write a complete sentence on your paper that begins as follows: "As I move from point P to point Q, the change in *x* is . . ." You should create a similar statement stating the slope of the line. Always use complete sentences to express your results.

3. For each of the following problems, set up a *separate* coordinate system on a sheet of graph paper, then carefully follow each of the following directions:

   i) Select two points P and Q on your graph. Label each point with its coordinates.

   ii) Clearly state the change in *x* and the change in *y*. Clearly state the slope of the line.

a)
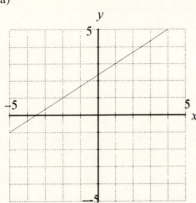

b)
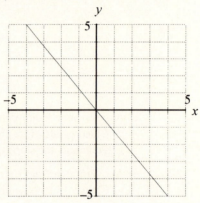

c)
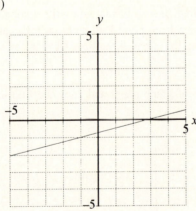

d)
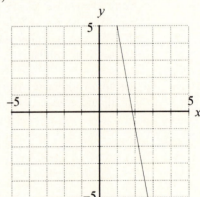

4. Set up a coordinate system on a sheet of graph paper. Plot the points P(−3,−2) and Q(7,5) on your coordinate system and draw a line through them.

   a) Use your sketch to find the change in *x* and the change in *y* and clearly state these changes on your homework paper. Use these results to compute the slope of the line.

   b) Use the slope formula to compute the slope of the line.

   c) Does your numerical solution from part (b) agree with your graphical solution from part (a)? If not, check your work for error.

5. Use the slope formula to compute the slope of the line through the given points. Include a sketch of the line with each of your calculations.

   a) (2,3) and (3,7)   b) (−5,1) and (7,2)   c) (−1,−1) and (5,−4)

   d) (2,−5) and (−5,8)   e) (5,4) and (8,4)   f) (9,2) and (9,7)

6. For each of the following problems, set up a *separate* coordinate system on a sheet of graph paper, then carefully follow each of the following directions:

   i)   Select two points P and Q on your graph. Label each point with its coordinates.

   ii) Use your sketch to find the change in x and the change in y and clearly state these changes on your homework paper. Use these results to compute the slope of the line.

   iii) Use the slope formula to compute the slope of the line. Make sure that this result agrees with the result from part (ii).

   a)

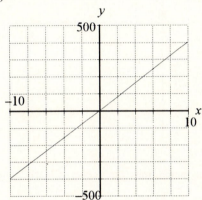

   b)

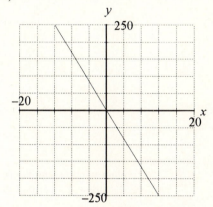

7. (TI-82) The following sketches were produced on a TI-82 graphing calculator. The WINDOW parameters for each sketch are included. Duplicate the graph on your homework paper and clearly state the slope of the line.

   a)

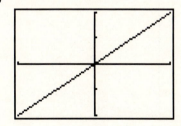

   WINDOW FORMAT
   Xmin=-1
   Xmax=1
   Xscl=1
   Ymin=-2
   Ymax=2
   Yscl=1

   b)

   WINDOW FORMAT
   Xmin=-2
   Xmax=2
   Xscl=1
   Ymin=-100
   Ymax=100
   Yscl=50

8. (TI-82) Use your calculator to compute the slope of the line through the following pairs of points. Include a sketch of the graph of the line and round your answer for the slope to the nearest hundredth (two decimal places).

   a) (1.2, 3.8) and (4.7, 5.3)   b) (−3.5, 4.5) and (5.7, −6.3)

9. Set up a coordinate system on a sheet of graph paper and duplicate the following graph on your coordinate system. Label each line with its slope. One of the lines has been labeled for you, but be sure to double-check the result.

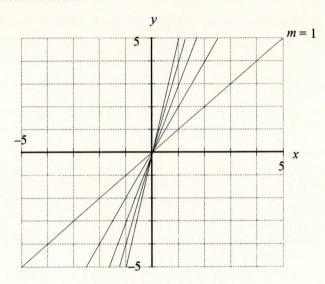

   Explain what is happening to the slope as the lines get steeper. Clearly state what you believe to be the slope of a vertical line.

10. Set up a coordinate system on a sheet of graph paper and duplicate the following graph on your coordinate system. Label each line with its slope. One of the lines has been labeled for you, but be sure to double-check the result.

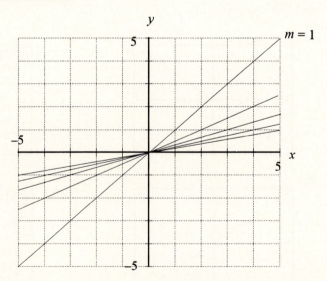

   Explain what is happening to the slope as the lines get closer to horizontal. Clearly state what you believe to be the slope of a horizontal line.

11. Complete the table of points that satisfy the equation $f(x) = 2x + 3$.

x	−4	−3	−2	−1	0	1	2	3	4
$f(x)$	−5								
Points to plot	(−4,−5)								

a) Set up a coordinate system on a sheet of graph paper and plot the points in the table on your coordinate system. Draw a line through your points.

b) What is the change in consecutive values of x in the table? What is the change in consecutive values of y in the table? Use these changes to compute the slope of the line.

c) What relationship do you see between the slope of the line and the equation of the line?

12. Complete the table of points that satisfy the equation $f(x) = -3x + 1$.

x	−4	−3	−2	−1	0	1	2	3	4
$f(x)$	13								
Points to plot	(−4,13)								

a) Set up a coordinate system on a sheet of graph paper and plot the points in the table on your coordinate system. Draw a line through your points.

b) What is the change in consecutive values of x in the table? What is the change in consecutive values of y in the table? Use these changes to compute the slope of the line.

c) What relationship do you see between the slope of the line and the equation of the line?

13. Suppose that the graph of the function f is a line. Furthermore, suppose that $f(-2) = 7$ and $f(3) = -5$.

a) Set up a coordinate system on a sheet of graph paper and plot the given data points. Draw a line through your plotted points.

b) What is the slope of the line through the given data points?

14. Complete the table of points that satisfy the equation $f(x) = x^2 - 2x - 3$.

x	−3	−1	0	1	2	3	4	5
$f(x)$								
Points to plot								

a) Set up a coordinate system on a sheet of graph paper and plot the points from the table on your coordinate system. Fill in the rest of the points that satisfy the equation of the function f.

b) Draw a line through the points $(1, f(1))$ and $(2, f(2))$. Clearly state the slope of this line on your homework paper.

c) Simplify the following expression as much as possible, then explain what your answer represents.

$$\frac{f(3) - f(1)}{3 - 1}$$

15. Consider the following graph.

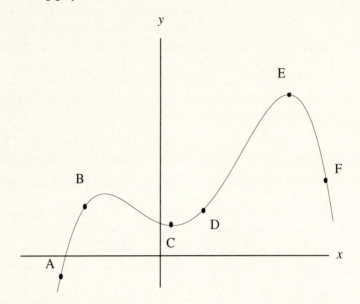

Arrange the following in ascending order, smallest number first, largest number last.

a) The slope of the line through points A and B.   b) The slope of the line through points B and C.

c) The slope of the line through points C and D.   d) The slope of the line through points D and E.

e) The slope of the line through points D and F.

16. Determine the slope of the line that passes through each of the following sets of data points.

a)

x	−10	−8	−6	−4	−2	0	2	4	6	8	10
y	35	29	23	17	11	5	−1	−7	−13	−19	−25

b)

x	−15	−12	−9	−6	−3	0	3	6	9	12	15
y	−19	−13	−7	−1	5	11	17	23	29	35	41

c)

x	2.0	2.1	2.2	2.3	2.4	2.5	2.6
y	−2.00	−2.31	−2.62	−2.93	−3.24	−3.55	−3.86

d)

x	3.0	3.2	3.4	3.6	3.8	4	4.2
y	8.00	8.44	8.88	9.32	9.76	10.20	10.64

17. Suppose that the public demand for a particular brand of toaster oven is a linear function of its unit price. If the price of the oven is set at $50, the demand is 100 toaster ovens. If the price of the oven is set at $60, the demand is 70 toaster ovens.

a) Set up a coordinate system on graph paper. Since the demand for toaster ovens depends on the price, place the unit price on the horizontal axis and the demand on the vertical axis. Scale each axis appropriately and plot the two data points. Draw a line through the two data points.

b) Use the slope formula to determine the rate at which the demand for toaster ovens is decreasing with respect to price. Write a sentence explaining the real-world significance of your number.

18. Jason is playing in his American Flyer, a brand-new red wagon. He pulls both feet up off the ground and tucks them into the wagon bed. As he starts down the hill, his mom screams for him to stop, but it's too late. With a smile of absolute glee, Jason rockets down the hill. Ten seconds after his mom's first warning shout, his wagon is traveling at a speed of 2 meters per second. Twenty seconds after his mom's first warning shout, his speed is 4 meters per second.

   a) Set up a coordinate system on a sheet of graph paper. Since Jason's speed depends on the amount of time that has passed since his mom's first warning shout, place the time on the horizontal axis and Jason's speed on the vertical axis. Scale each axis appropriately and plot the two data points. Assume that the relationship is linear and draw a line through your two data points.

   b) Use the slope formula to determine the rate at which Jason's speed is increasing with respect to time. Write a sentence explaining the real-world significance of your number.

19. A spring will stretch when a mass of some sort is attached to it. The length of the spring is a linear function of the amount of mass that is attached to it. The length of a spring is 4 centimeters when a mass of 100 grams is attached to it. The length of the spring is 6 centimeters when a mass of 200 grams is attached to it.

   a) Set up a coordinate system on a sheet of graph paper. Since the length of the spring depends on the amount of mass that is attached to it, place the amount of mass on the horizontal axis and the length of the spring on the vertical axis. Scale each axis appropriately and plot each of the data points. Draw a line through your two data points.

   b) At what rate is the length of the spring stretching with respect to the amount of mass attached to it? Attach units to your answer and write a sentence explaining the meaning of this number.

20. Felipe begins walking to the library to work on his history project. After 5 minutes of walking, he is 2000 feet from the library; after 10 minutes, he is 500 feet from the library.

   a) Set up a coordinate system on a sheet of graph paper. Place the distance from the library on the vertical axis and the amount of time that Felipe has been walking on the horizontal axis. Plot the two given data points. Assume that the relationship is linear and draw a line through your two data points.

   b) Find the rate at which Felipe's distance from the library is decreasing with respect to time. Attach units to your answer and write a short sentence explaining the meaning of your answer.

2.3 The Slope-Intercept Form of a Line

Sections 2.1 and 2.2 laid the foundation for introducing the equation of a line. In this section we derive the *slope-intercept* form of a line. We discuss how to find the equation of a line from its graph. Conversely, when given the equation of the line, we discuss how to draw the graph of the equation. Finally, we revisit some applications of linear functions.

Example 1   It is a windy March afternoon in Boise, Idaho. Rahki is trying to launch a kite from her garage roof, which is about 10 feet above ground level. She plays out the string attached to her kite, and the height of her kite above ground level begins to increase at a steady rate of 20 feet per second.

Assuming that the relation between height and time is linear, draw a graph of the height of Rahki's kite as a function of time.

Solution Since the height of the kite depends upon the amount of time that has passed, place the height on the vertical axis and the time on the horizontal axis. Let the variable h represent the height of the kite above ground level and let the variable t represent the amount of time that has passed since Rahki launched her kite. Let each tick mark on the t-axis represent 1 second and each tick mark on the h-axis represent 10 feet.

Because the kite is 10 feet above ground level at time $t = 0$, begin by plotting the point $(0,10)$. Because the rate of ascent is 20 feet per second, each time you move 1 second to the right, you move 20 feet upward, as follows:

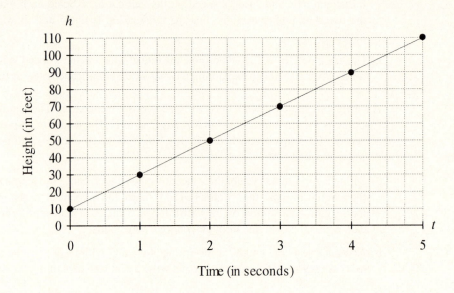

Example 2 Express the height, h, of Rahki's kite in Example 1 as a function of time, t.

Solution At time $t = 0$, the height of Rahki's kite is $h = 10 + 20(0)$ feet. At time $t = 1$, the height of Rahki's kite is $h = 10 + 20(1)$ feet, and so on. Arrange some of these results in a table.

t	h
0	10 + 20(0)
1	10 + 20(1)
2	10 + 20(2)
3	10 + 20(3)
4	10 + 20(4)
5	10 + 20(5)

A pattern emerges. It appears that the height, h, of Rahki's kite, as a function of time, t, is given by the following equation: $h = 10 + 20t$. This equation could also be written in the following form:

$$h = 20t + 10$$

The rate of ascent of Rahki's kite is 20 feet per second and the kite's initial height was 10 feet above the ground. It is important to note the position of these two numbers in the equation $h = 20t + 10$.

The Equation of a Line

We begin with a definition.

> *The point where the graph of a function intercepts the y-axis is called the y-intercept of the function.*

Find the equation of a line whose slope is the number m and whose y-intercept is the number b. You should begin with a sketch. Note that an arbitrary point (x, y) has been placed on the line.

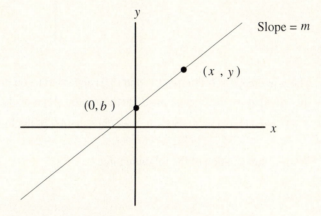

There are two points on the line, one given and one arbitrary. Use the slope formula from the previous section to compute the slope of the line.

$$\text{Slope} = \frac{\text{Change in } y}{\text{Change in } x}$$

$$m = \frac{y - b}{x - 0}$$

$$m = \frac{y - b}{x}$$

If you cross-multiply at this point, you arrive at the following result:

$$\frac{m}{1} = \frac{y - b}{x}$$

$$y - b = mx$$

Finally, if you add b to both sides of this equation, you arrive at the following result, called the *slope-intercept* form of a line.

$$y = mx + b$$

> *The equation of the line with y-intercept b and slope m is $y = mx + b$.*

Example 3   What is the equation of the following line?

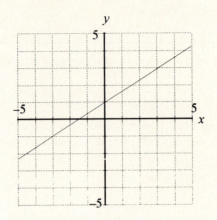

Solution   The y-intercept is 1. Points (0,1) and (3,3) appear to be on the line. The slope of the line is the change in y divided by the change in x. Therefore, the slope of the line is 2/3. If you substitute $m = \frac{2}{3}$ and $b = 1$ in the equation $y = mx + b$, then the equation of the line is $y = \frac{2}{3}x + 1$.

Example 4   What is the equation of the following line?

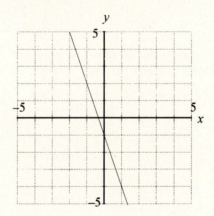

Solution   The y-intercept is -1. Points (0,-1) and (1,-4) appear to be on the line. The slope of the line is the change in y divided by the change in x. Therefore, the slope of the line is -3. If you substitute $m = -3$ and $b = -1$ in the equation $y = mx + b$, then the equation of the line is $y = -3x - 1$.

Example 5   Sketch the graph of the equation $y = -\frac{1}{2}x + 2$.

Solution   Since the equation is in the form $y = mx + b$, the graph of $y = -\frac{1}{2}x + 2$ will be a line with slope $-\frac{1}{2}$ and y-intercept 2. First, plot the point (0,2). Because the slope is $-\frac{1}{2}$, every time you move 2 units to the right, you must move 1 unit downward, as follows:

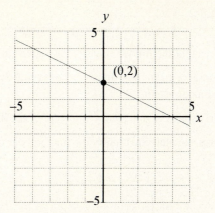

Example 6   Mandy's bike rental shop on Capistrano Beach rents bicycles to vacationers. Mandy charges a flat fee of $2.50 plus $5 for each hour of use. Let R represent the rental fee and t represent the time of use. Draw a graph of R versus t then find an equation that expresses R as a function of t.

Solution   Place the rental fee, R, on the vertical axis and the time of use, t, on the horizontal axis. Let each tick mark on the R-axis equal $2.50 and each tick mark on the t-axis equal 1 hour. The flat fee of $2.50 is the R-intercept of the line, so begin by plotting the point $(0, 2.50)$. The rental rate, which is $5 for each hour of use, is the slope of the line. Starting at point $(0, 2.50)$, each time you move 1 hour to the right, you move upward $5, as follows:

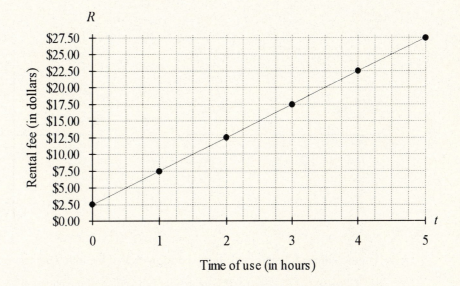

The slope-intercept form of a line is $y = mx + b$. However, y and x have been replaced with R and t, respectively. Therefore, the following changes must be made to the slope-intercept form of the line.

$$y = mx + b$$

$$R = mt + b$$

Finally, since the slope is $5 per hour and the R-intercept is $2.50, the equation of the line is as follows:

$$R = 5t + 2.50$$

Exercises for Section 2.3

1. Perform each of the following tasks for each of the following lines:

    i)   Set up a *separate* coordinate system on a sheet of graph paper and copy the line onto your coordinate system.

    ii) Clearly state the slope and *y*-intercept of the line on your paper.

    iii) Clearly state the equation of the line on your paper.

    a)   b)

    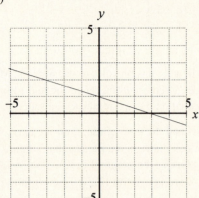   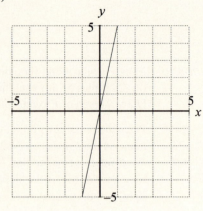

    c)   d)

    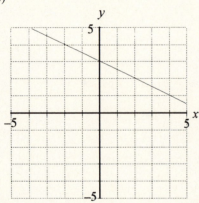   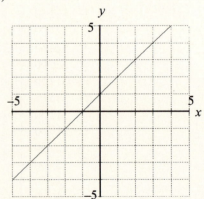

2. For each of the following equations, perform each of the following tasks:

    i)   Clearly state the slope and *y*-intercept of the graph of the equation.

    ii) For each line, set up a *separate* coordinate system on a sheet of graph paper and use the slope and *y*-intercept to draw the graph of the equation.

    a) $y = \dfrac{3}{4}x + 2$ b) $y = -\dfrac{3}{2}x + 1$ c) $y = -2x - 3$

    d) $y = 5x - 4$ e) $y = -\dfrac{1}{2}x + \dfrac{3}{2}$ f) $y = \dfrac{2}{3}x - \dfrac{5}{3}$

3. (TI-82) Enter each of the following functions in the Y= menu.

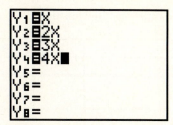

   a) Push the ZOOM button and select 6:ZStandard. Set up a coordinate system on a sheet of graph paper and clearly indicate the scale on each axis. Copy the image from your viewing window onto your coordinate system. Label each line with its equation.

   b) Explain, in your own words, why each of these lines goes through the origin.

4. (TI-82) Sketch the graph of $y = mx$ on your calculator for $m = 5, 6, 7, 8, 9,$ and 10. Copy these images onto a coordinate system on your homework paper.

   a) Explain, in your own words, what happens to the graph of the equation $y = mx$ as the value of m increases.

   b) What do you believe to be the slope of the y-axis?

5. (TI-82) Sketch the graph of $y = mx$ on your calculator for $m = 0.5, 0.4, 0.3, 0.2,$ and 0.1. Copy these images onto a coordinate system on your homework paper.

   a) Explain, in your own words, what happens to the graph of the equation $y = mx$ as the value of m approaches zero.

   b) What do you believe to be the slope of the x-axis?

6. (TI-82) Enter each of the following functions in the Y= menu.

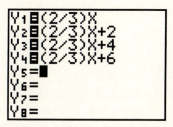

   a) Push the ZOOM button and select 6:ZStandard. Set up a coordinate system on a sheet of graph paper and clearly indicate the scale on each axis. Copy the image from your viewing window onto your coordinate system. Label each line with its equation.

   b) Explain, in your own words, why the lines are parallel.

7. Set up a coordinate system on a sheet of graph paper.

   a) Use the slope and y-intercept of the line whose equation is $y = \frac{1}{2}x - 2$ to draw the line on your coordinate system.

   b) Draw a line that so that the y-intercept is 4 and it is parallel to the line drawn in part (a). Clearly state the equation of this second line.

8. Consider the function f whose equation is $f(x) = \frac{2}{3}x + 1$. Complete the following table of points that satisfy the equation of the function f.

x	−12	−9	−6	−3	0	3	6	9	12
$f(x)$									
Points to plot									

a) Set up a coordinate system on a sheet of graph paper and plot the points from the table on your coordinate system. Draw a line through your points.

b) What is the change in consecutive x-values in the table? What is the change in consecutive y-values in the table? Use these changes to compute the slope of the line.

c) Explain, in your own words, how you can recognize that a function is linear from a table.

d) Which of the following tables were generated by linear functions? Explain how you made your decision for each table.

i)

x	1	2	3	4	5
y	2	5	8	11	14

ii)

x	2	5	8	11	14
y	15	13	11	9	7

iii)

x	0	5	10	15	20
y	1	2	4	8	16

9. Consider the following equations:

i) $y = 1.2x - 4.4$

ii) $y = 0.4x - 3.2$

iii) $y = -3.2x + 2.8$

iv) $y = -1.6x - 4.4$

Copy each of the following graphs on a *separate* coordinate system on your homework paper. Clearly label your graph with the equation from the above list of equations that is most likely the equation of the graph. Assume that the scale, though unknown, is the same on each of the four graphs.

Note: A calculator is not needed on this problem. A little understanding of the slope and y-intercept is all you will need to correctly match each equation with each graph.

a)

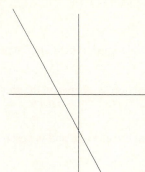

b)

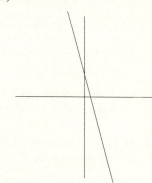

c)

d)

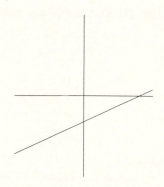

10. When creating a coordinate system for each of the following tasks, let each tick mark on each axis represent 1 unit.

   a) Set up a coordinate system on a sheet of graph paper. Sketch the graphs of $y = 2x$ and $y = -\frac{1}{2}x$ on your coordinate system.

   b) Set up a second coordinate system on your graph paper. Sketch the graphs of $y = -3x$ and $y = \frac{1}{3}x$ on this system.

   c) Set up a third coordinate system on your graph paper. Sketch the graphs of $y = 5x$ and $y = -\frac{1}{5}x$ on this system.

   d) Explain, in your own words, the geometrical relationship held in common by each pair of lines in parts (a), (b), and (c).

   e) Explain, in your own words, the numerical relationship between the slopes of each pair of lines in parts (a), (b), and (c).

11. When creating a coordinate system for each of the following tasks, let each tick mark on each axis represent 1 unit.

   a) Set up a coordinate system on a sheet of graph paper. Sketch the graphs of $y = \frac{2}{3}x$ and $y = -\frac{2}{3}x$ on your coordinate system.

   b) Set up a second coordinate system on your graph paper. Sketch the graphs of $y = -\frac{3}{4}x$ and $y = \frac{4}{3}x$ on this system.

c) Set up a third coordinate system on your graph paper. Sketch the graphs of $y = \frac{2}{5}x$ and $y = -\frac{5}{2}x$ on this system.

d) Explain, in your own words, the geometrical relationship held in common by each pair of lines in parts (a), (b), and (c).

e) Explain, in your own words, the numerical relationship between the slopes of each pair of lines in parts (a), (b), and (c).

12. What is the equation of the line that passes through the origin and is perpendicular to the graph of the equation $y = mx$?

Hint: See Exercises 10 and 11.

13. (TI-82) Load the equations $y = 2x$ and $y = -\frac{1}{2}x$ into the Y= menu in the following manner:

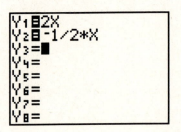

Caution! Do *not* enter the equations $y = 2x$ and $y = -\frac{1}{2}x$ as shown at the right. The equation entered in Y2 is actually the equation $y = -\frac{1}{2x}$, which is *not* the equation that should be graphed in this problem.	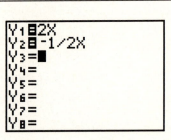

a) Press the ZOOM button and select 6:ZStandard. Copy the resulting image onto your homework paper. Note that the lines do not look perpendicular. This is due to distortion caused by the dimensions of the screen, which has more pixels horizontally than it does vertically.

b) Press the ZOOM button and select 5:ZSquare. This will set the WINDOW parameters so that the screen "squares up" and the lines will look perpendicular, as they should in this situation. Copy the image from this viewing window onto your homework paper. Label your axes with the WINDOW parameters.

c) Try 5:ZSquare on the following pairs of lines. Copy each image from the viewing window onto your homework paper. Label your axes with the WINDOW parameters.

i) $y = \frac{4}{5}x$ and $y = -\frac{5}{4}x$ ii) $y = -\frac{1}{4}x$ and $y = 4x$

14. Set up a coordinate system on a sheet of graph paper. Consider the line whose equation is $y = 3x + 2$. Use the slope and y-intercept to help draw the equation of this line.

    a) On the same coordinate system, draw a line through the origin that is perpendicular to the line whose equation is $y = 3x + 2$.

    b) Clearly state the equation of the line that you drew in part (a).

15. It's Christmas Eve at Macy's. At the beginning of the day, there were 100 teddy bears in stock. From the moment the store opened, the teddy bears were a popular item with the customers, being sold at a rate of 15 every 2 hours. Set up a coordinate system on a sheet of graph paper and place the number of teddy bears remaining in the store on the vertical axis and the amount of time that has passed since the store opened on the horizontal axis. Let N represent the number of teddy bears left in the store and let t represent the time that has passed since the store opened.

    a) Scale each axis appropriately. Assume that the relation between N and t is linear and draw the graph of N as a function of t on your coordinate system.

    b) Write an equation expressing N as a function of t.

16. Jasmijn loves to run each morning to school, a distance of about 10,000 feet. She leaves home at 7 A.M. and begins running toward school at a steady pace of 5 feet per second. Let D represent her distance from school and let t represent the amount of time that has passed since 7 A.M.

    a) Set up a coordinate system on a sheet of graph paper. Assuming that D is a linear function of t, sketch the graph of D as a function of t.

    Hint: It is somewhat challenging to scale the axes in this problem. However, note that 5 feet per second is equivalent to 50 feet every 10 seconds, which is equivalent to 500 feet every 100 seconds, and so on.

    b) Write an equation expressing D as a function of t.

17. Kyle is driving his jalopy at the local racetrack. He cranks his car up and reaches 150 miles per hour. At that point, he takes his foot off the accelerator and his car begins to lose speed at a steady 10 miles per hour every 20 seconds. Let S represent the speed of the car and let t represent the amount of time that has passed since Kyle took his foot off the accelerator.

    a) Set up a coordinate system on a sheet of graph paper and scale each axis appropriately. Assuming that S is a linear function of t, sketch the graph of S as a function of t.

    b) Express the slope of the line drawn in part (a) as a decimal. Include the appropriate units on your answer. Write a short sentence explaining the meaning of this number.

    c) Write an equation expressing S as a function of t.

18. Bob is in the gondola of his balloon at an altitude of 200 feet. In search of stronger winds, Bob works the controls of his balloon and begins to ascend at a steady rate of 100 feet per minute. Let H represent the height of the balloon above ground and let t represent the amount of time that has passed since Bob began his ascent.

    a) Set up a coordinate system on a sheet of graph paper. Assuming that H is a linear function of t, sketch the graph of H as a function of t.

    b) Write an equation expressing H as a function of t.

19. Suppose that the function f is linear. Furthermore, suppose that $f(0) = 3$ and $f(3) = 5$.

    a) Set up a coordinate system on a sheet of graph paper and plot the two given points. Draw a line through your points.

    b) Use the slope and y-intercept of your line to find the equation of the function f in the form $y = mx + b$.

    c) Remember, y and $f(x)$ are interchangeable. Use this fact to write the equation of the line in the form $f(x) = mx + b$. Use the equation of the function f to find $f(2)$.

2.4 The Point-Slope Form of a Line

The last section introduced the slope-intercept form of a line, $y = mx + b$. This form of the line is extremely useful, particularly when the slope and y-intercept are known. However, what happens if we do not know the y-intercept but are given a point on the line instead? We can still use the form $y = mx + b$, but it is a bit easier to use a form of the line called the *point-slope* form. In this section we'll discuss and use this new form of the line.

The Derivation of the Point-Slope Form of a Line

Suppose that you are given two pieces of information about a line: a point (x_0, y_0) that lies on the line and the slope m of the line. How can you find the equation of the line? First, draw a graph.

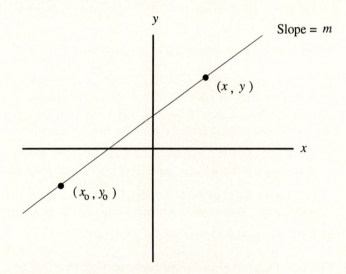

A line has been drawn through the given point (x_0, y_0) with slope m. In addition, an arbitrary point (x, y) has been selected on the line. Now use the slope formula to compute the slope of the line.

$$\text{Slope} = \frac{\text{Change in } y}{\text{Change in } x}$$

$$m = \frac{y - y_0}{x - x_0}$$

Now cross multiply.

$$\frac{m}{1} = \frac{y - y_0}{x - x_0}$$

$$y - y_0 = m\left(x - x_0\right)$$

This final result is known as the point-slope form of a line.

If you are given that the slope of the line is m and that the line passes through point $\left(x_0, y_0\right)$, *then the equation of the line is* $y - y_0 = m(x - x_0)$

You now have two forms of the line, $y = mx + b$ and $y - y_0 = m(x - x_0)$. You can use either one, but here is a good tip.

Tip | If you are given the slope and the *y*-intercept, use $y = mx + b$.
If you are given the slope and a point, use $y - y_0 = m\left(x - x_0\right)$.

Example 1   Draw the graph of the line that goes through (3,3) and has slope $\frac{2}{3}$. Use the point-slope form to find the equation of this line.

Solution   Plot the point (3,3). The slope is $\frac{2}{3}$. Each time you move 3 units to the right, you must move 2 units upward. Start at the point (3,3) and move 3 units to the right, then 2 units upward, arriving at the point (6,5). Draw a line through these two points.

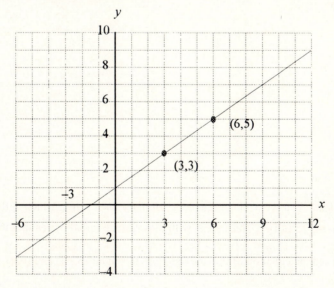

Use the point-slope form of a line to find the equation of the line. Insert $\frac{2}{3}$ for *m*, and then insert (3,3) for $\left(x_0, y_0\right)$.

$$y - y_0 = m\left(x - x_0\right)$$

$$y - 3 = \frac{2}{3}(x - 3)$$

Example 2   Use the graph from Example 1 to estimate the *y*-intercept of the line. Then use your equation to find the *exact* value of the *y*-intercept.

Solution   If you examine the graph, it appears that the *y*-intercept is about 1.

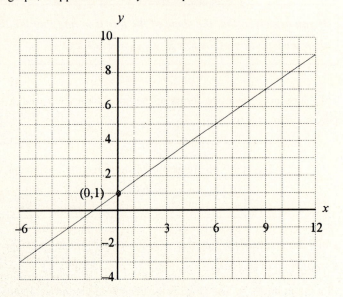

Put the equation into the form $y = mx + b$ by solving the equation for *y*.

$$y - 3 = \frac{2}{3}(x - 3)$$

$$y - 3 = \frac{2}{3}x - 2$$

$$y - 3 + 3 = \frac{2}{3}x - 2 + 3$$

$$y = \frac{2}{3}x + 1$$

This is now in slope-intercept form ($y = mx + b$), which tells you that the slope is $\frac{2}{3}$ and the *y*-intercept is 1. This agrees with the estimate found by reading the graph.

Example 3   Find the equation of the line that goes through (–3,–4) and (5,4).

Solution   Let's begin with a graph.

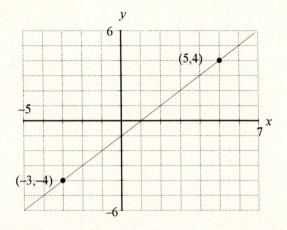

Use the slope formula to calculate the slope of the line that goes through $(-3,-4)$ and $(5,4)$.

$$\text{Slope} = \frac{\text{Change in } y}{\text{Change in } x} = \frac{4-(-4)}{5-(-3)} = \frac{8}{8} = 1$$

First Solution   If you use the point $\left(x_0, y_0\right) = (-3,-4)$ in the point-slope form of a line:

$$y - y_0 = m\left(x - x_0\right)$$

$$y - (-4) = 1(x - (-3))$$

$$y + 4 = 1(x + 3)$$

You can solve this equation for *y* to put it in slope-intercept form.

$$y + 4 = 1(x + 3)$$

$$y + 4 = x + 3$$

$$y + 4 - 4 = x + 3 - 4$$

$$y = x - 1$$

Second Solution   If you use the point $\left(x_0, y_0\right) = (5,4)$ in the point-slope form of a line:

$$y - y_0 = m\left(x - x_0\right)$$

$$y - 4 = 1(x - 5)$$

You can solve this equation for *y* to put it in slope-intercept form.

$$y - 4 = 1(x - 5)$$

$$y - 4 = x - 5$$

$$y - 4 + 4 = x - 5 + 4$$

$$y = x - 1$$

Note that it does not matter which point you use. Your final answer, in slope-intercept form, is $y = x - 1$, making the y-intercept of the line -1. When you examine your sketch, it does appear that the y-intercept is -1. This visualization, the ability to *see* and verify your results, develops confidence in your problem-solving ability.

Example 4   Suppose that the function f is linear. In addition, suppose that $f(2) = 1$ and $f(4) = 3$. Find $f(6)$.

Graphical Solution   Plot the point $(2, f(2))$, or $(2,1)$, and then plot the point $(4, f(4))$, or $(4,3)$. Because you know that the function is linear, draw a line through these two points, as follows:

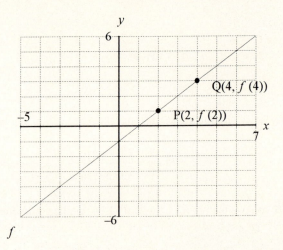

Next, locate the point on the line that has an x-value of 6.

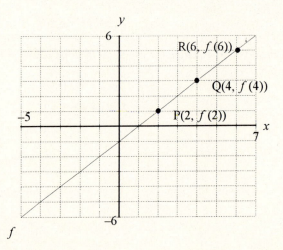

It appears that $f(6)$, the y-value of point R, is *approximately* equal to 5. Therefore, $f(6) \approx 5$. The symbol $\approx$ means "approximately equal to."

Analytical Solution   Find the equation of your line and use the equation to find $f(6)$. Begin by computing the slope of the line, using points P and Q.

$$\text{Slope} = \frac{\text{Change in } y}{\text{Change in } x} = \frac{f(4) - f(2)}{4 - 2} = \frac{3 - 1}{4 - 2} = \frac{2}{2} = 1$$

Now substitute $m = 1$ and $(x_0, y_0) = (2, f(2))$ in the point-slope form of a line.

$$y - y_0 = m\left(x - x_0\right)$$

$$y - f(2) = 1(x - 2)$$

$$y - 1 = 1(x - 2)$$

Now solve this equation for y.

$$y - 1 = 1(x - 2)$$

$$y - 1 = x - 2$$

$$y - 1 + 1 = x - 2 + 1$$

$$y = x - 1$$

Recall that y and $f(x)$ are interchangeable.

$$f(x) = x - 1$$

You can use this equation to compute the *exact* value of $f(6)$, as follows:

$$f(6) = 6 - 1$$

$$f(6) = 5$$

Note that this agrees nicely with your estimate from the graphical method. (Most of the time, you won't be so fortunate and your estimate will only closely agree with the exact solution.)

Example 5   Find the equation of the line that goes through point $(1,1)$ and is perpendicular to the line $y = 2x - 3$.

Solution   Begin by drawing the line whose equation is $y = 2x - 3$ and then plot the point $(1,1)$. The line has been labeled with the letter L and the point with the letter P for future reference.

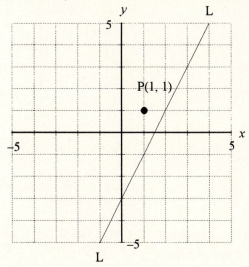

Next, draw a line through (1,1) that is perpendicular to line L.

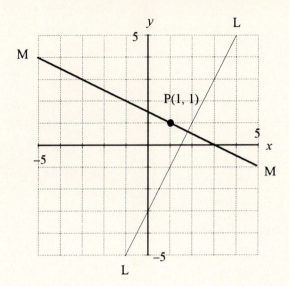

Select another point Q on the line M so that you can use your slope formula to find the slope of line M.

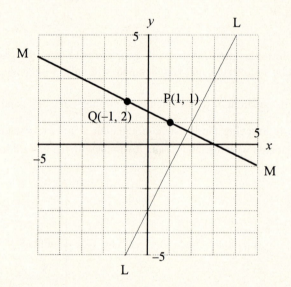

You can now use points P and Q to compute the slope of line M.

$$\text{Slope} = \frac{\text{Change in } y}{\text{Change in } x} = \frac{1-2}{1-(-1)} = -\frac{1}{2}$$

The slope of line L is 2 and the slope of line M is $-\frac{1}{2}$, the *negative reciprocal* of the slope of line L. This is not a coincidence.

> *If the slope of line L is m, then the slope of any line perpendicular to the line L is* $-\dfrac{1}{m}$, *provided that $m \neq 0$.*

Line M goes through point $(x_0, y_0) = (1,1)$ and has slope $m = -\frac{1}{2}$. You can use the point-slope form of a line to find the equation of line M.

$$y - y_0 = m(x - x_0)$$

$$y - 1 = -\frac{1}{2}(x - 1)$$

Place this equation in slope-intercept form by solving for y.

$$y - 1 = -\frac{1}{2}(x - 1)$$

$$y - 1 = -\frac{1}{2}x + \frac{1}{2}$$

$$y - 1 + 1 = -\frac{1}{2}x + \frac{1}{2} + 1$$

$$y = -\frac{1}{2}x + \frac{3}{2}$$

This last form implies that the y-intercept should be $\frac{3}{2}$ or 1.5. Examine the graph. Would you estimate that the y-intercept of the line M is 1.5?

Acceleration: The Runaway Truck Problem

Josephine Freeburg, the friendly California Highway Patrol officer, is waiting at the bottom of Loleta Hill on Highway 101 near College of the Redwoods. Josephine has just received a call from another officer near the top of the hill that a large semitrailer has lost its brakes and is accelerating down the hill. Ten seconds after receiving the call, Josephine spots the runaway truck, trains her radar gun it, and estimates its speed at 15 meters per second. After an additional 10 seconds pass, the truck draws nearer, and she estimates its speed to be 25 meters per second. Assume that the speed of the truck is a linear function of time. How fast will the truck be going when it passes Josephine, 30 seconds after the time that she received her initial warning?

Graphical Solution   First, as always, sketch a graph of the problem situation. The speed of the truck depends on the amount of time that has passed since Josephine's initial warning. Let S represent the speed of the truck and place it on the vertical axis. Let t represent the amount of time that has passed since Josephine received her first warning and place it on the horizontal axis. Ten seconds after her initial warning, the speed of the truck was 15 meters per second. Plot the point $(10,15)$ on your coordinate system. An additional 10 seconds pass and the radar captures the speed of the truck at 25 meters per second. Plot the point $(20,25)$ on your coordinate system. Because the relationship between speed and time is linear, draw a line through your two data points, as follows:

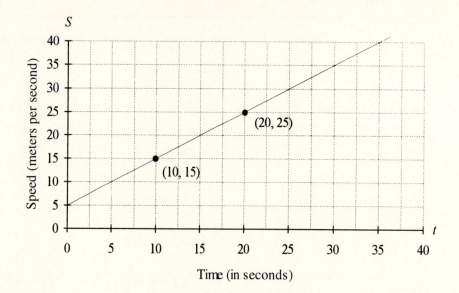

What is the speed of the truck when it passes Josephine, 30 seconds after her initial warning? Locate 30 seconds on the *t*-axis and draw a vertical line to the graph, followed by a horizontal line to the *S*-axis.

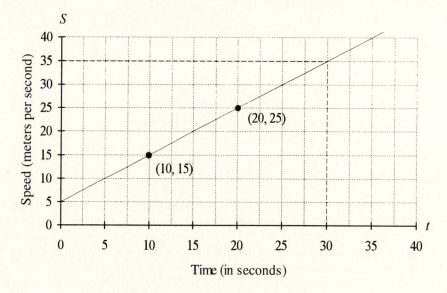

It appears that the answer is approximately 35 meters per second. This is read on the *S*-axis.

Analytical Solution   Begin by computing the slope of your line.

$$\text{Slope} = \frac{\text{Change in } y}{\text{Change in } x} = \frac{25-15}{20-10} = \frac{10}{10} = 1$$

Note: If you attach the appropriate units to your slope, then the slope is 1 meter per second per second. This is the rate at which the speed is changing with respect to time. Each second of time that elapses, the speed of the truck increases by 1 meter per second. This is commonly referred to as the *acceleration* of the truck.

Now substitute $m=1$ and $(x_0, y_0) = (10, 15)$ in the point-slope form of a line. Of course, you must remember that you have replaced *y* and *x* with *S* and *t*, respectively.

$$y - y_0 = m(x - x_0)$$

$$S - S_0 = m(t - t_0)$$

$$S - 15 = 1(t - 10)$$

Now, solve your equation for S:

$$S - 15 = 1(t - 10)$$

$$S - 15 = t - 10$$

$$S - 15 + 15 = t - 10 + 15$$

$$S = t + 5$$

The speed S is expressed as a function of the time t. Finally, if you want to determine the speed when $t = 30$ seconds, substitute $t = 30$ in the equation $S = t + 5$. You will arrive at an *exact* answer of $S = 30 + 5 = 35$ meters per second.

Exercises for Section 2.4

1. Perform each of the following tasks for each of the following problems:

   i) Set up a coordinate system on a sheet of graph paper and draw the line that goes through the given point with the given slope.

   ii) Clearly state the equation of the line in the form $y - y_0 = m(x - x_0)$.

   a) $(2,-3)$, $m = 2$ b) $(-3,4)$, $m = -3$ c) $(-4,-2)$, $m = \dfrac{2}{3}$

   d) $(5,-1)$, $m = -\dfrac{3}{2}$ e) $(-4,5)$, $m = 0$ f) $(3,-4)$, $m = 0$

2. Set up a coordinate system on a sheet of graph paper. Plot point P(2,3) and draw a line through point P with slope $\frac{1}{2}$.

   a) Use your graph to estimate the y-intercept. Clearly state this estimate on your paper.

   b) Use the point-slope form of a line to find the equation of your line.

   c) Place the equation of your line in the slope-intercept form by solving for y. Clearly state the *exact* value of the y-intercept on your paper. Does this answer agree with your estimate from part (a)? If not, check your work for error.

3. Set up a coordinate system on a sheet of graph paper. Plot point P(3,3) and draw a line through point P with slope $-\frac{2}{3}$.

   a) Use your graph to estimate the y-intercept. Clearly state this estimate on your paper.

   b) Use the point-slope form of a line to find the equation of your line.

   c) Place the equation of your line in the slope-intercept form by solving for y. Clearly state the *exact* value of the y-intercept on your paper. Does this answer agree with your estimate from part (a)? If not, check your work for error.

4. Find the exact value of the y-intercept of the line that goes through P($-1,2$) with slope $\frac{2}{5}$. Include a clearly labeled diagram with your work.

5. Find the equation of the line whose x-intercept is 4 and whose y-intercept is 5.

6. Set up a coordinate system on a sheet of graph paper and plot points P($1,1$) and Q($3,5$) on your coordinate system. Draw a line through points P and Q.

   a) Use the slope formula to find the slope of the line through points P and Q.

   b) Using the point-slope form of a line and the point P, find the equation of your line. Place this result in slope-intercept form by solving for the variable y.

   c) Using the point-slope form of a line and the point Q, find the equation of your line. Place this result in slope-intercept form by solving for the variable y. Does this solution agree with the solution found in part (b)? If not, check your work for error.

7. Set up a coordinate system on a sheet of graph paper and plot points P($-1,7$) and Q($1,1$) on your coordinate system. Draw a line through the points P and Q.

   a) Use the slope formula to find the slope of the line through points P and Q.

   b) Using the point-slope form of a line and the point P, find the equation of your line. Place this result in slope-intercept form by solving for the variable y.

   c) Using the point-slope form of a line and the point Q, find the equation of your line. Place this result in slope-intercept form by solving for the variable y. Does this solution agree with the solution found in part (b)? If not, check your work for error.

8. Perform each of the following tasks for each of the following problems:

   i) Set up a coordinate system on a sheet of graph paper and plot the given points. Draw a line through the two points.

   ii) Clearly state the equation of the line in the form $y - y_0 = m(x - x_0)$. You may use either point you wish for (x_0, y_0). (Ambitious students may try both points and compare the results.)

   a) ($-2,3$) and ($4,-1$)   b) ($-3,-2$) and ($4,5$)   c) ($-2,3$) and ($5,3$)   d) ($-3,-2$) and ($5,-4$)

9. Perform each of the following tasks for each of the following problems:

   i) Set up a coordinate system and plot each of the given points on your coordinate system. Draw a line through your points.

   ii) Clearly state an *estimate* of the y-intercept on your paper.

   iii) Use the point-slope form to find the equation of your line. Place your result in slope-intercept form by solving your result for the variable y. Clearly state the *exact* value of the y-intercept. If this exact value does not agree closely with your estimate from part (ii), check your work for error.

   a) ($-2,1$) and ($2,3$)   b) ($-1,5$) and ($2,4$)   c) ($-3,3$) and ($2,1$)   d) ($-7,5$) and ($3,7$)

10. Find the *y*-intercept of each of the following lines. The scale has been deliberately left off.

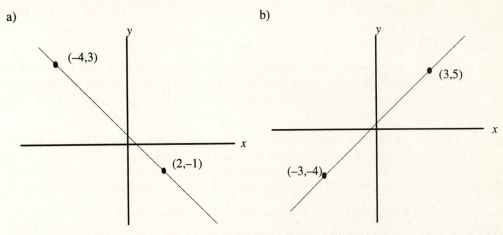

a)   b)

11. Suppose that *f* is a linear function. In addition, suppose that $f(1) = 3$ and $f(2) = 5$. Set up a coordinate system on a sheet of graph paper and plot the given points on your coordinate system. Draw a line through the points.

    a) Use your graph to *estimate* $f(-1)$. Clearly state this estimate on your paper.

    b) Use the slope formula to find the slope of the line.

    c) Use the point-slope form of a line to find the equation of this line. Place your answer in slope-intercept form by solving your equation for the variable *y*. Finally, replace *y* with $f(x)$.

    d) Use this last equation to find $f(-1)$. Does your answer agree closely with the estimate you found in part (a)? If not, check your work for error.

12. Suppose that *f* is a linear function. In addition, suppose that $f(-3) = 5$ and $f(4) = -1$. Set up a coordinate system on a sheet of graph paper and plot the given points on your coordinate system. Draw a line through the points.

    a) Use your graph to *estimate* $f(3)$. Clearly state this estimate on your paper.

    b) Use the slope formula to find the slope of the line.

    c) Use the point-slope form of a line to find the equation of this line. Place your answer in slope-intercept form by solving your equation for the variable *y*. Finally, replace *y* with $f(x)$.

    d) Use this last equation to find $f(3)$. Does your answer agree closely with the estimate you found in part (a)? If not, check your work for error.

13. Set up a coordinate system on a sheet of graph paper and use the slope and *y*-intercept of the line whose equation is $y = 3x + 2$ to draw the graph of the line on your coordinate system. Plot the point $(-2, 2)$ on your coordinate system. Label the line with the letter L and the point with the letter P for future reference.

    a) Draw a line M through point P that is perpendicular to line L.

    Hint: The slope of the line M is $-\frac{1}{3}$. Make sure that this is evident in your sketch.

    b) Use the point-slope form to find the equation of line M. Place this result in slope-intercept form by solving for the variable *y*.

14. Set up a coordinate system on a sheet of graph paper and draw the line M that goes through the point P$(-3,-2)$ and is perpendicular to the line $y = -\frac{3}{4}x + 4$. What is the *exact* value of the y-intercept of this line?

15. Water freezes at $32°$ Fahrenheit and $0°$ Celsius; it boils at $212°$ Fahrenheit and $100°$ Celsius. Use these two facts to express Celsius temperature as a function of Fahrenheit temperature. Earlier examples and exercises have shown that the relationship between Celsius and Fahrenheit temperature is linear.

   a) Set up a coordinate system on a sheet of graph paper. Place the Celsius temperature on the vertical axis and scale this axis appropriately. Let the variable C represent the Celsius temperature. Place the Fahrenheit temperature on the horizontal axis and scale this axis appropriately. Let the variable F represent the Fahrenheit temperature.

   b) Plot the two given data points and draw a line through them.

   c) Use the slope formula to calculate the slope of the line through your two points. What are the units associated with the slope? Explain, in your own words, the meaning of this number.

   d) Use the point-slope form of the line to find the equation of the line through your two data points. Be sure to replace the variables y and x in the form $y - y_0 = m(x - x_0)$ with the variables C and F, respectively.

   e) Solve the equation found in part (d) for the variable C. Use this equation to find the Celsius temperature if the Fahrenheit temperature is $80°$.

16. Suppose that the demand for a particular brand of teakettle is a function of its unit price and that this relationship is linear. When the unit price is fixed at \$30, the demand for teakettles is 100. This means that the public buys 100 teakettles. If the unit price is fixed at \$50, the demand for teakettles is 60.

   a) Set up a coordinate system on a sheet of graph paper. Place the demand for teakettles on the vertical axis and scale the axis appropriately. Use the variable D to represent the demand. Place the unit price on the horizontal axis and scale the horizontal axis appropriately. Use the variable p to represent the unit price.

   b) Plot the given data points and draw a line through them.

   c) Use the slope formula to find the slope of the line through your two data points. What are the units associated with the slope? Explain, in your own words, the meaning of this number.

   d) Use the point-slope form of a line to find the equation of the line through your two data points. Be sure to replace the variable y and x in the form $y - y_0 = m(x - x_0)$ with the variables D and p, respectively.

   e) Solve your equation for the variable D. You have now expressed the demand as a function of the unit price. Use this equation to compute the demand if the unit price is set at \$40.

17. It's perfect kite-flying weather on the coast of Oregon. Annie grabs her kite, climbs up on the roof of her two-story home, and begins playing out kite string. In 10 seconds, Annie's kite is 120 feet above the ground. After 20 seconds, it is 220 feet above the ground. Assume that the height of the kite above the ground is a linear function of the amount of time that has passed since Annie began playing out kite string.

   a) Set up a coordinate system on a sheet of graph paper. Place the height on the vertical axis and scale the axis appropriately. Let the variable h represent the height of the kite above the ground. Place the time on the horizontal axis and scale the axis appropriately. Let the variable t

represent the time that has passed since Annie began playing out the kite string. Plot the given data points on your coordinate system and draw a line through them.

b) Use the slope formula to calculate the slope of the line. What are the units associated with the slope? Explain, in your own words, the meaning of this number.

c) Use the point-slope form of the line to find the equation of the line through the two data points. Be sure to replace the variables y and x in the form $y - y_0 = m(x - x_0)$ with the variables h and t, respectively.

d) Solve your equation for the variable h. You have now expressed the height of the kite above the ground as a function of the amount of time that has passed since Annie began playing out kite string. Use your equation to find the height of the kite after 30 seconds have passed.

e) What is the height of Annie's garage?

18. Janelle runs a business that installs stereo systems in automobiles. If she installs 10 systems in a month, her costs are $1000. If she installs 20 systems in a month, her costs are $1500. Assume that Janelle's costs are a linear function of the number of systems she installs.

a) Set up a coordinate system on a sheet of graph paper. Let C represent Janelle's costs and place it on the vertical axis. Let x represent the number of systems she installs in a month and place it on the horizontal axis. Plot the given data points on your coordinate system and draw a line through them.

b) Use the slope formula to calculate the slope of your line. What are the units associated with the slope? Explain, in your own words, the meaning of this number.

c) Use the point-slope form of the line to find the equation of the line through your two data points. Be sure to replace the variables y and x in the form $y - y_0 = m(x - x_0)$ with the variables C and x, respectively.

d) Solve your equation for the variable C. You have now expressed Janelle's costs as a function of the number of systems she installs. Use this result to compute Janelle's costs if she installs 15 systems.

e) What are Janelle's *fixed costs*, the costs she incurs even if she installs no systems. Write a short paragraph explaining the possible meaning of these fixed costs.

19. (TI-82) Perform each of the following tasks for each of the following problems:

i) Set up a coordinate system on graph paper and draw the line that goes through the given points.

ii) Use the slope formula and your calculator to compute the slope of the line that goes through the given points. Round your answer to the nearest hundredth (two decimal places).

iii) Use the point-slope form to find the equation of the line. Place your answer in slope-intercept form by solving for the variable y. Again, use your calculator to assist you and round all numbers to the nearest hundredth (two decimal places).

a) $(-2.2, 4.5)$ and $(3.9, -1.7)$

b) $(-4.6, -5.3)$ and $(3.9, 4.0)$

c) $(-2.12, -3.12)$ and $(4.53, 5.15)$

d) $(-2.23, 4.56)$ and $(3.89, -5.99)$

20. Rudy has set up a spring in his lab. He attaches various weights to it, then measures its total length. Here are his data.

Weight in pounds	4	8	12	16	20
Length in inches	14.2	15.7	18.2	19.8	22.4

The weight is measured in pounds and the length of the spring is measured in inches. Set up a coordinate system on a sheet of graph paper. Let L represent the length of the spring and place it on the vertical axis. Let W represent the weight and place it on the horizontal axis. Scale each axis appropriately.

a) Plot the data points on your coordinate system and draw the line of best fit.

   Note: You might want to review Section 1.1 before continuing.

b) Select two points P and Q on your line of best fit *other than your original data points* and estimate the coordinates of your selected points. Use these two points and the slope formula to find the slope of your line of best fit. Use your calculator and round your answer to the nearest tenth (one decimal place).

c) Select one of two points P and Q and the slope computed in part (b) to find the equation of the line of best fit. Use the form $y - y_0 = m(x - x_0)$ but replace the variables y and x with the variables L and W, respectively.

d) Solve the equation found in part (c) for the variable L. Again, use your calculator to assist you and round all numbers to the nearest tenth (1 decimal place). Use your equation to predict the length of the spring, correct to the nearest tenth of an inch (one decimal place), if the weight attached is 18 pounds.

e) What is the length of the spring if no weight is attached to the spring?

2.5 Solving Linear Equations

In this section, we'll discuss how to solve *linear equations*, so-called because the graphs involved are lines. The techniques learned in earlier courses will be reviewed, but the *graphing approach* to solving linear equations will also be introduced. More work will also be done with graphing calculators.

Example 1   Sketch the graph of the function f whose equation is $f(x) = 4$.

Solution   Remember, y and $f(x)$ are interchangeable, so you are being asked to sketch the graph of the function whose equation is $y = 4$. Therefore, you want to graph all the points whose y-value equals 4. In the sketch that follows, a number of points whose y-value equals 4 have been plotted.

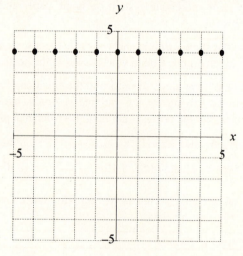

A pattern emerges. Plotting the remainder of the points that satisfy the equation $y = 4$ results in the following graph:

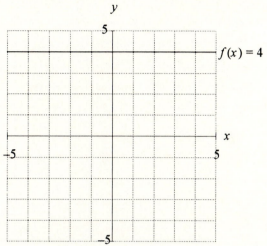

Note that $f(-5) = 4$, $f(-4) = 4$, $f(-3) = 4$, and so on. The y-values of the function f *constantly* equal 4, no matter what the x-value.

> **The function $f(x) = 4$ is called a constant function, because the function values constantly equal 4, no matter what the x-values.**

Necessary Fundamentals. You will need to review some important fundamentals before you begin to solve linear equations.

> **To solve the equation $f(x) = g(x)$ for x, draw the graphs of the functions f and g and note the x-value of the point of intersection of the graphs of f and g.**

There are several important ideas that you need to understand if you are going to successfully analyze equations. First, you may add or subtract any real number from both sides of an equation.

> **If $a = b$ and c is any real number, then $a + c = b + c$ and $a - c = b - c$.**

You may also multiply or divide both sides of an equation by any nonzero real number.

> **If $a = b$ and c is any nonzero real number, then $ac = bc$ and $\dfrac{a}{c} = \dfrac{b}{c}$.**

Example 2   Solve the following equation for x: $2x + 3 = 5$.

Graphical Solution   The equation $2x + 3 = 5$ is of the form $f(x) = g(x)$, where $f(x) = 2x + 3$ and $g(x) = 5$. Your first approach will be to draw the graphs of the functions f and g and note the x-value of the point of intersection. The graph of *f* is a line with slope 2 and *y*-intercept 3, and the graph of *g* is a horizontal line.

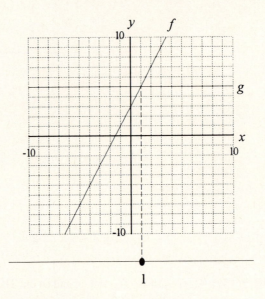

The *x*-value of the point of intersection of the two graphs is $x = 1$.

Analytical Solution   Let's begin with a tip for solving linear equations.

Tip	When solving a *linear* equation, try to isolate all of the terms containing the variable you are solving for on one side of the equation, leaving the rest of the terms on the other side of the equation.

Keeping the tip in mind, isolate all of the terms containing *x* on one side of the equation while gathering the rest of the terms on the other side of the equation. Add −3 to both sides of the equation and simplify:

$$2x + 3 = 5$$
$$2x + 3 - 3 = 5 - 3$$
$$2x = 2$$

Divide both sides of the equation by 2 and simplify.

$$\frac{2x}{2} = \frac{2}{2}$$
$$x = 1$$

It is important to note that the analytical solution matches the graphical solution. The approach used in this problem is one of the unifying themes of this book. We first look at a graphical solution, then at an analytical solution, and the solutions from each technique must agree.

Some Thoughts on Skipping Steps. Some readers may feel quite comfortable presenting each of the steps in the solution process, as done above. Others may feel that several steps in the previous analysis were unnecessary and that they can perform some of the work mentally, arranging their work as follows:

$$2x + 3 = 5$$

$$2x = 2$$

$$x = 1$$

When called upon to justify their work, they respond in the following manner: "I first subtracted 3 from both sides of the equation, simplified mentally and wrote $2x = 2$. I then divided both sides of the equation by 2, simplified mentally and wrote $x = 1$."

So, just what is the rule about skipping steps? Well, if you are skipping some of the algebraic steps, and you understand the underlying processes, and your problem is free of mistakes, that's great! If you feel that every time you skip steps you lose a bit of understanding and make errors, that's fine, too! Never hesitate to add an extra step for understanding or delete a step that you think is trivial. However, if you are called upon to justify your work, be sure that you are prepared to explain the underlying principles.

Some Thoughts on Organizing Your Work. We strongly believe the graphical solution and the analytic complement each other. *We expect our students to give both solutions*—the analytic solution on one half of the page and the graphical solution on the other half.

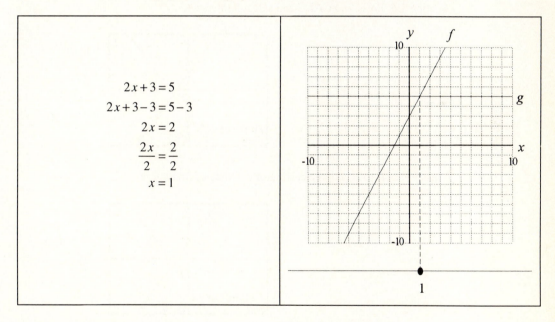

Example 3 Use the TI-82 graphing calculator to help solve the following equation for x:

$$3 - 2x = 2x - 5.$$

Graphical Solution   The equation $3 - 2x = 2x - 5$ is of the form $f(x) = g(x)$, where $f(x) = 3 - 2x$ and $g(x) = 2x - 5$. First draw the graphs of the functions f and g and note the x-value of the point of intersection. However, this time you may use your graphing calculator. Begin by loading the functions f and g into the Y= menu, as follows:

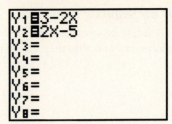

Push the ZOOM button and select 6:ZStandard, capturing the following image in your viewing window.

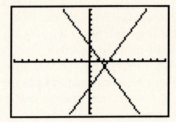

At this point you could push the TRACE button and use the arrow-keys to center the trace cursor over the point of intersection. However, it is better to use the intersection utility in the CALC menu of the TI-82. You might want to review the introduction to the intersection utility presented in Section 1.6. Press 2nd CALC and select 5:intersect.

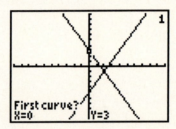

Press ENTER to accept the calculator's choice of first curve.

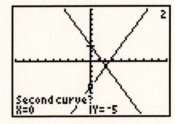

Press ENTER to accept the calculator's choice of second curve.

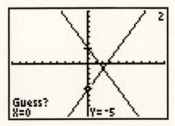

Press ENTER to send the current x-value of the cursor to the calculator as your guess.

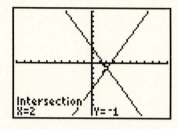

The solution of $3 - 2x = 2x - 5$ is $x = 2$, the x-value of the point of intersection.

Note: TI-81 users should consult the appendix for a program called INTERSCT which will enable them to emulate the intersection utility of the TI-82.

Analytical Solution   Isolate the terms with x in them on one side of the equation and the rest of the terms on the other side of the equation. Remove the $2x$ from the right side of the equation by subtracting $2x$ from both sides:

$$3 - 2x = 2x - 5$$
$$3 - 2x - 2x = 2x - 5 - 2x$$
$$3 - 4x = -5$$

Next, remove the 3 from the left side of the equation by subtracting 3 from both sides:

$$3 - 4x - 3 = -5 - 3$$
$$-4x = -8$$

Finally, divide both sides of the equation by -4:

$$\frac{-4x}{-4} = \frac{-8}{-4}$$
$$x = 2$$

Arrange your analytic solution and graphical solution side-by-side for easy comparison.

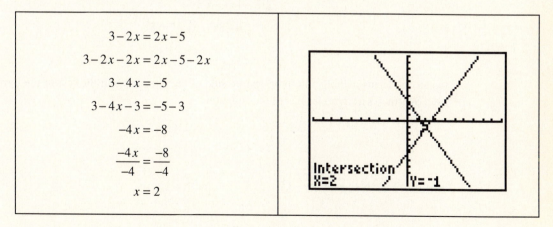

Note: When writing this textbook, we used TI-82 Link software to create the graph. When students use a calculator to find the solution, they simply duplicate the image from their viewing window onto their homework or examination paper. Then they draw a number line below their graph, drop a dotted line from the point of intersection, and shade their solution on their number line, as follows:

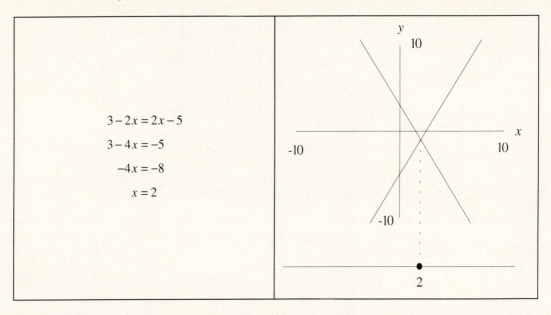

$$3 - 2x = 2x - 5$$
$$3 - 4x = -5$$
$$-4x = -8$$
$$x = 2$$

Note how this student included the WINDOW parameters on her sketch. Knowing the scale on the axes is essential and should always be included when possible. Also, note that this student skipped a few steps in her analytic method. Can you explain the steps she skipped?

Developing Analytical Skills

The analytic method combined with the graphing method is a union made in math heaven. One always accompanies and reinforces the other. However, for a little while, we will concentrate on the analytic method and place graphing method on hold. We want to concentrate on sharpening analytic skills. Soon, we will return to solving each problem twice, using both the graphing method and the analytic method.

Example 4   Solve the following equation for x: $2(3 - 4x) - 3(2x + 1) = 8$.

Solution   First, use the distributive property.

$$2(3 - 4x) - 3(2x + 1) = 8$$
$$6 - 8x - 6x - 3 = 8$$
$$-14x + 3 = 8$$

Isolate the terms with an x in them on one side of the equation and the rest of the terms on the other side. Subtract 3 from both sides of the equation.

$$-14x + 3 - 3 = 8 - 3$$
$$-14x = 5$$

Finally, divide both sides by -14.

$$\frac{-14x}{-14} = \frac{5}{-14}$$

$$x = -\frac{5}{14}$$

Example 5   Solve the following equation for x: $\frac{1}{2}x + \frac{1}{3} = \frac{3}{4}x - \frac{1}{2}$.

Solution   You can remove the fractions from this problem by multiplying both sides by the least common denominator.

$$12\left[\frac{1}{2}x + \frac{1}{3}\right] = 12\left[\frac{3}{4}x - \frac{1}{2}\right]$$

Next, use the distributive property to multiply.

$$12\left[\frac{1}{2}x\right] + 12\left[\frac{1}{3}\right] = 12\left[\frac{3}{4}x\right] - 12\left[\frac{1}{2}\right]$$

$$6x + 4 = 9x - 6$$

Isolate the terms with x on one side of the equation. Subtract $9x$ from both sides of the equation, effectively removing all of the terms with an x in them from the right side of the equation.

$$6x + 4 - 9x = 9x - 6 - 9x$$

$$-3x + 4 = -6$$

Subtract 4 from both sides of the equation, which will finally isolate the terms with an x in them on the left side of the equation.

$$-3x + 4 - 4 = -6 - 4$$

$$-3x = -10$$

Finally, divide both sides of the equation by -3.

$$\frac{-3x}{-3} = \frac{-10}{-3}$$

$$x = \frac{10}{3}$$

Example 6   (TI-82) Solve the following equation for x: $x - \frac{x-1}{2} = 3$.

Graphical Solution   Push the Y= button and enter the following functions into Y_1 and Y_2. Note the use of the parentheses.

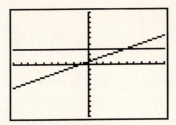

Push the ZOOM button and select 6:ZStandard to capture the following image in your viewing window:

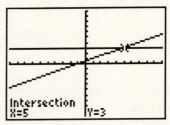

Press 2nd CALC and select 5:intersect. Press ENTER to select the default for first curve, ENTER to select the default for the second curve, and ENTER to select the current x-value of the trace cursor for a guess. The calculator will respond with the following image:

The solution of $x - \dfrac{x-1}{2} = 3$ is the x-value of the point of intersection, which is $x = 5$.

Analytical Solution   Multiply both sides of the equation by the least common denominator and use the distributive property.

$$2\left[x - \frac{x-1}{2}\right] = 2[3]$$

$$2[x] - 2\left[\frac{x-1}{2}\right] = 2[3]$$

Multiply, distribute the subtraction sign, then simplify.

$$2x - [x-1] = 6$$
$$2x - x + 1 = 6$$
$$x + 1 = 6$$

Subtract 1 from both sides of the equation and simplify.

$$x + 1 - 1 = 6 - 1$$

$$x = 5$$

Here is how one student arranged his work for this problem on his examination paper.

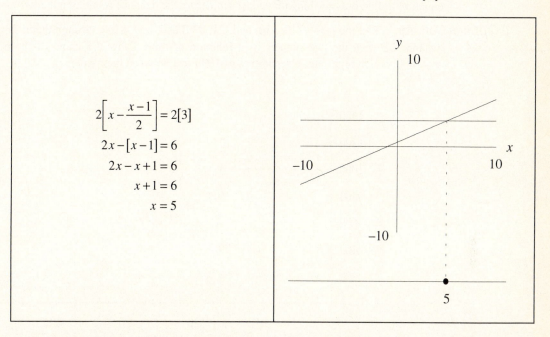

Notice that this student included the WINDOW parameters with his graph. It is essential to know the scale when interpreting a graph, so include the scale whenever possible. Also, note that this student skipped a few steps in his analytic method. Can you explain what he did?

Example 7   Solve the following equation for x: $0.15x - 0.03 = 0.45 + 0.02x$.

Solution   Isolate x on one side of the equation. Begin by subtracting $0.02x$ from both sides of the equation.

$$0.15x - 0.03 = 0.45 + 0.02x$$

$$0.15x - 0.03 - 0.02x = 0.45 + 0.02x - 0.02x$$

$$0.13x - 0.03 = 0.45$$

Add 0.03 to both sides of the equation.

$$0.13x - 0.03 + 0.03 = 0.45 + 0.03$$

$$0.13x = 0.48$$

Divide both sides of the equation by 0.13 and use a calculator to round the answer to the nearest hundredth (two decimal places).

$$\frac{0.13x}{0.13} = \frac{0.48}{0.13}$$

$$x = 3.69$$

Note: This solution is only an approximation. However, approximate answers can be quite useful and are perfectly acceptable in many situations.

Example 8   Brendan and Jennifer are two long-distance runners. After 10 minutes of running, they are separated by a distance of 550 feet. After 21 minutes of running, they are separated by a distance of 730 feet. Let D represent the distance separating the two runners. Let t represent the amount of time that they have been running. Assume that D is a linear function of t and express D as a function of t. Use this equation to find out how long they must run before the distance separating the two is 1000 feet.

Graphical Solution   Plot the points $(10, 550)$ and $(21, 730)$ and draw a line through them. This line shows the distance between the two runners as a function of time. Draw the graph of the constant function $y = 1000$, which represents a separation of 1000 feet between the runners.

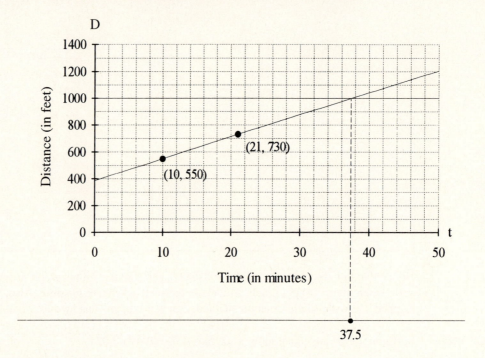

The separation between the runners appears to reach 1000 feet after approximately 37.5 minutes of running.

Analytical Solution   Use the slope formula to compute the slope of the line.

$$\text{Slope} = \frac{\text{Change in } D}{\text{Change in } t} = \frac{730 - 550}{21 - 10} \approx 16.4 \text{ feet per minute}$$

Replace y and x in the point-slope formula with D and t, respectively.

$$y - y_0 = m(x - x_0)$$

$$D - D_0 = m(t - t_0)$$

Place the point $(10, 550)$ and the slope 16.4 in this formula.

$$D - 550 = 16.4(t - 10)$$

Express D as a function of t by solving this equation for D.

$$D - 550 = 16.4t - 164$$

$$D - 550 + 550 = 16.4t - 164 + 550$$

$$D = 16.4t + 386$$

Use this formula to find the amount of time that must pass before the runners are separated by 1000 feet. Begin by substituting 1000 for D, then subtracting 386 from both sides of the equation.

$$1000 = 16.4t + 386$$

$$1000 - 386 = 16.4t + 386 - 386$$

$$614 = 16.4t$$

Exchange sides of the equation, divide both sides of the equation by 16.4, then use a calculator to round your answer to the nearest tenth of a minute (one decimal place).

$$16.4t = 614$$

$$\frac{16.4t}{16.4} = \frac{614}{16.4}$$

$$t = 37.4 \text{ minutes}$$

This agrees quite favorably with the estimate found from the graph.

Exercises for Section 2.5

1. (TI-82) Load each of the following *constant* functions into the Y= menu, as shown:

Press the ZOOM button and select 6:ZStandard. Set up a coordinate system on a sheet of graph paper and use the image in your viewing window to help draw *accurate* graphs of each of these *constant* functions on your coordinate system.

2. Set up a coordinate system on a sheet of graph paper. Consider the functions f and g whose equations are $f(x) = 2x - 1$ and $g(x) = 3$. Sketch the graphs of the functions f and g on your coordinate system.

   a) Draw a number line below your graph, drop a dotted line from the point of intersection of the two graphs to your number line, then shade and label the x-value of the point of intersection on your number line.

   b) Use an analytic method to solve the following equation for x: $2x - 1 = 3$. Does this analytic solution compare favorably with the solution found using the graphing method in part (a)? If not, check your work for error.

3. Set up a coordinate system on a sheet of graph paper. Consider the functions f and g whose equations are $f(x) = 3x + 2$ and $g(x) = 5$. Sketch the graphs of the functions f and g on your coordinate system.

   a) Draw a number line below your graph, drop a dotted line from the point of intersection of the two graphs to your number line, then shade and label the x-value of the point of intersection on your number line.

   b) Use an analytic method to solve the following equation for x: $3x + 2 = 5$. Does this analytic solution compare favorably with the solution found by using the graphing method in part (a)? If not, check your work for error.

4. Use an analytic method only to solve each of the following equations for x.

   a) $2x - 3 = 7$ b) $5 - 3x = -10$ c) $4 - 7x = 12$ d) $3x + 4 = -9$

5. Use *both* a graphing method and an analytic method to solve each of the following equations for x. Arrange your solutions side-by-side for comparison, as suggested in Example 2 in this section.

   a) $3 - 2x = 7$ b) $4x + 3 = 11$ c) $5 - x = -6$ d) $-2x + 9 = -11$

6. Set up a coordinate system on a sheet of graph paper. Consider the functions f and g whose equations are $f(x) = 2x + 1$ and $g(x) = x + 4$. Sketch the graphs of the functions f and g on your coordinate system.

   a) Draw a number line below your graph, drop a dotted line from the point of intersection of the two graphs to your number line, then shade and label the x-value of the point of intersection on your number line.

   b) Use an analytic method to solve the following equation for x: $2x + 1 = x + 4$. Does this analytic solution compare favorably with the solution found by using the graphing method in part (a)? If not, check your work for error.

7. Set up a coordinate system on a sheet of graph paper. Consider the functions f and g whose equations are $f(x) = 3 - x$ and $g(x) = 2x - 3$. Sketch the graphs of the functions f and g on your coordinate system.

   a) Draw a number line below your graph, drop a dotted line from the point of intersection of the two graphs to your number line, then shade and label the x-value of the point of intersection on your number line.

   b) Use an analytic method to solve the following equation for x: $3 - x = 2x - 3$. Does this analytic solution compare favorably with the solution found by using the graphing method in part (a)? If not, check your work for error.

8. Use an analytic method only to solve each of the following equations for x.

   a) $3 - x = 4 - 2x$ b) $2x + 5 = 3 - 2x$ c) $2(x + 3) - 3(3 - 2x) = 8$

   d) $4(2 - x) - 8(x + 1) = 11$ e) $2x - 3(x + 1) = 5x + 2$ f) $3x - (4 - 2x) = 2(x + 1)$

   g) $2(2x + 1) - 3(5 - x) = 4(x + 1)$ h) $2(1 - x) - (5 - x) = 3(4 - x)$

9. Use *both* a graphing method and an analytic method to solve each of the following equations for x. Arrange your solutions side-by-side for comparison, as suggested in Example 2 in this section.

   a) $3 - 2x = 2x + 7$ b) $1 - x = 2x + 4$ c) $5 - x = 11 - 4x$ d) $2x + 3 = 5 - x$

10. (TI-82) Load the following functions into the Y= menu.

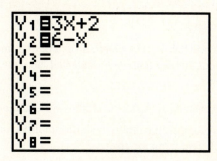

a) Set up a coordinate system on your homework paper. Push the ZOOM button and select 6:ZStandard. Copy the resulting image from your viewing window onto your coordinate system. Include the WINDOW parameters on your axes, as shown in Example 3 in this section.

b) Run the intersection utility in the CALC menu to find the point of intersection of the two graphs. Draw a number line under your coordinate system, drop a dotted line from your point of intersection to your number line, then shade and label the x-value of the point of intersection on your number line.

c) Use an analytic method to solve the following equation for x: $3x+2=6-x$. Does this analytic solution compare favorably with the solution found by using the graphing method in part (b)? If not, check your work for error.

11. (TI-82) Load the following functions into the Y= menu.

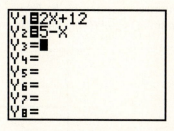

a) Set up a coordinate system on your homework paper. Push the ZOOM button and select 6:ZStandard. Copy the resulting image from your viewing window onto your coordinate system. Include the WINDOW parameters on your axes, as shown in Example 3 in this section.

b) Run the intersection utility in the CALC menu to find the point of intersection of the two graphs. Draw a number line under your coordinate system, drop a dotted line from your point of intersection to your number line, then shade and label the x-value of the point of intersection on your number line.

c) Use an analytic method to solve the following equation for x: $2x+12=5-x$. Does this analytic solution compare favorably with the solution found by using the graphing method in part (b)? If not, check your work for error.

12. Use an analytic method only to solve each of the following equations for x. In your first step, clear the fractions from the equation by multiplying both sides of the equation by the least common denominator.

a) $\dfrac{2}{3}x - \dfrac{3}{5} = \dfrac{1}{2}$

b) $\dfrac{1}{4} - \dfrac{2}{5}x = \dfrac{3}{2}$

c) $\dfrac{1}{2} - \dfrac{2}{3}x = \dfrac{3}{4}x - \dfrac{3}{2}$

d) $\dfrac{3}{4}x - \dfrac{1}{5} = \dfrac{1}{2} + \dfrac{2}{5}x$

e) $\dfrac{x}{3} - \dfrac{x}{5} = \dfrac{1}{2}$

f) $\dfrac{2x}{5} - \dfrac{3x}{2} = \dfrac{3}{4}$

13. Set up a coordinate system on a sheet of graph paper. Consider the functions f and g whose equations are $f(x) = -\frac{3}{2}x + 1$ and $g(x) = \frac{1}{2}x - 3$. Sketch the graphs of the functions f and g on your coordinate system.

   a) Draw a number line below your graph, drop a dotted line from the point of intersection of the two graphs to your number line, then shade and label the x-value of the point of intersection on your number line.

   b) Use an analytic method to solve the following equation for x: $-\frac{3}{2}x + 1 = \frac{1}{2}x - 3$. Does this analytic solution compare favorably with the solution found using the graphing method in part (a)? If not, check your work for error.

14. (TI-82) Press the CLEAR button on your calculator an appropriate number of times until your screen clears, then load the following functions in the Y= menu.

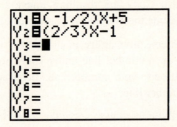

   a) Set up a coordinate system on your homework paper. Push the ZOOM button and select 6:ZStandard. Copy the resulting image from your viewing window onto your coordinate system. Include the WINDOW parameters on your axes, as shown in Example 3 in this section.

   b) Run the intersection utility in the CALC menu to find the point of intersection of the two graphs, capturing the following image on your viewing screen.

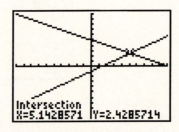

   c) You could be satisfied with this decimal approximation for x, but go a step further. Push the X,T,θ button on your keyboard, then the ENTER key, which will produce the following image in your viewing window.

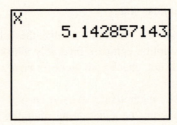

d) Press the MATH button to open the following menu:

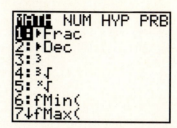

e) Select 1:▷Frac and press ENTER. Draw a number line under your coordinate system, drop a dotted line from your point of intersection to your number line, then shade and label the x-value of the point of intersection on your number line.

f) Use an analytic method to solve the following equation for x: $-\frac{1}{2}x+5 = \frac{2}{3}x-1$. Does this analytic solution compare favorably with the solution found by using the graphing method? If not, check your work for error.

Note: Exercise caution when using this routine. Keep in mind that not all numbers can be expressed in fractional form. For example, it is not possible to express $\sqrt{2}$ in fractional form. Also keep in mind that a finite precision machine like the TI-82 is bound to have difficulty finding the same fractional form that you can find with an analytic method.

15. (TI-82) Load the following functions into the Y= menu. Note the use of parentheses.

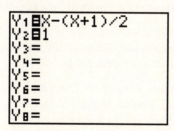

a) Set up a coordinate system on your homework paper. Push the ZOOM button and select 6:ZStandard. Copy the resulting image from your viewing window onto your coordinate system. Include the WINDOW parameters on your axes, as shown in Example 6 in this section.

b) Run the intersection utility in the CALC menu to find the point of intersection of the two graphs. Draw a number line under your coordinate system, drop a dotted line from your point of intersection to your number line, then shade and label the x-value of the point of intersection on your number line.

c) Use an analytic method to solve the following equation for x: $x - \frac{x+1}{2} = 1$. In your first step, be sure to multiply both sides of the equation by the least common denominator.

d) Does your analytic solution from part (c) compare favorably with the solution found by the graphing method in part (b)? If not, check your work for error.

16. Use an analytic method only to solve each of the following equations for x. In your first step, be sure to multiply both sides of the equation by the least common denominator.

a) $x - \frac{2x-3}{5} = 2$

b) $x - \frac{x+6}{7} = 2$

c) $x - \dfrac{10 - 5x}{7} = 5$

d) $3x - \dfrac{3 - 2x}{11} = 2$

17. Use *both* a graphing method and an analytic method to solve each of the following equations for x. Arrange your solutions side-by-side for comparison, as suggested in Example 6 in this section.

a) $x - \dfrac{x - 5}{4} = 2$

b) $x - \dfrac{2x + 1}{3} = 1$

18. Solve each of the following equations for x. Use your calculator to round your answer to the nearest hundredth (two decimal places).

a) $2.24x - 1.34 = 3.58$

b) $3.58 - 4.12x = 1.92$

c) $1.2 - 3.4x = 2.8x - 1.5$

d) $0.2x + 1.5 = 3.4x - 2.3$

e) $2.3(3 - x) - 1.8(2x + 1) = 5.3$

f) $1.12(8{,}000 - x) + 1.06x = 5{,}000$

19. Jimmy is paid a salary of $400 each month in addition to a 10 percent commission on the dollar value of all sales he makes during the month.

a) Let W represent Jimmy's total wages for the month. Let x represent the total dollar value of Jimmy's sales during the month. Express W as a function of x.

b) If Jimmy's wages are $1000 for the month, what is the total dollar value of Jimmy's sales? Express your answer to the nearest cent.

20. Raphael and Henrietta are in town for a computer show. They rent cars for their stay.

a) Raphael rents a car from World's Finest Rentals. He pays $50 plus $0.10 per mile. Let R represent the amount of his bill and let x represent the number of miles that he drives his rental car. Express R as a function of x.

b) Henrietta rents a car from Cheaper Is Better. Her bill is $40 plus $0.29 per mile. Let H represent the amount of her bill and let x represent the number of miles that she drives her rental car. Express H as a function of x.

c) If Raphael's bill is the same as Henrietta's, how far did each person drive the rental car? Express your answer to the nearest mile.

21. A water tank for a small community in Iowa holds 10,000 gallons of water. Suddenly, the tank springs a leak and water begins leaking from the tank at the rate of 7.8 gallons per minute.

a) Let V represent the volume of water remaining in the tank. Let t represent the amount of time in minutes that has passed since the tank sprung a leak. Express V as a function of t.

b) If the situation is not corrected, how long will it take the volume of the water in the tank to fall to 5000 gallons? Express your answer in hours and minutes, rounded to the nearest minute.

22. (TI-82) When the chase plane *Liberty One* spots the shuttle *Columbia* on its search radar, a distance of 150 miles separates the two aircraft. *Liberty One* is closing the distance between the two craft at a rate of 20.3 miles per minute. Let D represent the distance separating the two craft and let t represent the amount of time that has passed since *Columbia* showed up on *Liberty One's* search radar.

a) Express D as a function of t. Clearly state this result on your homework paper. Load this function in Y₁ in the Y= menu of the TI-82.

b) Find the amount of time required by *Liberty One* to close the gap between the two aircraft to 20 miles. Load the constant function $y = 20$ in Y2 in the Y= menu.

c) Adjust the WINDOW parameters so that the point of intersection of the two graphs is showing in your viewing window. Set up a coordinate system on your homework paper and duplicate the image from your viewing window on your coordinate system. Include the WINDOW parameters on the axes of your coordinate system.

d) Use the intersection utility in the CALC menu to find the time it takes *Liberty One* to close the gap to 20 miles. Draw a number line below your coordinate system on your homework paper and shade and label this solution on your number line in the usual manner.

e) Use an analytical method to find the amount of time required by *Liberty One* to close the gap between the aircraft to 20 miles. Round your answer to the nearest tenth (one decimal place). Include units with your answer.

23. Gitesh climbs onto his porch to fly his kite on a windy day in April. He launches his kite and after 30 seconds his kite is 224 feet in the air. After an additional 20 seconds, his kite is 360 feet in the air. Set up a coordinate system on a sheet of graph paper. Place the height of the kite on the vertical axis and the amount of time since the launch on the horizontal axis. Let H represent the height of the kite and let t represent the amount of time that has passed since the launch.

a) After scaling each axis appropriately, plot the two data points on your coordinate system. Assume that H is a linear function of t and draw a line through the two points.

b) Find the rate at which the kite is ascending. Round your answer to the nearest tenth (one decimal place) and include units with your answer.

c) Use your graph to *estimate* the time that it takes the kite to reach of height of 270 feet.

d) Use the point-slope form of a line to find the equation of the line. Be sure to replace y and x with H and t, respectively. Express the height H as a function of the time t by solving this equation for H.

e) Use an analytic method to find the amount of time that must pass for the kite to reach a height of 270 feet. Round your solution to the nearest tenth (one decimal place). Include units on your answer. Does your answer compare favorably with the estimate you found in part (c)? If not, check your work for error.

24. Set up a coordinate system on a sheet of graph paper. Suppose that f is a linear function whose graph has a y-intercept of 4 and whose slope is $-\frac{2}{3}$. Suppose that g is a linear function whose graph has a y-intercept of -2 and whose slope is $\frac{3}{4}$.

a) Sketch the graphs of the functions f and g on your coordinate system. Draw a number line below your graph, drop a dotted line from your point of intersection, then shade and label the solution of $f(x) = g(x)$ on your number line.

b) What are the equations of the functions f and g? Use these equations and an analytical method to find an *exact* solution of the equation $f(x) = g(x)$. Does this solution compare favorably with the solution found in part (a)? If not, check your work for error.

25. On a sheet of graph paper, set up a coordinate system and sketch the graph of the function f whose equation is $f(x) = 2x - 1$.

a) Plot the points P(1,4) and Q(3,2) and draw a line through them.

b) The graph of the function f and the line you just drew through the points P and Q intersect in a point. Use your graph to estimate the x-value of this point. Draw a number line below your graph, drop a dotted line from your point of intersection, and shade and label this estimate on your number line.

c) Use the point-slope form of a line to find the equation of the line through points P and Q. Place your answer in slope-intercept form by solving your equation for y. Replace y with $g(x)$.

d) Find the x-value of the point of intersection of the graphs of f and g by using an analytic method to solve the equation $f(x) = g(x)$ for x. Does this solution compare favorably with the solution found in part (b)? If not, check your work for error.

2.6 Solving Linear Inequalities

In this section we solve *linear inequalities*, so-called because the graphs involved are lines. When solving these linear inequalities, we use many of the same skills developed in the last section, with one major exception. We'll solve linear inequalities both graphically and analytically. We'll finish up this section with some nice applications of inequalities.

Important Fundamentals. Before beginning, you will need to review some important fundamentals, such as how to analyze a graph.

> *To solve the inequality $f(x) < g(x)$ for x, draw the graphs of the functions f and g and note where the graph of f is below the graph of g. To solve the inequality $f(x) > g(x)$ for x, draw the graphs of the functions f and g and note where the graph of f is above the graph of g.*

There are several important ideas that must be understood if you are going to successfully analyze inequalities. First, any real number may be added or subtracted from both sides of an inequality.

> *If $a < b$ and c is any real number, then $a + c < b + c$ and $a - c < b - c$.*

Second, both sides of an inequality may be multiplied or divided by any positive real number.

> *If $a < b$ and c is any positive real number, then $ac < bc$ and $\dfrac{a}{c} < \dfrac{b}{c}$.*

Reversing the Inequality Sign. There is one *major* difference between the techniques used for solving equations and inequalities. If you multiply both sides of the inequality $-2 < 3$ by -2, you get $4 > -6$. Note the reversal of the inequality sign. It was less than, but after multiplying both sides of the inequality by a negative number, it became greater than.

> *If $a < b$ and c is a negative real number, then $ac > bc$ and $\dfrac{a}{c} > \dfrac{b}{c}$. Note how the inequality sign is reversed.*

General Strategy. When solving linear equations and inequalities use the following strategy.

> *When solving a linear equation or inequality, try to isolate the terms containing the unknown on one side of the equation or inequality. For example, if you are solving for the variable x, try to isolate all of the terms containing an x on one side of the equation or inequality.*

Example 1   Solve the following inequality for x: $2x+1<3$.

Graphical Solution   First sketch the graph of functions f and g whose equations are $f(x)=2x+1$ and $g(x)=3$, then note where the graph of f is *below* the graph of g. The graph of f is a line with slope 2 and y-intercept 1. The graph of the *constant* function g is a horizontal line.

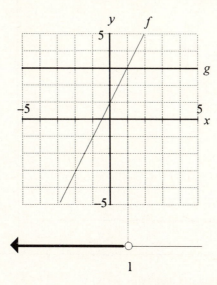

The graph of f is *below* the graph of g whenever x is to the left of 1. We have shaded the numbers to the left of 1 on the number line. Note the empty circle at 1. This is our way of conveying the idea that this point is *not* shaded and is *not* part of the solution. The approximate solution of $f(x)<g(x)$ is $\{x:x<1\}$, which is read "The set of all x such that x is less than 1."

Analytic Solution. Isolate the terms containing x on one side of the equation. Subtract 1 from both sides of the inequality, then divide both sides by 2.

$$2x+1<3$$
$$2x+1-1<3-1$$
$$2x<2$$
$$\frac{2x}{2}<\frac{2}{2}$$
$$x<1$$

When solving an inequality, the solution set $\{x:x<1\}$ must be shaded on a second number line, in the following manner:

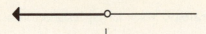

This is precisely the same solution arrived at using the graphing technique.

Organizing Your Work

It is important to note that the graphical solution and the algebraic or analytic solution complement each other, and *both are required* on your homework. Students should arrange their work on their homework or examination papers in the following manner:

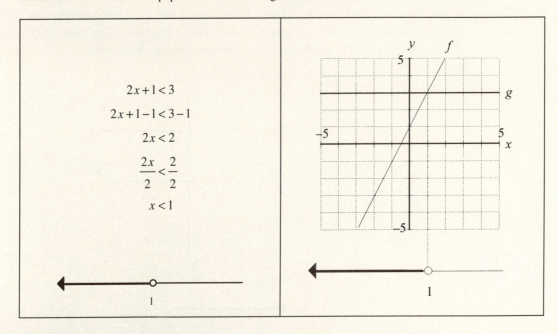

$$2x+1<3$$
$$2x+1-1<3-1$$
$$2x<2$$
$$\frac{2x}{2}<\frac{2}{2}$$
$$x<1$$

Example 2   Solve the following inequality for x: $x-2 \le 2x-1$.

Graphical Solution   This inequality is of the form $f(x) \le g(x)$, where $f(x)=x-2$ and $g(x)=2x-1$. Note the *less than or equal* symbol, $\le$. First sketch the graphs of f and g, then note where the graph of f is below *or* intersects the graph of g. Note how the slope and y-intercept is used to help draw the graph of each line.

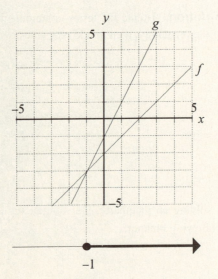

The graph of *f* is below the graph of *g* whenever *x* is to the right of −1. The graph of *f* intersects the graph of *g* at *x* equals −1. The numbers to the right of and including −1 on our number line have been shaded. Note the filled circle at −1. This is a way of indicating that −1 is shaded and is a solution. The approximate solution of $f(x) \leq g(x)$ is $\{x : x \geq -1\}$, which is read "The set of all *x* such that *x* is greater than or equal to −1."

Analytical Solution   Isolate *x* on one side of the inequality by adding 2 to both sides of the inequality and simplifying.

$$x - 2 \leq 2x - 1$$
$$x - 2 + 2 \leq 2x - 1 + 2$$
$$x \leq 2x + 1$$

Now subtract $2x$ from both sides of the inequality.

$$x - 2x \leq 2x + 1 - 2x$$
$$-x \leq 1$$

When multiplying or dividing by a negative number, remember to reverse the inequality sign. Multiply both sides of the inequality by −1, reversing the inequality sign the *moment you do so*. Then sketch your answer on a second number line.

$$-1(-x) \geq -1(1)$$
$$x \geq -1$$

Arrange the analytical solution and the graphical solution side-by-side for easy comparison.

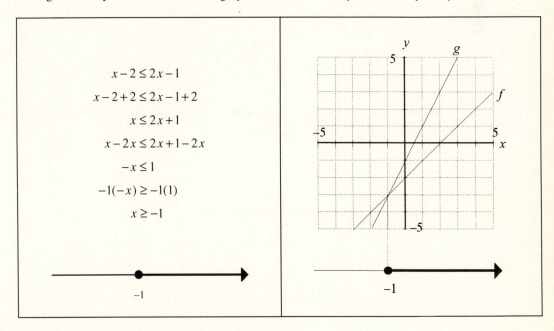

Example 3   Solve the following inequality for x: $-\frac{3}{4}x + 2 \geq \frac{3}{2}x - 1$.

Graphical Solution   This inequality is of the form $f(x) \geq g(x)$, where $f(x) = -\frac{3}{4}x + 2$ and $g(x) = \frac{3}{2}x - 1$. First draw the graphs of the functions f and g, then note where the graph of f is *above* the graph of g or where the graph of f *intersects* the graph of g.

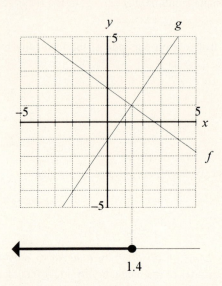

The x-value of the point of intersection of the graphs of f and g is *approximately* 1.4. The graph of f is *above* the graph of g whenever x is to the left of 1.4. The graph of f *intersects* the graph of g when x equals 1.4. The approximate solution of $f(x) \geq g(x)$ is $\{x : x \leq 1.4\}$, which is read: "The set of all x such that x is less than or equal to 1.4."

Analytical Solution   Begin by multiplying both sides of the inequality by the least common denominator.

$$4\left[-\frac{3}{4}x + 2\right] \geq \left[\frac{3}{2}x - 1\right]4$$

$$4\left[-\frac{3}{4}x\right] + 4[2] \geq 4\left[\frac{3}{2}x\right] - 4[1]$$

$$-3x + 8 \geq 6x - 4$$

Isolate the terms containing x on one side of the equation. Begin by subtracting $6x$ from both sides of the inequality.

$$-3x + 8 - 6x \geq 6x - 4 - 6x$$

$$-9x + 8 \geq -4$$

Now subtract 8 from both sides of the inequality.

$$-9x + 8 - 8 \geq -4 - 8$$

$$-9x \geq -12$$

Divide both sides by -9, reversing the inequality as you do so.

$$\frac{-9x}{-9} \le \frac{-12}{-9}$$

$$x \le \frac{4}{3}$$

Now shade your solution on a second number line.

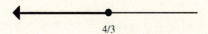

4/3

Does this *exact* solution agree favorably with your estimate from the graph? If you divide 4 by 3 on your calculator, you will find that $\frac{4}{3} \approx 1.3$, which agrees quite nicely with your graphical estimate. If you arrange your work side-by-side on our homework paper, you can easily compare the results of the analytic method and the graphing method.

$$4\left[-\frac{3}{4}x+2\right] \ge \left[\frac{3}{2}x-1\right]4$$

$$4\left[-\frac{3}{4}x\right]+4[2] \ge 4\left[\frac{3}{2}x\right]-4[1]$$

$$-3x+8 \ge 6x-4$$

$$-3x+8-6x \ge 6x-4-6x$$

$$-9x+8 \ge -4$$

$$-9x+8-8 \ge -4-8$$

$$-9x \ge -12$$

$$\frac{-9x}{-9} \le \frac{-12}{-9}$$

$$x \le \frac{4}{3}$$

4/3

1.4

Example 4   *(TI-82)* Solve the following inequality for x: $\dfrac{x-1}{2}-\dfrac{x+1}{3}<1$.

Graphical Solution   This inequality is of the form Y₁ < Y₂, where $Y_1 = \dfrac{x-1}{2}-\dfrac{x+1}{3}$ and Y₂ = 1. First draw the graphs of Y₁ and Y₂, then note where the graph of Y₁ is *below* the graph of Y₂. Load the following functions into the Y= menu:

Press the ZOOM button and select 6:ZStandard to produce the following image in your viewing window:

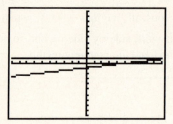

Make the following changes to the WINDOW parameters:

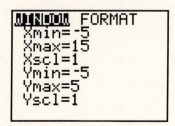

Press the GRAPH button to capture the following image in your viewing window:

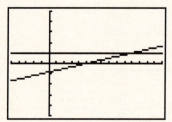

If you execute the intersection utility in the CALC menu (see Section 1.6 for a review of this technique), the following image will appear in your viewing window:

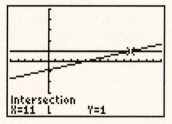

The graph of Y1 is *below* the graph of Y2 whenever *x* is to the left of 11. The approximate solution of Y1 < Y2 is $\{x : x < 11\}$, which is read "The set of all *x* such that *x* is less than 11." Duplicate this image from your viewing window onto your homework paper in the following manner:

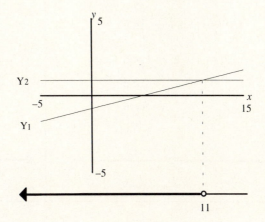

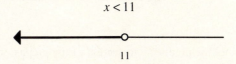

Note how this student has labeled his graphs with Y1 and Y2 and placed the WINDOW parameters on his axes.

Analytical Solution   Multiply both sides of the inequality by the least common denominator, which is 6. Note the use of parentheses.

$$6\left[\frac{x-1}{2} - \frac{x+1}{3}\right] < 6[1]$$

$$6\left(\frac{x-1}{2}\right) - 6\left(\frac{x+1}{3}\right) < 6$$

$$3(x-1) - 2(x+1) < 6$$

Next, distribute the 3 and the –2 and simplify.

$$3x - 3 - 2x - 2 < 6$$

$$x - 5 < 6$$

Add 5 to both sides of the inequality and sketch the solution on a second number line.

$$x < 11$$

When you arrange your work side-by-side, you can easily see that your analytical solution agrees with the solution found by using the graphing method.

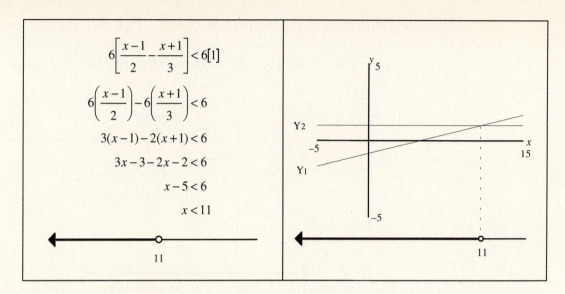

Example 5 Suppose that the function *f* is a linear function. In addition, suppose that $f(-3) = -4$ and $f(5) = 2$. For what values of *x* is $f(x) < 1$?

Graphical Solution Begin with a sketch. Because $f(-3) = -4$ and $f(5) = 2$, plot the points $(-3, -4)$ and $(5, 2)$ on the coordinate system. Because the function *f* is linear, draw a line through your two data points. Include the graph of the equation $y = 1$ in your sketch.

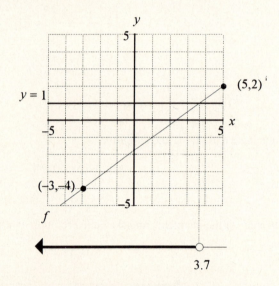

The *x*-value of the point of intersection of the graphs of *f* and $y = 1$ is approximately 3.7. The graph of *f* is *below* the graph of $y = 1$ whenever *x* is less than 3.7. The approximate solution of $f(x) < 1$ is $\{x : x < 3.7\}$, which is read: "The set of all *x* such that *x* is less than 3.7."

Analytical Solution Obtain the equation of the function *f*. You can use the slope formula to find the slope of the line through the points $(-3, -4)$ and $(5, 2)$.

$$\text{Slope} = \frac{\text{Change in } y}{\text{Change in } x} = \frac{2 - (-4)}{5 - (-3)} = \frac{6}{8} = \frac{3}{4}$$

Use the point-slope form of a line to find the equation of your line; then solve the equation for y.

$$y - y_0 = m(x - x_0)$$

$$y - 2 = \frac{3}{4}(x - 5)$$

$$y - 2 = \frac{3}{4}x - \frac{15}{4}$$

$$y - 2 + 2 = \frac{3}{4}x - \frac{15}{4} + 2$$

$$y = \frac{3}{4}x - \frac{15}{4} + \frac{8}{4}$$

$$y = \frac{3}{4}x - \frac{7}{4}$$

Finally, because y and $f(x)$ are interchangeable, replace y with $f(x)$.

$$f(x) = \frac{3}{4}x - \frac{7}{4}.$$

Now solve the inequality $f(x) < 1$ analytically. Begin by replacing $f(x)$ with $\frac{3}{4}x - \frac{7}{4}$.

$$f(x) < 1$$

$$\frac{3}{4}x - \frac{7}{4} < 1$$

$$4\left[\frac{3}{4}x - \frac{7}{4}\right] < [1]4$$

$$4\left[\frac{3}{4}x\right] - 4\left[\frac{7}{4}\right] < 4$$

$$3x - 7 < 4$$

$$3x - 7 + 7 < 4 + 7$$

$$3x < 11$$

$$x < \frac{11}{3}$$

11/3

If you divide 11 by 3 on your calculator, you will find that $\frac{11}{3} \approx 3.7$, which compares quite favorably with the solution found by the graphing method. By arranging the two methods of solution side-by-side on your homework paper, you can compare the results of the analytic and the graphing methods.

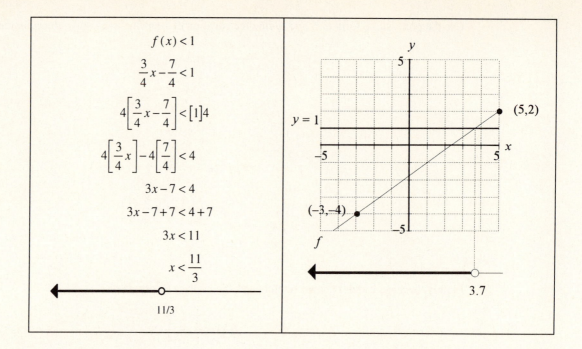

Exercises for Section 2.6

1. Set up a coordinate system on a sheet of graph paper. Consider the functions f and g whose equations are $f(x) = 2x - 1$ and $g(x) = 5$. Sketch the graphs of the functions f and g on your coordinate system.

   a) Draw a number line below your graph, drop a dotted line from the point of intersection of the two graphs to your number line, label the x-value of the point of intersection on your number line, then shade the solution of $f(x) < g(x)$ on your number line.

   b) Use an analytic method to solve the following inequality for x: $2x - 1 < 5$. Draw a second number line under your work and shade your analytic solution on your number line. Does the analytic solution compare favorably with the solution found by using the graphing method in part (a)? If not, check your work for error.

2. Set up a coordinate system on a sheet of graph paper. Consider the functions f and g whose equations are $f(x) = 3x + 2$ and $g(x) = 5$. Sketch the graphs of the functions f and g on your coordinate system.

   a) Draw a number line below your graph, drop a dotted line from the point of intersection of the two graphs to the number line, label the x-value of the point of intersection on the number line, then shade the solution of $f(x) \geq g(x)$ on the number line.

   b) Use an analytic method to solve the following inequality for x: $3x + 2 \geq 5$. Draw a second number line under your work and shade the analytic solution on the number line. Does your analytic solution compare favorably with the solution found by using the graphing method in part (a)? If not, check your work for error.

3. Set up a coordinate system on a sheet of graph paper. Consider the functions f and g whose equations are $f(x) = 3 - 2x$ and $g(x) = 7$. Sketch the graphs of the functions f and g on your coordinate system.

a) Draw a number line below your graph, drop a dotted line from the point of intersection of the two graphs to the number line, label the x-value of the point of intersection on your number line, then shade the solution of $f(x) < g(x)$ on the number line.

b) Use an analytic method to solve the following inequality for x: $3 - 2x < 7$. Draw a second number line under your work and shade the analytic solution on the number line. Does your analytic solution compare favorably with the solution found by using the graphing method in part (a)? If not, check your work for error.

4. Use an analytic method only to solve each of the following inequalities for x. Sketch the solution to each inequality on a number line. Use set-builder notation to describe each solution.

a) $2x - 11 < 5$ b) $3 - 4x \geq 15$ c) $8 - x > 4$ d) $2 - x \leq 5$

5. Use *both* a graphing method and an analytical method to solve each of the following equations for x. Arrange your solutions side-by-side for comparison, as suggested in Example 1 in this section.

a) $5x - 1 < 6$ b) $4 - x \geq 1$ c) $2x - 5 < 1$ d) $3 - 4x \geq 11$

6. Set up a coordinate system on a sheet of graph paper. Consider the functions f and g whose equations are $f(x) = 3 - x$ and $g(x) = x - 7$. Sketch the graphs of the functions f and g on your coordinate system.

a) Draw a number line below your graph, drop a dotted line from the point of intersection of the two graphs to the number line, label the x-value of the point of intersection on the number line, then shade the solution of $f(x) < g(x)$ on the number line.

b) Use an analytic method to solve the following inequality for x: $3 - x < x - 7$. Draw a second number line under your work and shade your analytic solution on the number line. Does your analytic solution compare favorably with the solution found by using the graphing method in part (a)? If not, check your work for error.

7. Set up a coordinate system on a sheet of graph paper. Consider the functions f and g whose equations are $f(x) = 2x - 3$ and $g(x) = 6 - x$. Sketch the graphs of the functions f and g on your coordinate system.

a) Draw a number line below your graph, drop a dotted line from the point of intersection of the two graphs to the number line, label the x-value of the point of intersection on the number line, then shade the solution of $f(x) \geq g(x)$ on the number line.

b) Use an analytic method to solve the following inequality for x: $2x - 3 \geq 6 - x$. Draw a second number line under your work and shade your analytic solution on the number line. Does your analytic solution compare favorably with the solution found by using the graphing method in part (a)? If not, check your work for error.

8. Use an analytic method only to solve each of the following inequalities for x. Sketch the solution to each inequality on a number line. Use set-builder notation to describe each solution.

a) $3 - x < x - 4$ b) $2x + 1 \geq 3 - 2x$ c) $2(3 - x) - 4(x + 1) > 5$

d) $2(5x + 1) - 4(3 - x) \leq 9$ e) $2x - (x - 1) < 8 - 5x$ f) $3x - (4 - 5x) \geq 5x + 1$

g) $3(4 - x) - (x - 5) \leq 2(x + 1)$ h) $2(3x + 1) - (6 - x) \leq 3(4 - 3x)$

9. Use *both* a graphing method and an analytical method to solve each of the following equations for x. Arrange your solutions side-by-side for comparison, as suggested in Example 2 in this section.

a) $2x - 2 > 4 - x$ b) $3x - 4 \leq x + 2$ c) $10 - x < 2x - 5$ d) $x - 4 \geq 3x + 2$

10. Set up a coordinate system on a sheet of graph paper. Consider the functions f and g whose equations are $f(x) = \frac{1}{3}x - 1$ and $g(x) = -\frac{1}{2}x + 2$. Sketch the graphs of the functions f and g on your coordinate system.

   a) Draw a number line below your graph, drop a dotted line from the point of intersection of the two graphs to the number line, label a decimal approximation of the x-value of the point of intersection on the number line, then shade the solution of $f(x) < g(x)$ on the number line.

   b) Use an analytic method to solve the following inequality for x: $\frac{1}{3}x - 1 < -\frac{1}{2}x + 2$. Draw a second number line under your work and shade your analytic solution on the number line. Use your calculator to obtain a decimal approximation of your answer. Does your analytic solution compare favorably with the solution found by using the graphing method in part (a)? If not, check your work for error.

11. Set up a coordinate system on a sheet of graph paper. Consider the functions f and g whose equations are $f(x) = -\frac{3}{4}x + 4$ and $g(x) = \frac{3}{2}x - 3$. Sketch the graphs of the functions f and g on your coordinate system.

   a) Draw a number line below your graph, drop a dotted line from the point of intersection of the two graphs to the number line, label a decimal approximation of the x-value of the point of intersection on the number line, then shade the solution of $f(x) \geq g(x)$ on the number line.

   b) Use an analytic method to solve the following inequality for x: $-\frac{3}{4}x + 4 \geq \frac{3}{2}x - 3$. Draw a second number line under your work and shade your analytic solution on the number line. Use your calculator to obtain a decimal approximation of your answer. Does your analytic solution compare favorably with the solution found by using the graphing method in part (a)? If not, check your work for error.

12. Use an analytic method only to solve each of the following inequalities for x. Sketch the solution to each inequality on a number line. Use set-builder notation to describe each solution.

   a) $\frac{3}{4}x + 1 > -\frac{2}{3}x - 3$

   b) $3 - \frac{2}{3}x \geq \frac{1}{5}x - 1$

   c) $\frac{1}{3}x + \frac{1}{5} > \frac{1}{6}$

   d) $\frac{1}{5} - \frac{3}{4}x \leq \frac{3}{2}$

   e) $\frac{1}{2}x + \frac{2}{3} \geq \frac{3}{4} - \frac{3}{2}x$

   f) $\frac{1}{8} + \frac{2}{3}x < \frac{3}{4} - \frac{1}{4}x$

13. Load the following functions into the Y= menu:

```
Y1B(X-2)/4-(X-2)
/3
Y2B1
Y3=
Y4=
Y5=
Y6=
Y7=
```

   a) Set up a coordinate system on your homework paper. Push the ZOOM button and select 6:ZStandard. Make the following changes to the WINDOW parameters:

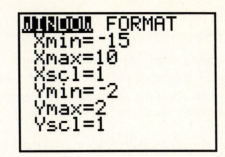

Push the GRAPH button and copy the resulting image from your viewing window onto your coordinate system. Label your graphs Y1 and Y2 and include the WINDOW parameters on the axes of your coordinate system, as shown in Example 4 in this section.

b) Use the intersect utility from the CALC menu to find the x-value of the point of intersection of Y1 and Y2. Draw a number line below your graph, drop a dotted line from the point of intersection of the two graphs to the number line, label the x-value of the point of intersection on the number line, then shade the solution of Y1 < Y2 on the number line.

c) Use an analytic method to solve the following inequality for x: $\dfrac{x-2}{4} - \dfrac{x-2}{3} < 1$. Draw a second number line under your work and shade your analytic solution on your number line. Does your analytic solution compare favorably with the solution found by using the graphing method in part (a)? If not, check your work for error.

14. Use an analytic method only to solve each of the following inequalities for x. Sketch the solution to each inequality on a number line. Use set-builder notation to describe each solution.

a) $x - \dfrac{x-1}{3} > 2$

b) $3x - \dfrac{x+1}{2} \le 1$

c) $\dfrac{x-2}{3} - \dfrac{x+1}{5} > 1$

d) $\dfrac{x-5}{2} - \dfrac{x+2}{7} \le 2$

e) $x - \dfrac{x+1}{2} < \dfrac{x-1}{3}$

f) $2x - \dfrac{x+5}{7} \le \dfrac{3-x}{4}$

15. Set up a coordinate system on a sheet of graph paper. Suppose that f is a linear function and $f(-3) = -1$ and $f(4) = 3$. Plot the given data points on the coordinate system and draw a line through them.

a) Draw a number line below your graph and shade the solution of the inequality $f(x) < 0$ on the number line. Use set-builder notation to describe your solution set.

b) Use the point-slope form to find the equation of the line. Place your answer in slope-intercept form by solving this equation for y, then replace y with $f(x)$.

c) Replace $f(x)$ in the inequality $f(x) < 0$ with the equation found in part (b) and use an analytic method to solve the resulting inequality for x. Shade your analytic solution on a second number line. Does this analytic solution compare favorably with the graphing solution found in part (a)? If not, check your work for error.

16. Set up a coordinate system on a sheet of graph paper. Sketch the line that goes through the point $(-3,3)$ and has slope $-\frac{2}{3}$ on your coordinate system. Draw the graph of the constant function $g(x) = 2$ on your coordinate system.

a) Draw a number line below your graph and shade the solution of the inequality $f(x) \ge g(x)$ on the number line. Use set-builder notation to describe your solution set.

b) Use the point-slope form of a line to find the equation of your line. Place this equation in the slope-intercept form by solving for the variable y, then replace y with $f(x)$.

c) Use an analytic method to solve the inequality $f(x) \geq g(x)$. Begin by replacing $f(x)$ with the equation you developed in part (b) and $g(x)$ with 2. Shade your analytic solution on a second number line. Does this analytic solution compare favorably with the graphing solution found in part (a)? If not, check your work for error.

17. For what values of x will the following rectangles have perimeters less than 100?

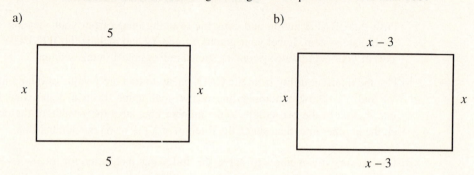

a)

b)

Note: Be sure that you exclude any x-values that cause either the width or length of the rectangle to be negative.

18. Find all values of x so that the area of the triangle remains under 100 square units.

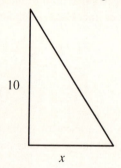

Note: Remember, length may not be negative. Be sure that your solution reflects this fact.

19. Big Frank runs a hot-dog cart on a street corner in New York. It costs Frank $10 just to open for business each day. In addition, for each hot-dog that he makes, his costs increase by an additional $0.50. Set up a coordinate system on a sheet of graph paper. Since Frank's costs depend on the number of hot-dogs he makes, place the number of hot-dogs on the horizontal axis, letting x represent the number of hot-dogs made by him. Place Frank's costs on the vertical axis, letting C represent his costs.

a) Scale each axis appropriately. Sketch the graph of Frank's costs versus the number of hot-dogs he makes. Draw a horizontal line representing costs of $100 on your coordinate system.

b) Draw a number line below your graph. Drop a vertical, dotted line from the point of intersection of the two graphs to the number line, and label the x-value of the point of intersection on the number line. On the number line shade all x-values so that Frank's costs remain under $100.

c) Express Frank's costs, C, as a function of the number of hot-dogs, x, that he makes.

d) Replace C in the inequality $C < 100$ with the equation found in part (c) and use an analytic method to solve the resulting inequality for x. Shade your analytic solution on a second number line. Does your analytic solution compare favorably with the graphing solution from part (b)?

Note: Can Big Frank make a negative number of hot-dogs? Does your solution reflect this fact?

20. Elizabeth's scores on her first four history exams are as follows: 90, 80, 92, 88. These are percentage scores on a 0 to 100 scale. A 0 represents a paper with no correct answers and a 100 represents a perfect paper, one with no errors at all. Use an analytic method to find all possible test scores that Elizabeth can score on her next exam and still have her average score on the five exams exceed 90.

Note: Does your solution reflect the fact that a score may not exceed 100?

21. Vanessa has 50 marbles in a bag. Whenever she shoots marbles with her friend Deirdre, she manages to win 3 marbles every hour, on the average. How long can Vanessa play and still keep the number of marbles in her bag under 100?

22. (TI-82) Mohammed is in town for a meeting of the National Council of Teachers of Mathematics, and he needs to rent a car for one day. He is trying to decide between two car rental companies.

a) Practical Transportation charges $29 per day plus 10 cents a mile. Express the daily cost, C, of renting a car from this company as a function of x, the number of miles driven before returning the car.

b) Thrifty Transportation charges $35 per day plus 8 cents a mile. Express the daily cost, C, of renting a car from this company as a function of x, the number of miles driven before returning the car.

c) Open the Y= menu and place the functions developed in parts (a) and (b) in Y1 and Y2, respectively. Experiment with the WINDOW parameters until you have a satisfactory viewing window that shows the point of intersection of the two graphs. Set up a coordinate system on your homework paper and copy the image in your viewing window onto the coordinate system. Label your graphs with Y1 and Y2 and include the window parameters on the axes of the coordinate system.

d) Use the intersection utility in the CALC menu to find the x-value of the point of intersection of the two functions. Draw a number line below your graph, drop a dotted line from the point of intersection, then shade and label this x-value on the number line.

e) Write a short paragraph explaining how Mohammed should decide which of the rental companies offers the better value. Be sure to explain exactly when each company offers the better value.

2.7 Elements of Logic

We will spend some time in this section talking about two small but difficult words: *and* and *or*. Let's begin with two statements about detecting the truth of *and* and *or* statements.

> ***An 'and' statement is true if and only if both parts of the 'and' statement are true.***

An 'or' statement is true if and only if either part of the 'or' statement is true.

Symbols Used in Logic. Here are the symbols that are used when working with the words *and* and *or*.

Symbol	Meaning
$\in$	is an element of
$\cap$	intersection; and
$\cup$	union; or

Some Important Definitions. First, let's define what is meant by $A \cap B$, the *intersection* of A and B.

$$A \cap B = \{x:\ x \in A\ \text{and}\ x \in B\}$$

This last set of symbols is read as follows: "The intersection of A and B is the set of all x such that x is an element of A *and x* is an element of B." We next define what is meant by $A \cup B$, the *union* of A and B.

$$A \cup B = \{x:\ x \in A\ \text{or}\ x \in B\}$$

This last set of symbols is read as follows: "The union of A and B is the set of all x such that x is an element of A *or x* is an element of B." At first look, it may appear that the two definitions are identical. On closer inspection, a major difference can be seen. The definition for intersection uses the word *and*; the definition for union uses the word *or*.

Intersection means and.
Union means or.

Visualizing 'and' and 'or'. Mathematicians use pictures, called Venn diagrams, to help explain the meaning of the words *and* and *or*. Consider the following diagram.

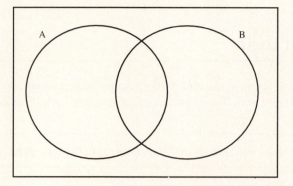

Everything within the boundary of the circle marked with the letter A is the set A. Everything within the boundary of the circle marked with the letter B is the set B.

***The Intersection of Sets* A *and* B.**   Remember, $A \cap B = \{x: x \in A \text{ and } x \in B\}$. If you are asked to shade the intersection of sets A and B, then all elements contained in set A *and* in set B will be shaded, as follows:

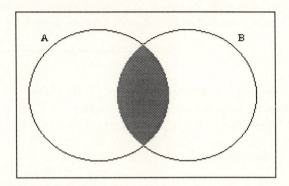

Tip	If you are asked to shade the objects that are in set A *and* in set B, then you shade where sets A and B *intersect*, or overlap.

***The Union of Sets* A *and* B.**   Remember, $A \cup B = \{x: x \in A \text{ or } x \in B\}$. If you are asked to shade the union of sets A and B, then all elements contained in set A *or* set B will be shaded, as follows:

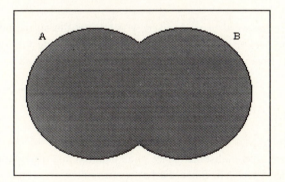

Tip	If you are asked to shade the objects that are in set A *or* in set B, then you shade everything from *either* set.

Using the Words 'and' and 'or'.   Because we will be using the words *and* and *or* in our work solving inequalities, we need to examine some special examples involving them.

Example 1   On a number line, sketch the set $\{x: x < 4 \text{ and } x < 5\}$.

Solution   First, do some scratch work. On one number line sketch the set of all real numbers that are less than 4. On a second number line, sketch the set of all real numbers that are less than 5.

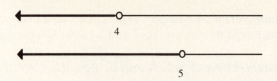

Because the word *and* is involved, you must shade the numbers that are shaded on *both* number lines in your scratch work. Remember, *and* means *intersection*, so you are searching for the region where the shaded regions in your scratch work intersect. Here is the final solution.

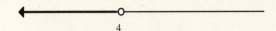

Therefore, $\{x : x < 4 \ \text{and} \ x < 5\} = \{x : x < 4\}$. Here is an analogy you might find useful. Suppose you have been asked to list the names of all students in a fifth grade class who are under 4 feet tall *and* under 5 feet. Would you not list the names of all the children who were under 4 feet in height?

Example 2   On a number line, sketch $\{x : x < 4 \ \text{or} \ x < 5\}$.

Solution   At first glance, this looks like the same problem you were asked to do in Example 1. However, on second glance, there is something distinctly different. The word *and* has been changed in the last example to the word *or* in the present example. Can that make a big difference? Let's find out. First, repeat your scratch work from Example 1.

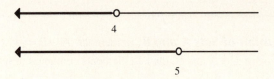

Because the word *or* is involved, you must shade all the numbers that are shaded on *either* of the number lines in our scratch work. In a sense, taking the *union* is a way of *combining* the shaded regions from your scratch work. Here is the final solution:

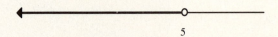

Therefore, $\{x : x < 4 \ \text{or} \ x < 5\} = \{x : x < 5\}$. Here is an analogy that might be useful. Suppose that you were asked to list the names of all children in a fifth grade class who were *either* under 4 feet tall *or* under 5 feet tall. Since the key word is now *or*, would you not include the children who were under 5 feet in height?

Important Note	If the solution to a problem requires that you write the word *and* in your solution and you have written the word *or* instead, you have not made a *minor* mistake. ***You have made a major mistake! You have completely changed the answer to the problem.***

Introduction to Interval Notation

In the examples that follow we will introduce *interval notation*. This notation is a useful way of describing the solution set of an inequality. Consider the solution for Example 1.

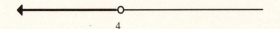

The following set-builder notation was used to describe this solution set: $\{x : x < 4\}$. If *interval notation* was used to describe this solution set, you would write $(-\infty, 4)$.

$$(-\infty, 4) = \{x : x < 4\}$$
The notation $(-\infty, 4)$ represents the set of all real numbers less than 4.

Let's look at another example. Suppose that the following set of real numbers are shaded on a number line.

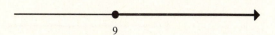

If set-builder notation is used, you would describe this set of real numbers with the following notation: $\{x : x \geq 9\}$. In *interval notation* you would describe this set of real numbers with the following notation: $[9, +\infty)$.

$$[9, +\infty) = \{x : x \geq 9\}$$
The notation $[9, +\infty)$ represents the set of all real numbers greater than or equal to 9.

Note that a parenthesis used in interval notation means the same as an empty circle on a number line. A parenthesis does not include the number. A bracket used in interval notation means the same as a filled-in number on a number line. A bracket includes the number. The symbols $-\infty$ and $+\infty$, since they represent a *concept* and not a number, are always enclosed in parentheses.

Example 3   Sketch the set $\{x : x > 4 \text{ and } x < 5\}$.

Solution   Do some preliminary scratch work, shading the real numbers that are greater than 4 and the real numbers that are less than 5 on separate number lines.

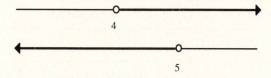

Because the word *and* is involved, the final solution will involve shading the real numbers that are shaded on *both* number lines. Show the *intersection* of the shaded regions in your scratch work.

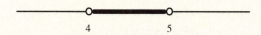

All the real numbers that are greater than 4 *and* less than 5 have been shaded. You might find the following analogy helpful. Suppose you were asked to list the names of all students in a fifth grade class who were over 4 feet tall *and* under 5 feet tall. Would you not list the names of the children who were between 4 and 5 feet in height?

If interval notation is used to describe the shaded region on the number line, you would write $(4, 5)$.

> $$(4,5) = \{x : x > 4 \text{ and } x < 5\}$$
> *The notation* **(4,5)** *represents the set of all real numbers between 4 and 5.*

Example 4 Sketch the set { $x : x > 4$ or $x < 5$ }.

Solution Is the same question being asked again? On careful inspection, you can see that the *and* in Example 3 has been changed to an *or* in this example. The graph of the final solution will probably be completely different from that of the last example. Repeat your scratch work from Example 3.

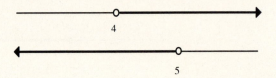

Remember, *or* means *union* and *union* is a way of combining sets. If all the numbers that belong to *either* of the sets pictured in your scratch work are shaded, then you will arrive at the following solution:

The entire line has been shaded. You might find the following analogy helpful. Suppose that you are asked to list the names of all children in a fifth grade class who are either over 4 feet tall *or* under 5 feet tall. Wouldn't that be everybody in the class?

If you want to describe this solution set using interval notation, you would write $(-\infty, +\infty)$.

> $$(-\infty, +\infty) = \{x : x \text{ is any real number}\}$$
> *The notation* $(-\infty, +\infty)$ *represents the entire set of real numbers.*

Compound Inequalities

There is an important notation meaning *and* that we will share with you now. Inequalities written with this notation are called *compound* inequalities.

> $a < x < b$ *is equivalent to* $x > a$ *and* $x < b$.

The notation $a < x < b$ is very compact and highly favored among mathematicians. There is no such compact notation for *or* statements. Remember, this notation means *and*.

Example 5 Sketch the set { $x : 0 \le x < 10$ }.

Solution This is equivalent to sketching the set $\{x : x \ge 0 \text{ and } x < 10\}$. Using the techniques of the previous examples, first draw some scratch work.

Because the word *and* is involved, you must shade all the real numbers that are shaded on *both* of the number lines in your scratch work. Notice that 0 is included in the final result, because it is shaded in *both* sets in the scratch work. The number 10 is not included in the final result because it is *not* shaded in both sets in the scratch work.

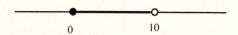

If interval notation is used to describe this solution set, you would write $[0,10)$.

$$[0,10) = \{x\colon 0 \le x < 10\}$$
The notation $[0,10)$ *represents the set of all real numbers between 0 and 10 and including 0.*

Reminder	A bracket is used to include a number and a parenthesis is used to exclude a number. In other words, the bracket is equivalent to a filled-in circle on a number line, and a parenthesis is equivalent to an empty circle on a number line.

Describing Betweenness. When the compound inequality is used properly, it is an excellent way to describe "betweenness." In the last example, the notation $\{x\colon 0 \le x < 10\}$ is an excellent way of describing the numbers between 0 and 10 and including zero. However, if used improperly, the compound inequality can give unexpected headaches.

Example 6   Sketch the set $\{x\colon 3 < x < 0\}$.

Solution   It would be a mistake to sketch all the points between 0 and 3 without first giving this problem a closer look. If 3 is smaller than x, and x is smaller than 0, shouldn't 3 be smaller than 0? Something is very wrong here! Perhaps we should break this problem down and analyze it properly. The set $\{x\colon 3 < x < 0\}$ is equivalent to the set $\{x\colon x > 3 \text{ and } x < 0\}$. Sketch each part of this *and* statement on separate number lines.

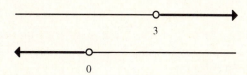

Because the word *and* is involved, you need to shade the real numbers that are shaded on *both* number lines. You need to shade where our two sets *intersect* or *overlap*. Here is the final answer.

There is nothing shaded on the line at all. Mathematicians refer to this set as the *empty set*. There are no numbers that are *both* greater than 3 *and* less than 0.

Example 7   Sketch the set $\{x\colon x < 2 \text{ or } x \ge 5\}$.

Solution   Begin with some scratch work. On one number line sketch the set of real numbers that are less than 2. On a second number line sketch the set of real numbers that are greater than or equal to 5.

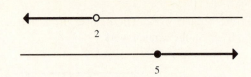

Because the word *or* is involved, you need to shade the real numbers that are shaded on *either* of the two number lines. Remember, *or* means *union*, and *union* is a way of combining everything from *either* set.

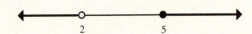

If interval notation is used to describe this solution set, you would write $(-\infty, 2) \cup [5, +\infty)$.

$$(-\infty, 2) \cup [5, +\infty) = \{x : x < 2 \text{ or } x \geq 5\}$$

The notation $(-\infty, 2) \cup [5, +\infty)$ represents the set of all real numbers that are less than 2 or greater than or equal to 5.

Note that the union symbol, $\cup$, is used between sets of numbers, but the word *or* is used between statements about numbers. To interchange their roles would be an abuse of the notation.

Example 8   Sketch the set $\{x : 0 < x \leq 3 \text{ or } x \geq 5\}$.

Solution   First present some scratch work. On one number line sketch the set of real numbers that are between 0 and 3, including 3. On a second number line, sketch the set of real numbers that are greater than or equal to 5.

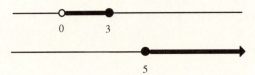

Because the word *or* is involved, you must shade the set of real numbers that are shaded on *either* number line in the scratch work. Remember, *or* means *union*, which is a way of combining all of the elements from either set.

Here is a description of the solution set in interval notation and set-builder notation:

$$(0, 3] \cup [5, +\infty) = \{x : 0 < x \leq 3 \text{ or } x \geq 5\}$$

Note that the union symbol, $\cup$, is used between sets of numbers, but the word *or* is used between statements about numbers. To reverse their roles would be an abuse of the notation.

Exercises for Section 2.7

1. Explain, in your own words, the meaning of the word *and*.

2. Explain, in your own words, the meaning of the word *or*.

3. Sketch each of the following sets on a number line.

   a) $\{x : x > 4\}$ b) $\{x : x \le 5\}$ c) $(5, +\infty)$ d) $(-\infty, 4]$

   e) $\{x : x \le -3\}$ f) $\{x : x > -9\}$ g) $(-\infty, -7)$ h) $[-5, +\infty)$

4. Use set-builder notation and interval notation to describe each of the following sets.

   a)
   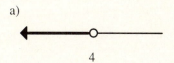

   b)

   c)
   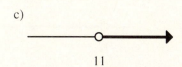

   d)
   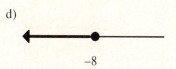

5. Sketch each of the following sets on a number line. If you wish, you may sketch intermediate results on number lines, but your final answer must be exhibited on a single, final number line. Use both set-builder notation and interval notation to describe your solution

   a) $\{x : x < 4 \text{ and } x < 8\}$ b) $\{x : x < 4 \text{ or } x < 8\}$

   c) $\{x : x < -3 \text{ and } x < -1\}$ d) $\{x : x < -3 \text{ or } x < -1\}$

   e) $\{x : x \ge 4 \text{ and } x \ge 2\}$ f) $\{x : x \ge 4 \text{ or } x \ge 2\}$

   g) $\{x : \le -1 \text{ and } x \le -5\}$ h) $\{x : \le -1 \text{ or } x \le -5\}$

   i) $\{x : x < 3 \text{ and } x \le 4\}$ j) $\{x : x < 3 \text{ or } x \le 4\}$

6. Break the following compound inequalities into two statements connected with the word *and*.

   a) $-2 < x < 5$ b) $3 \le x < 9$ c) $5 < x < 1$ d) $-4 \le x \le -8$

7. Use both set-builder notation and interval notation to describe each of the following solution sets.

   a)
   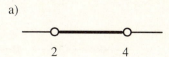

   b)
   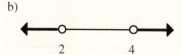

   c)
   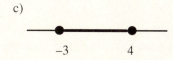

   d)
   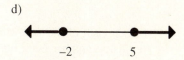

   e)
   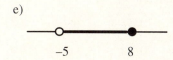

   f)
   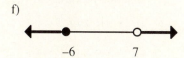

8. Sketch each of the following sets on a number line. If you wish, you may sketch intermediate results on number lines, but your final answer must be exhibited on a single, final number line. Use interval notation to describe your final solution, if possible.

a) $\{x:x<4 \text{ or } x>7\}$ b) $\{x:x<4 \text{ and } x>7\}$ c) $\{x:x>5 \text{ and } x<9\}$

d) $\{x:x>5 \text{ or } x<9\}$ e) $\{x:x\geq-1 \text{ and } x<8\}$ f) $\{x:x\geq-1 \text{ or } x<8\}$

g) $\{x:x\leq-4 \text{ and } x>10\}$ h) $\{x:x\leq-4 \text{ or } x>10\}$

9. Sketch each of the following sets on a number line. If you wish, you may sketch intermediate results on number lines, but your final answer must be exhibited on a single, final number line.

a) $\{x:-3<x<9\}$ b) $\{x:2<x<1\}$ c) $\{x:-4\leq x<-8\}$ d) $\{x:-3\leq x<8\}$

10. Use set-builder notation and a compound inequality to describe each of the following sets, then use interval notation to describe the sets.

a)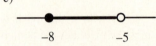

b)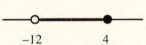

c)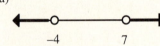

d)

11. Use set-builder and interval notation to describe each of the following solution sets.

a)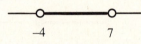

b)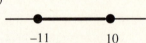

c)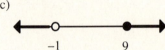

d)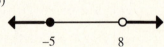

12. Use set-builder and interval notation to describe each of the following solution sets.

a)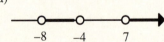

b)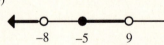

c)

d)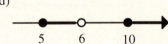

13. Sketch each of the following sets on a number line. If you wish, you may sketch intermediate results on number lines, but your final answer must be exhibited on a single, final number line. Describe your final solution using interval notation.

a) $\{x:x<0 \text{ or } 5<x\leq7\}$ b) $\{x:x<0 \text{ and } 5<x\leq7\}$

c) $\{x:x<3 \text{ or } 2<x<5\}$ d) $\{x:x<3 \text{ and } 2<x<5\}$

e) $\{x:x\leq4 \text{ and } 3<x\leq8\}$ f) $\{x:x\leq4 \text{ or } 3<x\leq8\}$

14. Sketch each of the following sets on a number line. If you wish, you may sketch intermediate results on number lines, but your final answer must be exhibited on a single, final number line. Use interval notation to describe your final answer.

a) $\{x : 0 < x < 6 \text{ or } 3 < x < 9\}$ b) $\{x : 0 < x < 6 \text{ and } 3 < x < 9\}$

c) $\{x : -3 \le x < 4 \text{ or } 5 \le x \le 9\}$ d) $\{x : -3 < x < 11 \text{ and } 5 \le x < 20\}$

2.8 Compound Inequalities

In this section we'll apply what we learned in Section 2.7 to aid us in solving compound inequalities. We'll be working with inequalities that involve *and* and *or* statements. Much of what we say will depend on the logic developed Section 2.7.

Example 1   It's a beautiful day to go fishing, so Alexei packs his gear and heads for the marina. After 10 minutes of rowing, Alexei is 1000 meters from his favorite fishing hole. After an additional 20 minutes of rowing, he is 400 meters from the fishing hole. If his distance from the fishing hole is a linear function of the amount of time that he has been rowing, during what time interval is Alexei between 900 meters and 500 meters from the fishing hole?

Graphical Solution   Begin by drawing a graph of Alexei's distance from the fishing hole as a function of the time that he has been rowing. Let D represent the distance *from* the fishing hole and place it on the vertical axis. Let t represent the amount of time that Alexei has been rowing and place it on the horizontal axis. Plot your data points and draw a line through them.

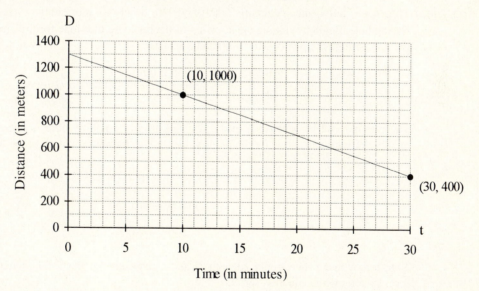

Find the interval of time when Alexei is between 500 meters and 900 meters from the fishing hole. If D represents the distance from the fishing hole, then you want to find when D is greater than 500 meters *and* less than 900 meters. You need to solve the inequality $D > 500$ and $D < 900$ for t. If you draw horizontal lines on the graph representing where $D = 500$ and $D = 900$, then you can find where the graph is above the line $D = 500$ and below the line $D = 900$.

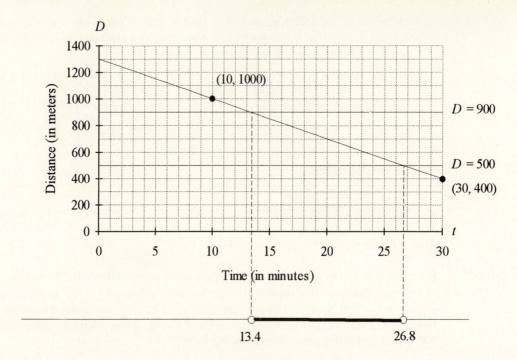

It appears that the graph of Alexei's distance from the fishing hole is above the line $D = 500$ *and* below the line $D = 900$ whenever the time t is approximately between 13.4 minutes and 26.8 minutes.

Analytical Solution   Solve the inequality $500 < D < 900$, where D represents Alexei's distance from the fishing hole. Begin by finding the equation of the line. Use the slope formula to calculate the slope of the line through the points (10, 1000) and (30, 400).

$$\text{Slope} = \frac{\text{Change in } D}{\text{Change in } t} = \frac{1000 - 400}{10 - 30} = -30 \text{ meters / minute}$$

The rate is negative because Alexei's distance from the fishing hole is *decreasing* at a rate of 30 meters every minute. Now use the point-slope form to find the equation of the line.

$$y - y_0 = m(x - x_0)$$

$$D - D_0 = m(t - t_0)$$

$$D - 1000 = -30(t - 10)$$

Note how y and x were replaced with D and t, respectively. Place this equation in slope-intercept form by solving the equation for D.

$$D - 1000 = -30t + 300$$

$$D - 1000 + 1000 = -30t + 300 + 1000$$

$$D = -30t + 1300$$

Now replace D in the inequality $500 < D < 900$ with the last result.

$$500 < D < 900$$

$$500 < -30t + 1300 < 900$$

Recall that compound inequalities of this form mean *and*. Break this compound inequality into two statements, separated by the word *and*, then solve each part separately for t.

$$-30t + 1300 > 500 \quad \text{and} \quad -30t + 1300 < 900$$

$$-30t + 1300 - 1300 > 500 - 1300 \quad \text{and} \quad -30t + 1300 - 1300 < 900 - 1300$$

$$-30t > -800 \quad \text{and} \quad -30t < -400$$

$$t < \frac{800}{30} \quad \text{and} \quad t > \frac{400}{30}$$

If you divide 800 by 30 on your calculator, $\dfrac{800}{30} \approx 26.7$. Similarly, $\dfrac{400}{30} \approx 13.3$.

$$t < 26.7 \quad \text{and} \quad t > 13.3$$

It is better to write this last statement in the following equivalent form.

$$t > 13.3 \quad \text{and} \quad t < 26.7$$

By shading this solution on a second number line, the following result is obtained:

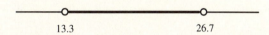

This agrees remarkably well with the graphical solution.

A Compact Technique for Solving Compound Inequalities

With the compound inequality, it is neater and more compact *not* to split the compound inequality into two separate inequalities separated with the word *and*. When working with compound inequalities, any real number may be added or subtracted from all three members of a compound inequality.

> *If $a < x < b$ and c is any real number, then $a + c < x + c < b + c$ and $a - c < x - c < b - c$.*

All three members of a compound inequality may be multiplied or divided by any *positive* real number.

> *If $a < x < b$ and c is any positive real number, then $ac < xc < bc$ and $\dfrac{a}{c} < \dfrac{x}{c} < \dfrac{b}{c}$.*

If you multiply or divide all three members by a negative real number, then you must remember to reverse the inequality signs.

> If $a < x < b$ and c is any negative real number, then $ac > xc > bc$ and $\dfrac{a}{c} > \dfrac{x}{c} > \dfrac{b}{c}$.

Example 2   Use a compact technique to solve the inequality in Example 1 for D.

Solution   Begin by subtracting 1300 from all three members.

$$500 < -30t + 1300 < 900$$

$$500 - 1300 < -30t + 1300 - 1300 < 900 - 1300$$

$$-800 < -30t < -400$$

Now divide all three members by -30. Note how the inequality signs reverse.

$$\frac{-800}{-30} > \frac{-30t}{-30} > \frac{-400}{-30}$$

$$\frac{800}{30} > t > \frac{400}{30}$$

It is best to put your answer in the following equivalent form.

$$\frac{400}{30} < t < \frac{800}{30}$$

If you now use your calculator to find decimal approximations for the fractions, you will obtain the following result:

$$13.3 < t < 26.7$$

If you organize your work in the following manner, it is easy to compare the results of the analytical method and the graphing method.

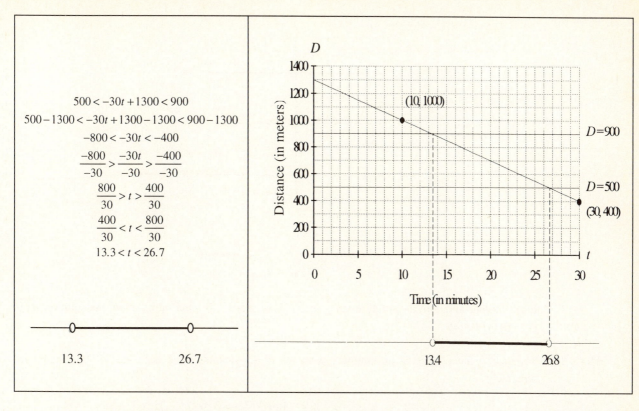

$$500 < -30t + 1300 < 900$$

$$500 - 1300 < -30t + 1300 - 1300 < 900 - 1300$$

$$-800 < -30t < -400$$

$$\frac{-800}{-30} > \frac{-30t}{-30} > \frac{-400}{-30}$$

$$\frac{800}{30} > t > \frac{400}{30}$$

$$\frac{400}{30} < t < \frac{800}{30}$$

$$13.3 < t < 26.7$$

13.3   26.7

13.4   26.8

In set-builder notation, the solution set is approximately $\{x : 13.3 < x < 26.7\}$. In interval notation, the solution set is approximately $(13.3, 26.7)$.

Example 3   Solve the following inequality for x: $3 - 2x < -3$ or $3 - 2x > 3$.

Graphical Solution   Begin by drawing the graph of the equation $f(x) = 3 - 2x$ on a coordinate system. The graph of this equation is a line, with slope -2 and y-intercept 3.

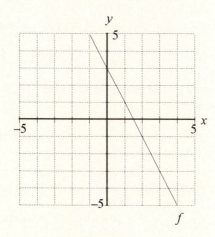

Because the inequality is of the form $f(x) < -3$ or $f(x) > 3$, draw the horizontal lines $y = -3$ and $y = 3$ on the graph and note where the graph of the function f is below the graph of $y = -3$ *or* above the graph of $y = 3$.

In set-builder notation, the solution set is $\{x : x < 0 \text{ or } x > 3\}$. In interval notation, the solution set is $(-\infty, 0) \cup (3, +\infty)$.

Analytical Solution Since there is no compact form for *or* statements, solve each half of this inequality separately.

$$3 - 2x < -3 \text{ or } 3 - 2x > 3$$

$$3 - 2x - 3 < -3 - 3 \text{ or } 3 - 2x - 3 > 3 - 3$$

$$-2x < -6 \text{ or } -2x > 0$$

$$\frac{-2x}{-2} > \frac{-6}{-2} \text{ or } \frac{-2x}{-2} < \frac{0}{-2}$$

$$x > 3 \text{ or } x < 0$$

The last line of the solution is equivalent to the following inequality.

$$x < 0 \text{ or } x > 3$$

It is best to write the inequality in this order, because it is now in the same order as the shaded solution on the number line.

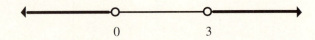

Here is the description of the solution set in interval notation and set-builder notation:

$$(-\infty, 0) \cup (3, +\infty) = \{x : x < 0 \text{ or } x > 3\}$$

If you arrange your work side-by-side, then it is easy to compare the results of the analytical method and the graphing method.

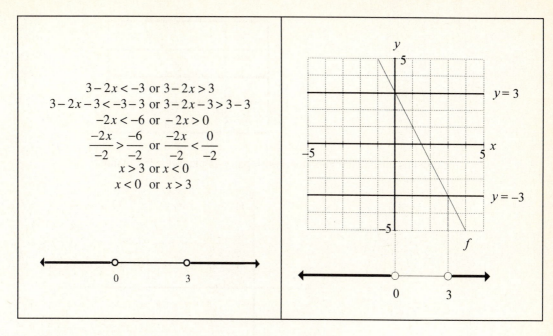

$$3 - 2x < -3 \text{ or } 3 - 2x > 3$$
$$3 - 2x - 3 < -3 - 3 \text{ or } 3 - 2x - 3 > 3 - 3$$
$$-2x < -6 \text{ or } -2x > 0$$
$$\frac{-2x}{-2} > \frac{-6}{-2} \text{ or } \frac{-2x}{-2} < \frac{0}{-2}$$
$$x > 3 \text{ or } x < 0$$
$$x < 0 \text{ or } x > 3$$

Example 4   *(TI-82)* Solve the following inequality for x: $-1 < x - \dfrac{2x+5}{3} < 1$.

Graphical Solution   Load the following functions into the Y= menu:

```
Y1█-1
Y2█X-(2X+5)/3
Y3█1
Y4=
Y5=
Y6=
Y7=
Y8=
```

Push the ZOOM button and select 6:ZStandard. Make the following adjustments to the WINDOW parameters:

```
WINDOW FORMAT
Xmin=-5
Xmax=15
Xscl=1
Ymin=-2
Ymax=2
Yscl=1
```

Push the GRAPH button to capture the following image in your viewing window:

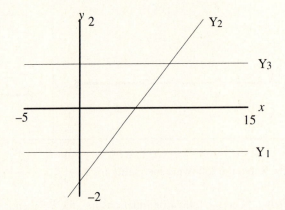

Copy this image onto your homework paper in the following manner:

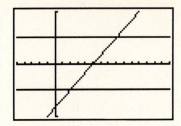

Note that each graph has been labeled with its name, Y1, Y2, or Y3. The axes have also been labeled with the WINDOW parameters. Use the intersect utility in the CALC menu to find the point of intersection of the graphs Y2 and Y3. Open the CALC menu by pressing 2nd CALC and select 5:intersect, capturing the following image in your viewing window:

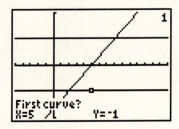

Note the 1 in the upper right corner of the viewing window. This indicates that the trace cursor is currently on the graph of the function Y1. The calculator is asking if this is one of the graphs to be used when finding a point of intersection. Because you want to find the point of intersection of the graphs of the functions Y2 and Y3, you do *not* want to use this graph. Press the up arrow on your keyboard to capture the following image in your viewing window:

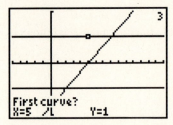

Note the 3 in the upper right corner. This indicates that the trace cursor is on the graph of the function Y3. Since this is one of the graphs that will be used, press ENTER to accept this curve. The calculator responds with the following image in the viewing window:

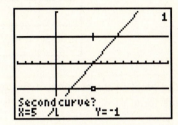

Note that the trace cursor has returned to the graph of Y1. Because Y2 will be used for the second curve, press the up arrow *twice* to capture the following image in your viewing window:

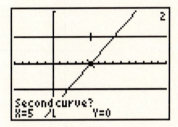

Because Y2 is one of the graphs that will be used, press ENTER to accept it. The calculator responds with the following image in the viewing window:

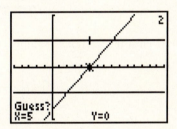

Move the cursor reasonably close to the point of intersection you wish to find. This is the guess. Press the ENTER key and the calculator will accept your guess and respond with the following image in the viewing window:

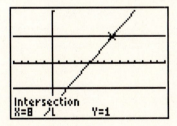

The *x*-value of the point of intersection of the graphs of the functions Y2 and Y3 is 8. You can find the *x*-value of the point of intersection of the graphs of the functions Y1 and Y2.

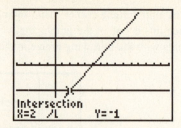

The *x*-value of the point of intersection of the graphs of the functions Y_1 and Y_2 is 2. Since the inequality is of the form $Y_1 < Y_2 < Y_3$, note where the graph of Y_2 is *above* the graph of Y_1 *and below* the graph of Y_3.

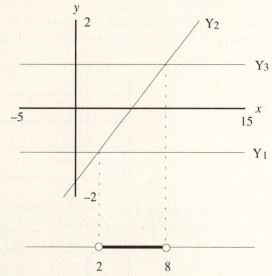

Analytical Solution First, to clear the fractions, multiply each member by the least common denominator.

$$3[-1] < 3\left[x - \frac{2x+5}{3} \right] < 3[1]$$

$$-3 < 3x - 3\left(\frac{2x+5}{3} \right) < 3$$

$$-3 < 3x - (2x+5) < 3$$

Note the effective use of parentheses. Next, distribute the minus sign and simplify.

$$-3 < 3x - 2x - 5 < 3$$

$$-3 < x - 5 < 3$$

Finally, add 5 to all three members and simplify.

$$-3+5 < x-5+5 < 3+5$$

$$2 < x < 8$$

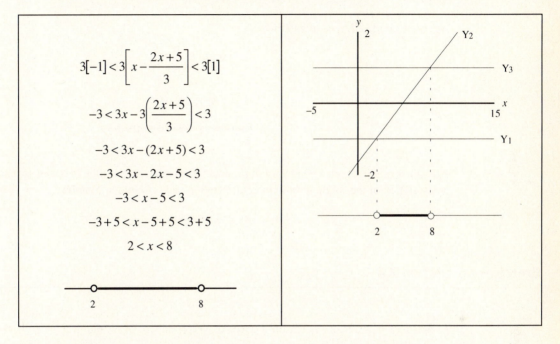

Place the analytical solution and graphing solution side-by-side for easy comparison.

$$3[-1] < 3\left[x - \frac{2x+5}{3}\right] < 3[1]$$

$$-3 < 3x - 3\left(\frac{2x+5}{3}\right) < 3$$

$$-3 < 3x - (2x+5) < 3$$

$$-3 < 3x - 2x - 5 < 3$$

$$-3 < x - 5 < 3$$

$$-3+5 < x-5+5 < 3+5$$

$$2 < x < 8$$

Since the graphical and the analytic solution agree, you can be confident that you have the correct solution. Here is a description of our solution set in interval notation and set-builder notation.

$$(2,8) = \{x : 2 < x < 8\}$$

Example 5   If you wish to keep the perimeter of the following rectangle between 100 and 200 units, where must the values of x lie?

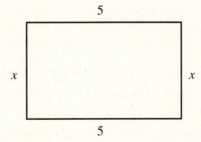

Solution   The perimeter must lie between 100 and 200 units. If you express this idea in symbols, you create the following compound inequality.

$$100 < P < 200$$

Of course, the perimeter, P, equals $x + 5 + x + 5$. Simplifying, you have $P = 2x + 10$. Now substitute this expression for P in the above inequality.

$$100 < 2x + 10 < 200$$

Subtract 10 from all three members of the compound inequality.

$$100 - 10 < 2x + 10 - 10 < 200 - 10$$
$$90 < 2x < 190$$

Next, we divide all three members by 2.

$$\frac{90}{2} < \frac{2x}{2} < \frac{190}{2}$$
$$45 < x < 95$$

All values of x between 45 and 95 will produce a perimeter that will lie between 100 and 200. Here is a description of the solution set in interval notation and set-builder notation.

$$(45, 95) = \{x : 45 < x < 95\}$$

Exercises for Section 2.8

1. Set up a coordinate system on a sheet of graph paper. Draw the graph of $f(x) = x + 1$ on the coordinate system. Draw the horizontal lines whose equations are $y = -2$ and $y = 3$ on the coordinate system.

   a) Draw a number line below the graph. Drop dotted, vertical lines to your number line from the points of intersection of the graphs of f and the horizontal lines $y = -2$ and $y = 3$. On the number line shade all values of x that are solutions of the inequality $-2 < f(x) < 3$.

   b) Use an analytic method to solve the following inequality for x: $-2 < x + 1 < 3$. Sketch the solution of this inequality on a second number line. Use set-builder notation and interval notation to describe the solution set. Does the solution from part (b) compare favorably with the solution from part (a)? If not, check your work for error.

2. Set up a coordinate system on a sheet of graph paper. Draw the graph of $f(x) = 2x$ on the coordinate system. Draw the horizontal lines whose equations are $y = -4$ and $y = 6$ on the coordinate system.

   a) Draw a number line below the graph. Drop dotted, vertical lines to the number line from the points of intersection of the graphs of f and the horizontal lines $y = -4$ and $y = 6$. On the number line shade all values of x that are solutions of the inequality $-4 < f(x) < 6$.

   b) Use an analytic method to solve the following inequality for x: $-4 < 2x < 6$. Sketch the solution of this inequality on a second number line. Use set-builder notation and interval notation to describe the solution set. Does the solution from part (b) compare favorably with the solution from part (a)? If not, check your work for error.

3. Set up a coordinate system on a sheet of graph paper. Draw the graph of $f(x) = 2x - 1$ on the coordinate system. Draw the horizontal lines whose equations are $y = -3$ and $y = 5$ on the coordinate system.

   a) Draw a number line below the graph. Drop dotted, vertical lines to the number line from the points of intersection of the graphs of f and the horizontal lines $y = -3$ and $y = 5$. On the number line shade all values of x that are solutions of the inequality $-3 < f(x) < 5$.

   b) Use an analytic method to solve the following inequality for x: $-3 < 2x - 1 < 5$. Sketch the solution of this inequality on a second number line. Use set-builder notation and interval notation to describe the solution set. Does the solution from part (b) compare favorably with the solution from part (a)? If not, check your work for error.

4. Set up a coordinate system on a sheet of graph paper. Draw the graph of $f(x) = 3 - 2x$ on the coordinate system. Draw the horizontal lines whose equations are $y = -3$ and $y = 7$ on the coordinate system.

   a) Draw a number line below the graph. Drop dotted, vertical lines to the number line from the points of intersection of the graphs of f and the horizontal line $y = -3$ and $y = 7$. On the number line shade all values of x that are solutions of the inequality $-3 < f(x) < 7$.

   b) Use an analytic method to solve the following inequality for x: $-3 < 3 - 2x < 7$. Sketch the solution of this inequality on a second number line. Use set-builder notation and interval notation to describe the solution set. Does the solution from part (b) compare favorably with the solution from part (a)? If not, check your work for error.

5. Use an analytic method only to solve each of the following compound inequalities for x. Use the compact technique demonstrated in Example 2. Shade your solution on a number line and use both set-builder notation and interval notation to describe the solution set.

   a) $-3 < x - 3 \le 5$
   b) $-5 \le x + 6 < 10$
   c) $-9 < 3x < 12$
   d) $-6 \le -3x < 9$
   e) $-3 < 2x + 5 < 3$
   f) $-3 \le 3 - x \le 3$
   g) $-2 \le 5x + 1 < 10$
   h) $-4 \le 4 - 5x \le 10$

6. For each of the following inequalities, set up a graphical solution on a sheet of graph paper, then present an analytical solution. Use set-builder notation and interval notation to describe the solution set. Present your graphical solution and analytical solution in a side-by-side format to compare the two methods.

   a) $-4 < 2x - 6 < 10$
   b) $-3 < 4 - x < 6$

7. Set up a coordinate system on a sheet of graph paper. Draw the graph of $f(x) = 2x + 3$ on the coordinate system. Draw the horizontal lines whose equations are $y = -7$ and $y = 7$ on the coordinate system.

   a) Draw a number line below your graph. Drop dotted, vertical lines to the number line from the points of intersection of the graph of f and the horizontal lines $y = -7$ and $y = 7$. On the number line shade all values of x that are solutions of the inequality $f(x) < -7$ or $f(x) > 7$.

   b) Use an analytic method to solve the following inequality for x: $2x + 3 < -7$ or $2x + 3 > 7$. Sketch the solution of this inequality on a second number line. Does the solution from part (b) compare favorably with the solution from part (a)? If not, check your work for error. Use set-builder notation and interval notation to describe the solution set.

8. Set up a coordinate system on a sheet of graph paper. Draw the graph of $f(x) = 4 - x$ on the coordinate system. Draw the horizontal lines whose equations are $y = -5$ and $y = 5$ on the coordinate system.

   a) Draw a number line below your graph. Drop dotted, vertical lines to the number line from the points of intersection of the graph of f and the horizontal lines $y = -5$ and $y = 5$. On the number line shade all values of x that are solutions of the inequality $f(x) < -5$ or $f(x) > 5$.

   b) Use an analytic method to solve the following inequality for x: $4 - x < -5$ or $4 - x > 5$. Sketch the solution of this inequality on a second number line. Use set-builder notation and interval notation to describe the solution set. Does the solution from part (b) compare favorably with the solution from part (a)? If not, check your work for error.

9. Use an analytic method only to solve each of the following inequalities for x. Remember, there is no compact form for *or* statements. Shade your solution on a number line and use both set-builder notation and interval notation to describe the solution set.

   a) $3x + 1 \leq -8$ or $3x + 1 \geq 8$

   b) $2 - x < -5$ or $2 - x > 5$

   c) $3 - 2x < -7$ or $3 - 2x > 7$

   d) $2x + 7 \leq -9$ or $2x + 7 \geq 9$

10. For each of the following inequalities, set up graphical solution on a sheet of graph paper, then present an analytical solution. Use set-builder notation and interval notation to describe the solution set. Present your graphical solution and analytical solution in a side-by-side format to compare the two methods.

    a) $2x - 1 < -5$ or $2x - 1 > 5$

    b) $3 - x \leq -5$ or $3 - x \geq 5$

11. Use an analytic method only to solve each of the following inequalities for x. Shade your solution on a number line and use set-builder notation and interval notation to describe the solution set.

    a) $-7 < 3x - 8 < 7$

    b) $2x - 3 < -11$ or $2x - 3 > 11$

    c) $-5 < 2x - (3 - x) \leq 8$

    d) $x - (2 - x) < -8$ or $x - (2 - x) > 8$

    e) $-3 \leq 2(x - 3) - 3(2 - x) \leq 3$

    f) $-4 < 2(4 - x) - (x - 5) \leq 4$

    g) $2(x - 3) - (1 - x) \leq -10$ or $2(x - 3) - (1 - x) \geq 10$

12. Set up a coordinate system on a sheet of graph paper. Draw the graph of $f(x) = \dfrac{2}{3}x + 1$ on the coordinate system. Draw the horizontal lines whose equations are $y = -3$ and $y = 5$ on the coordinate system.

    a) Draw a number line below your graph. Drop dotted, vertical lines to the number line from the points of intersection of the graphs of f and the horizontal lines $y = -3$ and $y = 5$. On the number line shade all values of x that are solutions of the inequality $-3 < f(x) < 5$.

    b) Use an analytic method to solve the following inequality for x: $-3 < \dfrac{2}{3}x + 1 < 5$. Be sure to *first* multiply all three members by the least common denominator. Sketch the solution of this inequality on a second number line. Use set-builder notation and interval notation to describe the solution set. Does the solution from part (b) compare favorably with the solution from part (a)? If not, check your work for error.

13. Use an analytic method only to solve each of the following inequalities for x. Be sure to multiply by the least common denominator in your *first* step. Shade your solution on a number line and use set-builder notation and interval notation to describe the solution set.

a) $-\dfrac{1}{2} < \dfrac{1}{2}x - \dfrac{1}{3} < \dfrac{1}{2}$

b) $-\dfrac{2}{5} \le \dfrac{3}{4} - \dfrac{1}{5}x < \dfrac{2}{5}$

c) $\dfrac{2}{3} < \dfrac{1}{2}x + \dfrac{1}{4} < \dfrac{7}{3}$

d) $-\dfrac{3}{5} \le \dfrac{2}{3} - \dfrac{1}{2}x \le \dfrac{4}{3}$

14. Use an analytic method only and your calculator to find approximate solutions of each of the following inequalities. Use a calculator to round your answers to the nearest hundredth (two decimal places). Shade your solution on a number line and use set-builder notation and interval notation to describe the solution set.

a) $-2.2 < 3.3x - 1.4 < 3.2$

b) $2.2x - 3.4 < -1.8$ or $2.2x - 3.4 > 1.8$

c) $-1.8 < 3.23 - 4.6x \le 4.9$

d) $-2.3 \le 2.2x - 3.4(x - 4) < 2.3$

e) $0.23 < 1.3x - 2.35 \le 4.6$

f) $4.3 - 3.8x < -9.2$ or $4.3 - 3.8x > 9.2$

15. (TI-82) Load the following functions into the Y= menu:

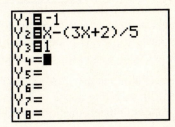

Push the ZOOM button and select 6:ZStandard. Make the following adjustments to the WINDOW parameters:

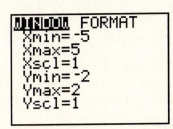

Set up a coordinate system on your homework paper. Push the GRAPH button and copy the image in the viewing window onto your homework paper. Label the appropriate graphs with Y1, Y2, or Y3. Include the WINDOW parameters on the axes.

a) Use the intersect utility in the CALC menu to find the points of intersection, first of the graphs of Y1 and Y2, then of the graphs of Y2 and Y3. Draw a number line below your graph and shade the solutions of the inequality Y1 <Y2 < Y3 on the number line.

b) Use an analytic method to solve the following inequality for x: $-1 < x - \dfrac{3x+2}{5} < 1$. Shade your solution on a second number line. Use both set-builder notation and interval notation to describe the solution set. Does the analytical solution compare favorably with the graphical solution of part (a)? If not, check your work for error.

16. (TI-82) Load the following functions into the Y= menu:

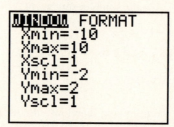

```
Y₁ ☐ -1
Y₂ ☐ (X+1)/2-(X+1)
    /3
Y₃ ☐ 1
Y₄ = ■
Y₅ =
Y₆ =
Y₇ =
```

Push the ZOOM button and select 6:ZStandard. Make the following adjustments to the WINDOW parameters:

```
WINDOW FORMAT
 Xmin=-10
 Xmax=10
 Xscl=1
 Ymin=-2
 Ymax=2
 Yscl=1
```

Set up a coordinate system on your homework paper. Push the GRAPH button and copy the image in the viewing window onto your homework paper. Label the appropriate graphs with Y1, Y2, or Y3. Include the WINDOW parameters on the axes.

a) Use the intersect utility in the CALC menu to find the points of intersection, first of the graphs of Y1 and Y2, then of the graphs of Y2 and Y3. Draw a number line below your graph and shade the solutions of the inequality Y1 < Y2 < Y3 on the number line.

b) Use an analytic method to solve the following inequality for x: $-1 < \dfrac{x+1}{2} - \dfrac{x+1}{3} < 1$. Shade your solution on a second number line. Use both set-builder notation and interval notation to describe the solution set. Does the analytical solution compare favorably with the graphical solution of part (a)? If not, check your work for error.

17. Use an analytic method only to solve each of the following inequalities for x. Be sure to multiply by the least common denominator in your *first* step. Shade your solution on a number line and use set-builder notation and interval notation to describe the solution set.

a) $-1 < x - \dfrac{2x-3}{7} < 1$

b) $-2 \le 3x - \dfrac{x-5}{4} \le 2$

c) $-1 < \dfrac{x+1}{3} - \dfrac{2x-5}{7} < 1$

d) $-1 \le \dfrac{3-x}{4} - \dfrac{x+1}{8} \le 1$

18. (TI-82) Present a graphing calculator solution and an analytical solution for each of the following inequalities. Use set-builder notation and interval notation to describe the solution set. Present your graphical solution and analytical solution in a side-by-side format to compare the two methods.

a) $-1 < x - \dfrac{x-4}{3} < 2$

b) $-2 \le \dfrac{x+1}{2} - \dfrac{x+1}{6} < 3$

19. Consider the following rectangle:

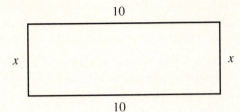

a) What values of x will keep the area of the rectangle between 100 and 200 square units?

b) What values of x will keep the perimeter of the rectangle between 100 and 200 units?

20. Assume that g is a linear function. In addition, suppose that $g(3) = 9$ and $g(7) = 17$. Set up a coordinate system on a sheet of graph paper, plot the given points, and draw a line through them.

a) Draw the lines whose equations are $y = 5$ and $y = 13$ on your sketch. Draw a number line below the graph. Drop dotted, vertical lines to your number line from the points of intersection of the graph of f and the horizontal lines $y = 5$ and $y = 13$. Shade the solution of $5 < g(x) < 13$ on the number line.

b) Find the equation of the function g. Express your answer in the form $g(x) = mx + b$.

c) Use an analytic method to solve the inequality $5 < g(x) < 13$ for x. Sketch your solution on a number line and use set-builder and interval notation to describe the solution set. Does the analytic solution agree favorably with the graphical solution from part (a)?

21. Set up a coordinate system on a sheet of graph paper. The graph of the function f is a line that goes through the point $(1,2)$ with slope $-\frac{2}{3}$. Draw the graph of f on your coordinate system as well as the graphs of the lines whose equations are $y = -5$ and $y = 5$.

a) Draw a number line below the graph. Shade and label the approximate solution of $-5 < f(x) < 5$ on the number line.

b) Find the equation of the function f. Express your answer in the form $f(x) = mx + b$.

c) Use an analytic method to find the solution of the inequality $-5 < f(x) < 5$. Shade your solution on a second number line. Use set-builder and interval notation to describe the solution set. Does the analytical solution compare favorably with the graphical solution from part (a)? If not, check your work for error.

22. Alfie is standing on a roof about 15 feet above ground level. He launches his kite and it begins to gain altitude at a rate of 10 feet per second. Set up a coordinate system on a sheet of graph paper and sketch the graph of the height, H, of the kite as a function of the amount of time, t, that it has been flying.

a) Draw the horizontal lines representing heights of 100 feet and 150 feet on your coordinate system. Draw a number line below the graph and shade and label the approximate solution of the inequality $100 < H < 150$ on the number line.

b) Express H as a function of t.

c) Use the equation you found for H in part (b) to help find an analytic solution of the inequality $100 < H < 150$. Shade the solution of this inequality on a second number line. Use set-builder and interval notation to describe this solution set. Does this analytic solution compare favorably with the graphing solution found in part (a)? If not, check your work for error.

23. Constantine runs a small shoe repair shop. It costs him $500 each month for operating expenses such as rent and utilities. In addition to these fixed costs, his costs increase by $10 every time he repairs a pair of shoes. Set up a coordinate system on a sheet of graph paper and sketch the graph of Constantine's costs, C, as a function of the number of pairs of shoes he repairs, x.

    a) Draw horizontal lines representing costs of $700 and $900 on your coordinate system. Draw a number line below the graph and shade and label the approximate solution of the inequality $700 < C < 900$ on the number line.

    b) Express C as a function of x.

    c) Use the equation you found for C in part (b) to help find an analytic solution of the inequality $700 < C < 900$. Shade the solution of this inequality on a second number line. Use set-builder and interval notation to describe this solution set. Does this analytic solution compare favorably with the graphing solution found in part (a)? If not, check your work for error.

24. Long-distance runners Jaime and Luis are running in a handicap race. Jaime is given a significant lead at the start of the race. After 30 minutes of running, the distance between Jaime and Luis is 500 meters. After 60 minutes, the distance between them is 200 meters. Set up a coordinate system on a sheet of graph paper. Let D represent the distance between the two runners and place it on the vertical axis. Let t represent the time that has passed since the race began and place it on the horizontal axis. Plot the two data points on your coordinate system. Assume that the distance between the two runners is a linear function of the time that they have been running and draw a line through the two data points.

    a) Draw horizontal lines representing distances between the runners of 400 meters and 300 meters on your coordinate system. Draw a number line below the graph and shade and label the approximate solution of the inequality $300 < D < 400$ on the number line.

    b) Express D as a function of t.

    c) Use the equation you found for D in part (b) to help find an analytic solution of the inequality $300 < D < 400$. Shade the solution of this inequality on a second number line. Use set-builder and interval notation to describe this solution set. Does this analytic solution compare favorably with the graphing solution found in part (a)? If not, check your work for error.

25. Racine is in Chicago to present a one-day workshop on a powerful spreadsheet program made by a well-known software company. While in town, she rents a car from Wheels Unlimited. She inquires about the rates and is told that if she drives 100 miles, her bill will be $39. If she drives 200 miles, her bill will be $49. Set up a coordinate system on a sheet of graph paper. Let A represent the amount of her bill and place it on the vertical axis. Let x represent the number of miles that she drives and place it on the horizontal axis. Plot the two data points on the coordinate system. Assume that the amount of her bill is a linear function of the number of miles that she drives and draw a line through your two data points.

    a) Draw horizontal lines representing charges of $42 and $45 on your coordinate system. Draw a number line below the graph and shade and label the approximate solution of the inequality $42 < A < 45$ on the number line.

    b) Express A as a function of x.

    c) Use the equation you found for A in part (b) to help find an analytic solution of the inequality $42 < A < 45$. Shade the solution of this inequality on a second number line. Use set-builder and interval notation to describe this solution set. Does this analytic solution compare favorably with the graphing solution found in part (a)? If not, check your work for error.

2.9 Inequalities Involving Absolute Value

In this section we learn how to solve inequalities involving absolute value. We begin by relating absolute value to distance on the number line. We'll use our definition of absolute value to draw the graph of $y = |x|$, a graph that will be fundamental to the work in this section. We'll use graphs to help solve inequalities involving absolute value. Finally, we'll use our definition of absolute value to introduce an analytical technique for solving inequalities involving absolute value.

Absolute Value as Distance. Let's begin with a definition.

$|x|$ = *the distance between x and 0 on the number line.*

Example 1   Simplify $|-3|$, $|0|$, and $|6|$.

Solution   Plot -3, 0, and 6 on a number line.

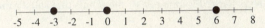

Because $|-3|$ represents the distance between -3 and 0 on the number line, $|-3| = 3$. Because $|0|$ represents the distance between 0 and 0 on the number line, $|0| = 0$. Because $|6|$ represents the distance between 6 and 0 on the number line, $|6| = 6$.

Important Note	Distance is never a negative quantity. Mathematicians prefer to say that distance is a *nonnegative* quantity.

Example 2   Sketch the graph of $f(x) = |x|$.

Solution   Create a table of points that satisfy the equation of the function f. Because $f(x) = |x|$, $f(x)$ represents the distance between x and 0 on the number line. Because $f(-5)$ represents the distance between -5 and 0 on the number line, $f(-5) = 5$. Because $f(-4)$ represents the distance between -4 and 0 on the number line, $f(-4) = 4$. You can complete the rest of the entries in the following table in a similar manner.

x	-5	-4	-3	-2	-1	0	1	2	3	4	5
$f(x)$	5	4	3	2	1	0	1	2	3	4	5

Plot the points in the table on a coordinate system.

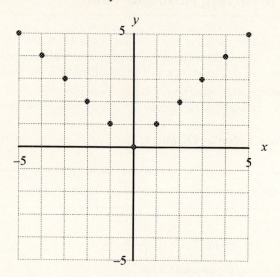

Do you see the pattern? Plot the rest of the points that satisfy the equation of the function f.

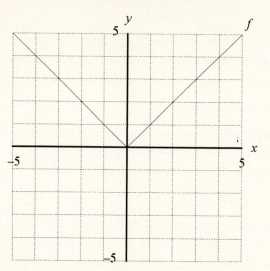

Example 3   Use the graph from Example 2 to solve the following inequality for x: $|x| < 3$.

Solution   The inequality is of the form $f(x) < g(x)$, where $f(x) = |x|$ and $g(x) = 3$. Draw the graphs of f and g on a coordinate system, then note where the graph of f is below the graph of g.

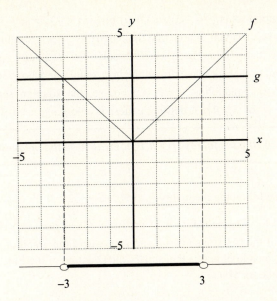

The solution set, in interval notation, is $(-3,3)$. In set-builder notation, the solution set is $\{x : -3 < x < 3\}$.

It is important to note that the inequality $|x| < 3$ is equivalent to the inequality $-3 < x < 3$. Suppose that you had been asked to solve $|x| < 4$. Instead of drawing the horizontal line whose equation is $y = 3$, you would have drawn the horizontal line whose equation is $y = 4$. Of course, the solution set would now be $(-4,4)$ or $\{x : -4 < x < 4\}$. A definite pattern is emerging.

> If $a > 0$, *the inequality* $|x| < a$ *is equivalent to the inequality* $-a < x < a$.

Example 4   Solve the following inequality for x: $|x| > 3$.

Solution   This inequality is of the form $f(x) > g(x)$, where $f(x) = |x|$ and $g(x) = 3$. Draw the graphs of f and g on a coordinate system, then note where the graph of f is above the graph of g.

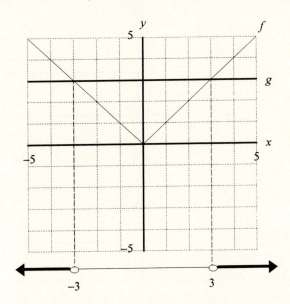

The solution set is $(-\infty, -3) \cup (3, +\infty)$ or $\{x : x < -3 \text{ or } x > 3\}$.

It is important to note that the inequality $|x| > 3$ is equivalent to the inequality $x < -3$ or $x > 3$. Suppose that you had been asked to solve the inequality $|x| > 4$ instead of the inequality $|x| > 3$. In that case, you would have drawn the horizontal line whose equation is $y = 4$ instead of the horizontal line whose equation is $y = 3$. Of course, the solution set would then be $(-\infty, -4) \cup (4, +\infty)$ or $\{x : x < -4 \text{ or } x > 4\}$. A definite pattern is emerging.

> **If $a > 0$, the inequality $|x| > a$ is equivalent to the inequality $x < -a$ or $x > a$.**

Example 5   Solve the following inequality for x: $|x - 3| < 2$.

Graphical Solution   The inequality is of the form $f(x) < g(x)$, where $f(x) = |x - 3|$ and $g(x) = 2$. Sketch the graphs of the functions f and g and note where the graph of f is below the graph of g. You will need to make a table of points that satisfy the equation of the function f. A few selected computations have been made for you.

$$f(-2) = |-2 - 3| = |-5| = 5$$

$$f(3) = |3 - 3| = |0| = 0$$

$$f(7) = |7 - 3| = |4| = 4$$

Continuing in this manner, you arrive at the following table of points that satisfy the equation of the function f.

x	-2	-1	0	1	2	3	4	5	6	7	8
$f(x)$	5	4	3	2	1	0	1	2	3	4	5

Plotting the points from this table and the rest of the points that satisfy the equation of the function f results in the following graph. The graph of the function g whose equation is $g(x) = 2$ is also included.

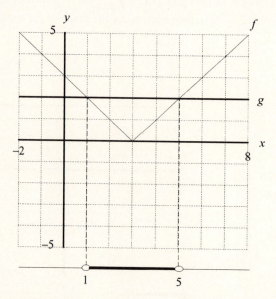

The graph of f is below the graph of g whenever x is between 1 and 5. In interval notation, the solution set is $(1, 5)$. In set-builder notation, our solution set is $\{x : 1 < x < 5\}$.

Analytical Solution   If $a > 0$, then the inequality $|x| < a$ is equivalent to the inequality $-a < x < a$. Consequently, the inequality $|x - 3| < 2$ is equivalent to the inequality $-2 < x - 3 < 2$. The techniques learned in Section 2.8 can be used to complete the solution.

$$|x - 3| < 2$$

$$-2 < x - 3 < 2$$

$$-2 + 3 < x - 3 + 3 < 2 + 3$$

$$1 < x < 5$$

Shade this solution on a second number line.

Place your analytic solution and your graphical solution side-by-side for easy comparison.

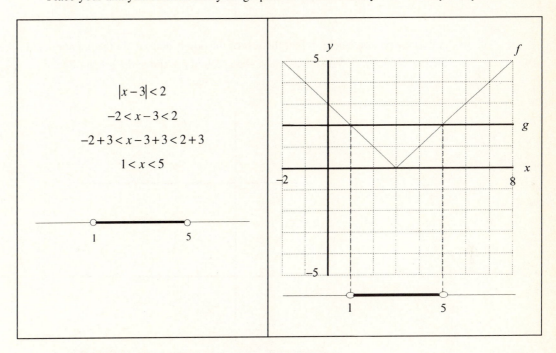

Example 6   Solve the following inequality for x: $|2x - 3| > 5$.

Graphical Solution   This inequality is of the form $f(x) > g(x)$, where $f(x) = |2x - 3|$ and $g(x) = 5$. Sketch the graphs of the functions f and g, then note where the graph of f is above the graph of g. Create a table of points that satisfy the equation of the function f. A few more points than usual have been added to the table in order to obtain the necessary accuracy. You should double-check the calculations before going on.

x	-3	-2.5	-2	-1.5	-1	-0.5	0	0.5	1
$f(x)$	9	8	7	6	5	4	3	2	1

x	1.5	2	2.5	3	3.5	4	4.5	5	5.5	6
$f(x)$	0	1	2	3	4	5	6	7	8	9

If you plot the points in this table, you arrive at the following sketch:

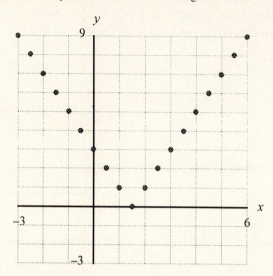

A definite pattern emerges. Plot the rest of the points that satisfy the equation $f(x) = |2x - 3|$. Include the graph of $g(x) = 5$ and note where the graph of f is above the graph of g.

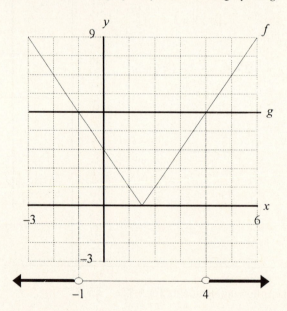

The graph of f is above the graph of g whenever x is less than -1 or x is greater than 4. In interval notation, the solution set is $(-\infty, -1) \cup (4, +\infty)$. In set-builder notation, the solution set is $\{x : x < -1 \text{ or } x > 4\}$

Analytical Solution   If $a > 0$, then the inequality $|x| > a$ is equivalent to the inequality $x < -a$ or $x > a$. Consequently, the inequality $|2x - 3| > 5$ is equivalent to the inequality $2x - 3 < -5$ or $2x - 3 > 5$. The techniques learned in Section 2.8 can now be used to complete the solution.

$$2x - 3 < -5 \text{ or } 2x - 3 > 5$$

$$2x - 3 + 3 < -5 + 3 \text{ or } 2x - 3 + 3 > 5 + 3$$

$$2x < -2 \text{ or } 2x > 8$$

$$\frac{2x}{2} < \frac{-2}{2} \text{ or } \frac{2x}{2} > \frac{8}{2}$$

$$x < -1 \text{ or } x > 4$$

Shade this solution set on a second number line.

Place the analytic solution and the graphical solution side-by-side for easy comparison.

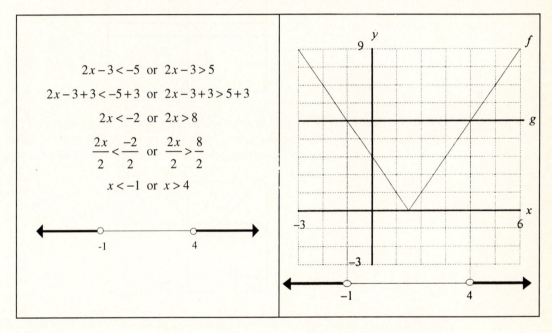

Example 7   *(TI-82)* Use your calculator to help solve the following inequality for x: $\left|\frac{1}{2}x + \frac{1}{3}\right| \le \frac{1}{4}$.

Solution   This inequality is of the form $f(x) \le g(x)$, where $f(x) = \left|\frac{1}{2}x + \frac{1}{3}\right|$ and $g(x) = \frac{1}{4}$. Use your calculator to sketch the graphs of the functions f and g, then note where the graph of f is below the graph of g or where the graph of f intersects the graph of g. Load the following functions into the Y= menu:

```
Y₁Babs (1/2*X+1/
3)
Y₂B1/4
Y₃=
Y₄=
Y₅=
Y₆=
Y₇=
```

Push the ZOOM button and select 6:ZStandard. After viewing the graph in the standard viewing window, make the following adjustments to the WINDOW parameters:

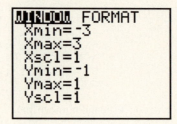

Push the GRAPH button to capture the following image in your viewing window:

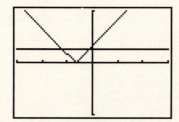

Use the intersect utility in the CALC menu to find the points of intersection of Y1 and Y2. Push 2nd CALC to open the CALC menu and select 5:intersect. Press ENTER *twice* to accept Y1 as the first curve and Y2 as the second curve, capturing the following image in your viewing window:

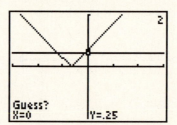

Find the *x*-value of the *leftmost* point of intersection of the graphs of Y1 and Y2. Move the cursor closer to this point of intersection than to the rightmost point of intersection, as shown:

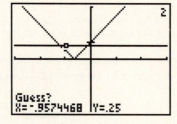

Press ENTER to instruct the calculator to accept the current *x*-value of the trace cursor as its initial guess in the algorithm used to find the desired point of intersection. Here is the resulting viewing window:

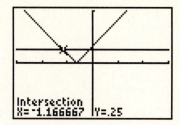

The x-value of the point of intersection is approximately -1.67. The x-value of the second point of intersection can be captured in a similar manner.

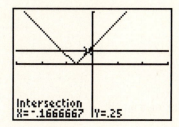

The x-value of the second point of intersection is approximately -0.17. Copy this image onto your homework paper in the following manner:

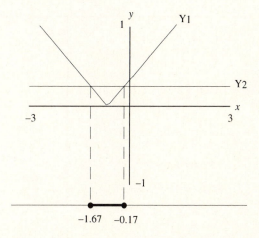

The appropriate graphs have been labeled with Y1 and Y2, respectively, and the WINDOW parameter settings have been included on the axes. The graph of Y1 is below the graph of Y2 whenever x is approximately between -1.67 and -0.17. The graph of Y1 intersects the graph of Y2 when x is approximately equal to -1.67 or -0.17. In interval notation, the solution set is approximately $[-1.67, -0.17]$. In set-builder notation, the solution set is approximately $\{x: -1.67 \le x \le -0.17\}$.

Analytical Solution   If $a > 0$, then the inequality $|x| \le a$ is equivalent to the inequality $-a \le x \le a$. Consequently, the inequality $\left| \frac{1}{2}x + \frac{1}{3} \right| \le \frac{1}{4}$ is equivalent to the inequality $-\frac{1}{4} \le \frac{1}{2}x + \frac{1}{3} \le \frac{1}{4}$. The techniques learned in Section 2.8 can be used to complete the solution of this inequality. Note that you should begin by multiplying all three members by the least common denominator.

$$-\frac{1}{4} \le \frac{1}{2}x + \frac{1}{3} \le \frac{1}{4}$$

$$12\left[-\frac{1}{4}\right] \le 12\left[\frac{1}{2}x + \frac{1}{3}\right] \le 12\left[\frac{1}{4}\right]$$

$$-3 \le 12\left[\frac{1}{2}x\right] + 12\left[\frac{1}{3}\right] \le 3$$

$$-3 \le 6x + 4 \le 3$$

$$-3 - 4 \le 6x + 4 - 4 \le 3 - 4$$

$$-7 \le 6x \le -1$$

$$\frac{-7}{6} \le \frac{6x}{6} \le \frac{-1}{6}$$

$$-\frac{7}{6} \le x \le -\frac{1}{6}$$

Shade this solution on a number line.

Arrange your work side-by-side for comparison.

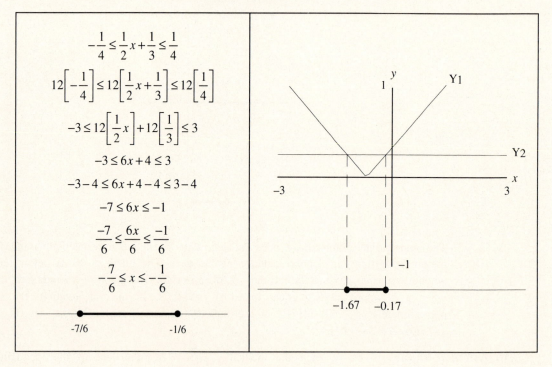

If you divide −7 by 6 on your calculator, $-\dfrac{7}{6} \approx -1.67$. Similarly, $-\dfrac{1}{6} \approx -0.17$. The graphing solution and the analytic solution compare quite nicely.

Exercises for Section 2.9

1. Draw a number line on a sheet of graph paper and plot points P, Q, and R on the number line at −4, −1, and 5, respectively. Label the point located at zero with the letter O.

   a) What is the distance from point P to point O? Simplify $|{-4}|$.

   b) What is the distance from point Q to point O? Simplify $|{-1}|$.

   c) What is the distance from point R to point O? Simplify $|5|$.

2. Draw a number line on a sheet of graph paper and plot points P, Q, and R on the number line at −6, 0, and 3, respectively. Label the point located at zero with the letter O.

   a) What is the distance from point P to point O? Simplify $|{-6}|$.

   b) What is the distance from point Q to point O? Simplify $|0|$.

   c) What is the distance from point R to point O? Simplify $|3|$.

3. Draw a number line on a sheet of graph paper. Label the point located at zero with the letter O.

   a) Choose an arbitrary point P to the *right* of zero whose coordinate is x. What is the distance between point P and point O? Simplify $|x|$.

   b) Choose an arbitrary point P to the *left* of zero whose coordinate is x. What is the distance between point P and point O? Simplify $|x|$.

   c) Explain, in your own words, why the following definition of absolute value makes perfectly good sense.

$$|x| = \begin{cases} x & \text{if } x > 0 \\ 0 & \text{if } x = 0 \\ -x & \text{if } x < 0 \end{cases}$$

4. Complete the following table of points that satisfy the equation $f(x) = |x - 4|$.

x	−1	0	1	2	3	4	5	6	7	8	9
$f(x)$											

   a) Set up a coordinate system on a sheet of graph paper. Plot each of the points in the table on your coordinate system. Use the pattern established to plot the rest of the points that satisfy the equation of the function f.

   b) The graph of f forms a sort of 'V'. What is the slope of the right-hand side of the 'V'? What is the slope of the left-hand side of the 'V'?

5. Complete the following table of points that satisfy the equation $f(x) = \left|\frac{1}{2}x + 1\right|$.

x	−12	−10	−8	−6	−4	−2	0	2	4	6	8
$f(x)$											

   a) Set up a coordinate system on a sheet of graph paper. Plot each of the points in the table on your coordinate system. Use the pattern established to plot the rest of the points that satisfy the equation of the function f.

b) The graph of f forms a sort of 'V'. What is the slope of the right-hand side of the 'V'? What is the slope of the left-hand side of the 'V'?

6. Complete the following table of points that satisfy the equation $f(x) = |2x - 4|$.

x	-3	-2	-1	0	1	2	3	4	5	6	7
$f(x)$											

a) Set up a coordinate system on a sheet of graph paper. Plot each of the points in the table on your coordinate system. Use the pattern established to plot the rest of the points that satisfy the equation of the function f.

b) The graph of f forms a sort of 'V'. What is the slope of the right-hand side of the 'V'? What is the slope of the left-hand side of the 'V'?

7. The graph of each of the following functions is a sort of 'V'. Perform the following tasks for each function.

i)   Set up a table of points that satisfy the equation of the function.

ii) Set up a *separate* coordinate system and plot the points from the table on your coordinate system. If the 'V' pattern is not present, add more points to the table and to the graph until the 'V' pattern is present. Plot the rest of the points that satisfy the equation of the function.

iii) Clearly state the slope of each side of the 'V' in your final sketch.

a) $f(x) = |x - 1|$ b) $f(x) = |x + 2|$ c) $f(x) = |3x - 9|$

d) $f(x) = |4x + 4|$ e) $f(x) = \left|\dfrac{1}{3}x + 2\right|$ f) $f(x) = \left|\dfrac{1}{2}x - 2\right|$

Hint: Use multiples of 3 for your x-values when working Exercise (e).

8. Complete the following table of points that satisfy the equation $f(x) = |x|$.

x	-5	-4	-3	-2	-1	0	1	2	3	4	5
$f(x)$											

a) Set up a coordinate system on a sheet of graph paper and plot each point from the table on your coordinate system. Plot the rest of the points that satisfy the equation of the function f.

b) Draw the graph of the equation $g(x) = 2$ on the coordinate system.

c) Draw a number line below the graph and shade and label the solution of $f(x) < g(x)$ on the number line. Use set-builder and interval notation to describe the solution set.

d) Explain, in your own words, why the inequality $|x| < 2$ is equivalent to the inequality $-2 < x < 2$.

9. Complete the following table of points that satisfy the equation $f(x) = |x|$.

x	-5	-4	-3	-2	-1	0	1	2	3	4	5
$f(x)$											

a) Set up a coordinate system on a sheet of graph paper and plot each point from the table on your coordinate system. Plot the rest of the points that satisfy the equation of the function f.

b) Draw the graph of the equation $g(x) = 2$ on your coordinate system.

c) Draw a number line below your graph and shade and label the solution of $f(x) > g(x)$ on the number line. Use set-builder and interval notation to describe the solution set.

d) Explain, in your own words, why the inequality $|x| > 2$ is equivalent to the inequality $x < -2$ or $x > 2$.

10. Match the inequality in column one with its equivalent inequality in column two.

a) $|x| < 8$

b) $|x| > 8$

c) $|x| \le 8$

d) $|x| \ge 8$

A) $-8 \le x \le 8$

B) $x < -8$ or $x > 8$

C) $-8 < x < 8$

D) $x \le -8$ or $x \ge 8$

11. Remember, if $a > 0$, then the inequality $|x| < a$ is equivalent to the inequality $-a < x < a$. Use this fact to transform each of the following inequalities into equivalent inequalities. If necessary, use the techniques of Section 2.8 to solve the equivalent inequality for x. Sketch the solution set on a number line and use set-builder and interval notation to describe the solution set.

a) $|x| < 9$

b) $|x| \le 7$

c) $|2x| < 6$

d) $|3x| \le 12$

e) $|x - 2| < 8$

f) $|x + 4| \le 11$

12. Remember, if $a > 0$, then the inequality $|x| > a$ is equivalent to the inequality $x < -a$ or $x > a$. Use this fact to transform each of the following inequalities into equivalent inequalities. If necessary, use the techniques of Section 2.8 to solve the equivalent inequality for x. Sketch the solution set on a number line and use set-builder and interval notation to describe the solution set.

a) $|x| > 12$

b) $|x| \ge 1$

c) $|5x| > 15$

d) $|4x| \ge 24$

e) $|x - 8| > 11$

f) $|x + 4| \ge 5$

13. Complete the following table of points that satisfy the equation $f(x) = |x - 5|$.

x	0	1	2	3	4	5	6	7	8	9	10
$f(x)$											

a) Set up a coordinate system on a sheet of graph paper and plot each of the points from the table on your coordinate system. Fill in the rest of the points that satisfy the equation of the function f.

b) Draw the graph of the function $g(x) = 4$ on your coordinate system. Draw a number line below the graph and shade and label the solution set of the inequality $f(x) < g(x)$.

c) Use an analytical method to solve the following inequality for x: $|x - 5| < 4$. Shade your solution on a second number line. Use set-builder and interval notation to describe the solution set. Does the analytical solution agree favorably with the graphical solution from part (b)? If not, check your work for error.

14. Complete the following table of points that satisfy the equation $f(x) = |x + 1|$.

x	-6	-5	-4	-3	-2	-1	0	1	2	3	4
$f(x)$											

a) Set up a coordinate system on a sheet of graph paper and plot each of the points from the table on the coordinate system. Fill in the rest of the points that satisfy the equation of the function f.

b) Draw the graph of the function $g(x) = 4$ on your coordinate system. Draw a number line below the graph and shade and label the solution set of the inequality $f(x) > g(x)$.

c) Use an analytical method to solve the following inequality for x: $|x+1| > 4$. Shade your solution on a second number line. Use set-builder and interval notation to describe the solution set. Does your analytical solution agree favorably with the graphical solution from part (b)? If not, check your work for error.

15. Complete the following table of points that satisfy the equation $f(x) = |2x|$.

x	−5	−4	−3	−2	−1	0	1	2	3	4	5
$f(x)$											

a) Set up a coordinate system on a sheet of graph paper and plot each of the points from the table on the coordinate system. Fill in the rest of the points that satisfy the equation of the function f.

b) Draw the graph of the function $g(x) = 6$ on your coordinate system. Draw a number line below the graph and shade and label the solution set of the inequality $f(x) \geq g(x)$.

c) Use an analytical method to solve the following inequality for x: $|2x| \geq 6$. Shade your solution on a second number line. Use set-builder and interval notation to describe the solution set. Does your analytical solution agree favorably with the graphical solution from part (b)? If not, check your work for error.

16. Transform each of the following inequalities into an equivalent inequality, then use an analytic method only to solve the equivalent inequality for x. Shade your solution set on a number line and use set-builder and interval notation to describe the solution set.

a) $|2x - 1| < 5$ b) $|3x + 1| > 8$ c) $|3 - 2x| \leq 5$

d) $|5 - x| \geq 11$ e) $|5x + 11| \leq 7$ f) $|4 - 3x| > 8$

17. Complete the following table of points that satisfy the equation $f(x) = |2x - 8|$.

x	−1	0	1	2	3	4	5	6	7	8	9
$f(x)$											

a) Set up a coordinate system on a sheet of graph paper and plot each of the points from the table on your coordinate system. Fill in the rest of the points that satisfy the equation of the function f.

b) Draw the graph of the function $g(x) = 6$ on your coordinate system. Draw a number line below the graph and shade and label the solution set of the inequality $f(x) \leq g(x)$ on the number line.

c) Use an analytical method to solve the following inequality for x: $|2x - 8| \leq 6$. Shade your solution on a second number line. Use set-builder and interval notation to describe the solution set. Does the analytical solution agree favorably with the graphical solution from part (b)? If not, check your work for error.

18. Complete the following table of points that satisfy the equation $f(x) = \left| \frac{1}{3}x - 3 \right|$.

x	−6	−3	0	3	6	9	12	15	18	21	24
$f(x)$											

a) Set up a coordinate system on a sheet of graph paper and plot each of the points from the table on your coordinate system. Fill in the rest of the points that satisfy the equation of the function f.

b) Draw the graph of the function $g(x) = 2$ on your coordinate system. Draw a number line below the graph and shade and label the solution set of the inequality $f(x) \geq g(x)$ on the number line.

c) Use an analytical method to solve the following inequality for x: $\left| \dfrac{1}{3}x - 3 \right| \geq 2$. Shade your solution on a second number line. Use set-builder and interval notation to describe the solution set. Does the analytical solution agree favorably with the graphical solution from part (b)? If not, check your work for error.

19. Transform each of the following inequalities into an equivalent inequality, then use an analytic method only to solve the equivalent inequality for x. Shade the solution set on a number line and use set-builder and interval notation to describe the solution set.

   a) $\left| \dfrac{1}{2}x - 1 \right| > 4$ b) $\left| \dfrac{1}{3}x - 4 \right| < 6$ c) $\left| \dfrac{3}{2}x - \dfrac{1}{3} \right| < 1$

   d) $\left| \dfrac{3}{4}x - \dfrac{1}{5} \right| > 1$ e) $\left| \dfrac{2}{3} - \dfrac{3}{4}x \right| \leq \dfrac{1}{5}$ f) $\left| \dfrac{3}{5} - \dfrac{3}{8}x \right| \geq \dfrac{1}{2}$

20. Transform each of the following inequalities into an equivalent inequality, then use an analytic method only to solve the equivalent inequality for x. Use a calculator to help you with your work and round the numbers in your solution to the nearest hundredth (two decimal places). Shade the solution set on a number line and use set-builder and interval notation to describe the solution set.

   a) $|2.3x - 1.4| < 3.8$ b) $|3.1 - 5.8x| > 4.2$

   c) $|3.12x - 4.53| \leq 1.12$ d) $|1.14 - 5.08x| \geq 4.5$

21. Transform each of the following inequalities into an equivalent inequality; then use an analytic method only to solve the equivalent inequality for x. Use a calculator to help you with your work, but do not round any numbers. Shade your solution set on a number line and use set-builder and interval notation to describe your solution set.

   a) $|x - 4| < 0.0001$ b) $|x + 5| < 0.00001$

   c) $|(2x - 3) - 5| < 0.001$ d) $|(3x - 5) - 10| < 0.0001$

22. (TI-82) Load the following functions into the Y= menu.

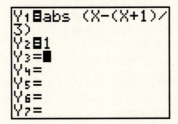

Push the ZOOM button and select 6:ZStandard. After viewing the graph in the standard viewing window, make the following adjustments to the WINDOW parameters:

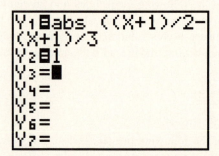

Push the GRAPH button. Set up a coordinate system on your homework paper and copy the resulting image onto the coordinate system. Label the graphs with Y1 and Y2, respectively, and include the WINDOW parameters on the coordinate axes.

a) Use the intersect utility in the CALC menu to find the *x*-values of the points of intersection of Y1 and Y2. Draw a number line below your graph and shade and label the solution set of the inequality Y1 <Y2 on the number line.

b) Use the methods discussed in this section to transform the inequality $\left| x - \dfrac{x+1}{3} \right| < 1$ into an equivalent inequality. Solve the equivalent inequality analytically. Shade your solution on a second number line. Use set-builder and interval notation to describe the solution set. Does this analytical solution agree favorably with the graphical solution in part (a)? If not, check your work for error.

23. (TI-82) Load the following functions into the Y= menu:

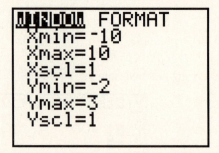

Push the ZOOM button and select 6:ZStandard. After viewing the graph in the standard viewing window, make the following adjustments to the WINDOW parameters:

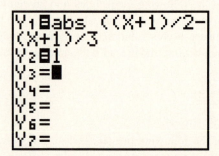

Push the GRAPH button. Set up a coordinate system on your homework paper and copy the resulting image onto your coordinate system. Label the graphs with Y1 and Y2, respectively, and include the WINDOW parameters on the coordinate axes.

a) Use the intersect utility in the CALC menu to find the *x*-values of the points of intersection of Y1 and Y2. Draw a number line below the graph and shade and label the solution set of the inequality Y1 <Y2 on the number line.

b) Use the methods of this section to transform the inequality $\left|\dfrac{x+1}{2} - \dfrac{x+1}{3}\right| < 1$ into an equivalent inequality. Solve the equivalent inequality analytically. Shade your solution on a second number line. Use set-builder and interval notation to describe the solution set. Does this analytical solution agree favorably with the graphical solution in part (a)? If not, check your work for error.

24. Transform each of the following inequalities into an equivalent inequality, then use an analytic method only to solve the equivalent inequality for x. Shade your solution set on a number line and use set-builder and interval notation to describe your solution set.

a) $\left|x - \dfrac{2x-3}{5}\right| < 1$

b) $\left|x - \dfrac{5-x}{7}\right| \le 1$

c) $\left|\dfrac{x-1}{2} - \dfrac{2x-1}{5}\right| \ge 1$

d) $\left|\dfrac{x+3}{5} - \dfrac{1-5x}{4}\right| \le 1$

25. Use both a graphing method and an analytical method to solve each of the following inequalities. Place your analytical and graphical solutions in the side-by-side format used in Example 7 in Section 2.9.

a) $|3x - 5| < 8$

b) $|2x + 3| > 5$

c) $\left|\dfrac{1}{2}x - \dfrac{1}{5}\right| \ge \dfrac{1}{10}$

d) $\left|2 - \dfrac{1}{5}x\right| < 1$

e) $\left|x - \dfrac{x-3}{2}\right| < 1$

f) $\left|\dfrac{x+1}{5} - \dfrac{x-1}{4}\right| \ge 1$

2.10 Summary and Review

Important Visualization Tools

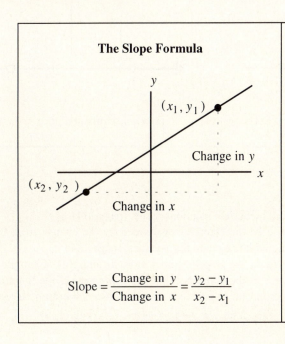

The Slope Formula

$$\text{Slope} = \frac{\text{Change in } y}{\text{Change in } x} = \frac{y_2 - y_1}{x_2 - x_1}$$

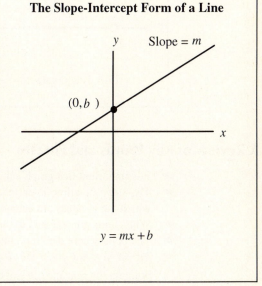

The Slope-Intercept Form of a Line

$$y = mx + b$$

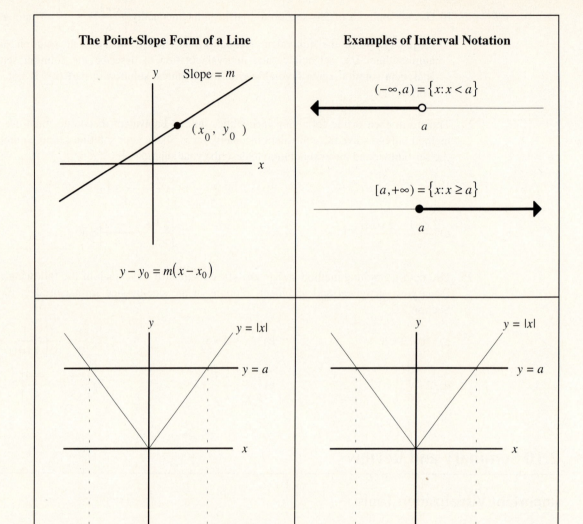

The Point-Slope Form of a Line

Slope $= m$

(x_0, y_0)

$y - y_0 = m(x - x_0)$

Examples of Interval Notation

$(-\infty, a) = \{x : x < a\}$

$[a, +\infty) = \{x : x \geq a\}$

$|x| < a$ is equivalent to $-a < x < a$

$|x| > a$ is equivalent to $x < -a$ or $x > a$

A Review of Key Words and Phrases

- It is important that graphical and numerical solutions agree. (page 114)
- The *slope* of a line is defined to be the change in y divided by the change in x. It is the rate at which the y-values are changing with respect to the x-values. (page 124)
- When units are attached to the slope, the slope becomes more meaningful. (Example 8, page 130)
- The point where the graph intercepts the y-axis is called the *y-intercept* of the function. (page 139)
- Tip: If you are given the slope and the y-intercept of a line, use $y = mx + b$. If you are given a point on the line and the slope of the line, use $y - y_0 = m(x - x_0)$. (page 149)

☛ If the slope of line L is m, then the slope of any line perpendicular to line L is $-\dfrac{1}{m}$. (page 155)

☛ The graph of a *constant* function is a horizontal line. (page 163)

☛ Important fundamentals to use when solving equations. (page 163)

　　1. If $a = b$ and c is any real number, then $a + c = b + c$ and $a - c = b - c$.

　　2. If $a = b$ and c is any nonzero real number, then $ac = bc$ and $\dfrac{a}{c} = \dfrac{b}{c}$.

☛ *Tip:* When solving a linear equation, try to isolate the terms containing x on one side of the equation. (page 164)

☛ When solving an equation with fractions, multiply both sides of the equation by the least common denominator in your *first* step. (page 169)

☛ Important fundamentals to use when solving inequalities. (page 180)

　　1. If $a < b$ and c is any real number, then $a + c < b + c$ and $a - c < b - c$.

　　2. If $a < b$ and c is any *positive* real number, then $ac < bc$ and $\dfrac{a}{c} < \dfrac{b}{c}$.

　　3. If $a < b$ and c is any *negative* real number, then $ac > bc$ and $\dfrac{a}{c} > \dfrac{b}{c}$.

　　Note the reversal of the inequality signs.

☛ $A \cap B = \{x : x \in A \text{ and } x \in B\}$. Intersection means *and*. (page 196)

☛ $A \cup B = \{x : x \in A \text{ or } x \in B\}$. Union means *or*. (page 196)

☛ Some further examples of interval notation. (Section 2.7)

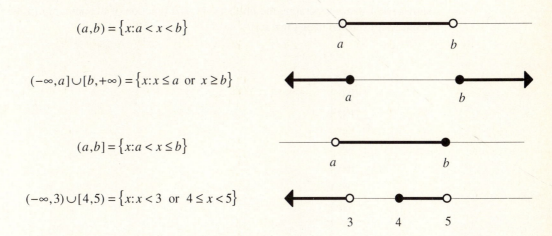

$$(a,b) = \{x : a < x < b\}$$

$$(-\infty, a] \cup [b, +\infty) = \{x : x \le a \text{ or } x \ge b\}$$

$$(a,b] = \{x : a < x \le b\}$$

$$(-\infty, 3) \cup [4,5) = \{x : x < 3 \text{ or } 4 \le x < 5\}$$

☛ For interval notation, a bracket includes the number and is the same as a filled-in circle. A parenthesis does not include the number and is the same as an empty circle. (page 201)

☛ $a < x < b$ is equivalent to $x > a$ and $x < b$. There is no compact form for *or* statements. (page 201)

☛ Technique for working with compound inequalities. (page 207)

　　1. If $a < x < b$ and c is any real number, then $a + c < x + c < b + c$ and $a - c < x - c < b - c$.

　　2. If $a < x < b$ and c is any *positive* real number, then $ac < xc < bc$ and $\dfrac{a}{c} < \dfrac{x}{c} < \dfrac{b}{c}$.

　　3. If $a < x < b$ and c is any *negative* real number, then $ac > xc > bc$ and $\dfrac{a}{c} > \dfrac{x}{c} > \dfrac{b}{c}$.

　　Note the reversal of the inequality signs.

☛ $|x| =$ the distance between x and 0 on a number line. (page 223)

TI-82 Keywords and Menus

☞   Using the TI-82 to solve linear equations. The intersection utility revisited. (page 165)
☞   Using the TI-82 to solve linear inequalities. The intersection utility revisited. (page (185)
☞   The fraction utility. (Problem 14, page 176).
☞   Finding points of intersection when 3 graphs are involved. (Example 4, page 211)

Chapter Review Exercises

1. Elizabeth and Jane are putting on their rollerblades. They are separated by a distance of 20 feet. They stand up simultaneously and begin to skate. With each passing minute, the distance between the two sisters increases by 5 feet. Assume that the distance, D, between the sisters is a linear function of the amount of time that they have been skating, t.

   a) Set up a coordinate system on a sheet of graph paper and sketch the graph of D as a function of t.

   b) Express D as a function of t. Use your equation to predict the distance between the sisters after 12.5 minutes of skating.

2. Jason is already running his morning workout when his brother Tim arrives at the track to join him. Of course, Tim is behind his brother and has some catching up to do. The graph that follows shows the distance separating the brothers, D, as a function of the amount of time that has passed, t, since Tim showed up at the track.

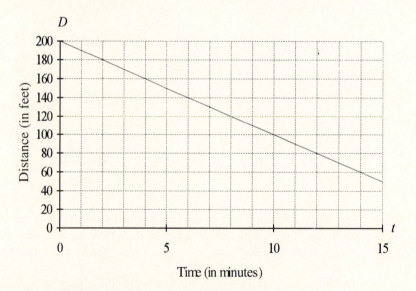

   a) What is the initial distance separating the brothers when Tim arrives?

   b) At what rate is the distance between the brothers decreasing? Include units with your answer to make the rate more meaningful.

   c) Express D as a function of t. Use this equation to find the distance between the brothers 8.5 minutes after Tim's arrival.

3. Use the slope formula to find the slope of the line that goes through the following pairs of points.

   a) $(2,3)$ and $(3,-4)$ b) $(-3,-5)$ and $(3,8)$
   c) $(4,3)$ and $(4,9)$ d) $(-3,4)$ and $(-6,4)$

4. For each of the following problems, perform the following tasks.

   i) Set up a separate coordinate system on a sheet of graph paper and sketch the line with the given slope and y-intercept.

   ii) Clearly state the equation of the line.

   a) $m = \dfrac{2}{3}, b = -1$ b) $m = -\dfrac{3}{2}, b = 2$

5. Suppose that $f(0) = -1$ and $f(2) = 5$. Furthermore, suppose that the graph of the function f is a line. Set up a coordinate system on a sheet of graph paper and sketch the graph of the line.

   a) Find the equation of the function in the form $f(x) = mx + b$.

   b) Use the equation of the function f that you found in part (a) to help find $f(10)$.

6. Initially, there are 50 gallons of water in the Driscolls' water tank. They turn on their pump and water begins to pour into the tank at the rate of 25 gallons per minute. Let V represent the volume of water in the tank and let t represent the time that the Driscolls' pump has been running. Assume that V is a linear function of t.

   a) Set up a coordinate system on a sheet of graph paper and sketch the graph of V as a function of t.

   b) Express V as a function of t. Use this equation to find the volume of water in the tank after 4.5 minutes.

7. For each of the following problems, perform the following tasks.

   i) Set up a separate coordinate system on a sheet of graph paper and draw the line that goes through the two data points. Clearly state the slope of the line on your paper.

   ii) Use the point-slope form of the line to find the equation of the line.

   iii) Place your answer in slope-intercept form by solving your equation for y. Clearly state the *exact* value of the y-intercept on your paper.

   a) $(-3, 4)$ and $(5, -8)$ b) $(-2, -2)$ and $(4, -9)$

8. Suppose that $f(-2) = -3$ and $f(3) = 4$. Furthermore, suppose that f is a linear function. Set up a coordinate system on a sheet of graph paper and sketch the equation of the line.

   a) Clearly state the slope of the line on your paper.

   b) Use the point-slope form to find the equation of your line. Solve your equation for y and find the equation of the function f in the form $f(x) = mx + b$.

   c) Use the equation in part (b) to find $f(10)$.

9. It's harvest time and Rachel begins picking apples at 6 A.M. on her family's farm in Washington. Rachel has a number of apples in her cart left over from the previous evening's work. At 6:10 A.M., there are 300 apples in her cart. At 6:20 A.M., there are 500 apples in her cart. Let N represent the number of apples in Rachel's cart and let t represent the amount of time that has passed since 6 A.M.. Assume that N is a linear function of t.

   a) Set up a coordinate system on a sheet of graph paper and sketch the graph of N as a function of t.

   b) At what rate is Rachel picking apples? Include units to make your answer more meaningful.

c) Use the point-slope form to find the equation of the line. Be sure to replace y and x with N and t, respectively.

d) Solve the equation you found in part (c) for N. How many apples were in Rachel's cart at 6 A.M.?

10. Use an analytic method only to solve each of the following equations for x.

a) $2x - 3 = 11$

b) $3 - 4x = -12$

c) $3x - 4 = 5 - 3x$

d) $5x + 2 = 12 - 3x$

e) $3(x - 4) - 2(5 - x) = 11$

f) $2(2x - 5) - (4 - x) = 3(5x + 1)$

g) $\dfrac{1}{2}x - \dfrac{3}{8} = \dfrac{2}{5}$

h) $\dfrac{1}{3} - \dfrac{3}{4}x = \dfrac{3}{5}x - \dfrac{3}{2}$

i) $x - \dfrac{x + 11}{13} = 1$

j) $\dfrac{x - 5}{8} - \dfrac{2x - 1}{9} = 3$

11. For each of the following problems, perform the following tasks.

i) Set up a coordinate system on a sheet of graph paper and sketch the graphs of the functions f and g on the coordinate system. Draw a number line below your graph and shade the approximate solution of $f(x) = g(x)$ on the number line in the usual manner.

ii) Use an analytic method to find the solution of the equation $f(x) = g(x)$. Make sure that this solution agrees with the graphical solution found in part (i).

a) $f(x) = 2x - 5$, $g(x) = 3$

b) $f(x) = 3 - \dfrac{1}{2}x$, $g(x) = -1$

c) $f(x) = 5 - x$, $g(x) = 2x - 1$

d) $f(x) = -\dfrac{5}{4}x + 4$, $g(x) = \dfrac{1}{2}x - 3$

12. Hans draws a fixed salary every month for polishing mirrors for telescopes. He also earns a certain amount for every mirror he polishes. If he polishes 20 mirrors in a month, his salary is $1200. If he polishes 40 mirrors in a month, his salary is $2200. Let S represent the total salary that Hans draws in a month and let x represent the number of mirrors that he polishes. Assume that S is a linear function of x.

a) Set up a coordinate system on a sheet of graph paper and sketch the graph of S as a function of x.

b) Use the slope formula to find the slope of the line. Attach units to your answer to make it more meaningful.

c) Use the point-slope form to find the equation of the line. Be sure to replace y and x with S and x, respectively. Place your answer in the slope-intercept form by solving this equation for S.

d) Use the equation in part (c) to find out how many mirrors Hans must polish if he wants to earn $1850 a month.

13. Use an analytic method to solve each of the following inequalities for x. Shade your solution set for each inequality on a number line and use set-builder and interval notation to describe your solution set.

a) $3 - 2x < -9$

b) $4x - 3 \geq -8$

c) $2(3 - x) - (4 - 5x) \leq 6$

d) $2(x + 1) - 3(4 - 3x) < -11$

e) $\dfrac{1}{3}x - \dfrac{5}{9} \leq \dfrac{5}{12}$

f) $\dfrac{2}{5}x - \dfrac{3}{2} \geq \dfrac{1}{20} + \dfrac{1}{2}x$

g) $\dfrac{x-8}{11} - \dfrac{x+9}{5} < 2$ h) $3x - \dfrac{4-5x}{13} > 1$

14. Use a calculator to help solve each of the following inequalities for x. Round the numbers in your final answer to the nearest hundredth (two decimal places). Shade this approximate solution set for each inequality on a number line and use set-builder and interval notation to describe this approximate solution set.

   a) $2.34(2x-7) - 3.12(5-x) > 1.23$ b) $3.12x - 0.05(4-2x) \geq 1.20$

15. For each of the following problems, perform the following tasks.

   i) Set up a coordinate system on a sheet of graph paper and sketch the graphs of the functions f and g on the coordinate system. Draw a number line below the graph and shade the approximate solution of $f(x) > g(x)$ on the number line in the usual manner.

   ii) Use an analytic method to find the solution of the equation $f(x) > g(x)$. Shade this analytical solution on a second number line. Use set-builder and interval notation to describe this solution set. Make sure that this solution agrees with the graphical solution found in part (i).

   a) $f(x) = 5 - 3x$, $g(x) = 2x - 5$ b) $f(x) = -\dfrac{3}{2}x + 4$, $g(x) = -2x - 3$

16. (TI-82) Load the following function into the Y= menu:

```
Y1=5X-(X+3)/4
Y2=1
Y3=
Y4=
Y5=
Y6=
Y7=
Y8=
```

   a) Press the ZOOM button and select 6:ZStandard. Set up a coordinate system on your homework paper and copy the image in the viewing window onto the coordinate system. Label the appropriate graphs with Y1 and Y2, respectively, and place your WINDOW parameters on the axes.

   b) Use the intersect utility to estimate the x-value of the point of intersection of Y1 and Y2. Draw a number line below the graph and shade the solution of Y1 < Y2 on the number line.

   c) Use an analytic method to find the solution of Y1 < Y2. Shade this analytic solution on a second number line. Use set-builder and interval notation to describe this solution set. Does this analytic solution compare favorably with the graphing solution found in part (b)? If not, check your work for error.

17. Use an analytic method to solve each of the following inequalities for x. Sketch your solution on a number line and use set-builder and interval notation to describe your solution set.

   a) $-3 < 3 - x < 8$ b) $3x + 2 < -8$ or $3x + 2 > 8$

   c) $-1 \leq x - \dfrac{x+1}{20} \leq 1$ d) $3 - \dfrac{1}{3}x \leq -2$ or $3 - \dfrac{1}{2}x \geq 2$

18. Suppose that $f(-2) = 0$ and $f(1) = 2$. Assume that f is a linear function. Set up a coordinate system on a sheet of graph paper and draw the graph of the function f.

    a) Draw the lines whose equations are $y = 4$ and $y = -2$ on the coordinate system. Draw a number line below your graph and shade and label the solution of $-2 < f(x) < 4$ on your number line in the usual manner.

    b) Find the equation of the function f in the form $f(x) = mx + b$. Use this result to find an analytical solution of the inequality $-2 < f(x) < 4$. Shade this analytical solution on a second number line and use set-builder and interval notation to describe this solution set. Does this analytic solution compare favorably with the graphical solution found in part (a)? If not, check your work for error.

19. Katerina tosses a baseball into the air. The velocity of the ball after 1 second is 88 feet per second. The velocity of the ball after 2 seconds is 56 feet per second. Let v represent the velocity of the ball and let t represent the time that has passed since the ball left Katerina's hand. Assume that v is a linear function of t. Set up a coordinate system on a sheet of graph paper and draw the graph of v as a function of t.

    a) Draw the lines whose equations are $v = 50$ and $v = 100$ on the coordinate system. Draw a number line below the graph and shade and label the times when the velocity of the ball is between 50 feet per second and 100 feet per second on the number line in the usual manner. When shading your solution on the number line, try to estimate it to the nearest tenth of a second (one decimal place).

    b) Express the velocity of the ball as a function of the time passing after it leaves Katerina's hand. Place your answer in the form $v = mt + b$. Use this equation to find an analytical solution of $50 < v < 100$. Shade this analytical solution on a second number line and use set-builder and interval notation to describe this solution set. Use your calculator to round the numbers in your answer to the nearest tenth of a second (one decimal place). Does this analytical solution compare favorably with the graphical solution found in part (a)? If not, check your work for error.

20. Consider the following rectangle. The length and width is measured in feet.

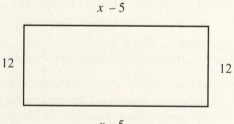

$x - 5$

12

12

$x - 5$

    a) For what values of x is the perimeter between 100 feet and 150 feet?

    b) For what values of x is the area between 200 square feet and 400 square feet?

    Hint: Does your answer take into consideration the fact that the length and width of a rectangle cannot be a negative number?

21. Complete the following table of points that satisfy the equation $f(x) = |x|$.

x	−5	−4	−3	−2	−1	0	1	2	3	4	5
$f(x)$											

a) Set up a coordinate system on a sheet of graph paper and plot each point in the table on the coordinate system. Plot the rest of the points that satisfy the equation of the function f.

b) Draw the line $y = 1$ on the coordinate system. Draw a number line below the graph and shade the solution set of the inequality $|x| < 1$ on the number line in the usual manner. Describe this solution using set-builder notation.

c) Draw a second number line below the graph and shade the solution set of the inequality $|x| > 1$ on this second number line in the usual manner. Use set-builder notation to describe this solution.

22. Remember, if $a > 0$, then $|x| < a$ is equivalent to $-a < x < a$, and $|x| > a$ is equivalent to $x < -a$ or $x > a$. Use these facts to write equivalent inequalities for each of the following problems. Then use an analytic method to complete the solution of the inequality. Shade your solution on a number line and use set-builder and interval notation to describe your solution set.

a) $|x - 8| < 12$

b) $|x + 11| > 15$

c) $|3 - 2x| \le 13$

d) $|5x - 9| \ge 11$

e) $\left|\dfrac{4}{7}x - \dfrac{1}{2}\right| > \dfrac{5}{14}$

f) $\left|1 - \dfrac{1}{8}x\right| \ge \dfrac{4}{5}$

g) $\left|x - \dfrac{x - 10}{9}\right| < 2$

h) $\left|\dfrac{2x - 5}{7} - \dfrac{3 - 4x}{5}\right| \le 1$

23. Complete the following table of points that satisfy the equation $f(x) = |3x - 18|$.

x	1	2	3	4	5	6	7	8	9	10	11
$f(x)$											

a) Set up a coordinate system on a sheet of graph paper and plot each point from the table on the coordinate system. Plot the rest of the points that satisfy the equation of the function f.

b) Draw the line whose equation is $y = 9$ on the coordinate system. Draw a number line below the graph and shade and label the solution of $f(x) < 9$ on the number line in the usual manner.

c) Use an analytic method to find the solution of $f(x) < 9$. Shade this analytic solution on a second number line and describe this solution using both set-builder and interval notation. Does this analytic solution compare favorably with the graphing solution found in part (b)? If not, check your work for error.

24. (TI-82) Load the following functions into the Y= menu:

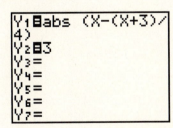

Push the ZOOM button and select 6:ZStandard. Set up a coordinate system and copy the image in the viewing window onto your coordinate system. Label the appropriate graphs with Y1 and Y2, respectively, and include the WINDOW parameters on the coordinate axes.

a) Use the intersect utility in the CALC menu to find the x-values of the points of intersection of Y1 and Y2. Draw a number line below the graph and shade the solution of Y1 > Y2 on the number line in the usual manner.

b) Use an analytic method to find the solution of Y1 > Y2. Shade this analytic solution on a second number line and describe this solution using both set-builder and interval notation. Does this analytic solution compare favorably with the graphing solution found in part (b)? If not, check your work for error.

Chapter 3
Systems of Linear Equations

In this chapter, we will work on methods of solving *systems* of *linear* equations. These are called *linear* systems because the graphs of the equations involved are lines. We will employ three different methods for solving such systems: the *graphing method,* the *substitution method*, and the *Gaussian elimination method.*

3.1 The Graphing Method

This section introduces the graphing method, but first we will discuss the *standard form* of a line.

The Standard Form of a Line

Solve the equation $Ax + By = C$ for y. Begin by subtracting Ax from both sides of the equation, then divide both sides of the result by B.

$$Ax + By = C$$
$$Ax + By - Ax = C - Ax$$
$$By = C - Ax$$
$$\frac{By}{B} = \frac{C - Ax}{B}$$

The *distributive property* is needed to finish the solution.

Division is distributive with respect to addition. In symbols, $\dfrac{a+b}{c} = \dfrac{a}{c} + \dfrac{b}{c}$.

In this case, divide each term by B, as follows.

$$\frac{By}{B} = \frac{C}{B} - \frac{Ax}{B}$$
$$y = \frac{C}{B} - \frac{A}{B}x$$

It is preferable to use the form $y = -\dfrac{A}{B}x + \dfrac{C}{B}$. Note that this equation is now in the form $y = mx + b$, so the graph of this equation is a line whose slope is $-\dfrac{A}{B}$ and whose y-intercept is $\dfrac{C}{B}$.

> *The graph of the equation $Ax + By = C$ is a line. The form $Ax + By = C$ is sometimes called the standard form of a line.*

Example 1   Find the slope and y-intercept of the line whose equation is $x + 2y = 8$. Use the slope and y-intercept to sketch the graph of the line.

Solution   First, solve the equation for y. This will place the equation in slope-intercept form.

$$x + 2y = 8$$
$$x + 2y - x = 8 - x$$
$$2y = -x + 8$$
$$\frac{2y}{2} = \frac{-x + 8}{2}$$
$$\frac{2y}{2} = \frac{-x}{2} + \frac{8}{2}$$
$$y = -\frac{1}{2}x + 4$$

You now know that the slope is $-\dfrac{1}{2}$ and the y-intercept is 4. You can use this information to draw the graph of the line.

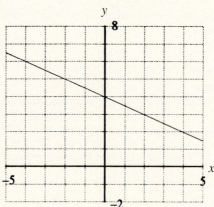

Finding the x- and y-Intercepts of a Line.   When your equation is in the standard form $Ax + By = C$, you can draw the graph by finding the x- and y-intercepts of the line.

Example 2   Sketch the graph of the line whose equation is $2x - 3y = 6$.

Solution   Because this equation is in the form $Ax + By = C$, you know that its graph will be a line. To draw a line, you need only plot *two* points, then draw a line through the two points. The easiest points to find are the x- and y-intercepts of the line.

> *To find the x-intercept of the line whose equation is* $Ax + By = C$, *let* $y = 0$ *in the equation and solve for x. To find the y-intercept of the line whose equation is* $Ax + By = C$, *let* $x = 0$ *in the equation and solve for y.*

To find the x-intercept of your line, let $y = 0$ in the equation of the line.

$$2x - 3y = 6$$
$$2x - 3(0) = 6$$
$$2x = 6$$
$$x = 3$$

The line crosses the x-axis at $(3, 0)$. To find the y-intercept of the line, let $x = 0$ in the equation of the line.

$$2x - 3y = 6$$
$$2(0) - 3y = 6$$
$$-3y = 6$$
$$y = -2$$

The line crosses the y-axis at $(0, -2)$. Plot the x- and y-intercepts and draw a line through them.

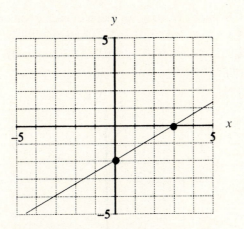

Finding the Solution of a Linear System by Guessing and Checking

Guessing and checking is a very effective problem-solving technique that is often passed over in favor of more sophisticated methods.

Example 3   Solve the following linear system for x and y.

$$x + y = 6$$
$$x - y = 4$$

Solution   Your task is to find an ordered pair (x, y) that satisfies *both* equations. Begin by guessing. Because $2 + 4 = 6$, the ordered pair $(2, 4)$ satisfies the first equation, $x + y = 6$. Here are some other ordered pairs that satisfy the equation $x + y = 6$: $(0,6)$, $(1,5)$, $(2,4)$, $(3,3)$, $(4,2)$, $(5,1)$, $(6,0)$. However, you must find an ordered pair that satisfies *both* equations. When you examine your list of points that satisfy the first equation, $x+y=6$, you note that $5-1=4$ and the ordered pair $(5,1)$ also satisfies the second equation, $x-y=4$. Therefore $(5,1)$ is a solution of the system of equations.

Tip	When solving a system of equations, you can always try to guess. The object is to find an ordered pair that satisfies *both* equations.

Using a Graph to Solve Systems of Equations

If you are to find an ordered pair (x,y) that satisfies both equations, then you must look for a point that lies on both graphs. Look for a point of intersection.

> *To solve a system of linear equations, first draw the graph of each equation. The point of intersection of the lines will be the solution of the system.*

Example 4   Solve the following linear system for x and y.

$$x+y=6$$
$$x-y=4$$

Graphical Solution   Begin by drawing the graph of each equation. The x- and y-intercepts of the line whose equation is $x+y=6$ are $(6,0)$ and $(0,6)$, respectively. The x- and y-intercepts of the line whose equation is $x-y=4$ are $(4,0)$ and $(0,-4)$, respectively. Plot each set of intercepts, then draw a line through them, as follows:

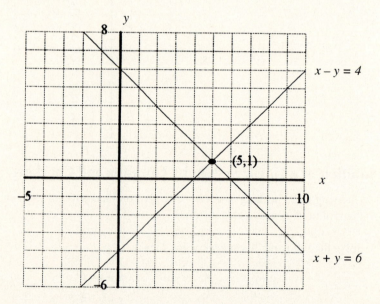

It appears that the approximate coordinates of the point of intersection of the two lines are $(5,1)$. Therefore $(5,1)$ is the solution of this system of linear equations. Note that this solution agrees with the guess and check solution found in Example 3.

Example 5   The sum of two numbers is 9. The larger number is 3 less than twice the smaller number. Find the numbers.

Graphical Solution   Begin by setting up a system of equations that models the problem situation. Let S represent the smaller number and let L represent the larger number.

Tip	Choose letters for your variables that will help you remember what each variable represents.

Because the sum of the two numbers is 9, the first equation is $S + L = 9$. Because the larger number is 3 less than twice the smaller number, $L = 2S - 3$. The following system of linear equations has been constructed:

$$S + L = 9$$
$$L = 2S - 3$$

Place L on the vertical axis and S on the horizontal axis. The S- and L-intercepts of the line whose equation is $S + L = 9$ are $(9,0)$ and $(0,9)$, respectively. Plot these two intercepts and draw a line through them, as follows:

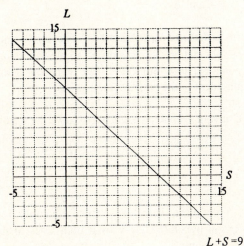

$L + S = 9$

The second equation, $L = 2S - 3$, is of the form $y = mx + b$, where L and S have taken the place of y and x, respectively. Therefore, the slope is 2 and the L-intercept is -3. Use the slope and the L-intercept to draw the graph of the line whose equation is $L = 2S - 3$.

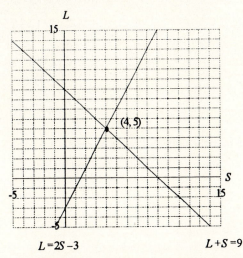

$L = 2S - 3$ $L + S = 9$

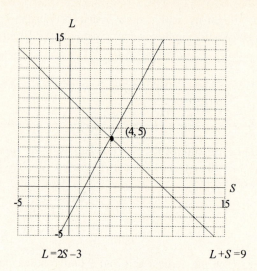

$$L = 2S - 3 \qquad\qquad L + S = 9$$

It appears that the point of intersection of the two lines is the point whose coordinates are $(4,5)$. Therefore, $S = 4$ and $L = 5$.

Does Your Solution Make Sense? The smaller number is 4 and the larger number is 5. The sum of these two numbers is certainly 9, so the first requirement of the problem is satisfied. Because the smaller number is 4, three less than twice this number is 5, which is the larger number. The second requirement of the problem is satisfied.

Example 6   *(TI-82)*   Use your graphing calculator to solve the following system of equations.

$$x + 2y = 12$$
$$y = \frac{1}{3}x + 1$$

Graphical Solution   You must solve the first equation for y before you can enter it in the Y= menu. Begin by subtracting x from both sides of the equation, then divide both sides of your result by 2, as follows:

$$x + 2y = 12$$
$$x + 2y - x = 12 - x$$
$$2y = 12 - x$$
$$\frac{2y}{2} = \frac{12 - x}{2}$$
$$y = \frac{12}{2} - \frac{x}{2}$$
$$y = 6 - \frac{1}{2}x$$

Of course, the preferred order is $y = mx + b$, which means you should write $y = -\frac{1}{2}x + 6$. Now enter your equations in the Y= menu of your calculator.

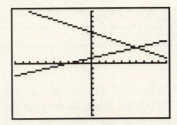

Push the ZOOM button and select 6:ZStandard to capture the following image in your viewing window:

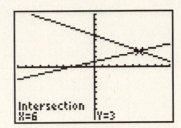

Select 5:intersect in the CALC menu, then press ENTER three consecutive times to capture the following image in your viewing window. (*Note:* If you wish to review how to use the intersect utility on the TI-82, refer to Example 10 in Section 1.6.)

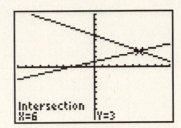

The calculator suggests that the solution of the system of equations is the ordered pair $(6,3)$.

Check Your Solution. Your solution must satisfy *both* equations. Substitute the solution $(6,3)$ in the first equation, $x + 2y = 12$.

$$x + 2y = 12$$
$$6 + 2(3) = 12$$
$$6 + 6 = 12$$
$$12 = 12$$

Your solution satisfies the first equation. Substitute the solution $(6,3)$ in the second equation, $y = \frac{1}{3}x + 1$.

$$y = \frac{1}{3}x + 1$$
$$3 = \frac{1}{3}(6) + 1$$
$$3 = 2 + 1$$
$$3 = 3$$

Your solution $(6,3)$ satisfies *both* equations and thus checks.

Example 7   Willie breaks open his piggy bank and heads for the ice cream parlor. He has 22 coins in his pocket, some of which are quarters, some of which are dimes. If the total value of the coins is $3.70, how many of each coin does Willie have in his pocket?

Graphical Solution   Let D represent the number of dimes in Willie's pocket and let Q represent the number of quarters. Now, D dimes, worth 10 cents apiece, have a monetary value of 10D cents. Similarly, Q quarters, worth 25 cents apiece, have a monetary value of 25Q cents. Willie has 22 coins in all, with a total monetary value of 370 cents. It's best to arrange this information in a table.

	Number of Coins	Value of Coins (Cents)
Dimes	D	$10D$
Quarters	Q	$25Q$
Total	22	370

Because Willie has 22 coins, $D+Q=22$. Because the monetary value of the coins is $3.70, or 370 cents, $10D+25Q=370$. The following system of equations has been created.

$$D+Q=22$$
$$10D+25Q=370$$

Place Q on the vertical axis and D on the horizontal axis. The D- and Q-intercepts of the line whose equation is $D+Q=22$ are $(22,0)$ and $(0,22)$, respectively.

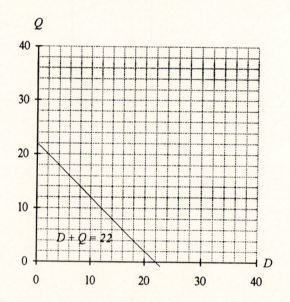

Note that the gridlines on both the Q and D axes are spaced 2 units apart. Let's find the D- and Q-intercepts of $10D+25Q=370$. To find the Q-intercept, let $D=0$.

$$10D + 25Q = 370$$
$$10(0) + 25Q = 370$$
$$25Q = 370$$
$$Q = \frac{370}{25}$$
$$Q = 14.8$$

The Q-intercept is 14.8. To find the D-intercept, let $Q = 0$.

$$10D + 25Q = 370$$
$$10D + 25(0) = 370$$
$$10D + 0 = 370$$
$$10D = 370$$
$$D = 37$$

The D-intercept is 37. Plot these intercepts and draw a line through them, as follows.

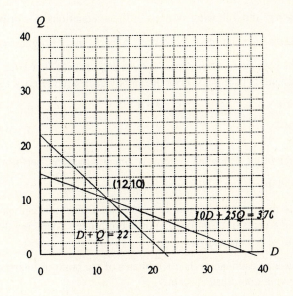

It appears that the point of intersection is approximately (12, 10). Willie has 12 dimes and 10 quarters.

Does Your Answer Make Sense?   You have determined that Willie has 12 dimes and 10 quarters. That is certainly a total of 22 coins, satisfying the first requirement of the problem. Next, 12 dimes is $1.20 and 10 quarters is $2.50. That is a total of $3.70, satisfying the second requirement of the problem.

One of the most important steps in the problem-solving process is the last one—checking to see if your answers are reasonable. Your problem is never complete until you've checked that your answer satisfies all the requirements of the problem.

Exercises for Section 3.1

1. Each of the following equations is in standard form $Ax + By = C$. Solve each equation for y to place it in slope-intercept form. Set up a separate coordinate system on a sheet of graph paper for each line and use the slope and the y-intercept to draw the graph of the line.

   a) $4x - 3y = 12$ b) $5x + 2y = 10$ c) $2x - 3y = -6$

   d) $2x + 5y = -10$ e) $\dfrac{1}{2}x + \dfrac{3}{4}y = \dfrac{4}{3}$ f) $\dfrac{3}{5}x + \dfrac{3}{8}y = \dfrac{3}{2}$

2. Each of the following equations is in standard form $Ax + By = C$. Find the x-intercept and y-intercept for each. Set up a separate coordinate system on a sheet of graph paper for each line and use the x-intercept and y-intercept to sketch the graph of each line.

   a) $x + 2y = 8$ b) $5x - 2y = 10$ c) $2x - 9y = -18$

   d) $5x + 4y = -20$ e) $2.2x - 1.8y = 12.3$ f) $4.2x - 3.8y = 24.1$

   Note: Use your calculator to estimate the x- and y-intercepts of problems (e) and (f).

3. For each of the following systems of linear equations, perform the following tasks:

   i) Guess the solution of the system. Clearly state this guess on your homework paper.

   ii) On your homework paper, clearly demonstrate that your guess satisfies each equation of the system. If your guess does not satisfy each equation, try again.

   a) $\begin{aligned} x + y &= 5 \\ x - y &= 3 \end{aligned}$ b) $\begin{aligned} x + y &= 8 \\ x - y &= 4 \end{aligned}$ c) $\begin{aligned} x + y &= -10 \\ x - y &= -2 \end{aligned}$ d) $\begin{aligned} x + y &= -6 \\ x - y &= 10 \end{aligned}$

4. For each of the following systems of linear equations, perform the following tasks:

   i) Find the x- and y-intercepts of each line.

   ii) Set up a separate coordinate system on a sheet of graph paper for each system. Plot the x- and y-intercepts of each line on your coordinate system and draw a line through them. Label each line with its equation.

   iii) Place an estimate of the coordinates of the point of intersection of the two lines near this point on your coordinate system.

   iv) Clearly demonstrate that your estimate satisfies each equation of the system.

   a) $\begin{aligned} x + y &= 4 \\ x - y &= 6 \end{aligned}$ b) $\begin{aligned} x + 2y &= 4 \\ x - y &= 1 \end{aligned}$ c) $\begin{aligned} x + y &= 7 \\ x - 2y &= 4 \end{aligned}$

   d) $\begin{aligned} x - y &= 5 \\ 3x + y &= 3 \end{aligned}$ e) $\begin{aligned} x - 2y &= -8 \\ x + 2y &= -4 \end{aligned}$ f) $\begin{aligned} x + y &= -3 \\ 2x - y &= 6 \end{aligned}$

5. For each of the following systems of equations, perform the following tasks:

   i) Set up a separate coordinate system on a sheet of graph paper for each system of equations. Find the x- and y-intercepts of the line represented by the first equation, plot them on your coordinate system, and draw a line through them. Label this line with its equation.

   ii) Use the slope and y-intercept to draw the line represented by the second equation on your coordinate system. Label this line with its equation.

iii) Place an estimate of the coordinates of the point of intersection of the two lines near this point on your coordinate system.

iv) Clearly demonstrate that your estimate satisfies each equation of the system.

a) $x + y = 4$

$y = x$

b) $x + y = -6$

$y = 2x$

c) $x - 2y = 6$

$y = -2x + 2$

d) $4x + y = 8$

$y = x + 3$

e) $x - 2y = -2$

$y = -\dfrac{3}{4}x - 4$

f) $x - 2y = -8$

$y = -\dfrac{5}{2}x - 2$

6. The sum of two numbers is 4 and their difference is 8.

a) Set up two equations in two unknowns to model this problem situation.

b) Set up a coordinate system on a sheet of graph paper and sketch the graph of each equation on your coordinate system. Label each graph with its equation. Clearly label your estimate of the point of intersection on your coordinate system.

c) Use complete sentences to explain why your solution satisfies the requirements of the problem. If your answer doesn't satisfy the requirements, check your work for error.

7. The sum of two numbers is 7. The larger number is 1 more than twice the smaller number.

a) Set up two equations in two unknowns to model this problem situation.

b) Set up a coordinate system on a sheet of graph paper and sketch the graph of each equation on your coordinate system. Label each graph with its equation. Clearly label your estimate of the point of intersection on your coordinate system.

c) Use complete sentences to explain why your solution satisfies the requirements of the problem. If your answer doesn't satisfy the requirements, check your work for error.

8. Consider the following rectangle.

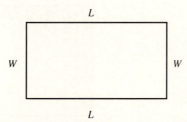

a) The length of the rectangle is twice its width. Express the length L as a function of the width W.

b) The perimeter of the rectangle is 24. Set up an equation in L and W that models this fact.

c) Set up a coordinate system on a sheet of graph paper and sketch the graph of each equation developed in parts (a) and (b). Label each graph with its equation. Place an estimate of the point of intersection of your two graphs near this point on your coordinate system.

d) Use complete sentences to explain why your solution satisfies the requirements of the problem. If your answer doesn't satisfy the requirements, check your work for error.

9. Satchel Paige loved to have a pocketful of change in his pocket when he walked. He'd place his hand in his pocket and jingle his change as he walked. He figured he wouldn't frighten people when he walked up behind them, for they could hear him coming before they turned around to say "Hello, Satch." Suppose that Satchel had 14 coins in his pocket, all in nickels and quarters, worth a total of $1.70.

   a) Set up two equations in the unknowns N and Q that model this problem, where N represents the number of nickels in Satchel's pocket and Q represents the number of quarters.

   b) Set up a coordinate system on a sheet of graph paper and draw the graph of each equation developed in part (a) on your coordinate system. Label each graph with its equation. Place your estimate of the point of intersection near this point on your coordinate system.

   c) Use complete sentences to explain why your solution satisfies the requirements of the problem. If your answer doesn't satisfy the requirements, check your work for error.

10. Antonio has been playing basketball all afternoon and he is hungry and thirsty. He walks to the corner grocery store to buy a snack and a drink. Antonio has a pocketful of change, all in dimes and quarters. Let Q represent the number of quarters in his pocket and D represent the number of dimes.

    a) The number of quarters in Antonio's pocket is one more than twice the number of dimes in his pocket. Express the number of quarters Q as a function of the number of dimes D.

    b) The total value of the coins in Antonio's pocket is $3.25. Set up an equation in Q and D that models this fact.

    c) Set up a coordinate system on a sheet of graph paper and draw the graphs of the equations developed in parts (a) and (b) on your coordinate system. Label each graph with its equation. Place your estimate of the point of intersection near this point on your coordinate system.

    d) Use complete sentences to explain why your solution satisfies the requirements of the problem. If your answer doesn't satisfy the requirements, check your work for error.

11. (TI-82) Load the following equations in the Y= menu.

```
Y₁■-1/3*X+4
Y₂■4/3*X-1
Y₃=
Y₄=
Y₅=
Y₆=
Y₇=
Y₈=
```

    a) Push the ZOOM button and select 6:ZStandard. Set up a coordinate system and copy the image in your viewing window onto your coordinate system. Label the graphs on your coordinate system with Y1 and Y2, and include the WINDOW parameters on your coordinate axes.

    b) Use the intersect utility in the CALC menu to find the point of intersection of the two lines. If you wish to review how to use the intersect utility on the TI-82, refer to Example 10 in Section 1.6. Place this estimate near the point of intersection on your coordinate system.

    c) Clearly demonstrate that this estimate satisfies the equations $Y_1 = -1/3 * X + 4$ and $Y_2 = 4/3 * X - 1$.

12. (TI-82) Consider the following system of equations.

$$2x + 3y = 6$$

$$y = \frac{1}{3}x - 4$$

a) Solve the equation $2x + 3y = 6$ for y.

b) Load both equations in the Y= menu, press the ZOOM button, and select 6:ZStandard. Set up a coordinate system and copy the image in your viewing window onto your coordinate system. Label the graphs on your coordinate system with Y1 and Y2, and include the WINDOW parameters on your coordinate axes.

c) Use the intersect utility in the CALC menu to find the point of intersection of the two lines. If you wish to review how to use the intersect utility on the TI-82, refer to Example 10 in Section 1.6. Place this estimate near the point of intersection on your coordinate system.

d) Clearly demonstrate that this estimate satisfies each equation in the system.

13. (TI-82) Consider the following system of equations.

$$7x - 3y = 6$$

$$2x + 3y = 21$$

a) Solve each equation of the system for y.

b) Load both equations in the Y= menu, press the ZOOM button, and select 6:ZStandard. Set up a coordinate system and copy the image in your viewing window onto your coordinate system. Label the graphs on your coordinate system with Y1 and Y2, and include the WINDOW parameters on your coordinate axes.

c) Use the intersect utility in the CALC menu to find the point of intersection of the two lines. If you wish to review how to use the intersect utility on the TI-82, refer to Example 10 in Section 1.6. Place this estimate near the point of intersection on your coordinate system.

d) Clearly demonstrate that this estimate satisfies each equation in the system.

14. The difference between two numbers is 9. One number is twice as large as the other. Set up a system of equations that models this situation and use the graphing method to solve your system. Use complete sentences to explain why your solution satisfies the requirements of the problem. If your answer doesn't satisfy the requirements, check your work for error.

    Note: Depending upon which way you subtract, this problem has two solutions.

15. The sum of two numbers is −5. One number is four more than twice the other number. Set up a system of equations that models this situation and use the graphing method to solve your system. Use complete sentences to explain why your solution satisfies the requirements of the problem. If your answer doesn't satisfy the requirements of the problem, check your work for error.

16. The perimeter of a rectangle is 26. The length of the rectangle is 1 foot longer than twice its width. Set up a system of equations that models this situation and use the graphing method to solve your system. Use complete sentences to explain why your solution satisfies the requirements of the problem. If your answer doesn't satisfy the requirements, check your work for error.

17. Linda reaches into her change purse. She discovers that she has only nickels and dimes worth a total of $1.20. There are three fewer dimes than there are nickels. Set up a system of equations that models this situation and use the graphing method to solve your system. Use complete

sentences to explain why your solution satisfies the requirements of the problem. If your answer doesn't satisfy the requirements, check your work for error.

18. Justin has just finished football practice, and he is hot and thirsty. He heads for the soda machine to buy a drink. He searches his pockets for change and finds 17 dimes and quarters, worth \$2.45. Set up a system of equations that models this situation and use the graphing method to solve your system. Use complete sentences to explain why your solution satisfies the requirements of the problem. If your answer doesn't satisfy the requirements, check your work for error.

19. Suppose that one line has x- and y-intercepts at $(4,0)$ and $(0,2)$, respectively, and that a second line has slope 1 and y-intercept -1. Drawing these lines on a coordinate system results in the following graph:

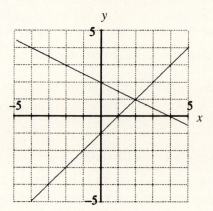

The coordinates of the point of intersection appear to be $(2,1)$. Find the equation of each line and show that this point satisfies each equation.

20. Suppose that one line goes through the points $(-3,-1)$ and $(3,2)$ and that a second line has slope -2 and y-intercept 3. Drawing these lines on a coordinate system results in the following graph:

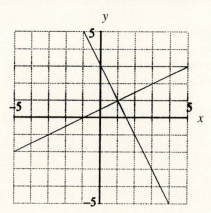

The coordinates of the point of intersection appear to be $(1,1)$. Find the equation of each line and show that this point satisfies each equation.

21. Consider the following system of equations.

$$2x + 3y = 6$$
$$2x + 3y = -12$$

a) Set up a coordinate system on a sheet of graph paper. Find the *x*- and *y*-intercepts of each line, plot them on your coordinate system, and draw a line through each set of intercepts. Label each graph with its equation.

b) Use complete sentences to explain why this system of equations has no solution.

22. Consider the following system of equations.

$$2x + 3y = 6$$
$$4x + 6y = 12$$

a) Set up a coordinate system on a sheet of graph paper. Find the *x*- and *y*-intercepts of each line, plot them on your coordinate system, and draw a line through each set of intercepts. Label each graph with its equation.

b) Use complete sentences to explain why this system of equations has an infinite number of solutions.

3.2 The Substitution Method

This section will develop an analytic method to find the exact solutions to systems of equations. However, we will not abandon the graphing method. In fact, we will depend on it a great deal. Our approach will be to first *estimate* a solution to our system by using the graphing method, then to find *exact* solutions of the system using the *substitution method*.

The substitution method is a two-step process:

1. Solve either equation for either variable.

2. Substitute the result of Step 1 into the other equation.

Example 1   Find the solution of the following system of linear equations.

$$x + 4y = 4$$
$$y = 2x - 3$$

Graphical Solution   The *x*- and *y*-intercepts of the line whose equation is $x + 4y = 4$ are $(4,0)$ and $(0,1)$, respectively. The line whose equation is $y = 2x - 3$ has slope 2 and *y*-intercept -3. Using this information, draw the graph of each equation on a coordinate system.

It appears that the approximate solution of the system of linear equations is $(1.7, 0.6)$. Place this estimate near the point of intersection of the two lines, then check it in each of the equations.

$$x + 4y = 4$$

$$1.7 + 4(0.6) = 4$$

$$4.1 = 4$$

The point $(1.7, 0.6)$ does not exactly satisfy the equation of $x + 4y = 4$, but it almost does. Now try your estimate in the other equation.

$$y = 2x - 3$$

$$0.6 = 2(1.7) - 3$$

$$0.6 = 0.4$$

The point $(1.7, 0.6)$ does not exactly satisfy the equation $y = 2x - 3$, but it almost does. This is one of the problems you must deal with when employing the graphing method. Your solutions will only be approximate solutions. If you want *exact* solutions, you must use the *substitution method*.

Analytical Solution   The equations will be numbered so that we can refer to them as we work. Try to emulate this technique in your own work.

$$x + 4y = 4 \qquad \text{Eq. (1)}$$
$$y = 2x - 3 \qquad \text{Eq. (2)}$$

Equation (2) is already solved for y, so it is not necessary to apply Step 1 of the substitution method. Proceed to Step 2.

Substitute Equation (2) into equation (1):

$$x + 4y = 4$$

$$x + 4(2x - 3) = 4$$

$$x + 8x - 12 = 4$$

$$9x - 12 = 4$$

$$9x = 16$$

$$x = \frac{16}{9}$$

To find the y-value of your solution, substitute $x = \dfrac{16}{9}$ in one of the equations. Because the point lies on both lines, you may substitute your x-value into either Equation (1) or Equation (2).

Substitute $x = \dfrac{16}{9}$ into Equation (2):

$$y = 2x - 3$$

$$y = 2\left(\frac{16}{9}\right) - 3$$

$$y = \frac{32}{9} - \frac{27}{9}$$

$$y = \frac{5}{9}$$

The *exact* solution of the system of equations is $\left(\dfrac{16}{9}, \dfrac{5}{9}\right)$. Use your calculator to find decimal approximations for these numbers.

$$\frac{16}{9} \approx 1.77777...$$

$$\frac{5}{9} \approx 0.55555...$$

The estimate found when employing the graphing method was (1.7, 0.6). It appears that the solutions from the substitution method agree quite favorably with this estimate. However, to be certain that you have found an *exact* solution of the system of linear equations, check the solution in each equation from your system.

Check Your Solution. Substitute $\left(\dfrac{16}{9}, \dfrac{5}{9}\right)$ into equation (1).

$$x + 4y = 4$$

$$\frac{16}{9} + 4\left(\frac{5}{9}\right) = 4$$

$$\frac{16}{9} + \frac{20}{9} = 4$$

$$\frac{36}{9} = 4$$

$$4 = 4$$

Point $\left(\dfrac{16}{9}, \dfrac{5}{9}\right)$ satisfies the first equation *exactly*. Substitute $\left(\dfrac{16}{9}, \dfrac{5}{9}\right)$ into Equation (2).

$$y = 2x - 3$$

$$\frac{5}{9} = 2\left(\frac{16}{9}\right) - 3$$

$$\frac{5}{9} = \frac{32}{9} - \frac{27}{9}$$

$$\frac{5}{9} = \frac{5}{9}$$

Point $\left(\frac{16}{9}, \frac{5}{9}\right)$ satisfies the second equation *exactly*. The solution satisfies *both* equations exactly, which means that this is the *exact* solution of this system of equations.

Example 2   Find the solution of the following system of linear equations.

$$x + 2y = 4$$
$$2x - 3y = 6$$

Graphical Solution   The x- and y-intercepts of the equation $x + 2y = 4$ are $(4,0)$ and $(0,2)$, respectively. The x- and y-intercepts of the equation $2x - 3y = 6$ are $(3,0)$ and $(0,-2)$, respectively. If you plot these pairs of intercepts, then draw lines through each pair, you arrive at the following graph.

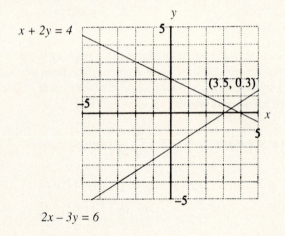

It appears that the solution to this system of equations is *approximately* (3.5, 0.3). Note that this estimate has been placed near the point of intersection of the two lines. Now check that this solution approximately satisfies each equation in the system.

Analytical Solution   Once again, the equations will be numbered so that we can refer to them in our work. Try to emulate this in your own work.

$$x + 2y = 4 \qquad \text{Eq. (1)}$$
$$2x - 3y = 6 \qquad \text{Eq. (2)}$$

Begin by solving either equation for either variable.

Tip	Since you can begin by solving either equation for either variable, it is beneficial to choose the equation and variable that will simplify the work as much as possible. Try to avoid fractions if at all possible.

Solve Equation (1) for x:

$$x + 2y = 4$$
$$x = -2y + 4 \qquad \text{Eq. (3)}$$

Now substitute this result into the *other* equation. Because this result came from Equation (1), you must substitute it into Equation (2).

Substitute Equation (3) into Equation (2).

$$2x - 3y = 6$$
$$2(-2y + 4) - 3y = 6$$
$$-4y + 8 - 3y = 6$$
$$-7y + 8 = 6$$
$$-7y = -2$$
$$y = \frac{2}{7}$$

Before proceeding, use your calculator to find a decimal approximation for $\frac{2}{7}$.

$$\frac{2}{7} = 0.285714285714...$$

The estimate produced by the graphing method was $y \approx 0.3$. The graphing method and the substitution method produce similar results. However, the graphing method produces *approximate* solutions, whereas the substitution method produces *exact* solutions. To find the exact solution for x, you can substitute the y-value into Equation (1), (2), or (3).

Substitute $y = \frac{2}{7}$ in Equation (3):

$$x = -2y + 4$$
$$x = -2\left(\frac{2}{7}\right) + 4$$
$$x = -\frac{4}{7} + \frac{28}{7}$$
$$x = \frac{24}{7}$$

Using your calculator to find the following approximation for $\frac{24}{7}$:

$$\frac{24}{7} = 3.4285714285714\ldots$$

The estimate produced by the graphing method was $x \approx 3.5$. You can be extremely confident that the system has been solved correctly, but the answers should be checked.

Use Your TI-82 to Check Your Solution. One can store numbers in variables on the TI-82. Note that the letters A-Z are printed on the calculator case above the keys. To access these letters, first push the ALPHA key, then push the letter of choice. Use the CLEAR button to clear your screen, then store the number $\frac{24}{7}$ in the variable X by entering 24/7 and pushing the STO▷ key. Press the ALPHA key and select the letter X, then press ENTER. Store $\frac{2}{7}$ in the variable Y by entering 2/7 and pushing the STO▷ key. Press the ALPHA key and select the letter Y. Press ENTER to produce the following image in your viewing window:

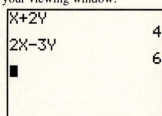

To check your solution, first press the CLEAR button to clear your screen. Enter the expression X + 2Y and press the ENTER key. Enter the expression $2X - 3Y$ and press the ENTER key. The following image should appear in your viewing window:

```
X+2Y
                       4
2X-3Y
                       6
■
```

Because the equations in this system are $x + 2y = 4$ and $2x - 3y = 6$, there is strong evidence that the solution is correct.

Example 3   Dave and Mary sold their home in Covina, California and are planning to use most of the proceeds to make a down payment on a home in Fortuna, California. However, they have held back $13,000 to pay taxes on capital gains. The tax is not due for a year, so Dave and Mary decide to invest part of the money in a certificate of deposit that pays 5 percent each year and they invest the rest of the money in a savings account that pays 3.2 percent per year. At the end of one year, the total interest from both investments is $560. How much did they invest in each account?

Graphical Solution (TI-82)   Let C represent the amount that they invested in the certificate of deposit and let S represent the amount they invested in a savings account. Because the certificate of deposit pays 5 percent per year, the amount of interest earned in this investment is $0.05C$. Because the savings account pays 3.2 percent per year, the amount of interest earned in this investment is $0.032S$. This information can be arranged in the following table.

	Amount of Investment	Interest
Certificate of deposit (5%)	C	$0.05C$
Savings account (3.2%)	S	$0.032S$
Totals	13,000	560

Because the total invested is \$13,000, the first equation is $C + S = 13,000$. Because the total interest for one year is \$560, the second equation is $0.05C + 0.032S = 560$. The following system of equations can be created:

$$C + S = 13,000 \qquad \text{Eq. (1)}$$
$$0.05C + 0.032S = 560 \qquad \text{Eq. (2)}$$

Place S on the vertical axis and C on the horizontal axis. In other words, replace y and x with S and C, respectively. If you wish to use your calculator to draw these equations, you must first solve each equation for S.

Solve Equation (1) for S:
$$C + S = 13,000$$
$$S = 13,000 - C$$

Solve Equation (2) for S:
$$0.05C + 0.032S = 560$$
$$0.032S = 560 - 0.05C$$
$$S = \frac{560 - 0.05C}{0.032}$$

When you enter these equations in your calculator, use Y and X instead of S and C.

```
Y₁▤13000-X
Y₂▤(560-0.05X)/0
.032
Y₃=█
Y₄=
Y₅=
Y₆=
Y₇=
```

Now try to find a suitable window to view your equations. Since S and C cannot be negative, you can concentrate your attention in the first quadrant. Since Dave and Mary could invest all of their money in the certificate of deposit, C could be as big as \$13,000. A similar statement can be made about S. Set your WINDOW parameters in the following manner:

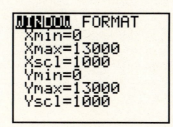

```
WINDOW FORMAT
Xmin=0
Xmax=13000
Xscl=1000
Ymin=0
Ymax=13000
Yscl=1000
```

Use the intersect utility in the CALC menu to capture the following image in your viewing window. If you need help reviewing how to use this utility, see Example 10 in Section 1.6.

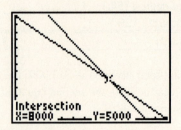

Set up a coordinate system on your homework paper and copy this image onto your coordinate system. Label each graph with its equation and include your WINDOW parameters on the axes, as follows:

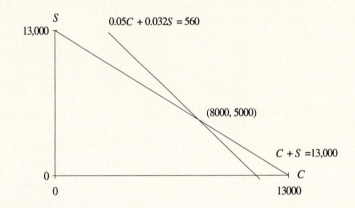

It appears that Dave and Mary invested $8,000 in the certificate of deposit and $5,000 in their savings account. Note how this estimate has been placed near the point of intersection of the two lines.

Analytical Solution   Use the substitution method to solve your system of equations.

$$C + S = 13,000 \qquad \text{Eq. (1)}$$
$$0.05C + 0.032S = 560 \qquad \text{Eq. (2)}$$

Solve Equation (1) for C:

$$C + S = 13,000$$
$$C = 13,000 - S \qquad \text{Eq. (3)}$$

Substitute Equation (3) into Equation (2):

$$0.05C + 0.032S = 560$$

$$0.05(13,000 - S) + 0.032S = 560$$

$$650 - 0.05S + 0.032S = 560$$

$$650 - 0.018S = 560$$

$$-0.018S = -90$$

$$S = \frac{-90}{-0.018}$$

$$S = 5,000$$

Plan to make extensive use of your calculator as you work.
Substitute $S = 5,000$ in Equation (3).

$$C = 13,000 - S$$

$$C = 13,000 - 5,000$$

$$C = 8,000$$

Dave and Mary invested $8,000 in a certificate of deposit and $5,000 in their savings account.

Is Your Answer Reasonable?   Dave and Mary's investment ($8,000 in the certificate of deposit and $5,000 in their savings account) totals $13,000, which satisfies the first requirement of the problem. The certificate of deposit pays 5 percent per year, so the interest earned on $8,000 would be $0.05(\$8,000)$, or $400. The savings account pays 3.2 percent, so the interest earned on $5,000 would be $0.032(\$5,000)$, or $160. The total interest earned from both investments is $560, satisfying the second requirement of the problem.

Tip	The last part of the problem-solving process is the most important. Always check to see if your answers are reasonable.

Example 4   *The Supply and Demand Problem*   Often, in the real world, the demand for a particular item is a function of its price. If the price is lowered, the demand for the article goes up. Conversely, if the price is raised, the demand for the article goes down. Suppose that the demand D as a function of its price p is given by the following equation:

$$D = 450 - 3p$$

The supply is also a function of the unit price. If the price of the article is high, the manufacturers are highly motivated to produce more of a particular item. However, if the price goes down, manufacturers are reluctant to produce the article, and the supply goes down. Suppose the supply, S, of an article as a function of its price, p, is given by the following equation:

$$S = 100 + 2p$$

When the supply equals the demand, a sort of equilibrium is reached. Therefore, the *equilibrium price* is the price per article that guarantees that the supply will equal the demand. In other words, if the price is set just right, consumers will buy everything that the manufacturers produce, and the manufacturers will be producing the exact amount that the buyers demand. Find the equilibrium price for the model.

Graphical Solution   Load the demand and supply functions into the Y= menu in the following manner:

The demand D and the supply S will not be negative, so you need only draw a graph in the first quadrant. Since the y-intercept of Y1 is 450 and the slope is negative, that will be the maximum y-value required. The y-intercept of Y2 is 100 and the slope is positive, so the graph will rise to meet the graph of Y1. With these thoughts in mind, and with a little experimenting, the following settings of the WINDOW parameters were obtained:

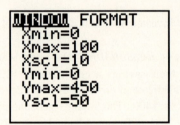

Use the intersect utility in the CALC menu to produce the following image in your viewing window:

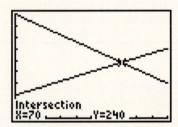

Set up a coordinate system on your homework paper and copy this image onto your coordinate system in the following manner. Note that each graph has been labeled with its equation and the WINDOW parameters have been included on the axes.

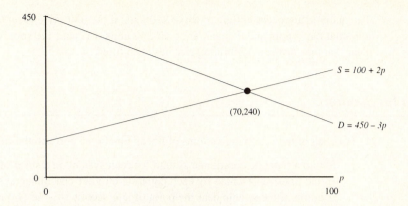

Note that the solution of this system appears to be $(70, 240)$, which has been placed near the point of intersection of the two lines. The supply S will equal the demand D where the graphs of S and D intersect. If the price per article is set at \$70, then both the demand and the supply will be 240. For example, the article might be toaster ovens. If the price of the toaster oven is set at \$70, then the manufacturers will produce 240 ovens and the consumers will buy 240 ovens. The supply and demand will be equal.

Analytical Solution   Since the equilibrium price is being determined, the supply must be set to be equal to the demand.

$$S = D \qquad \text{Eq. (1)}$$

However, $S = 100 + 2p$ and $D = 450 - 3p$. Substituting these into Equation (1) results in the following:

$$100 + 2p = 450 - 3p$$
$$2p + 3p = 450 - 100$$
$$5p = 350$$
$$p = 70$$

Note that you are being asked to do more and more of your work mentally when solving equations. The equilibrium price is exactly \$70. Now, substituting $p = 70$ in the demand equation $D = 450 - 3p$ results in the following:

$$D = 450 - 3p$$
$$D = 450 - 3(70)$$
$$D = 450 - 210$$
$$D = 240$$

The public will demand 240 of the article if the unit price is set at \$70. Substituting $p = 70$ in the supply equation $S = 100 + 2p$, results in the following:

$$S = 100 + 2p$$
$$S = 100 + 2(70)$$
$$S = 100 + 140$$
$$S = 240$$

The manufacturers are willing to make 240 units if the unit price is set at $70. If the unit price is $70, note that the supply and demand are both 240 units. Because the supply equals the demand, it seems clear that the correct equilibrium price has been found.

Exercises for Section 3.2

1. For each of the following systems of linear equations, perform the following tasks:

   i) Set up a separate coordinate system for each system on a sheet of graph paper. Draw the graph of the lines represented by each equation on your coordinate system. Place an estimate of the solution of the system near the point of intersection of your two lines.

   ii) Use the substitution method to find an exact solution of the system of equations. Use a calculator to find a decimal approximation of your solution and compare it with the estimate you found in part (i). If the solution from part (ii) does not compare favorably with your estimate from part (i), check your work for error.

   a) $x + y = -6$

       $y = x$

   b) $x + y = 6$

       $y = 2x$

   c) $2x - 5y = -10$

       $y = -x + 5$

   d) $6x - 5y = 30$

       $y = 2x + 1$

   e) $3x - 4y = -12$

       $y = -\dfrac{1}{2}x + 5$

   f) $x + 5y = -5$

       $y = \dfrac{3}{2}x + 1$

2. For each of the following systems of linear equations, perform the following tasks.

   i) Set up a separate coordinate system for each system on a sheet of graph paper. Draw the graph of the lines represented by each equation on your coordinate system. Place an estimate of the solution of the system near the point of intersection of your two lines.

   ii) Use the substitution method to find an exact solution of the system of equations. Use a calculator to find a decimal approximation of your solution and compare it with the estimate you found in part (i). If the solution from part (ii) does not compare favorably with your estimate from part (i), check your work for error.

   a) $2x + 7y = -14$

       $x - 2y = 4$

   b) $x - 4y = 8$

       $3x + 2y = 6$

   c) $5x - 8y = 40$

       $2x + y = 6$

   d) $4x + y = -4$

       $5x - 9y = 45$

   e) $2x + 8y = -16$

       $3x - 4y = -12$

   f) $2x + 5y = 10$

       $4x + 3y = 12$

3. (TI-82) Load the following functions into the Y= menu.

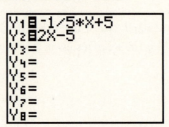

a) Push the ZOOM button and select 6:ZStandard. Set up a coordinate system on your homework paper and copy the image in your viewing window onto your coordinate system. Label the graphs with Y1 and Y2 and include the WINDOW parameters on your axes.

b) Use the intersect utility in the CALC menu to find the point of intersection. Place this estimate on your coordinate system near the point of intersection.

c) Use the substitution method to find exact solutions of this system of equations. Use your calculator to find decimal approximations for your solution and compare your solution to the estimate you found in part (b). If your analytic solution does not compare favorably with your estimate from part (b), check your work for error.

4. (TI-82) Consider the following system of equations.

$$4x+5y=-20 \quad \text{Eq. (1)}$$
$$y=x+1 \quad \text{Eq. (2)}$$

a) Solve Equation (1) for y. Enter your result in the Y= menu as Y1. Enter Equation (2) in the Y= menu as Y2.

b) Push the ZOOM button and select 6:ZStandard. Set up a coordinate system on your homework paper and copy the image in your viewing window onto your coordinate system. Label the graphs with Y1 and Y2 and include the WINDOW parameters on the axes.

c) Use the intersect utility in the CALC menu to find the point of intersection. Place this estimate on your coordinate system near the point of intersection.

d) Use the substitution method to find exact solutions of this system of equations. Use your calculator to find decimal approximations for your solution and compare your solution to the estimate you found in part (c). If your analytic solution does not compare favorably with your estimate from part (c), check your work for error.

5. (TI-82) Consider the following system of equations.

$$4x-5y=20 \quad \text{Eq. (1)}$$
$$2x+3y=6 \quad \text{Eq. (2)}$$

a) Solve Equation (1) for y. Enter your result in the Y= menu as Y1. Solve Equation (2) for y. Enter your result in the Y= menu as Y2.

b) Push the ZOOM button and select 6:ZStandard. Set up a coordinate system on your homework paper and copy the image in your viewing window onto your coordinate system. Label the graphs with Y1 and Y2 and include the WINDOW parameters on your axes.

c) Use the intersect utility in the CALC menu to find the point of intersection. Place this estimate on your coordinate system near the point of intersection.

d) Use the substitution method to find exact solutions of this system of equations. Use your calculator to find decimal approximations for your solution and compare your solution to the estimate you found in part (c). If your analytic solution does not compare favorably with your estimate from part (c), check your work for error.

6. The sum of two numbers is 8. One number is twice as big as the other. Let x and y represent these numbers and set up a system of equations in x and y that models this problem situation.

a) Set up a coordinate system on a sheet of graph paper. Draw the graph of each equation in your system on your coordinate system. Estimate the coordinates of the point of intersection.

b) Use the substitution method to find an exact solution of this system of equations. Compare this with the solution found in part (a).

c) Use complete sentences to explain why your solution satisfies the requirements of the problem. If your answer doesn't satisfy the requirements, check your work for error.

7. Emily, a waitress in a local restaurant, is sorting the tips she made during the evening's work shift. She has 50 coins, all in nickels and quarters. The monetary worth of the coins is $6.50. Let Q represent the number of quarters and N represent the number of nickels. Set up a system of equations in Q and N that models this problem situation.

a) Set up a coordinate system on a sheet of graph paper. Draw the graph of each equation in your system on your coordinate system. Estimate the coordinates of the point of intersection.

b) Use the substitution method to find an exact solution of this system of equations. Compare this with the solution found in part (a).

c) Use complete sentences to explain why your solution satisfies the requirements of the problem. If your answer doesn't satisfy the requirements, check your work for error.

8. Jason is the beneficiary of an inheritance of $10,000. He invests part of it in a savings account that pays 5 percent and part of it in a certificate of deposit that pays 6 percent. The total interest from both investments is $545. Let C represent the amount of money invested in the certificate of deposit and let S represent the amount in the savings account. Set up a system of equations in C and S that model the problem situation.

a) Set up a coordinate system on a sheet of graph paper. Draw the graph of each equation in your system on your coordinate system. Estimate the coordinates of the point of intersection.

b) Use the substitution method to find an exact solution of this system of equations. Compare this solution to the solution found in part (a).

c) Use complete sentences to explain why your solution satisfies the requirements of the problem. If your answer doesn't satisfy the requirements, check your work for error.

9. (TI-82) Suppose that the demand, D, for a particular brand of watch is given by the equation $D = 500 - 3.50p$, where p is the unit price. Suppose that the supply, S, of watches is given by the equation $S = 50 + 1.75p$, where p is the unit price.

a) Load these functions in the Y= menu. Adjust the WINDOW parameters so that the equilibrium point is displayed in the viewing window. Set up a coordinate system on your homework paper and copy the image in the viewing window onto your coordinate system. Label each graph with its equation and include the WINDOW parameters on the axes.

b) Use the intersect utility in the CALC menu to find the coordinates of the equilibrium point. Place this estimate on your coordinate system near the point of intersection of your two graphs.

c) Use the substitution method to find the coordinates of the point of intersection. Use your calculator to round this answer to the nearest cent. Compare this solution with the estimate found in part (b).

d) If the unit price is set at the value you found in part (c), what will be the demand and supply for this particular brand of watch? Round your answer to the nearest watch.

10. Use the substitution method only to solve each of the following systems of linear equations.

a) $\dfrac{1}{2}x + \dfrac{1}{3}y = \dfrac{1}{4}$

$\dfrac{2}{3}x - \dfrac{3}{4}y = \dfrac{4}{3}$

b) $\dfrac{2x}{5} - \dfrac{3y}{2} = \dfrac{5}{4}$

$\dfrac{3x}{4} - \dfrac{y}{5} = \dfrac{1}{2}$

c) $\dfrac{x-y}{2} - \dfrac{x+y}{3} = 1$

$\dfrac{2x+y}{5} - \dfrac{x-2y}{4} = 1$

d) $\dfrac{x-2y}{3} + \dfrac{x+y}{5} = 2$

$x - \dfrac{x-2y}{5} = 1$

11. Use the substitution method only to solve each of the following systems of linear equations. Use your calculator as you work. Round your solution to the nearest hundredth (two decimal places).

a) $2.3x - 4.5y = 1.8$

$y = 1.7x - 4$

b) $4.2x - 3.5y = 1.3$

$x + 1.2y = 4.3$

c) $1.08x + 1.05y = 525$

$x + y = 10{,}000$

d) $1.3(x - y) + 2.8(x + y) = 100$

$x - 2.5y = 50$

12. Apples sell for 40 cents each at Jerry the Greengrocer's. Oranges sell for 60 cents each. The total day's receipts for apples and oranges was \$190. Jerry sold 100 more apples than he did oranges. Set up two equations in two unknowns that model this problem situation and use the substitution method to solve your system of equations. Use complete sentences to explain why your solution satisfies the requirements of the problem. If your answer doesn't satisfy the requirements of the problem, check your work for error.

13. Jane wants to buy her mom a Mother's Day card. She raids her piggy bank and finds that she has \$2.40, all in nickels and dimes, 31 coins in all. Set up two equations in two unknowns that model this problem situation and use the substitution method to solve your system of equations. Use complete sentences to explain why your solution satisfies the requirements of the problem. If your answer doesn't satisfy the requirements, check your work for error.

14. Consider the following rectangle.

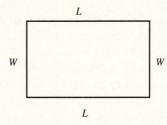

a) Suppose that the length of this rectangle is 1 foot longer than twice its width. Express the length of the rectangle as a function of its width.

b) If the perimeter of the rectangle is 20, set up an equation in L and W that models this fact.

c) Use the substitution method to solve your system of equations.

d) Use complete sentences to explain why your solution satisfies the requirements of the problem. If your answer doesn't satisfy the requirements, check your work for error.

15. Darnell presents the following problem as a challenge to his fellow students in an algebra class: "I'm thinking of two numbers. The sum of these two numbers is 10. The second number is three times larger than the first number. What numbers am I thinking of?" Let F and S represent the first number and second number, respectively. Set up two equations in the unknowns F and S that model this problem situation. Use the substitution method to solve your system of equations. Use complete sentences to explain why your solution satisfies the requirements of the problem. If your answer doesn't satisfy the requirements, check your work for error.

16. Set up a coordinate system on a sheet of graph paper. Sketch the line that goes through the points (1,5) and (6,−2) on the coordinate system. Sketch the line that has y-intercept 1 and slope 2 on the coordinate system. Place an estimate of the point of intersection of the two lines near the point of intersection on the coordinate system.

   a) Find an equation of each line. This will produce a system of linear equations.

   b) Use the substitution method to find the solution of the system of linear equations created in part (a). Use a calculator to find decimal approximations for this solution and compare your results with the estimate from your graph.

17. Suppose that the demand for television sets is a linear function of their unit price. If the unit price is $50, the demand is 600 television sets. If the unit price is $150, the demand is 400 television sets. Suppose that the supply of television sets is also a linear function of their unit price. If the unit price is $0 then the supply is 100 television sets. Thereafter the supply increases at a rate of three television sets per dollar increase in the unit price.

   a) Set up a coordinate system on a sheet of graph paper and sketch the demand function and the supply function on your coordinate system. Estimate the equilibrium point and place this estimate on your coordinate system near the point of intersection of the demand function and the supply function.

   b) Find the equation of the demand D as a function of the unit price p. Find the equation of the supply S as a function of the unit price p. Clearly state these equations on your homework paper.

   c) Use the substitution method and the equations you developed in part (b) to find the equilibrium price. If your solution does not agree with the estimate found in part (a), check your work for error.

18. Thaddeus invests money in two accounts, one that pays 5 percent per year and one that pays 4 percent per year. The interest from both accounts totals $69. The amount invested in the account that pays 4 percent interest is $100 more than twice the amount invested in the account that pays 5 percent. Set up two equations in two unknowns that models this problem situation. Use the substitution method to solve your system of equations. Use complete sentences to explain why your solution satisfies the requirements of the problem. If your answer doesn't satisfy the requirements of the problem, check your work for error.

19. Tricia has not cleaned out her change purse in some time. Consequently, she has quite a few coins. However, for some strange reason, all of the coins are either quarters or dimes. Also, the number of quarters is one more than twice the number of dimes. Tricia counts and discovers she has a total of $9.25 in her change purse. Set up two equations in two unknowns that model this problem situation. Use the substitution method to solve your system of equations. Use complete sentences to explain why your solution satisfies the requirements of the problem. If your answer doesn't satisfy the requirements, check your work for error.

20. Consider the following system of equations.

$$2x - y = 4$$
$$y = 2x - 7$$

   a) Set up a coordinate system on a sheet of graph paper. Sketch the graph of each equation from this system on your coordinate system. Examine your graph. Explain, in your own words, any suspicions you have.

b) Try to solve the system of equations using the substitution method. Explain how the weird behavior of the substitution method is in harmony with your suspicions from part (b).

c) What are the solutions of this system?

21. Consider the following system of equations.

$$6x + 2y = 24$$

$$y = -3x + 12$$

a) Set up a coordinate system on a sheet of graph paper. Sketch the graph of each equation from this system on your coordinate system. Examine your graph. Explain, in your own words, any suspicions you have.

b) Try to solve the system of equations using the substitution method. Explain how the weird behavior of the substitution method is in harmony with your suspicions from part (b).

c) What are the solutions of this system?

22. Here's a little move that is an eye-opener. Consider the following two functions:

$$y = f(x) \quad \text{Eq. (1)}$$
$$y = g(x) \quad \text{Eq. (2)}$$

What happens if you substitute Equation (1) into Equation (2)? You get the following statement:

$$f(x) = g(x)$$

What are you finding when you look for x-values that make these functions equal? That's right! You look at the point(s) of intersection, which is precisely what you are finding when you use the substitution method. Wow! Talk about things coming together.

Consider the following system of equations.

$$f(x) = 5 - 2x \quad \text{Eq. (1)}$$
$$g(x) = x - 3 \quad \text{Eq. (2)}$$

a) On a sheet of graph paper, set up a coordinate system and sketch the graphs of f and g. Estimate the point of intersection and record it on your graph paper, near the point of intersection.

b) Use the substitution method to solve the system of equations. Record this exact solution on your graph paper, near the point of intersection.

c) Does your solution from part (b) compare favorably with your estimate from part (a)? If not, go back and check your work for error.

3.3 The Gaussian Elimination Method

We will now look at another method to solve systems of equations. It is called the *Gaussian elimination method*, named after Karl Friedrich Gauss, one of the greatest mathematicians who ever lived. It is called an *elimination method* because the idea is to eliminate the x-terms and solve for y, then eliminate the y-terms and solve for x.

An Important Fundamental. Suppose you have two simple equations, as follows:

$$5 = 5$$
$$7 = 7$$

What happens when these equations are added together?

$$5 = 5$$
$$\underline{7 = 7}$$
$$12 = 12$$

If you have two equations, you can add them together, getting another equation. Let's state this idea in symbols:

$$a = b$$
$$\underline{c = d}$$
$$a + c = b + d$$

Example 1   Find the solution of the following system of linear equations.

$$x - y = 2$$
$$x + y = 4$$

Graphical Solution   Draw the graph of each equation and note the coordinates of the point of intersection.

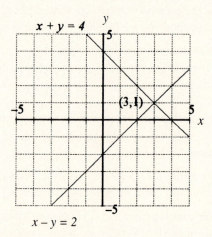

The solution of this system is approximately $(3,1)$.

Analytical Solution   Once again, the equations will be numbered so that we can reference them while we work.

$$x - y = 2 \quad \text{Eq. (1)}$$
$$x + y = 4 \quad \text{Eq. (2)}$$

If these equations are added together, the y terms will be eliminated.

$$x - y = 2 \qquad \text{Eq. (1)}$$
$$x + y = 4 \qquad \text{Eq. (2)}$$
$$\text{Eq. (1)} + \text{Eq. (2)} \qquad 2x \quad = 6$$
$$x = 3$$

You can now substitute $x = 3$ into either Equation (1) or Equation (2).
Substitute $x = 3$ into Equation (1):

$$x - y = 2$$
$$3 - y = 2$$
$$-y = -1$$
$$y = 1$$

The solution of this system is $(3,1)$. Note that this solution agrees exactly with the estimate from the graphing method. You should check this solution in each equation of the original system.

Example 2   Find the solution of the following system of equations:

$$2x + 5y = 10$$
$$3x - 2y = 6$$

Graphical Solution   Draw the graph of each equation and note the coordinates of the point of intersection.

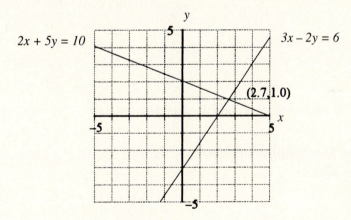

It appears that the solution of this system of equations is approximately $(2.7, 1.0)$.

Analytical Solution   The equations will be numbered so that we can reference them while we work. Try to emulate this in your own work.

$$2x + 5y = 10 \qquad \text{Eq. (1)}$$
$$3x - 2y = 6 \qquad \text{Eq. (2)}$$

The idea is to eliminate either x or y by adding the equations together. Multiply Equation (1) by 3 and Equation (2) by -2 and add the results together. Use the notation $3 \times$ Eq. (1) to remind yourself that you are multiplying Equation (1) by 3. The notation $-2 \times$ Eq. (2) is a reminder that you are multiplying Equation (2) by -2. These operations produce two equivalent equations, which will be numbered Equation (3) and Equation (4).

$$3 \times \text{Eq. (1)} \qquad 6x + 15y = 30 \qquad \text{Eq. (3)}$$
$$-2 \times \text{Eq. (2)} \qquad \underline{-6x + 4y = -12} \qquad \text{Eq. (4)}$$

Why is this done? Let's see what happens if these last two equations are added together. Use the notation Eq. (3) + Eq. (4) to remind yourself that you are adding Equation (3) and Equation (4) together.

$$6x + 15y = 30 \qquad \text{Eq. (3)}$$
$$\underline{-6x + 4y = -12} \qquad \text{Eq. (4)}$$
$$\text{Eq. (3) + Eq. (4)} \qquad 19y = 18$$
$$y = \frac{18}{19}$$

When Equation (3) and Equation (4) are added, the x-terms are eliminated. At this point, $y = \frac{18}{19}$ can be substituted into Equation (1) or Equation (2) and the x-value of the solution can be found. However, let's use the method of Gaussian elimination a second time. This time try to eliminate the y-terms in the system of equations. Multiply Equation (1) by 2 and Equation (2) by 5.

$$2 \times \text{Eq. (1)} \qquad 4x + 10y = 20 \qquad \text{Eq. (5)}$$
$$5 \times \text{Eq. (2)} \qquad \underline{15x - 10y = 30} \qquad \text{Eq. (6)}$$

This is good strategy. When you add these equations together, the y-terms will be eliminated.

$$4x + 10y = 20 \qquad \text{Eq. (5)}$$
$$\underline{15x - 10y = 30} \qquad \text{Eq. (6)}$$
$$\text{Eq. (5) + Eq. (6)} \qquad 19x \qquad = 50$$
$$x = \frac{50}{19}$$

The solution of this system of equations is the point whose coordinates are $\left(\dfrac{50}{19}, \dfrac{18}{19} \right)$.

Is This Solution Reasonable? Use your calculator to find decimal approximations for the exact solution.

$$\frac{50}{19} = 2.63157894...$$
$$\frac{18}{19} = 0.947368421...$$

These approximations agree quite favorably with the estimate (2.7, 1.0) that was found using the graphing method. However, if you want to be completely certain that your solution is correct, you will have to verify that the exact solution satisfies each of the equations in the system.

Check Your Solution. Substitute the solution $\left(\dfrac{50}{19}, \dfrac{18}{19}\right)$ into Equation (1).

$$2x + 5y = 10$$

$$2\left(\frac{50}{19}\right) + 5\left(\frac{18}{19}\right) = 10$$

$$\frac{100}{19} + \frac{90}{19} = 10$$

$$\frac{190}{19} = 10$$

$$10 = 10$$

Therefore the point $\left(\dfrac{50}{19}, \dfrac{18}{19}\right)$ satisfies Equation (1). Substitute the solution $\left(\dfrac{50}{19}, \dfrac{18}{19}\right)$ into Equation (2).

$$3x - 2y = 6$$

$$3\left(\frac{50}{19}\right) - 2\left(\frac{18}{19}\right) = 6$$

$$\frac{150}{19} - \frac{36}{19} = 6$$

$$\frac{114}{19} = 6$$

$$6 = 6$$

Therefore the point $\left(\dfrac{50}{19}, \dfrac{18}{19}\right)$ satisfies Equation (2). You can now be certain that $\left(\dfrac{50}{19}, \dfrac{18}{19}\right)$ is the exact solution of this system of equations.

Example 3   Find the solution of the following system of equations.

$$3x - 4y = 12 \qquad \text{Eq. (1)}$$
$$5x + y = 5 \qquad \text{Eq. (2)}$$

Graphical Solution   Draw the graph of each equation and note the coordinates of the point of intersection.

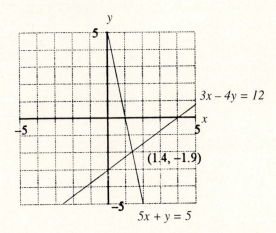

It appears that $(1.4, -1.9)$ is a reasonable *estimate* of the solution of this system of equations.

Analytical Solution   Multiply Equation (1) by 5 and multiply Equation (2) by −3. When the resulting equations are added together, the x-terms will be eliminated.

$$
\begin{array}{lrl}
5 \times \text{Eq. (1)} & 15x - 20y = 60 & \text{Eq. (3)} \\
-3 \times \text{Eq. (2)} & -15x - 3y = -15 & \text{Eq. (4)} \\
\text{Eq. (3) + Eq. (4)} & -23y = 45 & \\
& y = -\dfrac{45}{23} &
\end{array}
$$

You could substitute $y = -\frac{45}{23}$ in one of the equations to find the x-value of this solution. However, let's get a bit more practice with the elimination method. The goal is to eliminate the y-terms. Leave Equation (1) alone and multiply Equation (2) by 4. When the resulting equations are added together, the y-terms will be eliminated.

$$
\begin{array}{lrl}
 & 3x - 4y = 12 & \text{Eq. (1)} \\
4 \times \text{Eq. (2)} & 20x + 4y = 20 & \text{Eq. (5)} \\
\text{Eq. (1) + Eq. (5)} & 23x = 32 & \\
& x = \dfrac{32}{23} &
\end{array}
$$

Is Your Solution Reasonable?   Now, use your calculator to get decimal approximations of the exact solution.

$$
x = \frac{32}{23} \approx 1.391304348\ldots
$$

$$
y = -\frac{45}{23} \approx -1.956521739\ldots
$$

The estimate from the graphing method was $(1.4, -1.9)$. The solution from the elimination method agrees quite favorably with this estimate.

Use Your TI-82 to Check Your Solution.   Use the STO▷ key to store your solution in the variables X and Y on your calculator. For a review of this technique, see page 268.

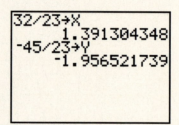

Next, evaluate $3X - 4Y$ and $5X + Y$.

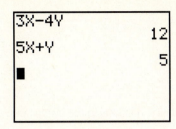

Since the equations in this system are $3x - 4y = 12$ and $5x + y = 5$, you have pretty striking evidence that you have found the correct solution to this system.

Example 4   Sweet Lou's Hotdogs is a hotdog stand downtown. You can buy two varieties of hotdogs: chili dogs and kraut dogs, both guaranteed to lead you to Heartburn City. Lou charges $2.25 for her chili dogs and $1.75 for her kraut dogs. On Tuesday, she sold 250 hotdogs, all of which were either chili dogs or kraut dogs. Lou's brother Ramone sells heartburn tablets in the pharmacy next door. It is rumored that Ramone does great business in heartburn tablets, but that's another problem. Anyway, Lou's Tuesday receipts for the chili dogs and kraut dogs totaled $487.50. How many of each type of hotdog did Lou sell?

Graphical Solution   Let C represent the number of chili dogs that Lou sold, and let K represent the number of kraut dogs sold. If the price of a chili dog is $2.25 and Lou sells C of them, then C chili dogs at $2.25 apiece yields a total of $2.25C$ in receipts. If the price of a kraut dog is $1.75 and Lou sells K of them, then K kraut dogs at $1.75 apiece yields a total of $1.75K$ in receipts. This information can be arranged in a table.

Type of Hotdog	Number Sold	Receipts
Chili dog	C	$2.25C$
Kraut dog	K	$1.75K$
Totals	250	487.50

The following two equations can be set up.

$$C + K = 250$$
$$2.25C + 1.75K = 487.50$$

You can arrive at the first equation by totaling the items in the column headed "Number Sold." The second equation can be found by totaling the items in the column headed "Receipts." You can remove the decimals in the second equation by multiplying through by 100, which moves the decimal point two places to the right in each term of the equation.

$$C + K = 250 \qquad \text{Eq. (1)}$$
$$225C + 175K = 48750 \qquad \text{Eq. (2)}$$

Use Your TI-82 to Draw the Graphs of the Equations in This System. Place K on the vertical axis and C on the horizontal axis and note that y and x have been replaced with K and C, respectively. You must solve each equation in your system for K before you can enter them in the Y= menu. Now solve the first Equation (1) for K.

$$C + K = 250$$
$$K = 250 - C$$

Solve the second Equation (2) for K.

$$225C + 175K = 48750$$

$$175K = 48750 - 225C$$

$$K = \frac{48750 - 225C}{175}$$

You must enter these equations in the Y= menu in the following manner:

You will need to find a good viewing window. Because the total number of hotdogs sold is 250, the number of kraut dogs must be a number between 0 and 250. The number of chili dogs must also be a number between 0 and 250. Based on these numbers, set your WINDOW parameters in the following manner:

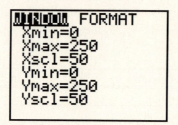

After pushing the GRAPH button to draw the graph, run the intersect utility in the CALC menu to capture the following image in your viewing window:

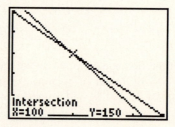

For a review of the intersect utility, see Example 10 in Section 1.6. Set up a coordinate system on your homework paper and copy this image onto the coordinate system in the following manner.

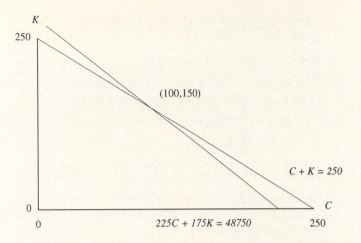

It appears that the approximate solution of this system of equations is (100,150).

Analytical Solution   Multiply Equation (1) by −225 and leave the second equation alone. Add the resulting equations together.

$$
\begin{array}{lll}
-225 \times \text{Eq. (1)} & -225C - 225K = -56250 & \text{Eq. (3)} \\
 & \underline{225C + 175K = 48750} & \text{Eq. (2)} \\
\text{Eq. (3)} + \text{Eq. (2)} & -50K = -7500 & \\
 & K = 150 &
\end{array}
$$

Next, eliminate the K-value in the system of equations. Multiply Equation (1) by −175 and leave Equation (2) alone. Then add the results.

$$
\begin{array}{lll}
-175 \times \text{Eq. (1)} & -175C - 175K = -43750 & \text{Eq. (4)} \\
 & \underline{225C + 175K = 48750} & \text{Eq. (5)} \\
\text{Eq. (4)} + \text{Eq. (5)} & 50C = 5000 & \\
 & C = 100 &
\end{array}
$$

Therefore, the exact solution of this system of equations is (100,150). Because this agrees with the approximate solution, you can be confident that you are correct.

Is Your Solution Reasonable?   You have determined that Sweet Lou sold 100 chili dogs and 150 kraut dogs. That total of 250 hotdogs satisfies the first requirement of the problem. Now, 100 chili dogs at $2.25 apiece will bring Sweet Lou receipts of $225, and 150 kraut dogs at $1.75 apiece will bring Lou receipts of $262.50. Receipts of $225 and $262.50 total $487.50 in receipts, satisfying the second condition of the problem. The solution is reasonable. Now you'll have to pay a visit to Sweet Lou's brother Ramone and get a few of those heartburn tablets.

Exercises for Section 3.3

1. For each of the following systems of linear equations, perform each of following tasks:

    i)   On a sheet of graph paper, set up a coordinate system. Find the x-intercept and y-intercept of Equation (1) and use this information to sketch the graph of Equation (1). Find the x-intercept and y-intercept of Equation (2) and use this information to sketch the graph of Equation (2).

ii) Estimate the y-value of the point of intersection of your two lines. Place this estimate on your graph, near the point of intersection of your two lines.

iii) Add Equations (1) and (2) together.

iv) Does this *exact* solution for y that you found in part (c) agree favorably with the *estimate* you found in part (b)? If not, check your work for error.

a) $x + 2y = 6$ Eq. (1)   b) $6x + 3y = 12$ Eq. (1)

  $-x - y = 2$ Eq. (2) $-6x - 2y = -6$ Eq. (2)

2. For each of the following systems of equations, perform the following tasks:

i)   Set up a separate coordinate system on a sheet of graph paper for each system. Find the x- and y-intercepts of the line represented by each equation, plot them on your coordinate system, and draw a line through them.

ii) Place your estimate of the solution of the system near the point of intersection of your line on your coordinate system.

iii) Use the Gaussian elimination method to find an *exact* solution of the system of equations. Compare this exact solution with your estimate from part (ii). If they do not agree favorably, check your work for error.

a) $2x - y = 8$ b) $x + 2y = 8$ c) $4x + 3y = 12$ d) $5x + 4y = 20$

  $-x + y = -4$ $-x - y = -6$ $-3x - 3y = -9$ $-5x - 3y = -15$

3. For each of the following systems of equations, perform the following tasks:

i)   Set up a separate coordinate system on a sheet of graph paper for each system. Find the x- and y-intercepts of the line represented by each equation, plot them on your coordinate system, and draw a line through them.

ii) Place your estimate of the solution of the system near the point of intersection of your line on your coordinate system.

iii) Use the Gaussian elimination method to find an *exact* solution of the system of equations. Use your calculator to find decimal approximations of your solution. Compare your solution with your estimate from part (ii). If they do not agree favorably, check your work for error.

a) $3x + 2y = 6$ b) $2x + 5y = -10$ c) $4x - 5y = -20$

  $x - y = 4$ $x - y = 4$ $3x + 4y = 12$

d) $2x - 7y = 14$ e) $\dfrac{x}{2} + \dfrac{y}{3} = 1$ f) $\dfrac{x}{5} + \dfrac{y}{3} = 1$

  $3x + 5y = 15$ $\dfrac{x}{4} - \dfrac{y}{3} = 1$ $\dfrac{x}{2} - \dfrac{y}{7} = 1$

4. (TI-82) For each of the following problems, perform the following tasks:

i)   Solve each equation in the system for y, then place each result in the Y= menu, one in Y1 and one in Y2.

ii) Set up a coordinate system on your homework paper. Press the ZOOM button on your calculator and select 6:ZStandard. Copy the resulting image onto your coordinate system. Label each curve with its equation and include the WINDOW parameters on your axes.

iii) Use the intersect utility in the CALC menu to find the coordinates of the point of intersection. Place this estimate on your coordinate system near the point of intersection of the two lines.

iv) Use the Gaussian elimination method to find the *exact* solution of the system. Use your calculator to find decimal approximations for your solution. If these do not compare favorably with the estimate you found in part (iii), check your work for error.

a) $2x + y = 4$ b) $-3x + y = -6$ c) $4x + 2y = -8$ d) $15x - 5y = -10$

$3x - y = 6$ $2x + y = 2$ $6x - 3y = 9$ $10x + 2y = 8$

5. The sum of two numbers is 14. Twice the larger of these numbers plus three times the smaller of these numbers totals 33. Let L represent the larger of the two numbers. Let S represent the smaller of the two numbers. Set up a system of equations in L and S that models the problem situation.

   a) Set up a coordinate system on a sheet of graph paper. Make the vertical axis the L axis and the horizontal axis the S axis. Sketch the graph of each line that is represented by your system of equations on your coordinate system. Place your estimate of the solution of your system of equations on your coordinate system near the point of intersection of your two lines.

   b) Use the Gaussian elimination method to solve your system of equations for L and S.

   c) Use complete sentences to explain why your solution satisfies the requirements of the problem. If your answer doesn't satisfy the requirements, check your work for error.

6. Jamal went to the post office to buy some stamps. He bought 20 stamps, priced at 15 cents and 25 cents apiece. He paid a total of $4.60 for the stamps. Let F represent the number of 15-cent stamps that Jamal purchased and let T represent the number of 25-cent stamps. Set up a system of equations in F and T that models the problem situation.

   a) Set up a coordinate system on a sheet of graph paper. Make the vertical axis the F axis and the horizontal axis the T axis. Sketch the graph of each line that is represented by your system of equations on your coordinate system. Place your estimate of the solution of your system of equations on your coordinate system near the point of intersection of your two lines.

   b) Use the Gaussian elimination method to solve your system of equations for F and T.

   c) Use complete sentences to explain why your solution satisfies the requirements of the problem. If your answer doesn't satisfy the requirements, check your work for error.

7. Regina loves to make ceramic pots. Every other Saturday, the craftspeople from her community set up booths at the park to sell their wares. Regina takes 30 of her ceramic pots and places them for sale in her booth. To her delight, she easily sells all of her pots. She sold two varieties of pots, a large size priced at $5.50 and a smaller priced at $2.50. Her total revenue for the day was $132. Let L represent the number of large pots she sold, and let S represent the number of small pots. Set up a system of equations in L and S that models the problem situation.

   a) Set up a coordinate system on a sheet of graph paper. Make the vertical axis the L axis and the horizontal axis the S axis. Sketch the graph of each line that is represented by your system of equations on your coordinate system. Place your estimate of the solution of your system of equations on your coordinate system near the point of intersection of your two lines.

   b) Use the Gaussian elimination method to solve your system of equations for L and S.

   c) Use complete sentences to explain why your solution satisfies the requirements of the problem. If your answer doesn't satisfy the requirements, check your work for error.

8. Use the Gaussian elimination method only to solve each of the following systems of linear equations.

a) $\dfrac{2x}{5} - \dfrac{3y}{2} = 1$

   $\dfrac{x}{2} + \dfrac{2y}{3} = 1$

b) $\dfrac{1}{4}x - \dfrac{2}{3}y = \dfrac{1}{2}$

   $\dfrac{3}{2}x + \dfrac{1}{3}y = -\dfrac{5}{6}$

c) $\dfrac{x-y}{3} + \dfrac{2x+y}{5} = 1$

   $\dfrac{x}{4} - \dfrac{x-2y}{3} = 1$

d) $\dfrac{x+y}{2} - \dfrac{x-y}{5} = 1$

   $\dfrac{x-2y}{4} + \dfrac{x-y}{3} = 1$

9. (TI-82) Consider the equation $2.35x - 4.72y = 8.36$. If you want to draw this graph on your calculator, you must solve this equation for y. Let your calculator do some of the work for you. This is how we solve for y:

$$2.35x - 4.72y = 8.36$$

$$-4.72y = 8.36 - 2.35x$$

$$y = \frac{8.36 - 2.35x}{-4.72}$$

Enter this function in the Y= menu in the following manner:

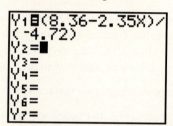

Note that both a subtraction sign and a negative sign are used when entering this expression. Solve each of the following equations for y as shown above. Enter them in the Y= menu. Press the ZOOM button and select 6:ZStandard. Set up a coordinate system on your homework paper and copy the resulting image in your viewing window onto your coordinate system. Label your graph with its equation and include the WINDOW parameters on your axes.

a) $2.32x + 4.12y = 10.83$

b) $1.04x - 2.34y = 8.62$

c) $1.235x - 8.442y = 24.112$

d) $2.201x - 4.003y = -9.001$

10. (TI-82) For each of the following systems of equations, perform the following tasks:

i)   Solve each equation for y.

ii) Enter your results in the Y= menu in Y1 and Y2, respectively. Press the ZOOM button and select 6:ZStandard. Set up a coordinate system on your homework paper and copy the image in your viewing window onto your coordinate system. Label each graph with its equation and include the WINDOW parameters on your coordinate axes.

iii) Run the intersect utility in the CALC menu to find the solution of the system of equations. Place this estimate on your coordinate system near the point of intersection of your two lines.

a) $2.32x - 4.78y = 9.12$

   $1.12x + 4.44y = -4.12$

b) $3.34x - 6.12y = 12.21$

   $8.12x + 5.13y = 12.23$

11. Suppose that your system of equations represents two lines that are parallel. Describe the solution of the system.

12. Suppose that your system contains two equations but both equations represent the same line. Describe the solution of the system.

13. Consider the following system of equations.

$$x + 3y = 6 \qquad \text{Eq. (1)}$$
$$2x + 6y = 18 \qquad \text{Eq. (2)}$$

a) Set up a coordinate system on a sheet of graph paper. Find the x-intercept and y-intercept of each of the equations and use this information to sketch the graph of each equation on your coordinate system. What can you say about your two lines? What is the solution of this system of equations?

b) What follows is an attempt to solve this system using Gaussian elimination.

$$-2 \times \text{Eq. (1)} \qquad -2x - 6y = -12 \qquad \text{Eq. (3)}$$
$$\underline{2x + 6y = 18} \qquad \text{Eq. (2)}$$
$$\text{Eq. (3) + Eq. (2)} \qquad 0 = 6$$

What should you suspect if this situation should occur again in your work with Gaussian elimination?

14. Consider the following system of equations:

$$3x - y = 3 \qquad \text{Eq. (1)}$$
$$6x - 2y = 6 \qquad \text{Eq. (2)}$$

a) Set up a coordinate system on a sheet of graph paper. Find the x-intercept and y-intercept of each of the equations and use this information to sketch the graph of each equation on your coordinate system. What can you say about the two lines? What is the solution of this system of equations?

b) What follows is an attempt to solve this system using Gaussian elimination.

$$-2 \times \text{Eq. (1)} \qquad -6x + 2y = -6 \qquad \text{Eq. (3)}$$
$$\underline{6x - 2y = 6} \qquad \text{Eq. (2)}$$
$$\text{Eq. (3) + Eq. (2)} \qquad 0 = 0$$

What should you suspect if this situation should occur again in your work with Gaussian elimination?

15. (TI-82) Racine recently moved to an area that has excellent public transportation and she feels that she no longer needs her truck with its accompanying financial worries. She sells her truck for $15,000 and takes part of the money and invests it in IBM stock and invests the rest in a certificate of deposit. Her investment with IBM pays 5 percent on the year and her certificate of deposit pays 6 percent on the year. The combined interest from her two investments at the end of the year is $865. Let I represent the amount invested in IBM stock and let C represent the amount invested in the certificate of deposit. Set up a system of equations in I and C that models this problem situation.

a) Solve both equations for C, then load them in the Y= menu of your calculator.

b) Set up a coordinate system on your homework paper. Because the total investment is $15,000, both I and C must be an amount between $0 and $15,000. Adjust your WINDOW parameters to reflect this fact. Push the GRAPH button and copy the resulting image in your viewing window onto your coordinate system. Label each graph with its equation and include the parameter settings on your coordinate axes.

c) Use the intersect utility on your calculator to find the solution of your system. Place this estimate on your coordinate system near the point of intersection of your two lines.

d) Use the Gaussian elimination method to find the solution of your system.

e) Use complete sentences to explain why your solution satisfies the requirements of the problem. If your answer doesn't satisfy the requirements of the problem, check your work for error.

16. Compact disks sell for $15 and record albums sell for $12 in Yolanda's Music Shoppe. Last week, a total of 200 CD's and record albums were sold. The revenue from these sales was $2,760. Let C represent the number of compact disks sold and let R represent the number of record albums sold. Set up a system of equations in C and R that models this problem situation. Use Gaussian elimination to solve your system of equations. Use complete sentences to explain why your solution satisfies the requirements of the problem. If your answer doesn't satisfy the requirements, check your work for error.

17. Constantine runs a catering service that frequently hosts large parties. Constantine mixes peanuts and cashews together as one of the snacks he serves during the social hour before dinner. Suppose that the total weight of the mix is 40 pounds and that peanuts cost $1.25 per pound and the cashews cost $2.50 per pound. The total cost of the mix is $87.50. Let P represent the number of pounds of peanuts used in the mix and let C represent the number of pounds of cashews in the mix. Set up a system of equations in P and C that models the problem situation. Use Gaussian elimination to solve your system of equations. Use complete sentences to explain why your solution satisfies the requirements of the problem. If your answer doesn't satisfy the requirements, check your work for error.

18. Jermaine has $4.35 in his pocket, all in nickels and quarters, 31 coins in all. Let N represent the number of nickels in his pocket and let Q represent the number of quarters in his pocket. Set up a system of equations in N and Q that models the problem situation. Use Gaussian elimination to solve this system of equations. Use complete sentences to explain why your solution satisfies the requirements of the problem. If your answer doesn't satisfy the requirements, check your work for error.

19. Maurice is a world class 800-meter runner. He runs a training race of 800 meters in 110 seconds (1 minute 50 seconds). He runs the first part of the race at a rate of 10 meters per second and the last part at 7 meters per second. Let t_1 represent the amount of time that he ran at the rate of 10 meters per second. Let t_2 represent the amount of time that he ran at 7 meters per second. Set up a system of equations in t_1 and t_2 that models the problem situation. Use Gaussian elimination to solve this system of equations. Use complete sentences to explain why your solution satisfies the requirements of the problem. If your answer doesn't satisfy the requirements, check your work for error.

20. Consider the lines L and M in the following graphs. Note that the lines L and M intersect at the point $(1, -1)$.

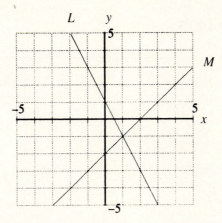

a) Note that the line L passes through the points $(-1, 3)$ and $(2, -3)$. Use this information and the point-slope form to find the equation of the line L. Place your answer in the form $Ax + By = C$.

b) Note that the line M has slope 1 and y-intercept -2. Use this information and the slope-intercept form to find the equation of the line M. Place your answer in the form $Ax + By = C$.

c) Use Gaussian elimination to solve the system of equations developed in parts (a) and (b). If your solution is not $(1, -1)$, check your work for error.

3.4 Summary and Review

Important Visualization Tools

The Substitution Method

$$2x + 3y = 6 \qquad \text{Eq. (1)}$$
$$\underline{x - 8y = 16} \qquad \text{Eq. (2)}$$

Solve Equation (2) for x:

$$x = 8y + 16 \qquad \text{Eq. (3)}$$

Substitute Equation (3) into Equation (1):

$$2(8y + 16) + 3y = 6$$
$$16y + 32 + 3y = 6$$
$$19y = -26$$
$$y = -\frac{26}{19}$$
$$y \approx -1.36842105263\ldots$$

Substitute $y = -\dfrac{26}{19}$ in Equation (3):

$$x = 8(-\frac{26}{19}) + 16$$
$$x = -\frac{208}{19} + \frac{304}{19}$$
$$x = \frac{96}{19}$$
$$x \approx 5.05263157895\ldots$$

The Gaussian Elimination Method

$$2x + 3y = 6 \qquad \text{Eq. (1)}$$
$$\underline{x - 8y = 16} \qquad \text{Eq. (2)}$$

$$\begin{array}{ll} & 2x + 3y = 6 \qquad \text{Eq. (1)} \\ -2 \times \text{Eq. (2)} & \underline{-2x + 16y = -32} \qquad \text{Eq. (3)} \\ \text{Eq. (1)} + \text{Eq. (3)} & 19y = -26 \end{array}$$
$$y = -\frac{26}{19}$$
$$y \approx -1.36842105263\ldots$$

$$\begin{array}{ll} 8 \times \text{Eq. (1)} & 16x + 24y = 48 \qquad \text{Eq. (4)} \\ 3 \times \text{Eq. (2)} & \underline{3x - 24y = 48} \qquad \text{Eq. (5)} \\ \text{Eq. (4)} + \text{Eq. (5)} & 19x = 96 \end{array}$$
$$x = \frac{96}{19}$$
$$x \approx 5.05263157895\ldots$$

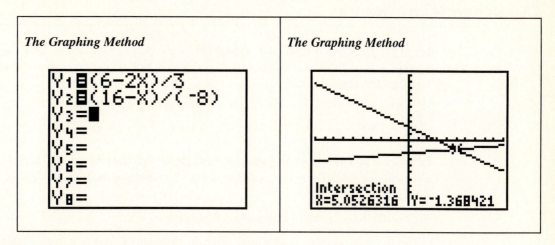

A Review of Key Words and Phrases

☞ Division is *distributive* with respect to addition. (page 249)

$$\frac{a+b}{c} = \frac{a}{c} + \frac{b}{c}$$

☞ The graph of the equation $Ax + By = C$ is a line. The form $Ax + By = C$ is sometimes called the *standard form* of a line. (page 250)

☞ To find the *x*-intercept of the line whose equation is $Ax + By = C$, let $y = 0$ in the equation and solve for *x*. To find the *y*-intercept of the line whose equation is $Ax + By = C$, let $x = 0$ in the equation and solve for *y*. (page 251)

☞ *The graphing method*: To solve a system of linear equations, first draw the graph of each equation. The point of intersection of the lines will be the solution of the system. (page 252)

☞ One of the most important steps in the problem-solving process is the last one, checking to see if your answers are reasonable. (page 257)

☞ Some systems have no solutions. (Section 3.1, Exercise 21; Section 3.3, Exercises 11 and 13)

☞ Some systems have more than one solution. (Section 3.1, Exercise 22; Section 3.3, Exercises 12 and 14)

☞ The *substitution method*: (1) Solve either equation for either variable. (2) Substitute the result of Step 1 into the *other* equation. (page 263)

☞ If $a = b$ and $c = d$, then $a + c = b + d$. (page 280)

$$a = b$$
$$\underline{c = d}$$
$$a + c = b + d$$

☞ The *Gaussian elimination method*: The idea behind this method is to eliminate either the *x*-terms and solve for *y*, or eliminate the *y*-terms and solve for *x*. (page 279)

TI-82 Keywords and Menus

☞ Using the TI-82 to solve a system of linear equations. (Example 6, page 254)

☞ Using the TI-82 to check the solution to a linear system. (pages 268, 284)

☞ Some thoughts on setting the WINDOW parameters. (Example 3, page 269; Example 4, page 286)

Chapter Review Exercises

1. Try to guess the solution of each of the following systems of equations. That's right, just guess, then check your guess in each equation. If your guess does not check, keep guessing until you find a solution that checks.

   a) $x + y = 5$

   $x - y = 1$

   b) $x + y = 10$

   $x - y = 4$

2. Set up a coordinate system on a sheet of graph paper. Plot the points $(-4, -1)$ and $(6, 3)$ and draw a line through them. Draw a second line on your coordinate system whose slope is 2 and whose y-intercept is -4.

   a) Estimate the coordinates of the point of intersection.

   b) Use the point-slope form of a line to find the equation of your first line and the slope-intercept form of a line to find the equation of your second line.

   c) Use the substitution method to find the solution of the system of equations you developed in part (b). Does this solution compare favorably with the estimate you found in part (a)? If not, check your work for error.

3. For each of the following systems of linear equations, perform each of the following tasks:

   i)   Set up a coordinate system on a sheet of graph paper and draw the graph of each equation on your coordinate system.

   ii) Estimate the solution of the system and place this solution on your coordinate system near the point of intersection of the two lines.

   iii) Use the *substitution method* to solve the system of equations. Use a calculator to find decimal approximations for your solution. If your solution does not compare favorably with the estimate found in part (ii), then check your work for error.

   a) $5x - y = -10$

   $y = -x + 4$

   b) $4x - 5y = -20$

   $y = x - 4$

   c) $-x - y = 7$

   $y = \dfrac{1}{2}x - 4$

   d) $2x - 9y = 18$

   $y = \dfrac{2}{3}x + 1$

   e) $8x - 3y = -24$

   $y = \dfrac{4}{3}x + 1$

   f) $6x + 5y = -30$

   $y = \dfrac{1}{3}x - 1$

4. For each of the following systems of linear equations, perform each of the following tasks.

   i)   Set up a coordinate system on a sheet of graph paper. Find the x- and y-intercepts of each line represented by an equation in the system. Plot each pair of intercepts on your coordinate system and draw a line through each pair of intercepts.

   ii) Estimate the solution of the system and place this solution on your coordinate system near the point of intersection of your two lines.

   iii) Use the *Gaussian elimination method* to solve the system of equations. Use a calculator to find decimal approximations for your solution. If your solution does not compare favorably with the estimate found in part (ii), then check your work for error.

   a) $2x - 9y = -18$

   $x + 8y = 16$

   b) $9x - y = 9$

   $x + 3y = -3$

   c) $2x + 5y = 5$

   $3x - 4y = 8$

d) $5x - 9y = 15$
      $2x + 7y = 14$

e) $\dfrac{x}{5} - \dfrac{y}{7} = 1$

      $\dfrac{x}{7} + \dfrac{y}{3} = 1$

f) $\dfrac{x}{9} + \dfrac{y}{8} = 1$

      $\dfrac{x}{5} - \dfrac{y}{3} = 1$

5. (TI-82) In each of the following problems, the system of equations is presented and the correct solution is also given. Use your calculator to check this solution in each equation using the method explained on page 296.

a) $2x + 3y = 7$

      $4x - y = 5$

      Solution $= \left(\dfrac{11}{7}, \dfrac{9}{7} \right)$

b) $5x - 8y = 11$

      $2x + 4y = 3$

      Solution $= \left(\dfrac{17}{9}, -\dfrac{7}{36} \right)$

6. (TI-82) For each of the following problems, perform each of the following tasks.

   i)   Load the functions in the Y= menu as shown. Press the ZOOM button and select 6:ZStandard. Set up a coordinate system on your homework paper and copy the resulting image onto your coordinate system. Label each graph with Y1 and Y2, respectively, and place your WINDOW parameters on your coordinate axes.

   ii) Use the intersect utility in the CALC menu to estimate the solution of the system of linear equations. Place this estimate on your coordinate system near the point of intersection of the two lines.

   iii) Use the substitution method to find the solution of the system of equations. Use your calculator to find decimal approximations of your solution. If these approximations of your solution do not compare favorably with the estimate from part (ii), check your work for error.

a)   b)

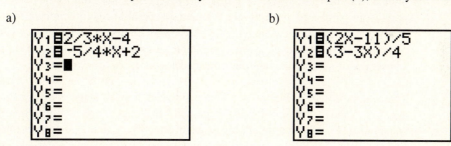

7. Use an analytic method of your choice to find a solution of each of the following systems of linear equations.

   a) $\dfrac{x-y}{4} - \dfrac{x+2y}{5} = 1$

         $\dfrac{x+y}{5} + \dfrac{2x-y}{3} = 1$

   b) $x - \dfrac{x+3y}{5} = 1$

         $\dfrac{y}{4} + \dfrac{x+5y}{3} = 1$

8. (TI-82) For each of the following systems of equations, perform each of the following tasks:

   i)   Solve each equation in the system for y. Enter your results in the Y= menu. Press the ZOOM button and select 6:ZStandard. Set up a coordinate system on your homework paper and copy the resulting image onto your coordinate system. Label each graph with Y1 and Y2, respectively, and place your WINDOW parameters on your coordinate axes.

ii) Use the intersect utility in the CALC menu to estimate the solution of the system of linear equations. Place this estimate on your coordinate system near the point of intersection of the two lines.

a) $1.28x - 3.44y = 9.995$

 $3.8x + 9.22y = 10.44$

b) $1.2x - 0.5y = 20$

 $0.3x + 4y = 12$

9. I am thinking of two numbers. If I multiply the smaller number by 2 and the larger number by 3 and add these results together, the sum is 24. The larger number is twice the smaller number. Let S and L represent the smaller number and larger number, respectively. Set up a system of linear equations in S and L that model the problem situation. Use the analytic method of your choice to find the solution of your system. Use complete sentences to explain why your solution satisfies the requirements of the problem. If your answer doesn't satisfy the requirements of the problem, check your work for error.

10. Mary goes to the post office to buy stamps for some postcards and letters that she wishes to mail. The stamps for postcards cost 19 cents each and the stamps for letters cost 29 cents. Mary is sending four more postcards than letters. Her total bill for the stamps is $6.04. Let L and P represent the number of 29-cent stamps and 19-cent stamps that Mary purchased. Set up a system of linear equations in L and P that model the problem situation. Use the analytic method of your choice to find the solution of your system. Use complete sentences to explain why your solution satisfies the requirements of the problem. If your answer doesn't satisfy the requirements, check your work for error.

11. Lily likes to stock up her freezer with meat at the beginning of each month so that she can avoid frequent trips to the market. Hamburger is priced at $2.89 per pound and pork chops are priced at $3.99 per pound. The combined weight of hamburger and pork chops in Lily's purchase is 25 pounds, and her bill is $88.75. Let H and P represent the number of pounds of hamburger and pork chops purchased, respectively. Set up a system of linear equations in H and P that models the problem situation. Use the analytic method of your choice to find the solution of your system. Use complete sentences to explain why your solution satisfies the requirements of the problem. If your answer doesn't satisfy the requirements, check your work for error.

12. The combined ages of Phyllis and Tina is 54 years. Phyllis's age is 3 years less than twice Tina's age. Let P and T represent Phyllis's age and Tina's ages, respectively. Set up a system of linear equations in P and T that model the problem situation. Use the analytic method of your choice to find the solution of your system. Use complete sentences to explain why your solution satisfies the requirements of the problem. If your answer doesn't satisfy the requirements, check your work for error.

13. (TI-82) The budget for the athletic department at a local college is $210,000. Part of the money in this budget comes from assessing a student athletic fee for each student that registers at the college. The rest of the budget is donated by the booster club. The amount of the budget donated by the booster club is $10,000 more than four times the amount that is collected in student athletic fees. Let S and B represent the amount of the fund that comes from student fees and the booster club, respectively.

a) Set up a system of linear equations in S and B that models the problem situation.

b) Set up a coordinate system on your homework paper. Place the variable B on the vertical axis and the variable S on the horizontal axis. Solve each equation in your system for the variable B. Enter each equation in the Y= menu.

c) When selecting an appropriate viewing window, consider that the combined budget is $210,000. The amount collected from student fees and the amount donated by the booster club cannot exceed this number.

d) Once you have found an appropriate viewing window, copy the image in your viewing window onto your coordinate system. Label each graph with its equation and include the WINDOW parameters on your axes. Use the intersect utility in the CALC menu to estimate the solution of your system of equations. Place this estimate on your coordinate system near the point of intersection of your two lines.

e) Use the analytic method of your choice to find the solution of your system. Use complete sentences to explain why your solution satisfies the requirements of the problem. If your answer doesn't satisfy the requirements, check your work for error.

14. (TI-82) Load the following function into the Y= menu.

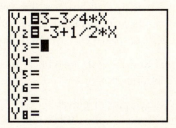

Press the ZOOM button and select 6:ZStandard. Set up a coordinate system on your homework paper and copy the image in your viewing window onto your coordinate system. Label each graph with its equation and include the WINDOW parameters on the coordinate axes.

a) Run the intersect utility in the CALC menu to capture the following image in your viewing window:

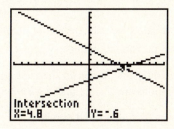

b) Press the X,T,Θ button on your keyboard. Press the MATH button and select 1:▷Frac and press the ENTER key to capture the following image in your viewing window:

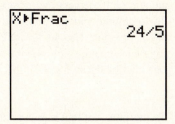

c) Press the ALPHA key, then the letter Y. Press the MATH button and select 1:▷Frac and press the ENTER key to capture the following image in your viewing window:

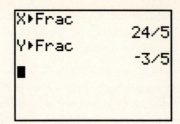

d) Use the substitution method to solve the system of equations that is loaded in the Y= menu. If your answer does not match that found in part (c), check your work for error.

Note: Trying to find exact fractional representations on a finite precision machine can be fruitless at times. Experiment with this utility, but be prepared for funny-looking results.

15. (TI-82) Roger runs a small business that specializes in making leather gloves. If he makes 100 pairs of gloves, his costs are $750. If he makes 400 pairs of gloves, his costs are $1,500. Let C represent Roger's costs and let x represent the number of pairs of gloves that he makes.

    a) Assume that Roger's costs are a linear function of the number of pairs of gloves that he makes. Express C as a function of x.

    b) Roger sells each pair of gloves for $4.75. Let R represent the revenue that he earns from the sales of his gloves. Express R as a function of x.

    c) The *break-even point* is the point where the revenue equals the cost. To break even, Roger must take in revenue equal to his costs. Use the intersect utility in the CALC menu to find the break-even point. Round your solution to the nearest glove. Use complete sentences to explain the meaning of this point.

    d) Use the substitution method to find the number of gloves that Roger must make to break even. Use your calculator to approximate your solution to the nearest glove.

16. Set up a coordinate system on a sheet of graph paper. Sketch the lines whose equations are $y = 2x$ and $x + 2y = 8$ on your coordinate system.

    a) Shade the triangular region in the first quadrant that is bordered by the x-axis and the lines whose equations are $y = 2x$ and $x + 2y = 8$.

    b) Find the area of your shaded region.

17. Jaime mixes some 10 percent hydrochloric acid solution with some 20 percent hydrochloric acid solution, producing 200 milliliters of solution, 12.5 percent of which is hydrochloric acid. Let x represent the amount of 10 percent hydrochloric acid solution used and let y represent the amount of 20 percent hydrochloric acid solution used. Set up a system of linear equations in x and y that model the problem situation. Use the analytic method of your choice to find the solution of your system. Use complete sentences to explain why your solution satisfies the requirements of the problem. If your answer doesn't satisfy the requirements of the problem, check your work for error.

18. Pedro begins his morning workout running at a steady pace of 2 miles per hour. During the second part of his run, he ups his pace to 3 miles per hour. Pedro's entire run takes 2.5 hours and he runs a total of 6 miles. Let t_1 represent the amount of time that Pedro runs at 2 miles per hour and let t_2 represent the amount of time that he runs at 3 miles per hour. Set up a system of linear equations in t_1 and t_2 that models the problem situation. Use the analytic method of your choice to find the solution of your system. Use complete sentences to explain why your solution satisfies

the requirements of the problem. If your answer doesn't satisfy the requirements, check your work for error.

19. Rachel receives a small inheritance and immediately invests part of it in a savings account that pays 4 percent interest per year. She invests the rest in a certificate of deposit that pays 5 percent per year. The interest at the end of 1 year from both investments totals $300. The amount invested in the savings account is $1,000 more than twice the amount invested in the certificate of deposit. Let S represent the amount invested in the savings account and let C represent the amount invested in the certificate of deposit. Set up a system of linear equations in S and C that models the problem situation. Use the analytic method of your choice to find the solution of your system. Use complete sentences to explain why your solution satisfies the requirements of the problem. If your answer doesn't satisfy the requirements, check your work for error.

20. Children's tickets at the Fortuna Theater cost $3.25 apiece and adult tickets cost $5.00. On a recent Saturday, ticket receipts totaled $1,400. There were 50 more children's tickets sold than adult tickets. Let C represent the number of children's tickets sold and let A represent the number of adult tickets. Set up a system of linear equations in C and A that models the problem situation. Use the analytic method of your choice to find the solution of your system. Use complete sentences to explain why your solution satisfies the requirements of the problem. If your answer doesn't satisfy the requirements, check your work for error.

21. The perimeter of a rectangle is 420 inches. The length of the rectangle is 10 inches longer than 3 times the width. Let L and W represent the length and width of the rectangle, respectively. Set up a system of linear equations in L and W that models the problem situation. Use the analytic method of your choice to find the solution of your system. Use complete sentences to explain why your solution satisfies the requirements of the problem. If your answer doesn't satisfy the requirements, check your work for error.

22. Aunt Rita is 4 years younger than Aunt Eileen. Their combined age is 158 years. Let R represent Rita's age and let E represent Eileen's age. Set up a system of linear equations that models the problem situation. Use the analytic method of your choice to find the solution of your system. Use complete sentences to explain why your solution satisfies the requirements of the problem. If your answer doesn't satisfy the requirements, check your work for error.

23. Harry mixes an amount of cement priced at 10 cents per pound with an amount of gravel priced at 8 cents per pound. The resulting mix weighs a total of 1,000 pounds and sells for 9 cents per pound. Let C represent the amount of cement used in the mix and let G represent the amount of gravel used in the mix. Set up a system of linear equations in C and G that models the problem situation. Use the analytic method of your choice to find the solution of your system. Use complete sentences to explain why your solution satisfies the requirements of the problem. If your answer doesn't satisfy the requirements, check your work for error.

Chapter 4
Introduction to Nonlinear Functions

Up to this point, most of the graphs that we have been working with have been lines. In this chapter, we will work with nonlinear equations, so-called because the graphs involved are not lines but curves. We'll introduce the laws of exponents and learn how to add, subtract, and multiply polynomials. Finally, we will work on factoring polynomials and apply our new skills to solving nonlinear equations.

4.1 The Laws of Exponents

We begin with a definition.

$$a^n = \underbrace{a \times a \times a \times \cdots \times a}_{n \text{ times}}$$
$$a^0 = 1, \, a \neq 0$$

In the expression a^n, the a is called the base and n is called the exponent. The exponent instructs us to use the base n times as a factor.

Example 1   Simplify the expressions 2^5 and $2+2+2+2+2$ and compare your results.

Solution   According to the definition, $2^5 = 2 \times 2 \times 2 \times 2 \times 2 = 32$. However, $2+2+2+2+2 = 10$. The product of five 2's is not the same as the sum of five 2's.

Example 2   Simplify the expression $a^5 a^3$.

Solution   Because $a^5 = a \cdot a \cdot a \cdot a \cdot a$ and $a^3 = a \cdot a \cdot a$, the following argument can be made:

$$a^5 a^3 = (a \cdot a \cdot a \cdot a \cdot a) \cdot (a \cdot a \cdot a)$$
$$= a \cdot a \cdot a \cdot a \cdot a \cdot a \cdot a \cdot a$$
$$= a^8$$

Note that it would be more efficient to add the exponents.

$$a^5 a^3 = a^{5+3}$$
$$= a^8$$

The First Law of Exponents. When multiplying powers of the same base, repeat the base and add the exponents.

303

$$a^m a^n = a^{m+n}$$

Remember, you are simply counting the total number of a's that are being multiplied together.

Example 3 Here are several applications of our first law of exponents.

$$a^3 a^7 = a^{10}$$
$$x^5 x^7 = x^{12}$$
$$y^{13} y^{15} = y^{28}$$

Note that the exponents were added.

Example 4 Simplify the expression $\left(a^5\right)^7$.

Solution Since a^5 is being raised to the seventh power, the following argument can be made:

$$\left(a^5\right)^7 = a^5 a^5 a^5 a^5 a^5 a^5 a^5$$
$$= a^{5+5+5+5+5+5+5}$$
$$= a^{35}$$

Because $5+5+5+5+5+5+5 = 5 \times 7$, it is much more efficient to multiply the exponents in the following manner.

$$\left(a^5\right)^7 = a^{5 \times 7}$$
$$= a^{35}$$

The Second Law of Exponents. When raising a power of a to another power, repeat the base a and multiply the exponents.

$$\left(a^m\right)^n = a^{mn}$$

Example 5 Here are several applications of the second law of exponents.

$$\left(a^7\right)^9 = a^{63}$$
$$\left(x^8\right)^7 = x^{56}$$
$$\left(y^{11}\right)^5 = y^{55}$$

Note that the exponents were multiplied.

Example 6   If $a \neq 0$, simplify the expression $\dfrac{a^8}{a^5}$.

Solution   Because $a^8 = a \cdot a \cdot a \cdot a \cdot a \cdot a \cdot a \cdot a$ and $a^5 = a \cdot a \cdot a \cdot a \cdot a$, the following argument can be made:

$$\frac{a^8}{a^5} = \frac{a \cdot a \cdot a \cdot a \cdot a \cdot a \cdot a \cdot a}{a \cdot a \cdot a \cdot a \cdot a}$$

$$= a \cdot a \cdot a$$

$$= a^3$$

It would be more efficient to subtract the exponents in the following manner.

$$\frac{a^8}{a^5} = a^{8-5}$$

$$= a^3$$

Example 7   If $a \neq 0$, simplify the expression $\dfrac{a^5}{a^8}$.

Solution   Because $a^5 = a \cdot a \cdot a \cdot a \cdot a$ and $a^8 = a \cdot a \cdot a \cdot a \cdot a \cdot a \cdot a \cdot a$, the following argument can be made:

$$\frac{a^5}{a^8} = \frac{a \cdot a \cdot a \cdot a \cdot a}{a \cdot a \cdot a \cdot a \cdot a \cdot a \cdot a \cdot a}$$

$$= \frac{1}{a \cdot a \cdot a}$$

$$= \frac{1}{a^3}$$

It would still be more efficient to subtract the exponents, but in the following manner.

$$\frac{a^5}{a^8} = \frac{1}{a^{8-5}}$$

$$= \frac{1}{a^3}$$

The Third Law of Exponents.   When dividing a larger power of a by a smaller power of a, repeat the a and subtract the exponents.

$$\boxed{\textit{If } a \neq 0 \textit{ and } m > n, \textit{ then } \frac{a^m}{a^n} = a^{m-n}}$$

When dividing a smaller power of a by a larger power of a, you would still subtract the exponents but in the following manner.

$$\text{If } a \neq 0 \text{ and } m < n, \text{ then } \frac{a^m}{a^n} = \frac{1}{a^{n-m}}$$

Example 8   Here are some examples of the third law of exponents. Assume each of the variables is a non-zero real number.

$$\frac{a^7}{a^3} = a^4 \qquad\qquad \frac{a^3}{a^7} = \frac{1}{a^4}$$

$$\frac{x^{11}}{x^6} = x^5 \qquad\qquad \frac{x^6}{x^{11}} = \frac{1}{x^5}$$

$$\frac{y^{23}}{y^{13}} = y^{10} \qquad\qquad \frac{y^{13}}{y^{23}} = \frac{1}{y^{10}}$$

Note that the exponents were subtracted in these examples.

Example 9   Simplify the expression $(ab)^5$.

Solution   The following argument can be made:

$$(ab)^5 = ab \cdot ab \cdot ab \cdot ab \cdot ab$$

$$= a \cdot a \cdot a \cdot a \cdot a \cdot b \cdot b \cdot b \cdot b \cdot b$$

$$= a^5 b^5$$

It would be more efficient to raise each factor to the fifth power.

$$(ab)^5 = a^5 b^5$$

The Fourth Law of Exponents.   When raising a product of two or more factors to a power, raise each factor to that power.

$$(ab)^n = a^n b^n$$

Example 10   Here are some examples of the fourth law of exponents.

$$(ab)^{11} = a^{11} b^{11}$$

$$(xy)^7 = x^7 y^7$$

$$(xyz)^9 = x^9 y^9 z^9$$

Example 11   If $b \neq 0$, simplify the expression $\left(\dfrac{a}{b}\right)^5$.

Solution   The following argument can be made:

$$\left(\frac{a}{b}\right)^5 = \frac{a}{b} \cdot \frac{a}{b} \cdot \frac{a}{b} \cdot \frac{a}{b} \cdot \frac{a}{b}$$

$$= \frac{a \cdot a \cdot a \cdot a \cdot a}{b \cdot b \cdot b \cdot b \cdot b}$$

$$= \frac{a^5}{b^5}$$

It would be more efficient to raise both the numerator and denominator to the fifth power.

$$\left(\frac{a}{b}\right)^5 = \frac{a^5}{b^5}.$$

The Fifth Law of Exponents.   If a quotient is raised to a power, then both the numerator and denominator are raised to that power.

$$\text{If } b \neq 0, \text{ then } \left(\frac{a}{b}\right)^n = \frac{a^n}{b^n}.$$

Example 12   Here are some examples of the fifth law of exponents. Assume all variables represent non-zero real numbers.

$$\left(\frac{a}{b}\right)^{11} = \frac{a^{11}}{b^{11}}$$

$$\left(\frac{x}{y}\right)^7 = \frac{x^7}{y^7}$$

A Summary of the Five Laws of Exponents

1. $a^m a^n = a^{m+n}$

2. $\left(a^m\right)^n = a^{mn}$

3. If $a \neq 0$ and $m > n$, then $\dfrac{a^m}{a^n} = a^{m-n}$. If $a \neq 0$ and $m < n$, then $\dfrac{a^m}{a^n} = \dfrac{1}{a^{n-m}}$.

4. $(ab)^n = a^n b^n$

5. If $b \neq 0$, then $\left(\dfrac{a}{b}\right)^n = \dfrac{a^n}{b^n}$.

Try Some Examples That Involve Combinations of the Laws of Exponents. We will now work some more complicated examples that require that more than one of the five laws of exponents be applied.

Example 13   Simplify the expression $\left(-3x^5y^7\right)^2$.

Solution   When raising a product to a power, raise each factor to that power.

$$\left(-3x^5y^7\right)^2 = (-3)^2\left(x^5\right)^2\left(y^7\right)^2$$
$$= 9x^{10}y^{14}$$

Note that the exponents in the last step were multiplied as required by the second law of exponents.

Example 14   Simplify the expression $\left(-3x^2y\right)^3\left(2x^3y^5\right)^2$.

Solution   When raising a product to a power, raise each factor to that power.

$$\left(-3x^2y\right)^3\left(2x^3y^5\right)^2 = (-3)^3\left(x^2\right)^3(y)^3(2)^2\left(x^3\right)^2\left(y^5\right)^2$$
$$= -27 \cdot x^6 \cdot y^3 \cdot 4 \cdot x^6 \cdot y^{10}$$
$$= -108x^{12}y^{13}$$

Note that the exponents in the last step were added, as required by the first law of exponents.

Example 15   Simplify the expression $\left(\dfrac{3x^4y^7}{2x^8y^4}\right)^2$.

Solution   Begin by raising the numerator and denominator to the second power, as required by the fifth law of exponents.

$$\left(\frac{3x^4y^7}{2x^8y^4}\right)^2 = \frac{9x^8y^{14}}{4x^{16}y^8}$$
$$= \frac{9y^6}{4x^8}$$

Note that exponents in the last step were subtracted, as required by the third law of exponents.

Simplifying Algebraic Expressions

We begin with two definitions.

> *A term is the product of a real number and one or more variables raised to powers. The real number is called the coefficient of the term.*

The coefficient of the term $-3x^4y$ is -3. The coefficient of the term $14a^5b^9$ is 14.

> *Two terms are called like terms if they are both constant numbers or if they have the same variables and those variables are raised to identical powers.*

The terms $-3x^4y$ and $5x^4y$ are like terms, but the terms $-3x^4y$ and $5x^4y^2$ are not like terms because of the different powers on the variable y. The terms $2a^3b^2$ and $5a^3b^2$ are like terms, but the terms $2a^3b^2$ and $5a^5b^2$ are not like terms because of the different powers on the variable a.

The Distributive Property.   We will make frequent use of a property known as the distributive property.

$$a(b+c) = ab + ac$$

Mathematicians like to say that multiplication is distributive over addition. When a sum is multiplied by a number, each term of the sum must be multiplied by the number.

Example 16   Verify the following: $2\cdot(3+4) = 2\cdot3 + 2\cdot4$.

Solution   Note that $2\cdot(3+4) = 2\cdot7 = 14$. Note that $2\cdot3 + 2\cdot4 = 6+8 = 14$. Therefore, $2\cdot(3+4) = 2\cdot3 + 2\cdot4$.

Combining Like Terms.   When adding like terms, you will find the distributive property extremely useful. Let's try a few examples.

Example 17   Simplify the expression $3x^2 + 4x^2$.

Solution   Using the distributive property, the following argument can be made:

$$3x^2 + 4x^2 = (3+4)x^2$$
$$= 7x^2$$

In practice, when adding like terms, you simply add the coefficients of the terms, and attach the common variables and powers: $3x^2 + 4x^2 = 7x^2$.

Example 18   Simplify the expression $-5x^{2n} - 3x^{2n}$.

Solution   Using the distributive property, the following argument can be made:

$$-5x^{2n} - 3x^{2n} = (-5-3)x^{2n}$$
$$= -8x^{2n}$$

In practice, when adding like terms, you simply add the coefficients of the terms, and attach the common variables and powers: $-5x^{2n} - 3x^{2n} = -8x^{2n}$.

Example 19   Simplify the expression $3a^2 + 2a$.

Solution   Since there are no like terms, this expression cannot be simplified any further.

Example 20   Simplify the expression $3x + 8x^2 - 7 + 4x^2 - 5x - 11$ and arrange your answer in ascending powers of x.

Solution   Rearrange the terms, then use the distributive property.

$$3x + 8x^2 - 7 + 4x^2 - 5x - 11 = 8x^2 + 4x^2 + 3x - 5x - 7 - 11$$
$$= (8+4)x^2 + (3-5)x - 18$$
$$= 12x^2 - 2x - 18$$

The answer is in *descending* powers of x. The powers of the variable x in our terms are decreasing. In practice, when adding like terms, simply add the coefficients of the terms, and attach the common variables and their powers.

$$3x + 8x^2 - 7 + 4x^2 - 5x - 11 = -18 - 2x + 12x^2$$

Note that this answer has been arranged in *ascending* powers of x. The powers on the variable x are increasing.

Example 21   Simplify the expression $3x^2y + 4xy^2 - 7x^2y + 8xy^2$.

Solution   Rearrange the terms, then use the distributive property.

$$3x^2y + 4xy^2 - 7x^2y + 8xy^2 = 3x^2y - 7x^2y + 4xy^2 + 8xy^2$$
$$= (3-7)x^2y + (4+8)xy^2$$
$$= -4x^2y + 12xy^2$$

In practice, the coefficients of the like terms are added, and the common variables and powers are attached:

$$3x^2y + 4xy^2 - 7x^2y + 8xy^2 = -4x^2y + 12xy^2.$$

Moving a Little Faster. In actual practice, we move a little faster. When simplifying expressions, simply add the coefficients of the like terms and attach the common variables with their common powers.

Example 22   Simplify the expression $3x^5 + 8x^5$.

Solution $3x^5 + 8x^5 = 11x^5$. Simply add the coefficients and attach the common variable with its common power.

Example 23   Simplify the expression $3z^3 - 9z^3$.

Solution $3z^3 - 9z^3 = -6z^3$. Simply add the coefficients and attach the common variable with its common power.

Example 24   Simplify the expression $3x^4 + 8x^3$.

Solution   There are no like terms. The expression $3x^4 + 8x^3$ is already simplified.

Example 25   Simplify the expression $8x^{2n} + 3x^n - 9 + 2x^{2n} - 5x^n - 12$.

Solution   Simply add the like terms.

$$8x^{2n} + 3x^n - 9 + 2x^{2n} - 5x^n - 12 = 10x^{2n} - 2x^n - 21$$

Note that the answer has been arranged in descending powers of x.

Exercises for Section 4.1

1. (TI-82) Enter the following expressions on your calculator and verify that the results are as shown.

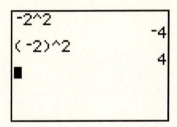

   Use your calculator to compute each of the following expressions. Record the problem and the answer on your homework paper.

   a) 2^{12} b) -3^{10} c) $(-4)^8$ d) -15^6 e) π^7 f) $(-3)^{10}$

2. (TI-82) Use your calculator to calculate each of the following expressions. Record both the problem and the answer on your homework paper.

   a) 5^0 b) -5^0 c) $(-5)^0$ d) 0^0 e) π^0 f) $10,000^0$

3. (TI-82) Copy each of the following expressions onto your homework paper. Use your calculator to compute each answer and place that on your homework paper as well.

   i) 10^0 ii) 10^1 iii) 10^2 iv) 10^3 v) 10^4 vi) 10^5

   a) Do you see a pattern? Use complete sentences to describe the pattern that is emerging.

   b) What does your pattern say about 10^{15}?

   c) Use your calculator to compute 10^{15}. Use complete sentences to explain the result in your viewing window.

4. (TI-82) Use your calculator to compute each of the following expressions. Copy the problem and the result onto your homework paper.

   a) 2^8 b) $2+2+2+2+2+2+2+2$ c) $2 \cdot 2 \cdot 2 \cdot 2 \cdot 2 \cdot 2 \cdot 2 \cdot 2$ d) 8×2

   Which of expressions (a), (b), (c), or (d) are equal?

5. (TI-82) Use your calculator to compute each of the following expressions. Copy the problem and the result onto your homework paper.

   a) 5^9 b) $5 \cdot 5 \cdot 5 \cdot 5 \cdot 5 \cdot 5 \cdot 5 \cdot 5 \cdot 5$ c) 9×5 d) $5+5+5+5+5+5+5+5+5$

   Which of expressions (a), (b), (c), or (d) are equal?

6. (TI-82) Use your calculator to compute $f(8)$ for each of the following functions.

   a) $f(x) = x^7$ b) $f(x) = 7^x$ c) $f(x) = 3x^6$ d) $f(x) = 3 \cdot 6^x$

7. Use the first law of exponents to simplify each of the following expressions:

   a) $x^8 x^{13}$

   b) $y^7 y^3$

   c) $\left(-3x^2 y^6\right)\left(2x^4 y^4\right)$

   d) $\left(-2xy^5\right)\left(-3x^6 y^6\right)$

   e) $\left(-2x^n\right)\left(3x^n\right)$

   f) $\left(-5x^{3n}\right)\left(2x^{5n}\right)$

8. Use the second law of exponents to simplify each of the following expressions.

   a) $\left(x^6\right)^7$ b) $\left(y^5\right)^8$ c) $\left(x^n\right)^2$ d) $\left(x^{2a}\right)^3$

9. Use the third law of exponents to simplify each of the following expressions.

   a) $\dfrac{x^8}{x^{10}}$ b) $\dfrac{a^{12}}{a^5}$ c) $\dfrac{x^5 y^9}{x^8 y^7}$ d) $\dfrac{a^7 b^3}{a^5 c^2}$ e) $\dfrac{z^{5n}}{z^{2n}}$ f) $\dfrac{x^{4a}}{x^{9a}}$

10. Use the fourth law of exponents to simplify each of the following expressions.

    a) $(xy)^9$

    b) $(ab)^4$

    c) $(-2x)^5$

    d) $(-3y)^4$

    e) $(-2xy)^4$

    f) $(5abc)^3$

11. Use the fifth law of exponents to simplify each of the following expressions.

    a) $\left(\dfrac{2}{y}\right)^4$ b) $\left(-\dfrac{x}{3}\right)^4$ c) $\left(\dfrac{x}{y}\right)^{2a}$ d) $\left(\dfrac{y}{z}\right)^{3b}$

12. Use the laws of exponents to simplify each of the following expressions.

    a) $\left(-2x^3\right)^2$

    b) $\left(3y^3\right)^4$

    c) $\left(\dfrac{2x^2}{5}\right)^3$

    d) $\left(-\dfrac{x^2 y^5}{4}\right)^2$

    e) $\left(2x^2 y\right)^2 \left(-3x^3 y^3\right)^3$

    f) $\left(-2a^7 b^2\right)^3 \left(3ab^5\right)^2$

    g) $\left(-2x^{3n}\right)^2$

    h) $\left(3x^n y^{2n}\right)^3$

    i) $\dfrac{\left(2x^2 y^3\right)^4}{\left(-3xy^5\right)^3}$

    j) $\left(\dfrac{3x^{5n}}{x^{3n}}\right)^2$

    k) $\left(x^{2n}\right)^n$

    l) $\left(y^{5n}\right)^{n^2}$

13. Use the *distributive property* to complete the blanks in each of the following problems. Write the completed problem on your homework paper.

a) $5x + 7x = \left(\underline{\quad} + \underline{\quad} \right)x$

   $= \underline{\quad}\, x$

b) $5a^2 - 8a^2 = \left(\underline{\quad} - \underline{\quad} \right)a^2$

   $= \underline{\quad}\, a^2$

c) $-4x^n + 7x^n = \left(\underline{\quad} + \underline{\quad} \right)x^n$

   $= \underline{\quad}\, x^n$

d) $10xy - 13xy = \left(\underline{\quad} - \underline{\quad} \right)xy$

   $= \underline{\quad}\, xy$

14. If possible, simplify each of the following expressions by adding coefficients.

a) $3x - 7x$

b) $-8y - 12y$

c) $2x^2 - 7x^2$

d) $7y^3 - 8y^2$

e) $7x^2y + 8xy^2$

f) $4x^{3n} - 9x^{3n}$

g) $-9x^4y + 18x^4y$

h) $-10x^{2a} - 12x^{2a}$

15. Simplify each of the following expressions by combining like terms. Arrange your answer in *descending* powers of x.

a) $x^2 + 3x - 11 + 4x^2 + x + 4$

b) $x^3 + 3x^2 - 11 + 5x^2 + 4x - 10$

c) $x^{3a} - 3 + 2x^{2a} - x^a + 4x^{2a} + 8x^{3a} - 11$

d) $3x^n + 4 - x^{2n} + 2x^n - 8 - 4x^{2n}$

16. Simplify each of the following expressions by combining like terms. Arrange your answer in *ascending* powers of y.

a) $4y^2 + 8 - 3y + 4 - y^2 + 9y$

b) $2y^4 - 3y^2 + 8 - 9y^3 - 7y^2 + y^3 - 11$

c) $y^{3n} - 5 + y^{2n} - 4y^n + 8 + 2y^{3n} + 4y^n$

d) $y^{3a} - 5 + y^a - 4y^{3a} - y^{2a} - 11 - 5y^a + y^{3a}$

17. Simplify each of the following expressions by combining like terms. Arrange your answer in some sort of reasonable order.

a) $x^2 + 4xy - 8y^2 + x^2 + 11xy - 9y^2$

b) $x^3 + x^2y - y^3 + 4x^2y + 2xy^2 - 8y^3 + 5x^3$

18. Simplify each of the following expressions.

a) $x^5 \cdot x^5$

b) $x^5 + x^5$

c) $x^2 \cdot x^3$

d) $x^2 + x^3$

e) $x^n \cdot x^n$

f) $x^n + x^n$

g) $\left(2x^{3n}\right)\left(-3x^{3n}\right)$

h) $2x^{3n} - 3x^{3n}$

19. Simplify each of the following expressions.

a) $\left(-3x^{n+1}\right)\left(2x^{2n+3}\right)$

b) $\left(-3x^{2n}\right)^3\left(2x^{n+1}\right)$

c) $\dfrac{x^{3n+2}}{x^{2n}}$

d) $\left(x^{3n}\right)^n\left(x^{3n+2}\right)^2$

20. (TI-82) Enter the following expressions on your calculator and verify that the results are as shown.

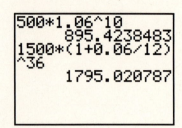

Copy the following expressions onto your homework paper. Use your calculator to find decimal approximations of each expression correct to the nearest hundredth (two decimal places).

a) $1,000 \cdot 1.02^{10}$

b) $250 \cdot 0.98^{20}$

c) $2,000 \cdot \left(1 + \dfrac{0.05}{4}\right)^{20}$

d) $900 \cdot \left(1 - \dfrac{0.05}{6}\right)^{12}$

21. A Superball is dropped from a ladder and the height it reaches on each bounce is given by the equation $h = 15(0.90)^n$, where h is the height the ball reaches on its nth bounce. What height does the ball reach on its tenth bounce? Round your answer to the nearest tenth of a foot (one decimal place).

22. The amount of money in Jenny's savings account is given by the equation $A = 100(1.06)^t$, where A is the number of dollars in the account and t is the number of years that the money has been in the account. How much is in Jenny's account after 10 years?

23. The amount of antibiotic in Sarah's bloodstream t hours after her injection is given by the equation $A = 500(0.85)^t$, where A is measured in milligrams. How much of the antibiotic is in Sarah's bloodstream 6 hours after her injection?

24. The amount of money in Jedediah's savings account t years after opening the account is given by $A = 12,000\left(1 + \dfrac{0.05}{6}\right)^{2t}$, where A is measured in dollars. How much is in Jedediah's account after 10 years?

25. Jenny keeps her retirement savings in a shoe box beneath the floorboards in her bedroom. The amount of money in the shoe box is given by the equation $A = 125,000(1 - 0.12)^t$, where A is the amount of money in the shoe box in dollars and t is the number of years that Jenny has been retired. How much money will be in the shoe box 5 years after Jenny retires?

26. Randolph has invested a portion of his savings in treasury bills. The value of the investment, in dollars, is given by the equation $V = 5,000\left(1 + \dfrac{0.055}{2}\right)^{2t}$, where t is the number of years that have passed since the money was first invested. What is the value of the investment 5 years after the money was first invested?

4.2 Operations with Polynomials

This section introduces an extremely important class of functions called polynomials. After some introductory definitions, some elementary operations involving polynomial functions will be performed.

Polynomials

The *terms* of an algebraic expression are the parts of the expression that are separated by addition. For example, the terms of the expression $-3x^2 + 5xy + 7y^2$ are $-3x^2$, $5xy$, and $7y^2$. The *coefficient* of the term $-3x^2$ is -3. The coefficient of the term $5xy$ is 5. The coefficient of the term $7y^2$ is 7.

Example 1   Identify the terms and coefficients of each term in the following expression:

$$-3a^3 - 4a^2b + 7ab^2 - 8b^3$$

Solution   This expression must be thought of as a sum of four terms, as follows:

$$-3a^3 - 4a^2b + 7ab^2 - 8b^3 = -3a^3 + (-4a^2b) + 7ab^2 + (-8b^3)$$

The coefficients of the terms $-3a^3$, $-4a^2b$, $7ab^2$, and $-8b^3$ are -3, -4, 7, and -8, respectively.

Defining Polynomials. The numbers $\{0, 1, 2, 3, 4, 5, \ldots\}$ are called the *nonnegative integers*. If all of the terms of an algebraic expression are of the form ax^k, where a is a real number and k is a nonnegative integer, then the expression is called a polynomial in x.

The expression $a_n x^n + a_{n-1} x^{n-1} + \cdots a_2 x^2 + a_1 x + a_0$, where each of the a_k is a real number and n is a nonnegative integer, is called a polynomial in x. The degree of this polynomial is n.

Tip	It is possible to have polynomials in more than one variable. For example, $a^3 + 3a^2b + 3ab^2 + b^3$ is a polynomial in a and b. However, most of our work will deal with polynomials of a single variable.

Example 2   Which of the following algebraic expressions are polynomials in x? If the algebraic expression is a polynomial, state its degree.

$$x^2 - 3x + 5$$

$$x^3 + 3x^2 - 4x^{-1} + 8$$

$$x^5 - 8x^3 + x - 7$$

$$4x^3 - 7x^{\frac{1}{2}} + 8x - 9$$

Solution   The expression $x^2 - 3x + 5$ is a polynomial because each of the terms is of the form ax^k, with a a real number and k a nonnegative integer. The degree of this polynomial is 2. The expression

$x^3 + 3x^2 - 4x^{-1} + 8$ is *not* a polynomial because the exponent in the term $-4x^{-1}$ is *not* a nonnegative integer. The expression $x^5 - 8x^3 + x - 7$ is a polynomial because each of the terms is of the form ax^k, with a a real number and k a nonnegative integer. The degree of this polynomial is 5. The expression $4x^3 - 7x^{\frac{1}{2}} + 8x - 9$ is *not* a polynomial because the exponent in the term $-7x^{\frac{1}{2}}$ is *not* a nonnegative integer.

Some Special Descriptive Names. If a polynomial has only one term, it is classified as a *monomial*. A polynomial with two terms is called a *binomial*. A polynomial with three terms is called a *trinomial*. If a polynomial has more than three terms, we'll simply call it a *polynomial*.

Adding and Subtracting Polynomials

When adding and subtracting polynomials, you simply combine the like terms. Let's begin with an example.

Example 3   If $p(x) = x^2 - 8x - 9$ and $q(x) = 2x^2 + 4x - 10$, find $p(x) + q(x)$ and $p(x) - q(x)$.

Solution   Adding the polynomials p and q is easy. Rearrange the terms and combine the like terms.

$$p(x) + q(x) = \left(x^2 - 8x - 9\right) + \left(2x^2 + 4x - 10\right)$$

$$= \left(x^2 + 2x^2\right) + (-8x + 4x) + (-9 - 10)$$

$$= 3x^2 - 4x - 19$$

When subtracting polynomials p and q, remember to distribute the subtraction sign.

$$p(x) - q(x) = \left(x^2 - 8x - 9\right) - \left(2x^2 + 4x - 10\right)$$

$$= x^2 - 8x - 9 - 2x^2 - 4x + 10$$

$$= \left(x^2 - 2x^2\right) + (-8x - 4x) + (-9 + 10)$$

$$= -x^2 - 12x + 1$$

Example 4   If $p(x) = 4x^3 - 8x^2 + 7$ and $q(x) = -2x^3 + 4x^2 - 7x - 11$, find $p(x) + q(x)$ and $p(x) - q(x)$.

Solution   When adding the polynomials p and q, rearrange the terms and combine the like terms.

$$p(x) + q(x) = \left(4x^3 - 8x^2 + 7\right) + \left(-2x^3 + 4x^2 - 7x - 11\right)$$

$$= \left(4x^3 - 2x^3\right) + \left(-8x^2 + 4x^2\right) - 7x + (7 - 11)$$

$$= 2x^3 - 4x^2 - 7x - 4$$

When subtracting the polynomials p and q, remember to distribute the subtraction sign.

$$p(x) - q(x) = \left(4x^3 - 8x^2 + 7\right) - \left(-2x^3 + 4x^2 - 7x - 11\right)$$

$$= 4x^3 - 8x^2 + 7 + 2x^3 - 4x^2 + 7x + 11$$

$$= \left(4x^3 + 2x^3\right) + \left(-8x^2 - 4x^2\right) + 7x + \left(7 + 11\right)$$

$$= 6x^3 - 12x^2 + 7x + 18$$

Moving a Little Faster. In the preceding examples, the terms were rearranged before combining them. In practice it is better to move a little faster and skip the rearrangement.

Example 5 Suppose that $p(x) = x^2 - 8x - 11$ and $q(x) = x^2 - 10x + 9$. Solve the following inequality for x: $p(x) - q(x) \le 0$. Sketch your solution on a number line.

Solution This time, when subtracting the polynomials p and q, combine the like terms without rearranging.

$$p(x) - q(x) \le 0$$

$$\left(x^2 - 8x - 11\right) - \left(x^2 - 10x + 9\right) \le 0$$

$$x^2 - 8x - 11 - x^2 + 10x - 9 \le 0$$

$$2x - 20 \le 0$$

Now add 20 to both sides of the inequality and divide both sides of the result by 2.

$$2x \le 20$$

$$x \le 10$$

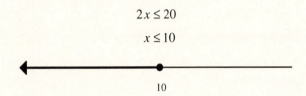

Example 6 Simplify the expression $\left(x^2 - 4xy - 8y^2\right) - \left(2x^2 - xy + y^2\right) - \left(4x^2 - 8xy - 5y^2\right)$.

Solution You can distribute the subtraction signs and combine like terms without rearranging the terms.

$$\left(x^2 - 4xy - 8y^2\right) - \left(2x^2 - xy + y^2\right) - \left(4x^2 - 8xy - 5y^2\right)$$

$$= x^2 - 4xy - 8y^2 - 2x^2 + xy - y^2 - 4x^2 + 8xy + 5y^2$$

$$= -5x^2 + 5xy - 4y^2$$

Note that the result is a polynomial in x and y.

Multiplying Polynomials

We are going to be making extensive use of the *distributive property*. Remember, multiplication is distributive over addition.

The distributive property: $a(b + c) = ab + ac$

Example 7 Simplify the expression $2(3x-5)$.

Solution When you multiply an expression by a number, you must multiply each term of the expression by that number.

$$2(3x-5) = 2(3x) - 2(5)$$
$$= 6x - 10$$

Example 8 Simplify the expression $2x(3x+7)$.

Solution When you multiply an expression by an expression, you must multiply each term of the expression by that expression.

$$2x(3x+7) = 2x(3x) + 2x(7)$$
$$= 6x^2 + 14x$$

Example 9 If $p(x) = 5x^2 - 3x - 8$, simplify the expression $3x^2 p(x)$.

Solution When you multiply an expression by an expression, you must multiply each term of the expression by that expression.

$$3x^2 p(x) = 3x^2(5x^2 - 3x - 8)$$
$$= 3x^2(5x^2) - 3x^2(3x) - 3x^2(8)$$
$$= 15x^4 - 9x^3 - 24x^2$$

Example 10 Simplify the expression $3ab(a^2 + 2ab + b^2)$.

Solution When you multiply an expression by an expression, you must multiply each term of the expression by that expression.

$$3ab(a^2 + 2ab + b^2) = 3ab(a^2) + 3ab(2ab) + 3ab(b^2)$$
$$= 3a^3b + 6a^2b^2 + 3ab^3$$

Example 11 Simplify the expression $3x^{3n}(3x^{2n} - 8x^n + 5)$.

Solution When you multiply an expression by an expression, you must multiply each term of the expression by that expression.

$$3x^{3n}(3x^{2n} - 8x^n + 5) = 3x^{3n}(3x^{2n}) - 3x^{3n}(8x^n) + 3x^{3n}(5)$$
$$= 9x^{5n} - 24x^{4n} + 15x^{3n}$$

Moving a Little Faster. In the last few examples, we have written down the distribution. In practice, most people like to perform the distribution mentally, writing the answer as they go. Let's try this in a few examples.

Example 12 If $p(x) = 3x^2 - 7x - 8$, simplify the expression $2p(x)$.

Solution Mentally distribute the 2.

$$2p(x) = 2\left(3x^2 - 7x - 8\right)$$

$$= 6x^2 - 14x - 16$$

Note how each term was multiplied by 2.

Example 13 Simplify the expression $5x^3(3x^2 - 2x - 7)$.

Solution Mentally distribute the $5x^3$.

$$5x^3(3x^2 - 2x - 7) = 15x^5 - 10x^4 - 35x^3$$

Note how each term was multiplied by $5x^3$.

Example 14 Simplify the expression $2xy\left(3x^3 - 2x^2y + 5xy^2 - 8y^3\right)$.

Solution Mentally distribute the $2xy$.

$$2xy\left(3x^3 - 2x^2y + 5xy^2 - 8y^3\right) = 6x^4y - 4x^3y^2 + 10x^2y^3 - 16xy^4$$

Note how each term was multiplied by $2xy$.

Up to this point, the problems have been restricted to multiplying a polynomial by a monomial. Let's try multiplying two arbitrary polynomials together.

Example 15 Simplify the expression $(x+5)(x-9)$.

Solution If you wish to multiply the expression $x+5$ by the expression $x-9$, then you must begin by multiplying each term of the expression $x+5$ by the expression $x-9$, as follows.

$$(x+5)(x-9) = x(x-9) + 5(x-9)$$

$$= x^2 - 9x + 5x - 45$$

$$= x^2 - 4x - 45$$

Note the second application of the distributive property and note how like terms were combined.

Example 16 Simplify the expression $(x+3)\left(x^2 + 2x + 5\right)$.

Solution If you wish to multiply the expression $x+3$ by the expression $x^2 + 2x + 5$, then you must begin by multiplying each term of the expression $x+3$ by the expression $x^2 + 2x + 5$, as follows.

$$(x+3)\left(x^2 + 2x + 5\right) = x\left(x^2 + 2x + 5\right) + 3\left(x^2 + 2x + 5\right)$$

$$= x^3 + 2x^2 + 5x + 3x^2 + 6x + 15$$

$$= x^3 + 5x^2 + 11x + 15$$

Note the second application of the distributive property and note how like terms were combined.

Example 17 If $p(x) = x^2 - 4x - 8$ and $q(x) = x^2 + x - 4$, simplify the expression $p(x)q(x)$.

Solution If you wish to multiply $p(x)$ by $q(x)$, you must begin by multiplying each term of $p(x)$ by $q(x)$, as follows.

$$\begin{aligned}
p(x)q(x) &= (x^2 - 4x + 8)(x^2 + x - 4) \\
&= x^2(x^2 + x - 4) - 4x(x^2 + x - 4) + 8(x^2 + x - 4) \\
&= x^4 + x^3 - 4x^2 - 4x^3 - 4x^2 + 16x + 8x^2 + 8x - 32 \\
&= x^4 - 3x^3 + 24x - 32
\end{aligned}$$

Note the second application of the distributive property and note how like terms were combined.

Example 18 The following diagram shows two concentric rectangles.

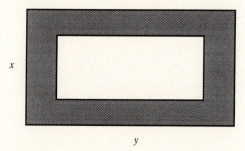

Let x represent the width of the outer rectangle and let y represent its length. Suppose the shaded region has a uniform width of 2 inches. Express the area of the shaded region as a function of x and y. Simplify your answer as much as possible.

Solution If the shaded region has a uniform width of 2 inches, then the width of the inner rectangle is $x - 4$ and the length of the inner rectangle is $y - 4$.

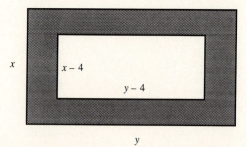

You can find the area of the shaded region by subtracting the area of the inner rectangle from the area of the outer rectangle. The area of the shaded region, as a function of x and y, is given by the equation $A = xy - (x - 4)(y - 4)$. The distributive property will now be used to simplify the answer.

$$A = xy - (x-4)(y-4)$$
$$A = xy - \left[x(y-4) - 4(y-4)\right]$$
$$A = xy - \left[xy - 4x - 4y + 16\right]$$
$$A = xy - xy + 4x + 4y - 16$$
$$A = 4x + 4y - 16$$

Exercises for Section 4.2

1. Explain why each of the following expressions is *not* a polynomial in x.

   a) $p(x) = x^{-3} + 2x^2 + 4x - 7$

   b) $p(x) = 4x^3 + 3x^{\frac{1}{2}} - 7x - 8$

   c) $p(x) = 2x^2 - 3x^{0.5} - 9$

   d) $p(x) = \dfrac{x^2 + 3x - 8}{x^2 - 4}$

2. Copy each of the following polynomials onto your homework paper and clearly state the degree.

   a) $p(x) = 2x^2 - 3x - 4$

   b) $p(x) = 2x + 5$

   c) $p(x) = x^4 - 3x^2 - 8x - 11$

   d) $p(x) = x^3 - 3x^2 - 2x + 8$

3. Arrange each of the following polynomials in *ascending* powers of x.

   a) $p(x) = 3 - 2x^2 + 4x$

   b) $p(x) = 3 - 2x^2 - 4x + x^3$

4. Arrange each of the following polynomials in *descending* powers of x.

   a) $p(x) = x + 5x^2 - 4$

   b) $p(x) = 2 + x^3 - 3x^2 + 4x$

5. Simplify each of the following expressions as much as possible. Arrange your answer in *descending* powers of x.

   a) $\left(2x^2 - 3x - 8\right) + \left(4x^2 - 4x + 5\right)$

   b) $\left(2x^2 + 3x - 6\right) - \left(x^2 - 4x + 9\right)$

   c) $\left(x^3 - 3x + 9\right) + \left(2x^2 - 4x - 10\right)$

   d) $\left(4x^3 + 3x^2 - 4x - 8\right) - \left(x^3 + 2x^2 - 8x + 9\right)$

6. Find and simplify $p(x) + q(x)$ and $p(x) - q(x)$ as much as possible for each of the following pairs of polynomial functions p and q.

   a) $p(x) = x + 1$
      $q(x) = x^2 - 8x - 8$

   b) $p(x) = 2x - 3$
      $q(x) = 5 - 4x - x^2$

   c) $p(x) = 3x^2 - 8x - 11$
      $q(x) = x^2 - 4x + 5$

   d) $p(x) = x^3 - 8x^2 - x - 1$
      $q(x) = -x^3 - 2x^2 + 5x + 7$

7. Simplify each of the following expressions as much as possible. Arrange your answer in *descending* powers of x.

a) $\left(7x - 8x^2 + x^3 - 9\right) + \left(2 - 3x + 5x^3 - 4x^2\right)$

b) $\left(3x - x^3 + 7x^2 - 5\right) - \left(9x^2 - 4x^3 - 2x^2\right)$

c) $\left(x^3 - 4x^2 + 5x - 8\right) + \left(2x^3 - 8x^2 - 9x + 11\right) - \left(4x^3 + 11x^2 - 9x - 9\right)$

d) $\left(x^4 - 2x^2 + 11\right) - \left(x^3 - 2x^2 - 9\right) - \left(x^4 - 3x^3 - 8x^2 - 19\right)$

8. Simplify each of the following expressions as much as possible. Arrange your answers in some sort of reasonable order.

a) $\left(x^2 + 2xy + 3y^2\right) + \left(2x^2 - 3xy - 9y^2\right)$

b) $\left(2x^2 - 3xy - 4y^2\right) - \left(4x^2 + 5xy + 9y^2\right)$

c) $\left(a^3 - 3a^2b + 3ab^2 - b^3\right) - \left(a^3 + 2a^2b - 4ab^2 + 5b^3\right) + \left(2a^3 + 3a^2b - 11ab^2 - 9b^3\right)$

d) $\left(x^4 - 4x^3y + 2xy^3 - y^4\right) - \left(2x^3y + 8x^2y^2 - 2y^4\right) - \left(2x^4 - x^2y^2 - 5xy^3 + 8y^4\right)$

9. Find the solution of the inequality $p(x) - q(x) \leq 0$ for each of the following pairs of polynomials p and q. Sketch your solution on a number line and use set-builder and interval notation to describe your solution set.

a) $p(x) = x^2 - 4x - 8$
   $q(x) = x^2 + 5x - 9$

b) $p(x) = x^3 - x^2 + 8x + 11$
   $q(x) = x^3 - x^2 - 5x - 9$

10. If $p(x) = x^2 + 2x + 4$ and $q(x) = x^2 - 3x - 1$, find the solution of each of the following inequalities. Sketch your solution on a number line and use set-builder and interval notation to describe your solution set.

a) $\left|p(x) - q(x)\right| \leq 3$

b) $\left|p(x) - q(x)\right| > 5$

11. Use the distributive property to help simplify each of the following expressions.

a) $2(x + 3)$

b) $x(2x + 5)$

c) $x\left(x^2 + 2x + 3\right)$

d) $x^2\left(x^2 - 2x - 5\right)$

e) $4x^2\left(3x^2 + 2x - 9\right)$

f) $3x^3\left(2x^3 - 3x - 9\right)$

g) $x^n\left(x^{2n} + 3x^n + 5\right)$

h) $x^{2n}\left(x^{2n} - 3x^n - 7\right)$

12. If $p(x) = x^2 - 3x - 8$, simplify each of the following expressions.

a) $3p(x)$

b) $-5p(x)$

c) $5x^2 p(x)$

d) $-7x^3 p(x)$

13. Use the distributive property to help simplify each of the following multiplication problems. Arrange your answer in descending powers of x.

a) $(x+2)(x+5)$

b) $(2x-5)(3x+7)$

c) $x+2(x-8)$

d) $(x+2)(x-8)$

e) $(x+2)(x^2-4x-5)$

f) $(2x-3)(3x^2-4x+1)$

g) $(x^2-4x-5)(x^2+x+4)$

h) $(2x^2-x-1)(x^2+4x+7)$

i) $(x+1)(x+2)(x+4)$

j) $(x-2)(2x-3)(x-5)$

k) $(x-1)(x^3+x^2+x+1)$

l) $(x-1)(x^2+x+1)(x+1)(x^2-x-1)$

14. If $p(x)=x+1$ and $q(x)=x^2-8x-8$, simplify each of the following expressions.

a) $p(x)+q(x)$

b) $p(x)-q(x)$

c) $p(x)q(x)$

d) $p(x)p(x)$

e) $(x+3)q(x)$

f) $x^2p(x)-3q(x)$

g) $(x+2)p(x)-xq(x)$ h) $q(x)q(x)$

15. Simplify each of the following expressions.

a) $(x^n+1)(x^{2n}-3x^n+2)$

b) $(x^{2n}+3x^n+1)(x^{2n}-4x^n+5)$

c) $(x^a-y^a)(x^{2a}+x^a y^a+y^{2a})$

d) $(x^n+y^n)(x^{3n}-x^{2n}y^n+x^n y^{2n}-y^{3n})$

16. (TI-82) The gold medal winning times in the women's Olympic 100-meter dash from the year 1928 through the year 1988 can be approximated by the following third-degree polynomial.

$$T = -0.00001354t^3 + 0.001290t^2 - 0.08093t + 12.11$$

The variable T represents the winning time in seconds and the variable t represents the number of years that have passed since 1928. Use this polynomial to predict the gold medal winning time in the year 1968.

Note: The actual winning time was 11.0 seconds, by Wyomia Tyus from the United States.

17. (TI-82) The gold medal winning time in the men's 1500-meter run in the Olympics from the year 1900 to the year 1992 can be approximated by the following third-degree polynomial.

$$T = 0.00007062t^3 - 0.006609t^2 - 0.2893t + 245.5$$

The variable T represents the gold medal winning time in seconds and the variable t represents the number of years that have passed since 1900. Use this polynomial to predict the winning gold medal time in the year 1968.

Note: The actual winning time was 214.9 seconds, by Kipchoge Keino from Kenya.

18. (TI-82) The number of new AIDS cases reported in California in each of the years from 1981 through 1991 can be approximated by the following third-degree polynomial.

$$N = -23.02t^3 + 316.6t^2 - 317.6t + 122.9$$

The variable N represents the number of AIDS cases reported in a particular year and t represents the number of years that have passed since 1981. Use this polynomial to predict the number of AIDS cases reported in the year 1985.

Note: The actual number of new cases reported in the year 1985 was 2298.

19. Consider the concentric squares in the following diagram:

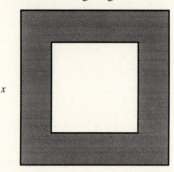

Suppose that the shaded region has a uniform width of 3 inches.

a) Express the area of the shaded region as a function of x.

b) Simplify the equation found in part (a) as much as possible.

c) If the area of the shaded region is 60 square inches, use the result developed in part (b) to find the length of the edge of the outer square.

20. Consider the concentric rectangles in the following diagram:

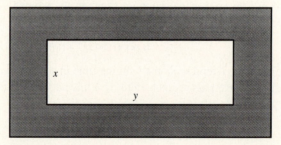

Let x and y represent the width and length of the inner rectangle, respectively.

a) If the shaded region has a uniform width of 4 inches, express the area of the shaded region as a function of x and y.

b) Simplify the equation found in part (a) as much as possible.

21. (TI-82) The minimum wage for workers in the United States from the year 1975 through the year 1981 can be approximated by the following fourth-degree polynomial.

$$W = 0.0001895t^4 - 0.006439t^3 + 0.05246t^2 + 0.09862t + 2.010$$

The variable W represents the minimum wage in dollars and the variable t represents the number of years that have passed since 1975. Use this polynomial to predict the minimum wage in the year 1977.

Note: The actual minimum wage in 1977 was $2.30 per hour.

22. (TI-82) Mathematicians have long been fascinated by prime numbers. A prime number is an integer whose only divisors are 1 and itself. Many mathematicians have tried to produce equations that will produce prime numbers. Consider the following second-degree polynomial.

$$p(x) = x^2 + x + 41$$

a) Use your calculator to compute $p(0)$, $p(1)$, $p(2)$, $p(3)$, $p(4)$, and $p(5)$. Are each of these values prime numbers?

b) Will this polynomial produce prime numbers for all nonnegative integers x?

4.3 Introduction to Factoring

In this section you will learn how to factor polynomials. It important to have a thorough understanding of the word factoring before you begin. In the last section, you were given two or more polynomials and were asked to find their product. This section introduces the *opposite* process. You will be given the product and you will be asked to find the original multiplication problem.

> ***In a certain sense, factoring is unmultiplying.***

Example 1   Factor 21.

Solution   You have been given the product and you must find the original multiplication problem. Note that this number is divisible by 3. To find the second factor or divisor, divide 21 by 3: 21 divided by 3 is 7. Therefore, 21 factors in the following manner:

$$21 = 3 \cdot 7$$

The numbers 3 and 7 are called *factors* or *divisors* of the number 21.

Example 2   A student is asked to factor 17. The student responds with $17 = 15 + 2$. Critique this solution.

Solution   The student has not expressed 17 as a *multiplication* problem. On the contrary, the student has expressed 17 as an *addition* problem. This is not factoring. When you factor, you are supposed to respond with a multiplication problem.

> ***When asked to factor, you must find the original multiplication problem.***

The First Rule of Factoring

When you are attempting to factor a polynomial, you must first look for common divisor of each term of the polynomial. Preferably, try to find the greatest common divisor of each term.

Example 3   Factor the expression $2x + 6$.

Solution   The terms of the polynomial are $2x$ and 6. They are both divisible by 2. To find the second factor, divide each term of $2x + 6$ by 2: $2x$ divided by 2 is x, and 6 divided by 2 is 3. Therefore, $2x + 6$ factors in the following manner:

$$2x + 6 = 2(x + 3)$$

Tip	When you factor, immediately multiply to check your work.

Check. Always check your work before moving on. Use the distributive property to multiply.

$$2(x+3) = 2(x) + 2(3)$$
$$= 2x + 6$$

The answer checks.

Example 4   Factor the expression $10x^2 + 15x + 20$.

Solution   The terms of the polynomial are each divisible by 5. To find the second factor, divide each term of $10x^2 + 15x + 20$ by 5: $10x^2$ divided by 5 is $2x^2$, $15x$ divided by 5 is $3x$, and 20 divided by 5 is 4. Therefore $10x^2 + 15x + 20$ factors in the following manner:

$$10x^2 + 15x + 20 = 5(2x^2 + 3x + 4)$$

Check. Multiply immediately to check the solution.

$$5(2x^2 + 3x + 4) = 5(2x^2) + 5(3x) + 5(4)$$
$$= 10x^2 + 15x + 20$$

The solution checks.

Example 5   Factor the expression $6x^2 + 9x$.

Solution   The terms of the polynomial are each divisible by $3x$. To find the second factor, divide each term of $6x^2 + 9x$ by $3x$: $6x^2$ divided by $3x$ is $2x$; $9x$ divided by $3x$ is 3. Therefore, $6x^2 + 9x$ factors in the following manner:

$$6x^2 + 9x = 3x(2x+3)$$

Check. Multiply immediately to check the solution.

$$3x(2x+3) = 3x(2x) + 3x(3)$$
$$= 6x^2 + 9x$$

The answer checks.

Example 6   Factor the expression $xz + 2z$.

Solution   The terms of the expression are both divisible by z. To find the second factor, divide each term of $xz + 2z$ by z: xz divided by z is x, $2z$ divided by z is 2. Therefore, $xz + 2z$ factors in the following manner:

$$xz + 2z = (x+2)z$$

Because $(x+2)z = z(x+2)$, it does not matter whether the factor of z is placed before or after $x+2$. Did you check this solution yet?

Example 7   Factor the expression $x(x-3)+2(x-3)$.

Solution   Note that there is no difference in structure between this example and the last one. The terms of the expression are both divisible by $x-3$. To find the second factor, divide each term of $x(x-3)+2(x-3)$ by $x-3$: $x(x-3)$ divided by $x-3$ is x, $3(x-3)$ divided by $x-3$ is 2. Therefore, $x(x-3)+2(x-3)$ factors in the following manner.

$$x(x-3)+2(x-3) = (x+2)(x-3)$$

Check. Multiply immediately to check the solution.

$$(x+2)(x-3) = x(x-3)+2(x-3)$$

The solution checks.

Factoring by Grouping

When you encounter an expression with four terms, you should try the method of grouping. Try to factor out a common factor from two of the terms and another common factor from the remaining two terms. Let's examine this technique in the next example.

Example 8   Factor the expression $2x^2 +2x-3x-3$.

Solution   This expression has four terms. Note that our first two terms are divisible by $2x$ and the second two terms are divisible by 3. Begin by factoring a $2x$ out of the first two terms and a -3 out of the second two terms.

$$2x^2 +2x-3x-3 = 2x(x+1)-3(x+1)$$
$$= (2x-3)(x+1)$$

Note that the problem was completed by factoring out the common factor $x+1$.

Check. Let's multiply to check the result.

$$(2x-3)(x+1) = 2x(x+1)-3(x+1)$$
$$= 2x^2 +2x-3x-3$$

The solution checks.

Example 9   Factor the expression $6x^2 -9x-16x+24$.

Solution   Because this is a four-term expression, try the method of grouping. The first two terms are divisible by $3x$ and the second two terms are divisible by 8. Begin by factoring a $3x$ out of the first two terms and a -8 out of the second two terms.

$$6x^2 -9x-16x+24 = 3x(2x-3)-8(2x-3)$$
$$= (3x-8)(2x-3)$$

Note that the problem was completed by factoring out the common factor $2x-3$.

Check. Multiply to check your result.

$$(3x - 8)(2x - 3) = 3x(2x - 3) - 8(2x - 3)$$

$$= 6x^2 - 9x - 16x + 24$$

The solution checks.

Solving Nonlinear Equations

Why do you need to learn how to factor algebraic expressions? There are several answers to this question. In this section you will find that factoring is a powerful tool to use when attempting to solve a class of equations called *nonlinear* equations. These equations are called nonlinear because the graphs involved in their solution are not lines. How do you spot a nonlinear equation?

The equations in this section will involve polynomials in x. Recall that a polynomial is an algebraic expression of the form $p(x) = a_n x^n + \cdots + a_2 x^2 + a_1 x + a_0$. The degree of this polynomial is n. Two cases will be considered: polynomials of degree 1 and polynomials of degree greater than 1.

Polynomials of Degree 1. If the polynomial is of degree 1, then the polynomial has the form $p(x) = a_1 x + a_0$. This is the same as the slope-intercept form of a line, $y = mx + b$. Therefore, if the degree of the polynomial is 1, the graph of the polynomial will be a line.

Polynomials of Degree Greater than 1. If the polynomial has degree greater than 1, it cannot have the form $y = mx + b$. Therefore the graph of the polynomial cannot be a line. The graph will be nonlinear. As you soon will see, the graph will usually be some sort of curve.

Degree	Equation Class
1	Linear equation
Greater than 1	Nonlinear equation

Before continuing with our discussion of nonlinear equations, let's pause to discuss an important problem-solving tool.

The Zero Product Property. Suppose that two or more numbers are multiplied together and the answer is zero? What must be true? Certainly, at least one of the numbers must be zero.

Zero product property: If ab = 0, then a = 0 or b = 0.

Note the use of the word *or*.

Tip	The zero product property is a powerful problem-solving tool.

Example 10 Use the zero product property to solve the following equation for x: $(x - 1)(x - 2) = 0$.

Solution The product is zero; therefore at least one of the factors must be zero.

$$(x-1)(x-2) = 0$$
$$x-1 = 0 \text{ or } x-2 = 0$$
$$x = 1 \text{ or } \quad x = 2$$

Check Your Solution. First, check the solution $x = 1$. Substituting 1 for x in the original equation results in the following:

$$(x-1)(x-2) = 0$$
$$(1-1)(1-2) = 0$$
$$(0)(-1) = 0$$
$$0 = 0$$

Therefore $x = 1$ checks. Now check the second solution, $x = 2$. Substituting $x = 2$ in the original equation results in the following:

$$(x-1)(x-2) = 0$$
$$(2-1)(2-2) = 0$$
$$(1)(0) = 0$$
$$0 = 0$$

Therefore $x = 2$ checks.

Equation-Solving Strategies. The last example illustrates that the zero product property is a powerful problem-solving tool. It is important to note that one side of the equation must equal zero and the other side of the equation must be in factored form before this tool can be applied. This suggests an important strategy for solving nonlinear equations.

> *If your equation is nonlinear, make one side of the equation zero and factor the remaining side of the equation. Finally, apply the zero product property.*

Example 11   *(TI-82)* Solve the following equation for x: $x^2 = -4x$.

Graphical Solution   The equation involves a power of x higher than one, so the equation is nonlinear. Make one side of the equation zero by adding $4x$ to each side of the equation.

$$x^2 + 4x = 0$$

This equation is of the form $f(x) = 0$. When you are asked to find where a function is equal to zero, you are being asked to find where the graph of that function crosses the x-axis. Load the equation into the Y= menu in the following manner: $Y_1 = X^2 + 4X$. Push the ZOOM button and select 6:ZStandard, capturing the following image in your viewing window:

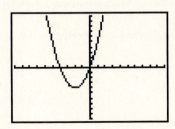

It appears that the graph of $f(x) = x^2 + 4x$ crosses the x-axis at -4 and at 0.

Copy this image onto your homework paper in the following manner.

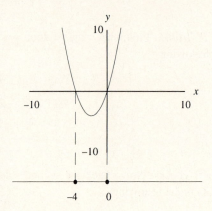

The solutions of this problem are $x = -4$ and $x = 0$.

Analytical Solution   First factor out the GCF, then use the zero product property.

$$x^2 + 4x = 0$$

$$x(x+4) = 0$$

$$x = 0 \ \text{ or } \ x + 4 = 0$$

$$x = 0 \ \text{ or } \qquad x = -4$$

Place your solutions side-by-side on your homework paper for comparison.

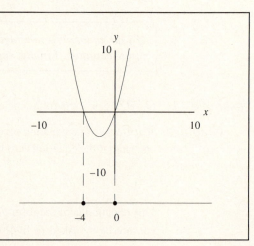

$$x^2 + 4x = 0$$

$$x(x+4) = 0$$

$$x = 0 \ \text{ or } \ x + 4 = 0$$

$$x = 0 \ \text{ or } \qquad x = -4$$

It's pretty clear these two solutions agree. However, you should check the solutions in the original equation to be absolutely certain you have found the correct answers to the equation.

Check Your Solution. First substitute $x = 0$ in the original equation.

$$x^2 = -4x$$

$$(0)^2 = -4(0)$$

$$0 = 0$$

Therefore, $x = 0$ checks. Now substitute $x = -4$ in the original equation.

$$x^2 = -4x$$

$$(-4)^2 = -4(-4)$$

$$16 = 16$$

Therefore, $x = -4$ checks.

Example 12   (TI-82) Solve the following equation for x: $6x(x+2) - 5(x+2) = 0$.

Graphical Solution   At first glance, you might think that this equation is linear. After all, there do not seem to be any powers of x present that are greater than 1. However, on closer inspection, if you were to begin multiplying, your first term would be $6x^2$. Therefore, this equation is nonlinear. Because one side of the equation is already zero, begin by loading the left side of the equation into the Y= menu in the following manner: $Y_1 = 6X(X+2) - 5(X+2)$. Push the ZOOM button and select 6:ZStandard, capturing the following image in your viewing window:

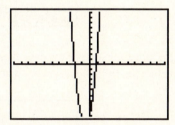

This graph is not what would be called a complete graph. For one thing, you cannot see the turning point at the bottom of the graph. Also, there is a great deal of white space that is not needed. Make the following changes to your WINDOW parameters:

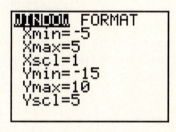

Push the GRAPH button to capture the following image in your viewing window:

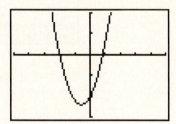

This viewing window is much better. It appears that the x-intercepts are approximately -2 and 1. To find more accurate estimates of these x-intercepts, use the root-finding utility in the CALC menu on your calculator. Begin by selecting 2:root in the CALC menu. Your calculator should respond with the following viewing window:

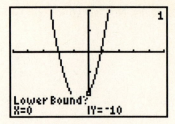

The 1 in the upper-right corner indicates that the trace cursor is currently on the graph of Y₁. The calculator is asking for a lower bound for the root. Find the root that is near 1. Move your cursor a little to the left of this root, as shown.

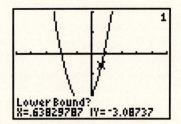

The *x*-value of the trace cursor is now less than the *x*-value of the root or *x*-intercept. Press the ENTER key to capture the following viewing window:

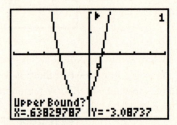

Note that the mark at the top of the viewing window represents the lower bound that was selected. The calculator is now asking for an upper bound for the root. Move the trace cursor a little to the right of the *x*-intercept or root, as shown.

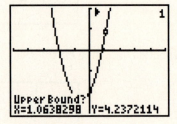

The *x*-value of the trace cursor is now greater than the *x*-value of the *x*-intercept. Press the ENTER key to capture the following viewing window:

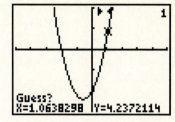

Note that the new mark at the top of the viewing window represents the upper bound that was just selected. The calculator now requires that you attempt a guess at the root or x-intercept. You must guess an x-value which is between the upper and lower bounds. Leave the trace cursor right where it is for your guess and press the ENTER key. The calculator responds with the following viewing window:

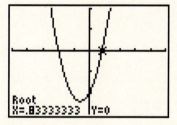

In a similar manner, you can find the second root.

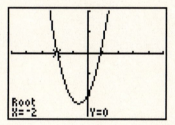

Copy this information onto your homework paper in the following manner:

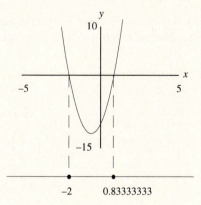

It appears that the solutions of the equation are approximately -2 and 0.83333333.

Analytical Solution   Factor out the common factor of $x+2$, then use the zero product property.

$$6x(x+2)-5(x+2)=0$$
$$(6x-5)(x+2)=0$$
$$6x-5=0 \text{ or } x+2=0$$
$$6x=5 \text{ or } \quad x=-2$$
$$x=\frac{5}{6} \text{ or } \quad x=-2$$

The solution $x=-2$ agrees quite nicely with the graphical estimate. Using your calculator, divide 5 by 6 to find a decimal approximation for 5/6.

$$\frac{5}{6}=0.8333333333...$$

This solution also compares quite favorably with the graphical estimate found with the root-finding utility of the TI-82.

Place your solutions side-by-side on your homework for comparison.

$$6x(x+2)-5(x+2)=0$$
$$(6x-5)(x+2)=0$$
$$6x-5=0 \text{ or } x+2=0$$
$$6x=5 \text{ or } \quad x=-2$$
$$x=\frac{5}{6} \text{ or } \quad x=-2$$

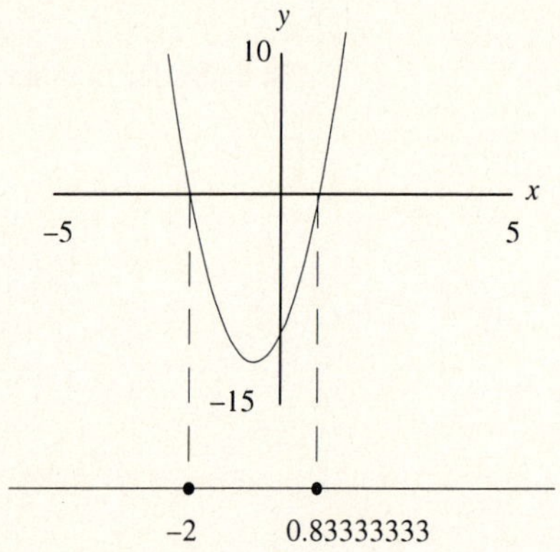

You should feel very confident about your solutions, but take a moment to check the solutions in the original equation. That way, you will be certain that your solution is correct.

Check Your Solution. Substitute $x=\frac{5}{6}$ in the original equation.

$$6x(x+2)-5(x+2)=0$$
$$6\left(\frac{5}{6}\right)\left(\frac{5}{6}+2\right)-5\left(\frac{5}{6}+2\right)=0$$
$$5\left(\frac{5}{6}+2\right)-5\left(\frac{5}{6}+2\right)=0$$
$$0=0$$

The solution $x=\frac{5}{6}$ checks. Substitute $x=-2$ in the original equation.

$$6x(x+2) - 5(x+2) = 0$$
$$6(-2)(-2+2) - 5(-2+2) = 0$$
$$-12(0) - 5(0) = 0$$
$$0 = 0$$

The solution $x = -2$ checks.

Example 13   *(TI-82)* Solve the following equation for x: $6x^2 + 15x - 14x - 35 = 0$.

Graphical Solution   This equation is of the form $f(x) = 0$. Sketch the graph of this function and note where it crosses the x-axis. Begin by loading the function into the Y= menu in the following manner: Y₁ = $6X^2 + 15X - 14X - 35$. Make the following changes to your WINDOW parameters.

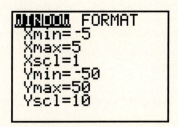

Use the root-finding utility in the CALC menu to find one of the roots.

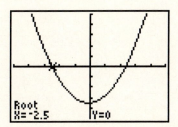

Use the root-finding utility in the CALC menu to find the other root.

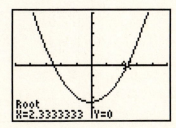

Place all of this information on your homework paper in the following manner:

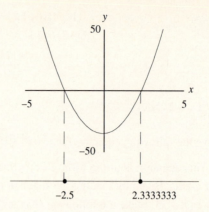

It appears that approximate solutions of the equation are $x = -2.5$ and $x = 2.33333333$.

Analytical Solution   Use the method of grouping to factor. Note that the first two terms are divisible by $3x$ and the second two terms are divisible by 7. Factor a $3x$ out of the first two terms and a -7 out of the second two terms.

$$6x^2 + 15x - 14x - 35 = 0$$

$$3x(2x+5) - 7(2x+5) = 0$$

Now factor out the common factor $2x+5$ and apply the zero product property.

$$(3x - 7)(2x + 5) = 0$$

$$3x - 7 = 0 \quad \text{or} \quad 2x + 5 = 0$$

$$3x = 7 \quad \text{or} \quad 2x = -5$$

$$x = \frac{7}{3} \quad \text{or} \quad x = -\frac{5}{2}$$

Use your calculator to divide 7 by 3 and -5 by 2 to find decimal approximations for these solutions. Note that they agree quite nicely with the estimates found with the graphing method.

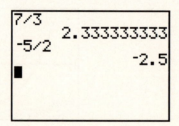

Place your solutions side-by-side on your homework for comparison.

$$6x^2 + 15x - 14x - 35 = 0$$

$$3x(2x+5) - 7(2x+5) = 0$$

$$(3x-7)(2x+5) = 0$$

$$3x - 7 = 0 \quad \text{or} \quad 2x + 5 = 0$$

$$3x = 7 \quad \text{or} \quad 2x = -5$$

$$x = \frac{7}{3} \quad \text{or} \quad x = -\frac{5}{2}$$

Use Your TI-82 to Check Your Solutions.

```
-5/2→X
              -2.5
6X²+15X-14X-35
               0
■
```

That's pretty good evidence that $x = -\dfrac{5}{2}$ satisfies the equation.

```
7/3→X
        2.333333333
6X²+15X-14X-35
               0
```

And that's pretty good evidence that $x = \dfrac{7}{3}$ satisfies the equation.

Note: There is a big difference between factoring an expression and solving an equation.

Example. Factor the expression $x^2 - 12x$.

Solution.

$$x^2 - 12x = x(x - 12)$$

Check. Multiply to check your answer.

$$x(x - 12) = x^2 - 12x$$

The solution checks.

Example. Solve for x: $x^2 - 12x = 0$.

Solution. After factoring, use the zero product property to complete the solution.

$$x^2 - 12x = 0$$
$$x(x - 12) = 0$$

$$x = 0 \text{ or } x - 12 = 0$$
$$x = 0 \text{ or } \qquad x = 12$$

Check. Check your solutions $x = 0$ and $x = 12$ in the original equation.

$$0^2 - 12(0) = 0 \quad 12^2 - 12(12) = 0$$
$$0 - 0 = 0 \qquad 144 - 144 = 0$$
$$0 = 0 \qquad\qquad 0 = 0$$

The solutions $x = 0$ and $x = 12$ both check.

Exercises for Section 4.3

1. Copy each of the following problems onto your homework paper and complete them. Check each problem by multiplying. Show this check work on your homework paper.

   a) $5x + 10 = 5(\underline{\quad} + \underline{\quad})$

   b) $6x + 9 = 3(\underline{\quad} + \underline{\quad})$

   c) $14x^2 + 21x - 35 = 7(\underline{\quad} + \underline{\quad} - \underline{\quad})$

   d) $24x^2 - 36x - 48 = 12(\underline{\quad} - \underline{\quad} - \underline{\quad})$

   e) $4x^{2a} - 8x^a = 4x^a(\underline{\quad} - \underline{\quad})$

   f) $x^{4n} - 5x^{3n} - 7x^{2n} = x^{2n}(\underline{\quad} - \underline{\quad} - \underline{\quad})$

   g) $6x^2y - 9xy^2 = 3xy(\underline{\quad} - \underline{\quad})$

   h) $14a^3b - 21a^2b^2 + 28ab^3 = 7ab(\underline{\quad} - \underline{\quad} + \underline{\quad})$

2. Copy each of the following expressions onto your homework paper and factor out the greatest common factor. Check your work by multiplying. Show this check work on your homework paper.

   a) $4x + 8$

   b) $10x + 15$

   c) $12x^2 - 8x - 4$

   d) $8y^2 - 16y + 24$

e) $z^2 - 9z$

f) $12x^2 + 16x$

g) $3x^3 + 9x^2 - 24x$

h) $8z^3 + 4z^2 - 8z$

i) $x^{2n} - 4x^n$

j) $3x^{3n} + 6x^{2n} - 9x^n$

Note: You are factoring expressions here, not solving equations.

3. Copy each of the following problems onto your homework paper and complete them. Check each problem by multiplying. Show this check work on your homework paper.

    a) $5x(x-9) - 7(x-9) = \left(\underline{\quad} - \underline{\quad} \right)(x-9)$

    b) $11x(x+8) + 3(x+8) = \left(\underline{\quad} + \underline{\quad} \right)(x+8)$

    c) $4x(2x-9) - (2x-9) = \left(\underline{\quad} - \underline{\quad} \right)(2x-9)$

    d) $7x(2x-7) + (2x-7) = \left(\underline{\quad} + \underline{\quad} \right)(2x-7)$

4. Copy each of the following expressions onto your homework paper and factor out the greatest common factor. Check your work by multiplying. Show this check work on your homework paper.

    a) $7x(x+2) - 5(x+2)$

    b) $8y(2y-3) - 9(2y-3)$

    c) $2x(2x+11) + (2x+11)$

    d) $3z(5z-1) - (5z-1)$

Note: You are factoring expressions here, not solving equations.

5. Copy each of the following problems onto your homework paper and complete them. Check each problem by multiplying. Show this check work on your homework paper.

    a) $\begin{aligned} 8x^2 + 36x - 10x - 45 &= 4x\left(\underline{\quad} + \underline{\quad} \right) - 5\left(\underline{\quad} + \underline{\quad} \right) \\ &= (4x-5)\left(\underline{\quad} + \underline{\quad} \right) \end{aligned}$

    b) $\begin{aligned} 6x^2 - 33x + 10x - 55 &= 3x\left(\underline{\quad} - \underline{\quad} \right) + 5\left(\underline{\quad} - \underline{\quad} \right) \\ &= (3x+5)\left(\underline{\quad} - \underline{\quad} \right) \end{aligned}$

    c) $\begin{aligned} 2x^2 - 4x - 9x + 18 &= \underline{\quad}(x-2) - \underline{\quad}(x-2) \\ &= \left(\underline{\quad} - \underline{\quad} \right)(x-2) \end{aligned}$

    d) $\begin{aligned} 4x^2 + 8x - 5x - 10 &= \underline{\quad}(x+2) - \underline{\quad}(x+2) \\ &= \left(\underline{\quad} - \underline{\quad} \right)(x+2) \end{aligned}$

6. Factor each of the following expressions by using the method of grouping. Check your work by multiplying. Show this check work on your homework paper.

    a) $5x^2 + 10x + 2x + 4$

    b) $3y^2 + 18y + 4y + 24$

    c) $2x^2 + 10x - 3x - 15$

    d) $3z^2 - 12z + 2z - 8$

e) $6x^2 - 8x + 15x - 20$

f) $16x^2 + 56x + 6x + 21$

g) $6y^2 + 27y - 2y - 9$

h) $12x^2 + 20x + 3x + 5$

Note: You are factoring expressions here, not solving equations.

7. Make a copy of the following table on your homework paper. Complete this table of points that satisfy the equation $f(x) = (x+2)(x-3)$.

x	−4	−3	−2	−1	0	1	2	3	4	5
$f(x)$										
Points to plot										

a) Set up a coordinate system on a sheet of graph paper and plot each of the points from the table on your coordinate system. Plot the rest of the points that satisfy the equation of the function f.

b) Draw a number line below your graph and shade and label the solutions of $f(x) = 0$ on your number line in the usual manner.

c) Use the zero product property to help solve the nonlinear equation $(x+2)(x-3) = 0$. Do these analytical solutions compare favorably with your estimates from part (b)? If not, check your work for error.

8. Copy the following nonlinear equations onto your homework paper, then use the zero product property to help find the solutions.

a) $(x-8)(x+10) = 0$

b) $(x-11)(x-12) = 0$

c) $(5x+11)(3x-17) = 0$

d) $(4x+7)(7x+11) = 0$

Note: You are solving equations here.

9. Make a copy of the following table on your homework paper. Complete this table of points that satisfy the equation $f(x) = x^2 - 4x$.

x	−2	−1	0	1	2	3	4	5	6
$f(x)$									
Points to plot									

a) Set up a coordinate system on a sheet of graph paper and plot each of the points from the table on your coordinate system. Plot the rest of the points that satisfy the equation of the function f.

b) Draw a number line below your graph and shade and label the solutions of $f(x) = 0$ on your number line in the usual manner.

c) Factor, then use the zero product property to help solve the nonlinear equation $x^2 - 4x = 0$. Do these analytic solutions compare favorably with your estimates from part (b)? If not, check your work for error.

10. (TI-82) Load the following function into the Y= menu: $Y_1 = X^2 + 6X$. Press the ZOOM button and select 6:ZStandard. Set up a coordinate system on your homework paper and copy the image in your viewing window onto your coordinate system. Label your graph with Y_1 and include the WINDOW parameters on your axes.

a) Draw a number line below your graph and shade and label the solutions of $Y_1 = 0$ on your number line in the usual manner.

b) Factor, then use the zero product property to help find the solutions of the nonlinear equation $x^2 + 6x = 0$. Do these analytical solutions compare favorably with your estimates from part (a)? If not, check your work for error.

11. Use an analytical method only to find the solutions of each of the following nonlinear equations.

a) $x^2 + 8x = 0$ b) $x^2 - 14x = 0$ c) $4x^2 = 3x$

d) $5x^2 = -8x$ e) $12x^2 = 24x$ f) $6x^2 = -9x$

Note: You are solving equations here.

12. Make a copy of the following table on your homework paper. Complete this table of points that satisfy the equation $f(x) = x(x+1) - 4(x+1)$.

x	−3	−2	−1	0	1	2	3	4	5	6
$f(x)$										
Points to plot										

a) Set up a coordinate system on a sheet of graph paper and plot each of the points from the table on your coordinate system. Plot the rest of the points that satisfy the equation of the function f.

b) Draw a number line below your graph and shade and label the solutions of $f(x) = 0$ on your number line in the usual manner.

c) Factor, then use the zero product property to help solve the nonlinear equation $x(x+1) - 4(x+1) = 0$. Do these analytical solutions compare favorably with your estimates from part (b)? If not, check your work for error.

13. (TI-82) Load the following function into the Y= menu: $Y_1 = X(X-3) + 2(X-3)$. Press the ZOOM button and select 6:ZStandard. Set up a coordinate system on your homework paper and copy the image in your viewing window onto your coordinate system. Label your graph with Y_1 and include the WINDOW parameters on your axes.

a) Draw a number line below your graph and shade and label the solutions of $Y_1 = 0$ on your number line in the usual manner.

b) Factor, then use the zero product property to help find the solutions of the nonlinear equation $x(x-3) + 2(x-3) = 0$. Do these analytical solutions compare favorably with your estimates from part (a)? If not, check your work for error.

14. Use an analytical method only to find the solutions of each of the following nonlinear equations.

a) $x(x+5) - 2(x+5) = 0$ b) $x(x-7) + 11(x-7) = 0$

c) $2x(3x+1) - 7(3x+1) = 0$ d) $9x(5x+8) - 7(5x+8) = 0$

Note: You are solving equations here.

15. Make a copy of the following table on your homework paper. Complete this table of points that satisfy the equation $f(x) = 2x^2 - 10x + x - 5$.

x	-2	-1	0	1	2	3	4	5	6	7
$f(x)$										
Points to plot										

a) Set up a coordinate system on a sheet of graph paper and plot each of the points from the table on your coordinate system. Plot the rest of the points that satisfy the equation of the function f.

b) Draw a number line below your graph and shade and label the solutions of $f(x) = 0$ on your number line in the usual manner.

c) Factor, then use the zero product property to help solve the nonlinear equation $2x^2 - 10x + x - 5 = 0$. Do these analytic solutions compare favorably with your estimates from part (b)? If not, check your work for error.

16. (TI-82) Load the following function into the Y= menu: $Y_1 = 6X^2 - 4X + 3X - 2$. Press the ZOOM button and select 6:ZStandard. Set up a coordinate system on your homework paper and copy the image in your viewing window onto your coordinate system. Label your graph with Y_1 and include the WINDOW parameters on your axes.

a) Use the root-finding utility in the CALC menu to find an estimate of each root of the function Y_1. Draw a number line below your graph and shade and label the solutions of $Y_1 = 0$ on your number line in the usual manner.

b) Factor, then use the zero product property to help find the solutions of the nonlinear equation $6x^2 - 4x + 3x - 2 = 0$. Do these analytical solutions compare favorably with your estimates from part (a)? If not, check your work for error.

17. Use an analytical method only to find the solutions of each of the following nonlinear equations.

a) $3x^2 + 5x + 6x + 10 = 0$ b) $2x^2 - 7x + 8x - 28 = 0$

c) $2x^2 + 6x - 5x - 15 = 0$ d) $15x^2 + 33x - 10x - 22 = 0$

Note: You are solving equations here.

18. (TI-82) For each of the following functions, perform each of the following tasks:

i) Load the function in the Y= menu and draw a complete graph of the function. If necessary, adjust the WINDOW parameters. Set up a coordinate system on your homework paper and copy the image in your viewing window onto your coordinate system. Label your graph and include the WINDOW parameters on your axes.

ii) Use the root-finding utility in the CALC menu to estimate each of the roots of the function. Use the fraction utility in the MATH menu to change each of these roots into fractional form. (See Exercise 14, in Section 2.5 for a review of this utility.) Draw a number line below your graph and shade and label your roots on your number line.

iii) Use an analytical method to find the roots of the function. If these analytical solutions do not compare favorably with those found in part (ii), check your work for error.

a) $f(x) = 5x^2 + 6x$ b) $f(x) = 2x(3x+7) - 7(3x+7)$

c) $f(x) = 15x^2 + 10x - 12x - 8$ d) $f(x) = 8x^2 - 6x + 20x - 15$

19. Use the appropriate method to factor each of the following expressions. Be sure to check your work.

a) $4y^2 - 12y$

b) $12z^2 - 18z$

c) $9x^2z - 6xz^2$

d) $6a^3b^2 + 9a^2b^3$

e) $6x^2 + 9x - 18$

f) $20y^3 - 40y^2 + 30y$

g) $6x^{2n} - 18x^n$

h) $9y^{3b} - 2y^{2b}$

i) $6x^3y - 15x^2y^2 + 3xy^2$

j) $6x^{4n} - 2x^{3n} + 14x^{2n}$

k) $11x(12x-5) - (12x-5)$

l) $21x(3x+11) - 4(3x+11)$

m) $6x^2 + 15x - 2x - 5$

n) $20x^2 + 8x - 15x - 6$

o) $x^{2n} - 5x^n + 6x^n - 10$

p) $18z^{2a} - 6z^a + 15z^a - 5$

Note: You are factoring expressions here, not solving equations.

4.4 Factoring Trinomials

This section introduces a method called the *ac* test that is used to factor trinomials of the form $ax^2 + bx + c$. You will find this new technique quite helpful when solving nonlinear equations. We'll continue to use the TI-82 graphing calculator to help compare graphical solutions and analytical solutions. This section finishes with some applications of the *ac* test. Let's begin with a description of the *ac* test.

The ac Test. The ac test is a process that is used to factor trinomials of the form $ax^2 + bx + c$. It is essentially a three-step process.

1. *If possible, find a pair of integers whose product is ac and whose sum is b.*
2. *Use this integer pair to express the middle term bx as a sum.*
3. *Factor by grouping.*

Example 1   Factor the expression $x^2 - 4x - 32$.

Solution   If you compare the trinomial $x^2 - 4x - 32$ with $ax^2 + bx + c$, you can see that $a = 1$, $b = -4$, and $c = -32$. The product ac equals -32. Your first job is to list all of the possible integer pairs whose product is -32.

$$-1,32 \quad\quad 1,-32$$
$$-2,16 \quad\quad 2,-16$$
$$-4,8 \quad\quad\ \ 4,-8$$

Tip	Try to be systematic when listing the integer pairs whose product is -32. List them in order as shown in the examples. Otherwise, you might miss the pair you are searching for.

Because the coefficient of the middle term is $b = -4$, choose the integer pair from your list whose sum is -4. The required pair of integers is $\{4, -8\}$ The product of these integers is -32 and the sum of these integers is -4. Use this pair $\{4, -8\}$ to write the middle term of the trinomial as a sum.

$$x^2 - 4x - 32 = x^2 + 4x - 8x - 32$$

Note that $4x - 8x$ equals $-4x$. Use the method of grouping to complete the factoring of this expression.

$$
\begin{aligned}
x^2 - 4x - 32 &= x^2 + 4x - 8x - 32 \\
&= x(x+4) - 8(x+4) \\
&= (x-8)(x+4)
\end{aligned}
$$

Check. As always, check immediately by multiplying.

$$
\begin{aligned}
(x-8)(x+4) &= x(x+4) - 8(x+4) \\
&= x^2 + 4x - 8x - 32 \\
&= x^2 - 4x - 32
\end{aligned}
$$

The problem checks.

Example 2 Factor the expression $2x^2 + x - 6$.

Solution If you compare the trinomial $2x^2 + x - 6$ with $ax^2 + bx + c$, you can see that $a = 2$, $b = 1$, and $c = -6$. The product ac equals -12. Your first job is to list all possible integer pairs whose product is -12.

$$
\begin{array}{ll}
-1, 12 & 1, -12 \\
-2, 6 & 2, -6 \\
-3, 4 & 3, -4
\end{array}
$$

Because the coefficient of the middle term is $b = 1$, you must choose an integer pair from the list whose sum is 1. The required pair of integers is $\{-3, 4\}$. The product of these integers is -12 and the sum of these integers is 1. Use the pair $\{-3, 4\}$ to write the middle term of the trinomial as a sum.

$$2x^2 + x - 6 = 2x^2 - 3x + 4x - 6$$

Note that $-3x + 4x$ equal x. Use the method of grouping to complete the factoring of this expression.

$$
\begin{aligned}
2x^2 + x - 6 &= 2x^2 - 3x + 4x - 6 \\
&= x(2x-3) + 2(2x-3) \\
&= (x+2)(2x-3)
\end{aligned}
$$

Check. Multiply immediately to check your result.

$$
\begin{aligned}
(x+2)(2x-3) &= x(2x-3) + 2(2x-3) \\
&= 2x^2 - 3x + 4x - 6 \\
&= 2x^2 + x - 6
\end{aligned}
$$

The solution checks.

Example 3   Factor the expression $6x^2 - x + 12$.

Solution   If you compare the trinomial $6x^2 - x + 12$ with $ax^2 + bx + c$, you can see that $a = 6$, $b = -1$, and $c = 12$. The product ac equals 72. Your first job is to list all integer pairs whose product is 72.

$$
\begin{array}{ll}
1,72 & -1,-72 \\
2,36 & -2,-36 \\
3,24 & -3,-24 \\
4,18 & -4,-18 \\
6,12 & -6,-12 \\
8,9 & -8,-9
\end{array}
$$

Note that there is no pair in the list whose sum is equal to the coefficient of the middle term, $b = -1$. Therefore, the trinomial $6x^2 - x + 12$ does not factor.

Example 4   Factor the expression $3x^2 + 5x - 12$.

Solution   Try to work a little faster by doing some of the work mentally. If you compare the trinomial $3x^2 + 5x - 12$ with $ax^2 + bx + c$, you can see that $a = 3$, $b = 5$, and $c = -12$. The product ac equals -36. You must find an integer pair whose product is -36 and whose sum is 5. The integer pair $\{-4, 9\}$ comes to mind.

Tip	When using the ac test, try to guess an integer pair whose product is ac and whose sum is b. However, if it is taking too long to guess, you should begin to systematically list the pairs whose product is ac.

Use the pair $\{-4, 9\}$ to write the middle term of the trinomial as a sum.

$$3x^2 + 5x - 12 = 3x^2 - 4x + 9x - 12$$

Note that $-4x + 9x$ equals $5x$. Use the method of grouping to complete the factoring of the expression.

$$
\begin{aligned}
3x^2 + 5x - 12 &= 3x^2 - 4x + 9x - 12 \\
&= x(3x - 4) + 3(3x - 4) \\
&= (x + 3)(3x - 4)
\end{aligned}
$$

Check.   Multiply immediately to check the result.

$$
\begin{aligned}
(x + 3)(3x - 4) &= x(3x - 4) + 3(3x - 4) \\
&= 3x^2 - 4x + 9x - 12 \\
&= 3x^2 + 5x - 12
\end{aligned}
$$

The solution checks.

Example 5   *(TI-82)* Solve the following equation for x: $x^2 + 3x = 10$.

Graphical Solution   Because this equation contains a power of x greater than one, the equation is nonlinear. Therefore, make one side of the equation zero by subtracting 10 from both sides.

$$x^2 + 3x - 10 = 0$$

Load this function in the Y= menu in the following manner: $Y_1 = X^2 + 3X - 10$. Pressing the ZOOM button and selecting 6:ZStandard produces the following image in your viewing window.

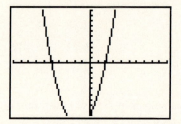

This viewing window fails to show the turning point of the curve. After some experimentation, the following settings of the WINDOW parameters are found to produce a satisfactory viewing window.

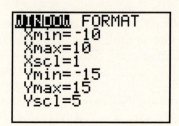

Push the GRAPH button to produce the following image in your viewing window:

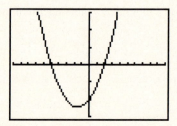

It appears that the graph crosses the x-axis at $x = -5$ and $x = 2$.

Place these results on your homework paper in the following manner:

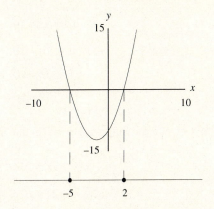

Analytical Solution   If you compare $x^2 + 3x - 10$ with $ax^2 + bx + c$, you can see that $a = 1$, $b = 3$, and $c = -10$. The product ac is -10. Find an integer pair whose product is -10 and whose sum is 3. The integer pair $\{5, -2\}$ comes to mind. Use the pair $\{5, -2\}$ to write the middle term of the trinomial as a sum.

$$x^2 + 3x - 10 = 0$$

$$x^2 + 5x - 2x - 10 = 0$$

Note that $5x - 2x$ equals $3x$. Use the method of grouping to complete the factoring of this expression.

$$x(x+5) - 2(x+5) = 0$$

$$(x-2)(x+5) = 0$$

Finish the solution of this equation by applying the zero product property.

$$x - 2 = 0 \text{ or } x + 5 = 0$$

$$x = 2 \text{ or } x = -5$$

Place your solutions side by side for comparison.

$x^2 + 3x - 10 = 0$ $x^2 + 5x - 2x - 10 = 0$ $x(x+5) - 2(x+5) = 0$ $(x-2)(x+5) = 0$ $x - 2 = 0 \text{ or } x + 5 = 0$ $x = 2 \text{ or } \quad x = -5$	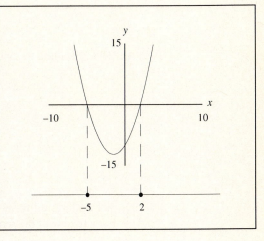

Check Your Solutions. Substitute $x = -5$ in the original equation $x^2 + 3x = 10$.

$$x^2 + 3x = 10$$
$$(-5)^2 + 3(-5) = 10$$
$$25 - 15 = 10$$
$$10 = 10$$

The solution $x = -5$ checks. Substitute $x = 2$ in the original equation $x^2 + 3x = 10$.

$$x^2 + 3x = 10$$
$$(2)^2 + 3(2) = 10$$
$$4 + 6 = 10$$
$$10 = 10$$

The solution $x = 2$ checks.

Example 6   *(TI-82)* Solve the following equation for x: $4x^2 + 6 = 25x$.

Solution   Because this equation contains a power of x greater than one, the equation is nonlinear. Make one side of the equation zero by subtracting $25x$ from both sides.

$$4x^2 - 25x + 6 = 0$$

Enter this function in the Y= menu in the following manner: $Y_1 = 4X^2 - 25X + 6$. Pushing the ZOOM button and selecting 6:ZStandard produces an unsatisfactory viewing window. After some experimenting, the following settings of the WINDOW parameters are found to produce a satisfactory viewing window.

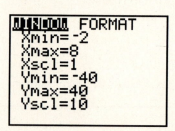

Push the GRAPH button to capture the following image in your viewing window:

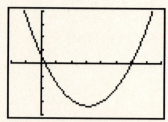

Use the root-finding utility in the CALC menu to produce the following image in your viewing window. For a review of the root-finding utility, see Example 12 in Section 4.3.

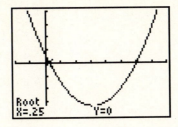

Use the root-finding utility in the CALC menu to produce the following image in your viewing window:

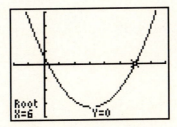

It appears that this graph crosses the x-axis at approximately $x = 0.25$ and $x = 6$. Arrange this information on your homework paper in the following manner:

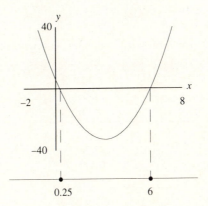

Analytical Solution   When the trinomial $4x^2 - 25x + 6$ is compared with $ax^2 + bx + c$, you can see that $a = 4$, $b = -25$, and $c = 6$. The product ac equals 24. You must find an integer pair whose product is 24 and whose sum is -25.

Tip	When factoring $ax^2 + bx + c$, you must find an integer pair whose product is ac and whose sum is b. If you can guess this integer pair, fine. If not, you can always create a list of integer pairs whose product is ac. Then the required integer pair should be apparent.

If nothing comes to mind, systematically list all the integer pairs whose product is 24.

1,24	−1,−24
2,12	−2,−12
3,8	−3,−8
4,6	−4,−6

The required integer pair is $\{-1,-24\}$. Use this integer pair to write the middle term of the trinomial as a sum.

$$4x^2 - 25x + 6 = 0$$
$$4x^2 - x - 24x + 6 = 0$$

Note that $-x - 24x$ equals $-25x$. Use the method of grouping to complete the factoring of this expression.

$$x(4x-1) - 6(4x-1) = 0$$
$$(x-6)(4x-1) = 0$$

The solution of this equation can be finished by applying the zero product property.

$$x - 6 = 0 \text{ or } 4x - 1 = 0$$
$$x = 6 \text{ or } \quad x = \frac{1}{4}$$

If you divide 1 by 4 on your calculator, you will find that $\frac{1}{4} = 0.25$. This solution, as well as $x = 6$, compare quite nicely with the approximate solutions found using the graphing method.

Place your solutions side-by-side for comparison.

$$4x^2 - 25x + 6 = 0$$
$$4x^2 - x - 24x + 6 = 0$$
$$x(4x-1) - 6(4x-1) = 0$$
$$(x-6)(4x-1) = 0$$

$$x - 6 = 0 \text{ or } 4x - 1 = 0$$
$$x = 6 \text{ or } x = \frac{1}{4}$$

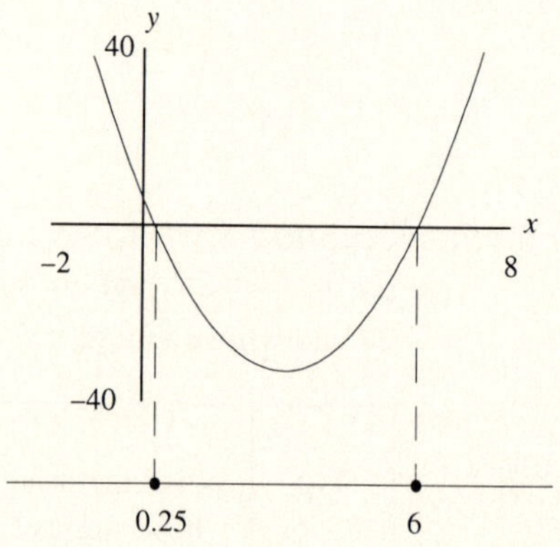

Use Your TI-82 to Check Your Solutions. Store 6 in X then evaluate both sides of the original equation $4x^2 + 6 = 25x$.

```
6→X
            6
4X²+6
         150
25X
         150
```

Because the left and right sides of the equation are equal, the solution $x = 6$ checks. You can check the solution $x = \frac{1}{4}$ in a similar manner.

```
1/4→X
          .25
4X²+6
         6.25
25X
         6.25
```

The solution $x = \dfrac{1}{4}$ checks.

Example 7   Consider the following rectangle, with width represented by the variable W and length represented by the variable L.

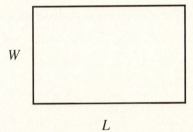

Suppose that you are given that the length of this rectangle is 1 foot longer than twice its width.

    a) Express the length of this rectangle as a function of W.

    b) Express the area of this rectangle as a function of W.

    c) If the area of the rectangle is 21 square feet, find the dimensions of the rectangle.

Solution

    a) Because the length of the rectangle is 1 foot longer than twice its width, you can express the length L as a function of the width W in the following manner.

$$L = 2W + 1 \qquad \text{Eq. (1)}$$

    b) Because the area of the rectangle is equal to the product of its width and length, you can express the area A as a function of length L and width W in the following manner.

$$A = LW \qquad \text{Eq. (2)}$$

However, you are asked to find the area as a function of W alone. If you substitute *Eq.* (1) into *Eq.* (2), the following result will be obtained:

$$A = (2W + 1)W$$

or

$$A = 2W^2 + W$$

You have now expressed the area A as a function of the width W.

c) Because the area is 21, you can let $A = 21$ in the equation for the area of the rectangle developed in part (b).

$$21 = 2W^2 + W$$

Because the powers of W are larger than 1, this last equation is nonlinear. Make one side of the equation zero by subtracting 21 from both sides.

$$0 = 2W^2 + W - 21$$

If you compare the trinomial $2W^2 + W - 21$ with $aW^2 + bW + c$, you can see that $a = 2$, $b = 1$, and $c = -21$. The product ac is -42. You must find an integer pair whose product is -42 and whose sum is 1. The integer pair $\{-6, 7\}$ comes to mind. Use this pair to write the middle term of the trinomial as a sum.

$$2W^2 + W - 21 = 0$$

$$2W^2 - 6W + 7W - 21 = 0$$

Note that $-6W + 7W = W$. Use the method of grouping to complete the factoring of this expression.

$$2W(W - 3) + 7(W - 3) = 0$$

$$(2W + 7)(W - 3) = 0$$

Apply the zero product to complete the solution of the nonlinear equation.

$$2W + 7 = 0 \quad \text{or} \quad W - 3 = 0$$

$$2W = -7 \quad \text{or} \quad W = 3$$

$$W = -\frac{7}{2} \quad \text{or} \quad W = 3$$

Practical Considerations. Because width cannot be a negative number, you can eliminate the solution $W = -\frac{7}{2}$ from further consideration. The width of this rectangle is 3 feet. Because the length of the rectangle is 1 foot longer than twice its width, $L = 2(3) + 1$ or $L = 7$ feet.

Is Your Solution Reasonable? Because the dimensions of the rectangle are 3 feet by 7 feet, the area would be 21 square feet. The solution seems to be reasonable.

Exercises for Section 4.4

1. Each of the following trinomials is of the form $ax^2 + bx + c$. Perform the following tasks for each trinomial:

i)   List all the integer pairs whose product is ac. Circle the integer pair whose sum equals b.

ii) Use your circled integer pair to express the term bx as a sum. Factor the resulting expression by grouping.

iii) Multiply to check your work. Show this check work on your homework paper.

a) $x^2 - 3x - 40$

b) $x^2 - 5x - 24$

c) $3x^2 - 11x - 20$

d) $3x^2 + 14x + 16$

e) $x^2 - 9x + 16$

f) $x^2 + 16x + 48$

g) $6x^2 + 5x - 6$

h) $2x^2 - 17x + 36$

i) $x^{2n} - 6x^n - 27$

j) $4x^{2n} + 4x^n - 15$

Note: You are factoring expressions here, not solving equations for x.

2. Try to guess an integer pair with the indicated product and sum. If you cannot guess the appropriate integer pair, systematically list all possible pairs until you find the appropriate pair.

a) Product = –54 and Sum = 3.

b) Product = –27 and Sum = 26.

c) Product = 81 and Sum = –30.

d) Product = 63 and Sum = 22.

e) Product = –21 and Sum = –4.

f) Product = 80 and Sum = –21.

3. Each of the following expressions is of the form $ax^2 + bx + c$. Perform the following tasks for each expression.

i) Try to guess an integer pair whose product is ac and whose sum is b. If you cannot guess the appropriate pair, systematically list all possible integer pairs until you find the appropriate pair.

ii) Use your integer pair to express the term bx as a sum. Factor the resulting expression by grouping.

iii) Multiply to check your work. Show this check work on your homework paper.

a) $x^2 + 14x - 32$

b) $x^2 - 9x - 36$

c) $2x^2 - 13x + 15$

d) $10x^2 + x - 2$

e) $x^2 + 14x - 72$

f) $x^2 - 3x - 60$

g) $5x^2 + 4x - 12$

h) $16x^2 + 22x - 3$

i) $x^{2a} + x^a - 90$

j) $15x^{2a} - 13x^a + 2$

Note: You are factoring expressions here, not solving equations for x.

4. Complete the following table of points that satisfy the equation $f(x) = x^2 - 4x - 5$.

x	–2	–1	0	1	2	3	4	5	6
$f(x)$									
Points to plot									

a) Set up a coordinate system on a sheet of graph paper and plot each point from the table on your coordinate system. Plot the remainder of the points on your coordinate system that satisfy the equation of the function f.

b) Draw a number line below your graph and shade and label estimates of the solutions of the equation $f(x) = 0$ on your number line in the usual manner.

c) Use the ac test and the zero product property to find the solutions of $x^2 - 4x - 5 = 0$. Do these analytical solutions compare favorably with your estimates from part (b)? If not, check your work for error.

5. Complete the following table of points that satisfy the equation $f(x) = 2x^2 + x - 10$.

x	−5	−4	−3	−2	−1	0	1	2	3	4
f(x)										
Points to plot										

a) Set up a coordinate system on a sheet of graph paper and plot each point from the table on your coordinate system. Plot the remainder of the points on your coordinate system that satisfy the equation of the function f.

b) Draw a number line below your graph and shade and label estimates of the solutions of the equation $f(x) = 0$ on your number line in the usual manner.

c) Use the *ac* test and the zero product property to find the solutions of $2x^2 + x - 10 = 0$. Do these analytical solutions compare favorably with your estimates from part (b)? If not, check your work for error.

6. Use the *ac* test and the zero product property to help find the solutions of each of the following nonlinear equations.

a) $x^2 + 2x = 8$ b) $2x^2 - 13x = 45$ c) $6x^2 + 5 = 17x$

d) $x^2 - 8x = 20$ e) $x^2 + 48 = 16x$ f) $15x^2 + 4x = 4$

Note: You are solving equations here.

7. (TI-82) Perform the following tasks for each of the following equations.

i)   Enter the equation in the Y = menu. Push the ZOOM button and select 6:ZStandard. Set up a separate coordinate system on your homework paper for each problem and copy the image in your viewing window onto your coordinate system. Label your graph with Y1 and include the WINDOW parameters on your coordinate axes.

ii) Use the root-finding utility in the CALC menu to estimate the roots of the function. Draw a number line below your graph and shade and label these estimates of the roots on your number line in the usual manner.

iii) Use the *ac* test and the zero product property to help find the solutions of Y1 = 0. Does this analytical solution compare favorably with your estimate from part (ii)? If not, check your work for error.

a) $Y1 = X^2 + 2X - 3$ b) $Y1 = 2X^2 - X - 1$ c) $Y1 = 4X^2 - 4X - 3$

d) $Y1 = -X^2 - 2X + 8$ e) $Y1 = -3X^2 + 5X + 2$ f) $Y1 = -2X^2 + 9X + 5$

8. Use the *ac* test and the zero product property to help solve the following equations for *x*:

a) $x^2 - 6x = 16$ b) $9 - 2x^2 = -7x$ c) $x = 12 - x^2$

d) $-13x = 7 - 2x^2$ e) $3x^2 + 5x = 22$ f) $-x^2 = 2x - 80$

Note: You are solving equations here.

9. Often, if both sides of an equation are divided by an appropriate number, the equation takes a much simpler form. For each of the following equations, perform the following tasks:

i)   Divide both sides of the equation by the greatest common divisor.

ii) Use the *ac* test and the zero product property to solve the resulting equation for *x*.

a) $2x^2 + 6x - 56 = 0$

b) $-4x^2 + 16x - 12 = 0$

c) $-5x^2 - 5x + 10 = 0$

d) $3x^2 - 27x + 54 = 0$

e) $20x^2 - 55x = 15$

f) $4x^2 - 28x = 240$

g) $20x^2 + 70x = 150$

h) $-30x^2 = 55x + 20$

Note: You are solving equations here.

10. Let the variables *L* and *W* represent the length and width of a rectangle, respectively.

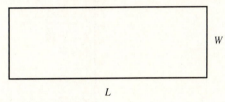

a) If the length of the rectangle is 3 feet longer than the width of the rectangle, express the length *L* as a function of its width *W*.

b) Express the area *A* of the rectangle as a function of *W*.

c) Suppose that the area of the rectangle is 40 square feet. Substitute this number in the equation you developed in part (b) and use the *ac* test and the zero product property to solve the resulting equation for *W*.

d) Eliminate any answers that are not reasonable and explain why you are eliminating them. Clearly state the length and width of the rectangle. Use complete sentences to explain why these dimensions satisfy the requirements of the problem.

11. Let the variable *b* represent the length of the base of a triangle and let *h* represent the height of the triangle.

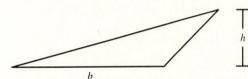

a) Suppose that the base of the triangle is 4 inches longer than its height. Express the base *b* of the triangle as a function of its height *h*.

b) Express the area *A* of the triangle as a function of *h*.

c) Suppose that the area of the triangle is 6 square inches. Substitute this number in the equation you developed in part (b) and use the *ac* test and the zero product property to solve the resulting equation for *h*.

d) Eliminate any answers that are not reasonable and explain why you are eliminating them. Clearly state the length of the base and the height of the triangle. Use complete sentences to explain why these dimensions satisfy the requirements of the problem.

12. Consider the functions whose equations are $f(x) = x^2$ and $g(x) = 2x + 3$.

a) Set up a table of points that satisfies the equation of the function *f*. Do the same for function *g*.

b) Set up a coordinate system on a sheet of graph paper. Plot the points from each table on your coordinate system. Plot the rest of the points that satisfy the equations of the functions f and g.

c) Draw a number line below your graph and shade and label the solutions of $f(x) = g(x)$ on your number line in the usual manner.

d) Use the ac test and the zero product property to help find the solutions of the nonlinear equation $x^2 = 2x + 3$. Do these analytical solutions compare favorably with your estimates from part (c)? If not, check your work for error.

13. (TI-82) Load the following functions into the Y = menu.

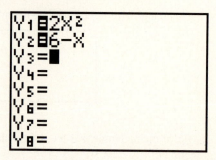

a) Push the ZOOM button and select 6:ZStandard. Set up a coordinate system on your homework paper and copy the image in your viewing window onto your homework paper. Label the appropriate graph with Y1 and the second graph with Y2. Include the WINDOW parameters on your axes.

b) Use the intersect utility in the CALC menu to find the points of intersection of the graphs of Y1 and Y2. For a review of this utility, see Example 10 in Section 1.6. Draw a number line below your graph and shade and label the x-values of the points of intersection on your number line in the usual manner.

c) Use the ac test and the zero product property to find the solutions of the equation $2x^2 = 6 - x$. Do these analytical solutions compare favorably with your estimated solutions found in part (b)? If not, check your work for error.

14. (TI-82) For the following pairs of equations, perform each of the following tasks.

i) Enter the functions in the Y = menu. Push the ZOOM button and select 6:ZStandard. Set up a coordinate system on your homework paper and copy the resulting image onto your coordinate system. Label the appropriate graph with the letter f and label the second graph with the letter g. Include your WINDOW parameters on your axes.

ii) Use the intersect utility in the CALC menu to find the solutions of the equation $f(x) = g(x)$. For a review of this utility, see Example 10 in Section 1.6. Draw a number line below your graph and shade and label the solutions on your number line in the usual manner.

iii) Use the ac test and the zero product property to find the solutions of the equation $f(x) = g(x)$. Do these analytical solutions compare favorably with the estimated solutions found in part (b)? If not, check your work for error.

a) $f(x) = x^2 - 4$, $g(x) = 2 - x$ b) $f(x) = 2 - x$, $g(x) = 4 - x^2$

c) $f(x) = x^2 - x$, $g(x) = 6 - x^2$ d) $f(x) = x^2 + 2x$, $g(x) = 4 - x^2$

15. Use the *ac* test and the zero product property to help find the solutions of each of the following nonlinear equations.

a) $x^2 + x - 38 = 10 - x$

b) $2x^2 - 2x = x^2 + x + 10$

c) $6x^2 - 5x = 2 - 6x^2$

d) $8x^2 - x = 6 - 7x^2$

16. The length of a rectangle is 5 feet longer than twice its width. Let *W* represent the width of the rectangle. Express the area *A* of the rectangle as a function of its width *W*.

a) Suppose that the area of the rectangle is 33 square feet. Substitute this value in the equation you developed in part (a). Use the *ac* test and the zero product property to solve the resulting equation for *W*.

b) Eliminate any solutions that are not practical and explain why you have eliminated them. Clearly state the dimensions of the rectangle. Use complete sentences to explain why these dimensions satisfy the requirements of the problem.

17. Consider the following concentric rectangles.

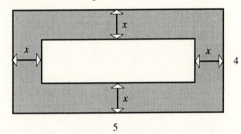

The dimensions of the outer rectangle are 5 feet by 4 feet. The shaded region is a border of uniform width. Let *x* represent this uniform width.

a) Express the area *A* of the shaded region as a function of *x*.

b) Suppose that the area of the shaded region is 8 square feet. Substitute this number in the equation you developed in part (a). Use the *ac* test and the zero product property to help find the solutions of the resulting equation.

c) Eliminate any solutions that are not reasonable and explain why you are doing so. Use complete sentences to explain why your solution satisfies the requirements of the problem.

18. The base of a triangle is 3 inches longer than twice the height of the triangle. Let *h* represent the height of the triangle.

a) Express the area *A* of the triangle as a function of its height *h*.

b) Suppose that the area of the triangle is 10 square inches. Substitute this number in the equation found in part (b) and use the *ac* test and the zero product property to help find the solutions of the resulting equation.

c) Eliminate any solutions that are not reasonable and explain why you are doing so. Use complete sentences to explain why your solution satisfies the requirements of the problem.

19. The demand for a certain brand of calculator is a function of its unit price. Usually, as the unit price goes up, the demand goes down. As the unit price goes down, the demand goes up. Suppose that the demand is related to the unit price by the following equation.

$$x = 54 - p$$

The variable x represents the number of calculators bought by consumers and the variable p represents the unit price of the calculator.

a) The revenue from sales of calculators is computed by multiplying the number of calculators sold by the unit price. Let R represent the revenue from sales of calculators. Express R as a function of x and p.

b) Use the fact that $x = 54 - p$ to express R as a function of the unit price p.

c) Suppose that the revenue made from the sales of calculators is \$200. Place this value in the equation developed in part (b) and use the *ac* test and the zero product property to help solve the resulting equation.

d) Eliminate any unreasonable answers and explain why you are doing so. Use complete sentences to explain why your solution satisfies each requirement of the problem.

20. A story is often told about Karl Friedrich Gauss, one of the great mathematicians of all time. When Gauss was a boy, his teacher tried to keep him busy by requiring that young Karl add the first 1000 positive integers. Karl was sent to his desk to find the following sum.

$$S = 1 + 2 + 3 + 4 + \ldots + 1000$$

Much to his teacher's chagrin, young Karl returned in moments with the answer. Today, mathematicians know that the sum of the first n positive integers is given by the following formula.

$$S = \frac{n(n+1)}{2}$$

a) Suppose that the sum of the first n integers is 55. Replace the variable S in the formula $S = \frac{n(n+1)}{2}$ with the number 55. Use the *ac* test and the zero product property to help find the solution of the resulting equation.

b) Use complete sentences to explain what your answer means and that it satisfies the requirements of the problem.

4.5 Useful Multiplication Patterns

This section introduces some patterns of multiplication that will be particularly useful in later chapters. The multiplication patterns will require that you do much of the work mentally. Once you have learned a multiplication pattern, you will immediately turn it around and use it as a tool for factoring.

The Difference of Squares Pattern

Use the distributive property to multiply $(a+b)(a-b)$.

$$(a+b)(a-b) = a(a-b) + b(a-b)$$
$$= a^2 - ab + ba - b^2$$
$$= a^2 - b^2$$

This multiplication pattern is called the *difference of squares*.

$$\textbf{\textit{Difference of squares:}}\ (a+b)(a-b) = a^2 - b^2$$

Two binomials that have the same first term and the same second term are being multiplied. One factor has a plus sign separating its terms, and the other factor has a minus sign separating its terms. Compute your answer by squaring the first term and second terms, then placing a minus sign between the squares. Hence the name, difference of squares.

Example 1   Multiply $(x+3)(x-3)$.

Solution   You are multiplying two binomials that have the same first term and the same second term. One factor has a plus sign separating its terms and the other has a minus sign separating its terms. Consequently, you can apply the difference of squares pattern. Square the first term, square the second term, and place a minus sign between the results.

$$(x+3)(x-3) = x^2 - 3^2$$
$$= x^2 - 9$$

Example 2   Use the difference of squares pattern to multiply $\left(x^{3n}+6\right)\left(x^{3n}-6\right)$.

Solution   Square the first term, square the second term, and place a minus sign between the squares.

$$\left(x^{3n}+6\right)\left(x^{3n}-6\right) = \left(x^{3n}\right)^2 - 6^2$$
$$= x^{6n} - 36$$

Example 3   Here are some more applications of the difference of squares pattern.

$$(5x+4y)(5x-4y) = (5x)^2 - (4y)^2 = 25x^2 - 16y^2$$

$$\left(2x^5+7\right)\left(2x^5-7\right) = \left(2x^5\right)^2 - 7^2 = 4x^{10} - 49$$

$$\left(x^{3a}+y^{3b}\right)\left(x^{3a}-y^{3b}\right) = \left(x^{3a}\right)^2 - \left(y^{3b}\right)^2 = x^{6a} - y^{6b}$$

Using the Difference of Squares Pattern to Factor. Once you have become proficient using the difference of squares pattern to multiply, then it should be easy to factor (unmultiply) using the same pattern.

Example 4   Factor the expression $25x^2 - 9y^2$.

Solution   If you square when you multiply, it makes sense that you take the *square root* when you unmultiply. Take the square root of each term, then separate the terms of one factor with a plus sign and the terms of the second factor with a minus sign.

$$25x^2 - 9y^2 = (5x+3y)(5x-3y)$$

Example 5   Here are some more examples of factoring that use the difference of squares pattern.

$$25a^2 - 36y^2 = (5a + 6y)(5a - 6y)$$

$$x^{14} - 100 = (x^7 + 10)(x^7 - 10)$$

$$x^{2n} - 9 = (x^n + 3)(x^n - 3)$$

In each case, take the square root of each term, then separate the terms of one factor with a plus sign and the terms of the second factor with a minus sign.

Example 6   *(TI-82)* Solve the following equation for x: $x^2 = 9$.

Graphical Solution   Because this equation contains a power of x greater than 1, this equation is nonlinear. Make one side of our equation zero by subtracting 9 from both sides.

$$x^2 - 9 = 0$$

Load the following function in the Y = menu: $Y_1 = X^2 - 9$. Push the ZOOM button and select 6:ZStandard to capture the following image in your viewing window:

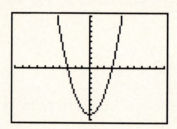

Arrange this information on your homework paper in the following manner:

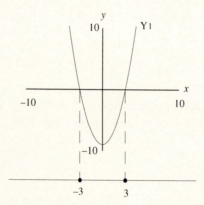

It appears that the graph of Y_1 crosses the x-axis at $x = -3$ and $x = 3$.

Example 9   Factor the expression $x^2 + 2x + 1$.

Solution   Note that the first and last terms of this trinomial are perfect squares. Taking the square root of the first and last terms, try the following:

$$x^2 + 2x + 1 = (x+1)^2$$

Check. Use the squaring a binomial pattern to check your result.

$$(x+1)^2 = x^2 + 2(x)(1) + (1)^2$$
$$= x^2 + 2x + 1$$

It checks. You have correctly factored.

Example 10   Factor the expression $4x^{2n} - 4x^n + 1$.

Solution   Note that the first and last terms of this trinomial are perfect squares. Taking the square root of the first and last terms, try the following:

$$4x^{2n} - 4x^n + 1 = \left(2x^n - 1\right)^2$$

Check. Use the squaring a binomial pattern to check your result.

$$\left(2x^n - 1\right)^2 = (2x^n)^2 + 2(2x^n)(-1) + (-1)^2$$
$$= 4x^{2n} - 4x^n + 1$$

The solution checks. You have correctly factored.

Example 11   Factor the expression $9x^2 - 50x + 25$.

Solution   Note that the first and last terms of this trinomial are perfect squares. Taking the square root of the first and last terms, try the following:

$$9x^2 - 50x + 25 = \left(3x - 5\right)^2$$

Check. Use the squaring a binomial pattern to check your result.

$$(3x - 5)^2 = 9x^2 + 2(3x)(-5) + (-5)^2$$
$$= 9x^2 - 30x + 25$$

The solution does *not* check. The middle term $-30x$ is incorrect.

One of the most important parts of the problem-solving process is the last part—checking your work to see if it is reasonable. This is often overlooked, usually with disastrous consequences.

Try Another Method. Try the ac test on this example. When you compare $9x^2 - 50x + 25$ with $ax^2 + bx + c$, you can see that $a = 9$, $b = -50$, and $c = 25$. The product ac equals 225. Begin listing integer pairs whose product is 225. Stop when you find a pair whose sum is -50.

$$1, 225 \qquad -1, -225$$
$$3, 75 \qquad -3, -75$$
$$5, 45 \qquad -5, -45$$

There are more pairs whose product is 225; for example, $\{15,15\}$ and $\{9,25\}$. However, your searching can stop when the required pair presents itself. The required pair is $\{-5, -45\}$. Write the middle term as a sum of this integer pair and use the method of grouping to finish the factoring of the expression.

$$9x^2 - 50x + 25 = 9x^2 - 5x - 45x + 25$$
$$= x(9x - 5) - 5(9x - 5)$$
$$= (x - 5)(9x - 5)$$

Check. Multiply immediately to check your result.

$$(x - 5)(9x - 5) = x(9x - 5) - 5(9x - 5)$$
$$= 9x^2 - 5x - 45x + 25$$
$$= 9x^2 - 50x + 25$$

The solution checks.

Example 12 *(TI-82)* Solve the following equation for x: $x^2 + 49 = 14x$.

Graphical Solution Because this equation contains a power of x that is greater than 1, the equation is nonlinear. Make one side of our equation zero by subtracting $14x$ from both sides.

$$x^2 - 14x + 49 = 0$$

Enter the following function into the Y = menu: $Y1 = X^2 - 14X + 49$. Make the following adjustments to the WINDOW parameters:

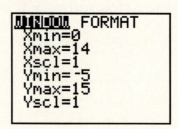

Push the GRAPH button to capture the following image in your viewing window:

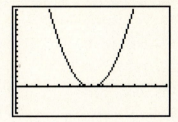

Arrange this information on your homework paper in the following manner:

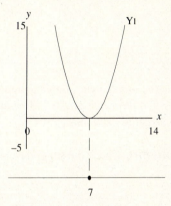

It appears that this graph comes down and touches the *x*-axis at $x = 7$, but it is difficult to be certain. Let's hope the analytical solution will reveal the true behavior of the graph near the *x*-axis.

Analytical Solution   The equation $x^2 + 49 = 14x$ has a power of *x* greater than 1, so it is nonlinear. Begin by making one side of the equation zero by subtracting $14x$ from both sides.

$$x^2 - 14x + 49 = 0$$

Noting that the first and last terms of this trinomial are perfect squares, you immediately attempt the following:

$$(x - 7)^2 = 0$$

Quickly multiply to see if this is correct. Because $(x - 7)^2 = x^2 - 14x + 49$, you can proceed. Use the zero product property to complete your solution.

$$(x - 7)^2 = 0$$
$$(x - 7)(x - 7) = 0$$
$$x - 7 = 0 \text{ or } x - 7 = 0$$
$$x = 7 \text{ or } \quad x = 7$$

Mathematicians call 7 a *double root* or a root of *multiplicity two*, meaning that the root is used twice.

Place the analytical solution and the graphical solution side-by-side for comparison.

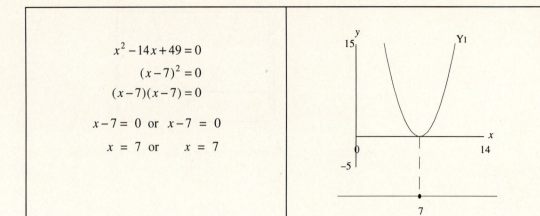

$$x^2 - 14x + 49 = 0$$
$$(x-7)^2 = 0$$
$$(x-7)(x-7) = 0$$

$$x-7 = 0 \text{ or } x-7 = 0$$
$$x = 7 \text{ or } \quad x = 7$$

It is now clear that the graph comes down and touches the x-axis exactly once, at $x = 7$. We say that this graph is *tangent to the x-axis* at $x = 7$. It is interesting to note that there is a double root at $x = 7$ and the graph is tangent to the x-axis at $x = 7$.

Check Your Solution.   Check the solution $x = 7$ in the original equation.

$$x^2 + 49 = 14x$$
$$(7)^2 + 49 = 14(7)$$
$$49 + 49 = 98$$
$$98 = 98$$

The solution $x = 7$ checks.

Example 13   A point P is selected on the line $3x + 2y = 24$. Horizontal and vertical lines are drawn from point P to the y-axis and x-axis, respectively, forming a rectangle in the first quadrant. Find the coordinates of the point P if the area of the rectangle is 24.

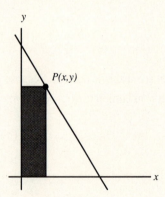

Solution   The sides of the rectangle must have lengths x and y.

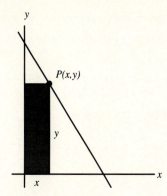

The area of the rectangle is given by the equation $A = xy$. Because the area of the rectangle is 24, this number can be substituted for A in the equation $A = xy$.

$$24 = xy \qquad \text{Eq. (1)}$$

Because Equation (1) has two unknowns, you will need a second equation involving x and y. The point (x, y) is on the line whose equation is $3x + 2y = 24$, so the point must satisfy the equation of the line.

$$3x + 2y = 24 \qquad \text{Eq. (2)}$$

A system of equations has been created in the unknowns x and y. Use the substitution method to solve them. First, solve Equation (2) for y.

$$3x + 2y = 24$$
$$2y = -3x + 24$$
$$y = -\frac{3}{2}x + 12 \qquad \text{Eq. (3)}$$

Substitute Equation (3) into Equation (1).

$$24 = xy$$
$$24 = x\left(-\frac{3}{2}x + 12\right)$$
$$24 = -\frac{3}{2}x^2 + 12x$$

Multiply both sides of this equation by 2.

$$48 = -3x^2 + 24x$$

Because this equation contains a power of x that is higher than 1, the equation is nonlinear. Make one side of the equation zero by adding $3x^2$ and $-24x$ to both sides.

$$3x^2 - 24x + 48 = 0$$

Divide both sides of this equation by 3.

$$x^2 - 8x + 16 = 0$$

Note that both x^2 and 16 are perfect squares, so you can immediately try the following:

$$(x-4)^2 = 0$$

Because $(x-4)^2 = x^2 - 8x + 16$, your work checks and you can use the zero product property to complete the solution of our equation.

$$(x-4)(x-4) = 0$$
$$x - 4 = 0 \text{ or } x - 4 = 0$$
$$x = 4 \text{ or } \quad x = 4$$

Note that $x = 4$ is a double root. To find the corresponding value of y, substitute $x = 4$ in Equation (3).

$$y = -\frac{3}{2}x + 12$$
$$y = -\frac{3}{2}(4) + 12$$
$$y = -6 + 12$$
$$y = 6$$

Therefore, $x = 4$ and $y = 6$.

Is Your Answer Reasonable? First of all, is the point (4,6) on the line whose equation is $3x + 2y = 24$? If it is, it should satisfy the equation of the line. Let's substitute the point (4,6) into the equation $3x + 2y = 24$.

$$3x + 2y = 24$$
$$3(4) + 2(6) = 24$$
$$12 + 12 = 24$$
$$24 = 24$$

That checks, so you can now be certain that (4,6) is on the line whose equation is $3x + 2y = 24$. Now, will the rectangle that is formed have an area of 24? Look at the diagram.

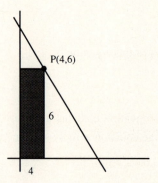

The area of a rectangle with dimensions 4 by 6 is most certainly 24. You can now be certain that you have done this problem correctly.

Exercises for Section 4.5

1. Use the difference of squares pattern to multiply each of the following pairs of binomials.

   a) $(x+8)(x-8)$

   b) $(9-x)(9+x)$

   c) $(11x-9)(11x+9)$

   d) $(15a-17b)(15a+17b)$

   e) $(5+3x)(5-3x)$

   f) $(x^3+1)(x^3-1)$

   g) $(11xy-13z)(11xy+13z)$

   h) $(x^5+2)(x^5-2)$

   i) $(x^{3n}+2)(x^{3n}-2)$

   j) $(x^a+y^a)(x^a-y^a)$

   k) $(x^{2n}+y^{2n})(x^{2n}-y^{2n})$

   l) $(7-2x^4)(7+2x^4)$

2. Factor each of the following using the difference of squares method.

   a) x^2-9

   b) $4x^2-1$

   c) $25x^2-49$

   d) $36x^2-121y^2$

   e) $x^{2n}-9$

   f) $y^{2b}-z^{2b}$

   g) x^4-9y^2

   h) $100x^2y^2-49z^4$

   i) x^4-16

   j) z^4-81

   k) $x^{6n}-4$

   l) $x^{8n}-1$

   Note: You are factoring expressions here, not solving equations for x.

3. Consider the function whose equation is $f(x)=x^2-4$. Copy and complete the following table on your homework paper.

x	-4	-3	-2	-1	0	1	2	3	4
$f(x)$									
Points to plot									

   a) Set up a coordinate system on a sheet of graph paper. Plot each point from the table on your coordinate system. Plot the rest of the points that satisfy the equation of the function f.

   b) Draw a number line below your graph and shade and label the solutions of $f(x)=0$ on your number line in the usual manner.

   c) Use an analytic method to find the solutions of the equation $f(x)=0$. Do these analytic solutions compare favorably with your estimates in part (b)? If not, check your work for error.

4. Use the difference of squares pattern and the zero product property to help find solutions of each of the following nonlinear equations.

   a) $x^2=16$

   b) $4x^2=9$

   c) $100=x^2$

   d) $25x^2=36$

   Note: You are solving equations for x here.

5. (TI-82) For each of the following equations, perform the following tasks:

   i) Load the equation in the Y= menu. Adjust the WINDOW parameters so that your viewing window shows the x-intercepts of the curve and the turning point of the curve. Set up a coordinate system on your homework paper and copy this image from your viewing screen onto your homework paper. Label the graph with its equation and include your WINDOW parameters on your coordinate axes.

ii) Draw a number line below your graph and shade and label the solutions of $f(x) = 0$ on your number line in the usual manner.

iii) Use an analytical method to find the solutions of the equation $f(x) = 0$. Do these analytical solutions compare favorably with your estimates in part (b)? If not, check your work for error.

a) $f(x) = x^2 - 25$ b) $f(x) = 9 - x^2$ c) $f(x) = x^2 - 81$ d) $f(x) = 121 - x^2$

6. (TI-82) Load the function $f(x) = 9 - 4x^2$ in the Y = menu like this: Y1 = 9 − 4X^2

a) Push the ZOOM button and select 6:ZStandard. Adjust the WINDOW parameters so that your viewing window shows the turning point of the curve. Set up a coordinate system on your homework paper and copy this image from your viewing screen onto your homework paper. Label the graph with its equation and include your WINDOW parameters on your coordinate axes.

b) Use the root utility in the CALC menu to find the roots (x-intercepts) of the function f. Draw a number line below your graph and shade and label these estimated solutions of $f(x) = 0$ on your number line in the usual manner.

c) Use an analytical method to find the solutions of the equation $f(x) = 0$. Do these analytical solutions compare favorably with your estimates in part (b)? If not, check your work for error.

7. Use the squaring a binomial pattern to multiply each of the following:

a) $(x+3)^2$ b) $(x-6y)^2$ c) $(3x+7)^2$ d) $\left(4x^2 - 9\right)^2$

e) $(3x^5 + 5y^3)^2$ f) $\left(x^n - 11\right)^2$ g) $\left(x^{2a} + y^{2a}\right)^2$ h) $\left(x^2 y^2 - 3z^2\right)^2$

8. Find $\left[f(x)\right]^2$ for each of the following.

a) $f(x) = x+5$ b) $f(x) = 2x-3$ c) $f(x) = 3x^3 + 8$ d) $f(x) = x^n + 2$

9. Consider the following concentric squares:

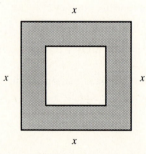

a) If the width of the shaded border is uniform and equal to 2 inches, express the area of the shaded region as a function of x.

b) Suppose that the area of the shaded region is 40 square inches. Substitute this number in the equation developed in part (a) and solve the resulting equation for x.

c) Use complete sentences to explain why your solution satisfies the requirements of the problem.

10. Factor each of the following expressions. First look to use the squaring a binomial pattern and check your solution. If your solution does not check, try the *ac* test and check that solution.

    a) $x^2 + 16x + 64$

    b) $x^2 - 12x + 36$

    c) $9x^2 - 6x + 1$

    d) $9x^2 - 15x + 4$

    e) $49x^2 - 14x + 1$

    f) $x^{2n} - 8x^n + 16$

    g) $x^{2a} - 20x^a + 64$

    h) $4x^{2b} - 4x^b y^b + y^{2b}$

    Note: You are factoring expressions here, not solving equations for *x*.

11. Consider the function whose equation is $f(x) = x^2 + 4x + 4$. Copy and complete the following table of points on your homework paper.

x	−5	−4	−3	−2	−1	0	1
f(x)							
Points to plot							

    a) Set up a coordinate system on a sheet of graph paper. Plot each point from the table on your coordinate system. Plot the rest of the points that satisfy the equation of the function *f*.

    b) Draw a number line below your graph and shade and label the solutions of $f(x) = 0$ on your number line in the usual manner.

    c) Use an analytical method to find the solutions of the equation $f(x) = 0$. Do these analytical solutions compare favorably with your estimates in part (b)? If not, check your work for error.

    d) Use complete sentences to explain the meaning of the phrase *double root*. What appears to happen to the graph at this double root?

12 (TI-82) For each of the following equations, perform the following tasks:

    i)   Load the equation in the Y= menu. Adjust the WINDOW parameters so that your viewing window shows the *x*-intercept of the curve. Set up a coordinate system on your homework paper and copy this image from your viewing screen onto your homework paper. Label the graph with its equation and include your WINDOW parameters on your coordinate axes.

    ii) Draw a number line below your graph and shade and label the solution of $f(x) = 0$ on your number line in the usual manner.

    iii) Use an analytical method to find the solution of the equation $f(x) = 0$. Do these analytical solutions compare favorably with your estimates in part (b)? If not, check your work for error.

    iv) Use complete sentences to explain what appears to happen to a function that has a double root.

    a) $f(x) = x^2 + 8x + 16$

    b) $f(x) = -x^2 + 4x - 4$

    c) $f(x) = x^2 + 16x + 64$

    d) $f(x) = -x^2 + 18x - 81$

13. Use an analytical method only to solve each of the following nonlinear equations for *x*. First attempt to use the squaring a binomial pattern to help factor. If that does not work, then give the *ac* test a try.

    a) $x^2 + 100 = 20x$

    b) $16x^2 = -40x - 25$

    c) $64 + 9x^2 = 48x$

    d) $25x^2 + 4 = 25x$

    e) $49x^2 + 4 = 28x$

    f) $4x^2 + 9 = 37x$

g) $x^2 = 10x - 16$ h) $144x^2 = -120x - 25$

Note: You are solving equations for x here.

14. The following diagram illustrates the graph of the line whose equation is $3x + y = 12$. A point P has been selected on the line, and horizontal and vertical lines have been drawn to the y-axis and x-axis, respectively. The area of the shaded rectangle is 12.

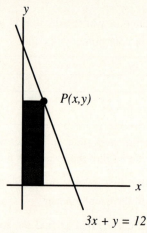

$3x + y = 12$

a) Express the area of the shaded rectangle as a function of x and y. Insert the number 12 for the area.

b) Since the point P(x,y) is on the line, it must satisfy the equation of the line, so $3x + y = 12$. This equation and the equation developed in part (a) give us a system of two equations in two unknowns. Use the substitution method to solve this system for x and y.

c) Use complete sentences to explain why your solution satisfies the requirements of the problem.

15. Consider the following concentric squares.

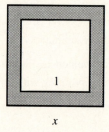

The variable x represents the length of the edge of the outer square. The length of the edge of the inner square is 1 foot.

a) Express the area of the shaded region as a function of x.

b) Suppose that the area of the shaded region is $\dfrac{25}{4}$ square feet. Insert this number in the equation developed in part (a) and use an analytical method to solve the resulting equation for x.

c) Use complete sentences to explain why your answer satisfies the requirements of the problem.

16. If a bowling ball is dropped from a height of 256 feet, its height above the ground is given by the equation $h = 256 - 16t^2$, where h represents the height of the ball and is measured in feet and t represents the amount of time that the ball has been falling and is measured in seconds.

   a) What is the height of the ball after 2 seconds?

   b) When the ball hits ground level, its height above the ground will be zero. Let $h = 0$ in the equation $h = 256 - 16t^2$ and solve the resulting equation for t.

   c) Use a complete sentence to explain the meaning of your answer.

17. Consider the circle below.

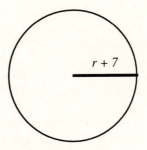

   a) Express the area of the circle as a function of r.

   b) Suppose that the area of the circle is 100π square inches. Substitute this number in the equation developed in part (a) and solve the resulting equation for r.

   c) Use complete sentences to explain why your answer satisfies the requirements of the problem.

4.6 Summary and Review

Important Visualization Tools

Place your analytical solutions and graphical solution side-by-side for comparison.

$$x^2 + 3x - 10 = 0$$
$$x^2 + 5x - 2x - 10 = 0$$
$$x(x+5) - 2(x+5) = 0$$
$$(x-2)(x+5) = 0$$

$$x - 2 = 0 \ \text{ or } \ x + 5 = 0$$
$$x = 2 \ \text{ or } \quad x = -5$$

There is a big difference between factoring an expression and solving an equation.

Example. Factor the expression $x^2 - 16$.

Solution.
$$x^2 - 16 = (x+4)(x-4)$$

Example. Solve for x: $x^2 - 16 = 0$.

Solution.
$$x^2 - 16 = 0$$
$$(x+4)(x-4) = 0$$
$$x+4 = 0 \quad \text{or} \quad x-4 = 0$$
$$x = -4 \qquad\qquad x = 4$$

A Review of Key Words and Phrases

☞ In the expression a^n, a is called the *base* and n is called the *exponent*. The exponent tells us how many times to write the base as a factor. (page 303)

$$a^n = \underbrace{a \times a \times a \times \cdots \times a}_{n \text{ times}}$$
$$a^0 = 1, a \neq 0$$

☞ The five laws of exponents are summarized below. (page 307)

1. $a^m a^n = a^{m+n}$

2. $\left(a^m\right)^n = a^{mn}$

3. If $a \neq 0$ and $m > n$, then $\dfrac{a^m}{a^n} = a^{m-n}$. If $a \neq 0$ and $m < n$, then $\dfrac{a^m}{a^n} = \dfrac{1}{a^{n-m}}$.

4. $\left(ab\right)^n = a^n b^n$

5. If $b \neq 0$, then $\left(\dfrac{a}{b}\right)^n = \dfrac{a^n}{b^n}$.

☞ A *term* is the product of a real number and one or more variables raised to a power. The real number is called the *coefficient* of the term. (page 308)

☞ *The distributive property*: $a(b+c) = ab + ac$. (page 309)

☞ The expression $a_n x^n + a_{n-1} x^{n-1} + \cdots a_2 x^2 + a_1 x + a_0$, where each of the a_k is a real number and n is a nonnegative integer, is called a *polynomial* in x. The *degree* of this polynomial is n. (page 315)

☞ A *monomial* is a polynomial with one term. A *binomial* is a polynomial with two terms. A *trinomial* is a polynomial with three terms. (page 316)

☞ When asked to factor, you must find the original multiplication problem. (page 325)

☞ *Tip:* When you factor, multiply immediately to check your work. (page 326)

☞ If the variable you are solving for is present in powers larger than 1, then the equation is *nonlinear*. The equation is called nonlinear because the graphs involved are curves, not lines. (page 328)

☞ *The zero product property:* If $ab = 0$, then either $a = 0$ or $b = 0$. (page 328)

☞ If your equation is nonlinear, make one side of the equation zero and factor the remaining side of the equation. Apply the zero product property to complete the solution. (page 329)

☞ The *ac* test is used to factor trinomials of the form $ax^2 + bx + c$. (page 343)

1. If possible, find a pair of integers whose product is *ac* and whose sum is *b*.
2. Use this integer pair to express the middle term *bx* as a sum.
3. Factor by grouping.

☞ The *difference of squares* multiplication pattern: $(a+b)(a-b) = a^2 - b^2$. (page 359)

☞ The *squaring a binomial* multiplication pattern: $(a+b)^2 = a^2 + 2ab + b^2$. (page 361)

TI-82 Keywords and Menus

☞ Using the root-finding utility on the TI-82. (Example 12, page 331)

☞ Checking solutions with the TI-82. (page 337)

Chapter Review Exercises

1. Use the laws of exponents to help simplify each of the following expressions.

a) $\left(3xy^3\right)^2$

b) $\left(-3x^{3a}\right)^2$

c) $\left(-3x^2y^3\right)^2\left(2x^2y^5\right)$

d) $\left(3x^n\right)^2\left(-2x^{2n}\right)^3$

e) $\left(\dfrac{3x^2y^5}{x^5y^3}\right)^2$

f) $\left(\dfrac{3x^{3a}}{x^{5a}}\right)^2$

2. Simplify each of the following expressions.

a) $\left(x^2 + 5xy - 9y^2\right) + \left(2x^2 - 6xy - y^2\right) - \left(x^2 + xy - 5y^2\right)$

b) $\left(x^{3n} - 2x^{2n} - 7x^n - 7\right) - \left(-x^{3n} - 5x^{2n} + 7x^n - 19\right)$

c) $\left(x^3 - 3x^2 + 7x - 11\right) - \left(x^3 - 4x - 9\right) - \left(2x^3 - 8x^2 - 11x + 4\right)$

3. Use the appropriate property or pattern to multiply.

a) $(2x-3)(4x-11)$

b) $(x-4)\left(x^2 + 4x + 16\right)$

c) $\left(x^2 + xy + y^2\right)\left(x^2 - xy + y^2\right)$

d) $(x+5)(x-3)(2x+1)$

e) $\left(x^n + 3y^n\right)\left(x^n - 3y^n\right)$

f) $\left(x^a + 2y^b\right)^2$

4. If $p(x) = x^2 - 3x - 5$ and $q(x) = x^2 + 2x - 1$, simplify each of the following expressions as much as possible.

a) $xp(x) + 2q(x)$

b) $p(x)q(x)$

c) $x^n q(x)$

d) $(x+3)p(x)$

e) $(x-1)p(x) - (x-2)q(x)$

f) $(x-1)p(x) - 2x + 3q(x)$

g) $x^3 p(x) - x^2 q(x)$

h) $x^2 p(x) - (x+1)q(x)$

5. If $p(x) = x^2 - x - 4$ and $q(x) = x^2 + 4x - 8$, solve each of the following inequalities for x. Draw a number line and shade your solution on it. Use set-builder and interval notation to describe your solution set.

a) $p(x) - q(x) \le 0$

b) $p(x) + 5 > -3x + q(x)$

c) $|p(x) - q(x)| \le 5$

d) $|p(x) - q(x)| > 3$

6. Factor each of the following expressions.

a) $21x^2 - 28x$

b) $2x(x-8) - (x-8)$

c) $x^2 - 10x - 2x + 20$

d) $x^2 - 12x + 27$

e) $40n^2 + 7n - 3$

f) $6x^{2a} - 7x^a - 24$

g) $x^{2n} - 144y^{2n}$

h) $100x^{2a} - 140x^a y^b + 49y^{2b}$

Note: You are being asked to factor here, not solve for x.

7. Use factoring and the zero product property to help you find the solutions of each of the following nonlinear equations.

a) $2x^2 = 13x$

b) $2x(3x-11) - (3x-11) = 0$

c) $x^2 - 84 = 5x$

d) $5x^2 + 36 = 63x$

e) $81x^2 = 100$

f) $100x^2 + 9 = 60x$

Note: You are being asked to solve for x here.

8. (TI-82) For each of the following equations, perform the following tasks.

i) Load the equation in the Y= menu. Adjust the WINDOW parameters so that the x-intercepts of the function and the turning point of the curve are visible in the viewing window. Set up a coordinate system on your homework paper and copy this image onto it. Label your graph with the function's equation and include your WINDOW parameters on the coordinate axes.

ii) Use the root-finding utility in the CALC menu to determine the roots of the function. Draw a number line below your graph and shade and label these estimated solutions of $f(x) = 0$ on your number line in the usual manner.

iii) Use factoring and the zero product property to find exact solutions of the equation $f(x) = 0$. Do these analytic solutions compare favorably with the estimates found in part (ii)? If not, check your work for error.

a) $f(x) = x^2 - 12x$

b) $f(x) = 2x(x+13) - 23(x+13)$

c) $f(x) = x^2 - 7x - 120$

d) $f(x) = 2x^2 - 19x - 33$

e) $f(x) = 81 - 4x^2$

f) $f(x) = x^2 - 22x + 121$

9. Consider the function whose equation is $f(x) = -x^2 + 8x - 16$. Copy and complete the following table of points that satisfy the equation of the function f on your homework paper.

x	1	2	3	4	5	6	7
$f(x)$							
Points to plot							

a) Set up a coordinate system on a sheet of graph paper. Plot each point from the table on your coordinate system. Plot the rest of the points on your coordinate system that satisfy the equation of the function f.

b) Draw a number line below your graph and shade and label the solution of the equation $f(x) = 0$ on your number line in the usual manner.

c) Use an analytical method to find the solutions of the equation $f(x) = 0$. Use a complete sentence to explain why your answer is called a double root. If your analytical solution does not compare favorably with the graphical solution found in part (b), check your work for error.

d) Use complete sentences to explain the relationship between the double root and the behavior of the graph of f near this root.

10. Jerome runs a small business making ladies' leather handbags. He has determined that his costs are given by the equation $C = x^2 - 13x + 80$, where C represents his costs in dollars and x represents the number of handbags that he makes.

a) What will Jerome's costs be if he makes five handbags?

b) Suppose that Jerome incurs costs of \$50. Substitute this value in the equation $C = x^2 - 13x + 80$ and use an analytical method to solve the resulting equation for x.

c) Use a complete sentence to explain the meaning of your answer for part (b).

11. Alice sells pot warmers in her shop. She has determined that her revenue from selling pot warmers is given by the equation $R = 40p - p^2$, where R represents her revenue in dollars and p represents the unit price for pot warmers.

a) What will Alice's revenue be if she sells the pot warmers for \$2 apiece?

b) Suppose that Alice's revenue from the sale of pot warmers is \$175. Substitute this value in the equation $R = 40p - p^2$ and use an analytic method to solve the resulting equation for p.

c) Use a complete sentence to explain the meaning of your answer for part (b).

12. Consider the following rectangle.

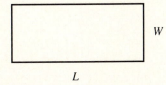

W

L

a) Express the perimeter of this rectangle as a function of L and W.

b) Suppose that the perimeter of the rectangle is 28 feet. Substitute this number into the equation you developed in part (a).

c) Express the area of the rectangle as a function of L and W.

d) Suppose that the area of the rectangle is 45 square feet. Substitute this number into the equation developed in part (c).

e) Use the substitution method to solve the system of equations developed in part (b) and (d).

f) Eliminate answers that are not practical. Clearly state the dimensions of the rectangle. Use complete sentences to explain why your solution satisfies the requirements of the problem.

13. The amount of money in Carlita's savings account is given by the equation $A = 1500\left(1 + \dfrac{0.05}{4}\right)^{4t}$,

    where A is the amount in the account in dollars and t is the number of years that have passed since Carlita put the money in the account. Use your calculator to estimate the amount in the account after 5 years. Round your answer to the nearest cent.

14. The size of the average farm in the United States can be predicted by the equation $S = -0.005333t^3 + 0.3250t^2 + 1.9833t + 167.6$, where S is the size of an average farm in acres and t is the number of years that have passed since 1940. Use this equation to predict the average size of a farm in acres in 1980.

    Note: The average farm size in the United States in 1980 was actually 426 acres.

Chapter 5

Radical Notation

We used radical notation in earlier chapters when the equation $f(x) = \sqrt{x}$ was introduced. However, no attempts were made at that point to provide rigorous definitions. This chapter will present a more formal introduction to radical notation. We will begin by discussing the real roots of the equation $x^n = b$ and by introducing the radical notation $\sqrt[n]{b}$. We will study several important properties of radical notation and use these properties to simplify expressions involving radical notation. We will introduce rational exponents and finish the chapter by solving equations containing radicals.

5.1 Introducing the *n*th Roots of *b*

Let's first examine the graphs of $y = x^n$ for $n = 2, 3, 4, \dots$. When the graph of the equation $y = x^n$ is understood, the real-valued solutions of the equation $x^n = b$ will be discussed, and the notation $\sqrt[n]{b}$ will be introduced.

The Graph of the Equation $y = x^n$. First, examine the graph of $y = x^n$ when n is an even integer.

Load the following functions into the Y= menu. Note that each of the exponents is an even integer.	Push the ZOOM button and select 6:ZStandard to capture the graphs of the equations $y = x^2$, $y = x^4$, and $y = x^6$.
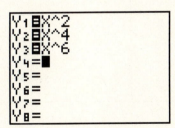	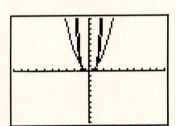

Next, consider the graph of $y = x^n$, where n is an odd integer.

Load the following functions into the Y= menu. Note that each of the exponents is an odd integer.	Push the ZOOM button and select 6:ZStandard to capture the graphs of the equations $y = x^3$, $y = x^5$, and $y = x^7$.
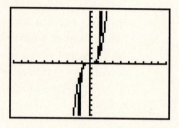	

The graph of $y = x^n$ assumes one of two basic shapes: one when the exponent is even, another when the exponent is odd.

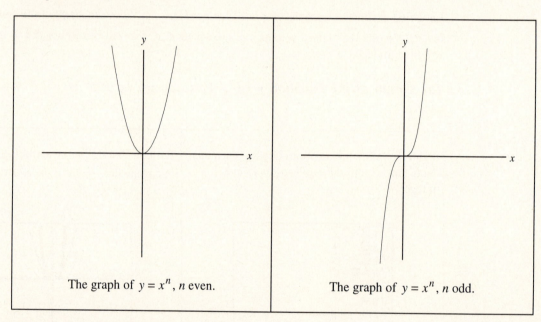

The graph of $y = x^n$, n even.	The graph of $y = x^n$, n odd.

The Real Solutions of $x^n = b$.

Let's begin with a definition. Don't expect to understand the definition thoroughly the first time through. The examples that follow the definition are designed to further help you understand this rather complicated and lengthy definition. Read the definition, understand what you can, then work through the examples that follow. After working the examples, read the definition again. Hopefully, it will be more meaningful this time.

The solutions of $x^n = b$ are called nth roots of b. Let's consider two cases.

Case 1. The exponent n is odd.

There is one real solution of the equation $x^n = b$—namely, $x = \sqrt[n]{b}$. Because $\sqrt[n]{b}$ is a solution of $x^n = b$, $\left(\sqrt[n]{b}\right)^n = b$.

Case 2. The exponent n is even. Let's consider two cases.

Case A. If $b < 0$, there are no real solutions.

Case B. If $b \geq 0$, then there are two real solutions of the equation $x^n = b$—namely, $x = \pm\sqrt[n]{b}$. The notation $-\sqrt[n]{b}$ calls for the nonpositive real solution of $x^n = b$ and the notation $\sqrt[n]{b}$ calls for the nonnegative real solution of $x^n = b$. Because $-\sqrt[n]{b}$ is a solution of $x^n = b$, $\left(-\sqrt[n]{b}\right)^n = b$. Because $\sqrt[n]{b}$ is a solution of $x^n = b$, $\left(\sqrt[n]{b}\right)^n = b$.

The n in the radical notation $\sqrt[n]{b}$ is called the index of the radical. If there is no index present, as in $\sqrt{b}$, square root is implied.

As you can see, this definition is not an easy one to absorb. The following examples are designed to help develop your understanding of it.

Example 1   Find all real-valued solutions of the equation $x^2 = 7$.

Graphical Solution   Since this equation is of the form $f(x) = g(x)$, begin by drawing the graphs of the functions $f(x) = x^2$ and $g(x) = 7$. The x-values of the points of intersection of the two graphs will be the real-valued solutions of the equation $x^2 = 7$.

Load the following equations into the Y= menu:	Push the ZOOM button and select 6:ZStandard to capture the following image in your viewing window:
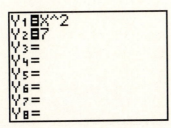	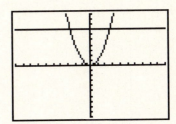

There are two points of intersection, so the equation $x^2 = 7$ has two real solutions. Use the intersect utility in the CALC menu to find the x-values of the points of intersection. For a review of this utility, see Example 10 in Section 1.6.

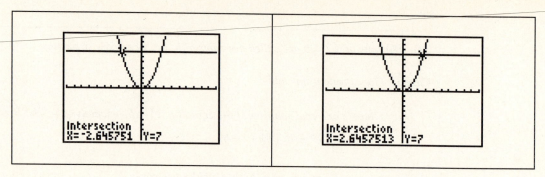

Analytical Solution

The solutions of $x^2 = 7$ are called square roots of 7. There are two real solutions.

$$x^2 = 7$$

$$x = \pm\sqrt{7}$$

Use your calculator to find decimal approximations for these solutions.

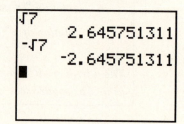

Place your analytical solution and graphical solution side-by-side on your homework paper for comparison.

The solutions of $x^2 = 7$ are called square roots of 7. There are two real solutions.

$$x^2 = 7$$

$$x = \pm\sqrt{7}$$

$$x \approx \pm 2.645751311\ldots$$

Arrange the results captured by the TI-82 in the following manner:

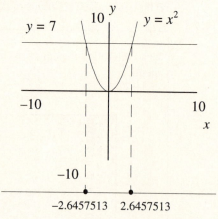

Check Your Solutions.

| Substitute the solution $\sqrt{7}$ for x in the original equation.

$$x^2 = 7$$
$$\left(\sqrt{7}\right)^2 = 7$$
$$7 = 7$$

The solution $\sqrt{7}$ checks. | Substitute the solution $-\sqrt{7}$ for x in the original equation.

$$x^2 = 7$$
$$\left(-\sqrt{7}\right)^2 = 7$$
$$7 = 7$$

The solution $-\sqrt{7}$ checks. |

Example 2   Find all real-valued solutions of $x^3 = -5$.

Graphical Solution

| Load the following equations into the Y= menu, then push the ZOOM button and select 6:ZStandard.

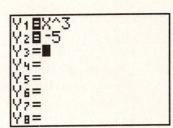 | Use the intersect utility in the CALC menu to find the coordinates of the point of intersection.

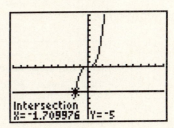 |

Analytical Solution

| The solutions of $x^3 = -5$ are called cube roots of -5. There is one real solution.

$$x^3 = -5$$
$$x = \sqrt[3]{-5}$$ | Use your calculator to find a decimal approximation for $\sqrt[3]{-5}$. Push the MATH button and select 4: $\sqrt[3]{\ }$, then enter -5.

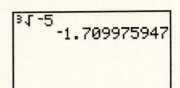 |

Place your analytical solution and graphical solution side-by-side on your homework paper for comparison.

The solutions of $x^3 = -5$ are called cube roots of -5. There is one real solution. $$x^3 = -5$$ $$x = \sqrt[3]{-5}$$ $$x \approx -1.709975947\ldots$$ **Check** $$x^3 = -5$$ $$\left(\sqrt[3]{-5}\right)^3 = -5$$ $$-5 = -5$$ The solution $\sqrt[3]{-5}$ checks.	Arrange the information captured by the TI-82 in the following manner. 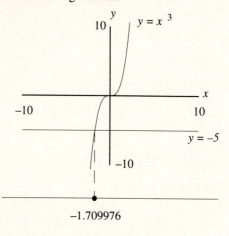 -1.709976

Example 3   Find all real solutions of $x^4 = -3$.

Graphical Solution

Load the following equations into the Y= menu: 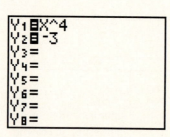	Push the ZOOM button and select 6:ZStandard to capture the following image in your viewing window: 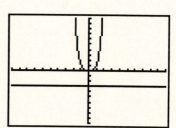 The graph of $y = x^4$ does not intersect the graph of $y = -3$, so the equation $x^4 = -3$ has no real solutions.

Analytical Solution   It is not possible to raise a real number to the fourth power and get a *negative* three.

$$x^4 = -3$$

There are no real solutions.

Place your analytical solution and graphical solution side-by-side on your homework paper for comparison.

It is not possible to raise a real number to the fourth power and get negative three. $$x^4 = -3$$ There are no real solutions.	The graph of $y = x^4$ does not intersect the graph of $y = -3$, so there are no real solutions of the equation $x^4 = -3$. 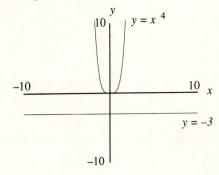

A Graphical Summary of the Solutions of $x^n = b$

If n is odd and b is positive, there is one real solution, $x = \sqrt[n]{b}$. 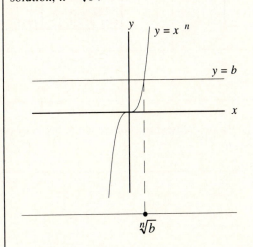	If n is odd and b is negative, there is one real solution, $x = \sqrt[n]{b}$. 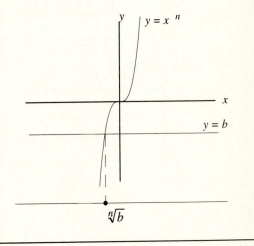

If n is even and b is positive, there are two real solutions of $x^n = b$, $x = \pm\sqrt[n]{b}$.

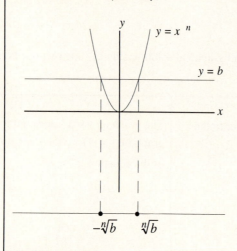

If n is even and b is negative, there are no real solutions of $x^n = b$.

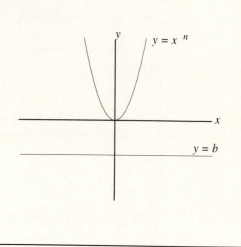

Example 4   Find all real solutions of the equation $2x^3 + 1 = 15$.

Graphical Solution   Begin by loading the following equations into the Y= menu: $Y_1 = 2X^3 + 1$ and $Y_2 = 15$.

Make the following adjustments to the WINDOW parameters.

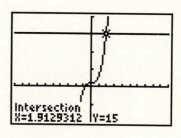

Use the intersect utility in the CALC menu to find the coordinates of the point of intersection.

Analytical Solution

| **Tip** | If there is just a single power of x present in your equation, begin by placing your equation in the form $x^n = b$. |

With this tip in mind, subtract 1 from both sides of the equation, then divide both sides of the resulting equation by 2.

$$2x^3 + 1 = 15$$ $$2x^3 = 14$$ $$x^3 = 7$$ Because the exponent is odd, this equation has one real solution. $$x = \sqrt[3]{7}$$	Use your calculator to find a decimal approximation of this solution. ³√7 1.912931183

Place your analytical solution and graphical solution side-by-side on your homework paper for comparison.

$$2x^3 + 1 = 15$$ $$2x^3 = 14$$ $$x^3 = 7$$ $$x \approx 1.912931183\ldots$$	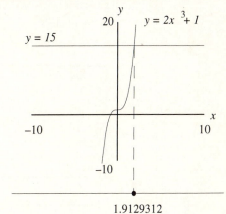 1.9129312

Some Reminders on Radical Notation

> *If n is odd, the notation $\sqrt[n]{b}$ calls for the one real solution of $x^n = b$.*

Example 5   If possible, find a real number simplification of the expression $\sqrt[3]{-8}$.

Solution   The notation $\sqrt[3]{-8}$ is calling for the one real solution of $x^3 = -8$. Because $(-2)^3 = -8$, $\sqrt[3]{-8} = -2$.

> *If n is even and b < 0, then the equation $x^n = b$ has no real solutions. If n is even and b < 0, then $\sqrt[n]{b}$ is not a real number.*

Example 6   If possible, find a real number simplification of the expression $\sqrt{-16}$.

Solution   The equation $x^2 = -16$ has no real solutions. Therefore $\sqrt{-16}$ is not a real number.

If n is even and $b \geq 0$, the notation $\sqrt[n]{b}$ calls for the nonnegative solution of $x^n = b$ and the notation $-\sqrt[n]{b}$ calls for the nonpositive solution of $x^n = b$.

Example 7   If possible, find real number simplifications of the expressions $\sqrt{25}$ and $-\sqrt{25}$.

Solution   The equation $x^2 = 25$ has two real solutions, $x = \pm 5$. The notation $\sqrt{25}$ calls for the nonnegative solution of the equation $x^2 = 25$, so $\sqrt{25} = 5$. The notation $-\sqrt{25}$ calls for the nonpositive solution of the equation $x^2 = 25$, so $-\sqrt{25} = -5$.

Example 8   Compare the graphs of $y = \sqrt{x^2}$ and $y = |x|$.

Solution

Enter the equation $y = \sqrt{x^2}$ in the Y= menu in the following manner: 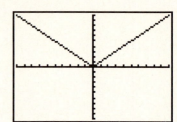	Push the ZOOM button and select 6:ZStandard to capture the graph of $y = \sqrt{x^2}$ in your viewing window.
Enter the equation $y = |x|$ in the Y= menu in the following manner: 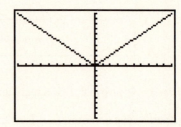	Push the ZOOM button and select 6:ZStandard to capture the graph of $y = |x|$ in your viewing window.

It appears that the graph of $y = \sqrt{x^2}$ and the graph of $y = |x|$ are identical. This visual evidence that $\sqrt{x^2} = |x|$ is even more striking if you draw the graphs of the equations $y = \sqrt{x^2}$ and $y = |x|$ simultaneously.

Enter the following functions in the Y= menu:	Push the ZOOM button and select 6:ZStandard to capture the graphs of $y = \sqrt{x^2}$ and $y = \lvert x \rvert$ in your viewing window.
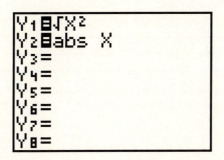	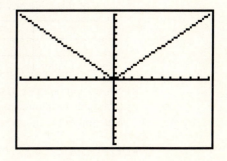

The graphs of the equations $y = \sqrt{x^2}$ and $y = \lvert x \rvert$ coincide. This is compelling evidence that $\sqrt{x^2} = \lvert x \rvert$.

How Would a Mathematician Formally Prove That $\sqrt{a^2} = \lvert a \rvert$?

It could be reasoned as follows. If a is a real number, then a^2 is nonnegative and the equation $x^2 = a^2$ has two real solutions. If $a \ge 0$, then $\lvert a \rvert = a$ and $\lvert a \rvert^2 = \lvert a \rvert \lvert a \rvert = aa = a^2$. If $a < 0$, then $\lvert a \rvert = -a$ and $\lvert a \rvert^2 = \lvert a \rvert \lvert a \rvert = (-a)(-a) = a^2$. In either case, $\lvert a \rvert^2 = a^2$. Substitute $\lvert a \rvert$ for x in the equation $x^2 = a^2$.

$$x^2 = a^2$$

$$\left(\lvert a \rvert \right)^2 = a^2$$

$$a^2 = a^2$$

Therefore $\lvert a \rvert$ is a nonnegative solution of the equation $x^2 = a^2$. By definition, $\sqrt{a^2}$ is the *unique* nonnegative solution of the equation $x^2 = a^2$. Logic demands that $\sqrt{a^2} = \lvert a \rvert$.

If a is a real number, then $\sqrt{a^2} = \lvert a \rvert$.

Note: It is important to remember that $\sqrt{a^2}$ is calling for a nonnegative square root of a^2. If you enclose a in absolute value bars, you ensure that the answer will be nonnegative. Therefore $\sqrt{a^2} = \lvert a \rvert$.

Exercises for Section 5.1

1. For each of the following equations, perform the following tasks:

   i) Complete the following table of points that satisfy the equation of the function f and place a copy of this table on a sheet of graph paper.

x	-3	-2	-1	0	1	2	3
$f(x)$							
Points to plot							

   ii) Set up a coordinate system on your graph paper. Scale the axes so that each point from your table can be plotted on your coordinate system. Plot the rest of the points that satisfy the equation of the function f.

   a) $f(x) = x^2$ b) $f(x) = x^3$ c) $f(x) = x^4$ d) $f(x) = x^5$

2. (TI-82) For each of the following equations, perform the following tasks:

   i) Make the following adjustments to the WINDOW parameters.

   ```
   WINDOW FORMAT
   Xmin=-10
   Xmax=10
   Xscl=1
   Ymin=-100
   Ymax=100
   Yscl=10
   ```

   ii) Load the equation into the Y= menu. Press the GRAPH button. Set up a coordinate system on your homework paper and copy the image in your viewing window onto it. Label your graph with its equation and include the WINDOW parameters on the coordinate axes.

   a) $f(x) = x^6$ b) $f(x) = x^7$ c) $f(x) = x^8$ d) $f(x) = x^9$

3. Set up a coordinate system on your homework paper and sketch the graph of each of the following equations.

   a) $f(x) = x^n$, n even   b) $f(x) = x^n$, n odd

4. (TI-82) For each of the following pairs of equations, perform the following tasks:

   i) Load the equations in the Y= menu, in Y1 and Y2, respectively.

   ii) Press the ZOOM button and select 6:ZStandard. Set up a coordinate system on your homework paper and copy the image in your viewing window onto your coordinate system. Label each graph with its equation and include the WINDOW parameters on your coordinate axes.

   iii) Use the intersect utility in the CALC menu to find the points of intersection of the graphs of the functions f and g. Draw a number line below your graph and shade and label these estimated real-valued solutions of $f(x) = g(x)$ on your number line in the usual manner.

   iv) Use an analytical method to find the *exact* real-valued solutions of the equation $f(x) = g(x)$ in radical form. Use your calculator to find decimal approximations for these solutions. If these solutions do not agree favorably with the graphical estimates found in part (iii), check your work for error.

a) $f(x) = x^2$, $g(x) = 3$ b) $f(x) = x^2$, $g(x) = -3$

c) $f(x) = x^3$, $g(x) = 3$ d) $f(x) = x^3$, $g(x) = -3$

5. (TI-82) Verify the following results by entering them into your calculator as shown. Copy these results onto your homework paper, including the result of the last computation.

Here is how to use your calculator to find $\sqrt{17}$, $-\sqrt{17}$, and $\sqrt{-17}$.
 	
``
√17
        4.123105626
-√17
       -4.123105626
√ -17■
``
 	
Explain your calculator's response to the last expression in this viewing window.
	Here is how to use your calculator to find $\sqrt[3]{10}$, $-\sqrt[3]{10}$, and $\sqrt[3]{-10}$.
	
	``
	³√10
	        2.15443469
	-³√10
	       -2.15443469
	³√ -10
	``
	
	Explain why your calculator's response on the last expression in this viewing window is so different from the response to the last expression in the window on the left.

6. (TI-82) Use radical notation to state the real-valued solutions of each of the following equations.

a) $x^2 = 17$ b) $x^3 = 13$ c) $x^4 = 11$ d) $x^5 = -9$

e) $x^6 = -15$ f) $x^7 = 19$ g) $x^8 = 21$ h) $x^9 = -10$

7. Verify the following results on your calculator.

If you wish to find evaluate radical expressions whose index is larger than 3, then you must select 5: $\sqrt[x]{}$ from the MATH menu.
 	
``
MATH NUM HYP PRB
1▶Frac
2:▶Dec
3:³
4:³√
5:ˣ√
6:fMin(
7↓fMax(
``
	Here is how you would find $\sqrt[4]{16}$, $\sqrt[5]{-32}$, $-\sqrt[6]{64}$, and $\sqrt[8]{-256}$.
	
	``
	4 ˣ√16
	              2
	5 ˣ√ -32
	             -2
	-6 ˣ√64
	             -2
	8 ˣ√ -256
	``

Explain the calculator's response with the last computation. Next, use this feature of your calculator to find decimal approximations of each of the following radical expressions. Show every digit from the viewing screen in your answer on your homework paper.

a) $\sqrt[4]{17}$ b) $\sqrt[5]{20}$ c) $-\sqrt[6]{31}$ d) $\sqrt[7]{-50}$ e) $\sqrt[8]{-40}$ f) $-\sqrt[8]{40}$

8. (TI-82) For each of the following pairs of equations, perform the following tasks:

i) Load the equations in the Y= menu, in Y1 and Y2, respectively.

ii) Press the ZOOM button and select 6:ZStandard. Set up a coordinate system on your homework paper and copy the image in your viewing window onto your coordinate system. Label each graph with its equation and include the WINDOW parameters on your coordinate axes.

iii) Use the intersect utility in the CALC menu to find the points of intersection of the graphs of the functions f and g. Draw a number line below your graph and shade and label these estimated real-valued solutions of $f(x) = g(x)$ on your number line in the usual manner.

iv) Use an analytical method to find the *exact* real-valued solutions of the equation $f(x) = g(x)$ in radical form. Use your calculator to find decimal approximations for these solutions. If these solutions do not agree favorably with the graphical estimates found in part (iii), check your work for error.

a) $f(x) = x^4$, $g(x) = 8$ b) $f(x) = x^5$, $g(x) = -8$ c) $f(x) = x^6$, $g(x) = 9$

d) $f(x) = x^7$, $g(x) = 3$ e) $f(x) = x^8$, $g(x) = -4$ f) $f(x) = x^9$, $g(x) = -4$

9. (TI-82) Load the following equations into the Y= menu.

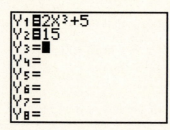

a) Adjust the WINDOW parameters so that both graphs, as well as their point of intersection, appear in the viewing window. Set up a coordinate system on a sheet of graph paper and copy the image in the viewing window onto your coordinate system. Label each graph with its equation and include your WINDOW parameters on the coordinate axes.

b) Use the intersect utility in the CALC menu to find the coordinates of the point of intersection. Draw a number line below your graph and shade and label this estimate on your number line in the usual manner.

c) Use an analytical method to find the solution of the equation $2x^3 + 5 = 15$ in radical form. Use your calculator to find a decimal approximation of this solution. If your solution does not agree favorably with the estimate found in part (b), check your work for error.

10. (TI-82) Load the following equations into the Y= menu.

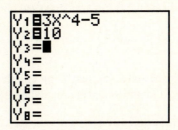

a) Adjust the WINDOW parameters so that both graphs, as well as their points of intersection, appear in the viewing window. Set up a coordinate system on a sheet of graph paper and copy the image in the viewing window onto your coordinate system. Label each graph with its equation and include your WINDOW parameters on the coordinate axes.

b) Use the intersect utility in the CALC menu to find the coordinates of the points of intersection. Draw a number line below your graph and shade and label these estimates on your number line in the usual manner.

c) Use an analytical method to find the solutions of the equation $3x^4 - 5 = 10$ in radical form. Use your calculator to find a decimal approximation of these solutions. If your solutions do not agree favorably with the estimates found in part (b), check your work for error.

11. Use an analytical method only to find the real-valued solutions of each of the following equations in radical form. Use your calculator to find decimal approximations for your solutions.

a) $2x^2 + 1 = 15$ b) $3 - 5x^3 = -7$ c) $2x^2 - 1 = x^2 + 4$

d) $3 - x^3 = x^3 + 7$ e) $x^4 - 7 = 11 - x^4$ f) $2x^5 - 3 = 5 - 2x^5$

12. (TI-82) Verify the following results by entering them into your calculator as shown. Copy these results onto your homework paper, including the result of the last computation.

Here is how to use your calculator to find $\left(\sqrt{5}\right)^2$, $\left(-\sqrt{5}\right)^2$, and $\left(\sqrt{-5}\right)^2$.	Here is how to use your calculator to find $\left(\sqrt[3]{5}\right)^3$, $\left(-\sqrt[3]{5}\right)^3$, and $\left(\sqrt[3]{-5}\right)^3$.
	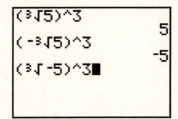
Explain your calculator's response on the last computation.	Why is the calculator's response on the last computation so different from the response on the last computation in the window on the left?

Mentally simplify each of the following expressions. Check your mental solutions with your calculator. Indicate the keystrokes you used on your homework paper.

a) $\left(\sqrt{17}\right)^2$ b) $\left(-\sqrt{17}\right)^2$ c) $\left(\sqrt{-17}\right)^2$ d) $\left(\sqrt[3]{11}\right)^3$ e) $\left(-\sqrt[3]{11}\right)^3$ f) $\left(\sqrt[3]{-11}\right)^3$

13. (TI-82) Verify the following results by entering them into your calculator as shown. Copy these results onto your homework paper, including the result of the last computation.

Here is how to use your calculator to find $\sqrt{5^2}$ and $\sqrt{(-5)^2}$.

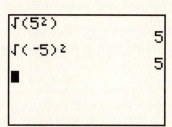

Here is how to use your calculator to find $\sqrt[3]{5^3}$ and $\sqrt[3]{(-5)^3}$.

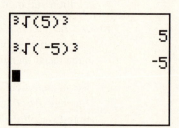

Use your calculator to evaluate each of the following pairs of radical expressions.

a) $\sqrt{7^2}$ and $\sqrt{(-7)^2}$ b) $\sqrt[3]{7^3}$ and $\sqrt[3]{(-7)^3}$

c) $\sqrt{11^2}$ and $\sqrt{(-11)^2}$ d) $\sqrt[3]{11^3}$ and $\sqrt[3]{(-11)^3}$

e) Use complete sentences to explain the differences you found between the even case and the odd case. Based on the evidence examined in this problem, simplify each of the following expressions.

i) If n is even, then $\sqrt[n]{x^n}$ = ?   ii) If n is odd, then $\sqrt[n]{x^n}$ = ?

14. (TI-82) Use your calculator to sketch the graphs of $y = \sqrt[3]{x^3}$, $y = x$, and $y = |x|$. Place each graph on a separate coordinate system on your homework paper. Does $\sqrt[3]{x^3} = x$ or does $\sqrt[3]{x^3} = |x|$? Explain how your made your decision.

15. (TI-82) Use your calculator to sketch the graphs of $y = \sqrt[4]{x^4}$, $y = x$, and $y = |x|$. Place each graph on a separate coordinate system on your homework paper. Does $\sqrt[4]{x^4} = x$ or does $\sqrt[4]{x^4} = |x|$? Explain how you made your decision.

16. (TI-82) Use your calculator to sketch the graphs of $y = \sqrt[5]{(x-2)^5}$, $y = x-2$, and $y = |x-2|$. Place each graph on a separate coordinate system on your homework paper. Does $\sqrt[5]{(x-2)^5} = x-2$ or does $\sqrt[5]{(x-2)^5} = |x-2|$? Explain how you made your decision.

17. (TI-82) Use your calculator to sketch the graphs of $y = \sqrt[6]{(x+2)^6}$, $y = x+2$, and $y = |x+2|$. Place each graph on a separate coordinate system on your homework paper. Does $\sqrt[6]{(x+2)^6} = x+2$ or does $\sqrt[6]{(x+2)^6} = |x+2|$? Explain how you made your decision.

18. Solve each of the following inequalities for x. Sketch your solution set on a number line and use interval notation to describe your solution set.

   a) $\sqrt{x^2} < 7$ b) $\sqrt{x^2} > 10$ c) $\sqrt{(x-2)^2} \leq 5$ d) $\sqrt{(2x+3)^2} \geq 8$

   Hint: Recall that $|x| < a$ if and only if $-a < x < a$ and $|x| > a$ if and only if $x < -a$ or $x > a$.

19. Consider the following concentric squares. The length of the edge of the inner square is 9 inches. Let the variable x represent the length of the edge of the outer square.

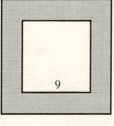

   x

   a) Express the area A of the shaded regions as a function of x.

   b) Suppose that the area of the shaded region is 100 square inches. Substitute this value in the equation developed in part (a) and solve the resulting equation for x. Use your calculator to find decimal approximations for your solutions, correct to the nearest tenth (one decimal place).

   c) Eliminate any answers that are not practical. Use complete sentences to explain why your solution satisfies the requirements of the problem.

20. A ball is dropped off the top of a building. The height of the ball above ground is given by the equation $h = 160 - 16t^2$, where h is the height of the ball above ground in feet and t is the time that the ball has been falling in seconds.

   a) How high is the top of the building above the ground?

   b) How high is the ball above ground level after 2 seconds?

   c) How long will it take the ball to fall to ground level? Give an answer in radical notation, then use your calculator to find a decimal approximation to the nearest tenth of a second (one decimal place).

5.2 Properties of Radicals

Like the laws of exponents, there are properties of radicals that allow the manipulation of expressions containing radical notation. This section discusses these properties of radicals and how to use them to put radical expressions into what is called *simple form*.

Example 1   Enter the following expressions into your TI-82 calculator and verify the results.

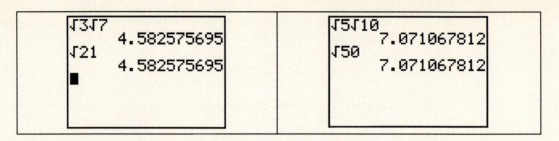

Note that $\sqrt{3}\sqrt{7} = \sqrt{21}$ and $\sqrt{5}\sqrt{10} = \sqrt{50}$. You could also use our calculator to show that $\sqrt{6}\sqrt{11} = \sqrt{66}$, $\sqrt{3}\sqrt{13} = \sqrt{39}$, etc. Although experimenting with a calculator does not constitute a proof, there is certainly some motivation here that $\sqrt{a}\sqrt{b} = \sqrt{ab}$.

How Would You Prove That $\sqrt{a}\sqrt{b} = \sqrt{ab}$? If a and b are nonnegative real numbers, then their product ab is a nonnegative real number. Because $ab \geq 0$, the equation $x^2 = ab$ has two real solutions, $\sqrt{ab}$ and $-\sqrt{ab}$. In particular, $\sqrt{ab}$ is the *unique* nonnegative real solution of the equation $x^2 = ab$.

Because $\sqrt{a}$ and $\sqrt{b}$ are nonnegative real numbers, $\sqrt{a}\sqrt{b}$ is also a nonnegative real number. Substitute $\sqrt{a}\sqrt{b}$ for x in the equation $x^2 = ab$.

$$x^2 = ab$$
$$\left(\sqrt{a}\sqrt{b}\right)^2 = ab$$
$$\left(\sqrt{a}\right)^2\left(\sqrt{b}\right)^2 = ab$$
$$ab = ab$$

Therefore, $\sqrt{a}\sqrt{b}$ is another nonnegative real solution of the equation $x^2 = ab$. Because the only nonnegative real solution of the equation $x^2 = ab$ is $\sqrt{ab}$, logic demands that $\sqrt{ab}$ and $\sqrt{a}\sqrt{b}$ are one and the same.

If $a \geq 0$ and $b \geq 0$, then $\sqrt{a}\,\sqrt{b} = \sqrt{ab}$.

In a similar fashion, you can also prove the following property of radicals.

If $a \geq 0$ and $b > 0$, then $\dfrac{\sqrt{a}}{\sqrt{b}} = \sqrt{\dfrac{a}{b}}$.

Example 2   Here are some sample applications of the properties of radicals.

$$\sqrt{5}\sqrt{2} = \sqrt{10}$$
$$\sqrt{7}\sqrt{11} = \sqrt{77}$$
$$\sqrt{3}\sqrt{22} = \sqrt{66}$$

$$\frac{\sqrt{12}}{\sqrt{7}} = \sqrt{\frac{12}{7}}$$
$$\frac{\sqrt{24}}{\sqrt{11}} = \sqrt{\frac{24}{11}}$$

Simple Form

Consider the following student solutions of the equation $\sqrt{2}x - 3 = 0$.

Julie proceeds by adding 3 to both sides of the equation, then divides both sides of the resulting equation by $\sqrt{2}$.

$$\sqrt{2}x - 3 = 0$$
$$\sqrt{2}x = 3$$
$$x = \frac{3}{\sqrt{2}}$$

Randolph proceeds by adding 3 to both sides of the equation and multiplying both sides of the resulting equation by $\sqrt{2}$. He then finishes by dividing both sides of the resulting equation by 2.

$$\sqrt{2}x - 3 = 0$$
$$\sqrt{2}x = 3$$
$$\sqrt{2}\sqrt{2}x = 3\sqrt{2}$$
$$2x = 3\sqrt{2}$$
$$x = \frac{3\sqrt{2}}{2}$$

Julie's answer does not look the same as Randolph's. Which student is correct? Use your calculator to find decimal approximations for each solution.

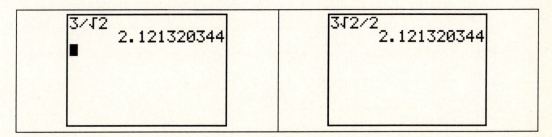

Although they don't look identical, Julie's and Randolph's solutions are precisely the same. It is important to agree on some sort of form for your answers so that you can compare answers with one another.

Finding a Standard Form for Your Answers. In this text, when working problems that involve radical notation, the form of your answer must comply with the following standard.

> *The rules for simple form*
>   *1. If you can factor out a perfect nth root, then you must.*
>   *2. No fractions are allowed under a radical.*
>   *3. No radicals are allowed in the denominator of a fraction.*

If your answer conforms with this standard, your answer is said to be in *simple form*.

Example 3   Place the expression $\sqrt{50}$ in simple form.

Solution   You can use the property $\sqrt{ab} = \sqrt{a}\sqrt{b}$ to help factor out a perfect square.

$$\sqrt{50} = \sqrt{25}\sqrt{2}$$
$$= 5\sqrt{2}$$

Example 4   In this example note how the property $\sqrt{ab} = \sqrt{a}\sqrt{b}$ is used to help factor out a perfect square. You can check your result on your calculator.

$$\sqrt{200} = \sqrt{100}\sqrt{2}$$
$$= 10\sqrt{2}$$

```
√200
           14.14213562
10√2
           14.14213562
```

Example 5   Place the expression $\sqrt{\dfrac{1}{12}}$ in simple form.

Tip	A good strategy is to try to make the denominator a perfect square.

Solution   Make your denominator a perfect square. Begin by multiplying the top and bottom of the fraction by 3, then apply the property $\sqrt{\dfrac{a}{b}} = \dfrac{\sqrt{a}}{\sqrt{b}}$.

$$\sqrt{\frac{1}{12}} = \sqrt{\frac{3}{36}}$$
$$= \frac{\sqrt{3}}{\sqrt{36}}$$
$$= \frac{\sqrt{3}}{6}$$

You no longer have a fraction under a radical, nor do you have a radical in the denominator.

Example 6   In this example the property $\sqrt{\dfrac{a}{b}} = \dfrac{\sqrt{a}}{\sqrt{b}}$ is used to remove the fraction from under the radical. Note how you can check your result on your calculator.

$$\sqrt{\frac{7}{20}} = \sqrt{\frac{35}{100}}$$

$$= \frac{\sqrt{35}}{\sqrt{100}}$$

$$= \frac{\sqrt{35}}{10}$$

```
√(7/20)
           .5916079783
√35/10
           .5916079783
```

Example 7   Place the expression $\dfrac{6}{\sqrt{8}}$ in simple form.

Tip	A good strategy is to try to make the denominator a perfect square.

Solution   Begin by multiplying the numerator and denominator of the fraction by $\sqrt{2}$.

$$\frac{6}{\sqrt{8}} = \frac{6}{\sqrt{8}} \cdot \frac{\sqrt{2}}{\sqrt{2}}$$

$$= \frac{6\sqrt{2}}{\sqrt{16}}$$

$$= \frac{6\sqrt{2}}{4}$$

$$= \frac{3\sqrt{2}}{2}$$

Example 8   Place the expression $\dfrac{12}{\sqrt{27}}$ in simple form. Check your result on your calculator.

Solution

$$\frac{12}{\sqrt{27}} = \frac{12}{\sqrt{27}} \cdot \frac{\sqrt{3}}{\sqrt{3}}$$

$$= \frac{12\sqrt{3}}{\sqrt{81}}$$

$$= \frac{12\sqrt{3}}{9}$$

$$= \frac{4\sqrt{3}}{3}$$

Check

```
12/√27
           2.309401077
4√3/3
           2.309401077
```

Example 9   Find all real solutions of the equation $x^2 - 2 = 7 - x^2$. Place your solutions in simple form.

Tip	To find a graphical solution of an equation of the form $f(x) = g(x)$, you could sketch the graphs of f and g and note where they intersect. However, you could also form a new equation $f(x) - g(x) = 0$ and note where the graph of the equation $f(x) - g(x)$ crosses the x-axis.

Graphical Solution   Make one side of the equation equal to zero.

$$x^2 - 2 = 7 - x^2$$
$$x^2 - 2 - 7 + x^2 = 7 - x^2 - 7 + x^2$$
$$2x^2 - 9 = 0$$

Load the equation $Y_1 = 2X^2 - 9$ in the Y= menu, then press the ZOOM button and select 6:ZStandard. Use the root-finding utility in the CALC menu to find where this graph crosses the x-axis. For a review of the root-finding utility, see Example 12 in Section 4.3.

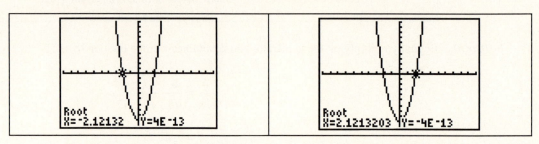

Tip	The notation $Y = 4E - 13$ means $Y = 4 \times 10^{-13}$ or $Y = 0.0000000000004$. Therefore, y is very close to zero, as it should be when finding a root.

Analytical Solution   It's easiest to place your equation in the form $x^2 = b$.

$$2x^2 - 9 = 0$$

$$2x^2 = 9$$

$$x^2 = \frac{9}{2}$$

$$x = \pm\sqrt{\frac{9}{2}}$$

$$x = \pm\frac{3}{\sqrt{2}}$$

$$x = \pm\frac{3}{\sqrt{2}} \cdot \frac{\sqrt{2}}{\sqrt{2}}$$

$$x = \pm\frac{3\sqrt{2}}{2}$$

You can use your calculator to find decimal approximations of your solution.

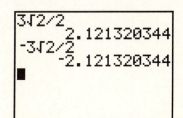

These agree with the graphical estimates.

Place your analytical solution and graphical solution side-by-side on your homework paper for comparison.

$$2x^2 - 9 = 0$$

$$2x^2 = 9$$

$$x^2 = \frac{9}{2}$$

$$x = \pm\sqrt{\frac{9}{2}}$$

$$x = \pm\frac{3}{\sqrt{2}}$$

$$x = \pm\frac{3}{\sqrt{2}} \cdot \frac{\sqrt{2}}{\sqrt{2}}$$

$$x = \pm\frac{3\sqrt{2}}{2}$$

$$x \approx \pm 2.121320344\ldots$$

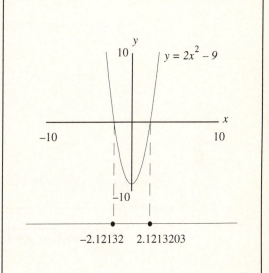

Simplifying Radical Expressions Containing Variables

Recall the following property.

> *If x is any real number, then* $\sqrt{x^2} = |x|$.

You can use this property to develop two properties involving absolute value.

$$|ab| = \sqrt{(ab)^2}$$
$$= \sqrt{a^2b^2}$$
$$= \sqrt{a^2}\sqrt{b^2}$$
$$= |a||b|$$

$$\left|\frac{a}{b}\right| = \sqrt{\left(\frac{a}{b}\right)^2}$$
$$= \sqrt{\frac{a^2}{b^2}}$$
$$= \frac{\sqrt{a^2}}{\sqrt{b^2}}$$
$$= \frac{|a|}{|b|}, \, b \neq 0$$

If a and b are any real numbers, then $|ab| = |a||b|$. If $b \neq 0$, then $\left|\dfrac{a}{b}\right| = \dfrac{|a|}{|b|}$.

Example 10   Place the expression $\sqrt{50x^2}$ in simple form.

Solution   Note that x can be any real number.

$$\sqrt{50x^2} = \sqrt{25x^2}\sqrt{2}$$
$$= \sqrt{(5x)^2}\sqrt{2}$$
$$= |5x|\sqrt{2}$$
$$= |5||x|\sqrt{2}$$
$$= 5|x|\sqrt{2}$$

Let's check this solution by comparing the graphs of the equations $y = 5|x|\sqrt{2}$ and $y = \sqrt{50x^2}$.

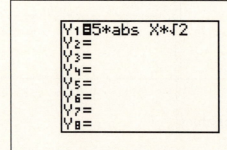
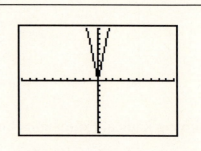

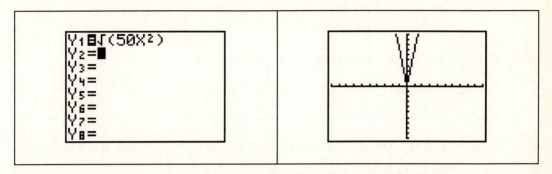

If one graph is superimposed on the other, you have striking visual evidence that you have captured the correct solution.

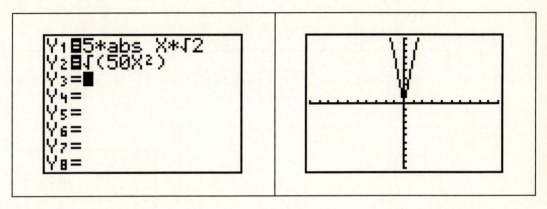

Example 11   Place the expression $\dfrac{6}{\sqrt{8x}}$ in simple form.

Solution   Note that the expression will not be a real number unless $x > 0$. Try to make the denominator a perfect square. Begin by multiplying numerator and denominator by $\sqrt{2x}$.

$$\frac{6}{\sqrt{8x}} = \frac{6}{\sqrt{8x}} \cdot \frac{\sqrt{2x}}{\sqrt{2x}}$$

$$= \frac{6\sqrt{2x}}{\sqrt{16x^2}}$$

$$= \frac{6\sqrt{2x}}{\sqrt{(4x)^2}}$$

$$= \frac{6\sqrt{2x}}{|4x|}$$

However, if $x > 0$, then $4x$ is a positive number and $|4x| = 4x$.

$$= \frac{6\sqrt{2x}}{4x}$$

$$= \frac{3\sqrt{2x}}{2x}$$

Therefore $\dfrac{6}{\sqrt{8x}} = \dfrac{3\sqrt{2x}}{2x}$, but only if $x > 0$.

Simplifying Expressions Involving *n*th Roots

It is not difficult to show that the properties of radicals hold for *n*th roots.

If $\sqrt[n]{a}$ and $\sqrt[n]{b}$ are real numbers, then $\sqrt[n]{a}\,\sqrt[n]{b} = \sqrt[n]{ab}$. If $b \neq 0$, then $\sqrt[n]{\dfrac{a}{b}} = \dfrac{\sqrt[n]{a}}{\sqrt[n]{b}}$.

Tip	These properties hold only if all radical expressions represent real numbers. You cannot use these properties with nonreal numbers such as $\sqrt{-4}$, $\sqrt[4]{-16}$, and $\sqrt[6]{-20}$.

Example 12   Use the property $\sqrt[3]{ab} = \sqrt[3]{a}\,\sqrt[3]{b}$ to place the expression $\sqrt[3]{16}$ in simple form. Use your calculator to check your result.

Solution

$$\sqrt[3]{16} = \sqrt[3]{8}\,\sqrt[3]{2}$$
$$= 2\sqrt[3]{2}$$

```
3√16
              2.5198421
2 3√2
              2.5198421
■
```

Example 13   Use the property $\sqrt[4]{\dfrac{a}{b}} = \dfrac{\sqrt[4]{a}}{\sqrt[4]{b}}$ to place the expression $\sqrt[4]{\dfrac{1}{2}}$ in simple form. Use your calculator to check your result.

Solution   Try to make the denominator a perfect fourth root. With this thought in mind, multiply the numerator and denominator of the fraction by 8.

$$\sqrt[4]{\dfrac{1}{2}} = \sqrt[4]{\dfrac{8}{16}}$$
$$= \dfrac{\sqrt[4]{8}}{\sqrt[4]{16}}$$
$$= \dfrac{\sqrt[4]{8}}{2}$$

```
4 ×√(1/2)
              .8408964153
4 ×√8/2
              .8408964153
```

Example 14   Place the expression $\dfrac{6}{\sqrt[5]{8}}$ in simple form.

Solution   Try to make the denominator a perfect fifth root.

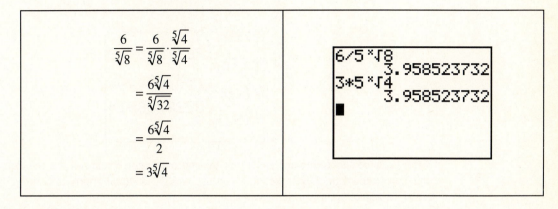

$$\frac{6}{\sqrt[5]{8}} = \frac{6}{\sqrt[5]{8}} \cdot \frac{\sqrt[5]{4}}{\sqrt[5]{4}}$$

$$= \frac{6\sqrt[5]{4}}{\sqrt[5]{32}}$$

$$= \frac{6\sqrt[5]{4}}{2}$$

$$= 3\sqrt[5]{4}$$

Exercises for Section 5.2

1. (TI-82) Enter the expressions $\sqrt{5}\sqrt{11}$, $\sqrt{55}$, and $\sqrt[3]{2}\sqrt[3]{7}$ in your calculator and verify the following results. What is the decimal approximation for the last expression, $\sqrt[3]{14}$?

Check each of the following equations by approximating each side of the equation on your calculator. Record the keystrokes that you use and the decimal approximation you find for each side of the equation on your homework paper.

a) $\sqrt{5}\sqrt{9} = \sqrt{45}$ b) $\sqrt{2}\sqrt{13} = \sqrt{26}$ c) $\sqrt[3]{6}\sqrt[3]{5} = \sqrt[3]{30}$ d) $\sqrt[3]{7}\sqrt[3]{11} = \sqrt[3]{77}$

2. (TI-82) Enter the expressions $\sqrt[4]{7}\sqrt[4]{2}$ and $\sqrt[4]{14}$ in your calculator and verify the following results.

Check each of the following equations by approximating each side of the equation on your calculator. Record the keystrokes that you use and the decimal approximation you find for each side of the equation on your homework paper.

a) $\sqrt[4]{8}\sqrt[4]{3} = \sqrt[4]{24}$ b) $\sqrt[5]{5}\sqrt[5]{12} = \sqrt[5]{60}$ c) $\sqrt[6]{7}\sqrt[6]{6} = \sqrt[6]{42}$ d) $\sqrt[7]{9}\sqrt[7]{12} = \sqrt[7]{108}$

3. (TI-82) Enter the expressions $\sqrt{\dfrac{7}{12}}$, $\dfrac{\sqrt{7}}{\sqrt{12}}$, and $\sqrt[3]{\dfrac{5}{9}}$ in your calculator and verify the following results. What is the decimal approximation for the last expression, $\dfrac{\sqrt[3]{5}}{\sqrt[3]{9}}$?

```
√(7/12)
        .7637626158
√7/√12
        .7637626158
³√(5/9)
        .8220706914
³√5/³√9
```

Check each of the following equations by approximating each side of the equation on your calculator. Record the keystrokes that you use and the decimal approximation you find for each side of the equation on your homework paper.

a) $\dfrac{\sqrt{5}}{\sqrt{8}} = \sqrt{\dfrac{5}{8}}$

b) $\dfrac{\sqrt{6}}{\sqrt{7}} = \sqrt{\dfrac{6}{7}}$

c) $\dfrac{\sqrt[3]{11}}{\sqrt[3]{2}} = \sqrt[3]{\dfrac{11}{2}}$

d) $\dfrac{\sqrt[3]{24}}{\sqrt[3]{5}} = \sqrt[3]{\dfrac{24}{5}}$

4. (TI-82) Enter the following expressions $\dfrac{\sqrt[5]{9}}{\sqrt[5]{7}}$ and $\sqrt[5]{\dfrac{9}{7}}$ in your calculator and verify the following results.

```
5ˣ√9/5ˣ√7
        1.051547497
5ˣ√(9/7)
        1.051547497
```

Check each of the following equations by approximating each side of the equation on your calculator. Record the keystrokes that you use and the decimal approximation you find for each side of the equation on your homework paper.

a) $\dfrac{\sqrt[4]{11}}{\sqrt[4]{2}} = \sqrt[4]{\dfrac{11}{2}}$

b) $\dfrac{\sqrt[5]{7}}{\sqrt[5]{22}} = \sqrt[5]{\dfrac{7}{22}}$

c) $\dfrac{\sqrt[6]{8}}{\sqrt[6]{7}} = \sqrt[6]{\dfrac{8}{7}}$

d) $\dfrac{\sqrt[7]{100}}{\sqrt[7]{49}} = \sqrt[7]{\dfrac{100}{49}}$

5. Place each of the following radical expressions in simple form. Check the result on your calculator. Record the keystrokes you used and decimal approximations you found on your homework paper.

a) $\sqrt{18}$

b) $\sqrt{75}$

c) $\sqrt{12}$

d) $\sqrt{72}$

e) $\sqrt[3]{24}$

f) $\sqrt[3]{81}$

g) $\sqrt[4]{32}$

h) $\sqrt[5]{64}$

6. Place each of the following radical expressions in simple form. Check the result on your calculator. Record the keystrokes you used and decimal approximations you found on your homework paper.

a) $\sqrt{\dfrac{5}{8}}$

b) $\sqrt{\dfrac{7}{32}}$

c) $\sqrt[3]{\dfrac{1}{25}}$

d) $\sqrt[3]{\dfrac{5}{2}}$

e) $\sqrt[4]{\dfrac{1}{2}}$

f) $\sqrt[5]{\dfrac{1}{16}}$

7. Place each of the following radical expressions in simple form. Check the result on your calculator. Record the keystrokes you used and decimal approximations you found on your homework paper.

a) $\dfrac{6}{\sqrt{12}}$
b) $\dfrac{5}{\sqrt{18}}$
c) $\dfrac{6}{\sqrt[3]{2}}$
d) $\dfrac{12}{\sqrt[3]{16}}$
e) $\dfrac{6}{\sqrt[4]{8}}$
f) $\dfrac{10}{\sqrt[5]{4}}$

8. For each of the following pairs of equations, perform the following tasks:

i) Load the equations in the Y= menu in Y1 and Y2, respectively. Adjust the WINDOW parameters so that both graphs appear in the viewing window as well as their points of intersection. Set up a coordinate system on your homework paper and copy the resulting image onto your coordinate system. Label each graph with its equation and include the WINDOW parameters on your coordinate axes.

ii) Use the intersect utility in the CALC menu to find the coordinates of the points of intersection. Draw a number line below your graph and shade and label these estimates of the real solutions of the equation $f(x) = g(x)$ in the usual manner.

iii) Use an analytical method to find the real solutions of the equation $f(x) = g(x)$ in radical form. Place your solutions in simple radical form. Use your calculator to find decimal approximations for your solutions. If these solutions do not compare favorably with the estimates found in part (ii), check your work for error.

a) $f(x) = 2x^2 + 1$, $g(x) = 6$
b) $f(x) = 3 - 2x^2$, $g(x) = -13$
c) $f(x) = 2x^3 - 1$, $g(x) = 2$
d) $f(x) = 8 - x^3$, $g(x) = -8$
e) $f(x) = 8x^4 - 1$, $g(x) = 1$
f) $f(x) = x^5 + 1$, $g(x) = 65$

9. Use the techniques presented in Example 9 in Section 5.2 to find both a graphical solution and an analytical solution for each of the following equations. Use the root utility in the CALC menu to assist you in finding your graphical solution. Place your analytical solutions in simple radical form and use your calculator to find decimal approximations for your answers. Place your analytical and your graphical solutions side-by-side on your homework paper for easy comparison. If your analytical solutions do not compare favorably with your graphical estimates, check your work for error.

a) $5x^2 - 10 = x^2 + 70$
b) $3x^3 - 5 = 11 + 2x^3$
c) $x^4 - 1 = 1 - 3x^4$
d) $x^5 = 128 - x^5$

10. Consider the following graph of the equation $y = \sqrt{4x^2}$.

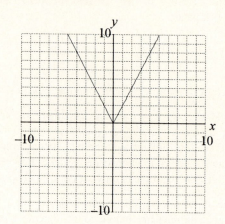

Set up a coordinate system on a sheet of graph paper and use a blue pencil or pen to make an *exact duplicate* of this graph on your coordinate system. Plot points if you have to.

a) On the same coordinate system, use a red pencil or pen to sketch the graph of $y = 2x$. Where does the blue graph of $y = \sqrt{4x^2}$ coincide with the red graph of $y = 2x$? In other words, for what values of x does $\sqrt{4x^2} = 2x$?

b) On the same coordinate system, use a green pencil or pen to sketch the graph of $y = -2x$. Where does the blue graph of $y = \sqrt{4x^2}$ coincide with the green graph of $y = -2x$? In other words, for what values of x does $\sqrt{4x^2} = -2x$?

11. Consider the function whose equation is $f(x) = \sqrt{x^2}$. Copy and complete the following table of points that satisfy the equation of the function f onto a sheet of graph paper.

x	−5	−4	−3	−2	−1	0	1	2	3	4	5
$f(x)$											
Points to plot											

a) Set up a coordinate system on a sheet of graph paper. Use a blue pencil to plot the points from your table on your coordinate system. Use a blue pencil to draw a complete graph of the function whose equation is $f(x) = \sqrt{x^2}$.

b) On the same coordinate system, use a red pencil or pen to draw the graph of the line whose equation is $y = x$. Where do the blue and red graphs coincide? In other words, for what values of x does $\sqrt{x^2} = x$?

c) On the same coordinate system, use a green pencil or pen to draw the graph of the line whose equation is $y = -x$. Where do the blue and green graphs coincide? In other words, for what values of x does $\sqrt{x^2} = -x$?

12. (TI-82) Load the following equations into the Y= menu:

Push the ZOOM button and select 6:ZStandard. Set up a coordinate system on a sheet of graph paper and copy the image in your viewing window onto your coordinate system. Use a blue pencil to draw the graph of Y1, a red pencil to draw the graph of Y2, and a green pencil to draw the graph of Y3. Label each graph with its equation and do this labeling with the appropriate colored pencil.

a) For what values of x does $\sqrt{25x^2} = 5x$?

b) For what values of x does $\sqrt{25x^2} = -5x$?

13. (TI-82) Use your calculator to draw the graphs of $f(x) = \sqrt{9x^2}$, $g(x) = 3x$, and $h(x) = |3x|$. Set up three coordinate systems on your homework paper. Place the graph of f on the first, the graph of g on the second, and the graph of h on the third coordinate system. Which of the graphs are identical? Which of the equations are identical?

14. (TI-82) Use your calculator to draw the graphs of $f(x) = \sqrt{16x^2}$, $g(x) = 4x$, and $h(x) = |4x|$. Set up three coordinate systems on your homework paper. Place the graph of f on the first, the graph of g on the second, and the graph of h on the third coordinate system. Which of the graphs are identical? Which of the equations are identical?

15. For what values of x are each of the following radical expressions real numbers?

a) $\sqrt{36x^2}$ b) $\sqrt{49x^3}$ c) $\dfrac{8x}{\sqrt{2x}}$ d) $\sqrt{\dfrac{-9}{10x^4}}$

16. Place each of the following radical expressions in simple form. Use set-builder notation to describe the set of x-values for which your solution is valid.

a) $\sqrt{45x^2}$ b) $\sqrt{27x^3}$ c) $\sqrt{54x^4}$ d) $\sqrt{72x^5}$ e) $\sqrt{\dfrac{1}{12x^2}}$

f) $\sqrt{\dfrac{12}{x^3}}$ g) $\dfrac{8}{\sqrt{12x^4}}$ h) $\dfrac{6}{\sqrt{24x^3}}$ i) $\sqrt{(x-4)^2}$ j) $\sqrt{(x-5)^3}$

17. (TI-82) Use your calculator to draw the graphs of $f(x) = \sqrt[3]{x^3}$ and $g(x) = x$. Set up a coordinate system on a sheet of graph paper. Use a blue pencil to draw the graph of the function f on your coordinate system. Use a red pencil to draw the graph of the function g on the same coordinate system. For what values of x do the graphs of f and g coincide? In other words, for what values of x is $\sqrt[3]{x^3} = x$?

18. (TI-82) Use your calculator to draw the graphs of $f(x) = \sqrt[4]{x^4}$, $g(x) = x$, and $h(x) = |x|$. Set up three coordinate systems on your homework paper. Place the graph of f on the first, the graph of g on the second, and the graph of h on the third coordinate system. Which of the graphs are identical? Which of the equations are identical?

19. (TI-82) Use your calculator to draw the graphs of $f(x) = \sqrt[5]{x^5}$ and $g(x) = x$. Set up a coordinate system on a sheet of graph paper. Use a blue pencil to draw the graph of the function f on your coordinate system. Use a red pencil to draw the graph of the function g on the same coordinate system. For what values of x do the graphs of f and g coincide? In other words, for what values of x is $\sqrt[5]{x^5} = x$?

20. (TI-82) Use your calculator to draw the graphs of $f(x) = \sqrt[6]{x^6}$, $g(x) = x$, and $h(x) = |x|$. Set up three coordinate systems on your homework paper. Place the graph of f on the first, the graph of g on the second, and the graph of h on the third coordinate system. Which of the graphs are identical? Which of the equations are identical?

5.3 Combining Radicals

There are a few more manipulative techniques involving radicals that need to be worked on. You will need to recall some properties that were already discussed and apply them to a selection of problems.

Useful Properties of Multiplication

With real numbers, multiplication is commutative. Changing the order of the factors does not affect the product.

> *The commutative property of multiplication: If a and b are real numbers, then ab = ba.*

With real numbers, multiplication is associative. When finding the product of three numbers, begin by taking the product of the first and second number or by taking the product of the second and third number. Then multiply this result by the remaining number.

> *The associative property of multiplication: If a and b are real numbers, then a(bc) = (ab)c.*

Example 1   Simplify the expression $5\left(2\sqrt{3}\right)$.

Solution

Multiplication is associative.
 	
$$5\left(2\sqrt{3}\right)=(5\cdot2)\sqrt{3}$$
$$=10\sqrt{3}$$
 	
It is helpful to note that $5\left(2\sqrt{3}\right)=10\sqrt{3}$ in much the same manner as $5(2x)=10x$.
	``5(2√3)
	        17.32050808
	10√3
	        17.32050808```

Example 2 Simplify the expression $-8\left(2\sqrt[3]{5}\right)$.

Solution In actual practice, when taking the product of three numbers, simply multiply the two numbers of your choice. Then multiply this result by the remaining number. Thus, $-8\left(2\sqrt[3]{5}\right)=-16\sqrt[3]{5}$.

Example 3 Simplify the expression $\left(2\sqrt{5}\right)\left(3\sqrt{7}\right)$.

Solution Because multiplication of real numbers is both commutative and associative, you are free to change the order or the grouping of any multiplication problem. In this example, four numbers are being multiplied. You may multiply in the order of your choice.

The numbers can be multiplied in the following order.
 	
$$\left(2\sqrt{5}\right)\left(3\sqrt{7}\right)=(2\cdot3)\cdot\left(\sqrt{5}\cdot\sqrt{7}\right)$$
$$=6\sqrt{35}$$
 	
It is helpful to note that $\left(2\sqrt{5}\right)\left(3\sqrt{7}\right)=6\sqrt{35}$ in much the same manner as $(2x)(3y)=6xy$.
	``(2√5)(3√7)
	         35.4964787
	6√35
	         35.4964787```

> ***If a and b are real numbers, then $(ab)^2=a^2b^2$.***

Example 4 Simplify the expression $\left(3\sqrt{6}\right)^2$.

Solution

The square of a product is the product of the squares of each factor.

$$\left(3\sqrt{6}\right)^2 = (3)^2 \cdot \left(\sqrt{6}\right)^2$$
$$= 9 \cdot 6$$
$$= 54$$

It is helpful to note that $\left(3\sqrt{6}\right)^2 = 9 \cdot 6$ in much the same manner as $(3x)^2 = 9x^2$.

Multiplication is distributive over addition. If a, b, and c are real numbers, then $a(b+c) = ab + ac$.

Example 5   Simplify the expression $5\sqrt{2} + 3\sqrt{2}$.

Solution

Use the distributive property to factor.

$$5\sqrt{2} + 3\sqrt{2} = (5+3)\sqrt{2}$$
$$= 8\sqrt{2}$$

It is helpful to note that $5\sqrt{2} + 3\sqrt{2} = 8\sqrt{2}$ in much the same manner as $5x + 3x = 8x$.

```
5√2+3√2
          11.3137085
8√2
          11.3137085
```

Example 6   Simplify the expression $-8\sqrt[3]{5} + 12\sqrt[3]{5}$.

Solution

Use the distributive property to factor.

$$-8\sqrt[3]{5} + 12\sqrt[3]{5} = (-8+12)\sqrt[3]{5}$$
$$= 4\sqrt[3]{5}$$

It is helpful to note that $-8\sqrt[3]{5} + 12\sqrt[3]{5} = 4\sqrt[3]{5}$ in much the same manner as $-8x + 12x = 4x$.

```
-8³√5+12³√5
          6.839903787
4³√5
          6.839903787
```

Example 7 Simplify the expression $2x\sqrt{2} + 3x\sqrt{2}$.

Solution In actual practice, this problem can be handled in the same manner that the problem $2y + 3y$ was handled. These are like terms, so you simply add the coefficients. Therefore, $2x\sqrt{2} + 3x\sqrt{2} = 5x\sqrt{2}$ in much the same manner as $2y + 3y = 5y$.

Example 8 Simplify the expression $\sqrt{8} + \sqrt{3} + \sqrt{2} + \sqrt{27}$.

Solution

Place each term in simple form, then add the like terms.
 	
$\sqrt{8} + \sqrt{3} + \sqrt{2} + \sqrt{27} = 2\sqrt{2} + \sqrt{3} + \sqrt{2} + 3\sqrt{3}$
 	
$\phantom{\sqrt{8} + \sqrt{3} + \sqrt{2} + \sqrt{27}} = 3\sqrt{2} + 4\sqrt{3}$
	`√8+√3+√2+√27`
	`        11.17084392`
	`3√2+4√3`
	`        11.17084392`

Example 9 Simplify the expression $\sqrt{243} + \dfrac{1}{\sqrt{3}} + \sqrt{\dfrac{1}{27}}$.

Solution

Place each term in simple form, then combine like terms.
 	
$\sqrt{243} + \dfrac{1}{\sqrt{3}} + \sqrt{\dfrac{1}{27}} = \sqrt{81}\sqrt{3} + \dfrac{\sqrt{3}}{\sqrt{9}} + \sqrt{\dfrac{3}{81}}$
 	
$ = 9\sqrt{3} + \dfrac{\sqrt{3}}{3} + \dfrac{\sqrt{3}}{9}$
 	
$ = \dfrac{81\sqrt{3}}{9} + \dfrac{3\sqrt{3}}{9} + \dfrac{\sqrt{3}}{9}$
 	
$ = \dfrac{85\sqrt{3}}{9}$
	`√243+1/√3+√(1/27`
	`)`
	`        16.35825763`
	`85√3/9`
	`        16.35825763`
	`■`

Example 10 Simplify the expression $\left(2\sqrt{5} + 3\right)\left(3\sqrt{5} - 7\right)$.

Solution

Use the distributive property to multiply.

$$\left(2\sqrt{5}+3\right)\left(3\sqrt{5}-7\right) = 2\sqrt{5}\left(3\sqrt{5}-7\right) + 3\left(3\sqrt{5}-7\right)$$

$$= 6\sqrt{25} - 14\sqrt{5} + 9\sqrt{5} - 21$$

$$= 30 - 5\sqrt{5} - 21$$

$$= 9 - 5\sqrt{5}$$

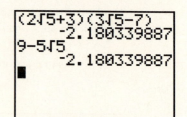

Squaring a binomial multiplication pattern: $(a+b)^2 = a^2 + 2ab + b^2$.

Example 11   Simplify the expression $\left(5\sqrt{2}+3\right)^2$.

Solution

Use the squaring a binomial multiplication pattern.

$$\left(5\sqrt{2}+3\right)^2 = \left(5\sqrt{2}\right)^2 + 2\left(5\sqrt{2}\right)(3) + (3)^2$$

$$= 25\cdot2 + 30\sqrt{2} + 9$$

$$= 50 + 30\sqrt{2} + 9$$

$$= 59 + 30\sqrt{2}$$

```
(5√2+3)²
         101.4264069
59+30√2
         101.4264069
■
```

Difference of squares pattern: If a and b are real numbers, then $(a+b)(a-b) = a^2 - b^2$

Example 12   Simplify the expression $\left(3\sqrt{6}+2\sqrt{5}\right)\left(3\sqrt{6}-2\sqrt{5}\right)$.

Solution

Use the difference of squares multiplication
pattern.

$$\left(3\sqrt{6}+2\sqrt{5}\right)\left(3\sqrt{6}-2\sqrt{5}\right)=\left(3\sqrt{6}\right)^2-\left(2\sqrt{5}\right)^2$$

$$=9\cdot 6-4\cdot 5$$

$$=54-20$$

$$=34$$

It is interesting to note that this answer is free of
radicals.

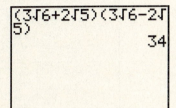

Example 13   Simplify the expression $\left(4\sqrt{3}+5\right)\left(4\sqrt{3}-5\right)$.

Solution

Use the difference of squares multiplication
pattern.

$$\left(4\sqrt{3}+5\right)\left(4\sqrt{3}-5\right)=\left(4\sqrt{3}\right)^2-\left(5\right)^2$$

$$=16\cdot 3-25$$

$$=48-25$$

$$=23$$

It is interesting to note that this answer is free
of radicals.

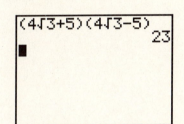

Example 14   Simplify the expression $\dfrac{\sqrt{3}+2}{4\sqrt{3}+5}$.

Solution   The solution to the last example provides a clue. Begin this problem by multiplying the numerator and denominator by $4\sqrt{3}-5$.

$$\frac{\sqrt{3}+2}{4\sqrt{3}+5} = \frac{\sqrt{3}+2}{4\sqrt{3}+5}\cdot\frac{4\sqrt{3}-5}{4\sqrt{3}-5}$$

$$= \frac{\sqrt{3}\left(4\sqrt{3}-5\right)+2\left(4\sqrt{3}-5\right)}{\left(4\sqrt{3}\right)^2-(5)^2}$$

$$= \frac{4\sqrt{9}-5\sqrt{3}+8\sqrt{3}-10}{16\cdot3-25}$$

$$= \frac{12+3\sqrt{3}-10}{48-25}$$

$$= \frac{2+3\sqrt{3}}{23}$$

```
(√3+2)/(4√3+5)
          .3128761923
(2+3√3)/23
          .3128761923
■
```

Example 15   Consider the following graph whose equation is $f(x) = x^2$. Points P and Q are on the graph of the function *f*. The *x*-values of points P and Q are $\sqrt{2}$ and $\sqrt{5}$, respectively. Find the slope of the line that passes through points P and Q. Place your answer in simple form.

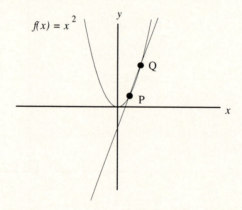

Solution   The coordinates of points P and Q are $\left(\sqrt{2}, f\left(\sqrt{2}\right)\right)$ and $\left(\sqrt{5}, f\left(\sqrt{5}\right)\right)$, respectively.

$$\text{Slope} = \frac{\text{change in } y}{\text{change in } x}$$

$$= \frac{f\left(\sqrt{5}\right) - f\left(\sqrt{2}\right)}{\sqrt{5} - \sqrt{2}}$$

$$= \frac{\left(\sqrt{5}\right)^2 - \left(\sqrt{2}\right)^2}{\sqrt{5} - \sqrt{2}}$$

$$= \frac{5 - 2}{\sqrt{5} - \sqrt{2}}$$

$$= \frac{3}{\sqrt{5} - \sqrt{2}}$$

$$\text{Slope} = \frac{3}{\sqrt{5} - \sqrt{2}} \cdot \frac{\sqrt{5} + \sqrt{2}}{\sqrt{5} + \sqrt{2}}$$

$$= \frac{3\left(\sqrt{5} + \sqrt{2}\right)}{\left(\sqrt{5}\right)^2 - \left(\sqrt{2}\right)^2}$$

$$= \frac{3\sqrt{5} + 3\sqrt{2}}{3}$$

$$= \frac{3\sqrt{5}}{3} + \frac{3\sqrt{2}}{3}$$

$$= \sqrt{5} + \sqrt{2}$$

Exercises for Section 5.3

1. Place each of the following expressions in simple radical form. Enter both the original expression and your solution into your calculator and find decimal approximations for each. Record these approximations on your homework paper.

   a) $5\left(6\sqrt{3}\right)$ b) $-3\left(-4\sqrt{7}\right)$ c) $12\left(4\sqrt[3]{7}\right)$ d) $-3\left(-5\sqrt[3]{5}\right)$

   e) $\left(2\sqrt{3}\right)\left(-5\sqrt{7}\right)$ f) $2\left(3\sqrt{2}\right)\left(5\sqrt{18}\right)$ g) $\left(2\sqrt[3]{11}\right)\left(5\sqrt[3]{2}\right)$ h) $\left(-3\sqrt[3]{4}\right)\left(2\sqrt[3]{2}\right)$

2. (TI-82) The expression $\left(2\sqrt[4]{5}\right)\left(3\sqrt[4]{4}\right)$ is equal to $6\sqrt[4]{20}$. Check this on your calculator in the following manner:

```
(2*4 ˣ√5)*(3*4 ˣ√4
)
            12.68845516
6*4 ˣ√20
            12.68845516
```

Place each of the following expressions in simple radical form. Enter both the original expression and the solution into your calculator and find decimal approximations for each. Record these approximations on your homework paper.

   a) $3\left(8\sqrt[4]{5}\right)$ b) $-4\left(-6\sqrt[5]{11}\right)$ c) $\left(2\sqrt[6]{8}\right)\left(-3\sqrt[6]{2}\right)$ d) $2\left(\sqrt[7]{32}\right)\left(\sqrt[7]{4}\right)$

3. Place each of the following expressions in simple radical form. Check your solution with your calculator.

a) $\left(5\sqrt{7}\right)^2$ b) $\left(-4\sqrt{11}\right)^2$ c) $\left(-2\sqrt[3]{2}\right)^3$ d) $\left(-4\sqrt[3]{7}\right)^3$

4. (TI-82) The expression $\left(2\sqrt[4]{5}\right)^4$ is equal to 80. Check this on your calculator in the following manner:

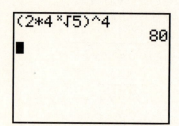

Place each of the following expressions in simple radical form. Check your solution with your calculator.

a) $\left(-2\sqrt[4]{7}\right)^4$ b) $\left(2\sqrt[5]{2}\right)^5$

5. Place each of the following expressions in simple radical form. Enter both the original expression and the solution into your calculator and find decimal approximations for each. Record these approximations on your homework paper.

a) $\left(2\sqrt[3]{5}\right)^2$ b) $\left(-3\sqrt[3]{4}\right)^2$ c) $\left(3\sqrt[3]{9}\right)^2$ d) $\left(-2\sqrt[4]{8}\right)^2$

6. Place each of the following expressions in simple radical form. Enter both the original expression and the solution into your calculator and find decimal approximations for each. Record these approximations on your homework paper.

a) $8\sqrt{3}+4\sqrt{3}$ b) $4\sqrt[3]{5}-8\sqrt[3]{5}$ c) $\sqrt{8}+\sqrt{32}$

d) $\sqrt[3]{2}+\sqrt[3]{16}$ e) $5\sqrt{12}-7\sqrt{27}$ f) $-5\sqrt{20}+12\sqrt{45}$

g) $\sqrt{2}+\sqrt{3}+\sqrt{50}-\sqrt{75}$ h) $\sqrt{6}+\sqrt[3]{6}+\sqrt{24}-2\sqrt[3]{48}$ i) $\sqrt{2}-\dfrac{4}{\sqrt{2}}+\sqrt{\dfrac{1}{2}}$

j) $\sqrt[3]{2}-\dfrac{1}{\sqrt[3]{4}}$

7. Use the distributive property to multiply. Place your answer in simple radical form. Enter both the original expression and the solution into your calculator and find decimal approximations for each. Record these approximations on your homework paper.

a) $\left(\sqrt{3}+2\right)\left(2\sqrt{3}-5\right)$ b) $\left(3-\sqrt{6}\right)\left(4+5\sqrt{6}\right)$

c) $\left(\sqrt{5}+\sqrt{3}\right)\left(2\sqrt{5}-\sqrt{3}\right)$ d) $\left(2\sqrt{3}+\sqrt{5}\right)\left(3\sqrt{3}-\sqrt{2}\right)$

e) $\left(\sqrt[3]{2}+1\right)\left(\sqrt[3]{4}-\sqrt[3]{2}+1\right)$ f) $\left(\sqrt[4]{3}+\sqrt[4]{2}\right)\left(\sqrt[4]{27}-\sqrt[4]{18}+\sqrt[4]{12}-\sqrt[4]{8}\right)$

8. Use the squaring a binomial pattern to multiply. Place your answer in simple radical form. Enter both the original expression and the solution into your calculator and find decimal approximations for each. Record these approximations on your homework paper.

a) $\left(\sqrt{2}+1\right)^2$

b) $\left(\sqrt{5}-\sqrt{3}\right)^2$

c) $\left(2\sqrt{5}-6\sqrt{3}\right)^2$

d) $\left(5-3\sqrt[3]{2}\right)^2$

9. Use the difference of squares pattern to multiply. Place your answer in simple radical form. Check your solution with your calculator.

a) $\left(\sqrt{5}+3\right)\left(\sqrt{5}-3\right)$

b) $\left(2\sqrt{3}+\sqrt{2}\right)\left(2\sqrt{3}-\sqrt{2}\right)$

c) $\left(\sqrt{11}-2\sqrt{2}\right)\left(\sqrt{11}+2\sqrt{2}\right)$

d) $\left(\sqrt[3]{5}+\sqrt[3]{2}\right)\left(\sqrt[3]{5}-\sqrt[3]{2}\right)$

10. Place each of the following expressions in simple radical form. Enter both the original expression and the solution into your calculator and find decimal approximations for each. Record these approximations on your homework paper.

a) $\dfrac{1}{2\sqrt{3}+1}$

b) $\dfrac{1}{\sqrt{5}-\sqrt{2}}$

c) $\dfrac{6}{2\sqrt{2}+1}$

d) $\dfrac{10}{\sqrt{5}-5}$

e) $\dfrac{2+\sqrt{2}}{3-2\sqrt{2}}$

f) $\dfrac{\sqrt{3}+1}{\sqrt{3}-1}$

g) $\dfrac{2\sqrt{2}+\sqrt{5}}{3\sqrt{2}-2\sqrt{5}}$

h) $\dfrac{\sqrt{7}+2\sqrt{5}}{\sqrt{7}-2\sqrt{5}}$

11. Each of the following figures is presented with its area formula. Use the area formula and the given information to find the area of the figure. Place your answer in simple radical form. Use your calculator to find a decimal approximation for your answer. Round your answer to the nearest tenth of a square inch (one decimal place).

a) $L=4+3\sqrt{2}$ inches, $W=1+\sqrt{2}$ inches

b) $x=2\sqrt{6}+5\sqrt{2}$ inches

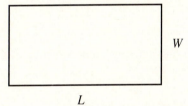

W

L

$A = LW$

x

x

$A = x^2$

c) $b=5+\sqrt{3}$ inches, $h=1+\sqrt{3}$ inches

d) $b=5+2\sqrt{6}$ inches, $h=5+\sqrt{6}$ inches

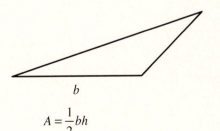

h

b

$A = \dfrac{1}{2}bh$

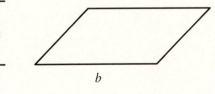

h

b

$A = bh$

e) $b_1 = \sqrt{2}$ inches, $b_2 = 2 + \sqrt{2}$ inches, $h = 3\sqrt{2}$ inches

f) $r = \sqrt{5 + 2\sqrt{2}}$ inches

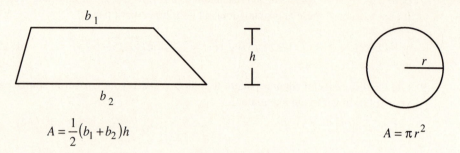

$$A = \frac{1}{2}(b_1 + b_2)h$$

$$A = \pi r^2$$

12. (TI-82) Load the equation $f(x) = \sqrt{2}x + 3$ into the Y= menu in the following manner:

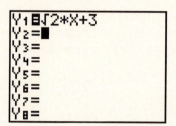

Press the ZOOM button and select 6:ZStandard. Set up a coordinate system on your homework paper and copy the image from your viewing window onto it.

a) Use the root-finding utility to find the x-intercept of your graph. Draw a number line below your graph and shade and label this estimated solution of $f(x) = 0$ on your number line.

b) Use an analytical method to find the x-intercept of your graph in simple radical form. Use your calculator to find a decimal approximation of this solution. If this solution does not agree favorably with the estimate found in part (a), check your work for error.

13. Find the slope of the line that goes through the points $\left(1, 1 + \sqrt{2}\right)$ and $\left(\sqrt{2}, 1\right)$. Place your answer in simple radical form. Use your calculator to find a decimal approximation of the slope, correct to the nearest ten-thousandth (four decimal places).

14. Consider the following graph of the function f whose equation is $f(x) = x^2$.

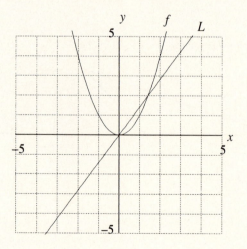

The line L passes through the origin and crosses the graph of f at the point $\left(\sqrt{2}, f\left(\sqrt{2}\right)\right)$. Find the equation of line L. Make sure all radicals are in simple form.

15. Consider the graph of the function f whose equation is $f(x) = \sqrt{x}$.

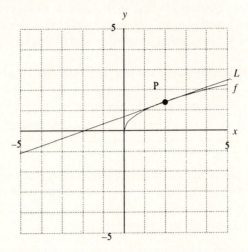

Line L touches the graph of f at point P. Mathematicians like to say that line L is tangent to the graph of f at point P. The coordinates of point P are $(2, f(2))$. The slope of line L is $\dfrac{\sqrt{2}}{4}$.

a) Set up a coordinate system on a sheet of graph paper and carefully copy this graph onto your coordinate system. Include an accurate reproduction of the line L. On your coordinate system, clearly state an estimate of the y-intercept of line L.

b) Use the point-slope form of a line to find the equation of line L.

c) Place your equation of line L in slope-intercept form by solving the equation found in part (b) for y. Clearly state the y-intercept of line L in simple radical form.

d) Use a calculator to estimate your answer from part (c). Compare this answer with the estimate made in part (a).

16. Find the x- and y-intercepts of the line whose equation is $\sqrt{2}x + \sqrt{3}y = 6$. Place your answers in simple radical form.

17. If you tie a string to a hook and attach a mass to the end of the string, you have created what physicists call a pendulum. If you pull the mass back and let it go, then the mass will swing back and forth at the end of the string. The time it takes to complete one swing is called the period of the pendulum. The expression $\sqrt{\dfrac{4\pi^2 L}{g}}$ can be used to compute the period of the pendulum, where L is the length of the string and g is the acceleration due to gravity. Place this expression in simple radical form.

18. Koh drops a rock into the Grand Canyon. The time that it takes to fall a distance of s feet is given by the expression $\sqrt{\dfrac{2s}{g}}$, where g is the acceleration due to gravity. Place this expression in simple radical form.

5.4 Solving Equations Containing Radical Expressions

In this section we will solve equations that contain radical expressions. To eliminate the radicals from an equation, the radical expression will have to be isolated on one side of the equation, then both sides of the equation will have to be raised to an appropriate power. You will find that raising both sides of an equation to an even power can change the solution set of an equation. The TI-82 will be used to predict the number of solutions and to find approximations for those solutions.

A Useful Technique.   You can raise both sides of an equation to a power of your choice.

$$\text{If } a = b, \text{ then } a^n = b^n.$$

Although this is an extremely useful technique, it can also introduce problems. For example, 3 is the only solution of the equation $x = 3$. If both sides of the equation $x = 3$ are squared, you get the equation $x^2 = 9$. However, both 3 and -3 are solutions of the equation $x^2 = 9$. Squaring both sides of the equation $x = 3$ has enlarged the solution set.

Tip	Raising both sides of an equation to an even power can add solutions that will not check in the original equation.

Example 1   *(TI-82)* Solve the following equation for x: $\sqrt{x+1} = -5$.

Graphical Solution   This equation is of the form $f(x) = g(x)$, where $f(x) = \sqrt{x+1}$ and $g(x) = -5$. Graph each function, then note where they intersect.

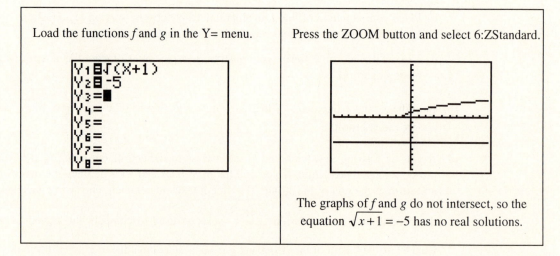

Load the functions f and g in the Y= menu.

Press the ZOOM button and select 6:ZStandard.

The graphs of f and g do not intersect, so the equation $\sqrt{x+1} = -5$ has no real solutions.

Note: Because the expression $\sqrt{x+1}$ is a real number only if $x \geq -1$, the graph of the equation $f(x) = \sqrt{x+1}$ exists only to the right of the point $(-1, 0)$.

If you take a second glance at the equation $\sqrt{x+1} = -5$, it is perfectly obvious that the equation can have no solutions. The left side of this equation calls for a *nonnegative* square root, but the right side of this equation is definitely negative. This is an impossible situation, so the equation cannot have any solutions.

Analytical Solution   Begin by squaring both sides of the equation.

$$\sqrt{x+1} = -5$$

$$\left(\sqrt{x+1}\right)^2 = (-5)^2$$

$$x+1 = 25$$

$$x = 24$$

The solution $x = 24$ is completely unexpected. Where did it come from?

Tip	If you square both sides of an equation, you may introduce wrong answers. All solutions should be checked, preferably in the original equation.

Check.   Check the solution $x = 24$ in the original equation.

$$\sqrt{x+1} = -5$$

$$\sqrt{24+1} = -5$$

$$\sqrt{25} = -5$$

$$5 = -5$$

Clearly, the solution does not check and must be discarded. Therefore, the equation has no solutions.

Place your analytical solution and graphical solution side-by-side on your homework paper for comparison.

<table>
<tr>
<td>

$$\sqrt{x+1} = -1$$

$$\left(\sqrt{x+1}\right)^2 = (-1)^2$$

$$x+1 = 1$$

$$x = 0$$

Check

$$\sqrt{x+1} = -1$$

$$\sqrt{0+1} = -1$$

$$\sqrt{1} = -1$$

$$1 = -1$$

The solution $x = 24$ does not check. The equation has no solutions.

</td>
<td>

There are no points of intersection. Therefore, the equation $\sqrt{x+1} = -5$ has no solutions.

</td>
</tr>
</table>

Example 2   *(TI-81)* Solve the following equation for *x*: $\sqrt{2x+3} = x$.

Graphical Solution

<table>
<tr>
<td>

Load each side of the equation into the Y= menu, then press the ZOOM button and select 6:ZStandard.

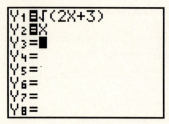

</td>
<td>

You can use the intersect utility in the CALC menu to find the point of intersection. For a review of the intersect utility, see Example 10 in Section 1.6.

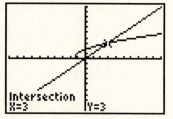

The graph indicates that the solution of the equation $\sqrt{2x+3} = x$ is $x = 3$.

</td>
</tr>
</table>

Note: If you want the expression $\sqrt{2x+3}$ to be a real number, then $2x+3$ must be a nonnegative real number. But $2x+3 \geq 0$ only if $x \geq -\dfrac{3}{2}$. Therefore, the graph of $f(x) = \sqrt{3x+2}$ only exists to the right of the point $\left(-\dfrac{3}{2}, 0\right)$.

Analytical Solution   First, square both sides of the equation, as follows:

$$\sqrt{2x+3} = x$$

$$\left(\sqrt{2x+3}\right)^2 = (x)^2$$

$$2x+3 = x^2$$

Because the preceding equation contains a power of *x* larger than 1, the equation is nonlinear. Make one side of the equation zero; then use the *ac* test and the zero product property to complete the solution.

$$0 = x^2 - 2x - 3$$

$$0 = x^2 + x - 3x - 3$$

$$0 = x(x+1) - 3(x+1)$$

$$0 = (x-3)(x+1)$$

$$x = 3, -1$$

Because of the graphical analysis, the solution $x = 3$ was expected but the solution $x = -1$ was not.

Check. Check the solution $x = 3$ in the original equation.

$$\sqrt{2x+3} = x$$
$$\sqrt{2(3)+3} = 3$$
$$\sqrt{9} = 3$$
$$3 = 3$$

The solution $x = 3$ checks. Now, check the solution $x = -1$. Remember, this solution is not expected to check.

$$\sqrt{2x+3} = x$$
$$\sqrt{2(-1)+3} = -1$$
$$\sqrt{1} = -1$$
$$1 = -1$$

Because it does not check in the original equation, discard the solution $x = -1$.

Place your analytical solution and graphical solution side-by-side on your homework paper for comparison.

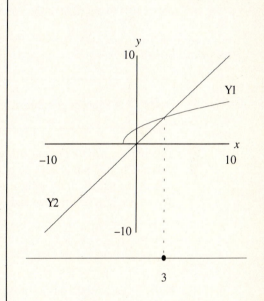

$$\sqrt{2x+3} = x$$
$$\left(\sqrt{2x+3}\right)^2 = (x)^2$$
$$2x+3 = x^2$$
$$0 = x^2 - 2x - 3$$
$$0 = x^2 + x - 3x - 3$$
$$0 = x(x+1) - 3(x+1)$$
$$0 = (x-3)(x+1)$$
$$x = 3, -1$$

Check

$$\sqrt{2x+3} = x \qquad \sqrt{2x+3} = x$$
$$\sqrt{2(3)+3} = 3 \quad \sqrt{2(-1)+3} = -1$$
$$\sqrt{9} = 3 \qquad\qquad \sqrt{1} = -1$$
$$3 = 3 \qquad\qquad 1 = -1$$

The solution $x = 3$ checks, the solution $x = -1$ does not.

The solution of the equation $\sqrt{2x+3} = x$ is $x = 3$.

Example 3   *(TI-81)* Solve the following equation for **x**: $1 + \sqrt{4x+13} = 2x$.

Graphical Solution

Load each side of the equation into the Y= menu, then press the ZOOM button and select 6:ZStandard. 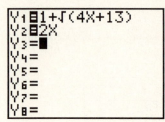	You can use the intersect utility in the CALC menu to find the coordinates of the point of intersection. 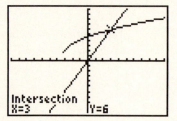 It appears that the solution of the equation $1+\sqrt{4x+13}=2x$ is $x=3$.

Note: In order for the expression $\sqrt{4x+13}$ to be a real number, the expression $4x+13$ must be a nonnegative real number. But $4x+13 \geq 0$ only if $x \geq -\dfrac{13}{4}$. Therefore the graph of $y=1+\sqrt{4x+13}$ only exists to the right of the point $\left(-\dfrac{13}{4}, 1\right)$.

Analytical Solution

Tip	Whenever possible, isolate the radical by itself on one side of the equation.

This phrase, *isolate the radical*, will become a strategy reminder when solving equations involving radicals. It may not always be possible to isolate the radical, but when the radical is isolated on one side of the equation, the radical will disappear when both sides of the equation are squared. Follow this advice in your first step and isolate the radical on one side of the equation.

$$1+\sqrt{4x+13}=2x$$
$$\sqrt{4x+13}=2x-1$$

Next, square both sides of the equation. Square $2x-1$ by using the squaring a binomial multiplication pattern, $(a-b)^2 = a^2 - 2ab + b^2$.

$$\left(\sqrt{4x+13}\right)^2 = (2x-1)^2$$
$$4x+13 = (2x)^2 - 2(2x)(1) + (1)^2$$
$$4x+13 = 4x^2 - 4x + 1$$

Since this equation has a power of x that is greater than 1, the equation is nonlinear. Make one side of the equation zero, then divide both sides of the resulting equation by 4.

$$0 = 4x^2 - 8x - 12$$

$$0 = x^2 - 2x - 3$$

Dividing both sides of the equation by the greatest common divisor is always a good thing to do because it makes the numbers smaller and the equation easier to deal with. Complete your solution by using the *ac* test and the zero product property.

$$0 = x^2 - 3x + x - 3$$

$$0 = x(x - 3) + (x - 3)$$

$$0 = (x + 1)(x - 3)$$

$$x = -1, 3$$

Because of the graphical analysis, the solution $x = 3$ was expected, but the solution -1 was not.

Check. Check the solution $x = -1$ in the original equation. Remember, this solution is not expected to check.

$$1 + \sqrt{4x + 13} = 2x$$

$$1 + \sqrt{4(-1) + 13} = 2(-1)$$

$$1 + \sqrt{9} = -2$$

$$1 + 3 = -2$$

$$4 = -2$$

This solution does not check and must be discarded. Now, check the solution $x = 3$ in the original equation.

$$1 + \sqrt{4x + 13} = 2x$$

$$1 + \sqrt{4(3) + 13} = 2(3)$$

$$1 + \sqrt{25} = 6$$

$$1 + 5 = 6$$

$$6 = 6$$

This solution checks.

Place your analytical solution and graphical solution side-by-side on your homework paper for comparison.

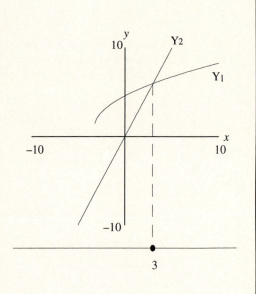

$$1+\sqrt{4x+13}=2x$$
$$\sqrt{4x+13}=2x-1$$
$$4x+13=4x^2-4x+1$$
$$0=x^2-3x+x-3$$
$$0=x(x-3)+(x-3)$$
$$0=(x+1)(x-3)$$
$$x=-1,3$$

Check

$$1+\sqrt{4x+13}=2x \qquad 1+\sqrt{4x+13}=2x$$
$$1+\sqrt{4(-1)+13}=2(-1) \quad 1+\sqrt{4(3)+13}=2(3)$$
$$1+\sqrt{9}=-2 \qquad\qquad 1+\sqrt{25}=6$$
$$1+3=-2 \qquad\qquad 1+5=6$$
$$4=-2 \qquad\qquad 6=6$$

The solution $x=-1$ does not check, the solution $x=3$ checks.

The solution of $1+\sqrt{4x+13}=2x$ is $x=3$.

Example 4 Solve the following equation for x: $\sqrt{2x}+\sqrt{2x+3}=3$.

Graphical Solution

Load each side of the equation into the Y= menu.

Push the ZOOM button and select 6:ZStandard.

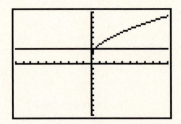

Note: In order for the expression $\sqrt{2x}$ to be a real number, x must be greater than or equal to zero. In order for the expression $\sqrt{2x+3}$ to be a real number, $2x+3$ must be a nonnegative real number. But $2x+3\geq 0$ only if $x\geq -\frac{3}{2}$. The expression $\sqrt{2x}+\sqrt{2x+3}$ will be a real number only if *both* of these conditions are satisfied. Therefore the graph of $y=\sqrt{2x}+\sqrt{2x+3}$ exists only to the right of the point $(0,\sqrt{3})$.

Make the following adjustments to the WINDOW parameters. 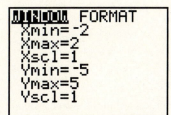	Push the GRAPH button then use the intersect utility in the CALC menu to produce the following image: 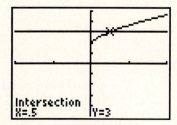 It appears that the solution of $\sqrt{2x}+\sqrt{2x+3}=3$ is $x=0.5$.

Analytical Solution   Since you cannot isolate both radicals at the same time, isolate the more complicated radical on one side of the equation.

$$\sqrt{2x}+\sqrt{2x+3}=3$$

$$\sqrt{2x+3}=3-\sqrt{2x}$$

Square both sides of the equation.

$$\left(\sqrt{2x+3}\right)^2=\left(3-\sqrt{2x}\right)^2$$

$$2x+3=(3)^2-2(3)\left(\sqrt{2x}\right)+\left(\sqrt{2x}\right)^2$$

$$2x+3=9-6\sqrt{2x}+2x$$

Isolate the radical on one side of the equation by subtracting 9 and $2x$ from both sides of the equation.

$$-6=-6\sqrt{2x}$$

Divide both sides of this equation by -6, then square both sides of the resulting equation.

$$1=\sqrt{2x}$$

$$(1)^2=\left(\sqrt{2x}\right)^2$$

$$1=2x$$

$$x=\frac{1}{2}$$

Check. Because of the graphical analysis and the fact that $\frac{1}{2}=0.5$, it is expected that $x=\frac{1}{2}$ will check in the original equation.

$$\sqrt{2x} + \sqrt{2x+3} = 3$$

$$\sqrt{2\left(\frac{1}{2}\right)} + \sqrt{2\left(\frac{1}{2}\right)+3} = 3$$

$$\sqrt{1} + \sqrt{4} = 3$$

$$1 + 2 = 3$$

$$3 = 3$$

The solution $x = \frac{1}{2}$ checks.

Place your analytical solution and graphical solutions side-by-side on your homework paper for comparison.

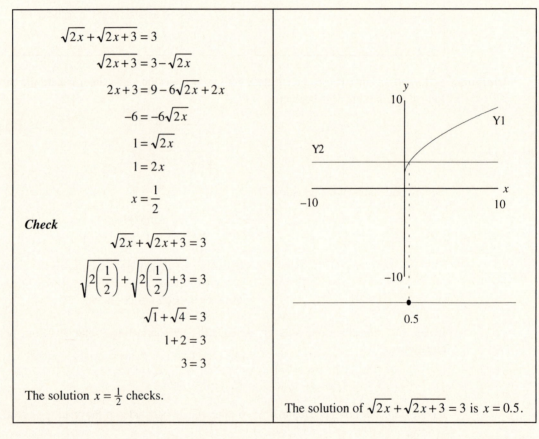

$\sqrt{2x} + \sqrt{2x+3} = 3$
$\sqrt{2x+3} = 3 - \sqrt{2x}$
$2x+3 = 9 - 6\sqrt{2x} + 2x$
$-6 = -6\sqrt{2x}$
$1 = \sqrt{2x}$
$1 = 2x$
$x = \frac{1}{2}$

Check

$$\sqrt{2x} + \sqrt{2x+3} = 3$$

$$\sqrt{2\left(\frac{1}{2}\right)} + \sqrt{2\left(\frac{1}{2}\right)+3} = 3$$

$$\sqrt{1} + \sqrt{4} = 3$$

$$1 + 2 = 3$$

$$3 = 3$$

The solution $x = \frac{1}{2}$ checks.

The solution of $\sqrt{2x} + \sqrt{2x+3} = 3$ is $x = 0.5$.

Example 5 If a mass is tied to the end of a string and allowed to swing back and forth, the result is an apparatus that physicists call a *pendulum*. The time to complete one swing is called the period of the pendulum and is given by the equation $T = 2\pi\sqrt{\dfrac{L}{g}}$. The variable T represents the period of the pendulum in seconds, the variable L represents the length of the string, and the variable g represents the acceleration due to gravity. Solve this equation for g and use your result to find the acceleration due to gravity if the length of the pendulum is 3.26 feet and the period is 2 seconds.

Solution Begin by squaring both sides of the equation.

$$T = 2\pi \sqrt{\frac{L}{g}}$$

$$(T)^2 = \left(2\pi \sqrt{\frac{L}{g}} \right)^2$$

$$T^2 = \frac{4\pi^2 L}{g}$$

Multiply both sides of this last equation by g.

$$g\left(T^2\right) = g\left(\frac{4\pi^2 L}{g} \right)$$

$$gT^2 = 4\pi^2 L$$

The highest power of g is one, so this equation is linear. All of the terms with a g in them should be placed on one side of the equation, and all of the remaining terms on the other side of the equation. This condition already exists. Complete your solution by dividing both sides of the equation by T^2

$$\frac{gT^2}{T^2} = \frac{4\pi^2 L}{T^2}$$

$$g = \frac{4\pi^2 L}{T^2}$$

Now substitute $L = 3.26$ feet and $T = 2$ seconds in this last equation.

$$g = \frac{4\pi^2 (3.26 \text{ ft})}{(2 \text{ s})^2}$$

$$g = 32.2 \text{ ft} / \text{s}^2$$

Note: What is the meaning of this solution? If a ball is dropped from a very tall building, its speed would change by 32.2 feet per second every second. After 1 second, the ball would be traveling at 32.2 feet per second. After 2 seconds, the ball would be traveling 64.4 feet per second, and so on.

Exercises for Section 5.4

1. Consider the function whose equation is $f(x) = \sqrt{x-2}$. Complete the following table of points that satisfy the equation of the function f. Copy this completed table onto a sheet of graph paper.

x	2	3	6	11	18
$f(x)$					
Points to plot					

   a) Set up a coordinate system on your graph paper and plot each point from the table on the coordinate system. Use these plotted points to help draw a complete graph of the function f. Sketch the graph of the constant function g whose equation is $g(x) = 1$ on this same coordinate system.

b) Use your graph to find the solution of the equation $f(x) = g(x)$. Draw a number line below your graph and shade and label your solution on your number line in the usual manner.

c) Use an analytical method to solve the equation $\sqrt{x-2} = 1$. Check your solution in the equation $\sqrt{x-2} = 1$. Does your analytic solution agree with the graphicalal estimate from part (b)? If not, check your work for error.

2. Consider the function whose equation is $f(x) = \sqrt{x+3}$. Complete the following table of points that satisfy the equation of the function f. Copy this completed table onto a sheet of graph paper.

x	−3	−2	1	6	13
$f(x)$					
Points to plot					

a) Set up a coordinate system on your graph paper and plot each point from the table on your coordinate system. Use these plotted points to help draw a complete graph of the function f. Sketch the graph of the constant function g whose equation is $g(x) = -2$ on this same coordinate system.

b) Use your graph to find the solution of the equation $f(x) = g(x)$. Draw a number line below your graph and shade and label your solution on your number line in the usual manner.

c) Use an analytical method to solve the equation $\sqrt{x+3} = -2$. Check your solution in the equation $\sqrt{x+3} = -2$. Does your analytical solution agree with your graphical estimate from part (b)? If not, check your work for error.

3. Consider the function whose equation is $f(x) = \sqrt{x+3}$. Complete the following table of points that satisfy the equation of the function f. Copy this completed table onto a sheet of graph paper.

x	−3	−2	1	6	13
$f(x)$					
Points to plot					

a) Set up a coordinate system on your graph paper and plot each point from the table on the coordinate system. Use these plotted points to help draw a complete graph of the function f. Sketch the graph of the function g whose equation is $g(x) = x+1$ on this same coordinate system.

b) Use your graph to find the solution of the equation $f(x) = g(x)$. Draw a number line below your graph and shade and label your solution on your number line in the usual manner.

c) Use an analytical method to solve the equation $\sqrt{x+3} = x+1$. Check your solutions in the equation $\sqrt{x+3} = x+1$. Does your analytical solution agree with your graphical estimate from part (b)? If not, check your work for error.

4. (TI-82) Load the following equations into the Y= menu:

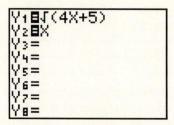

a) Press the ZOOM button and select 6:ZStandard. Set up a coordinate system on your homework paper and copy the image in your viewing window onto the coordinate system. Label each graph with its equation and include the WINDOW parameters on the coordinate axes.

b) Use the intersect utility in the CALC menu to find the solution of $\sqrt{4x+5} = x$. Draw a number line below the graph and shade and label this solution on your number line in the usual manner.

c) Use an analytical method to find the solution of the equation $\sqrt{4x+5} = x$. Check your solutions in the equation $\sqrt{4x+5} = x$. Does your analytical solution agree with the graphical estimate found in part (b)? If not, check your work for error.

5. (TI-82) Load the following equations into the Y= menu:

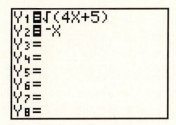

a) Press the ZOOM button and select 6:ZStandard. Set up a coordinate system on your homework paper and copy the image in your viewing window onto your coordinate system. Label each graph with its equation and include the WINDOW parameters on the coordinate axes.

b) Use the intersect utility in the CALC menu to find the solution of $\sqrt{4x+5} = -x$. Draw a number line below your graph and shade and label this solution on the number line in the usual manner.

c) Use an analytical method to find the solution of the equation $\sqrt{4x+5} = -x$. Check your solutions in the equation $\sqrt{4x+5} = -x$. Does your analytical solution agree with the graphical estimate found in part (b)? If not, check your work for error.

6. (TI-82) For each of the following equations, perform the following tasks:

i)   Load each side of the equation into the Y= menu, the left side in Y1 and the right side in Y2, respectively. Press the ZOOM button and select 6:ZStandard. Set up a coordinate system on your homework paper and copy the image in your viewing window onto the coordinate system. Label each graph with its equation and include the WINDOW parameters on your coordinate axes.

ii) Use the intersect utility in the CALC menu to find the solution of the equation Y1 = Y2. Draw a number line below your graph and shade and label this solution on your number line in the usual manner.

iii) Use an analytical method to find the solution of the equation Y1 = Y2. Check your solutions in the original equation. Does your analytical solution agree with the graphical estimate found in part (b)? If not, check your work for error.

a) $\sqrt{x+5} = 2$ b) $\sqrt{x+5} = -2$ c) $\sqrt{6x+7} = x$ d) $\sqrt{6x+7} = -x$

e) $x+1 = \sqrt{x+7}$ f) $x+1 = -\sqrt{x+7}$ g) $x - \sqrt{7-2x} = 2$ h) $x + \sqrt{7-2x} = 2$

7. (TI-82) For each of the following equations, perform the following tasks.

i)   Load each side of the equation into the Y= menu, the left side in Y1 and the right side in Y2, respectively. If necessary, adjust the WINDOW parameters so that all points of intersection of the two graphs are present in the viewing window. Set up a coordinate system on your homework paper and copy the image in your viewing window onto the coordinate system. Label each graph with its equation and include the WINDOW parameters on your coordinate axes.

ii) Use the intersect utility in the CALC menu to find the solution of the equation $Y_1 = Y_2$. Draw a number line below your graph and shade and label this solution on your number line in the usual manner.

iii) Use an analytical method to find the solution of the equation $Y_1 = Y_2$. Check your solutions in the original equation. Does your analytical solution agree with the graphical estimate found in part (b)? If not, check your work for error.

a) $\sqrt{x+5} = 4$
b) $\sqrt{4-x} = 4$
c) $\sqrt{2x+3} = 5$
d) $-\sqrt{2-5x} = 7$

e) $\sqrt{2x+14} = 5-x$
f) $\sqrt{3x+3} = x-5$
g) $x-\sqrt{4x+33} = 3$
h) $x-\sqrt{12-2x} = -6$

i) $\sqrt{x+6}-\sqrt{x+1} = 1$
j) $\sqrt{x+5}-\sqrt{x} = 1$
k) $\sqrt{2x+3}-\sqrt{x+1} = 1$
l) $\sqrt{x-4}-\sqrt{x+7} = 3$

8. Robina and her partner Dennis are working in Dr. Mills's physics lab on an experiment involving oscillating springs. They place a mass on a spring and measure the time required for the mass to oscillate up and down. This time is called the period of the motion and is given by the formula $T = 2\pi\sqrt{\dfrac{m}{k}}$, where T represents the period in seconds, m represents mass in kilograms, and k is the spring constant, a number that measures the stiffness of the spring.

a) Solve the equation $T = 2\pi\sqrt{\dfrac{m}{k}}$ for the spring constant k.

b) If the mass is 0.75 kilograms and the period of oscillation is 2.1 seconds, find the spring constant k. Round your answer to the nearest tenth (1 decimal place).

9. If an object is released from rest from a tall building, the time it takes for the object to strike the ground is given by the formula $t = \sqrt{\dfrac{2y}{g}}$, where t is the time in seconds, y is the height of the building in meters, and g is the acceleration due to gravity.

a) Solve the equation $t = \sqrt{\dfrac{2y}{g}}$ for y.

b) If it takes the object 3 seconds to strike the ground and the acceleration due to gravity is 9.8 meters per second per second, how tall is the building in meters? Round your answer to the nearest tenth of a meter (one decimal place).

10. Artificial gravity can be created in a space station by rotating the station. The number of rotations required is given by the formula $N = \dfrac{1}{2\pi}\sqrt{\dfrac{a}{r}}$, where N is measured in revolutions per second, a is the artificial acceleration produced and is measured in meters per second per second, and r is the radius of the space station in meters.

a) Solve the equation $N = \dfrac{1}{2\pi}\sqrt{\dfrac{a}{r}}$ for r.

b) If the space station is making 0.05 revolutions every second and the artificial acceleration produced is 9.8 meters per second per second, what is the radius of the space station in meters? Round your answer to the nearest tenth of a meter (one decimal place).

11. Solve each of the following equations for the indicated variable.

a) $r = \sqrt{\dfrac{A}{\pi}}$ for A.

b) $\omega = \sqrt{\dfrac{k}{m}}$ for m.

c) $T = 2\pi\sqrt{\dfrac{L}{g}}$ for L.

d) $d = \sqrt{2rh + h^2}$ for r.

5.5 Fractional Exponents

In this section you will learn how to raise a number to a fractional exponent. This should be an easy task, now that you understand radical notation. Some time will be spent changing expressions with fractional exponents into expressions with radical notation, and vice versa. It is important to be able to flow smoothly from one notation to the other. We begin with a very important statement.

In general, the laws of exponents hold true when the exponents are fractions.

The Meaning of a Fractional Exponent

The first goal of this section is to discover and define the meaning of the fractional exponent in the expression $a^{\frac{m}{n}}$. Let's begin with an example.

Example 1   *(TI-82)* Use your calculator to compare the expressions $\sqrt{7}$, $\sqrt{11}$, and $\sqrt{9}$ with the expressions $7^{\frac{1}{2}}$, $11^{\frac{1}{2}}$, and $9^{\frac{1}{2}}$.

Solution

Enter the expressions $\sqrt{7}$, $\sqrt{11}$, and $\sqrt{9}$ in the following manner:	Enter the expressions $7^{\frac{1}{2}}$, $11^{\frac{1}{2}}$, and $9^{\frac{1}{2}}$ in the following manner:
| ```
√7
          2.645751311
√11
          3.31662479
√9
                    3
■
``` | ```
7^(1/2)
          2.645751311
11^(1/2)
          3.31662479
9^(1/2)
                    3
``` |

It appears that $\sqrt{7} = 7^{\frac{1}{2}}$, $\sqrt{11} = 11^{\frac{1}{2}}$, and $\sqrt{9} = 9^{\frac{1}{2}}$.

Example 2   *(TI-82)* Use your calculator to compare the graphs of the functions f and g whose equations are

$$f(x) = \sqrt{x} \text{ and } g(x) = x^{\frac{1}{2}}.$$

Solution

Load the following equation in the Y= menu:	Push the ZOOM button and select 6:ZStandard.
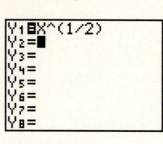	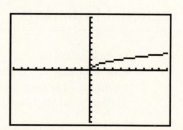

Load the following equation in the Y= menu:	Push the ZOOM button and select 6:ZStandard.
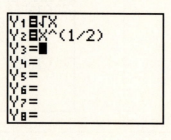	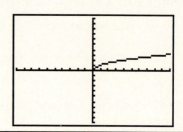

It appears that the graphs of $f(x) = \sqrt{x}$ and $g(x) = x^{\frac{1}{2}}$ are identical. This visual evidence that $\sqrt{x} = x^{\frac{1}{2}}$ is even more striking if you draw both graphs at the same time.

Load the following equations in the Y= menu:	Push the ZOOM button and select 6:ZStandard.
	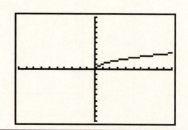

The graphs of $f(x) = \sqrt{x}$ and $g(x) = x^{\frac{1}{2}}$ coincide! This is compelling evidence that $\sqrt{x} = x^{\frac{1}{2}}$.

Can You Be Certain That $a^{\frac{1}{2}} = \sqrt{a}$? If $a \ge 0$, then $\sqrt{a}$ is the unique nonnegative real solution of $x^2 = a$. If $a \ge 0$, then $a^{\frac{1}{2}}$ is a nonnegative real number. Substitute $a^{\frac{1}{2}}$ in the equation $x^2 = a$.

$$x^2 = a$$

$$\left(a^{\frac{1}{2}} \right)^2 = a$$

$$a^1 = a$$

$$a = a$$

Therefore, $a^{\frac{1}{2}}$ is a nonnegative solution of $x^2 = a$. Because $\sqrt{a}$ is the unique nonnegative solution of $x^2 = a$, logic demands that $a^{\frac{1}{2}} = \sqrt{a}$.

$$\textbf{If } a \ge 0, \textbf{ then } a^{\frac{1}{2}} = \sqrt{a}.$$

Example 3   Compare the expressions $\sqrt[3]{7}$, $\sqrt[3]{-11}$, and $\sqrt[3]{-8}$ with the expressions $7^{\frac{1}{3}}$, $(-11)^{\frac{1}{3}}$, and $(-8)^{\frac{1}{3}}$.

Solution

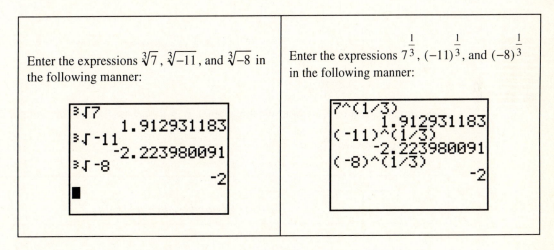

Enter the expressions $\sqrt[3]{7}$, $\sqrt[3]{-11}$, and $\sqrt[3]{-8}$ in the following manner:

```
³√7
            1.912931183
³√-11
           -2.223980091
³√-8
                      -2
■
```

Enter the expressions $7^{\frac{1}{3}}$, $(-11)^{\frac{1}{3}}$, and $(-8)^{\frac{1}{3}}$ in the following manner:

```
7^(1/3)
            1.912931183
(-11)^(1/3)
           -2.223980091
(-8)^(1/3)
                      -2
```

It appears that $\sqrt[3]{7} = 7^{\frac{1}{3}}$, $\sqrt[3]{-11} = (-11)^{\frac{1}{3}}$, and $\sqrt[3]{-8} = (-8)^{\frac{1}{3}}$.

Example 4   *(TI-82)* Use your calculator to compare the graphs of the functions f and g whose equations are $f(x) = \sqrt[3]{x}$ and $g(x) = x^{\frac{1}{3}}$.

Solution

Load the following equations in the Y= menu:

Push the ZOOM button and select 6:ZStandard.

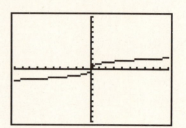

The graphs of $f(x) = \sqrt[3]{x}$ and $g(x) = x^{\frac{1}{3}}$ coincide. This is compelling visual evidence that $\sqrt[3]{x} = x^{\frac{1}{3}}$.

Can You Be Certain That $a^{\frac{1}{3}} = \sqrt[3]{a}$? If a is a real number, then $\sqrt[3]{a}$ is the unique real solution of $x^3 = a$. If a is a real number, then $a^{\frac{1}{3}}$ is a real number. Substitute the real number $a^{\frac{1}{3}}$ in the equation $x^3 = a$.

$$x^3 = a$$

$$\left(a^{\frac{1}{3}}\right)^3 = a$$

$$a^1 = a$$

$$a = a$$

Therefore, $a^{\frac{1}{3}}$ is a real solution of the equation $x^3 = a$. Because $\sqrt[3]{a}$ is the unique real solution of $x^3 = a$, logic demands that $a^{\frac{1}{3}} = \sqrt[3]{a}$.

If a is a real number, then $a^{\frac{1}{3}} = \sqrt[3]{a}$.

In general, the following definition can be made.

If a is a real number and n is odd, then $a^{\frac{1}{n}} = \sqrt[n]{a}$.

If $a \geq 0$ and n is even, then $a^{\frac{1}{n}} = \sqrt[n]{a}$. If $a < 0$ and n is even, then $a^{\frac{1}{n}}$ is not a real number.

Example 5   Simplify $25^{\frac{1}{2}}$, $(-27)^{\frac{1}{3}}$, and $(-16)^{\frac{1}{4}}$.

Solution

You can mentally compute $25^{\frac{1}{2}}$. $$25^{\frac{1}{2}} = \sqrt{25}$$ $$= 5$$	You can mentally compute $(-27)^{\frac{1}{3}}$. $$(-27)^{\frac{1}{3}} = \sqrt[3]{-27}$$ $$= -3$$	Use your calculator to check these solutions. `25^(1/2)` `                    5` `(-27)^(1/3)` `                   -3` `■`
$(-16)^{\frac{1}{4}}$ is not a real number.	Try this on your calculator. `(-16)^(1/4)`	Your calculator responds with an error message. `ERR:DOMAIN` `1█Goto` `2:Quit`

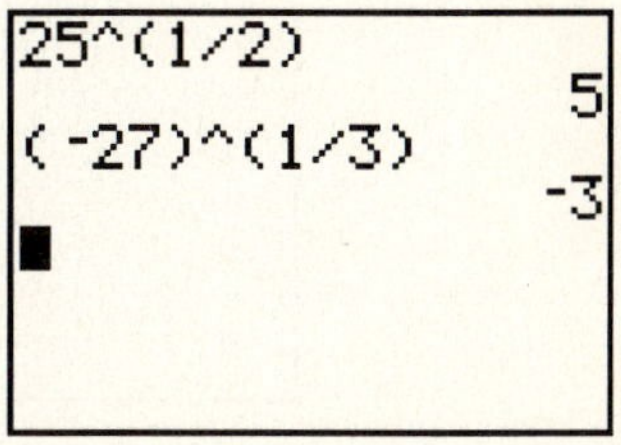

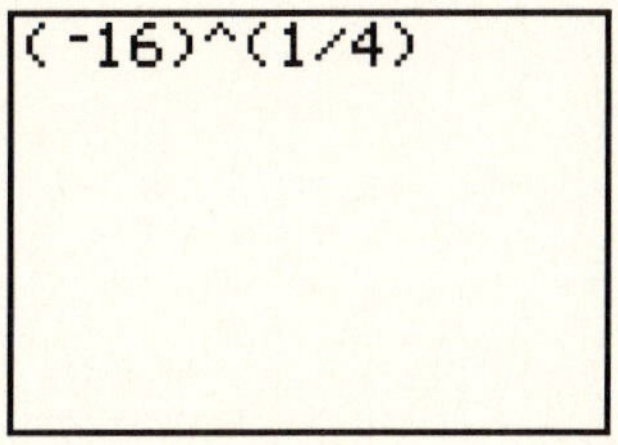

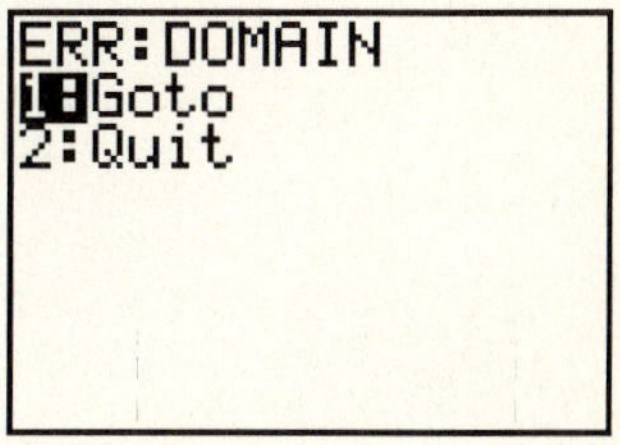

Example 6   Place the expressions $32^{\frac{1}{2}}$ and $\dfrac{2}{4^{\frac{1}{3}}}$ in simple radical form.

Solution

$$32^{\frac{1}{2}} = \sqrt{32}$$ $$= \sqrt{16}\sqrt{2}$$ $$= 4\sqrt{2}$$	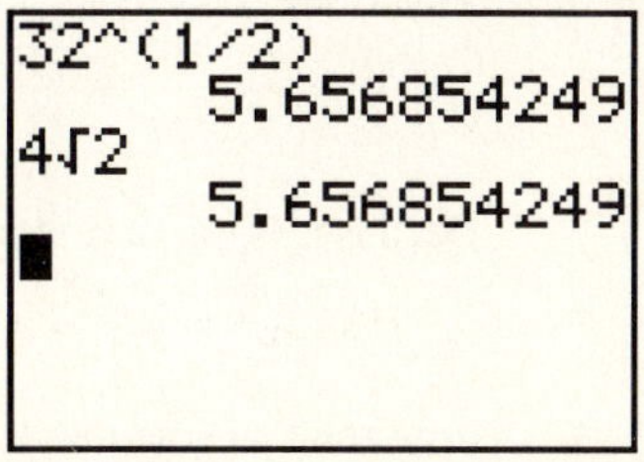 `32^(1/2)` `         5.656854249` `4√2` `         5.656854249` `■`

$$\frac{2}{4^{\frac{1}{3}}} = \frac{2}{\sqrt[3]{4}}$$

$$= \frac{2}{\sqrt[3]{4}} \cdot \frac{\sqrt[3]{2}}{\sqrt[3]{2}}$$

$$= \frac{2\sqrt[3]{2}}{\sqrt[3]{8}}$$

$$= \frac{2\sqrt[3]{2}}{2}$$

$$= \sqrt[3]{2}$$

```
2/4^(1/3)
          1.25992105
3√2
          1.25992105
```

Example 7   Place the expression $\dfrac{1}{x^{\frac{1}{2}}}$ in simple radical form.

Solution   The expression $\dfrac{1}{x^{\frac{1}{2}}}$ is a real number only if $x > 0$.

$$\frac{1}{x^{\frac{1}{2}}} = \frac{1}{\sqrt{x}}$$

$$= \frac{1}{\sqrt{x}} \cdot \frac{\sqrt{x}}{\sqrt{x}}$$

$$= \frac{\sqrt{x}}{\sqrt{x^2}}$$

$$= \frac{\sqrt{x}}{|x|}$$

$$= \frac{\sqrt{x}}{x}$$

Because $\dfrac{1}{x^{\frac{1}{2}}}$ is a real number only if $x > 0$, x is a positive number and $|x| = x$ in the last step. It is

important to note that $\dfrac{1}{x^{\frac{1}{2}}} = \dfrac{\sqrt{x}}{x}$ only if $x > 0$.

Example 8   Use the laws of exponents to simplify the expressions $x^{\frac{1}{2}} x^{\frac{1}{3}}$, and $\dfrac{x^{\frac{1}{2}}}{x^{\frac{1}{3}}}$, and $\left(x^{\frac{1}{2}} \right)^{\frac{1}{3}}$.

Solution

$x^{\frac{1}{2}}x^{\frac{1}{3}} = x^{\frac{1}{2}+\frac{1}{3}}$ $= x^{\frac{3}{6}+\frac{2}{6}}$ $= x^{\frac{5}{6}}$	$\dfrac{x^{\frac{1}{2}}}{x^{\frac{1}{3}}} = x^{\frac{1}{2}-\frac{1}{3}}$ $= x^{\frac{3}{6}-\frac{2}{6}}$ $= x^{\frac{1}{6}}$	$\left(x^{\frac{1}{2}}\right)^{\frac{1}{3}} = x^{\frac{1}{2}\cdot\frac{1}{3}}$ $= x^{\frac{1}{6}}$

In the first example, $x^{\frac{1}{2}}x^{\frac{1}{3}} = x^{\frac{5}{6}}$. How would you define $x^{\frac{5}{6}}$?

Defining $a^{\frac{m}{n}}$

When raising a power of a number to a power, the laws of exponents instruct us to repeat the base and multiply the exponents. Therefore, you can deal with $a^{\frac{m}{n}}$ in one of two ways.

$a^{\frac{m}{n}} = \left(a^{\frac{1}{n}}\right)^m$ $= \left(\sqrt[n]{a}\right)^m$	$a^{\frac{m}{n}} = \left(a^m\right)^{\frac{1}{n}}$ $= \sqrt[n]{a^m}$

In both cases, the denominator n of the fractional exponent becomes the index of the radical. You have the option of placing the numerator m inside or outside the radical.

Tip	This definition is valid only if the expression is a real number. For example, $\sqrt[4]{(-2)^2}$ is a valid real number, but $\left(\sqrt[4]{-2}\right)^2$ is not a real number. You will have to watch for this exception when applying the definition of $a^{\frac{m}{n}}$.

Example 9   Simplify $16^{\frac{3}{4}}$.

Solutions

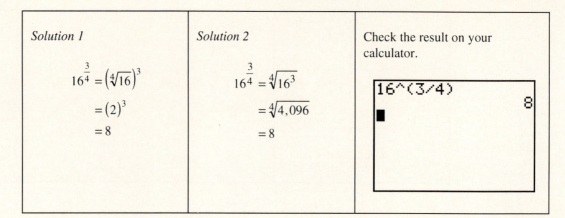

Solution 1	Solution 2	Check the result on your calculator.
$16^{\frac{3}{4}} = \left(\sqrt[4]{16}\right)^3$	$16^{\frac{3}{4}} = \sqrt[4]{16^3}$	
$= (2)^3$	$= \sqrt[4]{4,096}$	
$= 8$	$= 8$	

Although both methods yield the same answer, note that the first method is much easier, particularly if you are not allowed to use a calculator.

Example 10 Simplify $(-27)^{\frac{2}{3}}$.

Solutions

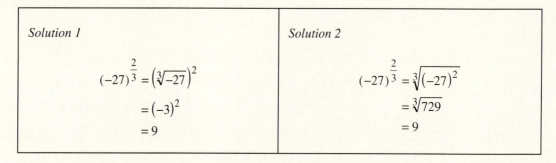

Solution 1	Solution 2
$(-27)^{\frac{2}{3}} = \left(\sqrt[3]{-27}\right)^2$	$(-27)^{\frac{2}{3}} = \sqrt[3]{(-27)^2}$
$= (-3)^2$	$= \sqrt[3]{729}$
$= 9$	$= 9$

Once again, the first solution is a lot easier.

Check Your Solution on Your Calculator.

Enter the following expression in your viewing window:	When you press the ENTER key, the following message appears in the viewing window:
(-27)^(2/3)	ERR:DOMAIN 1▮Goto 2:Quit

Unfortunately, the algorithm that the TI-82 uses to compute $a^{\frac{m}{n}}$ differs from the algorithm used to compute $a^{\frac{1}{n}}$. The TI-82 will not raise a negative base to the power $\frac{m}{n}$ unless $m = 1$ (nor will many other brands of calculators). At first, this seems to be a major inconvenience. However, you will not have much need for negative bases as most of the major applications in this text involve bases that are positive real numbers.

Example 11   Simplify $(-4)^{\frac{3}{2}}$.

Solution   Neither $\sqrt{(-4)^3}$ nor $\left(\sqrt{-4}\right)^3$ is a real number.

Example 12   Express $(x+2)^{\frac{2}{3}}$ in radical notation.

Solution   We have two options: $(x+2)^{\frac{2}{3}} = \sqrt[3]{(x+2)^2}$ or $(x+2)^{\frac{2}{3}} = \left(\sqrt[3]{x+2}\right)^2$.

Exercises for Section 5.5

1. (TI-82) Enter each of the following expressions on your calculator and verify the results.

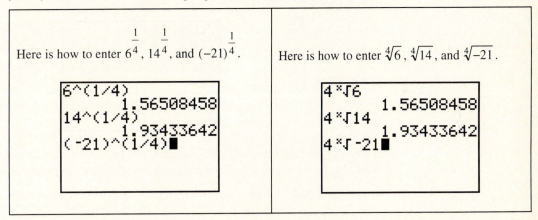

Here is how to enter $6^{\frac{1}{4}}$, $14^{\frac{1}{4}}$, and $(-21)^{\frac{1}{4}}$.

Here is how to enter $\sqrt[4]{6}$, $\sqrt[4]{14}$, and $\sqrt[4]{-21}$.

Use complete sentences to explain your calculator's response to the last expression in each window. Use your calculator to find decimal approximations for each of the following pairs of numbers. Place these approximations on your homework paper.

a) $8^{\frac{1}{4}}$ and $\sqrt[4]{8}$

b) $36^{\frac{1}{4}}$ and $\sqrt[4]{36}$

Use complete sentences to explain what you learned from this exercise.

2. (TI-82) Enter each of the following expressions on your calculator and verify the results.

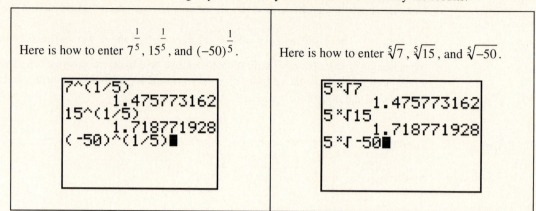

Here is how to enter $7^{\frac{1}{5}}$, $15^{\frac{1}{5}}$, and $(-50)^{\frac{1}{5}}$.

Here is how to enter $\sqrt[5]{7}$, $\sqrt[5]{15}$, and $\sqrt[5]{-50}$.

Use complete sentences to explain why your calculator's response to the last expression in each window is different from the calculator's response to the last item in each window in Exercise 1. Use your calculator to find decimal approximations for each of the following pairs of numbers. Place these approximations on your homework paper.

a) $20^{\frac{1}{5}}$ and $\sqrt[5]{20}$

b) $100^{\frac{1}{5}}$ and $\sqrt[5]{100}$

Use complete sentences to explain what you learned from this exercise.

3. (TI-82) Enter the following equations in the Y= menu.

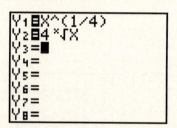

a) Push the ZOOM button and select 6:ZStandard. Set up a coordinate system on your homework paper and copy the image in your viewing window onto your coordinate system.

b) Use your calculator to draw a separate image for each equation. Set up two more coordinates systems on your homework paper. Place the graph of $y = x^{\frac{1}{4}}$ on one coordinate system and the graph of $y = \sqrt[4]{x}$ on the second coordinate system.

c) Use complete sentences to explain what you learned in this exercise.

4. (TI-82) Enter the following equations in the Y= menu.

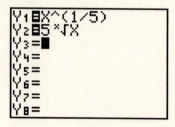

a) Push the ZOOM button and select 6:ZStandard. Set up a coordinate system on your homework paper and copy the image in your viewing window onto your coordinate system.

b) Use your calculator to draw a separate image for each equation. Set up two more coordinate systems on your homework paper. Place the graph of $y = x^{\frac{1}{5}}$ on one coordinate system and the graph of $y = \sqrt[5]{x}$ on the second coordinate system.

c) Use complete sentences to explain what you learned in this exercise.

5. Change each of the following expressions into radical notation and mentally simplify the result. When you have completed the problem, use your calculator to check your solution.

a) $8^{\frac{1}{3}}$ b) $36^{\frac{1}{2}}$ c) $(-8)^{\frac{1}{3}}$ d) $(-4)^{\frac{1}{2}}$ e) $16^{\frac{1}{4}}$ f) $32^{\frac{1}{5}}$

6. Place each of the following expressions in simple radical form. Enter both the original expression and your solution into your calculator and find decimal approximations for each. Record these approximations on your homework paper.

a) $50^{\frac{1}{2}}$ b) $\dfrac{6}{12^{\frac{1}{2}}}$ c) $16^{\frac{1}{3}}$ d) $\dfrac{12}{9^{\frac{1}{3}}}$ e) $\dfrac{8}{4^{\frac{1}{4}}}$ f) $\dfrac{6}{16^{\frac{1}{5}}}$

7. (TI-82) Enter each of the following expressions on your calculator and verify the results.

Here is how to enter $7^{\frac{2}{3}}$.

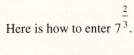

Here is how to enter $\sqrt[3]{7^2}$ and $\left(\sqrt[3]{7}\right)^2$.

Copy each expression on your homework, then use your calculator to find a decimal approximation for each expression.

a) $5^{\frac{2}{3}}$ and $\sqrt[3]{5^2}$ and $\left(\sqrt[3]{5}\right)^2$ b) $9^{\frac{2}{3}}$ and $\sqrt[3]{9^2}$ and $\left(\sqrt[3]{9}\right)^2$

Use complete sentences to explain what you learned in this exercise.

8. (TI-82) Enter each of the following expressions on your calculator and verify the results.

Here is how to enter $11^{\frac{2}{5}}$.

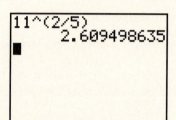

Here is how to enter $\sqrt[5]{11^2}$ and $\left(\sqrt[5]{11}\right)^2$.

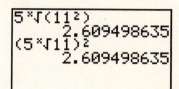

Copy each expression on your homework, then use your calculator to find a decimal approximation for each expression.

a) $18^{\frac{2}{5}}$ and $\sqrt[5]{18^2}$ and $\left(\sqrt[5]{18}\right)^2$

b) $3^{\frac{2}{5}}$ and $\sqrt[5]{3^2}$ and $\left(\sqrt[5]{3}\right)^2$

Use complete sentences to explain what you learned in this exercise.

9. Change each of the following expressions into radical notation and mentally simplify the result. When you have completed the problem, use your calculator to check your solution.

a) $4^{\frac{5}{2}}$ b) $16^{\frac{3}{2}}$ c) $8^{\frac{2}{3}}$ d) $125^{\frac{2}{3}}$ e) $16^{\frac{3}{4}}$ f) $(-32)^{\frac{3}{5}}$

10. Place each of the following expressions in simple radical form. Enter both the original expression and your solution into your calculator and find decimal approximations for each. Record these approximations on your homework paper.

a) $2^{\frac{3}{2}}$ b) $5^{\frac{4}{3}}$ c) $\dfrac{4}{2^{\frac{3}{2}}}$ d) $\dfrac{12}{3^{\frac{2}{3}}}$

11. Place each of the following expressions in simple radical form.

a) $x^{\frac{3}{2}}$ b) $x^{\frac{4}{3}}$ c) $\dfrac{4}{x^{\frac{1}{3}}}$ d) $\dfrac{12}{x^{\frac{3}{2}}}$

12. Use the laws of exponents to simplify each of the following expressions. Where appropriate, place your solution in simple radical form.

a) $2^{\frac{1}{2}} \cdot 2^{\frac{1}{4}}$ b) $3^{\frac{2}{3}} \cdot 3^{\frac{1}{4}}$ c) $\dfrac{3^{\frac{1}{4}}}{3^{\frac{1}{5}}}$ d) $\left(2^{\frac{1}{3}}\right)^{\frac{1}{4}}$

e) $x^{\frac{1}{2}} \cdot x^{\frac{2}{3}}$ f) $\left(x^{\frac{1}{2}}\right)^4$ g) $\left(2x^{\frac{1}{2}}y^{\frac{1}{3}}\right)^6$ h) $\left(\dfrac{x^{\frac{1}{2}}}{y^{\frac{2}{3}}}\right)^6$

13. Place each of the following radical expressions in fractional exponent form.

    a) $\sqrt{x+2}$ b) $\sqrt[3]{5-x}$ c) $\sqrt[3]{(x+1)^2}$ d) $\sqrt[3]{(1-x)^4}$

14. (TI-82) Because $1.5 = \dfrac{3}{2}$, $2^{1.5} = 2^{\frac{3}{2}}$. Thus, $2^{1.5}$ is well defined.

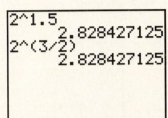

    Use your calculator to find decimal approximations for each of the following expressions.

    a) $3^{1.2}$ b) $10^{1.25}$ c) $125\left(8^{2.25}\right)$ d) $10,000\left(1.25^{0.1125}\right)$

15. The Parcel Post Rate is approximated by the function $R = 1.72w^{0.25}$, where R represents the rate for mailing the package in dollars and w represents the weight of the package in pounds. Use this formula and your calculator to approximate the rate for a package weighing in at 20 pounds.

    Note: The actual rate is $3.63 for zones 1 and 2, according to the *World Almanac Book of Facts,* 1994.

16. Jimmy is using the equation $P = 47.3t^{0.36}$ to predict the price of a gallon of gasoline. In this equation P represents the price of a gallon of regular unleaded gasoline in cents and t represents the number of years that have passed since 1980. Use this equation to find the price of a gallon of regular unleaded gasoline in 1991.

    Note: The actual average city price of a gallon of regular unleaded gasoline in 1991 was 114.0 cents.

17. Tim's mom kept a record of his height during his early childhood. She can use the equation $h = 30.0t^{0.31}$ to predict his height. In this equation, h represents Tim's height in centimeters and t represents his age in months. Use this equation to predict Tim's height at age 46 months.

    Note: Tim's actual height at age 46 months was 99.7 centimeters.

5.6 Summary and Review

Important Visualization Tools

If n is an odd integer and b is a real number, the one real solution of $x^n = b$ is $\sqrt[n]{b}$.

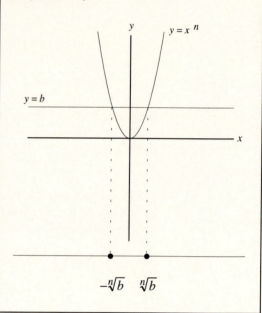

If n is an even integer and b is a nonnegative real number, the two real solutions of $x^n = b$ are $\sqrt[n]{b}$ and $-\sqrt[n]{b}$.

If n is an even integer and b is a negative real number, there are no real solutions of $x^n = b$.

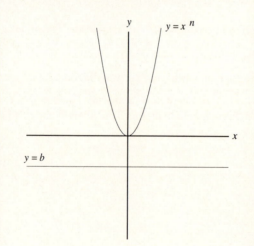

When solving an equation that involves only a single power of x, it is easiest to place the equation in the form $x^n = b$

$$3x^2 - 1 = x^2 + 15$$

$$3x^2 - x^2 = 1 + 15$$

$$2x^2 = 16$$

$$x^2 = 8$$

$$x = \pm\sqrt{8}$$

$$x = \pm 2\sqrt{2}$$

Place your analytical solution and graphical solution side-by-side on your homework paper for comparison.

Isolate the radical

$$x - \sqrt{x-1} = 3$$
$$x - 3 = \sqrt{x-1}$$
$$x^2 - 6x + 9 = x - 1$$
$$x^2 - 7x + 10 = 0$$
$$x^2 - 2x - 5x + 10 = 0$$
$$x(x-2) - 5(x-2) = 0$$
$$(x-5)(x-2) = 0$$
$$x = 5, 2$$

Check for extraneous roots

$$5 - \sqrt{5-1} = 3 \qquad 2 - \sqrt{2-1} = 3$$
$$5 - 2 = 3 \qquad\quad 2 - 1 = 3$$
$$3 = 3 \qquad\qquad 1 = 3$$

5 checks.   2 does not check.

Load $Y_1 = X - \sqrt{X-1}$ and $Y_2 = 3$ in the Y= menu.

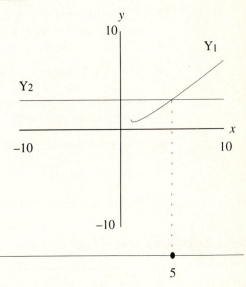

The solution of $x = \sqrt{x-1} = 3$ is 5.

A Review of Key Words and Phrases

☞ The solutions of $x^n = b$ are called nth roots of b. (page 381)

☞ If n is an odd integer and b is a real number, then $\sqrt[n]{b}$ is the one real solution of the equation $x^n = b$. Because $\sqrt[n]{b}$ is a solution of $x^n = b$, $\left(\sqrt[n]{b}\right)^n = b$. (page 381)

☞ If n is an even integer and b is a negative real number, then the equation $x^n = b$ has no real solutions. (page 381)

☞ If n is an even integer and b is a nonnegative real number, then the equation $x^n = b$ has two real solutions. The nonnegative real solution is $\sqrt[n]{b}$ and the nonpositive solution is $-\sqrt[n]{b}$. Because $-\sqrt[n]{b}$ is a solution of $x^n = b$, $\left(-\sqrt[n]{b}\right)^n = b$. Because $\sqrt[n]{b}$ is a solution of $x^n = b$, $\left(\sqrt[n]{b}\right)^n = b$. (page 381)

☞ If n is an even integer and b is a negative real number, then $\sqrt[n]{b}$ is not a real number. (page 387)

☞ If a is a real number, then $\sqrt{a^2} = |a|$. (page 389)

☞ If $a \geq 0$ and $b \geq 0$, then $\sqrt{a}\sqrt{b} = \sqrt{ab}$. Furthermore, if $b \neq 0$, then $\dfrac{\sqrt{a}}{\sqrt{b}} = \sqrt{\dfrac{a}{b}}$. (page 396)

☞ The rules for simple form are as follows (page 398):
   1. If you can factor out a perfect root, then you must.
   2. No fractions are allowed under a radical.
   3. No radicals are allowed in the denominator of a fraction.

- If a and b are real numbers, then $|ab| = |a||b|$. Furthermore, if $b \neq 0$, then $\left|\dfrac{a}{b}\right| = \dfrac{|a|}{|b|}$. (page 402)

- If $\sqrt[n]{a}$ and $\sqrt[n]{b}$ are real numbers, then $\sqrt[n]{a}\sqrt[n]{b} = \sqrt[n]{ab}$. Furthermore, if $b \neq 0$, then $\dfrac{\sqrt[n]{a}}{\sqrt[n]{b}} = \sqrt[n]{\dfrac{a}{b}}$. (page 404)

- Raising both sides of an equation to an even power can add solutions that will not check in the original equation. (page 422)

- Always try to isolate the radical before squaring both sides of the equation. (page 426)

- If a is a real number and n is an odd integer, then $a^{\frac{1}{n}} = \sqrt[n]{a}$. (page 438)

- If $a \geq 0$ and n is an even integer, then $a^{\frac{1}{n}} = \sqrt[n]{a}$. If $a < 0$ and n is an even integer, then $a^{\frac{1}{n}}$ is not a real number. (page 438)

- If a is a real number, then $a^{\frac{m}{n}} = \sqrt[n]{a^m}$ or $a^{\frac{m}{n}} = \left(\sqrt[n]{a}\right)^m$. You must be careful to apply this rule only when the expressions involved are real numbers. For example, the expression $\sqrt[4]{(-2)^2}$ is a real number, but the expression $\left(\sqrt[4]{-2}\right)^2$ is not a real number. (page 441)

TI-82 Keywords and Menus

The nth root button is found in the MATH menu. 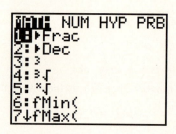	To enter $\sqrt[4]{7}$, first enter a 4, push the MATH button and select 5: $\sqrt[x]{\ }$, then enter a 7. ```4 ˣ√7 1.626576562```

Chapter Review Exercises

1. Simplify each of the following expressions mentally. Use your calculator to check your results.

   a) $\sqrt{36}$ b) $-\sqrt{36}$ c) $\sqrt{-36}$ d) $\sqrt[3]{8}$ e) $-\sqrt[3]{8}$ f) $\sqrt[3]{-8}$

   Use complete sentences to explain why the answer from part (c) is distinctly different from the answer in part (f).

2. Use your calculator to approximate each of the following roots. Round your answer to the nearest ten-thousandth (four decimal places).

   a) $\sqrt[4]{100}$ b) $\sqrt[5]{100}$ c) $-\sqrt[6]{100}$ d) $\sqrt[6]{-100}$ e) $\sqrt[7]{700}$ f) $\sqrt[7]{-700}$

   Use complete sentences to explain why the result from part (d) is distinctly different from the result in part (f).

3. (TI-82) For each of the following equations, perform the following tasks:

    i) Load each side of the equation in the Y= menu, in Y1 and Y2, respectively. Push the ZOOM button and select 6:ZStandard. Set up a coordinate system on your homework paper and copy the image in the viewing window onto your coordinate system. Label each graph with its equation and include the WINDOW parameter settings on your coordinate axes.

    ii) Use the intersect utility in the CALC menu to estimate the real solutions of the equation Y1 = Y2. For a review of this utility, see Example 10 in Section 1.6. Draw a number line below your graph and shade and label these solutions on your number line in the usual manner.

    iii) Use an analytical method to find the real solutions of the equation. Use your calculator to find decimal approximations for these solutions. If they do not compare favorably with the estimates found in part (ii), check your work for error.

    a) $x^2 = 7$ b) $x^3 = -7$ c) $x^4 = -7$ d) $x^5 = 7$ e) $x^6 = 7$ f) $x^7 = 7$

4. (TI-82) Use your calculator to draw the graphs of each of the following equations. For each equation, set up a separate coordinate system on your homework paper and copy the image in the viewing window onto your coordinate system.

    a) $y = \sqrt{x^2}$ b) $y = \sqrt[3]{x^3}$ c) $y = \sqrt[4]{x^4}$ d) $y = \sqrt[5]{x^5}$ e) $y = \sqrt[6]{x^6}$ f) $y = \sqrt[7]{x^7}$

    Use complete sentences to explain what you learned in this exercise.

5. Place each of the following expressions in simple radical form. Use a calculator to find a decimal approximation for the original problem and for your solution. Place these approximations on your homework paper. If they do not match, check your work for error.

    a) $\sqrt{200}$ b) $\sqrt{99}$ c) $\sqrt[3]{56}$ d) $\sqrt[4]{80}$ e) $\sqrt{\dfrac{7}{20}}$ f) $\sqrt{\dfrac{5}{6}}$

    g) $\sqrt[3]{\dfrac{5}{4}}$ h) $\sqrt[4]{\dfrac{5}{8}}$ i) $\dfrac{6}{\sqrt{3}}$ j) $\dfrac{6}{\sqrt{18}}$ k) $\dfrac{8}{\sqrt[3]{2}}$ l) $\dfrac{6}{\sqrt[4]{8}}$

6. Place each of the following radical expressions in simple radical form. Use set-builder notation to describe the set of x for which your solution is valid.

    a) $\sqrt{72x^2}$ b) $\sqrt{48x^2}$ c) $\sqrt{80x^3}$ d) $\sqrt{44x^4}$ e) $\sqrt[3]{24x^3}$ f) $\sqrt[3]{88x^4}$

    g) $\sqrt{\dfrac{1}{8x^2}}$ h) $\sqrt{\dfrac{1}{12x^3}}$ i) $\dfrac{6x}{\sqrt{24x^2}}$ j) $\dfrac{2x}{\sqrt[3]{24x^4}}$

7. Place each of the following expressions in simple radical form. Use a calculator to find a decimal approximation for the original problem and for your solution. Place these approximations on your homework paper. If they do not match, check your work for error.

    a) $\left(2\sqrt{3}+5\right)\left(3\sqrt{3}-6\right)$ b) $\left(\sqrt{2}+\sqrt{5}\right)^2$

    c) $\left(5\sqrt{5}+2\right)\left(5\sqrt{5}-2\right)$ d) $\sqrt{6}+\sqrt{5}+2\sqrt{24}-3\sqrt{20}$

    e) $\left(\sqrt[3]{5}+2\right)\left(\sqrt[3]{25}-2\sqrt[3]{5}+4\right)$ f) $\left(\sqrt[4]{2}+1\right)\left(\sqrt[4]{8}-\sqrt[4]{4}+\sqrt[4]{2}-1\right)$

    g) $\dfrac{2+\sqrt{5}}{3-\sqrt{5}}$ h) $\dfrac{2\sqrt{5}+\sqrt{2}}{\sqrt{5}-2\sqrt{2}}$

8. Find the slope of the line that goes through the points $\left(1+\sqrt{2},5\right)$ and $\left(2+3\sqrt{2},2\right)$. Place your answer in simple radical form.

9. Find the equation of the line that passes through the point (2,3) with slope $\sqrt{2}$. Find the x- and y-intercepts of this line and place your answers in simple radical form.

10. For each of the following equations, perform the following tasks:

   i)   Load each side of the equation in the Y= menu, in Y1 and Y2, respectively. Push the ZOOM button and select 6:ZStandard. Set up a coordinate system on your homework paper and copy the image in the viewing window onto your coordinate system. Label each graph with its equation and include the WINDOW parameter settings on your coordinate axes.

   ii) Use the intersect utility in the CALC menu to estimate the real solutions of the equation Y1 = Y2. For a review of this utility, see Example 10 in Section 1.6. Draw a number line below your graph and shade and label these solutions on your number line in the usual manner.

   iii) Use an analytical method to find the real solutions of the equation. If these analytical solutions do not compare favorably with the estimates found in part (ii), check your work for error.

   a) $\sqrt{x+10} = x-2$ b) $\sqrt{21-25x} = 3-2x$ c) $x-\sqrt{x-2} = 2$ d) $x-\sqrt{4x+4} = -1$

11. The period of a pendulum is given by the equation $T = 2\pi\sqrt{\dfrac{L}{g}}$. The variable T represents the period, the time taken for the pendulum to swing back and forth one time. The variable L represents the length of the pendulum and the variable g represents the acceleration due to gravity.

   a) Solve the equation $T = 2\pi\sqrt{\dfrac{L}{g}}$ for the variable L.

   b) If the period of the pendulum is 2.8 seconds and the acceleration due to gravity is 32.2 feet per second per second, substitute these numbers in the result found in part (a) to find the length L of the pendulum. Round your answer to the nearest tenth of a foot (one decimal place).

12. Solve each of the following equations for the indicated variable.

   a) $\omega = \sqrt{\dfrac{L}{g}}$ for g.   b) $I = \dfrac{I_0}{\sqrt{d}}$ for d.

13. Simplify each of the following expressions mentally. Use your calculator to check your results.

   a) $4^{\frac{1}{2}}$ b) $64^{\frac{1}{3}}$ c) $4^{\frac{3}{2}}$ d) $(-8)^{\frac{2}{3}}$ e) $16^{\frac{5}{4}}$ f) $(-32)^{\frac{4}{5}}$

14. Place each of the following expressions in simple radical form. Use a calculator to find a decimal approximation for the original problem and a decimal approximation for your solution. Place these approximations on your homework paper. If they do not match, check your work for error.

   a) $72^{\frac{1}{2}}$ b) $72^{\frac{1}{3}}$ c) $\dfrac{8}{2^{\frac{3}{2}}}$ d) $\dfrac{8}{2^{\frac{2}{3}}}$ e) $2^{\frac{2}{3}}\cdot 2^{\frac{1}{2}}$ f) $\left(2^{\frac{1}{2}}\right)^3$

15. Place each of the following radical expressions in fractional exponent form.

a) $\sqrt{4-5x}$

b) $\sqrt[3]{(x+6)^2}$

c) $\sqrt[4]{(x-4)^3}$

d) $\sqrt[4]{1+x^3}$

16. According to a recent *Scientific American* article on prostate cancer, the number of cases of prostate cancer in the United States is on the rise. The number of cases can be predicted with the equation $N = 61.9t^{0.2252}$, where the variable N represents the number of cases in thousands and the variable t represents the number of years that have passed since 1980. Use the equation $N = 61.9t^{0.2252}$ to predict the number of cases in 1990.

Note: The actual number of cases in 1990 was 106,000.

Chapter 6
Quadratic Functions

This chapter begins with a discussion of the powerful graphing techniques of translation and reflection. You will find that these new graphing techniques will enable you to draw the graphs of quadratic functions with ease. A powerful formula called the quadratic formula will be derived. This formula will be very useful when finding the x-intercepts of a parabola. Finally, you will find where the quadratic function has a maximum or minimum value and apply this knowledge to some interesting real-world situations.

6.1 Translations

This section begins with an introduction of four fundamental graphs. We will then investigate the graphs of the equations $y = f(x) + c$ and $y = f(x - c)$ and discover that they are *translations* or *shifts* of the graph of the equation $y = f(x)$.

Four Basic Graphs

Four fundamental graphs and their equations are shown below. They should be familiar as you have already encountered each of them in earlier work. Commit these graphs and equations to memory.

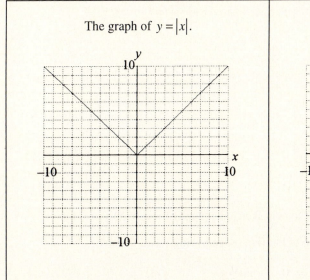

The graph of $y = |x|$.

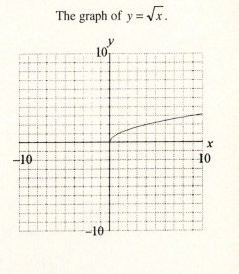

The graph of $y = \sqrt{x}$.

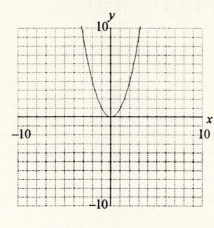

The graph of $y = x^2$.

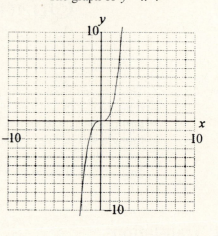

The graph of $y = x^3$.

Vertical Translations

A translation is a shift. A vertical translation is a shift in the vertical direction, either up or down.

Example 1   Consider the following graph of the equation $y = f(x)$.

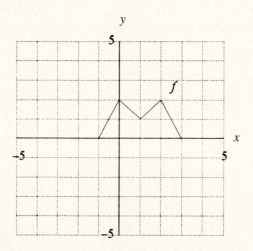

Use the graph of $y = f(x)$ to help sketch the graph of $y = f(x) + 1$.

Solution   Begin by reading the graph of f to find the following function values.

$$f(-1) = 0 \qquad f(0) = 2 \qquad f(1) = 1 \qquad f(2) = 2 \qquad f(3) = 0$$

You can arrange these results in a table.

x	-1	0	1	2	3
$y = f(x)$	0	2	1	2	0

Next, create a table of points that satisfy the equation $y = f(x) + 1$. Begin by substituting -1 for x in the equation $y = f(x) + 1$.

$$y = f(x) + 1$$
$$y = f(-1) + 1$$
$$y = 0 + 1$$
$$y = 1$$

Note how the fact that $f(-1) = 0$ was used to finish the calculation. In a similar manner, you can make the following calculations:

$$y = f(0) + 1 = 2 + 1 = 3$$
$$y = f(1) + 1 = 1 + 1 = 2$$
$$y = f(2) + 1 = 2 + 1 = 3$$
$$y = f(3) + 1 = 0 + 1 = 1$$

You can arrange these results in a table.

x	-1	0	1	2	3
$y = f(x) + 1$	1	3	2	3	1
Points to plot	$(-1,1)$	$(0,3)$	$(1,2)$	$(2,3)$	$(3,1)$

Note that each y-value in this table of points that satisfy the equation $y = f(x) + 1$ is 1 larger than the corresponding y-value in the table of points that satisfy the equation $y = f(x)$. Plot the points in this table.

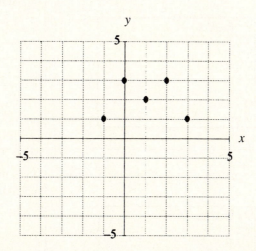

Because the y-values generated by the equation $y = f(x) + 1$ are 1 larger than the y-values generated by the equation $y = f(x)$, each point on the graph of the equation $y = f(x) + 1$ is 1 unit higher than the corresponding point on the graph of the equation $y = f(x)$. Therefore the graph of $y = f(x) + 1$ must look as follows:

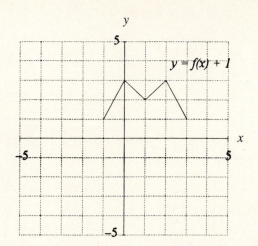

To draw the graph of $y = f(x) + 1$, shift the graph of $y = f(x)$ upward 1 unit.

Example 2   *(TI-82)* Use your calculator to draw the graphs of $y = |x|$ and $y = |x| - 2$ and discuss the relationship between the two graphs.

Solution

Load the equation $y = |x|$ into the Y= menu. ``` Y₁ᴮabs X Y₂= Y₃= Y₄= Y₅= Y₆= Y₇= Y₈= ```	Push the ZOOM button and select 6:ZStandard.
Load the equation $y = |x| - 2$ into the Y= menu. ``` Y₁ᴮabs X-2 Y₂= Y₃= Y₄= Y₅= Y₆= Y₇= Y₈= ```	Push the ZOOM button and select 6:ZStandard.

To draw the graph of $y = |x| - 2$, shift the graph of $y = |x|$ downward 2 units.

A pattern emerges.

To draw the graph of $y = f(x) + c$, shift the graph of $y = f(x)$ upward c units if c is positive and downward c units if c is negative.

This is a powerful graphing shortcut. Let's apply this shortcut to some of the basic graphs.

Example 3   Sketch the graph of $y = x^2 + 5$.

Solution

First sketch the graph of $y = x^2$.

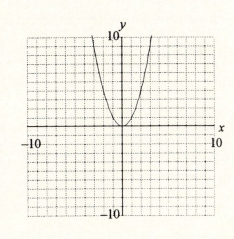

To draw the graph of $y = x^2 + 5$, shift the graph of $y = x^2$ upward 5 units.

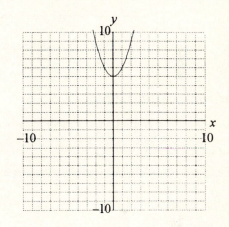

Note: The diagrams shown in the solution of Example 3 might indicate that you can pull precision graphs out of your head at this point. This is not the case. It is only expected that you capture the essence of the graph of $y = x^2 + 5$. Perhaps a solution presented by a student would be useful at this point.

Student Solution The following solution was presented by a student on an examination.

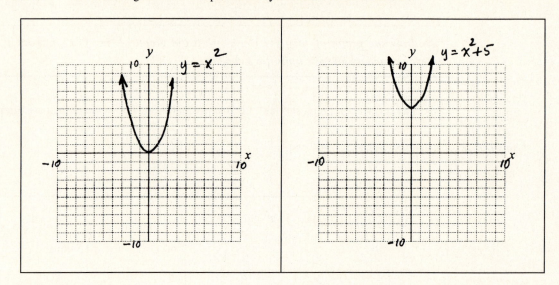

This student captured the essence of the graph of $y = x^2 + 5$ and received a perfect score for her work on this problem.

Horizontal Translations

A translation is a shift. A horizontal translation is a shift in the horizontal direction, either right or left.

Example 4 Consider the following graph whose equation is $y = f(x)$.

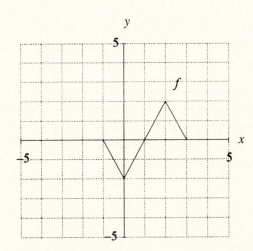

Use the graph of $y = f(x)$ to help sketch the graph of $y = f(x+2)$.

Solution Begin by reading the graph of f to find the following function values:

$$f(-1) = 0 \qquad f(0) = -2 \qquad f(1) = 0 \qquad f(2) = 2 \qquad f(3) = 0$$

You can arrange these results in a table.

x	−1	0	1	2	3
$y = f(x)$	0	−2	0	2	0

Next, create a table of points that satisfy the equation $y = f(x+2)$. Use the following x-values: −3, −2, −1, 0, and 1. Note that these x-values are 2 less than the x-values in the table of points satisfying the equation $y = f(x)$. Begin by substituting −3 for x in the equation $y = f(x+2)$.

$$y = f(x+2)$$

$$y = f(-3+2)$$

$$y = f(-1)$$

$$y = 0$$

Note how the fact that $f(-1) = 0$ was used to finish the calculation. In a similar manner, you can make the following calculations:

$$y = f(-2+2) = f(0) = -2$$

$$y = f(-1+2) = f(1) = 0$$

$$y = f(0+2) = f(2) = 2$$

$$y = f(1+2) = f(3) = 0$$

You can place these results in a table.

x	−3	−2	−1	0	1
$y = f(x+2)$	0	−2	0	2	0
Points to plot	(−3,0)	(−2,−2)	(−1,0)	(0,2)	(1,0)

Note that each x-value in the table of points that satisfy the equation $y = f(x+2)$ is 2 less than the corresponding x-value in the table of points that satisfy the equation $y = f(x)$. Plot the points in this table that satisfy the equation $y = f(x+2)$.

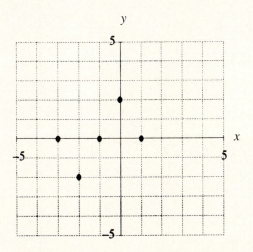

Because x-values that are 2 less than those used in the equation $y = f(x)$ will produce identical y-values when used in the equation $y = f(x+2)$, the graph of the equation $y = f(x+2)$ must be shifted 2 units to the left of the graph of the equation $y = f(x)$. The graph of $y = f(x + 2)$ must look as follows:

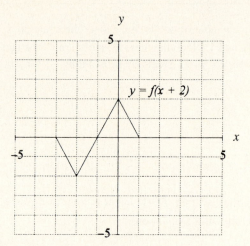

To draw the graph of $y = f(x + 2)$, shift the graph of $y = f(x)$ to the left 2 units.

Example 5 (TI-82) Use your calculator to draw the graphs of $y = \sqrt{x}$ and $y = \sqrt{x - 2}$ and discuss the relationship between them.

Solution

Load the equation $y = \sqrt{x}$ into the Y= menu. 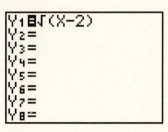	Push the ZOOM button and select 6:ZStandard. 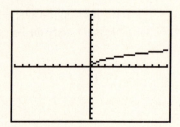
Load the equation $y = \sqrt{x - 2}$ into the Y= menu. 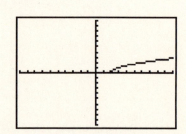	Push the ZOOM button and select 6:ZStandard.

> *To draw the graph of $y = \sqrt{x-2}$, shift the graph of $y = \sqrt{x}$ two units to the right.*

A pattern emerges.

> *To draw the graph of $y = f(x-c)$, shift the graph of $y = f(x)$ to the right c units if c is positive and to the left c units if c is negative.*

This is a powerful graphing shortcut. Let's apply this new shortcut to one of the basic graphs.

Example 6   Sketch the graph of $y = |x+3|$.

Solution

First sketch the graph of $y = |x|$.

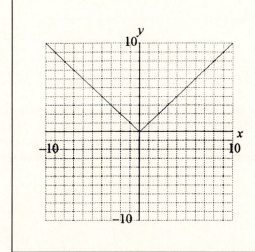

To draw the graph of $y = |x+3|$, shift the graph of $y = x^2$ to the left 3 units.

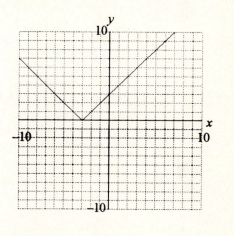

Student Solution   The following solution was presented by a student on an examination.

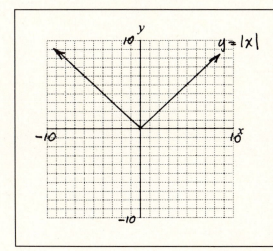

$y = |x|$

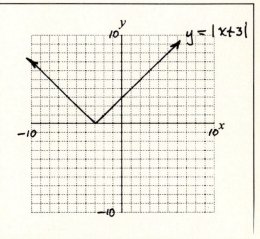

$y = |x+3|$

Combining Vertical and Horizontal Translations

We are going to sketch some graphs that involve both a horizontal and a vertical translation. Before beginning, review the rules that govern vertical and horizontal shifts.

> *To draw the graph of $y = f(x) + c$, shift the graph of $y = f(x)$ upward c units if c is positive and downward c units if c is negative.*
>
> *To draw the graph of $y = f(x - c)$, shift the graph of $y = f(x)$ to the right c units if c is positive and to the left c units if c is negative.*

Example 7 Simplify the equations $y = f(x) + 2$, $y = f(x + 2)$, and $y = f(x + 2) + 2$ for each of the following functions: $f(x) = x^2$, $f(x) = \sqrt{x}$, $f(x) = |x|$, and $f(x) = x^3$.

Solution The solution can be arranged in a table.

	$y = f(x) + 2$	$y = f(x + 2)$	$y = f(x + 2) + 2$
$f(x) = x^2$	$y = x^2 + 2$	$y = (x + 2)^2$	$y = (x + 2)^2 + 2$
$f(x) = \sqrt{x}$	$y = \sqrt{x} + 2$	$y = \sqrt{x + 2}$	$y = \sqrt{x + 2} + 2$
$f(x) = |x|$	$y = |x| + 2$	$y = |x + 2|$	$y = |x + 2| + 2$
$f(x) = x^3$	$y = x^3 + 2$	$y = (x + 2)^3$	$y = (x + 2)^3 + 2$

To draw the graph of each equation in column 2, shift the graph of the corresponding equation in column 1 upward 2 units. To draw the graph of each equation in column 3, shift the graph of the corresponding equation in column 1 to the left 2 units. To draw the graph of each equation in column 4, shift the graph of the corresponding equation in column 1 to the left 2 units and upward 2 units.

Example 8 Sketch the graph of $y = (x - 2)^2 - 3$.

Solution   If $f(x) = x^2$, then $y = (x-2)^2 - 3$ is equivalent to $y = f(x-2) - 3$.

First sketch the graph of $y = x^2$.

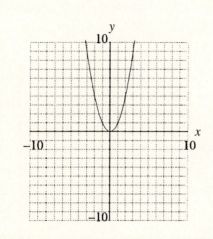

To draw the graph of $y = (x-2)^2 - 3$, shift the graph of $y = x^2$ to the right 2 units and down 3 units.

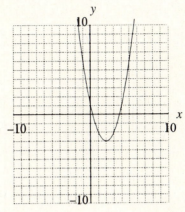

You can check this solution with your calculator.

Load the equation $y = (x-2)^2 - 3$ in the Y= menu.

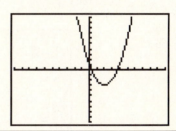

Push the ZOOM button and select 6:ZStandard.

| **Tip** | You should first draw the graph without your calculator. Use your calculator to check the problem, not to do the problem. |

Example 9   Sketch the graph of $y = (x+1)^3 + 1$.

Solution If $f(x) = x^3$, then $y = (x+1)^3 + 1$ is equivalent to $y = f(x+1) + 1$. You will need to shift the graph of f to the left 1 unit and upward 1 unit.

First sketch the graph of $y = x^3$.

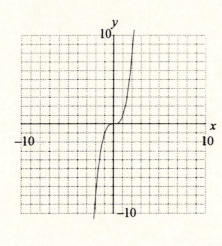

To draw the graph of $y = (x+1)^3 + 1$, shift the graph of $y = x^2$ to the left 1 unit and upward 1 unit.

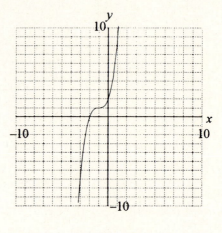

You can check the solution with your calculator.

Load the equation $y = (x+1)^3 + 1$ in the Y= menu.

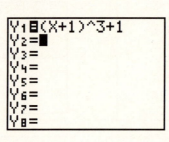

Push the ZOOM button and select 6:ZStandard.

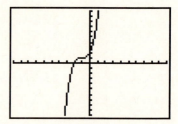

Exercises for Section 6.1

1. (TI-82) Sketch each of the following basic graphs on your calculator. Load the equation in Y₁ in the Y= menu, push the ZOOM button and select 6:ZStandard. Set up a coordinate system on your homework paper and copy the image in the viewing screen onto your coordinate system. Label the graph with its equation and include the WINDOW parameters on your coordinate axes.

   a) $y = x^2$
   b) $y = |x|$
   c) $y = \sqrt{x}$
   d) $y = x^3$

2. Set up one coordinate system on a sheet of graph paper. Place a copy of each of the following tables on your graph paper. Complete the missing entries in each table. Plot the points from each

table on the same coordinate system and use these points to help draw complete graphs of each equation. Label each graph with its equation.

x	−3	−2	−1	0	1	2	3
$y = x^2$							
Points to plot							

x	−3	−2	−1	0	1	2	3
$y = x^2 + 1$							
Points to plot							

x	−3	−2	−1	0	1	2	3
$y = x^2 + 2$							
Points to plot							

x	−3	−2	−1	0	1	2	3
$y = x^2 + 3$							
Points to plot							

Use complete sentences to explain what you learned in this exercise.

3. Set up one coordinate system on a sheet of graph paper. Place a copy of each of the following tables on your graph paper. Complete the missing entries in each table. Plot the points from each table on the same coordinate system and use these points to help draw complete graphs of each equation. Label each graph with its equation.

x	−3	−2	−1	0	1	2	3
$y = x^2$							
Points to plot							

x	−4	−3	−2	−1	0	1	2
$y = (x+1)^2$							
Points to plot							

x	−5	−4	−3	−2	−1	0	1
$y = (x+2)^2$							
Points to plot							

x	-6	-5	-4	-3	-2	-1	0
$y = (x+3)^2$							
Points to plot							

Use complete sentences to explain what you have learned from this exercise.

4. (TI-82) Load the equations $y = |x|$, $y = |x| - 1$, $y = |x| - 2$, and $y = |x| - 3$ in the Y= menu on your calculator in the following manner:

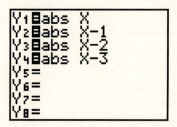

a) Set up a coordinate system on your homework paper. Push the ZOOM button and select 6:ZStandard, then copy the image from the viewing screen onto your coordinate system. Label each graph with its equation and include the WINDOW parameter settings on your coordinate axes.

b) Use complete sentences to explain what you learned from this exercise.

5. (TI-82) Load the equations $y = |x|$, $y = |x-1|$, $y = |x-2|$, and $y = |x-3|$ in the Y= menu on your calculator in the following manner:

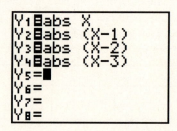

a) Set up a coordinate system on your homework paper. Push the ZOOM button and select 6:ZStandard, then copy the image from the viewing screen onto your coordinate system. Label each graph with its equation and include the WINDOW parameter settings on your coordinate axes.

b) Use complete sentences to explain what you learned from this exercise.

6. Set up a coordinate system on a sheet of graph paper and copy the following graph of the function *f* onto your coordinate system.

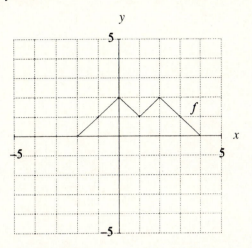

a) Make a copy of the following table on your graph paper. Examine the graph of *f* and complete the missing entries in the table.

x	−2	−1	0	1	2	3	4
$f(x)$							

b) Make a copy of the following table on your graph paper and complete the missing entries. Set up a second coordinate system on the same graph paper and plot each point from the table on your coordinate system. Plot the rest of the points that satisfy the equation $y = f(x) + 2$.

x	−2	−1	0	1	2	3	4
$y = f(x) + 2$							
Points to plot							

c) Make a copy of the following table on your graph paper and complete the missing entries. Set up a third coordinate system on the same graph paper and plot each point from the table on your coordinate system. Plot the rest of the points that satisfy the equation $y = f(x + 2)$.

x	−4	−3	−2	−1	0	1	2
$y = f(x + 2)$							
Points to plot							

d) Use complete sentences to explain what you have learned from this exercise.

7. Set up a coordinate system on a sheet of graph paper and copy the following graph of the function f onto your coordinate system.

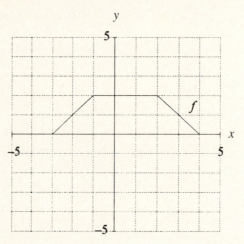

a) Make a copy of the following table on your graph paper. Examine the graph of f and complete the missing entries in the table.

x	-3	-2	-1	0	1	2	3	4
$f(x)$								

b) Make a copy of the following table on your graph paper and complete the missing entries. Set up a second coordinate system on the same graph paper and plot each point from the table on your coordinate system. Plot the rest of the points that satisfy the equation $y = f(x) - 2$.

x	-3	-2	-1	0	1	2	3	4
$y = f(x) - 2$								
Points to plot								

c) Make a copy of the following table on your graph paper and complete the missing entries. Set up a third coordinate system on the same graph paper and plot each point from the table on your coordinate system. Plot the rest of the points that satisfy the equation $y = f(x - 2)$.

x	-1	0	1	2	3	4	5	6
$y = f(x - 2)$								
Points to plot								

d) Use complete sentences to explain what you have learned from this exercise.

8. (TI-82) Load each of the equations $y = \sqrt{x}$, $y = \sqrt{x} + 2$, $y = \sqrt{x} + 4$, and $y = \sqrt{x} + 6$ into the Y= menu on your calculator in the following manner:

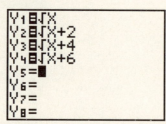

a) Set up a coordinate system on your homework paper. Push the ZOOM button and select 6:ZStandard, then copy the image from your viewing screen onto your coordinate system.

Label each graph with its equation and include the WINDOW parameter settings on your coordinate axes.

b) Use complete sentences to explain what you learned from this exercise.

9. (TI-82) Load each of the equations $y = \sqrt{x}$, $y = \sqrt{x+2}$, $y = \sqrt{x+4}$, and $y = \sqrt{x+6}$ into the Y= menu on your calculator in the following manner.

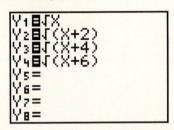

a) Set up a coordinate system on your homework paper. Push the ZOOM button and select 6:ZStandard, then copy the image from your viewing screen onto your coordinate system. Label each graph with its equation and include the WINDOW parameter settings on the coordinate axes.

b) Use complete sentences to explain what you learned from this exercise.

10. For each of the following equations, set up a coordinate system on your homework paper and use your knowledge of translations and the four fundamental graphs to sketch of the graph of the equation. For an example of how to proceed, refer to the student solution in Example 3 in this section.

a) $y = x^2 + 4$ b) $y = (x+4)^2$ c) $y = \sqrt{x} - 3$ d) $y = \sqrt{x-3}$

e) $y = |x| + 5$ f) $y = |x+5|$ g) $y = x^3 - 2$ h) $y = (x-2)^3$

11. For each of the following graphs, set up a coordinate system on a sheet of graph paper and copy the graph onto your coordinate system. Use your knowledge of translations and the four fundamental graphs to find the equation of the graph. Label the graph with its equation and use your calculator to check your solution.

a)

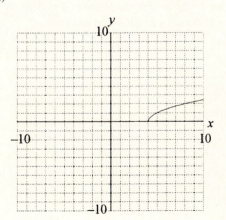

b)

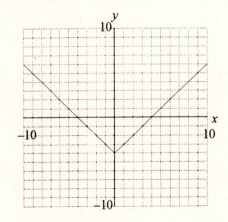

c)

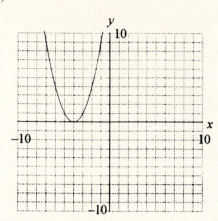

d)

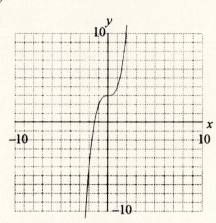

12. Set up a coordinate system on a sheet of graph paper and copy the following graph of the function f onto your coordinate system.

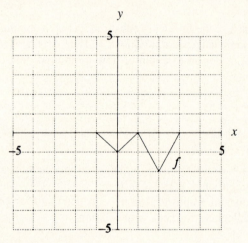

For each of the following equations, set up a separate coordinate system on your graph paper and use your knowledge of translations to sketch a graph of the equation.

a) $y = f(x) + 1$ b) $y = f(x + 1)$ c) $y = f(x) - 2$ d) $y = f(x - 2)$

13. Set up a coordinate system on a sheet of graph paper and copy the following graph of the function g onto your coordinate system.

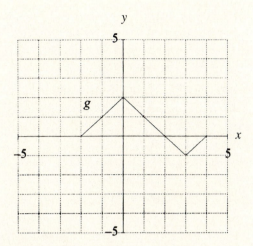

For each of the following equations, set up a separate coordinate system on your graph paper and use your knowledge of translations to sketch a graph of the equation.

a) $y = g(x) - 4$ b) $y = g(x-4)$ c) $y = g(x) + 3$ d) $y = g(x+3)$

14. For each of the following equations, set up a coordinate system on your homework paper and use your knowledge of translations and the four fundamental graphs to sketch of the graph of the equation. For an example of how to proceed, refer to the student solution in Example 3 of this section.

a) $y = (x-1)^2 - 3$ b) $y = \sqrt{x+2} + 1$ c) $y = |x-4| - 3$ d) $y = (x+2)^3 + 2$

e) $y = (x+4)^2 - 5$ f) $y = |x-7| + 5$ g) $y = \sqrt{x-5} - 3$ h) $y = (x+2)^3 + 3$

15. For each of the following graphs, set up a coordinate system on a sheet of graph paper and copy the graph onto your coordinate system. Use your knowledge of translations and the four fundamental graphs to find the equation of the graph. Label the graph with its equation and use your calculator to check your solution.

a)

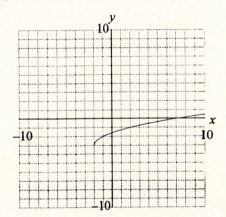

b)

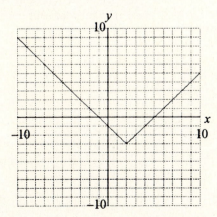

c)

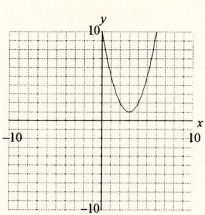

d)

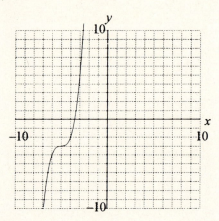

16. Set up a coordinate system on a sheet of graph paper and copy the following graph of the function f onto your coordinate system.

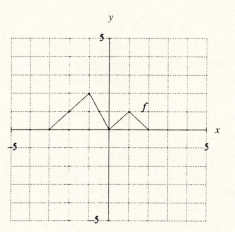

For each of the following equations, set up a separate coordinate system on your graph paper and use your knowledge of translations to sketch a graph of the equation.

a) $y = f(x+3) - 2$

b) $y = f(x-2) + 3$

17. In trigonometry, one period of the graph of $y = \sin x$ looks like the following.

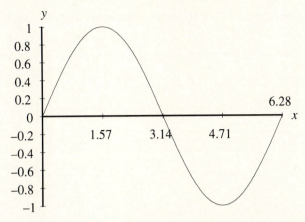

For each of the following equations, set up a separate coordinate system on your homework paper and use your knowledge of translations and the graph of $y = \sin x$ to sketch a graph of the given equation.

a) $y = \sin(x - 1.57)$

b) $y = \sin(x + 3.14)$

c) $y = \sin(x + 1.57) + 1$

d) $y = \sin(x - 4.71) - 2$

Note: If you set your calculator in radian mode, you can check your solutions with your calculator.

6.2 Reflections

This section will investigate the behavior of the graphs of the following two equations:

$$y = -f(x) \quad \text{and} \quad y = f(-x)$$

In the first of these two equations, the function f is applied, then the result is negated. Consequently, the y-values will be the negative of the y-values of the original function. In the second equation, the x-values are negated, then the function f is applied. This will cause quite different behavior in the graph.

Reflections Across the x-Axis

Example 1   Consider the following graph of the equation $y = f(x)$.

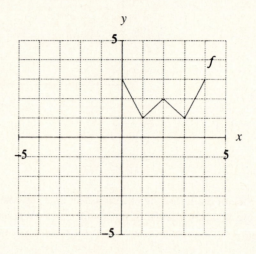

Use the graph of $y = f(x)$ to help sketch the graph of $y = -f(x)$.

Solution   Examine the graph of the function f to find the following function values:

$$f(0) = 3 \qquad f(1) = 1 \qquad f(2) = 2 \qquad f(3) = 1 \qquad f(4) = 3$$

These results can be arranged in a table.

x	0	1	2	3	4
$f(x)$	3	1	2	1	3

You can now create a table of points that satisfy the equation $y = -f(x)$. Use 0, 1, 2, 3, and 4 for your x-values. When you substitute these x-values into the equation, the following y-values will result:

$$y = -f(0) = -3 \qquad y = -f(1) = -1 \qquad y = -f(2) = -2 \qquad y = -f(3) = -1 \qquad y = -f(4) = -3$$

These results can be arranged in a table.

x	0	1	2	3	4
$y = -f(x)$	-3	-1	-2	-1	-3
Points to plot	(0,-3)	(1,-1)	(2,-2)	(3,-1)	(4,-3)

Plot the points in this table on a coordinate system.

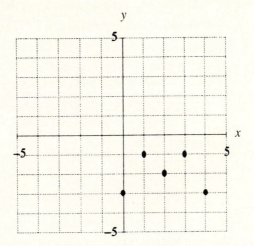

Because the y-values generated by the equation $y = -f(x)$ are the negative of the y-values generated by the equation $y = f(x)$, each point on the graph of the equation $y = -f(x)$ is the *reflection* of the corresponding point of $y = f(x)$ across the x-axis. Therefore the graph of $y = -f(x)$ must look as follows:

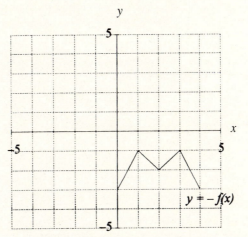

To draw the graph of $y = -f(x)$, reflect the graph of $y = f(x)$ across the x-axis.

A Great Experiment. Try the following experiment. With a dark pen or pencil, sketch the graph of the equation $y = f(x)$ on a sheet of graph paper. Now, hold the paper up to a strong light source, flip the paper across the x-axis, and observe the image of $y = -f(x)$ through the back of the paper.

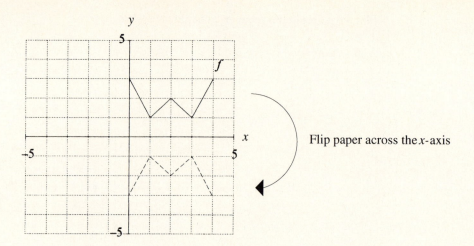

Flip paper across the x-axis

Observe the Symmetry. On the following coordinate system the graphs of $y = f(x)$ and $y = -f(x)$ have been sketched so that you can observe the symmetry with respect to the x-axis.

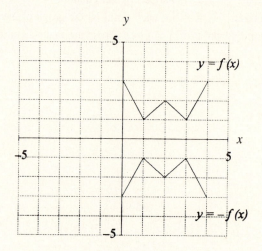

Example 2   *(TI-82)* Use your calculator to draw the graphs of $y = |x|$ and $y = -|x|$ and discuss the relationship between the two graphs.

Solution

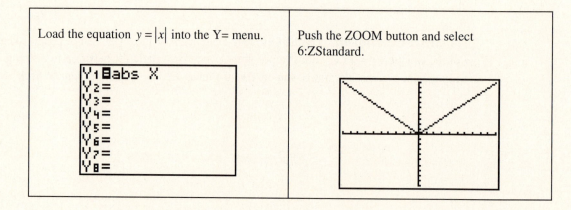

| Load the equation $y = |x|$ into the Y= menu. | Push the ZOOM button and select 6:ZStandard. |

Load the equation $y = -|x|$ into the Y= menu.

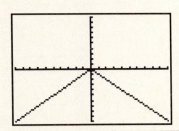

Push the ZOOM button and select 6:ZStandard.

To draw the graph of $y = -|x|$, reflect the graph of $y = |x|$ across the x-axis.

Example 3   Sketch the graph of $y = -\sqrt{x}$.

Solution   A student offered the following solution:

First sketch the graph of $y = \sqrt{x}$.

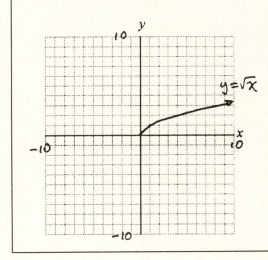

The graph of $y = -\sqrt{x}$ is the reflection of the graph of $y = \sqrt{x}$ across the x-axis.

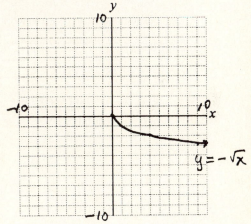

Reflections Across the *y*-Axis

Example 4   Consider again the graph from Example 1 whose equation is $y = f(x)$.

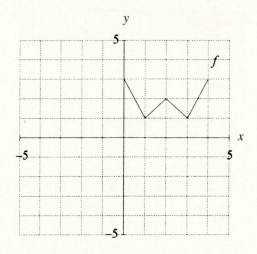

Use the graph of $y = f(x)$ to help sketch the graph of $y = f(-x)$.

Solution   Once again, the following function values can be read from the graph of *f*.

$$f(0) = 3 \qquad f(1) = 1 \qquad f(2) = 2 \qquad f(3) = 1 \qquad f(4) = 3$$

These results can be arranged in a table.

x	0	1	2	3	4
$f(x)$	3	1	2	1	3

Next, create a table of points that satisfy the equation $y = f(-x)$. Use $-4, -3, -2, -1$, and 0 for the x-values. Substitute each of these x-values into the equation $y = f(-x)$.

$$y = f(-(-4)) = f(4) = 3$$
$$y = f(-(-3)) = f(3) = 1$$
$$y = f(-(-2)) = f(2) = 2$$
$$y = f(-(-1)) = f(1) = 1$$
$$y = f(-0) = f(0) = 3·$$

These results can be arranged in a table.

x	-4	-3	-2	-1	0
$y = h(-x)$	3	1	2	1	3
Points to plot	$(-4,3)$	$(-3,1)$	$(-2,2)$	$(-1,1)$	$(0,3)$

Plot these points on a coordinate system.

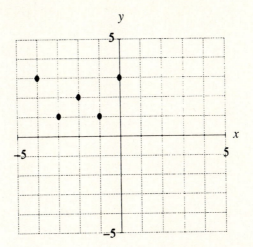

Each point on the graph of the equation $y = f(-x)$ is the reflection of the corresponding point of $y = f(x)$ across the y-axis. Therefore the graph of $y = f(-x)$ must look as follows:

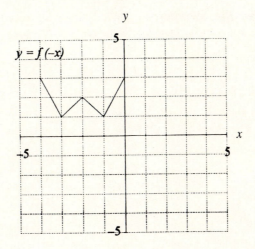

To draw the graph of $y = f(-x)$, reflect the graph of $y = f(x)$ across the y-axis.

Another Great Experiment. Try this experiment. With a dark pen or pencil, sketch the graph of the equation $y = f(x)$ on a sheet of graph paper. Hold the paper up to a strong light source, flip the paper across the y-axis, and observe the graph of $y = f(-x)$ through the back of the paper.

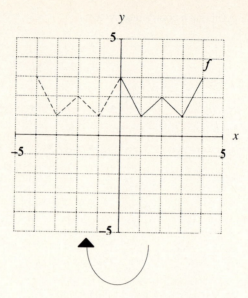

Flip paper across the y-axis.

Observe the Symmetry.   On the following coordinate system, the graphs of $y = f(x)$ and $y = f(-x)$ have been sketched, so that you can observe the symmetry with respect to the y-axis.

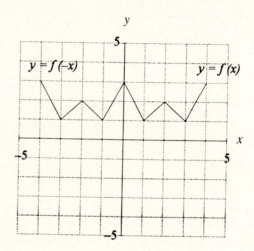

Example 5   *(TI-82)* Use your calculator to draw the graphs of $y = \sqrt{x}$ and $y = \sqrt{-x}$ and discuss the relationship between the two graphs.

Solution

Load the equation $y = \sqrt{x}$ in the Y= menu.

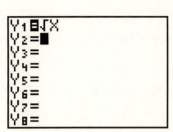

Push the ZOOM button and select 6:ZStandard.

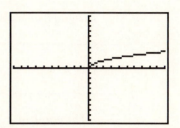

Load the equation $y = \sqrt{-x}$ in the Y= menu.

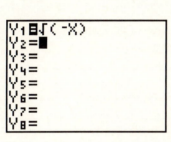

Push the ZOOM button and select 6:ZStandard.

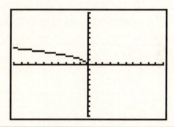

The graph of $y = \sqrt{-x}$ is a reflection of the graph of $y = \sqrt{x}$ across the y-axis.

Example 6   Sketch the graph of $y = \left|-x\right|$.

Solution   A student offered the following solution. The graph of $y = \left|-x\right|$ is a reflection of the graph of the graph of $y = \left|x\right|$ across the y-axis.

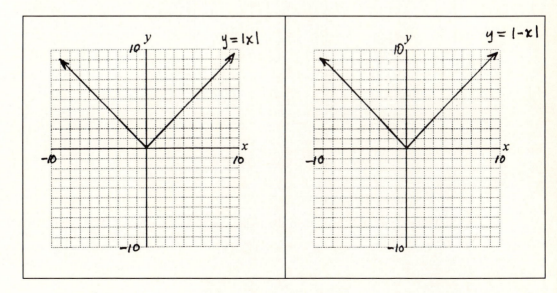

It is interesting to note that the graph of $y = |-x|$ is the same as the graph of $y = |x|$. When this happens, mathematicians say that the graph is *symmetric with respect to the y-axis*.

A Summary of Reflections.   Negating a function reflects the graph across the *x*-axis.

> **The graph of $y = -f(x)$ is a reflection of the graph of $y = f(x)$ across the x-axis.**

If the *x*-values are negated before applying the function, the graph will reflect across the *y*-axis.

> **The graph of $y = f(-x)$ is a reflection of the graph of $y = f(x)$ across the y-axis.**

Combining Reflections with Translations

The examples that follow combine one or more of the graphing shortcuts of translation and reflection to arrive at a final sketch. First an important tip.

Tip	It is recommended that you do the reflections first.

Example 7   Sketch the graph of $y = \sqrt{-(x-2)} - 3$.

Solution   It is best to present your work in a sequence of logical steps.

First, sketch the graph of $y = \sqrt{x}$.

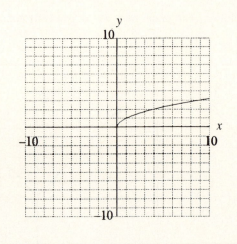

Reflect the graph of $y = \sqrt{x}$ across the *y*-axis to arrive at the graph of $y = \sqrt{-x}$.

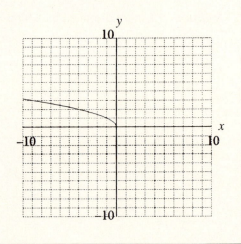

Slide the graph of $y = \sqrt{-x}$ to the right 2 units and downward 3 units to arrive at the graph of $y = \sqrt{-(x-2)} - 3$.

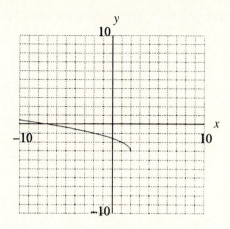

Example 8   Sketch the graph of $y = -|x-3| + 2$.

Solution   Here is a student solution.

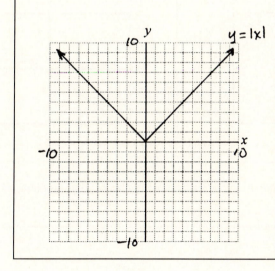

First sketch the graph of $y = |x|$.

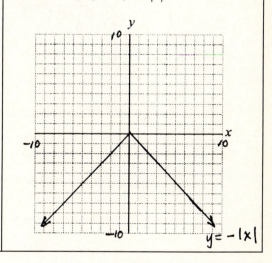

Reflect the graph of $y = |x|$ across the y-axis to arrive at the graph of $y = -|x|$.

Slide the graph of $y = -|x|$ right 3 units and upward 2 units to arrive at the graph of $y = -|x - 3| + 2$.

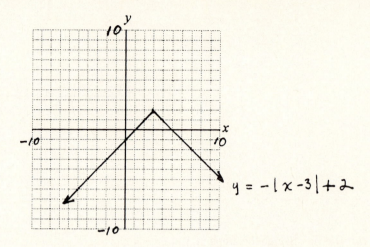

$$y = -|x - 3| + 2$$

Of course, you can check your solution on your calculator.

Load the equation $y = -|x - 3| + 2$ in the Y= menu.	Push the ZOOM button and select 6:ZStandard.
Y₁◻-abs (X-3)+2 Y₂=◼ Y₃= Y₄= Y₅= Y₆= Y₇= Y₈=	

Tip	It is important that you use your calculator to check your problem, not to do your problem.

Exercises for Section 6.2

1. Make a copy of the following table on a sheet of graph paper. Complete the missing entries in your table.

x	0	1	4	9	16	25
$y = \sqrt{x}$						
Points to plot						

a) Set up a coordinate system on your graph paper and plot each point from the table on your coordinate system. Plot the rest of the points that satisfy the equation $y = \sqrt{x}$.

10. This problem is designed to lead you through a sequence of sketches that will depict the graph of the equation $y = -|x| + 1$.

   a) Set up a coordinate system and sketch the graph of $y = |x|$.

   b) Set up a second coordinate system and sketch the graph of $y = -|x|$.

      Hint: Reflect the graph in part (a) across the *x*-axis.

   c) Set up a third coordinate system and sketch the graph of $y = -|x| + 1$.

      Hint: Translate the graph in part (b) up one unit.

   d) (TI-82) Enter the equation $y = -|x| + 1$ in the Y= menu of your calculator in the following manner.

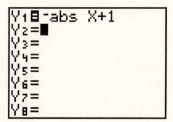

   Push the ZOOM button and select 6:ZStandard. If the image in your viewing window does not agree with the graph in part (c), check your work for error.

11. This problem is designed to lead you through a sequence of sketches that will depict the graph of the equation $y = -|x + 1|$.

   a) Set up a coordinate system and sketch the graph of $y = |x|$.

   b) Set up a second coordinate system and sketch the graph of $y = -|x|$.

      Hint: Reflect the graph in part (a) across the *x*-axis.

   c) Set up a third coordinate system and sketch the graph of $y = -|x + 1|$.

      Hint: Translate the graph in part (b) to the left one unit.

   d) (TI-82) Enter the equation $y = -|x + 1|$ in the Y= menu of your calculator in the following manner.

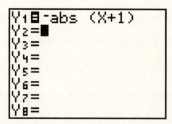

   Push the ZOOM button and select 6:ZStandard. If the image in your viewing window does not agree with the graph in part (c), check your work for error.

12. This problem is designed to lead you through a sequence of sketches that will depict the graph of the equation $y = \sqrt{-(x + 1)}$.

   a) Set up a coordinate system and sketch the graph of $y = \sqrt{x}$.

b) Set up a second coordinate system and sketch the graph of $y = \sqrt{-x}$.

   Hint: Reflect the graph in part (a) across the y-axis.

c) Set up a third coordinate system and sketch the graph of $y = \sqrt{-(x+1)}$.

   Hint: Translate the graph in part (b) to the left one unit.

d) (TI-82) Enter the equation $y = \sqrt{-(x+1)}$ in the Y= menu of your calculator in the following manner.

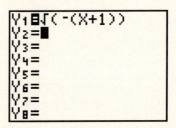

   Push the ZOOM button and select 6:ZStandard. If the image in your viewing window does not agree with the graph in part (c), check your work for error.

13. Set up a coordinate system on a sheet of graph paper and copy the following graph of the function f on your coordinate system.

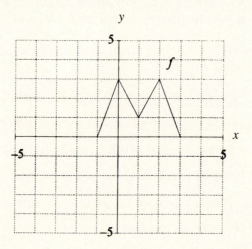

a) Set up a second coordinate system on your graph paper and sketch the graph of $y = -f(x)$.

   Hint: Reflect the graph of f across the x-axis.

b) Set up a third coordinate system on your graph paper and sketch the graph of $y = -f(x-1)$.

   Hint: Shift the graph from part (a) one unit to the right.

c) Set up a fourth coordinate system on your graph paper and sketch the graph of $y = -f(x-1)+2$.

   Hint: Shift the graph from part (b) up two units.

14. On graph paper, set up a sequence of graphs that lead to the graph of the given function. Check your solution with your calculator. If they do not match, check your work for error.

   a) $y = -|x| + 4$ b) $y = -x^2 - 1$ c) $y = -\sqrt{x-3}$

d) $y = -x^3 + 2$ e) $y = -|x - 3| + 1$ f) $y = -(x+2)^2 + 3$

g) $y = -|x + 1| - 1$ h) $y = -(x-2)^3 + 1$

Hint: Do the reflections first, the translations second.

15. Set up a coordinate system on a sheet of graph paper and copy the following graph of the function h on your coordinate system.

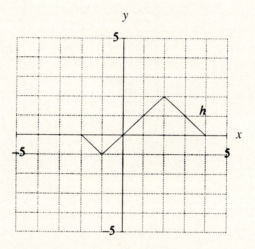

On graph paper, set up a sequence of graphs that lead to the graph of each of the following functions.

a) $y = -h(x - 1) + 2$ b) $y = -h(-x)$ c) $y = h(-x) - 2$ d) $y = -h(x + 2) - 1$

16. For each of the following graphs, set up a coordinate system on a sheet of graph paper and copy the graph onto your coordinate system. Use your knowledge of translations and reflection to find the equation of the graph. Label the graph with its equation and use your calculator to check your solution.

a)

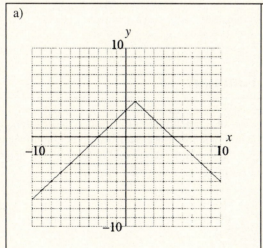

b)

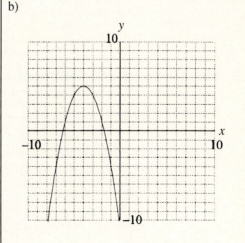

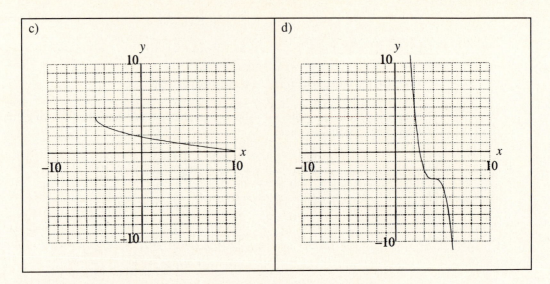

17. In trigonometry, one period of the *sine* function looks like the following:

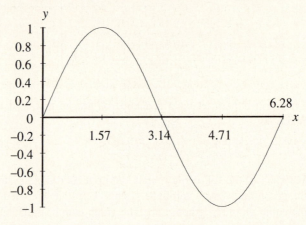

If the sketch is one period of the graph of $y = \sin(x)$, sketch one period of each of the following.

a) $y = -\sin(x)$ b) $y = \sin(-x)$ c) $y = -\sin(x - 3.14)$ d) $y = -\sin(x + 1.57)$

6.3 The Domain and Range of a Function

This section will begin with a review of the definition of a relation and its domain and range. It will also review the definition of a function and introduce a test for determining when a graph represents a function. With the powerful new graphing skills developed in Sections 6.1 and 6.2, the domain and range of a function can easily be determined by examining its graph.

Relations

Do you recall the following definition?

> *A relation is a set of ordered pairs. If the ordered pairs of a relation are denoted with the symbolism (x, y), then the domain of a relation is the set of all the x-values of the ordered pairs and the range of the relation is the set of all y-values of the ordered pairs.*

Example 1   Find the domain and range of the relation $R = \{(1,0),(2,3),(4,5)\}$.

Solution R is a set of ordered pairs. Therefore, by definition, R is a relation. Now, the domain is the set of x-values, and the range is the set of y-values.

$$\text{Domain} = \{1, 2, 4\}$$
$$\text{Range} = \{0, 3, 5\}$$

Mapping Diagrams.   Another way of indicating the ordered pairs of a relation is to use a construct called a mapping diagram. The domain of the relation is placed on the left, the range is placed on the right, and the ordered pairs are indicated with arrows.

Example 2   Construct a mapping diagram for the relation $R = \{(1,0),(2,3),(4,5)\}$.

Solution   List the domain on the left, the range on the right, and use arrows to indicate the ordered pairs.

$$R$$

$$1 \longrightarrow 0$$
$$2 \longrightarrow 3$$
$$4 \longrightarrow 5$$

The domain object 1 gets sent to 0, the domain object 2 gets sent to 3, and the domain object 4 gets sent to 5.

Functions

Do you recall the following definition?

> *A relation is a function if and only if each domain object is paired with (or sent to) exactly one range object.*

The relation R of Example 2 is a function. Each domain object is paired with (or sent to) exactly one range object.

Example 3   Is the relation $T = \{ (1,0), (1,2), (3,7) \}$ a function?

Solution   First, draw a mapping diagram. List the domain on the left and the range on the right, and use arrows to indicate the ordered pairs.

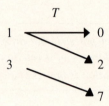

The relation *T* is *not* a function. The domain object 1 is paired with (or sent to) two different range objects; namely, 0 and 2.

Example 4   Is the relation $R = \{ (1,0), (2,0), (3,5) \}$ a function?

Solution   Draw a mapping diagram. List the domain on the left and the range on the right, and use arrows to indicate the ordered pairs.

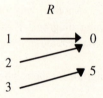

The domain object 1 gets sent to exactly one range object, namely 0. The domain object 2 gets sent to exactly one range object, namely 0. Finally, the domain object 3 gets sent to exactly one range object, namely 5. Therefore the relation *R* is a function.

Example 5   The following is the graph of a relation *f*. Is this relation a function?

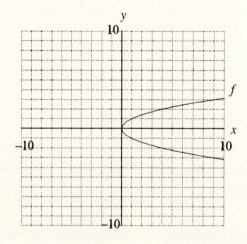

Solution   Certainly, f is a relation because it is a set of ordered pairs. But is it a function? Draw the vertical line whose equation is $x = 4$ on this graph.

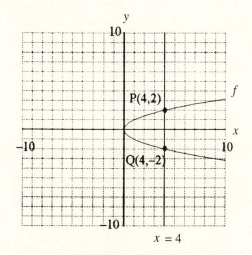

$$x = 4$$

This vertical line cuts the graph in two locations: at the points P(4,2) and Q(4,−2). Set up a sort of mapping diagram for f. You can't place all the points of this relation in the mapping diagram because there are an infinite number of points. However, you can at least place the two points P and Q in the mapping diagram.

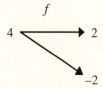

The domain object 4 is paired with (sent to) two different range objects. Therefore, f is *not* a function.

The Vertical Line Test: If any vertical line cuts the graph of f in more than one place, f will not be a function.

Example 6   Consider the following graph of the relation *R*. Is *R* a function?

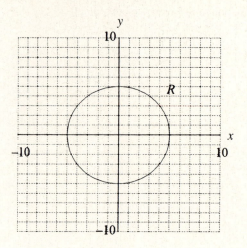

Solution

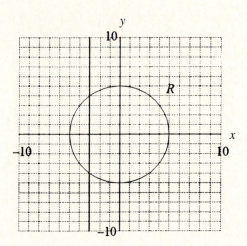

Because a vertical line cuts the graph of *R* in more than one location, *R* is *not* a function.

Example 7   Consider the following graph of the relation *f*. Is *f* a function?

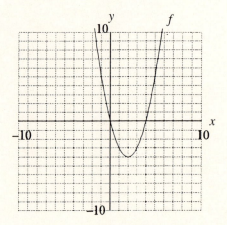

Solution   Because no vertical line will cut the graph of the relation *f* in more than one location, *f* is a function.

Finding the Domain and Range from a Graph

The most effective way to find the domain and range of a relation or of a function is to read the graph of that relation or function. Examine each point (x, y) on the graph. Add the *x*-value of that point to the domain and the *y*-value of that point to the range. Continue this process until you have examined each point on the graph.

Example 8   Find the domain and range of the function whose equation is $f(x) = \sqrt{x}$.

Solution   Begin by drawing the graph of the function *f*. Note that no vertical line will cut this graph more than once, so *f* is a function.

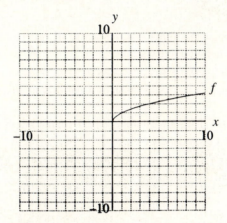

The Domain. Find the domain of *f*. Take an arbitrary point P on the graph of *f*. To find the *x*-value of point P, project point P onto the *x*-axis.

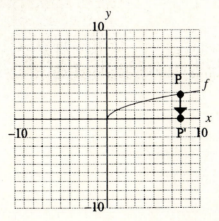

The point P' is called the *image* of point P under the projection. Imagine that it is a bright, sunny day, and the sun is shining directly overhead. Point P blocks the sunlight, causing a shadow P' to fall upon the *x*-axis. This shadow is the *x*-value of point P and is a domain object. To find the domain of the function *f*, you must project every point on the graph of *f* onto the *x*-axis. The part of the *x*-axis that lies in shadow will be the domain of the function *f*.

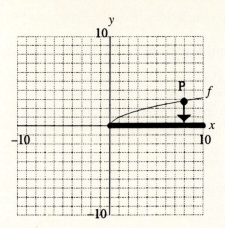

It appears that the domain is the set of all real numbers that are greater than or equal to zero. In set-builder notation, the domain equals $\{x : x \geq 0\}$. In interval notation, the domain equals $[0, +\infty)$. Of course, it is assumed that the graph of f continues to open to the right indefinitely. Because $\sqrt{x}$ is a real number for all x greater than or equal to zero, this is a reasonable assumption.

The Range. Take an arbitrary point P on the graph of f. To find the y-value of P, you will have to project point P onto the y-axis.

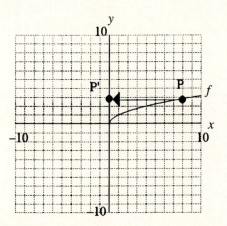

To find the range of function f, you must project every point on the graph of f onto the y-axis. Use your calculator to further investigate the behavior of the graph of f. Enter the equation $y = \sqrt{x}$ in the Y= menu as $Y_1 = \sqrt{X}$. The WINDOW parameter settings on the left will produce the graph on the right.

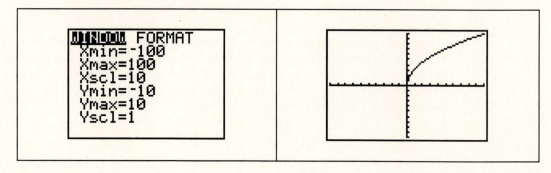

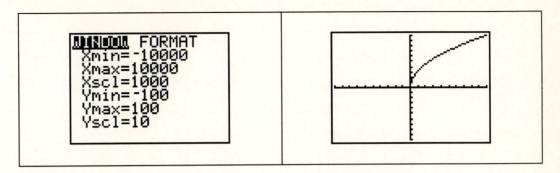

The first setting of WINDOW parameters indicates that the graph rises to $y = 10$. The second setting of WINDOW parameters indicates that the graph rises to $y = 100$. Further experimentation will show that the graph of $y = \sqrt{x}$ will rise indefinitely. Therefore, if every point on the graph of f is projected onto the y-axis, the following shadow will fall on the y-axis:

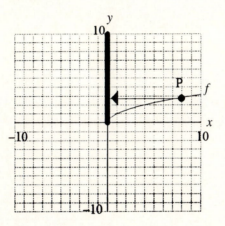

Therefore the range of this function is the set of all real numbers that are greater than or equal to zero. In set-builder notation, the range equals $\{y : y \geq 0\}$. In interval notation, the range equals $[0, +\infty)$.

Example 9   Find the domain and range of the function whose equation is $f(x) = -|x+1| + 2$.

Solution   First, use graphing shortcuts to sketch the graph of the function *f*.

First sketch the graph of $y = |x|$.

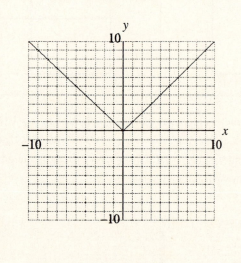

The graph of $y = -|x|$ is a reflection of the graph of $y = |x|$ across the *x*-axis.

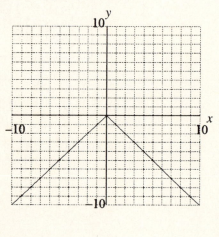

Slide the graph of $y = -|x|$ to the left 1 unit and upward 2 units to arrive at the graph of $y = -|x+1| + 2$. Note that *f* is a function because no vertical line can cut the graph of *f* more than once.

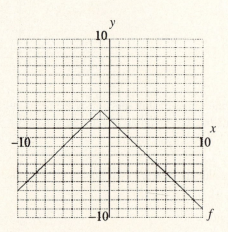

The Domain. If all of the points on the graph of the function f are projected onto the x-axis, the following shadow will fall on the x-axis:

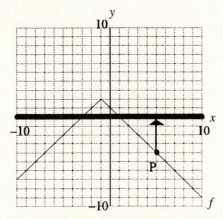

Further examination of the graph of f will show that it opens indefinitely to the right and left. Because the entire x-axis lies in shadow, the domain of f is the set of all real numbers. In set-builder notation, the domain equals $\{x: x \in R\}$, where R denotes the set of real numbers. In interval notation, the domain equals $(-\infty, +\infty)$.

The Range. To find the range of the function f, each point on the graph of f must be projected onto the y-axis.

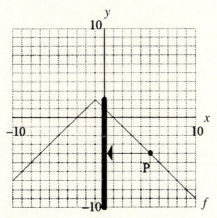

Further examination of the graph will show that the graph opens downward indefinitely. The shadow on the y-axis includes $y = 2$ and all points below. Therefore the range of the function f is the set of all real numbers that are less than or equal to 2. In set-builder notation, the range equals $\{y: y \le 2\}$. In interval notation, the range equals $(-\infty, 2]$.

Finding the Domain from an Equation

We begin with a remark.

> *The domain of a function is the set of real numbers for which the function values are real numbers.*

Example 10   What is the domain of the function whose equation is $f(x) = (x-2)^3 + 2$?

Solution   If x is any real number, the expression $(x-2)^3 + 2$ will also be a real number. Therefore the domain of f is $\{x : x \in R\}$. In interval notation, the domain is $(-\infty, +\infty)$.

Example 11   What is the domain of the function whose equation is $f(x) = \sqrt{5-2x}$?

Solution   The expression $\sqrt{5-2x}$ will be a real number only if $5-2x$ is a nonnegative real number.

$$5 - 2x \geq 0$$

$$-2x \geq -5$$

$$x \leq \frac{5}{2}$$

Therefore the domain of f is $\left\{ x : x \leq \dfrac{5}{2} \right\}$. In interval notation, the domain of f is $\left(-\infty, \dfrac{5}{2} \right]$.

Example 12   Find the domain of the function whose equation is $f(x) = \sqrt{x-3}$.

Analytical Solution   The expression $\sqrt{x-3}$ will be a real number only if $x-3$ is a nonnegative real number.

$$x - 3 \geq 0$$

$$x \geq 3$$

Therefore the domain of f is $\{x : x \geq 3\}$. In interval notation, the domain of f is $[3, +\infty)$.

Graphical Solution   A student offered the following solution.

First sketch the graph of $y = \sqrt{x}$.

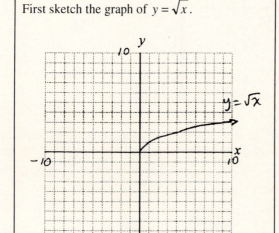

Slide the graph of $y = \sqrt{x}$ to the right 3 units to arrive at the graph of $y = \sqrt{x-3}$.

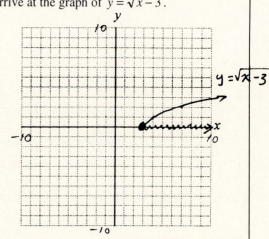

The domain is $\{x : x \geq 3\}$. In interval notation, the domain is $[3, +\infty)$. Note how the graphical solution and the analytical solution agree.

Exercises for Section 6.3

1. For each of the following relations, perform the following tasks:

   i)   Clearly list the domain and range of each relation.

   ii) Create a mapping diagram for each relation.

   iii) Clearly state whether or not the relation is a function. Use a complete sentence to justify your response.

   a) $R = \{(1,0),(2,3),(4,5)\}$ b) $R = \{(1,0),(2,0),(3,0)\}$

   c) $R = \{(1,0),(1,2),(2,3)\}$ d) $R = \{(1,0),(4,0),(1,5)\}$

   e) $R = \{(5,0),(5,1),(5,2)\}$ f) $R = \{(1,5),(2,3),(3,1)\}$

2. Set up a coordinate system on a sheet of graph paper and copy the following graph of the relation R onto your coordinate system.

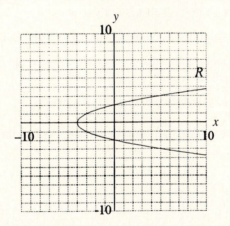

   a) Draw the graph of the line whose equation is $x = 5$ on your coordinate system. Label the points of intersection of the graph of R and the graph of $x = 5$ with the letters P and Q and include the coordinates of these points.

   b) Using only the coordinates of points P and Q, create a mapping diagram for the relation R.

   c) Clearly state whether or not the relation R is a function. Use a complete sentence to justify your response.

   Hint: Consider your mapping diagram.

3. Set up a coordinate system on a sheet of graph paper and copy the following graph of the relation *R* onto your coordinate system.

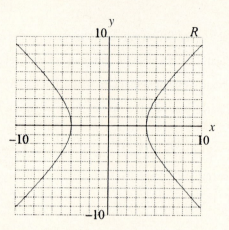

a) Draw the graph of the line whose equation is $x = -6$ on your coordinate system. Label the points of intersection of the graph of *R* and the graph of $x = -6$ with the letters P and Q and include the coordinates of these points.

b) Using only the coordinates of points P and Q, create a mapping diagram for the relation *R*.

c) Clearly state whether or not the relation *R* is a function. Use a complete sentence to justify your response.

   Hint: Consider your mapping diagram.

4. Use the vertical line test to determine which of the following graphs is the graph of a function.

a)

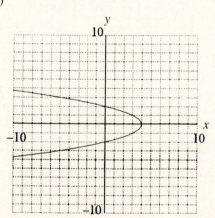

b)

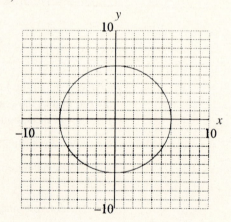

c)

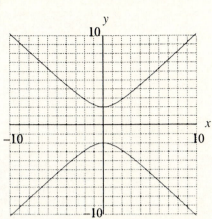

d)

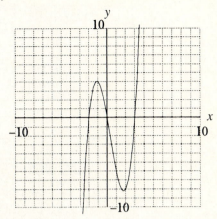

5. (TI-82) Load the equation $y = \sqrt{x}$ into the Y = menu in the following manner: Y1 = $\sqrt{}$ X. Enter each set of WINDOW parameters and push the GRAPH button to draw the graph. Set up a coordinate system and copy the image in the viewing window onto your coordinate system. Include the WINDOW parameters on your coordinate axes.

```
WINDOW FORMAT
 Xmin=-10
 Xmax=10
 Xscl=1
 Ymin=-10
 Ymax=10
 Yscl=1
```

```
WINDOW FORMAT
 Xmin=-100
 Xmax=100
 Xscl=10
 Ymin=-10
 Ymax=10
 Yscl=1
```

```
WINDOW FORMAT
 Xmin=-10000
 Xmax=10000
 Xscl=1000
 Ymin=-100
 Ymax=100
 Yscl=10
```

```
WINDOW FORMAT
 Xmin=-1000000
 Xmax=1000000
 Xscl=100000
 Ymin=-1000
 Ymax=1000
 Yscl=100
```

Based on the evidence discovered in this exploration, use both set-builder notation and interval notation to describe the domain and range of the function defined by the equation $y = \sqrt{x}$.

6. (TI-82) Load the equation $y = |x|$ into the Y = menu in the following manner: Y1 = abs X. Enter each set of WINDOW parameters and push the GRAPH button to draw the graph. Set up a coordinate system and copy the image in your viewing window onto your coordinate system. Include the WINDOW parameters on your coordinate axes.

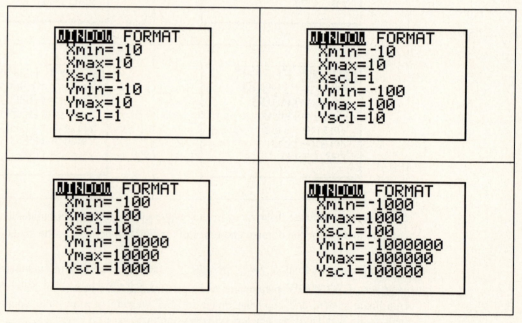

Based on the evidence discovered in this exploration, use both set-builder notation and interval notation to describe the domain and range of the function defined by the equation $y = |x|$.

7. (TI-82) Load the equation $y = x^2$ into the Y = menu in the following manner: $Y_1 = X^2$. Enter each set of WINDOW parameters and push the GRAPH button to draw the graph. Set up a coordinate system and copy the image in your viewing window onto your coordinate system. Include the WINDOW parameters on your coordinate axes.

```
WINDOW FORMAT          WINDOW FORMAT
 Xmin=-10               Xmin=-10
 Xmax=10                Xmax=10
 Xscl=1                 Xscl=1
 Ymin=-10               Ymin=-100
 Ymax=10                Ymax=100
 Yscl=1                 Yscl=10


WINDOW FORMAT          WINDOW FORMAT
 Xmin=-100              Xmin=-1000
 Xmax=100               Xmax=1000
 Xscl=10                Xscl=100
 Ymin=-10000            Ymin=-1000000
 Ymax=10000             Ymax=1000000
 Yscl=1000              Yscl=100000
```

Based on the evidence discovered in this exploration, use both set-builder notation and interval notation to describe the domain and range of the function defined by the equation $y = x^2$.

8. (TI-82) Load the equation $y = x^3$ into the Y = menu in the following manner: Y1 = X³. Enter each set of WINDOW parameters and push the GRAPH button to draw the graph. Set up a coordinate system and copy the image in your viewing window onto your coordinate system. Include the WINDOW parameters on your coordinate axes.

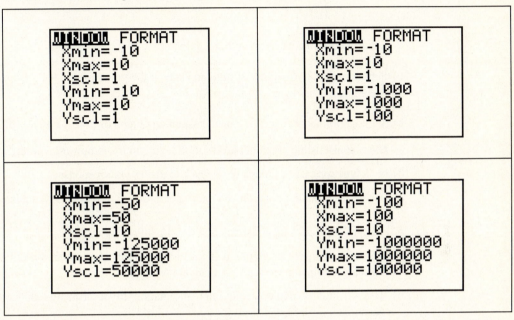

Based on the evidence discovered in this exploration, use both set-builder notation and interval notation to describe the domain and range of the function defined by the equation $y = x^3$.

9. The following sequence of steps is designed to aid you in sketching the graph of $f(x) = -|x-1| + 3$.

   a) Set up a coordinate system and sketch the graph of $y = |x|$.

   b) Set up a second coordinate system and sketch the graph of $y = -|x|$.

      Hint: Reflect the graph in part (a) across the x-axis.

   c) Set up a third coordinate system and sketch the graph of $y = -|x-1| + 3$.

      Hint: Translate the graph in part (b) to the right 1 unit and upward 3 units.

   d) Is f a function? Use a complete sentence to justify your response.

   e) Use both set-builder notation and interval notation to describe the domain and range of f.

10. The following sequence of steps is designed to aid you in sketching the graph of $g(x) = \sqrt{-(x-4)}$.

    a) Set up a coordinate system and sketch the graph of $y = \sqrt{x}$.

    b) Set up a second coordinate system and sketch the graph of $y = \sqrt{-x}$.

       Hint: Reflect the graph in part (a) across the y-axis.

    c) Set up a third coordinate system and sketch the graph of $y = \sqrt{-(x-4)}$.

       Hint: Translate the graph in part (b) to the right 4 units.

d) Is g a function? In your own words, justify your response.

e) Use both set-builder notation and interval notation to describe the domain and range of g.

11. The following sequence of steps is designed to aid you in sketching the graph of $f(x) = -(x+1)^2 + 5$.

a) Set up a coordinate system and sketch the graph of $y = x^2$.

b) Set up a second coordinate system and sketch the graph of $y = -x^2$. *Hint:* Reflect the graph in part (a) across the x-axis.

c) Set up a third coordinate system and sketch the graph of $y = -(x+1)^2 + 5$.

Hint: Translate the graph in part (b) to the left 1 unit and upward 5 units.

d) Is f a function? Use a complete sentence to justify your response.

e) Use both set-builder notation and interval notation to describe the domain and range of f.

12. For each of the following equations perform the following tasks:

i) Set up a sequence of graphs that lead to the graph of the function defined by the equation. Check your final graph with your calculator.

ii) Use set-builder notation and interval notation to describe the domain and range of the function.

a) $f(x) = \sqrt{-x} + 2$

b) $f(x) = -|-x|$

c) $g(x) = -x^2 + 5$

d) $h(x) = (-x)^3 + 1$

e) $f(x) = \sqrt{-(x+1)}$

f) $g(x) = -(x-3)^2 + 7$

h) $h(x) = -|x-5| + 5$

i) $f(x) = -(x+2)^3 - 2$

13. Consider the function f defined by the equation $f(x) = \sqrt{x-4}$.

a) You cannot take the square root of a negative number, so $x-4$ must be a nonnegative number—that is, $x-4 \geq 0$. Solve this inequality for x. Use set-builder notation to describe the domain of the function f.

b) Use graphing shortcuts to sketch the graph of the function f. Use the graph of f to determine the domain of the function f. Describe the domain using set-builder notation. If this solution does not agree with the solution from part (a), check your work for error.

14. Consider the function f defined by the equation $f(x) = \sqrt{x+5}$.

a) You cannot take the square root of a negative number, so $x+5$ must be a nonnegative number—that is, $x+5 \geq 0$. Solve this inequality for x. Use set-builder notation to describe the domain of the function f.

b) Use graphing shortcuts to sketch the graph of the function f. Use the graph of f to determine the domain of the function f. Describe the domain using set-builder notation. If this solution does not agree with the solution from part (a), check your work for error.

15. Find the domain of the function defined by each equation by examining the equation and determining the real numbers x for which the function is defined and a real number.

a) $f(x) = \sqrt{3x+6}$

b) $f(x) = \sqrt{10-2x}$

c) $f(x) = x^2 + 3$

d) $f(x) = -x^3 + 1$

e) $f(x) = \sqrt{x - \dfrac{x+1}{2}}$

f) $f(x) = \sqrt{2 - x - \dfrac{2-3x}{4}}$

6.4 The Parabola

The function f defined by the equation $f(x) = ax^2 + bx + c$ is called a *quadratic function*. The graph of this function is called a *parabola*. It is the intent of this section to give a thorough presentation of the graph of the parabola.

The Vertex of a Parabola

The turning point of a parabola is called its *vertex*.

The graph of $y = x^2$ is a parabola that opens upward.	The graph of $y = -x^2$ is a parabola that opens downward.
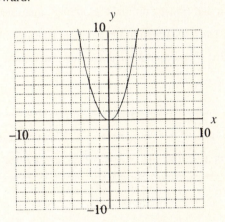	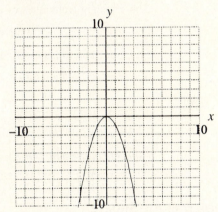
The vertex of this parabola is (0,0).	The vertex of this parabola is (0,0).

Slide the graph of $y = x^2$ to the right 1 unit and downward 2 units to arrive at the graph of $y = (x-1)^2 - 2$.

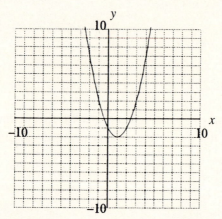

The vertex of this parabola is $(1, -2)$.

Slide the graph of $y = -x^2$ to the left 3 units and upward 5 units to arrive at the graph of $y = -(x+3)^2 + 5$.

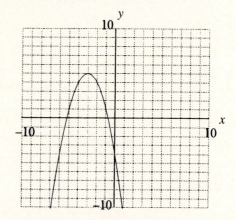

The vertex of this parabola is $(-3, 5)$.

The graphing techniques of reflection and translation make graphing equations like $y = (x-1)^2 - 2$ and $y = -(x+3)^2 + 5$ relatively easy. Once the graph of the parabola is drawn, it is a simple matter to find the coordinates of the vertex. Consequently, equations such as $y = (x-1)^2 - 2$ and $y = -(x+3)^2 + 5$ are said to be in *vertex form*.

> **The graph of the equation $y = a(x-h)^2 + k$ is a parabola. The vertex of the parabola is located at the point (h, k). The form $y = a(x-h)^2 + k$ is called vertex form.**

Completing the Square

Usually, the equation of the parabola will be presented in the form $y = ax^2 + bx + c$. A process called *completing the square* will enable you to place this equation in vertex form $y = a(x-h)^2 + k$.

The Squaring a Binomial Multiplication Pattern. Recall that $(a+b)^2 = a^2 + 2ab + b^2$ and $(a-b)^2 = a^2 - 2ab + b^2$.

Example 1   Use the squaring a binomial multiplication pattern to multiply $(x+3)^2$.

Solution

$$(x+3)^2 = (x)^2 + 2(x)(3) + (3)^2$$
$$= x^2 + 6x + 9$$

Example 2   Use the squaring a binomial multiplication pattern to multiply $(x-5)^2$.

Solution

$$(x-5)^2 = (x)^2 - 2(x)(5) + (5)^2$$
$$= x^2 - 10x + 25$$

Example 3   Find the vertex of the parabola whose equation is $y = x^2 + 2x - 3$.

Solution   A method called *completing the square* will be used to place this equation in vertex form. Take one-half of the coefficient of $2x$. Add and subtract the square of this result, as shown.

$$y = x^2 + 2x + (1)^2 - (1)^2 - 3$$

Regroup, as shown.

$$y = (x^2 + 2x + 1) - 1 - 3$$

Use the squaring a binomial multiplication pattern to factor the trinomial in parentheses, as shown.

$$y = (x+1)^2 - 4$$

This last equation is in vertex form. To draw the graph of $y = (x+1)^2 - 4$, slide the graph of $y = x^2$ to the left 1 unit and down 4 units.

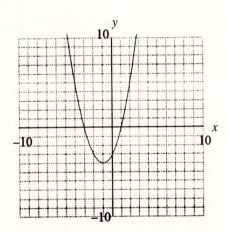

The vertex of this parabola is located at the point $(-1, -4)$. It is important to note that the parabola is *symmetric* with respect to a vertical line drawn through the vertex.

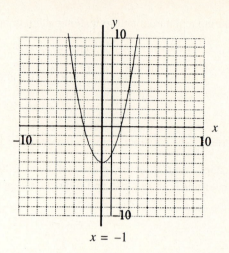

$$x = -1$$

The line $x = -1$ is called the *axis of symmetry* of the parabola. If you were to fold your paper in half along the axis of symmetry, one half of the parabola would fold right onto the other half.

Example 4 Find the vertex of the parabola whose equation is $y = x^2 - 4x - 5$.

Solution Take one-half of the coefficient of $-4x$. Add and subtract the square of this result, as shown.

$$y = x^2 - 4x + (-2)^2 - (-2)^2 - 5$$

Regroup, as shown.

$$y = (x^2 - 4x + 4) - 4 - 5$$

Use the squaring a binomial multiplication pattern to factor the trinomial in parentheses, as shown.

$$y = (x - 2)^2 - 9$$

This last equation is in vertex form. To draw the graph of $y = (x - 2)^2 - 9$, slide the graph of $y = x^2$ to the right 2 units and down 9 units.

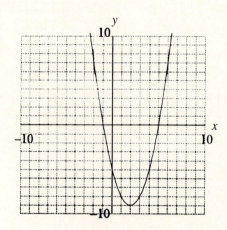

The vertex of this parabola is located at the point $(2, -9)$. The axis of symmetry is the line $x = 2$.

Note: Don't think that you should be able to draw precision parabolas at this point. Perhaps a student solution would be helpful.

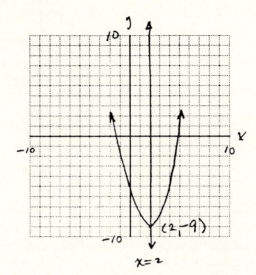

Note that it is rather difficult to draw a freehand parabola that is perfectly symmetric.

Example 5   Find the vertex of the parabola whose equation is $y = -x^2 + 12x - 30$ by using the method of completing the square to place the equation in vertex form.

Solution   Begin by factoring out a minus sign from the first two terms, as shown.

$$y = -(x^2 - 12x) - 30$$

Add and subtract the square of one half of the coefficient of $-12x$, as shown.

$$y = -(x^2 - 12x + (-6)^2 - (-6)^2) - 30$$

$$y = -(x^2 - 12x + 36 - 36) - 30$$

Distribute the negative sign, as shown.

$$y = -(x^2 - 12x + 36) - (-36) - 30$$

$$y = -(x^2 - 12x + 36) + 6$$

Use the squaring a binomial pattern to factor the trinomial in parentheses, as shown.

$$y = -(x - 6)^2 + 6$$

This last equation is in vertex form. To draw the graph of $y = -(x - 6)^2 + 6$, slide the graph of $y = -x^2$ to the right 6 units and up 6 units.

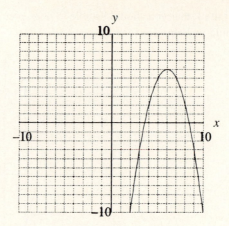

The vertex of this parabola is located at the point (6,6). The axis of symmetry is the line $x = 6$.

Check This Last Result with Your TI-82.   Load both the original equation and the vertex form of the equation into the Y= menu.

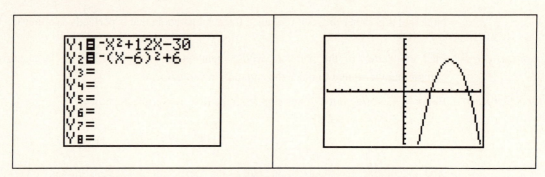

Because the graphs of Y1 and Y2 coincide, you can be confident that the vertex form $y = -(x-6)^2 + 6$ is equivalent to the original equation $y = -x^2 + 12x - 30$.

The Graph of $y = cf(x)$

What happens when a function is multiplied by a real number? Let's begin with an example.

Example 6   Consider the following graph, whose equation is $y = f(x)$.

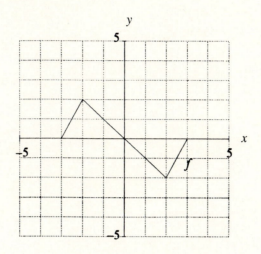

Use the graph of $y = f(x)$ to help sketch the graph of $y = 2f(x)$.

Solution   Begin by reading the graph of f to find the following function values.

$$f(-3) = 0 \qquad\qquad f(1) = -1$$
$$f(-2) = 2 \qquad\qquad f(2) = -2$$
$$f(-1) = 1 \qquad\qquad f(3) = 0$$
$$f(0) = 0$$

These results can be arranged in a table.

x	−3	−2	−1	0	1	2	3
$f(x)$	0	2	1	0	−1	−2	0

Now create a table of points that satisfy the equation $y = 2f(x)$. Use the following x-values: −3, −2, −1, 0, 1, 2, and 3. When these x-values are substituted into the equation $y = 2f(x)$, the following y-values are obtained:

$$y = 2f(-3) = 2(0) = 0 \qquad\qquad y = 2f(1) = 2(-1) = -2$$
$$y = 2f(-2) = 2(2) = 4 \qquad\qquad y = 2f(2) = 2(-2) = -4$$
$$y = 2f(-1) = 2(1) = 2 \qquad\qquad y = 2f(3) = 2(0) = 0$$
$$y = 2f(0) = 2(0) = 0$$

These results can be arranged in a table.

x	−3	−2	−1	0	1	2	3
$y = 2f(x)$	0	4	2	0	−2	−4	0
Points to plot	(−3,0)	(−2,4)	(−1,2)	(0,0)	(1,−2)	(2,−4)	(3,0)

Note that the y-values in this table are double the corresponding y-values in the previous table. Plot the points in this table on a coordinate system.

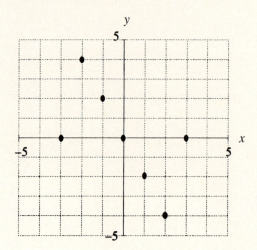

If you double the y-value of each point on the graph of $y = f(x)$, then the graph of $y = 2f(x)$ must look as follows:

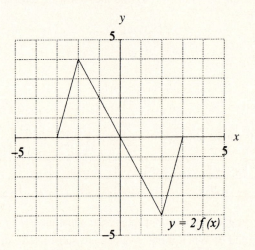

Because the y-values generated by the equation $y = 2f(x)$ are twice the y-values generated by the equation $y = f(x)$, the graph of $y = 2f(x)$ *stretches* in both vertical directions by a factor of 2.

Example 7   Compare the graphs of $y = x^2$ and $y = 2x^2$.

Solution   Create a table of points for each function.

x	-3	-2	-1	0	1	2	3
$y = x^2$	9	4	1	0	1	4	9
$y = 2x^2$	18	8	2	0	2	8	18

It is important to note that the y-values of the points satisfying the equation $y = 2x^2$ are twice the y-values of the points satisfying the equation $y = x^2$. To draw the graph of $y = 2x^2$, stretch the graph of $y = x^2$ in the vertical direction by a factor of 2.

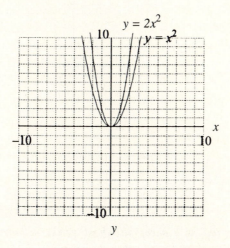

The vertex of the parabola $y = x^2$ is $(0,0)$. It is important to note that the vertex of the parabola $y = 2x^2$ is also $(0,0)$. This occurs because twice zero is still zero.

Example 8   Compare the graphs of $y = x^2$ and $y = -\frac{1}{2}x^2$.

Solution   Before beginning, try to guess. The y-values of $y = \frac{1}{2}x^2$ will be half the y-values of $y = x^2$. It's possible that the graph of $y = \frac{1}{2}x^2$ will be shrunk in the vertical direction by a factor of $\frac{1}{2}$. However, half of zero is still zero, so the vertex of $y = \frac{1}{2}x^2$ will remain at $(0,0)$. Finally, the graph of $y = -\frac{1}{2}x^2$ should be the reflection of the graph of $y = \frac{1}{2}x^2$ across the x-axis. Check your guess by creating a table of points that satisfy each equation.

x	-3	-2	-1	0	1	2	3
$y = x^2$	9	4	1	0	1	4	9
$y = -\dfrac{1}{2}x^2$	$-\dfrac{9}{2}$	-2	$-\dfrac{1}{2}$	0	$-\dfrac{1}{2}$	-2	$-\dfrac{9}{2}$

These points can be used to construct the graphs of $y = x^2$ and $y = -\frac{1}{2}x^2$.

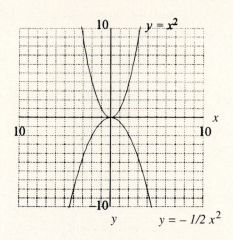

The graph of $y = ax^2$ is a parabola whose vertex is located at $(0,0)$. If a is positive, the parabola opens upward. If a is negative, the parabola opens downward.

Example 9   Sketch the graph of $y = 2x^2 - 4x - 6$.

Solution   Begin by placing the equation in vertex form. Factor a 2 from the first two terms, as shown.

$$y = 2(x^2 - 2x) - 6$$

Take one half of the coefficient of $-2x$. Add and subtract the square of this result, as shown.

$$y = 2(x^2 - 2x + (-1)^2 - (-1)^2) - 6$$
$$y = 2(x^2 - 2x + 1 - 1) - 6$$

Distribute the 2, as shown.

$$y = 2(x^2 - 2x + 1) - 2(1) - 6$$

Use the squaring a binomial pattern to factor the trinomial in parentheses, as shown.

$$y = 2(x - 1)^2 - 8$$

To draw the graph of $y = 2(x-1)^2 - 8$, slide the graph of $y = 2x^2$ to the right 1 unit and down 8 units. Therefore, the coordinates of the vertex would be $(1,-8)$.

The y-Intercept.   To find the coordinates of the y-intercept, let $x = 0$ in the original equation.

$$y = 2x^2 - 4x - 6$$

$$y = 2(0)^2 - 4(0) - 6$$

$$y = -6$$

The coordinates of the y-intercept are $(0, -6)$.

The x-Intercepts.   To find the coordinates of the x-intercepts, let $y = 0$ in the original equation.

$$y = 2x^2 - 4x - 6$$

$$0 = 2x^2 - 4x - 6$$

$$0 = x^2 - 2x - 3$$

$$0 = x^2 - 3x + x - 3$$

$$0 = x(x - 3) + 1(x - 3)$$

$$0 = (x + 1)(x - 3)$$

$$x = -1, 3$$

The coordinates of the x-intercepts are $(-1, 0)$ and $(3, 0)$.

The Graph.   Set up a coordinate system. Plot the vertex, the x- and y-intercepts, and the axis of symmetry.

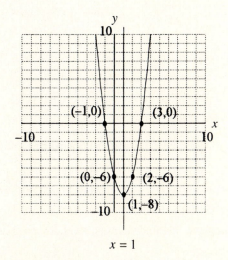

Because the parabola is symmetric with respect to the line $x = 1$ and $(0, -6)$ lies on the graph, the point $(2, -6)$ must also lie on the graph. Note that we have labeled each key point with its coordinates and labeled the axis of symmetry with its equation. If this labeling and the axis of symmetry were to be removed, the graph of $y = 2x^2 - 4x - 6$ would remain.

Example 10 Sketch the graph of $y = -\frac{1}{2}x^2 - 2x + 6$.

Solution First, use the completing the square method to place the equation in vertex form. Factor out a $-\frac{1}{2}$ from the first two terms, as shown.

$$y = -\frac{1}{2}(x^2 + 4x) + 6$$

Take one half of the coefficient of $4x$. Add and subtract the square of this result, as shown.

$$y = -\frac{1}{2}(x^2 + 4x + (2)^2 - (2)^2) + 6$$

$$y = -\frac{1}{2}(x^2 + 4x + 4 - 4) + 6$$

Distribute the $-\frac{1}{2}$, as shown.

$$y = -\frac{1}{2}(x^2 + 4x + 4) - \frac{1}{2}(-4) + 6$$

Use the squaring a binomial multiplication pattern to factor the trinomial in parentheses, as shown.

$$y = -\frac{1}{2}(x + 2)^2 + 8$$

The y-Intercept. To find the y-intercept, let $x = 0$ in the original equation.

$$y = -\frac{1}{2}x^2 - 2x + 6$$

$$y = -\frac{1}{2}(0)^2 - 2(0) + 6$$

$$y = 6$$

The coordinates of the y-intercept are $(0,6)$.

The x-Intercepts.   To find the coordinates of the x-intercepts, let $y = 0$ in the original equation.

$$y = -\frac{1}{2}x^2 - 2x + 6$$

$$0 = -\frac{1}{2}x^2 - 2x + 6$$

$$0 = x^2 + 4x - 12$$

Note how both sides of the equation were multiplied by -2. You can now use the *ac* test to factor.

$$0 = x^2 + 6x - 2x - 12$$

$$0 = x(x+6) - 2(x+6)$$

$$0 = (x-2)(x+6)$$

$$x = 2, -6$$

The coordinates of the x-intercepts are $(-6,0)$ and $(2,0)$.

The Graph.   Perhaps it would be helpful to see a student solution at this point. The student begins by plotting the vertex, x- and y-intercepts, and drawing the axis of symmetry.

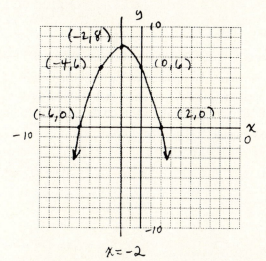

Because the parabola is symmetric with respect to the line $x = -2$ and $(0,6)$ lies on the graph, the point $(-4,6)$ must also lie on the graph. Note that the student has labeled each key point with its coordinates and labeled the axis of symmetry with its equation.

A summary of the steps required to draw the graph of $y = ax^2 + bx + c$.

1. Complete the square to place the equation in vertex form $y = a(x-h)^2 + k$.
2. Find the x- and y-intercepts.
3. Plot the vertex and the x- and y-intercepts, and draw the axis of symmetry.
4. Draw the parabola.
5. Label the vertex, x- and y-intercepts, and other points with their coordinates.
6. Label the axis of symmetry with its equation.

Example 11   Find the equation of the following parabola in the form $y = a(x-h)^2 + k$.

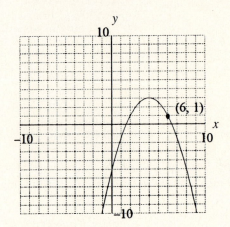

Solution   The parabola has been shifted 4 units to the right and 3 units upward. The coordinates of the vertex are (4,3). Therefore the equation of the parabola must be of the form $y = a(x-4)^2 + 3$. Because the parabola opens downward, you know that the value of a is negative. You need to find the numerical value of a. Note that the point (6,1) is on the graph, so (6,1) must satisfy the equation of the parabola. Substitute the point (6,1) into the equation $y = a(x-4)^2 + 3$.

$$y = a(x-4)^2 + 3$$
$$1 = a(6-4)^2 + 3$$
$$1 = 4a + 3$$
$$-2 = 4a$$
$$a = \frac{-2}{4}$$
$$a = -\frac{1}{2}$$

Thus the equation of the parabola is $y = -\frac{1}{2}(x-4)^2 + 3$.

Exercises for Section 6.4

1. For each of the following functions, perform the following tasks:

   i) Set up a separate coordinate system on a sheet of graph paper for each exercise. Clearly indicate the scale on each coordinate axis. Use the techniques of translation and reflection to draw a rough graph of the parabola represented by the equation.

   ii) Label the vertex of the parabola with its coordinates.

   iii) Use your calculator to check your result.

   a) $y = (x-4)^2 - 9$ b) $y = -(x+3)^2 + 5$

   c) $y = (x-8)^2 - 12$ d) $y = -(x+12)^2 + 15$

2. For each of the following parabolas, find the equation of the parabola in the form $y = a(x-h)^2 + k$. You may assume that $a = 1$ or $a = -1$ for each parabola.

a)

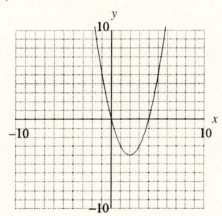

b)

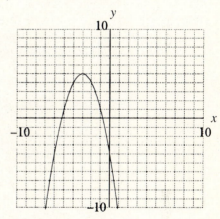

c)

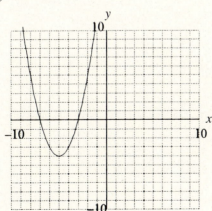

d)

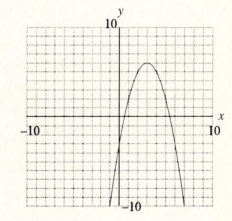

3. Use the squaring a binomial multiplication pattern to square each of the following binomials.

a) $(x+7)^2$ 　　　　b) $(x-3)^2$ 　　　　c) $\left(x-\dfrac{1}{2}\right)^2$ 　　　　d) $\left(x+\dfrac{7}{2}\right)^2$

4. Part of the completing the square process has been done for you in each of the following exercises. Complete the missing entries as you copy the exercise onto your homework paper.

a) 　$y = x^2 + 8x - 10$

　$y = x^2 + 8x + (\underline{\quad})^2 - (\underline{\quad})^2 - 10$

　$y = (x^2 + 8x + \underline{\quad}) - \underline{\quad} - 10$

　$y = (x + \underline{\quad})^2 - \underline{\quad}$

b) 　$y = x^2 - 10x + 12$

　$y = x^2 - 10x + (\underline{\quad})^2 - (\underline{\quad})^2 + 12$

　$y = (x^2 - 10x + \underline{\quad}) - \underline{\quad} + 12$

　$y = (x - \underline{\quad})^2 - \underline{\quad}$

5. Place each of the following equations in vertex form. Clearly state the coordinates of the vertex.

a) $y = x^2 - 12x - 20$ 　　　　　　　　b) $y = x^2 + 6x - 18$

c) $y = x^2 - 3x - 4$ 　　　　　　　　d) $y = x^2 + 5x - 6$

6. Part of the completing the square process has been done for you in each of the following exercises. Complete the missing entries as you copy the exercise onto your homework paper.

a) 　$y = -x^2 - 6x + 12$

　$y = -(x^2 + 6x + (\underline{\quad})^2 - (\underline{\quad})^2) + 12$

　$y = -(x^2 + 6x + \underline{\quad} - \underline{\quad}) + 12$

　$y = -(x^2 + 6x + \underline{\quad}) + \underline{\quad} + 12$

　$y = -(x + \underline{\quad})^2 + \underline{\quad}$

b) 　$y = -x^2 + 14x - 12$

　$y = -(x^2 - 14x + (\underline{\quad})^2 - (\underline{\quad})^2) - 12$

　$y = -(x^2 - 14x + \underline{\quad} - \underline{\quad}) - 12$

　$y = -(x^2 - 14x + \underline{\quad}) + \underline{\quad} - 12$

　$y = -(x - \underline{\quad})^2 + \underline{\quad}$

7. Place each of the following equations in vertex form. Clearly state the coordinates of the vertex.

a) $y = -x^2 + 8x + 9$ 　　　　　　　　b) $y = -x^2 - 10x + 11$

c) $y = -x^2 + 9x + 10$ 　　　　　　　　d) $y = -x^2 - 7x + 8$

8. For each of the following equations, perform the following tasks.

i)   Place the equation in vertex form.

ii) Find the x- and y-intercepts of the parabola represented by the equation.

iii) Set up a coordinate system on a sheet of graph paper. Clearly indicate the scale on each axis. Plot the vertex and the x- and y-intercepts and label each point with its coordinates. Draw the axis of symmetry and label this line with its equation.

iv) Draw the graph of the parabola on your coordinate system and check your solution with your calculator.

a) $y = x^2 + 6x - 7$ 　　　　b) $y = x^2 + 8x - 9$ 　　　　c) $y = x^2 - 9x + 8$

d) $y = x^2 + 11x + 24$ 　　　e) $y = -x^2 - 4x + 5$ 　　　f) $y = -x^2 + 10x + 11$

g) $y = -x^2 + 7x - 6$ 　　　h) $y = -x^2 - 9x - 8$

9. Set up a coordinate system on a sheet of graph paper and copy the following graph of the function
   f onto your coordinate system.

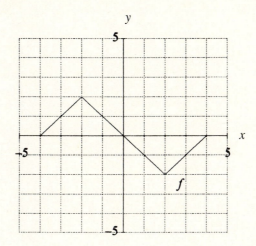

a) Make a copy of the following table on your graph paper. Examine the graph of *f* and complete
   the missing entries in the table.

x	-4	-3	-2	-1	0	1	2	3	4
$f(x)$									

b) Make a copy of the following table on your graph paper and complete the missing entries. Set
   up a second coordinate system on the same graph paper and plot each point from the table on
   your coordinate system. Plot the rest of the points that satisfy the equation $y = 2f(x)$.

x	-4	-3	-2	-1	0	1	2	3	4
$y = 2f(x)$									

c) Make a copy of the following table on your graph paper and complete the missing entries. Set
   up a second coordinate system on the same graph paper and plot each point from the table on
   your coordinate system. Plot the rest of the points that satisfy the equation $y = \frac{1}{2}f(x)$.

x	-4	-3	-2	-1	0	1	2	3	4
$y = \frac{1}{2}f(x)$									

10. Set up a coordinate system on a sheet of graph paper and copy the following graph of the function f onto your coordinate system.

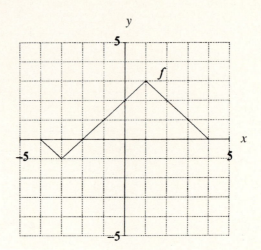

a) Make a copy of the following table on your graph paper. Examine the graph of f and complete the missing entries in the table.

x	-4	-3	-2	-1	0	1	2	3	4
$f(x)$									

b) Make a copy of the following table on your graph paper and complete the missing entries. Set up a second coordinate system on the same graph paper and plot each point from the table on your coordinate system. Plot the rest of the points that satisfy the equation $y = -2f(x)$.

x	-4	-3	-2	-1	0	1	2	3	4
$y = -2f(x)$									

c) Make a copy of the following table on your graph paper and complete the missing entries. Set up a second coordinate system on the same graph paper and plot each point from the table on your coordinate system. Plot the rest of the points that satisfy the equation $y = -\dfrac{1}{2}f(x)$.

x	-4	-3	-2	-1	0	1	2	3	4
$y = -\frac{1}{2}f(x)$									

11. The following is one period of the graph of the equation $y = \cos x$. Set up a coordinate system on a sheet of graph paper and copy this graph of $y = \cos x$ onto your coordinate system.

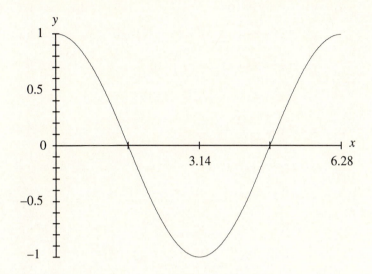

a) On the same coordinate system, use a blue pencil or pen to sketch one period of the function whose equation is $y = 2\cos x$.

b) On the same coordinate system, use a red pencil or pen to sketch one period of the function whose equation is $y = \frac{1}{2}\cos x$.

12. (TI-82) For each of the following exercises, perform the following tasks:

i) Load the indicated equations into the Y= menu.

ii) Push the ZOOM button and select 6:ZStandard. Set up a coordinate system on your homework paper and copy the image in your viewing window onto your coordinate system.

iii) Label each graph on your coordinate system with its equation.

a)

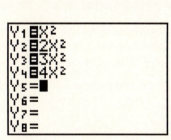

b)

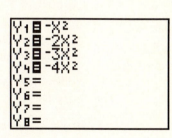

c)

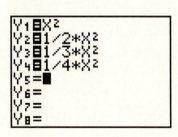

d)

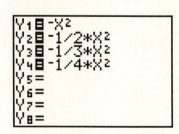

13. Part of the completing the square process has been done for you in each of the following exercises. Complete the missing entries as you copy the exercise onto your homework paper.

a) $y = 2x^2 + 8x - 10$

$y = 2\left(x^2 + 4x + (\underline{\quad})^2 - (\underline{\quad})^2\right) - 10$

$y = 2\left(x^2 + 4x + \underline{\quad} - \underline{\quad}\right) - 10$

$y = 2\left(x^2 + 4x + \underline{\quad}\right) - 2(\quad) - 10$

$y = 2\left(x + \underline{\quad}\right)^2 - \underline{\quad}$

b) $y = 3x^2 - 18x - 9$

$y = 3\left(x^2 - 6x + (\underline{\quad})^2 - (\underline{\quad})^2\right) - 9$

$y = 3\left(x^2 - 6x + \underline{\quad} - \underline{\quad}\right) - 9$

$y = 3\left(x^2 - 6x + \underline{\quad}\right) - 3(\quad) - 9$

$y = 3\left(x - \underline{\quad}\right)^2 - \underline{\quad}$

c) $y = -4x^2 + 8x - 12$

$y = -4\left(x^2 - 2x + (\underline{\quad})^2 - (\underline{\quad})^2\right) - 12$

$y = -4\left(x^2 - 2x + \underline{\quad} - \underline{\quad}\right) - 12$

$y = -4\left(x^2 - 2x + \underline{\quad}\right) + 4(\quad) - 12$

$y = -4\left(x - \underline{\quad}\right)^2 - \underline{\quad}$

d) $y = \dfrac{1}{2}x^2 - 3x - 2$

$y = \dfrac{1}{2}\left(x^2 - 6x + (\underline{\quad})^2 - (\underline{\quad})^2\right) - 2$

$y = \dfrac{1}{2}\left(x^2 - 6x + \underline{\quad} - \underline{\quad}\right) - 2$

$y = \dfrac{1}{2}\left(x^2 - 6x + \underline{\quad}\right) - \dfrac{1}{2}(\quad) - 2$

$y = \dfrac{1}{2}\left(x - \underline{\quad}\right)^2 - \underline{\quad}$

14. Place each of the following equations in vertex form. Clearly state the coordinates of the vertex.

a) $y = 2x^2 + 8x - 10$

b) $y = -3x^2 - 12x + 15$

c) $y = \dfrac{1}{2}x^2 - x - 2$

d) $y = -\dfrac{1}{3}x^2 - 2x - 2$

15. For each of the following equations, perform the following tasks.

i) Place the equation in vertex form.

ii) Find the x- and y-intercepts of the parabola represented by the equation.

iii) Set up a coordinate system on a sheet of graph paper. Clearly indicate the scale on each axis. Plot the vertex and the x- and y-intercepts and label each point with its coordinates. Draw the axis of symmetry and label this line with its equation.

iv) Draw the graph of the parabola on your coordinate system and check your solution with your calculator.

a) $y = 2x^2 + 4x - 16$

b) $y = -3x^2 + 6x + 45$

c) $y = \dfrac{1}{2}x^2 - 5x + 8$

d) $y = -\dfrac{1}{3}x^2 + 2x + 9$

16. Find the equation of each of the following parabolas in the form $y = a(x-h)^2 + k$.

a)

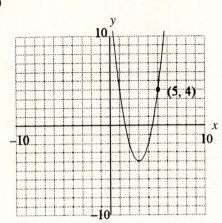

b)

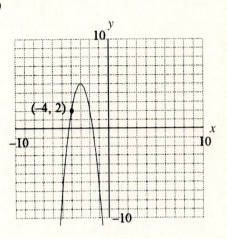

6.5 The Quadratic Formula

This section will continue our study of the parabola. A technique will be introduced that will enable you to find the coordinates of the vertex without completing the square. The quadratic formula will also be introduced. This will enable you to find the x-intercepts of a parabola with minimum effort.

Revisiting the Vertex of the Parabola

The graph of the equation $y = ax^2 + bx + c$ is a parabola. Place this equation in vertex form. Begin by factoring out a from the first two terms, as shown.

$$y = a\left(x^2 + \frac{b}{a}x\right) + c$$

Take one-half of the coefficient of $\frac{b}{a}x$. Add and subtract the square of this result, as shown.

$$y = a\left(x^2 + \frac{b}{a}x + \left(\frac{b}{2a}\right)^2 - \left(\frac{b}{2a}\right)^2\right) + c$$

$$y = a\left(x^2 + \frac{b}{a}x + \frac{b^2}{4a^2} - \frac{b^2}{4a^2}\right) + c$$

Distribute the a, as shown.

$$y = a\left(x^2 + \frac{b}{a}x + \frac{b^2}{4a^2}\right) - a\left(\frac{b^2}{4a^2}\right) + c$$

$$y = a\left(x^2 + \frac{b}{a}x + \frac{b^2}{4a^2}\right) + \left(c - \frac{b^2}{4a}\right)$$

Use the squaring a binomial multiplication pattern to factor the trinomial in the parentheses, as shown.

$$y = a\left(x + \frac{b}{2a}\right)^2 + \left(c - \frac{b^2}{4a}\right)$$

If you compare this last equation with $y = a(x-h)^2 + k$, you will find that the coordinates of the vertex are as follows:

$$\text{Vertex} = \left(-\frac{b}{2a}, c - \frac{b^2}{4a}\right)$$

The x-coordinate of the vertex of the parabola whose equation is $y = ax^2 + bx + c$ is $x = -\dfrac{b}{2a}$.

Procedure for finding the vertex of a parabola whose equation is $y = ax^2 + bx + c$.

1. Find the x-coordinate of the vertex by using the formula $x = -\dfrac{b}{2a}$.

2. Find the y-coordinate of the vertex by substituting the x-value in the equation $y = ax^2 + bx + c$.

Example 1   Find the coordinates of the vertex of the parabola whose equation is $y = x^2 - 4x - 3$.

Graphical Solution   Load the equation $y = x^2 - 4x - 3$ in the Y= menu in the following manner:

$\text{Y}_1 = \text{X}^2 - 4\text{X} - 3$.

Push the ZOOM button and select 6:ZStandard.	Push the TRACE button and use the arrow keys to move the trace cursor over the vertex.
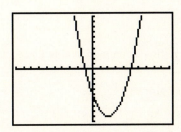	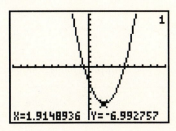

You may get a slightly different result, depending on your calculator settings and where you placed the trace cursor. In Section 6.8, we will explore a more efficient method for finding the coordinates of the vertex on your calculator.

Analytical Solution   If you compare the equation $y = x^2 - 4x - 3$ with $y = ax^2 + bx + c$, you will see that $a = 1$, $b = -4$, and $c = -3$. The x-coordinate of the vertex is found by using the formula $x = -\dfrac{b}{2a}$.

$$x = -\frac{b}{2a}$$

$$x = -\frac{-4}{2(1)}$$

$$x = 2$$

To find the y-coordinate of the vertex, substitute $x = 2$ in the equation $y = x^2 - 4x - 3$.

$$y = x^2 - 4x - 3$$

$$y = (2)^2 - 4(2) - 3$$

$$y = -7$$

The coordinates of the vertex are $(2, -7)$. This agrees quite nicely with the graphical estimate.

The *x*-Intercepts of the Parabola

To find the x-intercepts of the parabola, let $y = 0$ in the equation of the parabola. Let $y = 0$ in the vertex form of the equation $y = ax^2 + bx + c$.

$$y = a\left(x + \frac{b}{2a}\right)^2 + \left(c - \frac{b^2}{4a}\right)$$

$$0 = a\left(x + \frac{b}{2a}\right)^2 + \left(c - \frac{b^2}{4a}\right)$$

Now add $\dfrac{b^2}{4a} - c$ to both sides of this last equation.

$$\frac{b^2}{4a} - c = a\left(x + \frac{b}{2a}\right)^2$$

Simplify the left side of this equation by getting a common denominator and adding.

$$\frac{b^2}{4a} - \frac{4ac}{4a} = a\left(x + \frac{b}{2a}\right)^2$$

$$\frac{b^2 - 4ac}{4a} = a\left(x + \frac{b}{2a}\right)^2$$

Multiply both sides of this last equation by $\dfrac{1}{a}$.

$$\left(x + \frac{b}{2a}\right)^2 = \frac{b^2 - 4ac}{4a^2}$$

If $b^2 - 4ac < 0$, then the right side of this equation is negative and the equation has no real answers. If $b^2 - 4ac = 0$, then the equation will have one real answer. If $b^2 - 4ac > 0$, then the equation will have two real answers, as follows.

$$x + \frac{b}{2a} = \pm\sqrt{\frac{b^2 - 4ac}{4a^2}}$$

$$x + \frac{b}{2a} = \pm\frac{\sqrt{b^2 - 4ac}}{|2a|}$$

If $a < 0$, then $|2a| = -2a$. If $a > 0$, then $|2a| = 2a$. In either case, the following result is obtained:

$$x + \frac{b}{2a} = \pm\frac{\sqrt{b^2 - 4ac}}{2a}$$

$$x = -\frac{b}{2a} \pm \frac{\sqrt{b^2 - 4ac}}{2a}$$

$$x = \frac{-b \pm \sqrt{b^2 - 4ac}}{2a}$$

This last result is known as the *quadratic formula.*

> **The x-intercepts of the parabola whose equation is $y = ax^2 + bx + c$ are given by**
> $$x = \frac{-b \pm \sqrt{b^2 - 4ac}}{2a}.$$

Example 2   Find the x-intercepts of the parabola whose equation is $y = x^2 - 2x - 2$.

Graphical Solution   Load the equation $y = x^2 - 2x - 2$ in the Y= menu of the TI-82 in the following manner: $Y_1 = X^2 - 2X - 2$. Push the ZOOM button and select 6:ZStandard. Use the root-finding utility in the CALC menu to find the x-intercepts of the parabola. For a review of this utility, see Example 5 in Section 1.7.

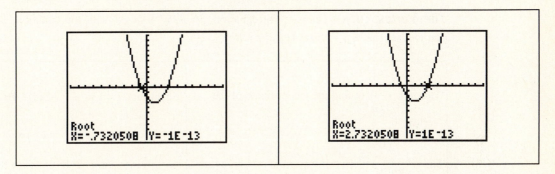

It appears that the x-intercepts of the parabola are approximately -0.7320508 and 2.7320508.

Analytical Solution   Use the quadratic formula to find the x-intercepts of the parabola. If you compare the equation $y = x^2 - 2x - 2$ with $y = ax^2 + bx + c$, you will see that $a = 1$, $b = -2$, and $c = -2$.

$$x = \frac{-b \pm \sqrt{b^2 - 4ac}}{2a}$$

$$x = \frac{-(-2) \pm \sqrt{(-2)^2 - 4(1)(-2)}}{2(1)}$$

$$x = \frac{2 \pm \sqrt{12}}{2}$$

Place your answer in simple form.

$$x = \frac{2 \pm 2\sqrt{3}}{2}$$

$$x = \frac{2}{2} \pm \frac{2\sqrt{3}}{2}$$

$$x = 1 \pm \sqrt{3}$$

Use your calculator to find decimal approximations for these exact answers so that you can compare them with the estimates found earlier.

```
1-√3
        -.7320508076
1+√3
         2.732050808
■
```

These decimal approximations agree quite nicely with estimates found by the graphing method.

Example 3   Use the techniques of this section to help sketch the graph of the parabola whose equation is $y = -2x^2 + 4x + 8$.

The Vertex. If you compare the equation $y = -2x^2 + 4x + 8$ with the equation $y = ax^2 + bx + c$, you will see that $a = -2$, $b = 4$, and $c = 8$.

Use $x = -\dfrac{b}{2a}$ to find the x-coordinate of the vertex. $$x = -\frac{b}{2a}$$ $$x = -\frac{4}{2(-2)}$$ $$x = 1$$	Substitute $x = 1$ in $y = -2x^2 + 4x + 8$. $$y = -2x^2 + 4x + 8$$ $$y = -2(1)^2 + 4(1) + 8$$ $$y = 10$$ The coordinates of the vertex are $(1,10)$.

The y-Intercept. To find the y-intercept, let $x = 0$ in the equation $y = -2x^2 + 4x + 8$.

$$y = -2x^2 + 4x + 8$$
$$y = -2(0)^2 + 4(0) + 8$$
$$y = 8$$

The x-Intercepts.   Use the quadratic formula to find the x-intercepts of $y = -2x^2 + 4x + 8$.

$$x = \frac{-b \pm \sqrt{b^2 - 4ac}}{2a}$$

$$x = \frac{-4 \pm \sqrt{(4)^2 - 4(-2)(8)}}{2(-2)}$$

$$x = \frac{-4 \pm \sqrt{80}}{-4}$$

$$x = \frac{-4 \pm 4\sqrt{5}}{-4}$$

$$x = \frac{-4}{-4} \pm \frac{4\sqrt{5}}{-4}$$

$$x = 1 \pm \sqrt{5}$$

Use your calculator to find estimates for decimal approximations for the x-intercepts.

```
1-√5
        -1.236067977
1+√5
        3.236067977
■
```

To the nearest tenth, $x \approx -1.2, 3.2$. You will need these approximations when you draw the graph.

The Graph.   Because $a = -2$, the parabola opens downward. Plot the vertex and the x- and y-intercepts, and draw the axis of symmetry.

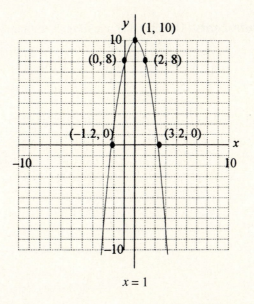

Because the parabola is symmetric with respect to the line $x = 1$ and the point $(0,8)$ is the y-intercept of the parabola, symmetry demands that the point $(2,8)$ also be on the parabola. It is important to note that -1.2 and 3.2 are only *approximate* values of the x-intercepts. The *exact* values of the x-intercepts are $1 - \sqrt{5}$ and $1 + \sqrt{5}$.

Example 4   Use the techniques of this section to help sketch the graph of the parabola whose equation is $y = x^2 - 2x + 1$.

The Vertex. If you compare the equation $y = x^2 - 2x + 1$ with the equation $y = ax^2 + bx + c$, you will see that $a = 1$, $b = -2$, and $c = 1$.

Use $x = -\dfrac{b}{2a}$ to find the x-coordinate of the vertex. $$x = -\frac{b}{2a}$$ $$x = -\frac{-2}{2(1)}$$ $$x = 1$$	Substitute $x = 1$ in $y = x^2 - 2x + 1$. $$y = x^2 - 2x + 1$$ $$y = (1)^2 - 2(1) + 1$$ $$y = 0$$ The coordinates of the vertex are $(1, 0)$.

The y-Intercept. To find the y-intercept, let $x = 0$ in the equation $y = x^2 - 2x + 1$.

$$y = x^2 - 2x + 1$$
$$y = (0)^2 - 2(0) + 1$$
$$y = 1$$

The x-Intercepts. Use the quadratic formula to find the x-intercepts of $y = x^2 - 2x + 1$.

$$x = \frac{-b \pm \sqrt{b^2 - 4ac}}{2a}$$
$$x = \frac{-(-2) \pm \sqrt{(-2)^2 - 4(1)(1)}}{2(1)}$$
$$x = \frac{2 \pm \sqrt{0}}{2}$$
$$x = \frac{2 \pm 0}{2}$$
$$x = 1$$

The Graph. Because $a = 1$, the parabola opens upward. Plot the vertex, the x-intercept, and the y-intercept. Draw the axis of symmetry.

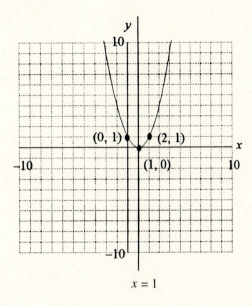

$x = 1$

There is exactly one intercept, at $(1,0)$. Note that $(1,0)$ is also the vertex of this parabola. Because the parabola is symmetric with respect to the line $x = 1$ and the point $(0,1)$ is on the y-intercept of the parabola, symmetry demands that the point $(2,1)$ also be on the parabola.

Example 5   Use the techniques of this section to help sketch the graph of the parabola whose equation is $y = x^2 - 2x + 2$.

The Vertex. If you compare the equation $y = x^2 - 2x + 2$ with the equation $y = ax^2 + bx + c$, you will see that $a = 1$, $b = -2$, and $c = 2$.

Use $x = -\dfrac{b}{2a}$ to find the x-coordinate of the vertex. $$x = -\frac{b}{2a}$$ $$x = -\frac{-2}{2(1)}$$ $$x = 1$$	Substitute $x = 1$ in $y = x^2 - 2x + 2$. $$y = x^2 - 2x + 2$$ $$y = (1)^2 - 2(1) + 2$$ $$y = 1$$ The coordinates of the vertex are $(1,1)$.

The y-Intercept. To find the y-intercept, let $x = 0$ in the equation $y = x^2 - 2x + 2$.

$$y = x^2 - 2x + 2$$
$$y = (0)^2 - 2(0) + 2$$
$$y = 2$$

The x-Intercepts. Use the quadratic formula to find the x-intercepts of $y = x^2 - 2x + 2$.

$$x = \frac{-b \pm \sqrt{b^2 - 4ac}}{2a}$$

$$x = \frac{-(-2) \pm \sqrt{(-2)^2 - 4(1)(2)}}{2(1)}$$

$$x = \frac{2 \pm \sqrt{-4}}{2}$$

Since $\sqrt{-4}$ is not a real number, this parabola does not have any x-intercepts.

The Graph. Because $a = 1$, the parabola opens upward. Plot the vertex and the y-intercept, and draw the axis of symmetry.

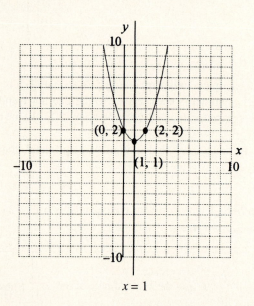

$$x = 1$$

Note that this parabola has no x-intercepts. Because the parabola is symmetric with respect to the line $x = 1$ and the point $(0, 2)$ is the y-intercept of the parabola, symmetry demands that the point $(2, 2)$ also lie on the parabola.

The Discriminant

The x-intercepts of the parabola $y = ax^2 + bx + c$ are given by the quadratic formula.

$$x = \frac{-b \pm \sqrt{b^2 - 4ac}}{2a}$$

The expression $b^2 - 4ac$ is called the discriminant.

   1. If $b^2 - 4ac > 0$, the parabola will have two x-intercepts.
   2. If $b^2 - 4ac = 0$, the parabola will have one x-intercept.
   3. If $b^2 - 4ac < 0$, the parabola will have no x-intercepts.

In Example 3, the discriminant was 12 and the parabola had two *x*-intercepts. In Example 4, the discriminant was 0 and the parabola had one *x*-intercept. In Example 5, the discriminant was −4 and the parabola had no *x*-intercepts.

Example 6   How many *x*-intercepts will the graph of $y = x^2 - 4x - 6$ have?

Solution   Use the discriminant as a tool for predicting the number of *x*-intercepts.

$$D = b^2 - 4ac$$

$$D = (-4)^2 = 4(1)(-6)$$

$$D = 40$$

Because the discriminant is positive, the parabola will have two *x*-intercepts.

Example 7   Find all values of k so that the parabola whose equation is $y = kx^2 - x - 4$ has exactly two *x*-intercepts.

Solution   To have two *x*-intercepts, the discriminant must be positive.

$$b^2 - 4ac > 0$$

$$(-1)^2 - 4(k)(-4) > 0$$

$$1 + 16k > 0$$

$$16k > -1$$

$$k > -\frac{1}{16}$$

If k is greater than $-\frac{1}{16}$, the parabola will have two *x*-intercepts.

Solving Quadratic Equations

Example 8   Solve the equation $-2x^2 + 2x = -12$.

Graphical Solution   The equation is nonlinear, so make one side of the equation zero by adding 12 to both sides of the equation.

$$-2x^2 + 2x + 12 = 0$$

Of course, you can make the solution of this equation a lot easier if you divide both sides of the equation by -2.

$$x^2 - x - 6 = 0$$

To find where an equation is equal to zero, draw its graph and note where the graph crosses the x-axis. Begin by loading the first equation in the Y= menu, like this: $Y_1 = -2X^2 + 2X + 12$. Push the ZOOM button and select 6:ZStandard. You can use the root-finding utility in the CALC menu to determine where the graph crosses the x-axis.

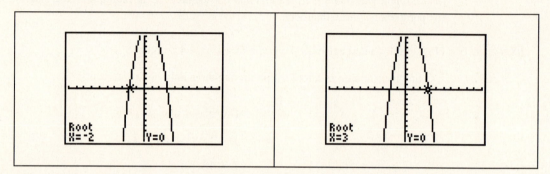

The solutions of the equation are -2 and 3. Now, load the second equation in the Y= menu in the following manner: $Y_1 = X^2 - X - 6$. Push the ZOOM button and select 6:ZStandard. You can use the root-finding utility in the CALC menu to find where this graph crosses the x-axis.

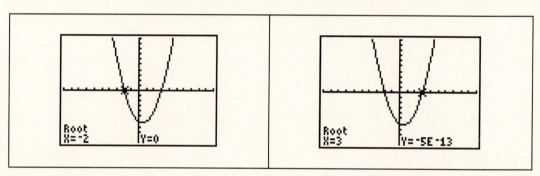

Even though the graph is different, the solutions are still -2 and 3. The equations $-2x^2 + 2x + 12 = 0$ and $x^2 - x - 6 = 0$ have the same solution set. Dividing both sides of an equation by -2 will change the graphs involved, but it will not affect where those graphs cross the x-axis.

Analytical Solution   Because you are finding where this graph crosses the x-axis, you are finding the x-intercepts. Because the graphs involved are parabolas, you can use the quadratic formula to solve the equations. Again, you can work with either equation.

$-2x^2 + 2x + 12 = 0$	$x^2 - x - 6 = 0$
$x = \dfrac{-b \pm \sqrt{b^2 - 4ac}}{2a}$	$x = \dfrac{-b \pm \sqrt{b^2 - 4ac}}{2a}$
$x = \dfrac{-2 \pm \sqrt{(2)^2 - 4(-2)(12)}}{2(-2)}$	$x = \dfrac{-(-1) \pm \sqrt{(-1)^2 - 4(1)(-6)}}{2(1)}$
$x = \dfrac{-2 \pm \sqrt{100}}{-4}$	$x = \dfrac{1 \pm \sqrt{25}}{2}$
$x = \dfrac{-2 - 10}{-4}, \dfrac{-2 + 10}{-4}$	$x = \dfrac{1 - 5}{2}, \dfrac{1 + 5}{2}$
$x = 3, -2$	$x = -2, 3$

Note that these solutions match the graphical solutions.

Tip	Dividing both sides of an equation by the greatest common divisor will greatly simplify the work and will not affect the solution set.

Exercises for Section 6.5

1. For each of the following equations, perform the following tasks:

   i) Use the method of completing the square to place the equation in vertex form $y = a(x - h)^2 + k$. Clearly state the coordinates of the vertex.

   ii) Use the formula $x = -\dfrac{b}{2a}$ to find the x-value of the vertex. Substitute this x-value into the equation to find the y-value of the vertex. If this answer does not match the answer in part (i), check your work for error.

   a) $y = x^2 + 4x - 8$ b) $y = -x^2 - 8x + 10$ c) $y = 2x^2 + 12x - 24$

   d) $y = -\dfrac{1}{2}x^2 - 3x + 5$ e) $y = x^2 - 3x - 5$ f) $y = -x^2 + 5x - 7$

2. (TI-82) For each of the following equations, perform the following tasks:

   i) Load the equation in the Y= menu, press the ZOOM button, and select 6:ZStandard. Adjust the WINDOW parameters so that the vertex of the parabola is visible in your viewing window. Set up a separate coordinate system on your homework paper and copy the image in your viewing window onto your coordinate system. Label your coordinate axes with your WINDOW parameters.

   ii) Use the TRACE capability of the calculator to estimate the coordinates of the vertex. Record this estimate on your coordinate system near the vertex of the parabola.

iii) Use the formula $x = -\dfrac{b}{2a}$ to find the x-value of the vertex. Substitute this x-value into the equation to find the y-value of the vertex. If this answer does not agree favorably with the estimate found in part (ii), check your work for error.

a) $y = x^2 - 2x - 5$ b) $y = -x^2 + 4x - 7$ c) $y = 2x^2 - x - 5$ d) $y = -3x^2 + 2x + 8$

3. For each of the following equations, perform the following tasks:

i) Substitute $y = 0$ in the equation and use the ac test to factor the resulting equation. Clearly state the x-intercepts of the parabola represented by the equation.

ii) Use the quadratic formula to find the x-intercepts of the parabola represented by the equation. If these intercepts do not agree exactly with those found in part (i), check your work for error.

a) $y = x^2 - 4x - 5$ b) $y = x^2 + 5x - 6$ c) $y = 2x^2 - x - 1$ d) $y = 3x^2 - 2x - 1$

4. (TI-82) For each of the following equations, perform the following tasks:

i) Load the equation in the Y= menu, press the ZOOM button, and select 6:ZStandard. Adjust the WINDOW parameters so that the x-intercepts and vertex of the parabola are visible in your viewing window. Set up a separate coordinate system on your homework paper and copy the image in the viewing window onto your coordinate system. Label your coordinate axes with your WINDOW parameters.

ii) Use the root-finding utility in the CALC menu to estimate the x-intercepts of the parabola. Record these estimates on your coordinate system, near the appropriate x-intercept on your coordinate system.

iii) Use the quadratic formula to find the exact value of the x-intercepts in simple radical form. Use your calculator to find decimal approximations of these exact answers. If your decimal approximations do not closely agree with the estimates found in part (ii), check your work for error.

a) $y = x^2 - 4x - 4$ b) $y = x^2 + 6x - 9$ c) $y = -x^2 + 5x + 7$

d) $y = -x^2 - 3x + 9$ e) $y = 2x^2 - x - 2$ f) $y = 3x^2 - 6x - 7$

5. Use the quadratic formula only to find the exact value of the x-intercepts of each of the following parabolas in simple radical form.

a) $y = x^2 - 2x - 4$ b) $y = 2x^2 - 4x - 8$ c) $y = 4x^2 - 2x - 3$ d) $y = \dfrac{1}{2}x^2 - x - 1$

6. Consider the function f whose equation is $f(x) = x^2 - 4x - 6$. On a sheet of graph paper, make a copy of the following table of points that satisfy the equation of the function f and complete the missing entries.

x	−2	−1	0	1	2	3	4	5	6
$f(x)$									
Points to plot									

a) Set up a coordinate system on your graph paper. Clearly indicate the scale on each coordinate axis. Plot each of the points in your table on your coordinate system and use these points to help draw a complete graph of the function f.

b) Place a decimal estimate of each x-intercept on your graph near the appropriate x-intercept.

c) Use the quadratic formula to find the exact values of the x-intercepts in simple radical form. Use your calculator to find decimal approximations of these exact values, correct to the nearest tenth (one decimal place). If these approximation do not agree favorably with the estimates from part (b), check your work for error.

7. Consider the function f whose equation is $f(x) = -x^2 + 6x + 10$. On a sheet of graph paper, make a copy of the following table of points that satisfy the equation of the function f and complete the missing entries.

x	-2	-1	0	1	2	3	4	5	6	7	8
$f(x)$											
Points to plot											

a) Set up a coordinate system on your graph paper. Clearly indicate the scale on each coordinate axis. Plot each of the points in your table on your coordinate system and use these points to help draw a complete graph of the function f.

b) Place a decimal estimate of each x-intercept on your graph near the appropriate x-intercept.

c) Use the quadratic formula to find the exact values of the x-intercepts in simple radical form. Use your calculator to find decimal approximations of these exact values, correct to the nearest tenth (one decimal place). If these approximation do not agree favorably with the estimates from part (b), check your work for error.

8. For each of the following equations perform the following tasks:

   i)   Use the formula $x = -\dfrac{b}{2a}$ to help find the coordinates of the vertex.

   ii) Find the y-intercept.

   iii) Use the quadratic formula to find the x-intercepts.

   iv) Set up a separate coordinate system on a sheet of graph paper for each equation. Clearly indicate the scale on each axis. Plot the vertex, the x- and y-intercepts, and draw the axis of symmetry. Plot any points provided by symmetry. Draw the graph of the parabola.

   a) $y = x^2 + 4x - 5$ b) $y = -x^2 - 8x + 9$ c) $y = x^2 - 4x + 4$

   d) $y = -x^2 - 8x - 16$ e) $y = x^2 - 2x + 4$ f) $y = -x^2 - 4x - 8$

9. (TI-82) For each of the following equations, perform the following tasks:

   i)   Load the equation in the Y= menu, push the ZOOM button, and select 6:ZStandard. Adjust the WINDOW parameters so that the vertex and intercepts are visible in the viewing window. Set up a separate coordinate system for each equation and include the WINDOW parameter settings on each coordinate axis. Copy the image in your viewing window onto your coordinate system.

   ii) Calculate the discriminant. Clearly state the value of the discriminant on your coordinate system.

   a) $y = x^2 - 2x - 10$ b) $y = -x^2 + 6x - 9$ c) $y = x^2 - 3x + 3$

   d) $y = -x^2 + 3x + 11$ e) $y = x^2 - 10x + 25$ f) $y = -x^2 + 4x - 7$

10. For each of the following equations, perform the following tasks:

   i) Use the formula $x = -\dfrac{b}{2a}$ to help find the coordinates of the vertex.

   ii) Find the y-intercept.

   iii) Use the quadratic formula to find the exact value of the x-intercepts in simple radical form. Use your calculator to find decimal approximations of these exact values, correct to the nearest tenth (one decimal place).

   iv) Set up a separate coordinate system on a sheet of graph paper for each equation. Clearly indicate the scale on each axis. Plot the vertex and the x- and y-intercepts, and draw the axis of symmetry. Plot any points provided by symmetry. Draw the graph of the parabola.

   a) $y = x^2 - 6x - 8$ b) $y = -x^2 + 5x - 5$ c) $y = 2x^2 - 8x - 2$ d) $y = -2x^2 - 10x + 1$

11. (TI-82) Use your calculator to perform the following tasks.

Load the equation $y = kx^2 - 4x + 1$ in the Y= menu.	Store the number 1 in the variable K in the following manner.
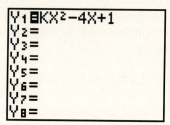	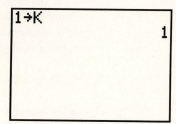
Press the ALPHA key followed by the variable K to enter the K.	Enter a 1, press the STO▷ key, press the ALPHA key followed by the variable K, and press the ENTER key.

   i) Press the ZOOM button and select 6:ZStandard to view the graph of the equation $y = kx^2 - 4x + 1$.

   ii) Store various numbers for K until you find a value for K that causes the graph of $y = kx^2 - 4x + 1$ to have exactly one x-intercept. When you have found such a value, set up a coordinate system on your paper and copy the graph from your viewing window onto your coordinate system. Record the value of K that produced this graph on your coordinate system.

   iii) Use the discriminant to find the exact value of k so that the graph of $y = kx^2 - 4x + 1$ has exactly one x-intercept.

12. Find all values of k so that the parabola represented by the given equation has two x-intercepts.

   a) $y = kx^2 - 2x + 4$ b) $y = 2x^2 - 4x - 4k$

13. Find all values of k so that the parabola represented by the given equation has no x-intercepts.

   a) $y = 2kx^2 - x - 8$ b) $y = 4 - 4x - kx^2$

14. Find all values of k so that the parabola represented by the equation $y = (k+1)x^2 - kx - 1$ has exactly one x-intercept.

15. Use the quadratic formula to find exact solutions of each of the following equations in simple radical form.

    a) $x^2 = 4x + 8$

    b) $2x^2 + 3 = 8x$

    c) $x^2 = 4(2x + 3)$

    d) $4 = x(2x - 5)$

    e) $x^2 - (2x + 3) = x(2 - x)$

    f) $x(x + 1) - 3(4 - x) = -10$

16. (TI-82) Perform the following tasks.

    Load the equations $y = x^2 - 5x - 6$ and $y = -2x^2 + 10x + 12$ in the Y= menu.

    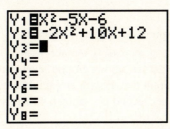

    Make the following adjustments to the WINDOW parameters.

    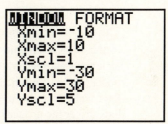

    i)   Push the GRAPH button on your calculator. Set up a coordinate system on your paper and copy the image in the viewing window onto your coordinate system. Label each graph with its equation and include the WINDOW parameter settings on each coordinate axis.

    ii) Although their equations are different, the parabolas share the same x-intercepts. Use the graph in the viewing window to help estimate these intercepts. Record these estimates on your coordinate system, near the appropriate x-intercept.

    iii) Use the quadratic formula to solve the equation $x^2 - 5x - 6 = 0$.

    iv) Use the quadratic formula to solve the equation $-2x^2 + 10x + 12 = 0$. Work with the existing coefficients.

17. Use the quadratic equation to find solutions of the following equations in simple radical form.

    Hint: First clear the equation of fractions by multiplying both sides by the least common denominator.

    a) $\dfrac{1}{2}x^2 - x - \dfrac{1}{2} = 0$

    b) $\dfrac{2}{3}x^2 - \dfrac{1}{2}x - 1 = 0$

    c) $\dfrac{1}{2}x(x - 3) = 2$

    d) $\dfrac{2}{3}x - \dfrac{1}{2}x(x - 1) = 1$

18. Use the quadratic equation to help find approximate solutions of each of the following equations. Round your answers to the nearest hundredth (two decimal places).

    a) $5.23x^2 - 3.44x - 1.22 = 0$

    b) $5.62 - 3.04x - 0.50x^2 = 0$

19. Solve each of the following equations for x in terms of r.

a) $rx^2 - (r+1)x - 2 = 0$

b) $x^2 - 2rx - (r+1) = 0$

6.6 Applications of the Quadratic Formula

This section presents some fascinating applications of the quadratic formula. You will have a chance to try your hand at some applications in the exercises at the end of the section.

Example 1 Professor David Mills is discussing projectile motion with his physics class. He explains to his students that the height of an object, as a function of time, is given by the following equation.

$$y = y_0 + v_0 t - \frac{1}{2}gt^2$$

The variable y_0 represents the initial height of the object, the variable v_0 represents its initial velocity in the vertical direction, and the variable g represents the acceleration due to the gravity. Suppose Helen is standing on top of a building, 120 feet above ground level. She takes a baseball out of her pocket and throws the baseball straight up in the air with an initial velocity of 80 feet per second. If the acceleration due to gravity is 32 feet per second per second, how many seconds will pass until the ball strikes the ground?

Graphical Solution The initial height y_0 is 120 feet. The initial velocity v_0 in the vertical direction is 80 feet per second. The acceleration g due to gravity is 32 feet per second per second. Substitute these numbers into the equation of motion.

$$y = y_0 + v_0 t - \frac{1}{2}gt^2$$

$$y = 120 + 80t - \frac{1}{2}(32)t^2$$

$$y = 120 + 80t - 16t^2$$

When the object strikes the ground, y will equal zero. Enter this function in the Y= menu in the following manner: $Y_1 = 120 + 80X - 16X^2$.

After some experimentation, the following adjustments to the WINDOW parameters were made.	Use the root-finding utility in the CALC menu to determine when the object strikes the ground.
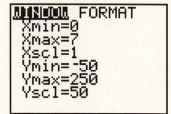	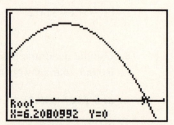

Analytical Solution   When the object returns to the ground, its height y will equal zero.

$$y = 120 + 80t - 16t^2$$

$$0 = 120 + 80t - 16t^2$$

Divide both sides of this equation by -8.

$$0 = 2t^2 - 10t - 15$$

Use the quadratic formula to solve this equation.

$$t = \frac{-b \pm \sqrt{b^2 - 4ac}}{2a}$$

$$t = \frac{-(-10) \pm \sqrt{(-10)^2 - 4(2)(-15)}}{2(2)}$$

$$t = \frac{10 \pm \sqrt{220}}{4}$$

Because the problem calls for an approximation of the time required for the ball to return to the ground, you should dispense with simple radical form and find decimal approximations of your answers.

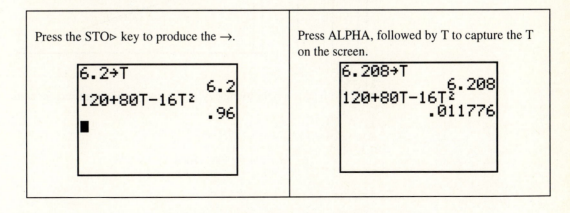

The first number makes no sense in this problem and should be discarded. Note that the second number agrees with the estimate found with the graphing method. Rounding this solution to the nearest tenth of a second, you can determine that it will take approximately 6.2 seconds for the ball to return to the ground.

Is Your Answer Reasonable? Use your calculator to check your solution.

Press the STO▷ key to produce the →.	Press ALPHA, followed by T to capture the T on the screen.
6.2→T 6.2 120+80T-16T² .96 ■	6.208→T 6.208 120+80T-16T² .011776

The first screen indicates that the height of the ball is 0.96 feet after 6.2 seconds. The second screen indicates that the height of the ball will be 0.011776 feet after 6.208 seconds. This is closer to the required height, $y = 0$. You may find this next screen interesting.

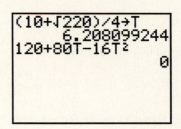

The Pythagorean Theorem

If an angle contains $90°$, then the angle is called a *right angle*. If one of the angles of a triangle is a right angle, then the triangle is called a *right triangle*. A classic formula from antiquity, called the Pythagorean theorem, relates the three sides of a right triangle in a very special way.

The Pythagorean Theorem: If the following triangle is a right triangle, then $c^2 = a^2 + b^2$.

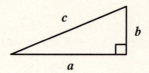

The side c, opposite the right angle, is called the *hypotenuse* of the right triangle. The sides a and b are called the *legs* of the right triangle. The square of the hypotenuse is equal to the sum of the squares of the legs.

Example 2   Kara's rescue team is practicing a maneuver that one day might save the life of a drowning person. Kara, dressed in a lifejacket, is lowered into the current of the Trinity River by means of a rope attached to a harness that she wears. The current takes her downstream. The other end of the rope is anchored to the edge of the dock, which is 10 feet above the level of the river. If 50 feet of line is played out, how far is Kara downstream?

Solution   Begin by drawing a sketch of the problem situation.

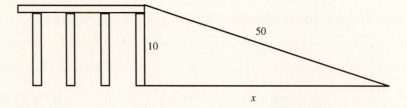

Because the triangle involved is a right triangle, you may use the Pythagorean theorem.

$$50^2 = 10^2 + x^2$$

$$2500 = 100 + x^2$$

You could use the quadratic formula to solve this equation. However, because you only have a single power of x, it will be easier to place this equation in the form $x^n = b$ and proceed from there.

$$2500 = 100 + x^2$$

$$2400 = x^2$$

$$x = \pm\sqrt{2400}$$

$$x = \pm\sqrt{400}\sqrt{6}$$

$$x = \pm 20\sqrt{6}$$

Use your calculator to approximate these results.

```
-20√6
        -48.98979486
20√6
         48.98979486
```

Because Kara's distance downstream is not a negative quantity, you can discard the first answer. Her distance downstream, rounded to the nearest tenth of a foot, is 49.0 feet.

Is Your Answer Reasonable?

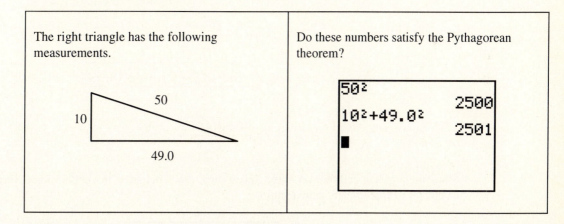

The right triangle has the following measurements.

Do these numbers satisfy the Pythagorean theorem?

```
50²
         2500
10²+49.0²
         2501
```

Because your measurements very nearly satisfy the Pythagorean theorem, you can be confident that Kara is about 49.0 feet downstream.

Example 3   Joe and Louise attend the same school in the Mojave Desert, a flat section of land where you can see for miles in all directions. Both students are members of the cross-country team, and their afternoon workout entails running to their homes after school. Louise leaves first, running due north at a steady 5 kilometers per hour. Joe spends 1 hour stretching and rubbing in the Ben-Gay because he is nursing a sore hamstring. One hour after Louise has started, Joe heads due west, running at a steady 4 kilometers per hour. If Louise left at 3:00 P.M., at what time will they be precisely 10 kilometers apart?

Solution   Let t represent the time Louise runs until she and Joe are 10 kilometers apart. If Louise is running at a steady 5 kilometers per hour for t hours, then she will run a distance of $5t$ kilometers. If Joe starts out 1 hour after Louise leaves, by the time they are 10 kilometers apart, he will have been running for 1 hour less than has Louise. Consequently, he has been running for $t-1$ hours. If he is running at a steady rate of 4 kilometers per hour for $t-1$ hours, then he will run a distance of $4(t-1)$ or $4t-4$ kilometers. Draw a diagram of the problem situation.

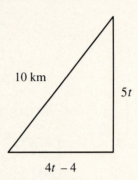

This is a right triangle, so you can use the Pythagorean theorem.

$$10^2 = (5t)^2 + (4t-4)^2$$

$$100 = 25t^2 + 16t^2 - 32t + 16$$

$$0 = 41t^2 - 32t - 84$$

Now use the quadratic formula.

$$t = \frac{-b \pm \sqrt{b^2 - 4ac}}{2a}$$

$$t = \frac{-(-32) \pm \sqrt{(-32)^2 - 4(41)(-84)}}{2(41)}$$

$$t = \frac{32 \pm \sqrt{14,800}}{82}$$

Since the problem calls for an approximate time, don't bother with simple radical form. Use your calculator to approximate these solutions.

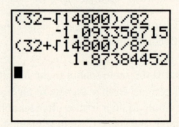

Because you know that Louise has not been running for a negative amount of time, eliminate the negative answer from consideration. Recall that the variable t represents the time that Louise has been running. If you round to two decimal places, she has been running for approximately 1.87

hours. You must convert the fraction of an hour (0.87) to minutes. Using unit dimensional analysis, a great technique used by science teachers, the following result is obtained.

$$0.87 \text{ hour} \times 60 \frac{\text{minutes}}{\text{hour}} \approx 52 \text{ minutes}$$

Louise ran for approximately 1 hour and 52 minutes. Because she left at 3 P.M., the two runners were 10 kilometers apart at approximately 4:52 P.M.

Is Your Answer Reasonable?   If Louise runs at 5 kilometers per hour for 1.87 hours, she will run approximately 9.35 kilometers. Remember, Joe remained behind for an hour to loosen up, so he ran for only 0.87 hours. If Joe runs at 4 kilometers per hour for 0.87 hours, he will run approximately 3.48 km.

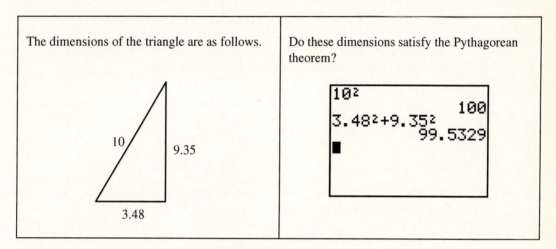

The dimensions of the triangle are as follows.

Do these dimensions satisfy the Pythagorean theorem?

Because the Pythagorean theorem is nearly satisfied, you can be confident that your approximate solution is correct.

Example 4   Henry is a tinsmith. He takes a square piece of tin and cuts out 2-inch squares from each corner. He folds up the sides to form an open box with a volume of 80 cubic inches. Find the length of a side of the original piece of tin.

Solution   Let x represent the length of the edge of the square piece of tin. If Henry removes 2-inch squares from each corner, the remaining edge has a length of $x - 4$ inches, as shown.

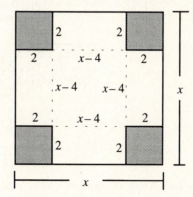

Discard the material in the corners and fold up the sides to create an open box.

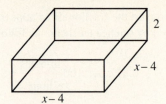

You can express the volume V of the box as a function of x.

$$V = (\text{length})(\text{width})(\text{height})$$

$$V = (x-4)(x-4)(2)$$

$$V = 2(x^2 - 8x + 16)$$

$$V = 2x^2 - 16x + 32$$

The volume of Henry's box is 80 cubic inches. Because the resulting equation is nonlinear, make one side zero by subtracting 80 from both sides of the equation.

$$80 = 2x^2 - 16x + 32$$

$$0 = 2x^2 - 16x - 48$$

$$0 = x^2 - 8x - 24$$

Note that both sides of the equation were divided by 2 in the last step. Use the quadratic formula to find the solutions of this last equation.

$$x = \frac{-b \pm \sqrt{b^2 - 4ac}}{2a}$$

$$x = \frac{-(-8) \pm \sqrt{(-8)^2 - 4(1)(-24)}}{2(1)}$$

$$x = \frac{8 \pm \sqrt{160}}{2}$$

Because Henry needs an approximate measurement before he can begin cutting, dispense with simple radical form and find decimal approximations for these solutions.

```
(8-√160)/2
           -2.32455532
(8+√160)/2
           10.32455532
```

Since you cannot have a side of negative length, discard the negative answer. The length of the edge of the original square piece of tin is approximately 10.3 inches.

Is Your Solution Reasonable?   Suppose that you begin with a square piece of tin that is 10.3 inches on a side. If you cut 2-inch squares from each corner, you will be left with an edge of 6.3 inches, as shown.

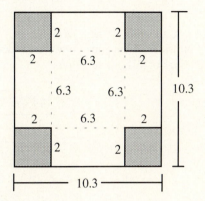

If you discard the cut corners and fold up the sides, you will have an open box with the following dimensions:

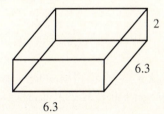

The volume of this box will be $V = (6.3)(6.3)(2)$ or 79.38 cubic inches. This volume agrees favorably with the required volume of 80 cubic inches.

You Try It	Try a better approximation for $\dfrac{8+\sqrt{160}}{2}$ inches, perhaps one that is rounded to 4 decimal places. Use this new approximation to compute the volume of the open box. How much better is this solution?

Example 5   The sum of the squares of two consecutive odd integers is 107,650. Find the integers.

Solution   Examples of consecutive integers are 5,6,7,8, ... Examples of consecutive even integers are 8, 10, 12, 14, ... Examples of consecutive odd integers are 9, 11, 13, 15 . . . . Let *k* represent an odd integer. The next consecutive odd integer would be represented by the expression $k + 2$. The sum of the squares of these consecutive odd integers is given by the expression $k^2 + (k+2)^2$. This last expression must equal 107,650.

$$k^2 + (k+2)^2 = 107,650$$

This equation is nonlinear. Simplify and make one side of the equation zero.

$$k^2 + k^2 + 4k + 4 = 107,650$$

$$2k^2 + 4k - 107,646 = 0$$

$$k^2 + 2k - 53,823 = 0$$

Note that both sides of the equation were divided by 2 in our last step. Use the quadratic formula to find the solutions of this last equation.

$$k = \frac{-b \pm \sqrt{b^2 - 4ac}}{2a}$$

$$k = \frac{-2 \pm \sqrt{(2)^2 - 4(1)(-53,823)}}{2(1)}$$

$$k = \frac{-2 \pm \sqrt{215,296}}{2}$$

$$k = \frac{-2 - 464}{2}, \frac{-2 + 464}{2}$$

$$k = -233, 231$$

If $k = -233$, then $k + 2 = -231$. The integer pair $\{-233, -231\}$ is a solution. If $k = 231$, then $k + 2 = 233$. The integer pair $\{231, 233\}$ is also a solution.

Is Your Solution Reasonable?

Check the sum of the squares of the first integer pair.	Check the sum of the squares of the second integer pair.

```(-233)²+(-231)²```
```   107650```
■
	```231²+233²```
	``` 107650```

In each case, you have consecutive odd integers and the sum of their squares is 107,650.

Tip	One of the most important steps in the problem-solving process is the last. Is your answer reasonable? Never fail to address this question.

Exercises for Section 6.6

1. Janine throws a ball into the air. The height of the ball in feet is given by the equation $y = 12 + 40t - 16t^2$, where t represents the number of seconds that have passed since the ball left Janine's hand.

   a) Find the initial height of the ball.

   b) Find the height of the ball after 1.5 seconds have elapsed.

   c) (TI-82) Load the equation into the Y= menu in the following manner:

   $Y_1 = 12 + 40X - 16X^2$. Adjust the WINDOW parameters in the following manner:

   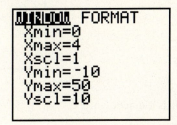

   Push the GRAPH button. Set up a coordinate system on your paper and copy the image in the viewing window onto your coordinate system. Include the WINDOW parameters on the coordinate axes.

   d) Use the root-finding utility in the CALC menu to find the time when the ball returns to the ground. Record this number on your graph, near the appropriate point.

   e) Use the quadratic formula to find the time required for the ball to return to the ground. Use your calculator to find decimal approximations for your solutions. Discard solutions that are not practical and explain why you are doing so. If your final solution does not compare favorably with the estimate found in part (d), check your work for error.

   f) Use complete sentences and calculations to explain why your solution is reasonable.

2. Use the Pythagorean theorem to find the length of the missing side of each of the following right triangles in simple radical form.

   a)

   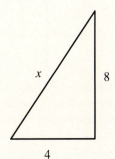

   b)

   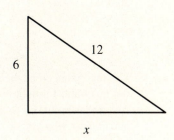

c)   d)

3. Consider the following right triangle.

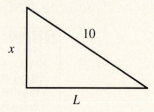

a) Express L as a function of x.

b) Express the area A of the right triangle as a function of x.

4. Two high-powered racing boats, built for ocean racing, leave Catalina Island 1 hour apart. The first boat, the *Older But Wiser,* heads due west at 50 kilometers per hour. The second boat, the *Mae West,* has to have a fuel line repaired and leaves 1 hour later. The *Mae West* heads due north at 80 kilometers per hour. You are to find the time that must pass until they are 500 kilometers apart.

a) Make a copy of the following chart on your paper and complete the missing entries.

	Distance	Speed	Time
Older But Wiser		50	t
Mae West		80	$t-1$

b) Make a copy of the following right triangle on your paper. On each leg of the right triangle, place the expression representing the distance each boat travels.

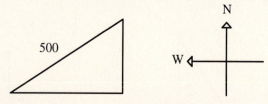

c) Use the Pythagorean theorem to set up an equation involving the three sides of the triangle. Use the quadratic formula to help find solutions of the resulting equation. Use your calculator to round your solutions to the nearest tenth of an hour. Discard any solutions that are not practical and explain why you are doing so.

d) Use complete sentences and calculations to explain why your solution is reasonable.

5. Maria and Ernie run a small packaging business. One project involves taking a square piece of cardboard, cutting three-inch squares out of each of the four corners, then folding up the sides of the cardboard to make an open box, as follows.

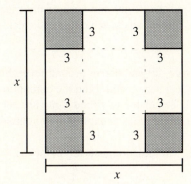
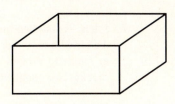

a) Express the volume V of the box as a function of x.

b) Suppose that the volume of the open box is 120 cubic inches. Let V equal 120 in the equation you developed in part (a). Use the quadratic formula to solve the resulting equation for x.

c) Use your calculator to round the solutions found in part (b) to the nearest tenth of an inch (1 decimal place). Discard any solutions that are not practical and explain why you are doing so.

d) Use complete sentences and calculations to explain why your answer is reasonable.

6. The sum of the squares of two consecutive integers is 24,865. Let k represent the first of these integers.

a) Set up an equation in k that models this problem situation. Use the quadratic formula to find the solutions of this equation.

b) Use complete sentences and calculations to explain why your solution is reasonable.

7. Following are two concentric squares. The side of the outer square is 1 inch longer than twice the side of the inner square. Let x represent the length of the side of the inner square.

a) Express the length L of the side of the outer square as a function of x.

b) Express the area A of the shaded region as a function of x.

c) Suppose that the area of the shaded region is 40 square inches. Let A equal 40 in the function in part (b) and use the quadratic formula to solve the resulting equation for x. Find decimal approximations of your solutions correct to the nearest tenth of an inch (1 decimal place). Discard any solutions that are not practical and explain why you are doing so.

d) Use complete sentences and calculations to explain why your solution is reasonable.

8. The area of a rectangle is 100 square feet. If the length of the rectangle is 3 feet more than twice its width, find the dimensions of the rectangle, correct to the nearest tenth of a foot (one decimal place). Use complete sentences and calculations to explain why your solution is reasonable.

9. The area of a triangle is 200 square inches. If the height of the triangle is 5 inches more than the length of the base of the triangle, find the height and base of the triangle, correct to the nearest tenth of an inch (one decimal place). Use complete sentences and calculations to explain why your solution is reasonable.

10. Jolene is standing on top of her garage, 20 feet above ground level. She hurls her Superball into the air with an initial velocity of 80 feet per second. Find the amount of time required for the ball to return to ground level. Use your calculator to round your answer to the nearest tenth of a second. Use complete sentences and calculations to explain why your solution is reasonable.

11. Consider the following right triangle. All measurements are in inches.

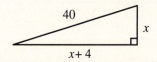

a) Use the Pythagorean theorem to set up an equation that models this problem situation.

b) Use the quadratic formula to solve the resulting equation for x. Use your calculator to round your solutions to the nearest tenth of an inch (one decimal place). Discard any solutions that are not practical and explain why you are doing so.

c) Use complete sentences and calculations to explain why your solution is reasonable.

12. What follows is the graph of $f(x) = 5 - x$. The point P is on the graph of f.

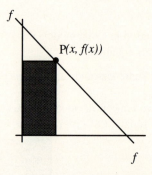

a) Express the area A of the shaded rectangle as a function of x.

b) Suppose that the area of the shaded rectangle is 3 square feet. Let A equal 3 in the equation developed in part (a) and use the quadratic formula to solve the resulting equation for x.

c) Use your calculator to estimate your solutions to the nearest tenth of a foot. Discard any solutions that are not practical and explain why you are doing so. Remember, your rectangle is in the first quadrant.

d) Find the coordinates of the point P, correct to the nearest tenth of a foot.

e) Use complete sentences and calculations to explain why your solution is reasonable.

13. (TI-82) The revenue earned from the sale of a particular article is given by the equation $R = 36p - \frac{1}{2}p^2$, where R represents the revenue earned in dollars and p represents the unit price of the article in dollars. Suppose that you wish to find the unit price that will bring in a revenue of $500.

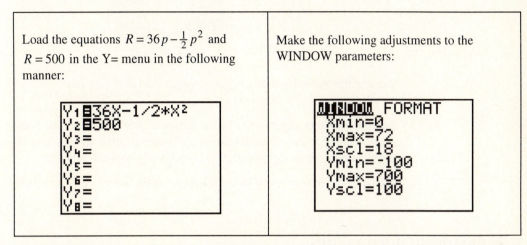

> Load the equations $R = 36p - \frac{1}{2}p^2$ and $R = 500$ in the Y= menu in the following manner:
>
> Make the following adjustments to the WINDOW parameters:

a) Set up a coordinate system on your paper. Push the GRAPH button and copy the image in the viewing window onto your coordinate system. Label each graph with its equation and include the WINDOW parameters on your coordinate axes.

b) Use the intersect utility in the CALC menu to determine how to set the unit price so the incoming revenue is $500. Draw a number line below your graph and shade and label these solutions on your number line in the usual manner.

c) Use the quadratic formula to help find the solutions of the equation $500 = 36p - \frac{1}{2}p^2$. Use your calculator to find decimal approximations for these solutions, correct to the nearest cent (two decimal places). If these approximations do not agree favorably with the estimates found in part (b), check your work for error.

14. The demand for solar powered calculators, as a function of the unit price, is given by the equation $x = 50 - p$, where x represents the number of calculators demanded by the public and p represents the unit price of the calculator.

a) Let R represent the revenue resulting from the sale of solar powered calculators. The revenue equals the number of articles sold times the unit price, so $R = xp$. Express the revenue R as a function of p only.

b) Suppose that the revenue resulting from the sale of solar powered calculators is $600. Let R equal 600 in the function you developed in part (a) and use the quadratic formula to solve the resulting equation for p.

c) Discard any solutions that are not practical and explain why you discarded them. How many solar powered calculators will the public demand at each unit price you found?

d) Use complete sentences and calculations to explain why your solutions are reasonable.

15. Following are two concentric circles. Let *r* represent the radius of the inner circle.

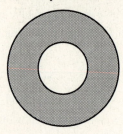

a) Suppose that the outer radius is 1 inch longer than twice the inner radius. Express the area *A* of the shaded region as a function of *r*.

b) Suppose that the area of the shaded region is 65π. Let *A* equal 65π in the equation you developed in part (a). Use the quadratic formula to solve the resulting equation for *r*. Discard any solutions that are not practical and explain why you are doing so.

c) Use complete sentences and calculations to explain why your solution is reasonable.

6.7 Quadratic Inequalities

This section discusses how to solve quadratic inequalities of the form $ax^2 + bx + c > 0$ — or perhaps how to solve a quadratic inequality that uses the symbols <, ≤, or ≥. It is heartening to know that you have already learned everything necessary to solve these inequalities. You must simply note where the graph of the parabola is above, below, or on the *x*-axis.

Example 1   Solve the following inequality for *x*: $x^2 - 2x - 3 < 0$.

Solution   The graph of the equation $y = x^2 - 2x - 3$ is a parabola that opens upward. The formula $x = -\dfrac{b}{2a}$ will show that the coordinates of the vertex are $(1, -4)$ and the axis of symmetry is the line $x = 1$. The *y*-intercept is -3 and the quadratic formula will show that the *x*-intercepts are -1 and 3.

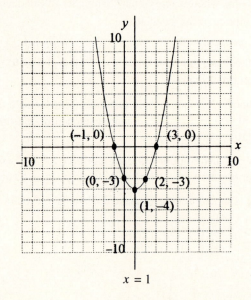

To solve the inequality $x^2 - 2x - 3 < 0$, note where the graph of $y = x^2 - 2x - 3$ is below the x-axis.

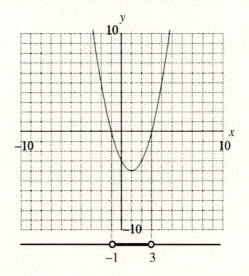

In set-builder notation, the solution of the inequality $x^2 - 2x - 3 < 0$ is $\{x: -1 < x < 3\}$. In interval notation, the solution set is $(-1, 3)$.

Was All of That Work Necessary?   To solve the inequality $x^2 - 2x - 3 < 0$ you only need to know the x-intercepts and the fact that the parabola opens upward. Nothing else is needed. Perhaps a student solution would be helpful at this point.

$$x = \frac{-b \pm \sqrt{b^2 - 4ac}}{2a}$$

$$x = \frac{-(-2) \pm \sqrt{(-2)^2 - 4(1)(-3)}}{2(1)}$$

$$x = \frac{2 \pm \sqrt{16}}{2}$$

$$x = \frac{2 - 4}{2}, \frac{2 + 4}{2}$$

$$x = -1, 3$$

$y = x^2 - 2x - 3$

This student did not even attempt to find the vertex and the y-intercept. She used only the x-intercepts and the fact that the parabola opens upward.

> *Procedure for solving quadratic inequalities.*
>
>   *1. Use the quadratic formula or the ac test to find the x-intercepts.*
>   *2. Plot the x-intercepts correctly. Use the fact that the parabola opens upward or downward. It is not necessary to accurately locate the vertex and the y-intercept. Draw a rough outline of the parabola with this minimal information.*
>   *3. Draw a number line below your graph and shade the solution of the inequality in the usual manner.*
>   *4. Use set-builder notation or interval notation to describe your solution set.*

Example 2   Solve the following inequality for x: $-x^2 + 2x + 2 \le 0$.

Solution   The graph of $y = -x^2 + 2x + 2$ is a parabola that opens downward. Use the quadratic formula to find the x-intercepts.

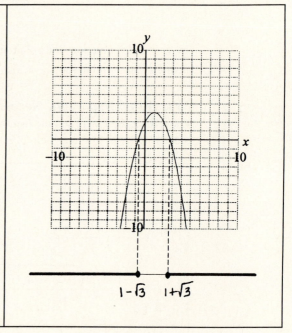

$$x = \frac{-b \pm \sqrt{b^2 - 4ac}}{2a}$$

$$x = \frac{-2 \pm \sqrt{(2)^2 - 4(-1)(2)}}{2(-1)}$$

$$x = \frac{-2 \pm \sqrt{12}}{-2}$$

$$x = \frac{-2 \pm 2\sqrt{3}}{-2}$$

$$x = \frac{-2}{-2} \pm \frac{2\sqrt{3}}{-2}$$

$$x = 1 \pm \sqrt{3}$$

$$x \approx -0.7, 2.7$$

The solution set is $\left\{ x : x \le 1 - \sqrt{3} \text{ or } x \ge 1 + \sqrt{3} \right\}$ or $\left(-\infty, 1 - \sqrt{3} \right] \cup \left[1 + \sqrt{3}, +\infty \right)$.

Example 3   Use the TI-82 graphing calculator to solve the following inequality for x: $x^2 - 3x + 5 > 0$.

Solution

<table>
<tr>
<td>

Load the equation $y = x^2 - 3x + 5$ into the Y= menu.

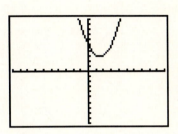

</td>
<td>

Push the ZOOM button and select 6:ZStandard.

</td>
</tr>
</table>

Because the graph of $y = x^2 - 3x + 5$ is above the x-axis for all values of x, the solution set of the inequality $x^2 - 3x + 5 > 0$ is $\{x : x \in R\}$ or $(-\infty, +\infty)$.

Example 4   Solve the following inequality for x: $-2x^2 + x - 1 \geq 0$.

Solution   The graph of $y = -2x^2 + x - 1$ is a parabola that opens downward. Use the quadratic formula to find the x-intercepts.

$$x = \frac{-b \pm \sqrt{b^2 - 4ac}}{2a}$$

$$x = \frac{-1 \pm \sqrt{(1)^2 - 4(-2)(-1)}}{2(-2)}$$

$$x = \frac{-1 \pm \sqrt{-7}}{-4}$$

Because $\sqrt{-7}$ is not a real number, the parabola $y = -2x^2 + x - 1$ does not have any x-intercepts. If the parabola opens downward and has no x-intercepts, the parabola must lie entirely below the x-axis. Check this with our calculator.

<table>
<tr>
<td>

Load the equation $y = -2x^2 + x - 1$ into the Y= menu.

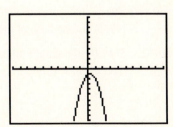

</td>
<td>

Push the ZOOM button and select 6:ZStandard.

</td>
</tr>
</table>

Because the parabola lies entirely below the x-axis, the solution set of $-2x^2 + x - 1 \geq 0$ is the empty set. There are no solutions.

Example 5   Find the domain of the function $h(x) = \sqrt{9 - x^2}$.

Solution   In order for the expression $\sqrt{9 - x^2}$ to be a real number, the expression $9 - x^2$ must be a nonnegative real number. Therefore, to find the domain of the function $h(x) = \sqrt{9 - x^2}$, you must solve the inequality $9 - x^2 \geq 0$. First find the x-intercepts of this parabola by letting $y = 0$ in the equation $y = 9 - x^2$.

$$y = 9 - x^2$$
$$0 = 9 - x^2$$
$$x^2 = 9$$
$$x = \pm 3$$

The x-intercepts are -3 and 3. The parabola whose equation is $y = 9 - x^2$ opens downward. Use these facts to draw the graph of $y = 9 - x^2$. To solve the inequality $9 - x^2 \geq 0$, note where the graph of $y = 9 - x^2$ is above or intercepts the x-axis.

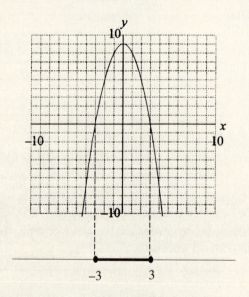

The domain of the function h is $\{x : -3 \leq x \leq 3\}$ or $[-3, 3]$.

Example 6   The altitude of a triangle is 6 feet less than the length of the base of the triangle. Find all possible lengths of the base of the triangle, if it is known that the area of the triangle is *less than* 8 square feet.

Solution   Start by drawing a picture. Let x represent the length of the base. Because the altitude is 6 feet less than the base of the triangle, $x - 6$ will represent the altitude of the triangle.

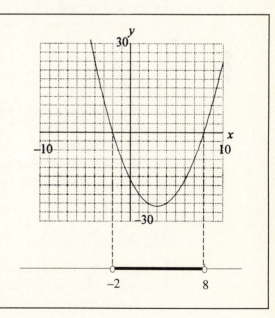

The area is less than 8 square feet. The area of a triangle is found by multiplying $\dfrac{1}{2}$ times the base times the height of the triangle.

$$\text{Area} < 8$$

$$\frac{1}{2}x(x-6) < 8$$

This inequality is nonlinear, so you should make one side equal to zero by multiplying both sides of the inequality by 2, then subtracting 16 from both sides of the resulting inequality.

$$\frac{1}{2}x^2 - 3x < 8$$

$$x^2 - 6x < 16$$

$$x^2 - 6x - 16 < 0$$

To solve the inequality $x^2 - 6x - 16 < 0$, note where the graph of $y = x^2 - 6x - 16$ is below the x-axis.

Use the ac test to help find the x-intercepts of the parabola whose equation is $y = x^2 - 6x - 16$.

$$y = x^2 - 6x - 16$$

$$0 = x^2 - 6x - 16$$

$$0 = x^2 - 8x + 2x - 16$$

$$0 = x(x-8) + 2(x-8)$$

$$0 = (x+2)(x-8)$$

$$x = -2, 8$$

The solution set $\{x : -2 < x < 8\}$ or $(-2, 8)$.

Practical Considerations. There are two practical considerations that must be dealt with.

1. The base of a triangle cannot have negative length. You must ensure that the base of the triangle is greater than zero. Because x represents the length of the base of the triangle, x must be greater than zero.

2. The altitude of the triangle must have a positive length. Consequently, $x - 6 > 0$ or $x > 6$.

It can be concluded that x must be a number greater than 6 and less than 8.

Exercises for Section 6.7

1. Copy the following table of points that satisfy the equation $f(x) = x^2 - 4x - 5$ onto a sheet of graph paper. Complete the missing entries.

x	−1	0	1	2	3	4	5
$f(x)$							
Points to plot							

   a) Set up a coordinate system on your graph paper and clearly indicate the scale on each axis. Plot each point in the table on your coordinate system and use these points to draw a complete graph of the function f.

   b) Draw a number line below your graph and shade and label the solution of the inequality $x^2 - 4x - 5 > 0$ on your number line in the usual manner.

   c) Use both set-builder and interval notation to describe your solution set.

2. Copy the following table of points that satisfy the equation $f(x) = -x^2 + 2x + 3$ onto a sheet of graph paper. Complete the missing entries.

x	−2	−1	0	1	2	3	4
$f(x)$							
Points to plot							

   a) Set up a coordinate system on your graph paper and clearly indicate the scale on each axis. Plot each point in the table on your coordinate system and use these points to draw a complete graph of the function f.

   b) Draw a number line below your graph and shade and label the solution of the inequality $-x^2 + 2x + 3 > 0$ on your number line in the usual manner.

   c) Use both set-builder and interval notation to describe your solution set.

3. (TI-82) Load the equation $y = x^2 + x - 6$ in the Y= menu, as follows: Y1 = X^2 + X − 6.

   a) Push the ZOOM button and select 6:ZStandard. Set up a coordinate system on your paper and copy the image in your viewing window onto your coordinate system. Label the graph with its equation and include the WINDOW parameters on the coordinate axes.

   b) Use the root-finding utility in the CALC menu to find the x-intercepts of the parabola. Place these results on your coordinate system, near the appropriate point.

   c) Draw a number line below your graph and shade and label the solution of the inequality $x^2 + x - 6 \geq 0$ on your number line in the usual manner. Use both set-builder and interval notation to describe your solution set.

4. (TI-82) Load the equation $y = -x^2 - x + 12$ in the Y= menu, as follows: $Y_1 = -X^2 - X + 12$.

a) Adjust the WINDOW parameters so that the vertex and x-intercepts are visible in your viewing window. Set up a coordinate system on your paper and copy the image in your viewing window onto your coordinate system. Label the graph with its equation and include the WINDOW parameters on your coordinate axes.

b) Use the root-finding utility in the CALC menu to find the x-intercepts of the parabola. Place these results on your coordinate system, near the appropriate point.

c) Draw a number line below your graph and shade and label the solution of the inequality $-x^2 - x + 12 \geq 0$ on your number line in the usual manner. Use both set-builder and interval notation to describe your solution set.

5. For each of the following equations, perform the following tasks:

i)   Place the equation in vertex form.

ii) Find the y-intercept.

iii) Use the ac test to help find the x-intercepts.

iv) Set up a coordinate system on a sheet of graph paper. Clearly indicate the scale on each axis. Plot the vertex, the x- and y-intercepts, and draw the axis of symmetry. Use this information to draw the parabola.

v)   Draw a number line below your graph and shade and label the solution set of the inequality $f(x) > 0$ on your number line in the usual manner. Use both set-builder and interval notation to describe your solution set.

a) $f(x) = -x^2 + 4x + 12$

b) $f(x) = x^2 - 2x - 8$

c) $f(x) = 2x^2 + 4x - 6$

d) $f(x) = \frac{1}{2}x^2 - 2x - 16$

6. For each of the following equations, perform the following tasks:

i)   Use the formula $x = -\frac{b}{2a}$ to help find the coordinates of the vertex.

ii) Find the y-intercept.

iii) Use the quadratic formula to help find the x-intercepts in simple radical form. Use your calculator to find decimal approximations for these solutions, correct to the nearest tenth (one decimal place).

iv) Set up a coordinate system on a sheet of graph paper. Clearly indicate the scale on each axis. Plot the vertex, the x- and y-intercepts, and draw the axis of symmetry. Use this information to draw the parabola.

v) Draw a number line below your graph and shade and label the solution set of the inequality $f(x) \leq 0$ on your number line in the usual manner. Use both set-builder and interval notation to describe your solution set.

Note: Use the exact values of the x-intercepts, in simple radical form, when answering this question.

a) $f(x) = -x^2 + 2x + 2$

b) $f(x) = x^2 - 8x - 8$

c) $f(x) = 2x^2 - 8x - 12$

d) $f(x) = \frac{1}{2}x^2 - 2x - 2$

7. For each of the following inequalities, perform the following tasks:

i) Use the quadratic formula to find the x-intercepts of the associated parabola in simple radical form. Use a calculator to find decimal approximations of the x-intercepts, correct to the nearest tenth (one decimal place).

ii) Set up a coordinate system on graph paper. Clearly indicate the scale on each axis. Carefully locate the x-intercepts on your coordinate system. Using only the x-intercepts and the fact that the parabola opens up or down, sketch a rough graph of the parabola on your coordinate system.

iii) Draw a number line below your graph and shade and label the solution of the inequality on your number line in the usual manner. Use both set-builder and interval notation to describe your solution set.

Note: Use the exact values of the x-intercepts, in simple radical form, when answering this question.

a) $x^2 - 8x - 6 > 0$

b) $-x^2 + 4x + 6 \leq 0$

c) $5x^2 - 2x - 5 \leq 0$

d) $x^2 - 3x + 5 > 0$

e) $-2x^2 + x + 4 < 0$

f) $-x^2 + 2x - 5 \geq 0$

8. (TI-82) Janine and Tommy are studying together. They are trying to solve the inequality $-2x^2 + 8x + 42 > 0$.

Janine loads the equation $y = -2x^2 + 8x + 42$ into the Y= menu.	Janine makes the following adjustments to the WINDOW parameters:
	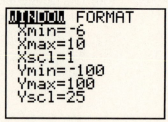

a) Push the GRAPH button. Set up a coordinate system on your paper and copy the image in Janine's viewing window onto your coordinate system. Draw a number line below your graph and shade and label the solution set of $-2x^2 + 8x + 42 > 0$ on your number line in the usual manner.

Tommy begins by dividing both sides of the inequality by -2, as follows.

$$-2x^2 + 8x + 42 > 0$$

$$x^2 - 4x - 21 < 0$$

Tommy loads the equation $y = x^2 - 4x - 21$ in the Y= menu. 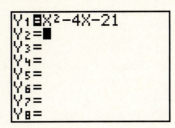	Tommy makes the following adjustments to his WINDOW parameters: 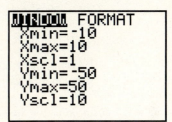

b) Push the GRAPH button. Set up a coordinate system on your paper and copy the image in Tommy's viewing window onto your coordinate system. Draw a number line below your graph and shade and label the solution set of $x^2 - 4x - 21 < 0$ on your number line in the usual manner.

c) Whose solution is correct, Janine's or Tommy's?

9. Use both set-builder and interval notation to describe the solution sets of each of the following inequalities.

a) $x^2 < 8x + 9$

b) $6 \geq x(x+5)$

c) $x(x+3) - 2(x+1) < 2$

d) $x(x+2) \geq 4x^2 - (2x+5)$

10. Find the domain of each of the following functions.

a) $f(x) = \sqrt{25 - x^2}$

b) $f(x) = \sqrt{x^2 - 16}$

c) $g(x) = \sqrt{4 - 3x - x^2}$

d) $f(x) = \sqrt{2x^2 - x - 1}$

11. The following is the graph of the function whose equation is $f(x) = \sqrt{x^2 - 9}$

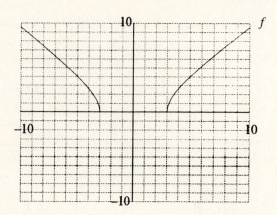

a) Examine the graph to determine the domain of the function f. Use set-builder and interval notation to describe the domain of f.

b) In order for the expression $\sqrt{x^2 - 9}$ to be a real number, $x^2 - 9$ must be a nonnegative real number. Determine the domain of the function $f(x) = \sqrt{x^2 - 9}$ by solving the inequality $x^2 - 9 \geq 0$. Use set-builder and interval notation to describe the domain of f.

12. For each of the following equations, perform the following tasks:

   i) (TI-82) Load the function into the Y= menu, push the ZOOM button and select 6:ZStandard. Set up a coordinate system on your paper and copy the image in your viewing window onto your coordinate system. Label the graph with its equation and include the WINDOW parameters on your coordinate axes.

   ii) Use your graph to determine the domain of the function. Use set-builder and interval notation to describe the domain of f.

   iii) Use the technique of Example 5 in this section to find the domain of the function. Use set-builder and interval notation to describe the domain of f. If this description does not agree with the description from part (ii), check your work for error.

   a) $f(x) = \sqrt{3 - 2x - x^2}$ b) $f(x) = \sqrt{x^2 + 4x - 5}$

13. Consider the following rectangle:

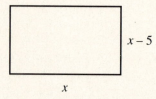

a) Express the area A of the rectangle as a function of x.

b) Find all values of x such that the area of the rectangle is less than 14. Start with the inequality $A < 14$ and replace A with the expression involving x that you found in part (a). Shade and label the solution of this inequality on a number line.

c) Discard any solutions that are not practical and explain why you are doing so. Once you've done this, shade and label your final solution to this problem on a second number line.

14. Gene is making open boxes out of square pieces of cardboard by cutting 1-inch squares out of each of the four corners and folding up the sides.

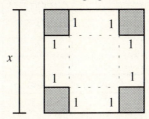

a) Express the volume V of the open box as a function of x.

b) Gene would like to keep the volume under 49 cubic inches. Start with the inequality $V < 49$ and replace V with the expression containing x that you developed in part (a). Shade and label the solution of this inequality on a number line.

c) Discard any solutions that are not practical and explain why you are doing so. Once you've done this, shade and label your final solution to this problem on a second number line.

d) Use complete sentences to explain the meaning of our solution to our friend Gene.

15. Sandra owns an appliance store in Santa Rosa. She prides herself on being a pretty good amateur mathematician, and she has determined that her weekly revenue from the sale of Zenith TVs, expressed as a function of the unit price, is given by $R = -0.1p^2 + 100p$, where R is her weekly revenue and p is the unit price. How should Sandra set the unit price so that her weekly revenue will *exceed* $9,000?

16. An object is propelled upward with an initial velocity of 128 feet per second from ground level. The object's height as a function of the amount of time that has passed since the object's release is given by $h = -16t^2 + 128t$, where h is the height of the object in feet and t is the number of seconds that have passed since the object's release. During what interval of time is the object *over* 192 feet in the air?

6.8 Max-Min Problems

This section introduces problems that involve the words *maximum* and *minimum*. A class of max-min problems that involve the parabola is easily solved using the basics that you have already learned.

> *The graph of the function defined by the equation $f(x) = ax^2 + bx + c$ is a parabola. If the parabola opens upward, then the function will have a minimum function value at $x = -\dfrac{b}{2a}$. If the parabola opens downward, then the function will have a maximum function value at $x = -\dfrac{b}{2a}$. The maximum or minimum value of the function is $f\left(-\dfrac{b}{2a}\right)$.*

Example 1 Find the minimum value of the function $f(x) = x^2 - 2x - 3$.

Intuitive Solution Try to develop some intuition by guessing. Substitute a few x-values into the equation $f(x) = x^2 - 2x - 3$.

x	$f(x)$
−2	5
−1	0
0	−3
1	−4
2	−3
3	0
4	5

Thus far, −4 is the minimum function value. This minimum value occurs when $x = 1$. However, if you were to substitute more values of x into this equation, is it possible you might find a smaller function value? Are you certain that you have the minimum function value?

Graphical Solution (TI-82) The graph of the equation $f(x) = x^2 - 2x - 3$ is a parabola that opens upward. Therefore, the minimum function value will have to occur at the vertex.

Load the equation into the Y= menu.	Press the ZOOM button and select 6:ZStandard.
Y₁◼X²−2X−3 Y₂=◼ Y₃= Y₄= Y₅= Y₆= Y₇= Y₈=	

You could use the TRACE and ZOOM capability of the TI-82 to find the coordinates of the vertex. However, the TI-82 has a max-min utility in the CALC menu that you will find useful.

Press 2nd CALC to open the following menu:	Select 3:minimum to capture the following image:
CALCULATE 1:value 2:root 3:minimum 4:maximum 5:intersect 6:dy/dx 7:∫f(x)dx	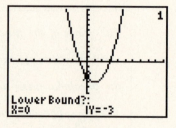 Lower Bound?: X=0 Y=-3

The calculator is asking for a lower bound. Because the cursor is already a little to the left of the vertex, push ENTER to capture the following image:

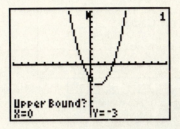

The calculator is asking for an upper bound. Move the cursor a little to the right of the vertex and push ENTER to capture the following image:

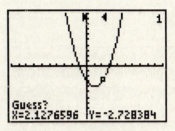

Note: Your screen may differ slightly depending upon the location of your cursor before you pushed the ENTER key. The calculator is now asking that you guess the location of the vertex. Note the lower and upper bound marks at the top of the viewing screen. As long as the *x*-value of the cursor is located between or on these two marks, it is a valid guess. At this point, we will stay right where we are and push the ENTER key to capture the following image:

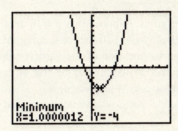

Your screen may look different if you have chosen a different lower bound, or upper bound, or a different guess. All of these affect the outcome of the result. However, all results will indicate that the location of the vertex is probably (1,4).

Analytical Solution   Find the coordinates of the vertex of the parabola whose equation is $f(x) = x^2 - 2x - 3$.

$$x = -\frac{b}{2a}$$ $$x = -\frac{-2}{2(1)}$$ $$x = 1$$	$$f(x) = x^2 - 2x - 3$$ $$f(1) = (1)^2 - 2(1) - 3$$ $$f(1) = -4$$ The minimum function value is −4, which occurs when $x = 1$.

Example 2   Mary lives in a rural area where the deer love to raid vegetable gardens. She will have to build a fence to keep the deer out of her garden. Suppose that she has enough material to build a fence 60 feet in length. She wants a rectangular plot, one side of which will be bounded by her garage wall. The fencing material will be used to build a fence around the remaining three sides of her garden. What should be the dimensions of her garden so that it will have maximum area?

Intuitive Solution   First, draw a picture of the problem situation. You need a rectangular garden, with fence on three sides and the garage wall on the remaining side. Two of the sides of the garden will have equal length, which have been denoted by the variable x. The length of the remaining side of the rectangular garden is represented by the variable y.

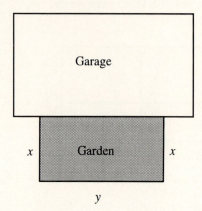

Try to develop some intuition by guessing. Suppose that $x = 5$ feet. The opposite side of the garden will also be 5 feet, which means that you've already used 10 feet of fencing material. Since Mary has 60 feet of fencing material available, that leaves 50 feet of fencing for the side of the garden represented by the variable y. The dimensions of the garden will be 5 feet by 50 feet and the garden area will be 250 square feet.

Try another guess. Suppose that $x = 10$ feet. Since the opposite side of the garden will also be 10 feet, you've already used 20 feet of fencing. Mary has 60 feet of fencing available, so that leaves 40 feet of fencing for the side of the garden represented by the variable y. The dimensions of the garden will be 10 feet by 40 feet and the area of the garden will be 400 square feet. These dimensions lead to an area larger than your first guess.

Some additional guesses have been arranged in a table.

x	y	Area
5	50	250
10	40	400
11	38	418
12	36	432
13	34	442
14	32	448
15	30	450
16	28	448

When you examine this table, it appears that Mary can have a garden with maximum area if she chooses dimensions of 15 feet by 30 feet, creating a rectangular garden with an area of 450 square feet.

Analytical Solution   Draw the picture of the problem situation again.

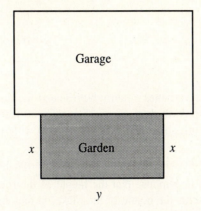

Mary has enough material available for a fence of length 60 feet. Since the garage bounds the fourth side of her garden, the sum of the three remaining sides must total 60 feet in length.

$$x + y + x = 60$$

or

$$2x + y = 60 \qquad \text{Eq. (1)}$$

Mary wishes to maximize the area of her garden, so you must now find a formula for the area of the garden. This area is given by the following equation:

$$A = xy \qquad \text{Eq. (2)}$$

The area A of Mary's garden has been expressed as a function of x and y. Express the area A as a function of a single variable, and solve Equation (1) for y, as follows:

$$y = 60 - 2x \qquad \text{Eq. (3)}$$

Substitute Equation (3) into Equation (2), as follows:

$$A = xy$$

$$A = x(60 - 2x)$$

$$A = -2x^2 + 60x$$

You now have the area A of Mary's garden as a function of the single variable x. The graph of the equation $A = -2x^2 + 60x$ is a parabola that opens downward, so A will have a maximum value and it will occur at the vertex of the parabola.

$x = -\dfrac{b}{2a}$ $x = -\dfrac{60}{2(-2)}$ $x = 15$	$A = -2x^2 + 60x$ $A = -2(15)^2 + 60(15)$ $A = 450$

To find the value of y, substitute $x = 15$ in Equation (3).

$$y = 60 - 2x$$

$$y = 60 - 2(15)$$

$$y = 30$$

Remember, the variables x and y represented the dimensions of Mary's garden. Therefore, if Mary creates a garden with dimensions 15 feet by 30 feet, she will have a maximum area of 450 square feet. Note that this solution agrees quite nicely with your intuitive solution.

Example 3   Melissa has received a beautiful model rocket for her birthday, and she and her dad go to the local park to launch it. Suppose that the height of the rocket is given by the equation $h = -16t^2 + 128t$, where h represents the height of the rocket in feet and t represents the number of seconds that have passed since the rocket's launch. What is the maximum height that Melissa's rocket will reach, and how much time will have elapsed when the it attains this height?

Graphical Solution (TI-82)   Load the equation in the Y= menu, like this: $Y_1 = -16X^2 + 128X$.

Make the following adjustments to the WINDOW parameters:	Use 4:maximum in the CALC menu to capture the following image:
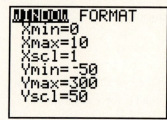	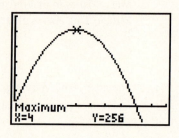

It appears that the maximum height of the rocket will be 256 feet and that height will be attained at $t = 4$ seconds.

Analytical Solution   Find the coordinates of the vertex of the parabola represented by the equation $h = -16t^2 + 128t$.

$t = -\dfrac{b}{2a}$ $t = -\dfrac{128}{2(-16)}$ $t = 4$	$h = -16t^2 + 128t$ $h = -16(4)^2 + 128(4)$ $h = 256$

You can now be certain. The maximum height of the rocket will be 256 feet and that height will occur at $t = 4$ seconds.

Exercises for Section 6.8

1. What is the maximum value of each of the following functions? What is the *x*-value that produces this maximum function value?

   a)   b)

   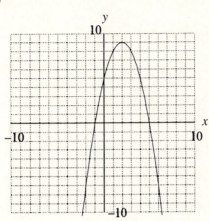 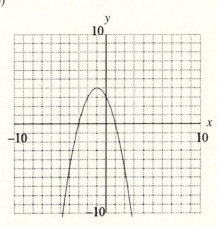

2. What is the minimum value of each of the following functions? What is the *x*-value that produces this minimum function value?

   a)   b)

   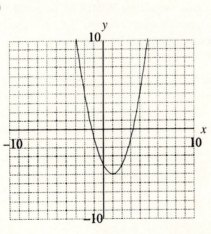 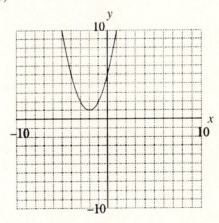

3. Consider the function whose equation is $f(x) = x^2 - 6x - 4$. Make a copy of the following table on your homework paper and complete the missing entries.

x	0	1	2	3	4	5	6
$f(x)$							

   a) What is the minimum function value in the table? At what *x*-value does it occur?

   b) Use the formula $x = -\dfrac{b}{2a}$ to help find the minimum function value.

4. Consider the function whose equation is $f(x) = -x^2 - 4x + 6$. Make a copy of the following table on your homework paper and complete the missing entries.

x	-5	-4	-3	-2	-1	0	1
$f(x)$							

a) What is the maximum function value in the table? At what x-value does it occur?

b) Use the formula $x = -\dfrac{b}{2a}$ to help find the maximum function value.

5. For each of the following equations, perform the following tasks:

i) Clearly state whether the function represented by the equation has a maximum or minimum and explain how you arrived at your decision.

ii) Use the formula $x = -\dfrac{b}{2a}$ to help find this maximum or minimum function value and the x-value at which it occurs.

a) $f(x) = x^2 - 8x - 12$ b) $f(x) = -x^2 + 4x + 4$ c) $g(x) = -3x^2 - 4x - 18$

d) $h(x) = \dfrac{2}{3}x^2 - x - 2$ e) $h(x) = -16x^2 + 256x$ f) $f(x) = -0.01x^2 + 900x$

6. (TI-82) For each of the following equations, perform the following tasks:

i) Load the equation into the Y= menu. Adjust the WINDOW parameters so that the vertex and both x-intercepts are visible in your viewing window. Set up a coordinate system on your paper and copy the image in your viewing window onto your coordinate system. Include the WINDOW parameters on your coordinate axes.

ii) Use 3:minimum or 4:maximum in the CALC menu to find the coordinates of the vertex. Record these coordinates on your coordinate system near the vertex of the parabola.

iii) Use the formula $x = -\dfrac{b}{2a}$ to help find this maximum or minimum function value and the x-value at which it occurs. If this result does not agree favorably with the result found in part (ii), check your work for error.

a) $f(x) = -x^2 + 6x + 11$ b) $g(x) = -x^2 - 10x + 17$ c) $h(x) = 2x^2 - 12x - 8$

d) $f(x) = -\dfrac{1}{4}x^2 + 2x + 2$ e) $g(x) = -16x^2 + 512x$ f) $h(x) = 0.01x^2 - 100x$

7. Jose Canseco hits a towering fly toward the left-field seats. The height of the ball above ground level is given by the equation $h = -16t^2 + 192t$, where h represents the height of the ball in feet and t represents the number of seconds that have passed since the ball left Jose's bat.

a) (TI-82) Load the equation into the Y= menu. Adjust the WINDOW parameters so that the vertex and both t-intercepts are visible in your viewing window. Set up a coordinate system on your paper and copy the image in the viewing window onto your coordinate system. Include the WINDOW parameters on your coordinate axes.

b) Use the max-min utility in the CALC menu to find the maximum height of the ball and the time at which it occurs.

c) Use the formula $t = -\dfrac{b}{2a}$ to help find the maximum height of the ball and the time that it occurs. If this solution does not agree favorably with the estimate found in part (b), check your work for error.

8. Find two numbers whose sum is 12 and whose product is a maximum.

a) Begin by guessing. Let x and y represent the two numbers. Let P represent their product. Make a copy of the following table on your paper and complete the missing entries.

x	y	P
0	12	0
1	11	11
2	10	20
3		
4		
5		
6		
7		
8		
9		
10		
11		
12		

Based on these tabular results, what is the maximum product P? What numbers x and y produce this maximum product?

b) Express the sum S of the numbers as a function of x and y. The sum of the two numbers is 12, so replace the variable S with the number 12.

c) Express the product P of the numbers as a function of x and y. Use the equation developed in part (b) to express the product P as a function of x only.

d) The graph of the equation developed in part (c) is a parabola that opens downward. Use the formula $x = -\dfrac{b}{2a}$ to find the coordinates of the vertex of this parabola.

e) What is the maximum product? What two numbers will produce this maximum product?

9. Mary wants to fence all four sides of a rectangular garden and she has 100 feet of fencing available for the job. Find the dimensions of the garden of maximum area. Begin by letting x and y represent the width and length of the rectangular garden.

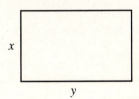

a) Try a few guesses first. Make a copy of the following table on your paper and complete the missing entries.

x	y	Perimeter	Area
5	45	100	225
10			
15			
20			
25			
30			

Based on these tabular results, what appears to be the dimensions of the rectangle that produces a maximum area?

b) Express the perimeter P of the rectangle as a function of x and y. Because the perimeter is 100, replace P with 100.

c) Express the area A of the rectangle as a function of x and y. Use your result from part (b) to express the area A as a function of x only.

d) If you have progressed correctly to this point, the graph of the last equation developed in part (c) is a parabola that opens downward. Use the formula $x = -\dfrac{b}{2a}$ to find the coordinates of the vertex.

e) What are the dimensions of the rectangle that produce a rectangle of maximum area?

10. Willie owns a small curio shop on the boardwalk near Venice Beach. He sells and rents all sorts of items to beachgoers. Willie finds that the demand for a particular brand of sunglasses, as a function of their unit price, is given by the equation $x = 100 - 2p$, where x is the number of sunglasses bought by the public and p is the unit price. Let R represent Willie's revenue from the sale of sunglasses.

a) Express R as a function of x and p.

b) Use the demand function $x = 100 - 2p$ to help express Willie's revenue as a function of p only.

c) If you have proceeded correctly to this point, the equation developed in part (b) is a parabola that opens downward. Use the formula $t = -\dfrac{b}{2a}$ to help find the coordinates of the vertex of the parabola.

d) What is the maximum value of R and for what p-value does it occur?

e) How many sunglasses are bought by the public at this unit price?

11. The sum of two numbers is 10. Let the variable x represent one of these numbers. Let the variable P represent the product of these numbers.

a) Express P as a function of x.

b) What is the maximum possible product and what numbers produce this maximum product?

12. The sum of two numbers is 100. Let the variable x represent one of these numbers. Let the variable S represent the sum of the squares of these two numbers.

a) Express S as a function of x.

b) Find the minimum value of S and the numbers that produce this minimum value.

13. Jaime stands on the roof of his apartment building, 30 feet above ground level. He fires his toy rocket into the air with an initial velocity of 256 feet per second.

a) Use the equation $y = y_0 + v_0 t - \frac{1}{2} g t^2$ to express the height y of the rocket as a function of t, the number of seconds that have passed since its launch.

b) Use the formula $t = -\dfrac{b}{2a}$ to find the coordinates of the vertex of the parabola represented by the equation in part (a).

c) What maximum height will Jaime's rocket reach and at what time will this height be attained?

6.9 Summary and Review

Important Visualization Tools

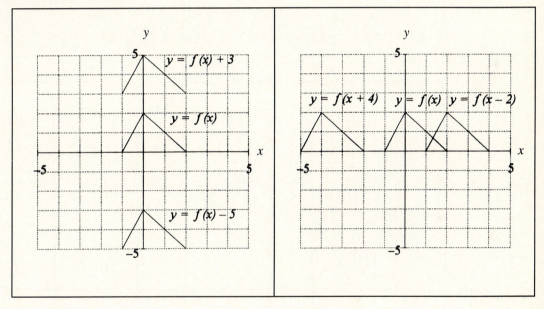

The graphs of $y = -f(x)$ and $y = f(x)$ are symmetric with respect to the x-axis.

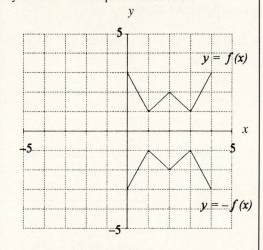

The graphs of $y = f(-x)$ and $y = f(x)$ are symmetric with respect to the y-axis.

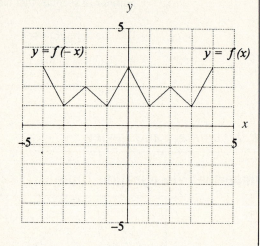

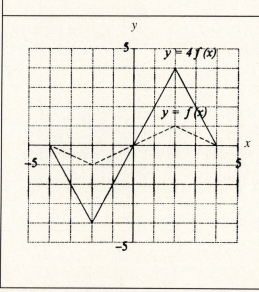

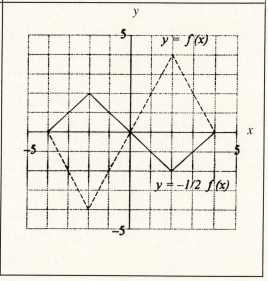

A Review of Key Words and Phrases

☛ To draw the graph of $y = f(x) + c$, shift the graph of $y = f(x)$ upward c units if c is positive or downward c units if c is negative. (page 459)

☛ To draw the graph of $y = f(x - c)$, shift the graph of $y = f(x)$ to the right c units if c is positive or to the left c units if c is negative. (page 463)

☛ To draw the graph of $y = -f(x)$, reflect the graph of $y = f(x)$ across the x-axis. (page 477)

☛ To draw the graph of $y = f(-x)$, reflect the graph of $y = f(x)$ across the y-axis. (page 481)

☛ Tip: Remember to do the reflections first. (page 484)

☛ A relation is a set of ordered pairs. If the ordered pairs of the relation are denoted with the symbolism (x, y), then the domain of the relation is the set of all x-values of the ordered pairs and the range of the relation is the set of all y-values. (page 495)

☛ A relation is a function if and only if each domain object is paired with (or sent to) exactly one range object. (page 495)

☛ The vertical line test: If any vertical line cuts the graph of f in more than one place, f is not a function. (page 497)

☛ To find the domain of a function from its graph, project all the points of the graph onto the x-axis. (page 499)

☛ To find the range of a function from its graph, project all the points of the graph onto the y-axis. (page 500)

☛ To find the domain of a function from its equation, find the set of all real numbers for which the function values are real numbers. (page 503)

☛ The graph of $y = ax^2 + bx + c$ is a parabola. (page 511)

☛ The graph of $y = a(x-h)^2 + k$ is a parabola. The coordinates of the vertex of the parabola are (h,k). The form $y = a(x-h)^2 + k$ is called vertex form. (page 512)

☛ The parabola whose equation is $y = ax^2 + bx + c$ is symmetric with respect to a vertical line drawn through its vertex. This line is called the axis of symmetry. (page 513)

☛ The graph of $y = ax^2$ is a parabola whose vertex is located at $(0,0)$. If a is positive, the parabola opens upward. If a is negative, the parabola opens downward. (page 520)

☛ Use a process called completing the square to place the equation of the parabola in vertex form. (page 520)

$$y = 2x^2 - 4x - 6$$

$$y = 2(x^2 - 2x) - 6$$

$$y = 2(x^2 - 2x + 1 - 1) - 6$$

$$y = 2(x^2 - 2x + 1) - 2 - 6$$

$$y = 2(x-1)^2 - 8$$

☛ The following steps are required to draw the graph of $y = ax^2 + bx + c$. (page 524)

1. Complete the square to place the equation in vertex form $y = a(x-h)^2 + k$.
2. Find the x- and y-intercepts.
3. Plot the vertex and the x- and y-intercepts, and draw the axis of symmetry.
4. Draw the parabola.
5. Label the vertex, x- and y-intercepts, and other points with their coordinates.
6. Label the axis of symmetry with its equation.

☛ The x-coordinate of the vertex of the parabola $f(x) = ax^2 + bx + c$ is given by the formula $x = -\dfrac{b}{2a}$. The y-coordinate of the vertex is given by the formula $f\left(-\dfrac{b}{2a}\right)$. (page 531, 573)

☛ The x-intercepts of the parabola whose equation is $y = ax^2 + bx + c$ are given by the formula

$$x = \frac{-b \pm \sqrt{b^2 - 4ac}}{2a}$$

(page 534)

☛ The expression under the radical, $b^2 - 4ac$, is called the discriminant. (page 540)

1. If $b^2 - 4ac > 0$, the parabola will have two x-intercepts.
2. If $b^2 - 4ac = 0$, the parabola will have one x-intercept.
3. If $b^2 - 4ac < 0$, the parabola will have no x-intercepts.

☛ If an object is hurled into the air, its height above ground is given by the equation $y = y_0 + v_0 t - \frac{1}{2}gt^2$, where y represents the height, y_0 represents the initial height, v_0 represents the initial velocity, g represents the acceleration due to gravity, and t represents the amount of time since the object's launch. (page 548)

☛ The Pythagorean theorem: In a right triangle, the square of the hypotenuse equals the sum of the squares of the legs. (page 550)

☛ Procedure for solving quadratic inequalities. (page 564)

1. Find and plot the x-intercepts.
2. Use the x-intercepts and the fact that the parabola opens upward or downward to draw a rough graph of the parabola.
3. Draw a number line below your graph and shade the solution of the inequality in the usual manner.
4. Use set-builder and interval notation to describe the solution set.

☛ The graph of the function defined by the equation $f(x) = ax^2 + bx + c$ is a parabola. If the parabola opens upward, then the function will have a minimum function value at $x = -\frac{b}{2a}$. If the parabola opens downward, then the function will have a maximum function value at $x = -\frac{b}{2a}$. The maximum or minimum value of the function is $f\left(-\frac{b}{2a}\right)$. (page 573)

TI-82 Keywords and Menus

☛ Use the max-min utility in the CALC menu to help find the coordinates of the vertex of a parabola. (page 574)

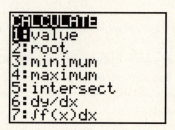

Chapter Review Exercises

1. Use graphing shortcuts to sketch the graph of each of the following equations. Check your result on your calculator. *Note*: Use your calculator to check the problem, not to do the problem.

   a) $y = -\sqrt{x-3}$ b) $y = -x^3 + 2$ c) $y = -|x+2| + 4$ d) $y = \sqrt{-(x+1)}$

2. Consider the following graph of the function f.

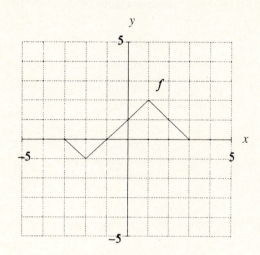

Sketch the graphs of each of the following equations. Place each graph on its own coordinate system on a sheet of graph paper.

a) $y = f(x-1) + 2$ b) $y = f(x+3) - 4$ c) $y = -f(x)$ d) $y = f(-x)$

e) $y = -f(x+2) + 5$ f) $y = -f(-x)$ g) $y = 2f(x)$ h) $y = -\dfrac{1}{2} f(x)$

3. For each of the following graphs, set up a coordinate system on a sheet of graph paper and make a copy of the graph on your coordinate system. Label the graph with its equation . Check your result with your calculator.

a)
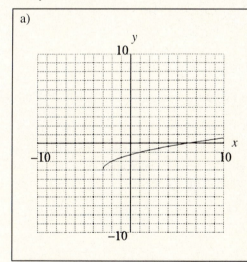

b)
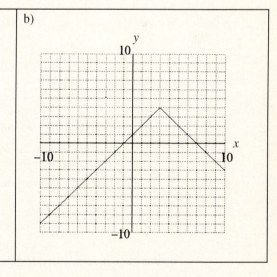

4. Use the graphing techniques of translation and reflection to help sketch the graphs of the following equations. Use both set-builder and interval notation to describe the domain and range of the function described by the equation.

a) $f(x) = -(x+3)^2 - 4$ b) $f(x) = -\sqrt{x-5}$

5. Use set-builder notation and interval notation to describe the domain of each of the functions defined by the following equations.

   a) $f(x) = \sqrt{2-x}$

   b) $f(x) = \sqrt{2x+11}$

6. For each of the following equations, perform the following tasks:

   i) Use the method of completing the square to place the equation in vertex form. Use the *ac* test to help find the *x*-intercepts.

   ii) Set up a coordinate system on a sheet of graph paper. Clearly indicate the scale on each coordinate axis. Plot the vertex and the *x*- and *y*-intercepts, and draw the axis of symmetry. Use this information to draw the graph of the parabola.

   a) $y = x^2 - 10x - 24$

   b) $y = -x^2 - 6x + 40$

7. Find the equation of the following parabola.

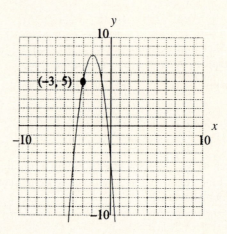

8. For each of the following equations, perform the following tasks:

   i) Use the formula $x = -\dfrac{b}{2a}$ to find the coordinates of the vertex. Use the quadratic formula to find the *x*-intercepts in simple radical form.

   ii) Set up a coordinate system on a sheet of graph paper. Clearly indicate the scale on each coordinate axis. Plot the vertex, the *x*- and *y*-intercepts, and draw the axis of symmetry. Use this information to draw the graph of the parabola.

   a) $y = -x^2 + 8x + 12$

   b) $y = \dfrac{1}{2}x^2 - 3x - 5$

9. Find all values of k so that the equation $x^2 - kx + 4 = 0$ has exactly one solution.

10. Find all values of k so that the equation $(k+1)x^2 - 2x - 4 = 0$ has two real solutions.

11. Use the quadratic formula to find the real solutions of each of the following equations in simple radical form.

   a) $x(x+4) - 3(x+2) = x$

   b) $x^2 - (x-4) = 2x(x-1)$

12. Lucille launches a rocket from atop her roof, which is 20 feet above ground level. The initial velocity of the rocket in the vertical direction is 256 feet per second. Let h represent the height of the rocket and let t represent the number of seconds that have passed since launch time.

   a) Express h as a function of t.

   b) What is the maximum height of the rocket? At what time will this maximum height occur?

   c) During what interval of time is the rocket more than 100 feet in the air? Use interval notation to describe your answer and round all numbers to the nearest tenth of a second.

   d) When will the rocket return to ground level? Round your answer to the nearest tenth of a second.

13. Use set-builder notation and interval notation to describe the solution set of each of the following inequalities.

   a) $x^2 - 5x > 24$

   b) $6 - 4x \le x^2$

14. Margaret's garden is shaped like a rectangle. Its length is 5 feet longer than its width. A gas line runs underground diagonally from one corner of the garden to the opposite corner and is 20 feet in length. Find the dimensions of the garden, correct to the nearest tenth of a foot.

15. Two concentric squares are shown below:

   The length of the edge of the outer square is 3 feet longer than twice the length of the edge of the inner square. The area of the shaded region is 100 square feet. Find the length of the edge of the inner square, correct to the nearest tenth of a foot.

16. Use set-builder notation and interval notation to describe the domain of the functions defined by the following equations:

   a) $f(x) = \sqrt{50 - x^2}$

   b) $f(x) = \sqrt{50 + 5x - x^2}$

17. The sum of two numbers is 20. Let x represent one of these numbers. Let P represent their product.

   a) Express P as a function of x.

   b) What is the maximum value of P? What numbers x and y will produce this maximum value of P?

Chapter 7
The Graph of a Polynomial

In this chapter we will study the graphs of polynomial functions. The definition of a polynomial was first introduced in Chapter 4.

> *The expression $a_nx^n + a_{n-1}x^{n-1} + \cdots + a_2x^2 + a_1x + a_0$—where each of the a_k's is a real number and n is a nonnegative integer—is called a polynomial in x. The a_k's are called the coefficients of the polynomial. The degree of the polynomial is n, the highest power of x. The term a_nx^n is called the leading term of the polynomial.*

We will begin by looking at the behavior of the graph for very large or very small values of x and at the relationship between the factors and roots of a polynomial. We will learn how to draw the graphs of polynomials and use the graph to help solve polynomial inequalities.

7.1 End-Behavior

In this section we will discuss the *end-behavior* of the graphs of polynomials.

> *The behavior of the graph for values of x that have large absolute value is called the end-behavior of the graph.*

How does the graph behave for large values of x such as 10, 100, 1000, ...? How does the graph behave for values of x such as -10, -100, -1000,...? This behavior is referred to as the end-behavior of the graph.

Recalling the Graph of $y = x^n$. The graph of $y = x^n$ was first introduced in Chapter 5. If n is even, the graph of $y = x^n$ falls from positive infinity and completes its journey by rising to positive infinity. If n is odd, the graph of $y = x^n$ rises from negative infinity and completes its journey by rising to positive infinity. Understanding the end-behavior of the graph of $y = x^n$ is essential to understanding the end-behavior of the graph of a polynomial.

The graph of $y = x^n$, n even

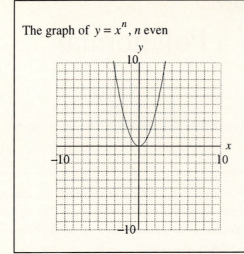

The graph of $y = x^n$, n odd

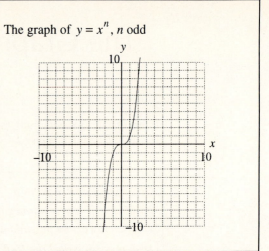

Example 1   Discuss the end-behavior of the graph of $y = -2x^5$.

Solution

The graph of $y = x^5$

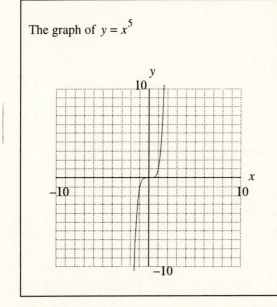

To draw the graph of $y = 2x^5$, double the y-values of each point on the graph of $y = x^5$.

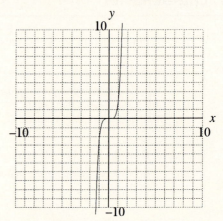

To draw the graph of $y = -2x^5$, reflect the graph of $y = 2x^5$ across the x-axis.

From left to right, the graph falls from positive infinity, wiggles a bit, then falls to negative infinity.

Example 2   *(TI-82)* Compare the end-behavior of the graph of $y = x^3$ with that of the graph of $y = x^3 - 2x^2 - 11x + 12$.

Solution   Enter the equations $y = x^3$ and $y = x^3 - 2x^2 - 11x + 12$ into the Y= menu.

```
Y₁⬛X^3
Y₂⬛X^3-2X^2-11X+
12
Y₃=
Y₄=
Y₅=
Y₆=
Y₇=
```

Make the following adjustments to the WINDOW parameters:	Push the GRAPH button to capture the following image:
| ```
WINDOW FORMAT
Xmin=-10
Xmax=10
Xscl=1
Ymin=-50
Ymax=50
Yscl=10
``` | 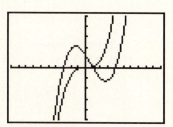 |

For values of x such as −4, −3, −2, −1, 0, 1, 2, 3, and 4, there is quite a difference between the graph of $y = x^3 - 2x^2 - 11x + 12$ and the graph of $y = x^3$. However, as the x-values move toward the left or right sides of the viewing window, the graphs begin to exhibit similar behavior.
```

Make the following adjustments to the WINDOW parameters:	Push the GRAPH button to capture the following image:
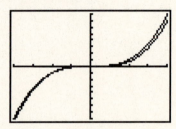	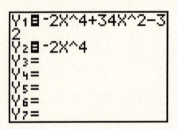

For values of x such as 10, 15, and 20, or for values of x such as -10, -15, and -20, there is not a great deal of difference between the graphs of $y = x^3$ and $y = x^3 - 2x^2 - 11x + 12$. At the left and right ends of the viewing window, the graph of the polynomial $y = x^3 - 2x^2 - 11x + 12$ behaves as if it were the graph of $y = x^3$. We like to say that the end-behavior of the graph of $y = x^3 - 2x^2 - 11x + 12$ is the same as the end-behavior of the graph of $y = x^3$.

Example 3   *(TI-82)* Compare the end-behavior of the graph of $y = -2x^4 + 34x^2 - 32$ with that of the graph of $y = -2x^4$.

Solution   Enter the equations $y = -2x^4 + 34x^2 - 32$ and $y = -2x^4$ into the Y = menu.

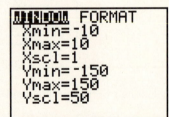

Make the following adjustments to the WINDOW parameters:	Push the GRAPH button to capture the following image:
	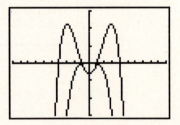

For values of x such as -4, -3, -2, -1, 0, 1, 2, 3, and 4, there is quite a difference between the graph of $y = -2x^4 + 34x^2 - 32$ and the graph of $y = -2x^4$. However, as the x-values move toward the left or right sides of the viewing window, the graphs begin to exhibit similar behavior.

Make the following adjustments to the WINDOW parameters:

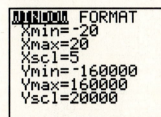

Push the GRAPH button to capture the following image:

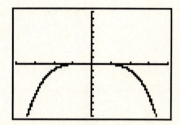

For values of x such as 10, 15, and 20, or for values of x such as -10, -15, and -20, there is not a great deal of difference between the graphs of $y = -2x^4 + 34x^2 - 32$ and $y = -2x^4$. At the left and right ends of the viewing window, the graph of the polynomial $y = -2x^4 + 34x^2 - 32$ behaves as if it were the graph of $y = -2x^4$. You can say that the end-behavior of the graph of $y = -2x^4 + 34x^2 - 32$ is the same as the end-behavior of the graph of $y = -2x^4$.

> *In general, the end-behavior of the polynomial $p(x) = a_n x^n + \cdots + a_2 x^2 + a_1 x + a_0$ is identical to the end-behavior of its leading term. The end-behavior of $p(x) = a_n x^n + \cdots + a_2 x^2 + a_1 x + a_0$ is identical to the end-behavior of $y = a_n x^n$.*

Example 4   (TI-82) Predict the end-behavior of the polynomial $p(x) = (5-x)(x+2)(x+6)$. Check your solution with your calculator.

Solution   Begin by multiplying the factors of $p(x)$ together.

$$p(x) = (5-x)(x+2)(x+6)$$
$$= (5-x)(x^2 + 6x + 2x + 12)$$
$$= (5-x)(x^2 + 8x + 12)$$
$$= 5x^2 + 40x + 60 - x^3 - 8x^2 - 12x$$
$$= -x^3 - 3x^2 + 28x + 60$$

The leading term of this polynomial is $-x^3$. Therefore the end-behavior of the polynomial $y = -x^3 - 3x^2 + 28x + 60$ will be the same as the end-behavior of $y = -x^3$. To draw the graph of $y = -x^3$, reflect the graph of $y = x^3$ across the x-axis.

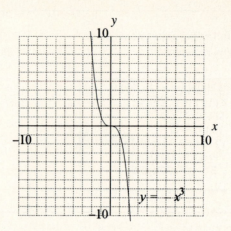

The graph of $y = -x^3 - 3x^2 + 28x + 60$ will have the same end-behavior as the graph of $y = -x^3$. The graph of $y = -x^3 - 3x^2 + 28x + 60$ will fall from positive infinity, wiggle a bit, then fall to negative infinity.

Check with the TI-82. Load the equation $y = -x^3 - 3x^2 + 28x + 60$ in the Y= menu in the following manner: Y1 = -X^3 - 3X^2 + 28X + 60.

Make the following adjustments to the WINDOW parameters:	Push the GRAPH button to capture the following image:
WINDOW FORMAT Xmin=-10 Xmax=10 Xscl=1 Ymin=-150 Ymax=150 Yscl=50	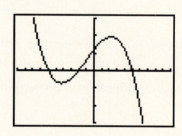

As predicted, the graph of $y = -x^3 - 3x^2 + 28x + 60$ falls from positive infinity, wiggles a bit, then falls to negative infinity.

Exercises for Section 7.1

1. (TI-82) Make the following adjustments to the WINDOW parameters:

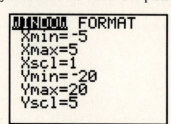

WINDOW FORMAT
Xmin=-5
Xmax=5
Xscl=1
Ymin=-20
Ymax=20
Yscl=5

Load each of the following sets of equations into the Y= menu. Push the GRAPH button. Set up a coordinate system and copy the image in the viewing window onto your coordinate system. Label each graph with its equation and include the WINDOW parameter settings on the coordinate axes.

a)

b)

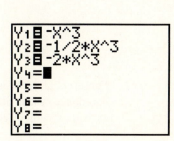

c)

d)

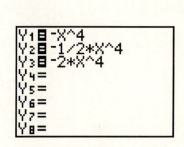

e)

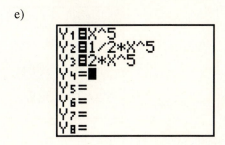

f)

2. Without the use of a calculator, sketch a graph of the following equations, each on its own coordinate system.

a) $y = -4x^5$ b) $y = 3x^6$ c) $y = -\frac{1}{3}x^7$ d) $y = \frac{3}{2}x^8$

Note: Use your calculator to check your solution, but do not use it until you have completed your sketch.

3. Sketch the graph of the following equations, each on a separate coordinate system.

a) $y = ax^n$, $a > 0$, n even   b) $y = ax^n$, $a < 0$, n even

c) $y = ax^n$, $a > 0$, n odd   d) $y = ax^n$, $a < 0$, n odd

4. (TI-82) Load the equations $y = x^3 + x^2 - 4x - 4$ and $y = x^3$ into the Y= menu.

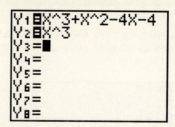

For each of the following WINDOW settings, push the GRAPH button. Set up a separate coordinate system and copy the image in the viewing window onto your coordinate system. Label each graph with its equation and include the WINDOW parameters on the coordinate axes.

a)

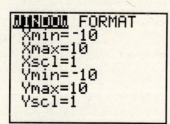

b)

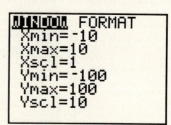

c)

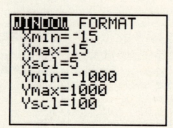

d)

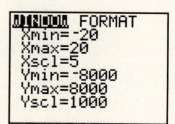

Use complete sentences to explain the purpose of this problem.

5. (TI-82) Load the equations $y = -2x^4 + 35x^2 - 8x - 72$ and $y = -2x^4$ into the Y= menu.

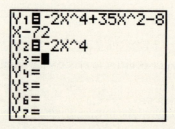

For each of the following WINDOW settings, push the GRAPH button. Set up a separate coordinate system and copy the image in the viewing window onto your coordinate system. Label each graph with its equation and include the WINDOW parameters on your coordinate axes.

a)

b)

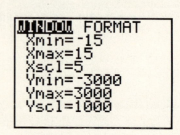

c)

```
WINDOW FORMAT
 Xmin=-20
 Xmax=20
 Xscl=5
 Ymin=-10000
 Ymax=10000
 Yscl=1000
```

d)

```
WINDOW FORMAT
 Xmin=-30
 Xmax=30
 Xscl=10
 Ymin=-800000
 Ymax=800000
 Yscl=100000
```

Use complete sentences to explain the purpose of this problem.

6. (TI-82) For each of the following equations, perform the following tasks:

   i) Load the equation into the Y= menu. Press the ZOOM button and select 6:ZStandard.

   ii) Set up a separate coordinate system and copy the image in the viewing window onto your coordinate system. Label the graph with its equation and include the WINDOW parameter settings on your coordinate axes.

   iii) Use complete sentences to explain what the graph has in common with the graph of the equation $y = x^3$.

   a) $y = x^3 - x - 7$ b) $y = x^3 - 4x^2 + 2x + 2$ c) $y = x^3 + 2x^2 - 5x - 3$

7. (TI-82) For each of the following equations, perform the following tasks:

   i) Load the equation into the Y= menu. Press the ZOOM button and select 6:ZStandard.

   ii) Set up a separate coordinate system and copy the image in the viewing window onto your coordinate system. Label the graph with its equation and include the WINDOW parameter settings on your coordinate axes.

   iii) Use complete sentences to explain what the graph has in common with the graph of the equation $y = x^4$.

   a) $y = x^4 - 5x^2 + 4$ b) $y = x^4 - x - 5$ c) $y = x^4 - 3x^3 + x - 2$

8. (TI-82) For each of the following equations, perform the following tasks:

   i) Load the equation into the Y= menu. Press the ZOOM button and select 6:ZStandard.

   ii) Set up a separate coordinate system and copy the image in the viewing window onto your coordinate system. Label the graph with its equation and include the WINDOW parameter settings on your coordinate axes.

   iii) Use complete sentences to explain what the graph has in common with the graph of the equation $y = -2x^3$.

   a) $y = -2x^3 - x - 1$ b) $y = -2x^3 + 4x^2 - 7$ c) $y = -2x^3 + 8x^2 - 4x - 6$

9. For each of the following polynomials, perform each of the following tasks:

   i) Set up a coordinate system on a sheet of graph paper. Without the use of a calculator, sketch a graph of the polynomial that exhibits the proper end-behavior.

   ii) (TI-82) Load the equation into the Y= menu. Experiment with the WINDOW parameters until you have a viewing window that shows a complete graph, one that shows all of the important features of the graph. Set up a second coordinate system on your graph paper and copy the image in your viewing window onto your coordinate system. Label the graph with its equation and include the WINDOW parameters on your coordinate axes.

a) $y = x^3 - 8x^2 + 11x + 12$

b) $y = -x^4 + 6x^3 - 7x^2 + 6x + 11$

c) $y = x^5 - 2x^4 - 13x^3 + 14x^2 + 24x$

d) $y = -2x^3 + 2x^2 + 16x - 24$

10. Set up a separate coordinate system for each graph and make a copy of the graph on your coordinate system. Assume that the equation of each of the following graphs is a polynomial. Clearly state whether the degree of the polynomial is even or odd.

a)

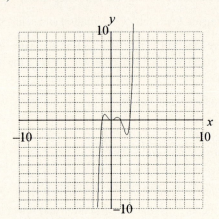

b)

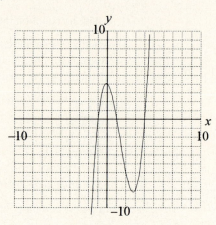

c)

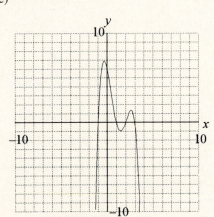

d)

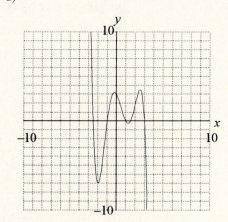

e)

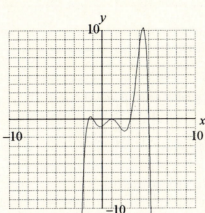

f)

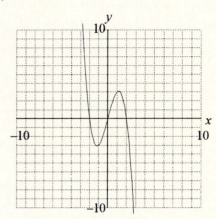

11. Eloise cuts four congruent squares, of undetermined dimension, out of the four corners of a 10-inch square piece of tin. The squares are discarded and the sides of the remaining material are folded upward to make an open box, as shown.

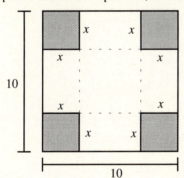
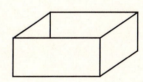

a) Let V represent the volume of the box. Express V as a function of x.

b) Place the result found in part (a) in the form $V = a_3x^3 + a_2x^2 + a_1x + a_0$.

c) (TI-82) Load the function in part (b) into the Y= menu. Experiment with the WINDOW parameter settings until you have a complete graph showing all of the important behavior of the graph in your viewing window. Set up a coordinate system on your paper and copy the image in the viewing window onto your coordinate system. Include the WINDOW parameter settings on the coordinate axes.

12. (TI-82) The number of new AIDS cases reported in California in each of the years from 1981 through 1991 can be approximated by the following third-degree polynomial.

$$N = -23.02t^3 + 316.6t^2 - 317.6t + 122.9$$

The variable N represents the number of AIDS cases reported in a particular year and t represents the number of years that have passed since 1981.

a) (TI-82) Load the function into Y1 in the Y= menu. Experiment with the WINDOW parameter settings until you have a complete graph showing all of the important behavior of the graph in your viewing window. Set up a coordinate system on your paper and copy the image in the viewing window onto your coordinate system. Include the WINDOW parameter settings on the coordinate axes.

7.2 Roots and Factors

This section will explain an important relationship between the *zeros* of a polynomial and the *factors* of a polynomial. Let's begin with the definition of a zero of a polynomial.

---

r is a zero of a polynomial if and only if p(r) = 0.

If $p(r) = 0$, then the graph of p will cross the x-axis at $(r, 0)$.

Example 1   Find the zeros of the polynomial function whose equation is $p(x) = x^3 - 2x^2 - 3x$. Use the zeros of the polynomial and your knowledge of the end-behavior of the polynomial to help draw a rough graph of the polynomial.

Solution   To find the zeros of the polynomial, you must solve the equation $p(x) = 0$.

$$p(x) = 0$$

$$x^3 - 2x^2 - 3x = 0$$

First factor out the greatest common factor, then use the *ac* test.

$$x(x^2 - 2x - 3) = 0$$

$$x(x^2 + x - 3x - 3) = 0$$

$$x(x(x+1) - 3(x+1)) = 0$$

$$x(x-3)(x+1) = 0$$

$$x = 0, 3, -1$$

The zeros are 0, 3, and −1. The graph of the polynomial must cross the x-axis at $(-1, 0)$, $(0, 0)$, and $(3, 0)$. The end-behavior of the polynomial $p(x) = x^3 - 2x^2 - 3x$ is the same as the end-behavior of its leading term, which is x^3. From right to left, the graph must rise from negative infinity, cross the x-axis at $(-1, 0)$, $(0, 0)$, and $(3, 0)$, then rise to positive infinity. A rough sketch follows.

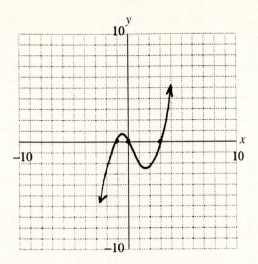

Check. Check your solution with your calculator. Load the equation $y = x^3 - 2x^2 - 3x$ into the Y= menu in the following manner: Y₁ = X³ − 2X² − 3X. Push the ZOOM button and select 6:ZStandard to capture the following image in your viewing window:

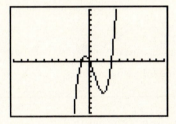

Although the rough graph was not as precise as this calculator-generated graph, it did capture much of the essential behavior of the polynomial.

Example 2   Find the zeros of the polynomial function whose equation is $p(x) = x^3 + 5x^2 - 4x - 20$. Use the zeros of the polynomial and your knowledge of the end-behavior of the polynomial to help draw a rough graph of the polynomial.

Solution   To find the zeros of the polynomial, you must solve the equation $p(x) = 0$.

$$p(x) = 0$$

$$x^3 + 5x^2 - 4x - 20 = 0$$

When trying to factor an expression with four terms, first try the method of grouping.

$$x^2(x+5) - 4(x+5) = 0$$

$$(x^2 - 4)(x+5) = 0$$

Use the difference of squares multiplication pattern to factor further.

$$(x+2)(x-2)(x+5) = 0$$

$$x = -2, 2, -5$$

The zeros are -5, -2, and 2. The graph of the polynomial must cross the x-axis at $(-5,0)$, $(-2,0)$, and $(2,0)$. The end-behavior of the polynomial $p(x) = x^3 + 5x^2 - 4x - 20$ is the same as the end-behavior of its leading term, which is x^3. As we move from right to left, the graph must rise from negative infinity, cross the x-axis at $(-5,0)$, $(-2,0)$, and $(2,0)$, then rise to positive infinity. A rough sketch follows.

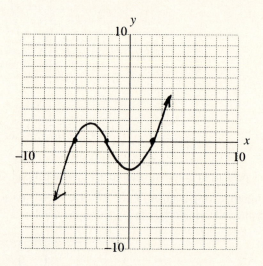

Check. Check your solution with your calculator. Load the equation $p(x) = x^3 + 5x^2 - 4x - 20$ into the Y= menu in the following manner: $Y_1 = X^3 + 5X^2 - 4X - 20$.

Make the following adjustments to the WINDOW parameters:	Push the GRAPH button to capture the following image in your viewing window:
WINDOW FORMAT Xmin=-10 Xmax=10 Xscl=1 Ymin=-30 Ymax=30 Yscl=5	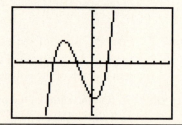

The scale on the y-axis in the rough graph was not correct and the turning points of the graph were not correctly plotted. However, the rough graph did capture most of the essential characteristics.

Example 3   Find the zeros of the polynomial function whose equation is $p(x) = -x^4 + 5x^2 - 4$. Use the zeros of the polynomial and your knowledge of the end-behavior of the polynomial to help draw a rough graph of the polynomial.

Solution   To find the zeros of the polynomial, you must solve the equation $p(x) = 0$.

$$p(x) = 0$$

$$-x^4 + 5x^2 - 4 = 0$$

Multiply both sides of this equation by -1 and use the ac test to factor the trinomial.

$$x^4 - 5x^2 + 4 = 0$$

$$x^4 - x^2 - 4x^2 + 4 = 0$$

$$x^2(x^2 - 1) - 4(x^2 - 1) = 0$$

$$(x^2 - 4)(x^2 - 1) = 0$$

Use the difference-of-squares multiplication pattern to complete the factoring.

$$(x + 2)(x - 2)(x + 1)(x - 1) = 0$$

$$x = -2, 2, -1, 1$$

The zeros are -2, -1, 1, and 2. The graph of the polynomial must cross the x-axis at $(-2, 0)$, $(-1, 0)$, $(1, 0)$, and $(2, 0)$. The end-behavior of the polynomial $p(x) = -x^4 + 5x^2 - 4$ is the same as the end-behavior of its leading term, which is $-x^4$. As we move from right to left, the graph must rise from negative infinity, cross the x-axis at $(-2, 0)$, $(-1, 0)$, $(1, 0)$, and $(2, 0)$, then return to negative infinity. A rough sketch follows.

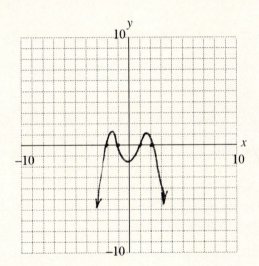

Check. Check your solution with your calculator. Load the equation $p(x) = -x^4 + 5x^2 - 4$ into the Y= menu in the following manner: $Y_1 = -X^4 + 5X^2 - 4$. Push the ZOOM button and select 6:ZStandard to capture the following image in your viewing window:

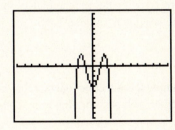

Though the turning points of the rough graph are not located as precisely as they should be, the sketch did capture most of the essential behavior of the graph.

The Factor Theorem

The examples of this section seem to indicate the following.

The factor theorem: If $x - r$ is a factor of the polynomial p, then r is a zero of p.

Example 4   Find the zeros of $p(x) = (x+3)(x-2)(x-5)$.

Solution   Because $x+3$ is a factor of $p(x)$, -3 is a zero of the polynomial. Because $x-2$ is a factor of $p(x)$, 2 is a zero of the polynomial. Because $x-5$ is a factor of $p(x)$, 5 is a zero of the polynomial.

Example 5   Find the zeros of $p(x) = (x-3)(2x+1)(5x-2)$.

Solution   You could set each factor equal to zero and find the zeros, but try to change the form of the factors so that you can use the factor theorem to spot the zeros of the polynomial. Begin by factoring a 2 from the second factor and a 5 from the third factor.

$$p(x) = (x-3)2\left(x+\frac{1}{2}\right)5\left(x-\frac{2}{5}\right)$$

The product of 2 and 5 is 10, which is moved to the front.

$$p(x) = 10(x-3)\left(x+\frac{1}{2}\right)\left(x-\frac{2}{5}\right)$$

Because $x-3$ is a factor of $p(x)$, 3 is a zero of the polynomial. Because $x+\frac{1}{2}$ is a factor of $p(x)$, $-\frac{1}{2}$ is a zero of the polynomial. Because $x-\frac{2}{5}$ is a factor of $p(x)$, $\frac{2}{5}$ is a factor of the polynomial.

What If You Know the Zeros? If you know the factors, you know the zeros. Conversely, if you know the zeros, you should know the factors.

The factor theorem: If r is a zero of the polynomial p, then $x - r$ is a factor of p.

Example 6   Find a third-degree polynomial with the following roots: -1, 2, and 5.

Solution   If -1, 2, and 5 are zeros of the polynomial p, then $x+1$, $x-2$, and $x-5$ must be factors of the polynomial p. Here is a good candidate for p.

$$p(x) = (x+1)(x-2)(x-5)$$

If you were to multiply these factors together, the highest power of x would be x^3. This results in the requested third-degree polynomial. However, consider the following polynomials, which are also third-degree.

$$p(x) = 2(x+1)(x-2)(x-5)$$

$$p(x) = \frac{1}{2}(x+1)(x-2)(x-5)$$

Because the factors of each of these third-degree polynomials are $x+1$, $x-2$, and $x-5$, the zeros of each of these polynomials are -1, 2, and 5.

Check. Load the equations $p(x)=(x+1)(x-2)(x-5)$, $p(x)=2(x+1)(x-2)(x-5)$, and $p(x)=\frac{1}{2}(x+1)(x-2)(x-5)$ into the Y= menu.

Load the equations in the Y= menu, as follows:	Make the following adjustments to the WINDOW parameters:
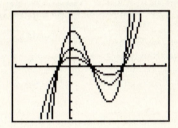	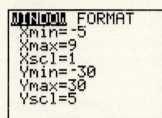

Push the GRAPH button to capture the following image in your viewing window:

Although each of the graphs is different, each one exhibits the end-behavior of a third-degree polynomial and each crosses the x-axis at -1, 2, and 5.

Example 7   Find a third-degree polynomial with zeros -3, -1, and 3 so that $p(5)=5$.

Solution   Because the zeros of this polynomial are -3, -1, and 3, you know that $x+3$, $x+1$, and $x-3$ are its factors.

$$p(x)=(\text{something})(x+3)(x+1)(x-3)$$

The (something) is there because there may be more zeros, and consequently more factors, that have not been stated. However, the degree of $p(x)$ is already at least 3, so the degree of the (something) must be zero and the (something) must be a constant. Consequently, the third-degree polynomial must have the following form:

$$p(x)=a(x+3)(x+1)(x-3)$$

By changing the value of the constant a, you can find many different third-degree polynomials with the required zeros. However, because the problem requests that $p(5)=5$, the polynomial must pass through the point (5,5). Try to draw a third-degree polynomial with zeros -3, -1, and 3 that passes through the point (5,5).

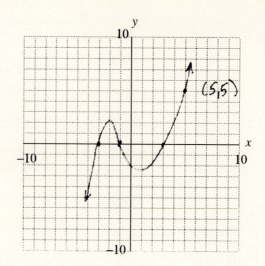

The graph has the same end-behavior as the graph of $y = ax^3$, $a > 0$. Use the fact that $p(5) = 5$ to determine the value of a.

$$p(x) = a(x+3)(x+1)(x-3)$$

$$5 = p(5)$$

$$5 = a(5+3)(5+1)(5-3)$$

$$5 = 96a$$

$$a = \frac{5}{96}$$

Note that $a > 0$, as expected. The required third-degree polynomial is $p(x) = \dfrac{5}{96}(x+3)(x+1)(x-3)$.

Check. Load the polynomial in the Y= menu like this: $Y_1 = 5/96*(X+3)(X+1)(X-3)$.

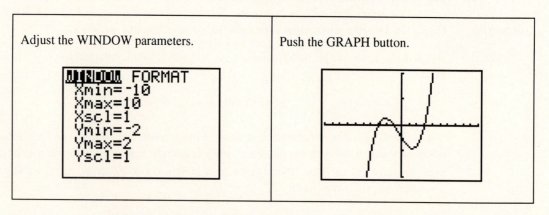

The polynomial apparently passes through the x-axis at -3, -1, and 3, but does it pass through the point $(5,5)$? Push the ZOOM button and select 8:ZInteger, then press the ENTER key. Push the TRACE button and use the arrow keys to move the trace cursor until you capture the following image in your viewing window.

This is not a good window for viewing the *x*-intercepts, but it does show that the polynomial passes through the point (5,5), as required.

Exercises for Section 7.2

1. Multiply the factors of each of the following equations together to place the equation in the form $p(x) = a_n x^n + \cdots + a_2 x^2 + a_1 x + a_0$. Clearly state the leading term of the polynomial and the degree of the polynomial.

   a) $p(x) = (x+1)(x-1)(x-2)$

   b) $p(x) = -(x+2)(x-1)(x-5)$

   c) $p(x) = \frac{1}{2}x(x+3)(x+2)(x-6)$

   d) $p(x) = -2(x+3)(x+1)(x-3)(x-5)$

2. Clearly state the zeros of each of the following polynomials.

   a) $p(x) = (x+3)(x-4)(x-6)$

   b) $p(x) = 2(x+4)(x+1)(x-6)$

   c) $p(x) = -3(x+5)(x+3)(x-1)(x-2)$

   d) $p(x) = (x+3)(2x+1)(3x-4)(x-5)$

3. For each of the following polynomials, perform the following tasks:

   i) Set up a coordinate system on a sheet of graph paper. Clearly indicate the scale on each coordinate axis. Plot the zeros of the function. Use the zeros and your knowledge of the polynomial's end-behavior to draw a rough graph of the polynomial.

   ii) (TI-82) Load the polynomial's equation into the Y= menu. Adjust the WINDOW parameters until you have a complete graph in the viewing window, one that shows all of the zeros of the polynomial and all of the turning points of the graph. Set up a second coordinate system on your graph paper and copy the image in the viewing window onto your coordinate system as accurately as possible. Include the WINDOW parameters on each coordinate axis.

   a) $p(x) = (x+2)(x-3)(x-5)$

   b) $p(x) = -(x+2)(x+1)(x-3)(x-6)$

   c) $p(x) = 2x(x+1)(x-2)(x-4)$

   d) $p(x) = -\frac{1}{2}x(x+4)(x+2)(x-3)(x-6)$

4. For each of the following polynomials, perform the following tasks:

   i) Express the polynomial as a product of linear (first degree) factors. Clearly state the zeros of the polynomial.

   ii) Set up a coordinate system on a sheet of graph paper. Clearly indicate the scale on each coordinate axis. Plot the zeros of the function. Use the zeros and your knowledge of the polynomial's end-behavior to draw a rough graph of the polynomial.

   iii) (TI-82) Load the polynomial's equation into the Y= menu. Adjust the WINDOW parameters until you have a complete graph in the viewing window, one that shows all of the zeros of the polynomial and all of the turning points of the graph. Set up a second coordinate system on your graph paper and copy the image in the viewing window onto your coordinate system as accurately as possible. Include the WINDOW parameters on each coordinate axis.

a) $p(x) = x^3 - 4x^2 - 5x$ b) $p(x) = 2x^3 - 6x^2 - 36x$

c) $p(x) = -2x^3 + 7x^2 + 15x$ d) $p(x) = -4x^3 - 14x^2 + 30x$

Hint: First factor out the greatest common factor, then use the *ac* test, as in Example 1 of this section.

5. For each of the following polynomials, perform the following tasks:

    i)   Express the polynomial as a product of linear (first-degree) factors. Clearly state the zeros of the polynomial.

    ii)   Set up a coordinate system on a sheet of graph paper. Clearly indicate the scale on each coordinate axis. Plot the zeros of the function. Use the zeros and your knowledge of the polynomial's end-behavior to draw a rough graph of the polynomial.

    iii)   (TI-82) Load the polynomial's equation into the Y= menu. Adjust the WINDOW parameters until you have a complete graph in the viewing window, one that shows all of the zeros of the polynomial and all of the turning points of the graph. Set up a second coordinate system on your graph paper and copy the image in the viewing window onto your coordinate system as accurately as possible. Include the WINDOW parameters on each coordinate axis.

a) $p(x) = x^3 + 5x^2 - 9x - 45$ b) $p(x) = 4x^3 + 16x^2 - x - 4$

c) $p(x) = -x^3 + 6x^2 + x - 6$ d) $p(x) = -x^3 - 5x^2 + 4x + 20$

Hint: Try the method of grouping, then use the difference of squares multiplication pattern, as in Example 2 of this section.

6. For each of the following polynomials, perform the following tasks:

    i)   Express the polynomial as a product of linear (first-degree) factors. Clearly state the zeros of the polynomial.

    ii)   Set up a coordinate system on a sheet of graph paper. Clearly indicate the scale on each coordinate axis. Plot the zeros of the function. Use the zeros and your knowledge of the polynomial's end-behavior to draw a rough graph of the polynomial.

    iii)   (TI-82) Load the polynomial's equation into the Y= menu. Adjust the WINDOW parameters until you have a complete graph in the viewing window, one that shows all of the zeros of the polynomial and all of the turning points of the graph. Set up a second coordinate system on your graph paper and copy the image in the viewing window onto your coordinate system as accurately as possible. Include the WINDOW parameters on each coordinate axis.

a) $p(x) = x^4 - 17x^2 + 16$ b) $p(x) = -x^4 + 13x^2 - 36$

c) $p(x) = 4x^4 - 17x^2 + 4$ d) $p(x) = -4x^4 + 29x^2 - 25$

Hint: Use the *ac* test to factor the trinomial, then use the difference of squares multiplication pattern, as in Example 3 of this section.

7. (TI-82) For each of the following sets of equations, perform the following tasks:

   i) Load the equations into the Y= menu. Adjust the WINDOW parameters so that the zeros of each equation and the turning points of each graph are visible in your viewing window. Set up a coordinate system and copy the image in the viewing window onto your coordinate system. Label each graph with its equation and include the WINDOW parameters on your coordinate axes.

   ii) Use complete sentences to describe everything that each graph in your viewing window has in common.

   a)

   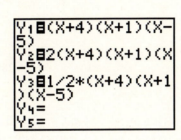

   b)

   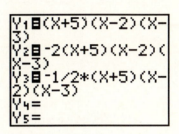

8. Find at least three third-degree polynomials that have the following zeros.

   a) Zeros: $-5, -1$, and 5 b) Zeros: $-4, 2$, and 5

9. Find at least three fourth-degree polynomials that have the following zeros.

   a) Zeros: $-5, -1, 0$, and 6 b) Zeros: $-7, -2, 3$, and 8

10. For each of the following graphs, perform the following tasks:

    i)   Set up a coordinate system on a sheet of graph paper. Clearly indicate the scale on each coordinate axis, then make a copy of the graph on your coordinate system.

    ii) Find the equation of the graph.

    iii) (TI-82) Use your calculator to check your solution, as in Example 7 of this section.

    a)

    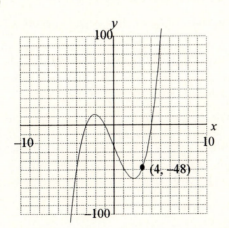

    b)

    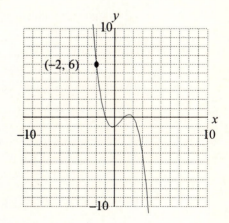

c)

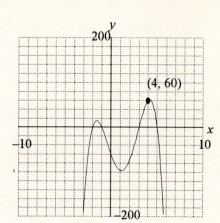

d)

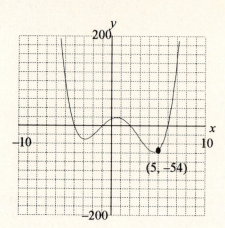

11. Set up a coordinate system on graph paper with the coordinate axes clearly scaled. Draw a rough graph of a third-degree polynomial whose zeros are −4, 1, and 3 so that the graph goes through the point (2, 3). Find the equation of this graph.

12. Set up a coordinate system on graph paper with the coordinate axes clearly scaled. Draw a rough graph of a third-degree polynomial whose zeros are −5, −1, and 3 so that the graph goes through the point (2, −3). Find the equation of this graph.

13. Set up a coordinate system on graph paper with the coordinate axes clearly scaled. Draw a rough graph of a fourth-degree polynomial whose zeros are −3, −1, 3, and 7 so that the graph goes through the point (4, 3). Find the equation of this graph.

14. Set up a coordinate system on graph paper with the coordinate axes clearly scaled. Draw a rough graph of a fifth-degree polynomial whose zeros are −2, −1, 0, 1, and 4 so that the graph goes through the point (3, 2). Find the equation of this graph.

15. For each of the following polynomials, perform the following tasks:

   i) Express the polynomial as a product of linear (first-degree) factors. Clearly state the zeros of the polynomial.

   ii) Set up a coordinate system on a sheet of graph paper. Clearly indicate the scale on each coordinate axis. Plot the zeros of the function. Use the zeros and your knowledge of the polynomial's end-behavior to draw a rough graph of the polynomial.

   iii) (TI-82) Load the polynomial's equation into the Y= menu. Adjust the WINDOW parameters until you have a complete graph in your viewing window, one that shows all of the zeros of the polynomial and all of the turning points of the graph. Set up a second coordinate system on your graph paper and copy the image in the viewing window onto your coordinate system as accurately as possible. Include the WINDOW parameters on each coordinate axis.

   a) $p(x) = 2x^3 - 21x^2 + 40x$

   b) $p(x) = -4x^3 + 24x^2 + 9x - 54$

   c) $p(x) = x^5 - 13x^3 + 36x$

   d) $p(x) = -4x^3 - 4x^2 + 15x$

   e) $p(x) = -x^4 + 101x^2 - 100$

   f) $p(x) = -8x^3 + 12x^2 + 50x - 75$

7.3 Division of Polynomials

Let's pause for a moment and focus on some division skills that will be needed when finding the zeros of polynomials. We'll examine a technique called *long division*, and use words such as *divisor*, *dividend*, *quotient*, and *remainder*. Let's begin with a review of these words:

$$\text{Divisor} \enspace \overline{)\,\dfrac{\text{Quotient}}{\text{Dividend}}}$$

$$\overline{}$$

$$\text{Remainder}$$

The Division Algorithm

When a polynomial $f(x)$ is divided by a nonzero polynomial $d(x)$ of less or equal degree, the quotient $q(x)$ is a polynomial, and the remainder $r(x)$ is a zero polynomial or a polynomial with degree less than the degree of the divisor.

$$\boxed{f(x) = d(x)q(x) + r(x), \textit{ where } r(x)=0 \textit{ or degree } r(x) < \textit{degree } d(x)}$$

The following arrangement is equivalent to $f(x)=d(x)q(x)+r(x)$, but it may be more meaningful to some readers.

$$d(x)\overline{)\,\dfrac{q(x)}{f(x)}}$$

$$\overline{}$$

$$r(x)$$

Example 1   Use the method of long division to divide $x^2 - 4x - 5$ by $x - 2$.

Solution   It is important to arrange both the divisor and the dividend in descending powers of x.

$$x-2\overline{)\,x^2 - 4x - 5}$$

Divide the first term of $x^2 - 4x - 5$ by the first term of $x - 2$. Because $\dfrac{x^2}{x} = x$, place an x in the quotient.

$$x-2\overline{)\,\dfrac{x}{x^2 - 4x - 5}}$$

Multiply $x - 2$ by x. Place the product $x^2 - 2x$ below $x^2 - 4x$.

$$\begin{array}{r} x \phantom{{}^2 - 4x - 5} \\ x-2{\overline{\smash{\big)}\,x^2-4x-5}} \\ \underline{x^2-2x\phantom{{}-5}} \end{array}$$

Subtract x^2-2x from x^2-4x.

$$\begin{array}{r} x \phantom{{}^2 - 4x - 5} \\ x-2{\overline{\smash{\big)}\,x^2-4x-5}} \\ \underline{x^2-2x\phantom{{}-5}} \\ -2x\phantom{{}-5} \end{array}$$

Note: $-4x-(-2x)=-2x$. Divide $-2x$ by the first term of $x-2$. Because $\dfrac{-2x}{x}=-2$, place a -2 in the quotient.

$$\begin{array}{r} x-2 \phantom{{}^2 - 4x - 5} \\ x-2{\overline{\smash{\big)}\,x^2-4x-5}} \\ \underline{x^2-2x\phantom{{}-5}} \\ -2x\phantom{{}-5} \end{array}$$

Multiply $x-2$ by -2. Place the product $-2x+4$ below $-2x$.

$$\begin{array}{r} x-2 \phantom{{}^2 - 4x - 5} \\ x-2{\overline{\smash{\big)}\,x^2-4x-5}} \\ \underline{x^2-2x\phantom{{}-5}} \\ -2x\phantom{{}-5} \\ \underline{-2x+4} \end{array}$$

Bring down the -5 and subtract $-2x+4$ from $-2x-5$.

$$\begin{array}{r} x-2 \phantom{{}^2 - 4x - 5} \\ x-2{\overline{\smash{\big)}\,x^2-4x-5}} \\ \underline{x^2-2x\phantom{{}-5}} \\ -2x-5 \\ \underline{-2x+4} \\ -9 \end{array}$$

Note: $-5-4=-9$, so the remainder is -9. The degree of the remainder is 0 and the degree of the divisor is 1. Because the degree of the remainder is strictly less than the degree of the divisor, the division is finished.

Check. If you take the product of the divisor and the quotient, then add the remainder, you should get the dividend.

$$(x-2)(x-2)-9 = x^2 - 2x - 2x + 4 - 9$$

$$= x^2 - 4x - 5$$

Because $x^2 - 4x - 5$ is the dividend, the solution checks.

Example 2 Use the method of long division to divide $x^3 + 3x - 8$ by $1 - x$.

Solution Arrange the divisor and dividend in descending powers of x. Insert $0x^2$ as a place holder. This will keep the columns aligned properly.

$$-x+1\overline{)x^3 + 0x^2 + 3x - 8}$$

Divide the first term of $x^3 + 0x^2 + 3x - 8$ by $-x$. Because $\dfrac{x^3}{-x} = -x^2$, place $-x^2$ in the quotient.

$$\begin{array}{r} -x^2 \\ -x+1\overline{)x^3 + 0x^2 + 3x - 8} \end{array}$$

Multiply $-x+1$ by $-x^2$. Place the product $x^3 - x^2$ below $x^3 + 0x^2$ and subtract.

$$\begin{array}{r} -x^2 \\ -x+1\overline{)x^3 + 0x^2 + 3x - 8} \\ \underline{x^3 - x^2 } \\ x^2 \end{array}$$

Note: $0x^2 - (-x^2) = x^2$. The degree of the remainder is 2 and the degree of the divisor is 1. Because the degree of the remainder is not smaller than the degree of the divisor, you must continue. Divide x^2 by the first term of $-x+1$. Because $\dfrac{x^2}{-x} = -x$, place $-x$ in the quotient. Multiply $-x+1$ by $-x$. Place the product $x^2 - x$ below $x^2 + 3x$ and subtract.

$$
\begin{array}{r}
-x^2 - x \\
-x+1 \overline{)\, x^3 + 0x^2 + 3x - 8} \\
\underline{x^3 - x^2} \\
x^2 + 3x \\
\underline{x^2 - x} \\
4x
\end{array}
$$

Note: $3x - (-x) = 4x$. The degree of the remainder is 1 and the degree of the divisor is 1. Since the degree of the remainder is not less than the degree of the divisor, you must continue. Divide $4x$ by the first term of $-x+1$. Because $\dfrac{4x}{-x} = -4$, place -4 in the quotient. Multiply -4 times $-x+1$. Place the product $4x - 4$ below $4x - 8$ and subtract.

$$
\begin{array}{r}
-x^2 - x - 4 \\
-x+1 \overline{)\, x^3 + 0x^2 + 3x - 8} \\
\underline{x^3 - x^2} \\
x^2 + 3x \\
\underline{x^2 - x} \\
4x - 8 \\
\underline{4x - 4} \\
-4
\end{array}
$$

Note: $-8 - (-4) = -4$. The degree of the remainder is zero, which is less than the degree of the divisor, so the division is finished.

Check. To check a division problem, you must multiply the divisor and quotient together, then add the remainder.

$$
(-x+1)(-x^2 - x - 4) - 4 = x^3 + x^2 + 4x - x^2 - x - 4 - 4
$$
$$
= x^3 + 3x - 8
$$

Because $x^3 + 3x - 8$ is the dividend, the solution checks.

Example 3   Use long division to divide $x^4 - x^3 + 2x^2 - 8$ by $3 - 2x + x^2$.

Solution   Arrange the divisor and dividend in descending powers of x. Insert $0x$ as a place holder.

$$\begin{array}{r} x^2 + x + 1 \\ x^2 - 2x + 3 \overline{\smash{\big)}\, x^4 - x^3 + 2x^2 + 0x - 8} \\[4pt] \underline{x^4 - 2x^3 + 3x^2} \\[4pt] x^3 - x^2 + 0x \\[4pt] \underline{x^3 - 2x^2 + 3x} \\[4pt] x^2 - 3x - 8 \\[4pt] \underline{x^2 - 2x + 3} \\[4pt] - x - 11 \end{array}$$

The degree of the remainder is 1 and the degree of the divisor is 2. Because the degree of the remainder is less than the degree of the divisor, the division is finished.

Check. Multiply the quotient by the divisor and add the remainder.

$$(x^2 - 2x + 3)(x^2 + x + 1) + (-x - 11) = x^4 + x^3 + x^2 - 2x^3 - 2x^2 - 2x + 3x^2 + 3x + 3 - x - 11$$

$$= x^4 - x^3 + 2x^2 - 8$$

Because $x^4 - x^3 + 2x^2 - 8$ is the dividend, the solution checks.

Example 4   Use long division to divide $x^3 + 27$ by $3 + x$.

Solution   Arrange the divisor and dividend in descending powers of x. Insert $0x^2$ and $0x$ as place holders.

$$\begin{array}{r} x^2 - 3x + 9 \\ x + 3 \overline{\smash{\big)}\, x^3 + 0x^2 + 0x + 27} \\[4pt] \underline{x^3 + 3x^2} \\[4pt] -3x^2 + 0x \\[4pt] \underline{-3x^2 - 9x} \\[4pt] 9x + 27 \\[4pt] \underline{9x + 27} \\[4pt] 0 \end{array}$$

Check. Multiply the divisor by the quotient and add the remainder.

$$(x + 3)(x^2 - 3x + 9) + 0 = x^3 - 3x^2 + 9x + 3x^2 - 9x + 27$$

$$= x^3 + 27$$

Because $x^3 + 27$ is the dividend, the solution checks.

Note: Because the remainder in the last example was 0, $x^3 + 27$ is *divisible* by $x+3$. Therefore $x+3$ is a factor of $x^3 + 27$. In fact, $x^3 + 27$ factors in the following manner:

$$x^3 + 27 = (x+3)(x^2 - 3x + 9)$$

Applying the Division Algorithm

Now that you know how to divide one polynomial by another, you can apply this newfound skill to help find the zeros of polynomials. At first you will need a few hints, but soon you will be able to uncover your own hints.

Example 5 Given that -5 is a zero of $p(x) = 2x^3 + 3x^2 - 39x - 20$, find the remaining zeros of $p(x)$. Use the zeros and your knowledge of the polynomial's end-behavior to sketch a rough graph of the polynomial.

Solution Because -5 is a zero of $p(x)$, you immediately know that $x+5$ is a factor of $p(x)$. If $p(x)$ is divided by $x+5$, the remainder should be zero.

$$
\begin{array}{r}
2x^2 - 7x - 4 \\
x+5 \overline{\smash{)}2x^3 + 3x^2 - 39x - 20} \\
\underline{2x^3 + 10x^2} \\
-7x^2 - 39x \\
\underline{-7x^2 - 35x} \\
-4x - 20 \\
\underline{-4x - 20} \\
0
\end{array}
$$

You can use the results of this division to express $p(x)$ as a product of a linear (first-degree) factor and a quadratic (second-degree) factor .

$$p(x) = (x+5)(2x^2 - 7x - 4)$$

You can use the *ac* test to express $2x^2 - 7x - 4$ as a product of two linear (first-degree) factors .

$$p(x) = (x+5)(2x^2 - 7x - 4)$$
$$= (x+5)(2x^2 - 8x + x - 4)$$
$$= (x+5)(2x(x-4) + 1(x-4))$$
$$= (x+5)(2x+1)(x-4)$$

At this point you could set each factor equal to zero to find the zeros, but you can also factor a 2 out of the second factor.

$$p(x) = 2(x+5)\left(x + \frac{1}{2}\right)(x-4)$$

Because the factors of $p(x)$ are $x+5$, $x+\dfrac{1}{2}$, and $x-4$, the zeros of $p(x)$ are -5, $-\dfrac{1}{2}$, and 4. Draw a rough sketch of the polynomial.

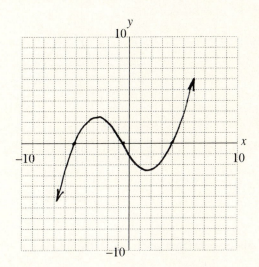

Check. Use a TI-82 calculator to check. Load the equation $p(x) = 2x^3 + 3x^2 - 39x - 20$ into the Y= menu in the following manner: $Y_1 = 2X^3 + 3X^2 - 39X - 20$. Experiment until you find a viewing window that will show the zeros and the turning points of the polynomial.

Make the following adjustments to the WINDOW parameters:	Push the GRAPH button to capture the following image in your viewing window:
WINDOW FORMAT Xmin=-10 Xmax=10 Xscl=1 Ymin=-100 Ymax=100 Yscl=10	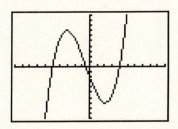

The scale on the y-axis of the rough graph was all wrong. The turning points of the polynomial were not accurately located. However, the rough graph did capture most of the essential character of the polynomial.

Example 6   Use the graph from Example 5 to help solve the following inequality:

$$2x^3 + 3x^2 - 39x - 20 > 0$$

Solution   You are being asked to locate where the graph of $p(x) = 2x^3 + 3x^2 - 39x - 20$ is above the x-axis. Use your rough graph to find the solution.

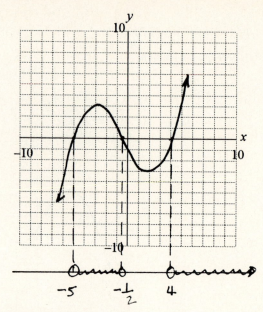

Use a computer-generated graph to find the solution.

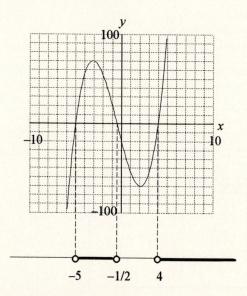

It is important to note that the rough graph produced the same solution set as the computer-generated graph. When solving an inequality involving a polynomial, it's the x-intercepts—not the turning points of the graph—that are important. In set-builder notation, the solution set is $\left\{x\!:\!-5 < x < -\dfrac{1}{2} \text{ or } x > 4\right\}$. In interval notation, the solution set is $\left(-5, -\dfrac{1}{2}\right) \cup (4, +\infty)$.

Exercises for Section 7.3

1. Check each of the following long-division problems.

   a)
   $$x+1{\overline{\smash{\big)}\,x^2-4x-6}}$$
   quotient $x-5$
   $$\underline{x^2+\ x}$$
   $$-5x-6$$
   $$\underline{-5x-5}$$
   $$-1$$

   b)
   $$x-2{\overline{\smash{\big)}\,x^2+8x-10}}$$
   quotient $x+10$
   $$\underline{x^2-2x}$$
   $$10x-10$$
   $$\underline{10x-20}$$
   $$10$$

   Check: $(x+1)(x-5)-1=$

   Check: $(x-2)(x+10)+10=$

2. In each of the following problems, the long-division process has been almost completed. Find the remainder, then check your solution.

   a)
   $$x-2{\overline{\smash{\big)}\,x^2+4x-8}}$$
   quotient $x+6$
   $$\underline{x^2-2x}$$
   $$6x-8$$
   $$6x-12$$

   b)
   $$x+3{\overline{\smash{\big)}\,x^2-\ x-7}}$$
   quotient $x-4$
   $$\underline{x^2+3x}$$
   $$-4x-7$$
   $$-4x-12$$

3. Perform each of the following long-division problems, then check your result when you are done.

   a) $x-3{\overline{\smash{\big)}\,x^2-5x-7}}$

   b) $x+2{\overline{\smash{\big)}\,x^2+5x-9}}$

   c) $x-4{\overline{\smash{\big)}\,2x^3-7x^2+5x-9}}$

   d) $x+3{\overline{\smash{\big)}\,3x^3-8x^2-7x+10}}$

   e) $x+1{\overline{\smash{\big)}\,x^3+0x^2-8x-11}}$

   f) $x-2{\overline{\smash{\big)}\,x^4+0x^3-4x^2+0x-9}}$

4. Perform each of the following divisions and check your result.

   a) Divide $5-4x-x^2$ by $x+3$.

   b) Divide x^2-4x-8 by $1-x$.

   c) Divide $11-5x-x^3$ by $x+1$.

   d) Divide $2x^3+5x^2-7x-4$ by $2-x$.

   e) Divide x^4-3x^2+9 by $x+4$.

   f) Divide $7-5x^3-x^4$ by $x+2$.

5. Perform each of the following long-division problems, then check your result when you are done.

   a) $x^2-x-1{\overline{\smash{\big)}\,x^3-4x^2-8x+15}}$

   b) $-x^2+3x-4{\overline{\smash{\big)}\,x^3-5x^2+0x-13}}$

   c) $-x^2+2x-3{\overline{\smash{\big)}\,2x^4+5x^3-3x^2+0x-10}}$

   d) $x^2-4x-1{\overline{\smash{\big)}\,x^4-3x^3+0x^2-5x-11}}$

e) $x^2 - 4 \overline{)x^3 - 8x^2 + 8x - 9}$ f) $x^2 + 1 \overline{)x^3 + 0x^2 - 8x - 9}$

6. Perform each of the following divisions and check your result.

   a) Divide $x^3 - 3x^2 + 11$ by $2 - x - x^2$.   b) Divide $x^3 - 9x - 19$ by $x^2 - 2x - 2$.

   c) Divide $x^4 - 3x^2 - 8$ by $x^2 - 1$ d) Divide $8 - x - 7x^2 - x^4$ by $x^2 + 2$

7. Perform each of the following tasks.

   a) Divide $x^3 - 1$ by $x - 1$.

   b) Divide $x^4 - 1$ by $x - 1$.

   c) Predict what the quotient will be when you divide $x^5 - 1$ by $x - 1$. Use long division to verify your prediction.

   d) Predict what the quotient will be when you divide $x^n - 1$ by $x - 1$.

8. Perform each of the following tasks.

   a) Divide $x^3 - a^3$ by $x - a$.

   b) Divide $x^4 - a^4$ by $x - a$.

   c) Predict what the quotient will be when you divide $x^5 - a^5$ by $x - a$. Use long division to verify your prediction.

   d) Predict what the quotient will be when you divide $x^n - a^n$ by $x - a$.

9. For the following polynomials, you are given a zero r of the polynomial. Perform each of the tasks below.

   i)   Use long division to divide the polynomial by $x - r$. Express the result of your division to express $p(x)$ as a product of $x - r$ and a quadratic (second-degree) factor.

   ii) Use the ac test to express the quadratic factor as a product of two linear (first-degree) factors.

   iii) Clearly state the zeros of the polynomial.

   iv) Set up a coordinate system on a sheet of graph paper. Clearly indicate the scale on each axis. Use your knowledge of the zeros and end-behavior of the polynomial to draw a rough sketch of the polynomial on your coordinate system.

   v)   (TI-82) Load the equation of the polynomial into the Y= menu. Adjust the WINDOW parameters so that the zeros and the turning points of the parabola are visible in the viewing window. Set up a second coordinate system on your graph paper. Clearly indicate the scale on each axis. Make an accurate copy of the image in your viewing window on this second coordinate system.

   a) $p(x) = x^3 - 6x^2 - x + 30$, zero $= -2$.   b) $p(x) = x^3 - 5x^2 - 34x + 80$, zero $= 2$.

   c) $p(x) = x^3 - 9x^2 - x + 105$, zero $= -3$.   d) $p(x) = x^3 - 7x^2 - 10x + 16$, zero $= 1$.

   e) $p(x) = x^3 - 6x^2 - 19x + 24$, zero $= -3$.   f) $p(x) = x^3 - 10x^2 + 3x + 54$, zero $= -2$.

10. For the following polynomials, you are given a zero r of the polynomial. Perform each of the tasks below.

    i)   Find the remaining zeros.

    ii) Set up a coordinate system on a sheet of graph paper. Clearly indicate the scale on each axis. Use your knowledge of the zeros and the end-behavior of $p(x)$ to draw a rough sketch of the polynomial on your coordinate system.

    iii) Draw a number line below your graph. Shade and label the solution set of the inequality $p(x) > 0$ on your number line in the usual manner. Use set-builder and interval notation to describe your solution set.

    a) $p(x) = 4x^3 - 16x^2 + 5x + 7$, zero = 1.   b) $p(x) = -3x^3 + 8x^2 + 33x + 10$, zero = -2.

11. For the following polynomials, you are given a zero r of the polynomial. Perform each of the tasks below.

    i)   Find the remaining zeros.

    ii) Set up a coordinate system on a sheet of graph paper. Clearly indicate the scale on each axis. Use your knowledge of the zeros and the end-behavior of $p(x)$ to draw a rough sketch of the polynomial on your coordinate system.

    iii) Draw a number line below your graph. Shade and label the solution set of the inequality $p(x) \leq 0$ on your number line in the usual manner. Use set-builder and interval notation to describe your solution set.

    a) $p(x) = 5x^3 + 14x^2 - 93x + 18$, zero = 3.   b) $p(x) = 6x^3 - 35x^2 + 21x + 20$, zero = 5.

7.4 The Graph/Reduce Method

Up until now, you have either been able to factor the given polynomial using conventional methods or you've been given a hint, usually in the form of a zero to try. What happens if you do not know how to factor the polynomial and you are given no hints at all? You can draw the graph on your calculator and use the graph to estimate a zero. Let's begin with an example.

Example 1   Find the *exact* zeros of the polynomial $p(x) = x^3 - 3x^2 - 10x + 24$.

Solution   Load the polynomial into the Y= menu as follows: $Y_1 = X^3 - 3X^2 - 10X + 24$.

Make the following adjustments to the WINDOW parameters:	Push the GRAPH button to capture the following image in your viewing window:
WINDOW FORMAT Xmin=-10 Xmax=10 Xscl=1 Ymin=-50 Ymax=50 Yscl=10	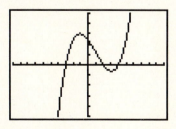

It appears that the graph of the polynomial crosses the x-axis at -3, 2, and 4. Therefore -3 is a serious *candidate* for a zero of the polynomial. However, -3 is only an *estimate* or *approximation;* the graph could just as easily cross the x-axis at -2.99 or -3.01. How can you be certain that -3 is a zero? You can begin by dividing $p(x)$ by $x+3$.

$$
\begin{array}{r}
x^2 - 6x + 8 \\
x+3 \overline{\smash{)}\ x^3 - 3x^2 - 10x + 24} \\
\underline{x^3 + 3x^2} \\
-6x^2 - 10x \\
\underline{-6x^2 - 18x} \\
8x + 24 \\
\underline{8x + 24} \\
0
\end{array}
$$

Because the remainder is 0, $x+3$ is a factor of $p(x)$.

$$p(x) = (x+3)(x^2 - 6x + 8)$$

It is now certain that -3 is a zero of the polynomial. The *ac* test can be used to factor $x^2 - 6x + 8$.

$$
\begin{aligned}
p(x) &= (x+3)(x^2 - 6x + 8) \\
&= (x+3)(x^2 - 4x - 2x + 8) \\
&= (x+3)(x(x-4) - 2(x-4)) \\
&= (x+3)(x-2)(x-4)
\end{aligned}
$$

Because $x+3$ is a factor of $p(x)$, -3 is a zero. Because $x-2$ is a factor of $p(x)$, 2 is a zero. Because $x-4$ is a factor of $p(x)$, 4 is a zero. These zeros are no longer estimates. They are *exact* zeros of the polynomial.

Example 2   Find the *exact* zeros of the polynomial $p(x) = -x^3 - 3x^2 + 12x + 10$.

Solution   Load the polynomial into the Y= menu as follows: $Y_1 = -X^3 - 3X^2 + 12X + 10$.

Make the following adjustments to the WINDOW parameters:

Push the GRAPH button to capture the following image in your viewing window:

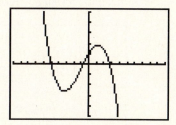

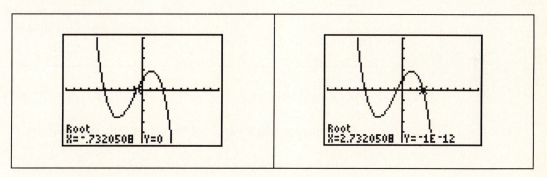

It appears that −5 will be a serious candidate for one of the zeros. The root-finding utility in the CALC menu can be used to find estimates of the remaining zeros. For a review of this utility, see Example 5 in Section 1.7.

Although −5 appears to be a zero, this is only an estimate. The graph could easily cross the x-axis at −5.01 or −4.99. Divide $p(x)$ by $x+5$.

$$
\begin{array}{r}
-x^2 + 2x + 2 \\
x+5 \overline{\smash{\big)}\,-x^3 - 3x^2 + 12x + 10} \\
\underline{-x^3 - 5x^2} \\
2x^2 + 12x \\
\underline{2x^2 + 10x} \\
2x + 10 \\
\underline{2x + 10} \\
0
\end{array}
$$

Because the remainder is 0, $x+5$ is a factor of $p(x)$ and it is now certain that −5 is an exact zero of the polynomial.

$$p(x) = (x+5)(-x^2 + 2x + 2)$$

Although it is not necessary, it's better to factor out a −1 from the quadratic factor before continuing.

$$p(x) = -(x+5)(x^2 - 2x - 2)$$

A quick check, using the *ac* test, reveals that the quadratic factor will not factor. However, the quadratic formula will find the zeros of this quadratic factor.

$$x = \frac{-b \pm \sqrt{b^2 - 4ac}}{2a}$$ $$x = \frac{-(-2) \pm \sqrt{(-2)^2 - 4(1)(-2)}}{2(1)}$$ $$x = \frac{2 \pm \sqrt{12}}{2}$$ $$x = \frac{2 \pm 2\sqrt{3}}{2}$$ $$x = 1 \pm \sqrt{3}$$	Use your calculator to approximate these zeros. 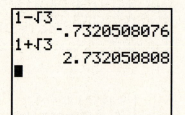 Note that these compare nicely with the estimates found using the root-finding utility in the CALC menu.

The *exact* zeros of the polynomial $p(x) = -x^3 - 3x^2 + 12x + 10$ are -5, $1 - \sqrt{3}$, and $1 + \sqrt{3}$.

Example 3 Use the results of Example 2 to find the *exact* solution of the following inequality :

$$-x^3 - 3x^2 + 12x + 10 \le 0$$

Solution Take a look at the following student solution. Note the different scales on the coordinate axes. Note also that this student attempted to create an accurate copy of the image in the viewing window of her calculator.

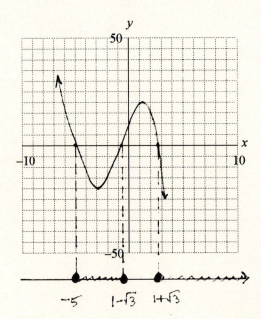

In set-builder notation, the solution set is $\left\{x : -5 \le x \le 1 - \sqrt{3} \text{ or } x \ge 1 + \sqrt{3}\right\}$. In interval notation, the solution set is $\left[-5, 1 - \sqrt{3}\right] \cup \left[1 + \sqrt{3}, +\infty\right)$.

Applications of Polynomials

Mathematicians such as Abel, Tartaglia, Cardan, Gauss, Galois, and countless others struggled for many years to find *exact* zeros of polynomials. Much of what is studied today in algebra is a result of the efforts of these and other great mathematicians. However, when a real-world application involves a polynomial, it is not usually possible to find an exact zero. In fact, it is more than likely that the coefficients of the polynomial are not exact in the first place, making any pursuit of an exact zero somewhat silly at best. Perhaps an example will help make this point.

Example 4   The number of Japanese cars sold in the United States from 1980 through 1992 can be approximated by the polynomial $S = -0.00147t^3 + 0.0109t^2 + 0.0388t + 1.814$. The variable S represents the number of Japanese cars sold (in millions of cars), and the variable t represents the number of years that have passed since 1980. In what years were there more than 1,950,000 Japanese cars sold in the United States?

Solution   Someone has probably taken the actual data on imported car sales in the United States and found a polynomial that approximately fits the data points. The graph of the polynomial $S = -0.00147t^3 + 0.0109t^2 + 0.0388t + 1.814$ will not pass through the actual data points, but it will pass reasonably close to most of them. Consequently, trying to find an exact solution of this problem would be stretching reality. Try to find an approximate solution. Because S is in millions of cars, you want to find when S is more than 1.95.

$$S > 1.95$$

Substitute $S = -0.00147t^3 + 0.0109t^2 + 0.0388t + 1.814$ into this inequality.

$$-0.00147t^3 + 0.0109t^2 + 0.0388t + 1.814 > 1.95$$

You want to find when the polynomial $S = -0.00147t^3 + 0.0109t^2 + 0.0388t + 1.814$ is above the horizontal line $y = 1.95$. Load the polynomial $S = -0.00147t^3 + 0.0109t^2 + 0.0388t + 1.814$ and the horizontal line $y = 1.95$ into the Y= menu.

```
Y1目-0.00147X3+0.
0109X2+0.0388X+1
.814
Y2目1.95
Y3=■
Y4=
Y5=
Y6=
```

Because t represents the number of years that have passed since 1980, and the equation presents a model of Japanese car sales from 1980 through 1992, you need only view the graph for t-values that run from 0 through 12. Make the following adjustments to the WINDOW parameters:

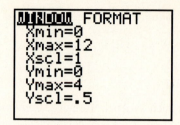

You can use the intersect utility in the CALC menu to find the points of intersection of the graphs represented by the equations loaded in Y1 and Y2. For a review of this utility, see Example 10 in Section 1.6.

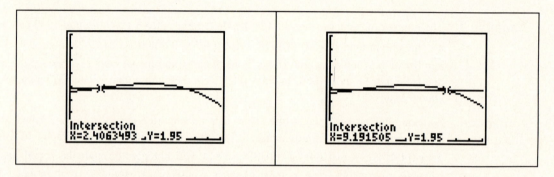

Therefore the graph of the polynomial in Y1 is above 1.95 whenever t is between approximately 2.4 years and 9.2 years. Finally, data are probably collected at the end of each year. The graph of the polynomial is above 1.95 for $t = 3, 4, 5, 6, 7, 8,$ and 9. Therefore the sales of Japanese cars were more than 1,950,000 during the years 1983 through 1989.

Note: According to the *World Almanac*, sales of Japanese cars were more than 1,950,000 between the years 1985 and 1988. Hopefully, this example demonstrates that trying to find an exact solution would be pointless.

> ***There is nothing wrong with finding approximate solutions.***

Example 5   Squares of equal dimension are cut from the four corners of a rectangular piece of tin whose width is 6 inches and whose length is 12 inches. The sides are folded to form an open box, as shown below.

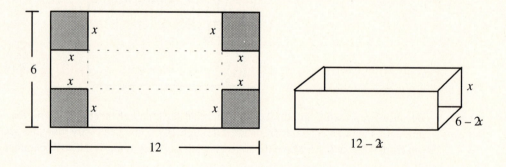

What size squares should be cut from the corners of the tin material to produce a box of *maximum* volume?

Solution   Let *V* represent the volume of the box. You can find the volume of a box by multiplying the length, width, and height of the box together. Therefore, $V = x(6-2x)(12-2x)$. The leading term of this polynomial is $4x^3$. If each of the factors of $V = x(6-2x)(12-2x)$ is set equal to zero, the zeros are 0, 3, and 6. It would not be difficult to draw the graph of this polynomial by hand, but your TI-82 will draw the graph. Load $V = x(6-2x)(12-2x)$ into the Y= menu like this: Y1 = X(6-2X)(12-2X).

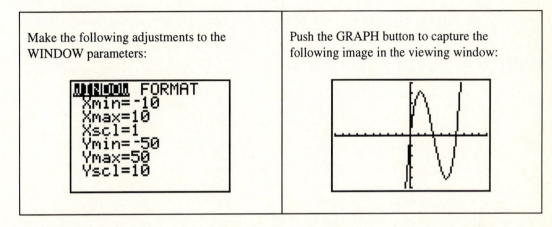

Make the following adjustments to the WINDOW parameters:

Push the GRAPH button to capture the following image in the viewing window:

Practical Considerations.   The width of the original piece of material is 6 inches and *x* represents the edge of the square that will be cut from each corner. Therefore *x* must be larger than 0 inches and smaller than 3 inches. Furthermore, the volume of the box must be greater than or equal to zero. With these thoughts in mind, adjust the WINDOW parameters in the following manner:

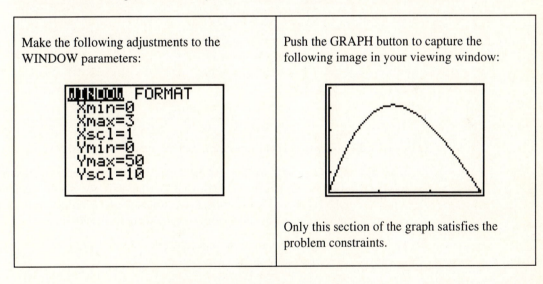

Make the following adjustments to the WINDOW parameters:

Push the GRAPH button to capture the following image in your viewing window:

Only this section of the graph satisfies the problem constraints.

You can use the maximum utility to find the maximum value of the polynomial in the viewing window. For a review of this utility, see Example 1 in Section 6.8.

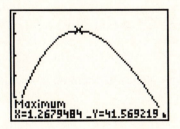

The maximum volume will be approximately 41.57 cubic inches, which will occur if you cut squares with an edge of approximate length 1.27 inches from each corner.

Exercises for Section 7.4

1. In each of the following problems, the graph of the polynomial has been drawn for you.

   i)   Examine the graph and determine a candidate for a zero r.

   ii) Use long division to verify that $x - r$ is a factor of $p(x)$. Express $p(x)$ as a product of a $x - r$ and a quadratic factor.

   iii) Use the ac test to factor the quadratic factor.

   iv) Clearly state the exact zeros of the polynomial.

   a) $p(x) = 6x^3 - 43x^2 + 66x + 27$ b) $p(x) = -6x^3 - x^2 + 82x - 40$

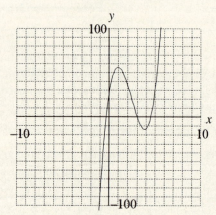
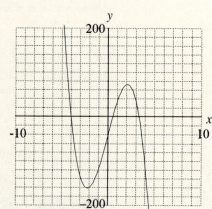

2. (TI-82) For the polynomials below, perform each of the following tasks:

   i)   Load the equation in the Y= menu. Adjust the WINDOW parameters so that the zeros and the turning points of the graph are visible in your viewing window. Set up a coordinate system on a sheet of graph paper and clearly indicate the scale on each coordinate axis. Make an accurate copy of the image in the viewing window on your coordinate system.

   ii) Examine the graph and determine a candidate zero r. Divide $p(x)$ by $x - r$. Express $p(x)$ as a product of a $x - r$ and quadratic factor.

   iii) Use the ac test to factor the quadratic factor. Clearly state the exact zeros of the polynomial.

   a) $p(x) = 2x^3 - 9x^2 - 11x + 30$ b) $p(x) = -3x^3 + 2x^2 + 63x - 110$

   c) $p(x) = 4x^3 - 12x^2 - 37x + 45$ d) $p(x) = -4x^3 + 4x^2 + 41x - 21$

3. In each of the following problems, the graph of the polynomial has been drawn for you.

   i)   Examine the graph and determine a candidate for a zero r.

ii) Use long division to verify that $x - r$ is a factor of $p(x)$. Express $p(x)$ as a product of a

$x - r$ and a quadratic factor.

iii) Use the quadratic formula to find the exact zeros of the quadratic factor in simple radical form.

a) $p(x) = x^3 - 5x^2 + 4$

b) $p(x) = -x^3 - 2x^2 + 14x - 12$

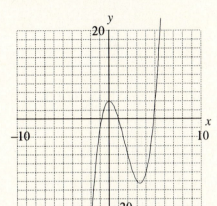

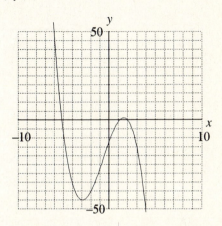

4. (TI-82) For each of the following polynomials, perform each of the following tasks:

i) Load the equation in the Y= menu. Adjust the WINDOW parameters so that the zeros and the turning points of the graph are visible in the viewing window. Set up a coordinate system on a sheet of graph paper and clearly indicate the scale on each coordinate axis. Make an accurate copy of the image in the viewing window on your coordinate system.

ii) Examine the graph and determine a candidate zero r. Divide $p(x)$ by $x - r$. Express $p(x)$ as a product of a $x - r$ and quadratic factor.

iii) Use the quadratic formula to find the zeros of the quadratic factor. Clearly state the exact zeros of the polynomial in simple radical form.

a) $p(x) = x^3 - 5x^2 + 2x + 12$

b) $p(x) = -x^3 - 3x^2 + 14x + 40$

c) $p(x) = 2x^3 - 11x^2 - 5x + 50$

d) $p(x) = -x^3 + 5x + 2$

5. In each of the following problems, the graph of the polynomial has been drawn for you.

i) Show all work necessary to find the *exact* roots of the polynomial.

ii) Set up a coordinate system on a sheet of graph paper and make a copy of the graph on your coordinate system. Draw a number line below the graph and shade and label the *exact* solution of the inequality $p(x) \geq 0$ on the number line.

iii) Use set-builder and interval notation to describe the *exact* solution set.

a) $p(x) = 4x^3 - 8x^2 - 51x + 27$

b) $p(x) = -5x^3 - 16x^2 + 67x + 78$

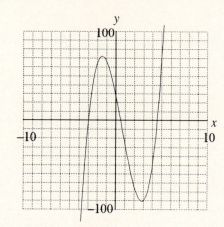

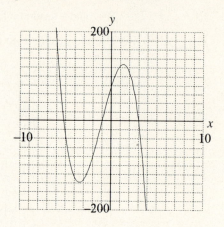

6. (TI-82) For the inequalities below, perform each of the following tasks:

   i)   Show all work necessary to find the *exact* zeros of the polynomial.

   ii) Set up a coordinate system on a sheet of graph paper. Clearly indicate the scale on each axis. Draw a rough graph of the polynomial on your coordinate system. Draw a number line below the graph and shade and label the solution set of the inequality on the number line.

   iii) Use set-builder and interval notation to describe your solution set.

   a) $x^3 + x^2 - 22x - 40 > 0$

   b) $-x^3 - 4x^2 + 7x + 10 \geq 0$

   c) $4x^3 + 4x^2 - 55x - 28 < 0$

   d) $-16x^3 - 24x^2 + 247x + 255 \leq 0$

7. In each of the following problems, the graph of the polynomial has been drawn for you.

   i)   Show all necessary work to find the *exact* roots of the polynomial in simple radical form.

   ii) Set up a coordinate system on a sheet of graph paper and make a copy of the graph on your coordinate system. Draw a number line below your graph and shade and label the *exact* solution of the inequality $p(x) < 0$ on the number line.

   iii) Use set-builder and interval notation to describe the *exact* solution set.

   a) $p(x) = x^3 - 6x^2 + 10x - 4$

   b) $p(x) = -x^3 + 30x - 36$

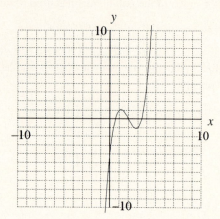

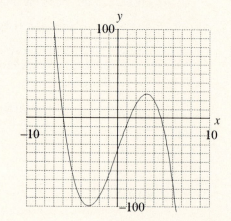

8. (TI-82) For the inequalities below, perform each of the following tasks:

   i)   Show all work necessary to find the *exact* zeros of the polynomial in simple radical form.

   ii) Set up a coordinate system on a sheet of graph paper. Clearly indicate the scale on each axis. Draw a rough graph of the polynomial on your coordinate system. Draw a number line below the graph and shade and label the solution set of the inequality on the number line.

   iii) Use set-builder and interval notation to describe your solution set.

   a) $x^3 - 11x^2 + 30x - 20 > 0$ b) $-x^3 + 4x^2 + 30x + 36 \geq 0$

   c) $x^3 - 4x^2 - 16x + 24 < 0$ d) $-x^3 + 13x^2 - 42x + 30 \leq 0$

9. (TI-82) For the inequalities below, perform each of the following tasks:

   i)   Load the polynomial into the Y= menu. Adjust the WINDOW parameters until the zeros and the turning points of the polynomial are visible in the viewing window. Set up a coordinate system on a sheet of graph paper and clearly indicate the scale on each axis. Make an accurate copy of the image in the viewing window on the coordinate system.

   ii) Use the root-finding utility in the CALC menu to find the zeros of the polynomial, correct to the nearest hundredth. Draw a number line below your graph and shade and label the solution set of the inequality on the number line.

   iii) Use set-builder and interval notation to describe your solution set.

   a) $x^3 - 0.13x^2 - 22.9888x + 26.986 > 0$ b) $-3x^3 - 8.91x^2 + 72.3315x + 87.163083 \geq 0$

10. (TI-82) The United States' silver production for 1930 through 1992 can be approximated by the polynomial $P = 0.00002452t^3 - 0.001628t^2 + 0.01244t + 1.5665$, where S is measured in thousands of metric tons of silver produced and t represents the number of years that have passed since 1930. In what year was silver production 2008 metric tons?

   Note: 2008 metric tons of silver were produced in the year 1989.

11. The gold production in South Africa can be approximated by the polynomial $P = -0.002976t^3 + 0.1317t^2 - 2.054t + 31.94$, where P is measured in millions of troy ounces and t represents the number of years that have passed since 1970. In what year were 19,965,611 troy ounces of gold produced in South Africa?

   Note: 19,965,611 troy ounces of gold were produced in South Africa in the year 1988.

12. Four congruent square corners of undetermined size are cut from a piece of tin that measures 12 inches by 16 inches, as shown below.

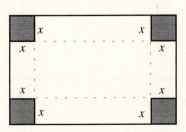

   The squares are discarded and the remaining sides are folded along the dotted lines to form an open box. Let *V* represent the volume of the box.

a) Express V as a function of x. What are the zeros of this function? Set up a coordinate system on a sheet of graph paper and sketch a rough graph of this function.

b) Because x represents the length of the edge of the square, x must be positive. What is the largest value possible for x?

c) (TI-82) Adjust Xmin and Xmax to reflect your findings in part (b). Because the volume of the box must be positive, set Ymin equal to 0. Experiment with Ymax until the maximum value of the polynomial is visible in your viewing window. Set up a second coordinate system on your graph paper and clearly indicate the scale on each coordinate axis. Make an accurate copy of the image in the viewing window on your second coordinate system.

d) Use the maximum utility in the CALC menu to find the maximum value of your function in the viewing window.

e) Clearly state the size of the squares that should be cut to create a box of maximum volume.

13. (TI-82) Four congruent square corners are cut from a square piece of tin that measures 12 inches on a side.

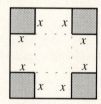

The squares are discarded and the remaining edges are folded upward to form an open box. What value of x will produce a box of volume 100 cubic inches?

14. (TI-82) Assume that the point P is in the first quadrant and lies on the graph of the parabola whose equation is $y = 36 - x^2$.

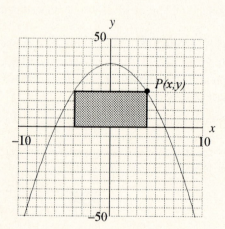

a) Let A represent the area of the inscribed rectangle. Express A as a function of x.

b) For what values of x is the area of the inscribed rectangle larger than 50 square units?

7.5 Summary and Review

Important Visualization Tools

The graph of $y = x^n$, n even

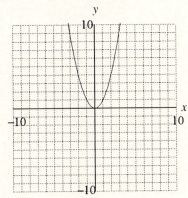

The graph of $y = x^n$, n odd

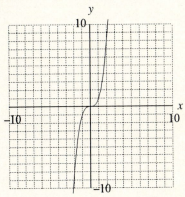

The end-behavior of $y = x^3 - 3x^2 - 18x + 40$ is the same as the end-behavior of $y = x^3$.

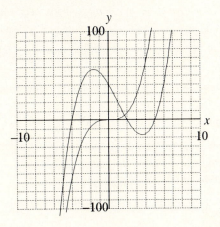

The end-behavior of $y = -x^4 + 26x^2 - 25$ is the same as the end-behavior of $y = -x^4$.

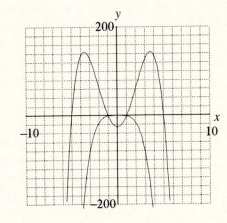

The factors of $p(x) = (x+5)(x+1)(x-3)$ are $x+5$, $x+1$, and $x-3$. Therefore the zeros are $-5, -1$, and 3.

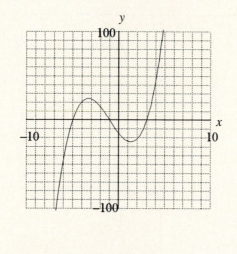

You can use the graph of $p(x)$ to solve the inequality $p(x) > 0$.

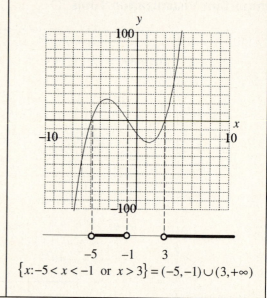

$$\{x : -5 < x < -1 \text{ or } x > 3\} = (-5, -1) \cup (3, +\infty)$$

A Review of Key Words and Phrases

☛ The expression $a_n x^n + a_{n-1} x^{n-1} + \cdots + a_2 x^2 + a_1 x + a_0$, where each of the a_k's is a real number and n is a nonnegative integer, is called a polynomial in x. The a_k's are called the coefficients of the polynomial. The degree of the polynomial is n, the highest power of x. The term $a_n x^n$ is called the leading term of the polynomial. (page 591)

☛ The behavior of the graph for values of x that have large absolute value is called the end-behavior of the graph. (page 591)

☛ In general, the end-behavior of the polynomial $p(x) = a_n x^n + a_{n-1} x^{n-1} + \cdots + a_2 x^2 + a_1 x + a_0$ is identical to the end-behavior of its leading term, or the same as $y = a_n x^n$. (page 595)

☛ r is a zero of the polynomial p if and only if $p(r) = 0$. (page 602)

☛ If $p(r) = 0$, then the graph of p will cross the x-axis at $(r, 0)$. (page 602)

☛ According to the factor theorem, r is a zero of the polynomial p if and only if $x - r$ is a factor of $p(x)$. (page 606)

☛ According to the division algorithm, when a polynomial $f(x)$ is divided by a nonzero polynomial $d(x)$ of lesser or equal degree, the quotient $q(x)$ is a polynomial and the remainder $r(x)$ is either zero or a polynomial of degree less than the degree of the divisor. (page 613)

$$f(x) = d(x)q(x) + r(x), \text{ where } r(x) = 0 \text{ or degree } r(x) < \text{degree } d(x)$$

☛ Most real-world applications of polynomials will require that you find approximate solutions. (page 627)

TI-82 Keywords and Menus

☛ Review the intersect utility and the root-finding utility in the CALC menu in Example 10 in Section 1.6 and Example 5 in Section 1.7.

Chapter Review Exercises

1. Load the equations $y = -4x^3 + 4x^2 + 71x - 126$ and $y = -4x^3$ into the Y= menu in the following manner: Y1 = -4X^3 + 4X^2 + 71X - 126 and Y2 = -4X^3. Set up a separate coordinate system for each of the following viewing windows and sketch the image in the viewing window on your coordinate system. Clearly indicate the scale on each coordinate axis.

   a)
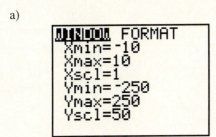

   b)
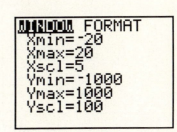

   c)
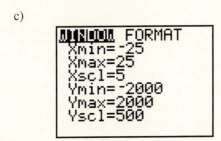

   d)
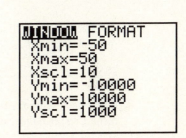

2. For the polynomials below, perform each of the following tasks:

   i)   Express the polynomial as a product of linear (first-degree) factors. Each of these polynomials can be factored by using the techniques learned in Section 7.2. There should be no need for long division. Clearly state the zeros of the polynomial.

   ii) Set up a coordinate system on a sheet of graph paper. Clearly indicate the scale on each coordinate axis. Plot the zeros of the function. Use the zeros and your knowledge of the polynomial's end-behavior to draw a rough graph of the polynomial.

   iii) (TI-82) Load the polynomial's equation into the Y= menu. Adjust the WINDOW parameters until you have a complete graph in the viewing window, one that shows all of the zeros of the polynomial and all of the turning points of the graph. Set up a second coordinate system on your graph paper and copy the image in the viewing window onto the coordinate system as accurately as possible. Include the WINDOW parameters on each coordinate axis.

   a) $p(x) = 4x^3 + 4x^2 - 35x$

   b) $p(x) = 4x^4 - 53x^2 + 49$

   c) $p(x) = 18x^3 + 27x^2 - 8x - 12$

   d) $p(x) = -4x^3 - 16x^2 + 33x$

   e) $p(x) = -4x^4 + 29x^2 - 25$

   f) $p(x) = -2x^3 + 3x^2 + 72x - 108$

3. Set up a coordinate system on a sheet of graph paper and sketch a rough graph of a third-degree polynomial with zeros -1, 3, and 6 that goes through the point $(4, -2)$. Find the equation of this polynomial.

4. Set up a coordinate system on a sheet of graph paper and sketch a rough graph of a fourth-degree polynomial with zeros -5, -3, 2, and 6 that goes through the point $(3, 2)$. Find the equation of this polynomial.

5. Find the zeros of the given polynomial. Place any radicals in simple form.

   a) $p(x) = x^3 + 3x^2 - 15x - 25$ b) $p(x) = -x^3 + 20x + 16$

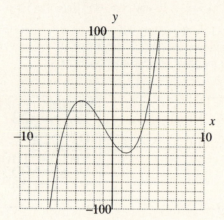 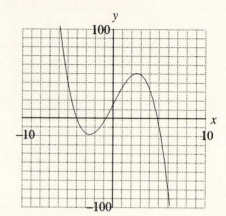

6. Use the results of Exercise 5 to solve each of the following inequalities. Use set-builder and interval notation to describe your solution set.

   a) $x^3 + 3x^2 - 15x - 25 > 0$ b) $-x^3 + 20x + 16 \geq 0$

7. Find exact solutions of each of the following inequalities. Place any radicals in simple radical form. Use set-builder and interval notation to describe your solution set.

   a) $-x^3 - 12x^2 - 27x + 40 < 0$ b) $x^3 + 14x^2 + 58x + 60 \leq 0$

   c) $-x^3 + 7x^2 + 10x - 60 > 0$ d) $x^3 + 15x^2 + 47x - 63 \geq 0$

8. Use long division to divide $p(x) = x^4 - 4x^3 - 10x^2 + 28x + 24$ by $x^2 - 2x - 2$.

   a) Express $p(x)$ as a product of two quadratic factors. Use the quadratic formula to find the zeros of each of these quadratic factors in simple radical form.

   b) Use your calculator to estimate these zeros, correct to the nearest tenth.

   c) Set up a coordinate system on a sheet of graph paper and clearly indicate the scale on each coordinate axis. Use the zeros and your knowledge of the end-behavior of $p(x)$ to draw a rough sketch of the graph of $p(x)$. Check the graph with your TI-82.

   d) Draw a number line below your graph and shade and label the solution set of the inequality $x^4 - 4x^3 - 10x^2 + 28x + 24 \leq 0$ on the number line. Use set-builder and interval notation to describe your solution set.

9. The average monthly precipitation for the 30-year period from 1951 to 1980 in Seattle can be approximated by the polynomial $P = 0.003949t^3 + 0.09683t^2 - 1.805t + 7.695$, where P is the precipitation in inches. The variable t represents the month, $t = 1$ being January, $t = 2$ being February, etc. During which months is the average monthly rainfall below 3 inches?

10. Assume that the point P is in the first quadrant and on the graph of the parabola whose equation is $y = 9 - x^2$.

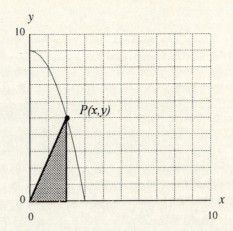

a) Let A represent the area of the shaded triangle. Express A as a function of x. What are the zeros of this function? Set up a coordinate system on a sheet of graph paper. Use the zeros and your knowledge of the end-behavior to draw a rough sketch of this function.

b) Because the point P is in the first quadrant, x must be greater than zero. What is the maximum value that x can attain? *Hint:* What is the x-intercept of the parabola?

c) Use your TI-82 to draw the graph of the area function over the interval established in part (b). Set up a second coordinate system on your graph paper and draw an accurate sketch of the image in your viewing window on your second coordinate system.

d) Use your calculator to estimate the value of x that will maximize the area of the shaded triangle. Clearly state this answer on your second coordinate system.

Chapter 8
Rational Functions

This chapter discusses an important class of functions called *rational* functions.

> If p and q are polynomials, then the function f defined by $f(x) = \dfrac{p(x)}{q(x)}$ is called a rational function.

Our study will begin with a drawing of the graph of $f(x) = \frac{1}{x}$. Once you understand this fundamental graph, you will be able to draw the graphs of more complicated rational functions. Some algebraic manipulations involving rational functions will then be reviewed, and equations with rational functions in them will be solved.

8.1 The Graph of $f(x) = \frac{1}{x}$

The goal of this section is to thoroughly understand the behavior of the graph of the equation $f(x) = \frac{1}{x}$. Begin by creating a table of points that satisfy this equation.

x	-5	-4	-3	-2	-1	0	1	2	3	4	5
$f(x) = \dfrac{1}{x}$	$-\dfrac{1}{5}$	$-\dfrac{1}{4}$	$-\dfrac{1}{3}$	$-\dfrac{1}{2}$	-1	Undefined	1	$\dfrac{1}{2}$	$\dfrac{1}{3}$	$\dfrac{1}{4}$	$\dfrac{1}{5}$

Because division by zero is meaningless, $f(0) = \frac{1}{0}$ is undefined. Plot the points from this table on a coordinate system.

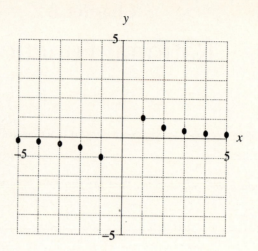

Two important questions remain: (1) What is the end-behavior of the graph? (2) How does the graph behave for x-values between -1 and 1?

The End-Behavior

Tables will be used to help investigate the end-behavior of the function f.

x	10	100	1,000	10,000	100,000
$f(x) = \dfrac{1}{x}$	0.1	0.01	0.001	0.0001	0.00001

As the x-values approach positive infinity, the function values decrease toward zero. Therefore, as you move farther and farther to the right, the graph of $f(x) = \frac{1}{x}$ must fall closer and closer to the x-axis.

x	-10	-100	$-1,000$	$-10,000$	$-100,000$
$f(x) = \dfrac{1}{x}$	-0.1	-0.01	-0.001	-0.0001	-0.00001

As the x-values approach negative infinity, the function values increase toward zero. Therefore, as you move farther and farther to the left, the graph of $f(x) = \frac{1}{x}$ must rise closer and closer to the x-axis.

The end-behavior indicated in the tables is easily seen in the graph. The farther you move to the right or left, the closer the graph gets to the x-axis.

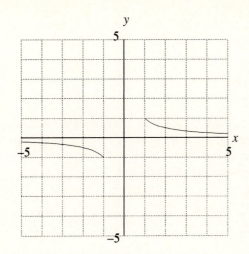

Mathematicians say that the graph of $f(x) = \dfrac{1}{x}$ approaches the x-axis *asymptotically* and that the x-axis is a *horizontal asymptote*.

Investigating the Behavior of the Graph for x-Values Between –1 and 1.   Create another table using x-values between -1 and 1.

x	$-\dfrac{1}{2}$	$-\dfrac{1}{3}$	$-\dfrac{1}{4}$	$-\dfrac{1}{5}$	0	$\dfrac{1}{5}$	$\dfrac{1}{4}$	$\dfrac{1}{3}$	$\dfrac{1}{2}$
$f(x) = \dfrac{1}{x}$	-2	-3	-4	-5	Undefined	5	4	3	2

Plot these additional points on the coordinate system.

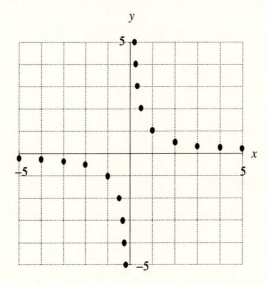

Create two final tables, the first to examine the behavior of the function as the *x*-values approach zero from the right.

x	0.1	0.01	0.001	0.0001	0.00001
f(*x*)	10	100	1,000	10,000	100,000

As the *x*-values approach zero from the right, the function values increase without bound. Therefore, as you approach zero from the right, the graph will rise to positive infinity, getting closer and closer to the *y*-axis as it moves upward. The second table examines the behavior of the function as the *x*-values approach zero from the left.

x	−0.1	−0.01	−0.001	−0.0001	−0.00001
f(*x*)	−10	−100	−1,000	−10,000	−100,000

As the *x*-values approach zero from the left, the function values decrease without bound. Therefore, as you approach zero from the left, the graph will fall to negative infinity, getting closer and closer to the *y*-axis as it moves downward.

The behavior indicated in these last two tables is exhibited in the final graph. As the graph moves closer to the *y*-axis from the left, the graph falls to negative infinity. As the graph moves closer to the *y*-axis from the right, the graph rises to positive infinity.

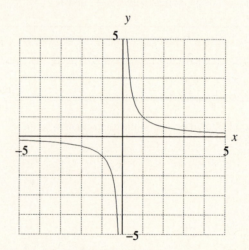

Mathematicians say that the graph approaches the *y*-axis *asymptotically* and that the *y*-axis is a *vertical asymptote*.

Applying Graphing Shortcuts

Now that you know what the basic graph of $f(x) = \dfrac{1}{x}$ looks like, try to apply some graphing shortcuts.

Example 1   Sketch the graph of $y = \dfrac{1}{x-3} + 2$.

Solution

Begin by sketching the graph of $y = \dfrac{1}{x}$.

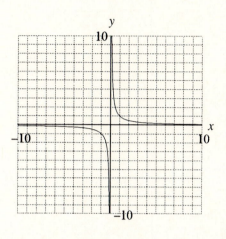

To draw the graph of $y = \dfrac{1}{x-3} + 2$, slide the

graph of $y = \dfrac{1}{x}$ to the right 3 units and up 2 units.

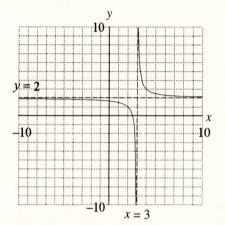

Dotted lines have been drawn to represent the translated horizontal and vertical asymptotes. The equation of the horizontal asymptote is $y = 2$ and the equation of the vertical asymptote is $x = 3$. Once your asymptotes are drawn, it is a simple matter to draw the graph so that it approaches the asymptotes in the prescribed manner.

Note: The function defined by the equation $y = \dfrac{1}{x-3} + 2$ is undefined when $x = 3$ (division by zero is meaningless) and the equation of its vertical asymptote is also $x = 3$.

Example 2   Sketch the graph of $y = -\dfrac{1}{x+2} - 3$.

Solution Take a look at the following student solution.

The graph of $y = -\dfrac{1}{x}$ is a reflection of the graph of $y = \dfrac{1}{x}$ across the *x*-axis.

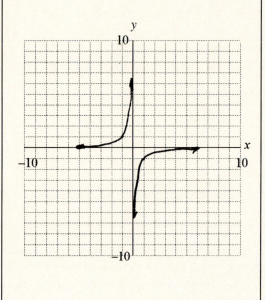

Slide the graph of $y = -\dfrac{1}{x}$ to the left 2 units and downward 3 units to arrive at the graph of $y = -\dfrac{1}{x+2} - 3$.

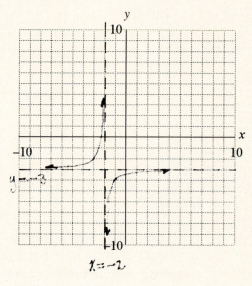

The equations of the horizontal and vertical asymptotes of the graph of $y = -\dfrac{1}{x+2} - 3$ are $y = -3$ and $x = -2$, respectively. Once this student placed the asymptotes in the proper position, it was a simple matter to draw the graph so that it approaches the asymptotes in the prescribed manner.

Note: The function defined by the equation $y = -\dfrac{1}{x+2} - 3$ is undefined when $x = -2$ (division by zero is meaningless) and the equation of the vertical asymptote is also $x = -2$.

Check Your Solution with a TI-82 Calculator.

Load the equation $y = -\dfrac{1}{x+2} - 3$ into the Y= menu.

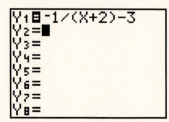

Push the ZOOM button and select 6:ZStandard to capture the following image in your viewing window:

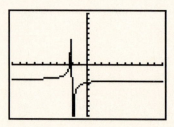

The image in the viewing screen somewhat resembles the graph, but it has some bothersome features.

How Does the TI-81 Draw the Graph of $y = -\dfrac{1}{x+2} - 3$?   The TI-82 begins at the left side of the viewing window, plots a point that satisfies the equation, then goes a certain incremental distance in the x-direction and plots the next point that satisfies the equation. These two points are connected with a line segment. The calculator goes the same incremental distance in the x-direction, plots the next point that satisfies the equation, then connects this point to the last point plotted with a line segment. This process is repeated over and over until the graph is completed: Go an incremental distance in the x-direction, plot the new point, then connect it to the last point plotted with a line segment.

This algorithm performs brilliantly on continuous functions like polynomials but not so well on functions with equations like $y = -\dfrac{1}{x+2} - 3$. A problem can occur near the vertical asymptote.

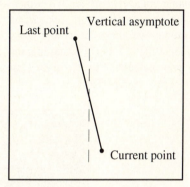

Take another look at the graph of the equation $y = -\dfrac{1}{x+2} - 3$ in the TI-82 viewing window.

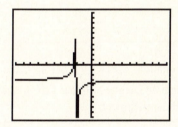

From the left to the right toward $x = -2$, the graph begins to rise upward toward positive infinity. The graph should continue to rise, getting closer and closer to the asymptote $x = -2$ as it travels upward. However, all the TI-82 knows is "plot the next point, then connect it to the last point with a line segment." The incremental distance has carried you to the other side of the vertical asymptote and a false line is drawn, connecting the last point on the left of the vertical asymptote to current point on the right of the vertical asymptote. *This line segment is false and should not be drawn.* It is up to you to recognize that this is a false line that should be omitted from the final picture.

There is a feature on the TI-82 that will turn off this algorithm in favor of another.

Push the MODE key. Use the arrow keys to highlight Dot. Press the ENTER key.	Push the ZOOM button and select 6:ZStandard to capture the following image:
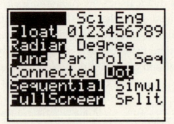	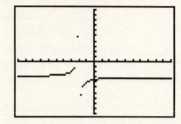

The TI-82 no longer connects consecutive points with line segments. As long as the points are fairly close together, the graph assumes the appearance of a solid curve, but soon only a few scattered points appear as the curve jumps larger and larger distances in the vertical direction. Notice that the false line is no longer there. It is now clear that this particular line segment did not belong.

To return your calculator to its original mode, push the MODE key. Use the arrow keys to select Connected. Press the ENTER key.

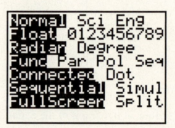

Exercises for Section 8.1

1. Consider the equation $f(x) = -\frac{2}{x}$. Set up a coordinate system on a sheet of graph paper. Clearly indicate the scale on each coordinate axis. Make a copy of the following table on the graph paper and complete the missing entries.

x	-3	-2	-1	$-\dfrac{1}{2}$	$-\dfrac{1}{4}$	0	$\dfrac{1}{4}$	$\dfrac{1}{2}$	1	2	3
$f(x)$											
Points to plot											

a) Plot each point in the table on your coordinate system.

b) Make a copy of each of the following tables on your graph paper and complete the missing entries.

x	10	100	$1,000$	$10,000$
$f(x)$				
Points to plot				

x	-10	-100	$-1,000$	$-10,000$
$f(x)$				
Points to plot				

Use complete sentences to explain the end-behavior of the function f. Include the equation of the horizontal asymptote in your explanation. Draw a dotted line representing the horizontal asymptote on the coordinate system.

c) Make a copy of each of the following tables on your graph paper and complete the missing entries.

x	0.1	0.01	0.001	0.0001
$f(x)$				
Points to plot				

x	-0.1	-0.01	-0.001	-0.0001
$f(x)$				
Points to plot				

Use complete sentences to explain the behavior of the function f near $x = 0$. Include the equation of the vertical asymptote in your explanation. Draw a dotted line representing the vertical asymptote on the coordinate system.

d) Draw a complete graph of the equation $f(x) = -\dfrac{2}{x}$ on your coordinate system.

2. Complete the following steps to help draw the graph of $y = \dfrac{1}{x+1} - 3$.

a) Set up a coordinate system on a sheet of graph paper. Clearly indicate the scale on each coordinate axis. Draw a rough sketch of the graph of the equation $y = \dfrac{1}{x}$.

b) Set up a second coordinate system on your graph paper. Draw a rough sketch of the graph of the equation $y = \dfrac{1}{x+1} - 3$.

Hint: Translate the graph of $y = \dfrac{1}{x}$ to the left 1 unit and down 3 units. Draw dotted lines representing the horizontal and vertical asymptotes and label each asymptote with its equation.

3. For the equations below, perform the following tasks:

i)   Set up a coordinate system on a sheet of graph paper. Clearly indicate the scale on each axis.

ii) Draw dotted lines representing the horizontal and vertical asymptotes. Label each asymptote with its equation.

iii) Draw a rough sketch of the equation. Check your solution with your calculator.

a) $y = \dfrac{1}{x-3} + 4$ b) $y = \dfrac{1}{x+2} - 5$ c) $y = \dfrac{1}{x-4} - 2$ d) $y = \dfrac{1}{x+3} + 4$

4. Complete the following steps to help draw the graph of $y = -\dfrac{1}{x-2} + 4$.

 a) Set up a coordinate system on a sheet of graph paper. Clearly indicate the scale on each coordinate axis. Draw a rough sketch of the graph of the equation $y = \dfrac{1}{x}$.

 b) Set up a second coordinate system on your graph paper. Draw a rough sketch of the graph of $y = -\dfrac{1}{x}$.

 Hint: Reflect the graph of $y = \dfrac{1}{x}$ across the x-axis.

 c) Set up a third coordinate system on your graph paper. Draw a rough sketch of the graph of $y = -\dfrac{1}{x-2} + 4$.

 Hint: Translate the graph of $y = -\dfrac{1}{x}$ to the right 2 units and upward 4 units. Draw dotted lines representing the horizontal and vertical asymptotes and label each asymptote with its equation.

5. For the equations below, perform the following tasks:

 i) Set up a coordinate system on a sheet of graph paper. Clearly indicate the scale on each axis.

 ii) Draw dotted lines representing the horizontal and vertical asymptotes. Label each asymptote with its equation.

 iii) Draw a rough sketch of the equation. Check your solution with your calculator.

 a) $y = -\dfrac{1}{x+4}$ b) $y = -\dfrac{1}{x} + 2$ c) $y = -\dfrac{1}{x-3} - 2$ d) $y = -\dfrac{1}{x+5} + 1$

6. For the graphs below, perform each of the following tasks:

 i) Set up a coordinate system on a sheet of graph paper. Clearly indicate the scale on each coordinate axis. Draw dotted lines on your coordinate system that represent what you feel are the asymptotes of the graph. Make an accurate copy of the graph on your coordinate system.

 ii) Clearly state a possible equation of your graph. Check the result with your calculator.

 a)

 b)

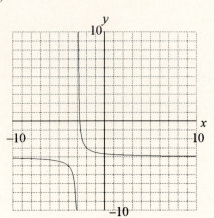
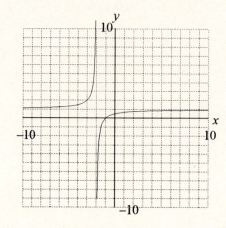

c)

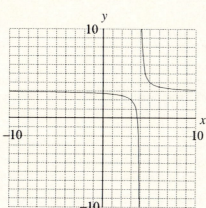

d)

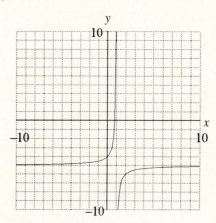

7. For the equations below, perform each of the following tasks:

   i)   Set up a coordinate system on a sheet of graph paper. Clearly indicate the scale on each axis.

   ii) Draw dotted lines representing the horizontal and vertical asymptotes. Label each asymptote with its equation.

   iii) Draw a rough sketch of the equation. Check the solution with your calculator.

   a) $y = \dfrac{1}{x+4} + 500$

   b) $y = -\dfrac{1}{x+8} + 1000$

   c) $y = -\dfrac{1}{x-20} + 400$

   d) $y = \dfrac{1}{x-2} - 40$

8. (TI-82) Suppose that a number of rabbits is placed in a particular area of desert. Because food is scarce, the rabbit population begins to decrease. Suppose that the number of rabbits in the area is given by the equation $N = \dfrac{1000}{t+4} + 200$, where N represents the number of rabbits and t represents the number of months that have passed since the rabbits were first placed in the area.

Load the equation $N = \dfrac{1000}{t+4} + 200$ into the Y= menu.
 	
`Y₁■1000/(X+4)+20`
`0`
`Y₂=`
`Y₃=`
`Y₄=`
`Y₅=`
`Y₆=`
`Y₇=`
	Make the following adjustments to the WINDOW parameters.
	
	`WINDOW FORMAT`
	`Xmin=0`
	`Xmax=50`
	`Xscl=10`
	`Ymin=0`
	`Ymax=500`
	`Yscl=100`
	
	Push the GRAPH button to view the graph.

   a) Set up a coordinate system on a sheet of graph paper. Clearly indicate the scale on each axis. Make an accurate copy of the image in your viewing window on your coordinate system.

b) How many rabbits were first introduced to the area? Plot this data point on your coordinate system.

   Hint: Let $t = 0$ in the equation $N = \dfrac{1000}{t+4} + 200$.

c) Draw a dotted line on your coordinate system representing what you feel is the horizontal asymptote of this function. Clearly state what you feel will be the eventual size of the rabbit population.

9. (TI-82) At 7 P.M. people began waiting in line to see a popular movie that will be shown at 9 P.M. As time passes, the number of people in line begins to increase. Later in the evening, people are discouraged by the length of the line and decide to seek their entertainment elsewhere. Suppose that the number of people waiting in line is given by the equation $N = \dfrac{-60}{t+1} + 100$, where N represents the number of people standing in line and t represents the number of minutes that have passed since 7 P.M.

Load the equation $N = \dfrac{-60}{t+1} + 100$ into the Y= menu.

```
Y₁ ∎ -60/(X+1)+100

Y₂=
Y₃=
Y₄=
Y₅=
Y₆=
Y₇=
```

Make the following adjustments to the WINDOW parameters:

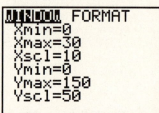

Push the GRAPH button to view the graph.

a) Set up a coordinate system on a sheet of graph paper. Clearly indicate the scale on each axis. Make an accurate copy of the image in your viewing window on your coordinate system.

b) How many people were waiting in line at 7 P.M.? Plot this data point on your coordinate system.

   Hint: Let $t = 0$ in the equation $N = \dfrac{-60}{t+1} + 100$.

c) Draw a dotted line on your coordinate system representing what you feel is the horizontal asymptote of this function. Clearly state what you feel will be the eventual number of people waiting in line.

10. A potato is placed in an oven. At first, the temperature of the potato rises very quickly, but eventually the rate at which the temperature is rising begins to slow. Suppose that the temperature of the potato is given by the equation $T = \dfrac{-160}{t+0.4} + 450$, where T is the temperature of the potato in degrees Fahrenheit and t is the number of minutes that have passed since the potato was placed in the oven.

a) (TI-82) Draw a graph of the potato's temperature during the first 10 minutes.

b) What is the initial temperature of the potato?

c) What do you feel is the eventual temperature of the potato?

11. A steel rod is taken from a furnace. It begins to cool rapidly at first, but the rate at which it cools begins to slow. Suppose that the temperature of the rod is given by the equation $T = \dfrac{1000}{t+1} + 70$, where T is the temperature of the rod in degrees Fahrenheit and t is the number of minutes that have passed since the rod was removed from the furnace.

a) (TI-82) Draw a graph of the rod's temperature during the first 30 minutes.

b) What is the initial temperature of the rod?

c) What do you feel is the eventual temperature of the rod?

8.2 Graphing Rational Functions

The preceding section introduced the equation $f(x) = \dfrac{1}{x}$. The function defined by this equation is an example of what is called a *rational* function.

> **If $p(x)$ and $q(x)$ are polynomials, then the function f defined by $f(x) = \dfrac{p(x)}{q(x)}$ is called a rational function.**

Here are some examples of rational functions.

$$f(x) = \frac{x+1}{x-2}$$

$$g(x) = \frac{x+1}{x^2 - 5x + 6}$$

$$h(x) = \frac{x^2 - 2x - 3}{x + 2}$$

The purpose of this section is to introduce techniques that are used by mathematicians to draw the graphs of rational functions. The examples will be kept simple so that you can concentrate on the ideas involved. The more sophisticated techniques and examples will be left to your college algebra and calculus teachers. Let's first establish a fundamental concept.

> **Assume that N and D are real numbers.**
>
> **If $D = 0$, then $\dfrac{N}{D}$ is undefined. Division by zero is meaningless.**
>
> **If $N = 0$ and $D \neq 0$, then $\dfrac{N}{D} = 0$.**

These principles can be applied to rational functions.

Suppose that you have a rational function f with equation $f(x) = \dfrac{p(x)}{q(x)}$.

1. If $p(x) = 0$ and $q(x) \neq 0$, then $f(x) = 0$ and x is a zero of the function f. The graph of the function f will cross the x-axis at $(x, 0)$.

2. If $p(x) \neq 0$ and $q(x) = 0$, then the graph of the function f will have a vertical asymptote at x.

Example 1   Sketch the graph of $f(x) = \dfrac{x+2}{x-3}$.

Solution   If $x = -2$, the numerator of $f(x) = \dfrac{x+2}{x-3}$ equals zero and the denominator is nonzero. Therefore $f(-2) = 0$ and the graph of f will cross the x-axis at $(-2, 0)$. If $x = 3$, the denominator of $f(x) = \dfrac{x+2}{x-3}$ equals zero and the numerator is nonzero. Therefore the graph of f will have a vertical asymptote at $x = 3$.

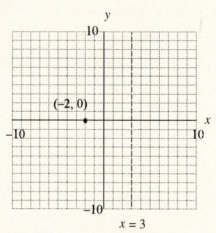

$x = 3$

As you gain experience graphing rational functions, you will see that this preliminary information is incredibly powerful. Now plot three points on each side of the vertical asymptote.

Tip	A program called EVAL on the TI-82 is being used to produce the tables in this section. Th program listing is in the appendix.

x	0	1	2	3	4	5	6
$f(x)$	−0.7	−1.5	−4.0	Undefined	6.0	3.5	2.7

The function values have been rounded to the nearest tenth. Now plot each point in the table on your coordinate system.

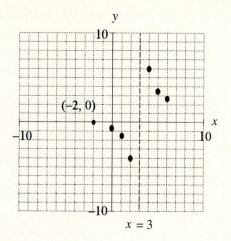

$x = 3$

Investigating the End-Behavior

x	10	100	1,000	10,000	100,000
$f(x)$	1.7142857	1.0515464	1.005015	1.0005002	1.00005

As the x-values approach positive infinity, the function values appear to be decreasing toward 1. Therefore, as we move farther and farther to the right, the graph of $f(x) = \dfrac{x+2}{x-3}$ must fall closer and closer to the horizontal asymptote $y = 1$.

x	-10	-100	$-1,000$	$-10,000$	$-100,000$
$f(x)$	0.6153846	0.9514563	0.995015	0.9995001	0.99995

As the x-values approach negative infinity, the function values appear to be increasing toward 1. Therefore, as we move farther and farther to the left, the graph of $f(x) = \dfrac{x+2}{x-3}$ must rise closer and closer to the horizontal asymptote $y = 1$.

The end-behavior indicated in these tables is easily seen in the graph. The farther you move to the right or the left, the closer the graph of $f(x) = \dfrac{x+2}{x-3}$ gets to the horizontal asymptote $y = 1$.

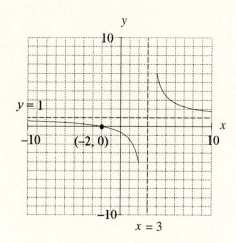

$x = 3$

Investigating the Behavior Near the Vertical Asymptote

x	2.7	2.8	2.9	2.99	2.999
$f(x)$	−15.666667	−24	−49	−499	−4999

As the x-values approach 3 from the left, the function values decrease without bound. Therefore, as you approach 3 from the left, the graph of f will fall to negative infinity, getting closer and closer to the vertical asymptote $x = 3$ as it moves downward.

x	3.3	3.2	3.1	3.01	3.001
$f(x)$	17.666667	26	51	501	5001

As the x-values approach 3 from the right, the function values increase without bound. Therefore, as you approach 3 from the right, the graph of f will rise to positive infinity, getting closer and closer to the vertical asymptote $x = 3$ as it moves upward.

The behavior near the vertical asymptote $x = 3$ indicated in the last two tables is exhibited in the final graph. As the graph moves closer to the vertical asymptote $x = 3$ from the left, the graph falls to negative infinity. As the graph moves closer to the vertical asymptote from the right, the graph rises to positive infinity.

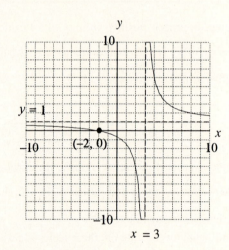

Example 2   Use the graph of the function from Example 1 to help find the solution set of the following inequality.

$$\frac{x+2}{x-3} \ge 0$$

Solution   You will need to find where the graph of $f(x) = \dfrac{x+2}{x-3}$ is above or on the x-axis.

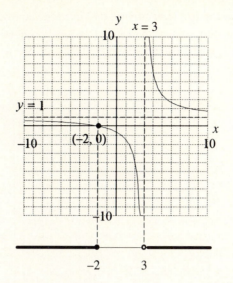

In set-builder notation, the solution set is $\{x : x \le -2 \ \text{or} \ x > 3\}$. In interval notation, the solution set is $(-\infty, -2] \cup (3, +\infty)$.

Example 3   Use the TI-82 to help draw the graph of $f(x) = \dfrac{2-x}{x+1}$.

Solution   If $x = 2$, the numerator of $f(x) = \dfrac{2-x}{x+1}$ is zero and the denominator is nonzero. Therefore $f(2) = 0$ and the graph of f will cross the x-axis at $(2,0)$. If $x = -1$, the denominator of $f(x) = \dfrac{2-x}{x+1}$ is zero and the numerator is nonzero. Therefore the graph of f will have a vertical asymptote at $x = -1$.

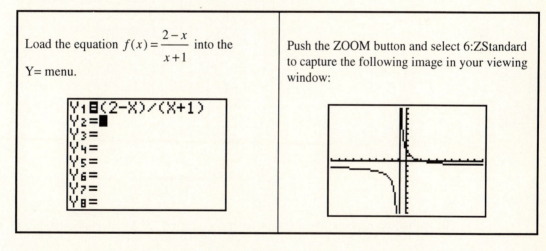

Load the equation $f(x) = \dfrac{2-x}{x+1}$ into the Y= menu.

Push the ZOOM button and select 6:ZStandard to capture the following image in your viewing window:

If you examine the image in the viewing window, it appears that the equation of the horizontal asymptote might be $y = -1$. However, the equation could just as well be $y = -1.1$. To be certain of the equation of the horizontal asymptote, you will have to examine the behavior of the function as the x-values approach positive infinity.

x	10	100	1,000	10,000	100,000
$f(x)$	−0.7272727	−0.970297	−0.997003	−0.9997	−0.99997

As x approaches positive infinity, the function values decrease toward -1. A similar table will reveal that the function values approach -1 as your x-values approach negative infinity. Therefore the line $y = -1$ will be a horizontal asymptote.

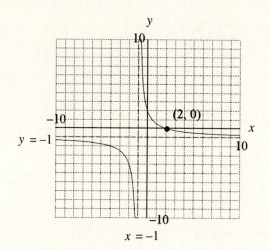

Example 4   Use the graph from Example 3 to help find the solution set of the inequality $\dfrac{2-x}{x+1} \geq 0$.

Solution   Determine where the graph of f is above or on the x-axis.

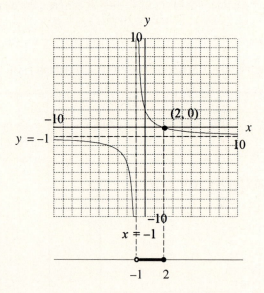

In set-builder notation, the solution set is $\{x : -1 < x \leq 2\}$. In interval notation, the solution set is $(-1, 2]$.

Exercises for Section 8.2

1. Consider the equation $f(x) = \dfrac{x+3}{x-2}$. Set up a coordinate system on a sheet of graph paper and clearly indicate the scale on each axis.

   a) Plot and label the *x*-intercept of the graph of *f*. Draw a dotted line representing the vertical asymptote of the function *f*. Label the vertical asymptote with its equation.

   b) Make a copy of the following table on your graph paper and complete the missing entries.

x	−1	0	1	2	3	4	5
f (*x*)							

   Plot each point in the table on your coordinate system.

   c) Make a copy of each of the following tables on your graph paper and complete the missing entries.

x	10	100	1,000	10,000	100,000
f (*x*)					

x	−10	−100	−1,000	−10,000	−100,000
f (*x*)					

   Draw a dotted line representing the horizontal asymptote of the function *f*. Label the horizontal asymptote with its equation.

   d) Make a copy of each of the following tables on your graph paper and complete the missing entries.

x	1.7	1.8	1.9	1.99	1.999
f (*x*)					

x	2.3	2.2	2.1	2.01	2.001
f (*x*)					

   Use the information in these last two tables to complete the graph of *f*. Use your TI-82 to check your solution.

   e) Draw a number line below your graph of the function *f* and shade the solution set of the inequality $\dfrac{x+3}{x-2} > 0$ on your number line in the usual manner. Use set-builder and interval notation to describe your solution set.

2. Consider the function $f(x) = \dfrac{x-2}{x+1}$.

   a) Plot and label the *x*-intercept of the graph of *f*. Draw a dotted line representing the vertical asymptote of the function *f*. Label the vertical asymptote with its equation.

   b) Make a copy of the following table on your graph paper and complete the missing entries.

x	−4	−3	−2	−1	0	1	2
f (*x*)							

   Plot each point in the table on your coordinate system.

c) Make a copy of each of the following tables on your graph paper and complete the missing entries.

x	10	100	1,000	10,000	100,000
$f(x)$					

x	−10	−100	−1,000	−10,000	−100,000
$f(x)$					

Draw a dotted line representing the horizontal asymptote of the function f. Label the horizontal asymptote with its equation.

d) Make a copy of each of the following tables on your graph paper and complete the missing entries.

x	−0.7	−0.8	−0.9	−0.99	−0.999
$f(x)$					

x	−1.3	−1.2	−1.1	−1.01	−1.001
$f(x)$					

Use the information in these last two tables to complete the graph of f. Use your TI-82 to check your solution.

e) Draw a number line below your graph of the function f and shade the solution set of the inequality $\dfrac{x-2}{x+1} < 0$ on the number line in the usual manner. Use set-builder and interval notation to describe your solution set.

3. For the graphs below, perform each of the following tasks:

i) Set up a coordinate system on graph paper and clearly indicate the scale on each coordinate axis. Make an accurate copy of the graph on your coordinate system.

ii) Given that the equation of the graph is of the form $y = \dfrac{x+a}{x+b}$, clearly state the equation of the graph. Use a TI-82 calculator to check your solution.

a)

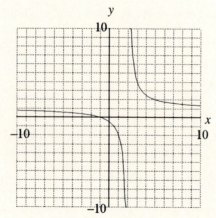

b)

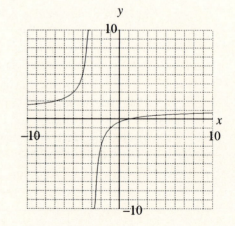

4. For the equations below, perform each of the following tasks:

i) Set up a coordinate system on a sheet of graph paper. Clearly indicate the scale on each coordinate axis. Plot and label the x-intercept of the graph of the equation. Draw a dotted line on your coordinate system representing the vertical asymptote. Label this asymptote with its equation.

ii) Make a copy of the following table on your graph paper and complete the missing entries.

x	10	100	1,000	10,000	100,000
$f(x)$					

Draw a dotted line on your coordinate system representing the horizontal asymptote. Label this asymptote with its equation.

iii) (TI-82) Load the equation into the Y= menu, push the ZOOM button, and select 6:ZStandard. Use the image in the viewing window to help draw the graph of f on your coordinate system.

iv) Draw a number line below your graph and shade and label the solution set of the inequality $f(x) \leq 0$ on the number line in the usual manner. Use set-builder and interval notation to describe your solution set.

a) $f(x) = \dfrac{x}{x+3}$

b) $f(x) = \dfrac{4-x}{x+1}$

c) $f(x) = \dfrac{4-2x}{x-6}$

d) $f(x) = \dfrac{2x+1}{x-3}$

e) $f(x) = \dfrac{3-x}{2x+1}$

f) $f(x) = \dfrac{5x+15}{2x-3}$

5. Solve each of the following inequalities for x. Use both set-builder and interval notation to describe your solution set.

a) $\dfrac{x-7}{x+2} \geq 0$

b) $\dfrac{2-x}{x+4} < 0$

c) $\dfrac{2x+5}{3-x} > 0$

d) $\dfrac{2x+1}{3-2x} \leq 0$

6. For the equations below, perform each of the following tasks:

i) Set up a coordinate system on a sheet of graph paper. Clearly indicate the scale on each coordinate axis. Plot and label the x-intercept of the graph of the equation. Draw a dotted line on your coordinate system representing the vertical asymptote. Label this asymptote with its equation.

ii) Make a copy of the following table on your graph paper and complete the missing entries.

x	10	100	1,000	10,000	100,000
$f(x)$					

Draw a dotted line on your coordinate system representing the horizontal asymptote. Label this asymptote with its equation.

iii) (TI-82) Load the equation into the Y= menu, push the ZOOM button, and select 6:ZStandard. Use the image in your viewing window to help draw the graph of f on your coordinate system.

a) $f(x) = \dfrac{350x+850}{x+3}$

b) $f(x) = \dfrac{200x+150}{x+4}$

7. (TI-82) A potato is placed into an oven. The temperature of the potato is given by the equation $T = \dfrac{450t+180}{t+3}$, where T is the temperature of the potato in degrees Fahrenheit and t is the number of minutes that have passed since the potato was placed in the oven.

Load the equation $T = \dfrac{450t + 180}{t + 3}$ into the Y= menu.

Make the following adjustments to the WINDOW parameters.

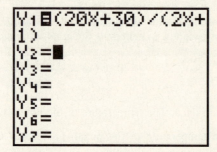

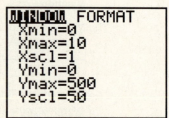

a) Set up a coordinate system on a sheet of graph paper. Clearly indicate the scale on each axis. Make an accurate copy of the image in the viewing window on your coordinate system.

b) What is the initial temperature of the potato?

   Hint: Let $t = 0$ in the equation $T = \dfrac{450t + 180}{t + 3}$.

c) Make a copy of the following table on your graph paper and complete the missing entries.

t	10	100	1,000	10,000	100,000
T					

   Draw a dotted line on your coordinate system representing the horizontal asymptote of the graph. Clearly state what you feel is the eventual temperature of the potato.

d) Do you think that this model is realistic? How long do you think it would take the potato to warm to the same temperature as the oven?

8. (TI-82) Wheelsburg has a large water tank located in the hills above the town. The tank develops a leak and water begins to pour from the tank. Initially, the water shoots out at a tremendous rate, but as the water level and pressure drop, the rate begins to slow. Suppose that the height of the water in the tank is given by the equation $h = \dfrac{20t + 30}{2t + 1}$, where h represents the water level in feet and t represents the number of hours that have passed since the leak started.

Load the equation $h = \dfrac{20t + 30}{2t + 1}$ into the Y= menu.

Make the following adjustments to the WINDOW parameters.

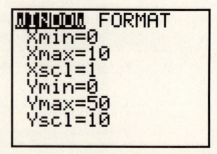

a) Set up a coordinate system on a sheet of graph paper. Clearly indicate the scale on each axis. Make an accurate copy of the image in the viewing window on your coordinate system.

b) What was the initial height of the water in the tank at the time the leak started?

Hint: Let $t = 0$ in the equation $h = \dfrac{20t + 30}{2t + 1}$.

c) Make a copy of the following table on your graph paper and complete the missing entries.

t	5	10	15	20	25
h					

Draw a dotted line on your coordinate system representing the horizontal asymptote of the graph. Clearly state what you feel is the eventual water level in the tank.

9. A glass of ice water with a thermometer in it is placed in the open air. Suppose that the temperature of the water in the glass is given by the equation $T = \dfrac{700t + 290}{10t + 7}$, where T represents the water in the glass in degrees Fahrenheit and t represents the number of hours that have passed since the glass was placed in the open air.

a) Draw a graph of the temperature of the water in the glass during the first 10 hours.

b) What is the initial temperature of the water in the glass?

c) What will be the eventual temperature in the glass?

d) Do you think that this model is realistic? How long do you think it would take for the water in the glass to warm to room temperature?

10. A sudden thaw starts a snowpack melting at a rapid rate. Consequently, the water in a nearby lake begins to rise at a rapid rate. As the weather cools, the rate at which the snowpack melts begins to slow. Suppose that the water level, measured from the base of a dam at the lake, is given by the equation $h = \dfrac{150t + 1300}{t + 10}$, where h is the height of the water in feet and t is the number of hours that have passed since the beginning of the thaw.

a) Draw a graph of the height of the lake over the first 120 hours of the thaw.

b) What was the initial height of the lake?

c) What will be the eventual height of the lake?

11. Suppose that the concentration of soot falling to the ground from a smokestack is given by the equation $C = \dfrac{100x + 11,400}{x + 14}$, where C is the concentration of soot in parts per million and x is the distance from the smokestack in feet.

a) Draw a graph of the concentration of soot falling to the ground for distances of anywhere from 0 to 500 feet from the smokestack.

b) What is the concentration of soot falling to the ground at the base of the smokestack?

c) What will be the concentration of soot falling to the ground at great distances from the smokestack?

d) Do you think this model is realistic? How well do you think the model mirrors reality for distances between 0 and 500 feet? How well does the model mirror reality for distances beyond 500 feet?

8.3 Reducing Rational Expressions

This section will first concentrate on techniques used in reducing rational expressions. Once the art of reducing rational expressions has been perfected, some time will be spent multiplying and dividing rational expressions. The skills for doing this are identical to the skills required for reducing rational expressions.

Let's begin by defining multiplication of two rational expressions.

$$\textit{Definition of multiplication: If } b \neq 0 \textit{ and } d \neq 0, \textit{ then } \frac{a}{b} \cdot \frac{c}{d} = \frac{ac}{bd}.$$

Example 1   Simplify the expression $\dfrac{2}{3} \cdot \dfrac{5}{7}$.

Solution   According to the definition of multiplication, $\dfrac{2}{3} \cdot \dfrac{5}{7} = \dfrac{2 \cdot 5}{3 \cdot 7} = \dfrac{10}{21}$.

Example 2   If $b \neq 0$ and $c \neq 0$, simplify the expression $\dfrac{ac}{bc}$.

Solution   Use the definition of multiplication to remove a factor of 1.

$$\frac{ac}{bc} = \frac{a}{b} \cdot \frac{c}{c}$$

$$= \frac{a}{b} \cdot 1$$

$$= \frac{a}{b}$$

This last example is such an important example that it is repeated here for emphasis.

$$\textit{If } b \neq 0 \textit{ and } c \neq 0, \textit{ then } \frac{ac}{bc} = \frac{a}{b}.$$

Tip   It is important to note that the numerator and denominator must have a *common factor* to remove a factor of 1. For example, $\dfrac{a+c}{b+c} \neq \dfrac{a}{b}$ because c is *not* a common factor but a *common addend*.

Example 3   Simplify the expression $\dfrac{x^5 y^4}{x^2 y^2}$.

Solution   Use the definition of multiplication to remove a factor of 1. Note that the highest power of x and y that is common to both the numerator and denominator has been removed.

$$\frac{x^5 y^4}{x^2 y^7} = \frac{x^3}{y^3} \cdot \frac{x^2 y^4}{x^2 y^4}$$

$$= \frac{x^3}{y^3} \cdot 1$$

$$= \frac{x^3}{y^3}$$

This process can be streamlined by removing a factor of 1 in the following manner:

$$\frac{x^5 y^4}{x^2 y^7} = \frac{x^5 y^4}{x^2 y^7} \qquad\qquad \frac{x^5 y^4}{x^2 y^7} = \frac{\overset{3}{\cancel{x^5}}\,\cancel{y^4}}{\cancel{x^2}\,\cancel{y^7}\,_3}$$

$$= \frac{x^3}{y^3} \qquad\qquad\qquad\qquad\quad = \frac{x^3}{y^3}$$

Note: Because division by zero is meaningless, this argument is valid only if x does not equal 0 and y does not equal zero.

Example 4   Simplify the expression $\dfrac{18}{24}$.

Solution   Factor the numerator and denominator, then use the definition of multiplication to remove a factor of 1.

$$\frac{18}{24} = \frac{2 \cdot 3 \cdot 3}{2 \cdot 2 \cdot 2 \cdot 3}$$

$$= \frac{3}{2 \cdot 2} \cdot \frac{2 \cdot 3}{2 \cdot 3}$$

$$= \frac{3}{4} \cdot 1$$

$$= \frac{3}{4}$$

This process can be streamlined by removing a factor of 1 in the following manner:

$$\frac{18}{24} = \frac{\cancel{2} \cdot 3 \cdot \cancel{3}}{\cancel{2} \cdot 2 \cdot 2 \cdot \cancel{3}}$$

$$= \frac{3}{4}$$

Example 5   Simplify the expression $\dfrac{x^2 - 9}{x^2 + 3x}$.

Solution Factor the numerator and denominator and use the definition of multiplication to remove a factor of 1.

$$\frac{x^2-9}{x^2+3x}=\frac{(x+3)(x-3)}{x(x+3)}$$

$$=\frac{x-3}{x}\cdot\frac{x+3}{x+3}$$

$$=\frac{x-3}{x}\cdot 1$$

$$=\frac{x-3}{x}$$

This process can be streamlined by removing a factor of 1 in the following manner:

$$\frac{x^2-9}{x^2+3x}=\frac{\cancel{(x+3)}(x-3)}{x\cancel{(x+3)}}$$

$$=\frac{x-3}{x}$$

Note: Division by zero is meaningless, so this argument will not hold if any of the denominators are zero. Therefore x must not equal 0 or -3.

Example 6 Simplify the expression $\dfrac{5x^3-x^2-5x+1}{5x^3+4x^2-x}$.

Solution Factor the numerator and denominator, then use the definition of multiplication to remove a factor of 1.

$$\frac{5x^3-x^2-5x+1}{5x^3+4x^2-x}=\frac{x^2(5x-1)-1(5x-1)}{x(5x^2+4x-1)}$$

$$=\frac{(x^2-1)(5x-1)}{x(5x^2+5x-x-1)}$$

$$=\frac{(x+1)(x-1)(5x-1)}{x(5x(x+1)-1(x+1))}$$

$$=\frac{(x+1)(x-1)(5x-1)}{x(5x-1)(x+1)}$$

$$=\frac{x-1}{x}\cdot\frac{(x+1)(5x-1)}{(x+1)(5x-1)}$$

$$= \frac{x-1}{x} \cdot 1$$

$$= \frac{x-1}{x}$$

This presentation can be streamlined by removing a factor of 1 in the following manner.

$$\frac{5x^3 - x^2 - 5x + 1}{5x^3 + 4x^2 - x} = \frac{\cancel{(x+1)}(x-1)\cancel{(5x-1)}}{x\cancel{(5x-1)}\cancel{(x+1)}}$$

$$= \frac{x-1}{x}$$

Note: This argument is meaningless if any of the denominators equals zero. Therefore the argument is valid only if x does not equal 0, $\frac{1}{5}$, or -1.

Example 7   The following is the graph of the equation $f(x) = x^2 - 2x - 3$.

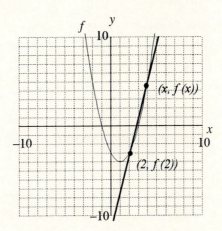

Find the slope of the line that passes through the points $(2, f(2))$ and $(x, f(x))$.

Solution   The slope of a line is found by dividing the change in y by the change in x.

$$m = \frac{\text{Change in } y}{\text{Change in } x}$$

$$m = \frac{f(x) - f(2)}{x - 2}$$

$$= \frac{\left(x^2 - 2x - 3\right) - \left((2)^2 - 2(2) - 3\right)}{x - 2}$$

$$= \frac{x^2 - 2x}{x - 2}$$

$$= \frac{x(x - 2)}{x - 2}$$

$$= x$$

Note: Because division by zero is meaningless, this argument is valid only if x does not equal 2.

A Sign Change Rule for Fractions

Every fraction has three signs: one in the numerator, one in the denominator, and one on the fraction bar. In the fraction $-\frac{2}{3}$, positive signs in both the numerator and denominator are understood.

$$-\frac{2}{3} = -\frac{+2}{+3}$$

The sign change rule: If any two parts of a fraction are negated, the fraction will remain the same.

Making two negations is the same as multiplying by -1, then multiplying by a second -1. Because $(-1)(-1) = 1$, making two negations is the same as multiplying by 1, which leaves the fraction unchanged.

Example 8 Consider the fraction $-\frac{-2}{3}$. Two negations will be made. If the numerator and fraction bar are negated, and the denominator is left alone, then $-\frac{-2}{3} = \frac{2}{3}$.

Example 9 Consider the fraction $-\frac{-2}{-3}$. We will make two negations. If we negate the numerator and denominator, leaving the fraction bar alone, then $-\frac{-2}{-3} = -\frac{2}{3}$.

Example 10 Simplify the expression $\frac{x(x - 3)}{3 - x}$.

Solution Begin by making two negations, the fraction bar and the denominator. When the denominator is negated, note that $-(3 - x) = x - 3$.

$$\frac{x(x-3)}{3-x} = -\frac{x(x-3)}{x-3}$$

$$= -x$$

Note: Because division by zero is meaningless, this argument is valid only if x does not equal 3.

Example 11   Suppose that $f(x) = x^4 - 13x^2 + 36$ and $g(x) = 6 + x - x^2$. Simplify the expression $\dfrac{f(x)}{g(x)}$.

Solution

$$\frac{f(x)}{g(x)} = \frac{x^4 - 13x^2 + 36}{6 + x - x^2}$$

Note that the numerator is arranged in descending powers of x but the denominator is arranged in ascending powers of x. You could simply rearrange the denominator in descending powers of x, but the leading term would be negative. It is better to invoke the sign change rule for fractions. The fraction bar and denominator are negated. When the denominator is negated, note that $-(6 + x - x^2)$ equals $x^2 - x - 6$.

$$\frac{f(x)}{g(x)} = \frac{x^4 - 13x^2 + 36}{x^2 - x - 6}$$

Factor the numerator and denominator and remove a factor of 1.

$$\frac{f(x)}{g(x)} = \frac{x^4 - 9x^2 - 4x^2 + 36}{x^2 - 3x + 2x - 6}$$

$$= \frac{x^2(x^2 - 9) - 4(x^2 - 9)}{x(x-3) + 2(x-3)}$$

$$= \frac{(x^2 - 4)(x^2 - 9)}{(x+2)(x-3)}$$

$$= \frac{(x+2)(x-2)(x+3)(x-3)}{(x+2)(x-3)}$$

$$= (x-2)(x+3)$$

Note: Because division by zero is meaningless, this argument is valid only if x does not equal -2 or 3.

Example 12   Suppose that $f(x) = \dfrac{x^2 - 4x}{x^2 - 2x - 8}$ and $g(x) = \dfrac{x^2 - 4}{x^2 - 4x + 4}$. Simplify the expression $f(x)g(x)$.

Solution   First use the definition of multiplication.

$$f(x)g(x) = \frac{x^2-4x}{x^2-2x-8} \cdot \frac{x^2-4}{x^2-4x+4}$$

$$= \frac{(x^2-4x)(x^2-4)}{(x^2-2x-8)(x^2-4x+4)}$$

Factor the numerator and denominator and remove a factor of 1.

$$f(x)g(x) = \frac{x(x-4)(x+2)(x-2)}{(x^2-4x+2x-8)(x-2)^2}$$

$$= \frac{x(x-4)(x+2)(x-2)}{(x(x-4)+2(x-4))(x-2)^2}$$

$$= \frac{x(x-4)(x+2)(x-2)}{(x+2)(x-4)(x-2)^2}$$

$$= \frac{x}{x-2}$$

Note: Because division by zero is meaningless, this argument is valid only if x does not equal -2, 4, or 2.

Definition of division. If $b \neq 0$, $c \neq 0$, and $d \neq 0$, then $\dfrac{a}{b} \div \dfrac{c}{d} = \dfrac{a}{b} \cdot \dfrac{d}{c}$ or $\dfrac{\frac{a}{b}}{\frac{c}{d}} = \dfrac{a}{b} \cdot \dfrac{d}{c}$.

Example 13 If $f(x) = \dfrac{x^2-3x-10}{x^2+2x}$ and $g(x) = \dfrac{x^2-9x+20}{x+6}$, simplify the expression $\dfrac{f(x)}{g(x)}$.

Solution Begin with the definition of division, then use the definition of multiplication.

$$\frac{f(x)}{g(x)} = \frac{\dfrac{x^2-3x-10}{x^2+2x}}{\dfrac{x^2-9x+20}{x+6}}$$

$$= \frac{x^2-3x-10}{x^2+2x} \cdot \frac{x+6}{x^2-9x+20}$$

$$= \frac{(x^2-3x-10)(x+6)}{(x^2+2x)(x^2-9x+20)}$$

Factor the numerator and denominator and remove a factor of 1.

$$\frac{f(x)}{g(x)} = \frac{(x^2 - 5x + 2x - 10)(x+6)}{x(x+2)(x^2 - 5x - 4x + 20)}$$

$$= \frac{(x(x-5) + 2(x-5))(x+6)}{x(x+2)(x(x-5) - 4(x-5))}$$

$$= \frac{(x+2)(x-5)(x+6)}{x(x+2)(x-4)(x-5)}$$

$$= \frac{x+6}{x(x-4)}$$

Note: Because division by zero is meaningless, none of the denominators in this argument may equal zero. Therefore, this argument is valid only if x does not equal 0, −2, −6, 4, or 5.

Exercises for Section 8.3

1. Reduce each of the following rational expressions by removing a factor of 1 as in Example 3 in this section.

   a) $\dfrac{x^4 y^6}{x^2 y^{10}}$

   b) $\dfrac{a^8 b^3}{a^5 b^{10}}$

   c) $\dfrac{x^5 (x^2 + 3)^7}{x^7 (x^2 + 3)^5}$

   d) $\dfrac{x^2 (x^2 + 4)^5 (x^2 + 9)^3}{x^5 (x^2 + 4)^2 (x^2 + 9)^7}$

2. Consider the rational expression $\dfrac{14}{21}$.

   a) Reduce $\dfrac{14}{21}$ by dividing both the numerator and denominator by 7.

   b) Factor the numerator and denominator of $\dfrac{14}{21}$, then remove a factor of 1, as in Example 4 in this section. If your result is not the same as the result in part (a), check your work for error.

3. Use the method of Example 4 in this section to reduce each of the following rational expressions.

   a) $\dfrac{15}{75}$

   b) $\dfrac{35}{42}$

   c) $\dfrac{60}{96}$

   d) $\dfrac{34}{51}$

4. Use the method of Example 5 in this section to reduce each of the following rational expressions. Clearly state the x-values for which your argument is not valid.

   a) $\dfrac{x^2 + 2x}{x^2 - 4}$

   b) $\dfrac{x^2 - 4x + 4}{x^2 - 2x}$

   c) $\dfrac{x^2 - 2x - 3}{x^2 + x}$

   d) $\dfrac{2x^2 - x - 1}{2x^2 - x}$

   e) $\dfrac{2x^2 - 16x}{x^3 - 3x^2 - 40x}$

   f) $\dfrac{x^4 - 10x^2 + 9}{x^3 + 4x^2 + 3x}$

   g) $\dfrac{4x^3 + 20x^2 - 9x - 45}{2x^3 + 13x^2 + 15x}$

   h) $\dfrac{6x^2 + x - 40}{4x^2 - 4x - 15}$

5. If $f(x) = x(x+3) - 2(x+3)$ and $g(x) = x(x+3)$, simplify the rational expression $\dfrac{f(x)}{g(x)}$. Clearly state the x-values for which your argument is not valid.

6. The following is the graph of $f(x) = x^2 - 4x - 5$.

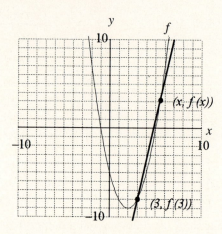

Find the slope of the line that goes through the points $(x, f(x))$ and $(3, f(3))$. Reduce the expression and clearly state the x-values for which the slope does not exist.

7. Use the sign change rule for fractions to help reduce each of the following rational expressions. Clearly state all x-values for which your argument is not valid.

a) $\dfrac{3-x}{x(x-3)}$

b) $\dfrac{-x-5}{x^2(x+5)}$

c) $\dfrac{4-x^2}{x^2-2x}$

d) $\dfrac{3+2x-x^2}{x^2-3x}$

e) $\dfrac{4x^3-12x^2-x+3}{3+5x-2x^2}$

f) $\dfrac{4x^4-5x^2+1}{x+x^2-2x^3}$

8. If $f(x) = 3x+5$, reduce the expression $\dfrac{f(x)-f(5)}{5-x}$. Clearly state all x-values for which your argument is not valid.

9. If $f(x) = 2x^2+3x$, reduce the expression $\dfrac{f(x)-f(3)}{3-x}$. Clearly state all x-values for which your argument is not valid.

10. The following is the graph of $f(x) = x^2$.

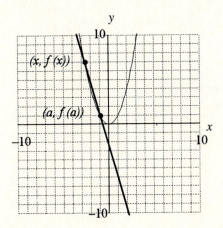

Find the slope of the line that passes through the points $(x, f(x))$ and $(a, f(a))$. Reduce the resulting expression and clearly state all x-values for which the slope of the line is undefined.

11. Simplify each of the following rational expressions. Clearly state all x-values for which your argument is not valid.

a) $\dfrac{x^2+4x}{2-x} \cdot \dfrac{x^2-11x+18}{x^2-5x-36}$

b) $\dfrac{-x-1}{x} \div \dfrac{x^2-1}{4x^2}$

c) $\dfrac{x^2+x}{x^2-1} \cdot \dfrac{x^2-2x+1}{x^2-3x+2}$

d) $\dfrac{x^2-x-12}{x^2+5x+6} \cdot \dfrac{x^2+8x+16}{x^2-16}$

e) $\dfrac{2x^2-7x-15}{x^2-2x-15} \div \dfrac{6x^2+7x-3}{x^2-3x}$

f) $\dfrac{x(x+5)-2(x+5)}{x^2-4} \div \dfrac{x^2+8x+15}{x^2+5x+6}$

g) $\dfrac{x^3+3x^2-25x-75}{x^2+9x+20} \cdot \dfrac{24+2x-x^2}{x^3-2x^2-15x}$

h) $\dfrac{x^4-2x^2+1}{x-x^3} \div (x^2+x)$

12. Suppose that you have two functions $f(x) = \dfrac{x}{x+1}$ and $g(x) = \dfrac{x^2-1}{x^3-x^2}$. Find and simplify both $f(x)g(x)$ and $\dfrac{f(x)}{g(x)}$. Clearly state all x-values for which each argument is not valid.

13. Simplify each of the following rational expressions. Clearly state all x-values for which your argument is not valid.

a) $\dfrac{-x-8}{x^2+8x}$

b) $\dfrac{x^3+x^2+x+1}{x^4-1}$

c) $\dfrac{x^2+2x-a^2-2a}{x-a}$

d) $\dfrac{x}{-x-3} \cdot \dfrac{x^2-9}{x^3-3x^2}$

e) $\dfrac{\dfrac{x}{x+6}}{\dfrac{x^2+2x}{30-x-x^2}}$

f) $\dfrac{\dfrac{x}{-x-5}}{x^2+3x} \cdot \dfrac{x+3}{5x-x^2}$ $\dfrac{}{x^2-25}$

14. Suppose that $f(x) = x^2$. Simplify the expression $\dfrac{f(x+h) - f(x)}{h}$. Clearly state all values of h for which your argument is not valid.

15. If $f(x) = 2x^2 + x$, simplify the expression $\dfrac{f(x) - f(a)}{x - a}$. Clearly state all values of x for which your argument is not valid.

16. (TI-82) It is possible to reduce $\dfrac{x^2 - 4}{x - 2}$ in the following manner:

$$\frac{x^2 - 4}{x - 2} = \frac{(x+2)(x-2)}{x - 2}$$
$$= x + 2$$

Because division by zero is meaningless, this argument is valid only if x does not equal 2. You can check this result on a TI-82 calculator in the following manner.

Load the equations $y = \dfrac{x^2 - 4}{x - 2}$ and $y = x + 2$ into the Y= menu.

```
Y18(X²-4)/(X-2)
Y28X+2
Y3=■
Y4=
Y5=
Y6=
Y7=
Y8=
```

Press 2nd TblSet to open the following window. You will find the TblSet command above the WINDOW button.

```
TABLE SETUP
 TblMin=0
 △Tbl=1
Indpnt: Auto Ask
Depend: Auto Ask
```

Use the arrow keys and the ENTER key to duplicate these settings.

Press 2nd TABLE to open the following viewing window. You will find the TABLE command above the GRAPH button.

Note that the function values in the Y1 column and the Y2 column are identical, except at $x = 2$. You can now be confident that you have correctly simplified this rational expression.

Note: Try using the arrow keys to scroll up and down through the table.

8.4 Adding and Subtracting Rational Expressions

This section will illustrate how to add and subtract rational expressions. We will build on familiar and useful elementary methods and techniques.

> *When adding or subtracting rational expressions:*
>
> 1. *Find a least common denominator (LCD).*
> 2. *Make equivalent fractions that have the LCD as their denominator.*
> 3. *Add or subtract the numerators and place this result over our LCD.*
> 4. *Reduce.*

Example 1 Simplify the expression $\dfrac{1}{2}+\dfrac{2}{3}$.

Solution The least common denominator is 6. Make equivalent fractions with a denominator of 6 by multiplying each fraction by the appropriate form of the number 1. Add the results.

$$\frac{1}{2}+\frac{1}{3}=\frac{1}{2}\cdot\frac{3}{3}+\frac{1}{3}\cdot\frac{2}{2}$$

$$=\frac{3}{6}+\frac{2}{6}$$

$$=\frac{5}{6}$$

The numerator and denominator of $\dfrac{5}{6}$ have no common factors, so $\dfrac{5}{6}$ is reduced.

Example 2 Simplify the expression $\dfrac{5}{24}+\dfrac{11}{36}$.

Solution You must first find a common denominator. Begin by factoring each of the denominators.

$$24 = 2\cdot2\cdot2\cdot3$$
$$36 = 2\cdot2\cdot3\cdot3$$
$$\text{LCD} = ?$$

In order that the LCD be a multiple of 24, or $2\cdot2\cdot2\cdot3$, it must contain at least three 2's and one 3.

$$24 = 2\cdot2\cdot2\cdot3$$
$$36 = 2\cdot2\cdot3\cdot3$$
$$\text{LCD} = 2\cdot2\cdot2\cdot3\cdot?$$

In order that the LCD be a multiple of 36, or $2\cdot2\cdot3\cdot3$, it must contain at least two 2's and two 3's. Another 3 will need to be added to the LCD.

$$24 = 2\cdot2\cdot2\cdot3$$
$$36 = 2\cdot2\cdot3\cdot3$$
$$\text{LCD} = 2\cdot2\cdot2\cdot3\cdot3$$

Therefore the LCD is $2 \cdot 2 \cdot 2 \cdot 3 \cdot 3$, or 72. Multiply each fraction by the appropriate form of 1 to produce an equivalent fraction whose denominator is the LCD, $2 \cdot 2 \cdot 2 \cdot 3 \cdot 3$.

$$\frac{5}{24} + \frac{11}{36} = \frac{5}{2 \cdot 2 \cdot 2 \cdot 3} + \frac{11}{2 \cdot 2 \cdot 3 \cdot 3}$$

$$= \frac{5}{2 \cdot 2 \cdot 2 \cdot 3} \cdot \frac{3}{3} + \frac{11}{2 \cdot 2 \cdot 3 \cdot 3} \cdot \frac{2}{2}$$

$$= \frac{5 \cdot 3}{2 \cdot 2 \cdot 2 \cdot 3 \cdot 3} + \frac{11 \cdot 2}{2 \cdot 2 \cdot 2 \cdot 3 \cdot 3}$$

$$= \frac{15}{72} + \frac{22}{72}$$

$$= \frac{37}{72}$$

Because the numerator and denominator of $\dfrac{37}{72}$ contain no common factors, $\dfrac{37}{72}$ is reduced.

Example 3 Simplify the expression $\dfrac{4}{x^2 + 2x} - \dfrac{2}{x^2 + 3x + 2}$.

Solution First factor the denominators.

$$x^2 + 2x = x(x+2)$$

$$x^2 + 3x + 2 = (x+1)(x+2)$$

$$\text{LCD} = ?$$

In order that the LCD be a multiple of $x^2 + 2x$, or $x(x+2)$, the LCD must contain factors x and $x+2$.

$$x^2 + 2x = x(x+2)$$

$$x^2 + 3x + 2 = (x+1)(x+2)$$

$$\text{LCD} = x(x+2)?$$

In order that the LCD be a multiple of $x^2 + 3x + 2$, or $(x+1)(x+2)$, the LCD must contain factors $x+1$ and $x+2$. A factor of $x+1$ must be added to the LCD.

$$x^2 + 2x = x(x+2)$$

$$x^2 + 3x + 2 = (x+1)(x+2)$$

$$\text{LCD} = x(x+2)(x+1)$$

Multiply each fraction by the appropriate form of 1 to produce an equivalent fraction whose denominator is the LCD, $x(x+2)(x+1)$.

$$\frac{4}{x^2+2x}-\frac{2}{x^2+3x+2}=\frac{4}{x(x+2)}-\frac{2}{(x+1)(x+2)}$$

$$=\frac{4}{x(x+2)}\cdot\frac{x+1}{x+1}-\frac{2}{(x+1)(x+2)}\cdot\frac{x}{x}$$

$$=\frac{4x+4}{x(x+2)(x+1)}-\frac{2x}{x(x+2)(x+1)}$$

Subtract the numerators and place the result over the LCD.

$$=\frac{4x+4-2x}{x(x+2)(x+1)}$$

$$=\frac{2x+4}{x(x+2)(x+1)}$$

This example can be reduced by removing a factor of 1.

$$=\frac{2(x+2)}{x(x+2)(x+1)}$$

$$=\frac{2}{x(x+1)}$$

Note: Because division by zero is meaningless, this argument is valid only if x does not equal 0, -2, or -1.

Example 4   Simplify the expression $\dfrac{x-5}{x^2-x-6}+\dfrac{x-3}{10+3x-x^2}$.

Solution   Note that all numerators and denominators are arranged in descending powers of x except for the denominator of the second fraction. Invoke the sign change rule, and negate the fraction bar and denominator of the second fraction.

$$\frac{x-5}{x^2-x-6}+\frac{x-3}{10+3x-x^2}=\frac{x-5}{x^2-x-6}-\frac{x-3}{x^2-3x-10}$$

Factor each denominator.

$$=\frac{x-5}{(x-3)(x+2)}-\frac{x-3}{(x-5)(x+2)}$$

The LCD is $(x-3)(x+2)(x-5)$. Multiply each fraction by the appropriate form of 1 to produce an equivalent fraction whose denominator is the LCD, $(x-3)(x+2)(x-5)$.

$$=\frac{x-5}{(x-3)(x+2)}\cdot\frac{x-5}{x-5}-\frac{x-3}{(x-5)(x+2)}\cdot\frac{x-3}{x-3}$$

$$=\frac{x^2-10x+25}{(x-3)(x+2)(x-5)}-\frac{x^2-6x+9}{(x-3)(x+2)(x-5)}$$

Note that the squaring-a-binomial multiplication shortcut was used in each numerator. Next, subtract the numerators and place the result over the LCD.

$$= \frac{(x^2 - 10x + 25) - (x^2 - 6x + 9)}{(x-3)(x+2)(x-5)}$$

$$= \frac{x^2 - 10x + 25 - x^2 + 6x - 9}{(x-3)(x+2)(x-5)}$$

$$= \frac{-4x + 16}{(x-3)(x+2)(x-5)}$$

Can this be reduced? The numerator will factor as $-4(x-4)$. However, this will not provide a common factor, so the expression is reduced.

Note: Because division by zero is meaningless, this argument is valid only if x does not equal 3, -2, or 5.

Example 5   If $f(x) = \dfrac{x+1}{x-3}$ and $g(x) = \dfrac{x-2}{x+5}$, simplify the expression $f(x) - g(x)$.

Solution

$$f(x) - g(x) = \frac{x+1}{x-3} - \frac{x-2}{x+5}$$

The LCD is $(x-3)(x+5)$. Multiply each fraction by the appropriate form of 1 to produce an equivalent fraction whose denominator is the LCD, $(x-3)(x+5)$.

$$f(x) - g(x) = \frac{x+1}{x-3} \cdot \frac{x+5}{x+5} - \frac{x-2}{x+5} \cdot \frac{x-3}{x-3}$$

$$= \frac{x^2 + 5x + x + 5}{(x-3)(x+5)} - \frac{x^2 - 3x - 2x + 6}{(x-3)(x+5)}$$

$$= \frac{x^2 + 6x + 5}{(x-3)(x+5)} - \frac{x^2 - 5x + 6}{(x-3)(x+5)}$$

Subtract the numerators and place the result over the LCD.

$$f(x) - g(x) = \frac{(x^2 + 6x + 5) - (x^2 - 5x + 6)}{(x-3)(x+5)}$$

$$= \frac{x^2 + 6x + 5 - x^2 + 5x - 6}{(x-3)(x+5)}$$

$$= \frac{11x - 1}{(x-3)(x+5)}$$

Because the numerator and denominator have no common factors, this last rational expression is reduced.

Note: Because division by zero is meaningless, this argument is valid only if x does not equal 3 or -5.

Example 6   If $f(x) = \dfrac{x+2}{x+5}$, simplify the expression $\left[f(x) - f(2) \right] \cdot \dfrac{1}{x-2}$. Clearly state all values of x for which this argument is not valid.

Solution

$$\left[f(x) - f(2) \right] \cdot \frac{1}{x-2} = \left[\frac{x+2}{x+5} - \frac{2+2}{2+5} \right] \cdot \frac{1}{x-2}$$

$$= \left[\frac{x+2}{x+5} - \frac{4}{7} \right] \cdot \frac{1}{x-2}$$

Begin by adding the fractions within the grouping symbols.

$$\left[f(x) - f(2) \right] \cdot \frac{1}{x-2} = \left[\frac{x+2}{x+5} \cdot \frac{7}{7} - \frac{4}{7} \cdot \frac{x+5}{x+5} \right] \cdot \frac{1}{x-2}$$

$$= \left[\frac{7x+14}{7(x+5)} - \frac{4x+20}{7(x+5)} \right] \cdot \frac{1}{x-2}$$

$$= \left[\frac{(7x+14) - (4x+20)}{7(x+5)} \right] \cdot \frac{1}{x-2}$$

$$= \left[\frac{7x+14-4x-20}{7(x+5)} \right] \cdot \frac{1}{x-2}$$

$$= \frac{3x-6}{7(x+5)} \cdot \frac{1}{x-2}$$

Now multiply.

$$\left[f(x) - f(2) \right] \cdot \frac{1}{x-2} = \frac{3(x-2)}{7(x+5)(x-2)}$$

$$= \frac{3}{7(x+5)}$$

Note: Because division by zero is meaningless, this argument is valid only if x does not equal -5 or 2.

Exercises for Section 8.4

1. Use the method shown in Example 2 in this section to find the least common denominator for each of the following.

   a) $24 =$ b) $54 =$ c) $56 =$ d) $99 =$ e) $18 =$ f) $27 =$

        $18 =$ $36 =$ $21 =$ $66 =$ $24 =$ $36 =$

     LCD $=$ LCD $=$ LCD $=$ LCD $=$ $36 =$ $81 =$

                                                                        LCD $=$ LCD $=$

2. Use the method shown in Example 3 in this section to find the least common denominator for each of the following.

   a) $x^2 + 4x =$ b) $x^2 - 6x =$ c) $x^2 - 4x - 5 =$

        $x^2 - 4 =$ $x^2 - 5x - 6 =$ $x^2 + 3x + 2 =$

        LCD $=$ LCD $=$ LCD $=$

   d) $x^2 - 6x - 16 =$ e) $x^3 - 3x^2 - 4x + 12 =$ f) $x^4 - 5x^2 + 4 =$

        $2x^2 - 15x - 8 =$ $x^3 - x^2 - 6x =$ $x^3 + 3x^2 + 2x =$

        LCD $=$ LCD $=$ LCD $=$

3. Multiply each of the following fractions by an appropriate form of 1 to produce an equivalent fraction with the given denominator.

   a) $\dfrac{5}{2 \cdot 2 \cdot 2 \cdot 3} \cdot \dfrac{?}{?} = \dfrac{?}{2 \cdot 2 \cdot 2 \cdot 3 \cdot 3}$
       b) $\dfrac{7}{2 \cdot 3 \cdot 3 \cdot 5} \cdot \dfrac{?}{?} = \dfrac{?}{2 \cdot 2 \cdot 3 \cdot 3 \cdot 5}$

   c) $\dfrac{11}{2 \cdot 2 \cdot 5 \cdot 7} \cdot \dfrac{?}{?} = \dfrac{?}{2 \cdot 2 \cdot 2 \cdot 5 \cdot 5 \cdot 7}$
       d) $\dfrac{7}{2 \cdot 3 \cdot 3 \cdot 11} \cdot \dfrac{?}{?} = \dfrac{?}{2 \cdot 3 \cdot 3 \cdot 3 \cdot 11 \cdot 11}$

   e) $\dfrac{x}{x-2} \cdot \dfrac{?}{?} = \dfrac{?}{(x-2)(x-3)}$
       f) $\dfrac{x+1}{x(x-5)} \cdot \dfrac{?}{?} = \dfrac{?}{x(x+2)(x-5)}$

   g) $\dfrac{x-3}{(x+1)(x-5)} \cdot \dfrac{?}{?} = \dfrac{?}{(x-1)(x+1)(x-5)}$
       h) $\dfrac{x^2-4}{x(x+5)} \cdot \dfrac{?}{?} = \dfrac{?}{x(x+5)(2x-3)}$

4. Use the method shown in Example 2 in this section to simplify each of the following expressions.

   a) $\dfrac{11}{21} + \dfrac{5}{28}$
       b) $\dfrac{5}{32} + \dfrac{1}{24}$
       c) $\dfrac{11}{20} - \dfrac{5}{28}$

   d) $\dfrac{11}{18} - \dfrac{5}{27}$
       e) $\dfrac{5}{24} + \dfrac{5}{36} - \dfrac{11}{48}$
       f) $\dfrac{5}{44} - \dfrac{7}{66} + \dfrac{4}{99}$

5. Simplify each of the following expressions. Clearly state the values of all variables for which your argument is not valid.

   a) $\dfrac{x}{x+1} - \dfrac{2}{x}$
       b) $\dfrac{x+1}{x} - \dfrac{x-3}{x+1}$
       c) $\dfrac{1}{x^2+x} - \dfrac{1}{x^2-1}$

d) $1 - \dfrac{1}{t} + \dfrac{1}{t^2}$

e) $\dfrac{t+1}{t-3} - \dfrac{t-3}{t+1}$

f) $\dfrac{1}{t^2} - \dfrac{t}{(t+1)^2}$

g) $\dfrac{1}{r} - \dfrac{1}{r+1} - \dfrac{1}{r-1}$

h) $\dfrac{x+1}{x^2-5x+6} - \dfrac{x-2}{x^2-2x-3}$

i) $\dfrac{1}{x^2} + \dfrac{2}{(10-x)^2}$

j) $\dfrac{x+3}{x^2+9x} - \dfrac{x}{x^2+12x+27}$

6. Suppose that $f(x) = \dfrac{x+1}{x+2}$ and $g(x) = \dfrac{x+3}{x+1}$. Simplify each of the following expressions. Clearly state all values of x for which your argument is not valid.

   a) $f(x)g(x)$

   b) $\dfrac{f(x)}{g(x)}$

   c) $f(x) + g(x)$

   d) $f(x) - g(x)$

7. Suppose that $f(x) = \dfrac{1}{x}$. Simplify each of the following expressions. Clearly state all values of x for which your argument is not valid.

   a) $f(x) - f(a)$

   b) $f(x+h) - f(x)$

8. Suppose that $f(x) = \dfrac{x-2}{x^2-x-12}$ and $g(x) = \dfrac{x+3}{x^2-6x+8}$. Find and simplify each of the following. Clearly state all values of x for which your argument is not valid.

   a) $f(x)g(x)$

   b) $f(x) - g(x)$

9. Simplify each of the following expressions. Clearly state all values of x for which your argument is not valid.

   a) $\dfrac{2x}{x+3} - \dfrac{6}{-x-3}$

   b) $\dfrac{3x}{x-3} + \dfrac{9}{3-x}$

   c) $\dfrac{1}{x^2+4x} - \dfrac{1}{16-x^2}$

   d) $\dfrac{1}{-x^2-2x} - \dfrac{1}{4-x^2}$

   Hint: First use the sign change rule for fractions.

10. Simplify each of the following expressions. Clearly state all values of all variables for which your argument is not valid.

   a) $\dfrac{3}{x^2-6x} + \dfrac{x}{x^2-36}$

   b) $\dfrac{x+2}{x^2-4x+3} + \dfrac{x-3}{2-x-x^2}$

   c) $\dfrac{-x-3}{7+6x-x^2} - \dfrac{x+2}{x^2+5x+4}$

   d) $\dfrac{x-5}{x^2-3x-4} - \dfrac{x+2}{x^2-10x+24}$

   e) $\dfrac{x+1}{x^2+x-6} + \dfrac{x+2}{x^2-6x+8} - \dfrac{x+5}{x^2-x-12}$

   f) $\dfrac{x+1}{x^2+4x} - \dfrac{x}{x^2+5x+4} + \dfrac{x+2}{4x-x^2}$

11. Simplify each of the following expressions. Clearly state all values of x for which your argument is not valid.

   a) $\left[\dfrac{1}{x} + \dfrac{1}{x+1} \right] \div \left[\dfrac{1}{x} - \dfrac{1}{x+1} \right]$

   b) $\left[\dfrac{x+1}{x-2} - \dfrac{x-2}{x+1} \right] \cdot \dfrac{x-2}{2x-1}$

   c) $\left[\dfrac{x}{x+1} - \dfrac{2}{3} \right] \div (x-2)$

d) $\dfrac{1}{h}\cdot\left[\dfrac{1}{x+h}-\dfrac{1}{x}\right]$

e) $\dfrac{x}{x+1}-\dfrac{x+1}{x}\cdot\dfrac{x}{x-3}$

f) $\dfrac{x}{x+1}\cdot\dfrac{x-1}{x-3}+\dfrac{1}{x-1}$

12. If $f(x)=\dfrac{x}{x-5}$, simplify each of the following expressions. Clearly state all values of x for which your argument is not valid.

   a) $\left[f(x)-f(2)\right]\cdot\dfrac{1}{x-2}$

   b) $\left[f(7+h)-f(7)\right]\cdot\dfrac{1}{h}$

13. If $f(x)=\dfrac{2}{x^2}$, simplify each of the following expressions. Clearly state all values of x for which your argument is not valid.

   a) $\left[f(x)-f(a)\right]\cdot\dfrac{1}{x-a}$

   b) $\left[f(x+h)-f(x)\right]\cdot\dfrac{1}{h}$

8.5 Complex Fractions

Maybe you're thinking that this material has been hard enough—and now we're going to do complex fractions? Weren't they complex enough before? Rest easy. The word *complex* does not necessarily mean difficult. A complex fraction is essentially a fraction problem over another fraction problem, as will be shown in the examples that follow.

Example 1   Simplify $\dfrac{\dfrac{1}{2}+\dfrac{1}{3}}{\dfrac{1}{3}+\dfrac{1}{4}}$.

Solution   The first attempt will go as follows: Simplify the numerator, simplify the denominator, then divide.

$$\frac{\dfrac{1}{2}+\dfrac{1}{3}}{\dfrac{1}{3}+\dfrac{1}{4}}=\frac{\dfrac{3}{6}+\dfrac{2}{6}}{\dfrac{4}{12}+\dfrac{3}{12}}$$

$$=\frac{\dfrac{5}{6}}{\dfrac{7}{12}}$$

$$=\frac{5}{6}\cdot\frac{12}{7}$$

$$=\frac{60}{42}$$

$$=\frac{10}{7}$$

A More Sophisticated Method. Sophisticated does not mean more difficult. In fact, this second method is easier than the first method used. Begin by selecting a least common denominator for *all* of the fractions in the numerator and denominator, which is 12 in this case. Multiply both the numerator and denominator of this complex fraction by 12.

$$\frac{\dfrac{1}{2}+\dfrac{1}{3}}{\dfrac{1}{3}+\dfrac{1}{4}}=\frac{12\left(\dfrac{1}{2}+\dfrac{1}{3}\right)}{12\left(\dfrac{1}{3}+\dfrac{1}{4}\right)}$$

$$=\frac{12\left(\dfrac{1}{2}\right)+12\left(\dfrac{1}{3}\right)}{12\left(\dfrac{1}{3}\right)+12\left(\dfrac{1}{4}\right)}$$

$$=\frac{6+4}{4+3}$$

$$=\frac{10}{7}$$

If you perform a bit of the work mentally, this second method is quick and concise.

Example 2   Simplify $\dfrac{\dfrac{1}{x}-\dfrac{1}{y}}{x-y}$.

Solution   Multiply the numerator and denominator by xy, invoke the sign change rule for fractions, then reduce.

$$\frac{\dfrac{1}{x}-\dfrac{1}{y}}{x-y}=\frac{xy\left(\dfrac{1}{x}-\dfrac{1}{y}\right)}{xy(x-y)}$$

$$=\frac{xy\left(\dfrac{1}{x}\right)-xy\left(\dfrac{1}{y}\right)}{xy(x-y)}$$

$$=\frac{y-x}{xy(x-y)}$$

$$=-\frac{x-y}{xy(x-y)}$$

$$=-\frac{1}{xy}$$

Note: Because division by zero is meaningless, this argument is valid only if $x, y \neq 0$ and $x \neq y$.

Example 3   Simplify $\dfrac{1-\dfrac{1}{x^2}}{1-\dfrac{1}{x}}$.

Solution   Multiply the numerator and denominator of this fraction by x^2. That should be enough to clear the fractions from both the numerator and denominator of this fraction. Once the fractions are cleared, factor and reduce.

$$\frac{1-\dfrac{1}{x^2}}{1-\dfrac{1}{x}} = \frac{x^2\left(1-\dfrac{1}{x^2}\right)}{x^2\left(1-\dfrac{1}{x}\right)}$$

$$= \frac{x^2(1)-x^2\left(\dfrac{1}{x^2}\right)}{x^2(1)-x^2\left(\dfrac{1}{x}\right)}$$

$$= \frac{x^2-1}{x^2-x}$$

$$= \frac{(x+1)(x-1)}{x(x-1)}$$

$$= \frac{x+1}{x}$$

Note: Because division by zero is meaningless, this argument is valid only if x does not equal 0 or 1.

Example 4   If $f(x) = \dfrac{1}{x}+1$ and $g(x) = x^2-1$, simplify the expression $\dfrac{f(x)}{g(x)}$.

Solution   Begin by multiplying the numerator and denominator by x.

$$\frac{f(x)}{g(x)} = \frac{\dfrac{1}{x}+1}{x^2-1}$$

$$= \frac{x\left(\dfrac{1}{x}+1\right)}{x(x^2-1)}$$

$$= \frac{x\left(\dfrac{1}{x}\right)+x(1)}{x(x^2-1)}$$

$$= \frac{1+x}{x(x^2-1)}$$

Arrange the numerator in descending powers of x, factor the denominator, then reduce.

$$= \frac{x+1}{x\left(x^2 - 1\right)}$$

$$= \frac{x+1}{x(x+1)(x-1)}$$

$$= \frac{1}{x(x-1)}$$

Note: Because division by zero is meaningless, this argument is valid only if x does not equal 0, −1, or 1.

Example 5   If $f(x) = \dfrac{1}{x}$, simplify the expression $\dfrac{f(x) - f(3)}{x - 3}$.

Solution

$$\frac{f(x) - f(3)}{x - 3} = \frac{\dfrac{1}{x} - \dfrac{1}{3}}{x - 3}$$

Multiply the numerator and denominator of this complex fraction by $3x$ to clear the numerator of fractions.

$$= \frac{3x\left(\dfrac{1}{x} - \dfrac{1}{3}\right)}{3x(x-3)}$$

$$= \frac{3x\left(\dfrac{1}{x}\right) - 3x\left(\dfrac{1}{3}\right)}{3x(x-3)}$$

$$= \frac{3 - x}{3x(x-3)}$$

Negate the numerator and the fraction bar, but leave the denominator alone. Reduce the resulting fraction.

$$= -\frac{x-3}{3x(x-3)}$$

$$= -\frac{1}{3x}$$

Note: Because division by zero is meaningless, this argument is valid only if x does not equal 0 or 3.

Example 6   The following is the graph of the equation $f(x) = \dfrac{1}{x}$.

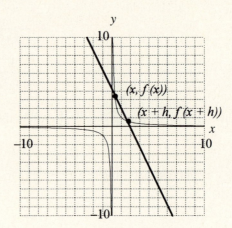

Find the slope of the line through the points $(x+h, f(x+h))$ and $(x, f(x))$.

Solution   The slope is found by dividing the change in y by the change in x.

$$m = \frac{\text{Change in } y}{\text{Change in } x}$$

$$m = \frac{f(x+h) - f(x)}{(x+h) - x}$$

$$= \frac{\dfrac{1}{x+h} - \dfrac{1}{x}}{h}$$

Multiply both the numerator and denominator of this fraction by $x(x+h)$.

$$m = \frac{x(x+h)\left(\dfrac{1}{x+h} - \dfrac{1}{x}\right)}{hx(x+h)}$$

$$= \frac{x(x+h)\left(\dfrac{1}{x+h}\right) - x(x+h)\left(\dfrac{1}{x}\right)}{hx(x+h)}$$

$$= \frac{x - (x+h)}{hx(x+h)}$$

$$= \frac{x - x - h}{hx(x+h)}$$

$$= \frac{-h}{hx(x+h)}$$

$$= \frac{-1}{x(x+h)}$$

Note: Because division by zero is meaningless, this argument is valid only if $x, h \neq 0$ and $x \neq -h$.

Exercises for Section 8.5

1. Simplify each of the following expressions twice. Use each of the two methods presented in Example 1 in this section.

a) $\dfrac{\dfrac{1}{2} + \dfrac{1}{3}}{\dfrac{1}{4} + \dfrac{1}{3}}$

b) $\dfrac{\dfrac{1}{4} - \dfrac{1}{5}}{\dfrac{1}{2} + \dfrac{3}{4}}$

c) $\dfrac{\dfrac{1}{4} + \dfrac{1}{3}}{\dfrac{1}{8}}$

d) $\dfrac{\dfrac{1}{5}}{\dfrac{1}{3} + \dfrac{1}{6}}$

e) $\dfrac{\dfrac{3}{4} + \dfrac{5}{8}}{\dfrac{3}{2} - \dfrac{5}{4}}$

f) $\dfrac{\dfrac{7}{10} + \dfrac{3}{2}}{\dfrac{5}{4} - \dfrac{11}{10}}$

2. Simplify the expression $\dfrac{\dfrac{1}{2} - \dfrac{1}{x}}{x - 2}$ by multiplying both the numerator and denominator by $2x$. Clearly state all values of x for which your argument is not valid.

3. Simplify each of the following expressions. Use the technique presented in Example 2 in this section. Clearly state all values of all variables for which your argument is not valid.

a) $\dfrac{\dfrac{1}{5} - \dfrac{1}{x}}{x - 5}$

b) $\dfrac{\dfrac{1}{z} - \dfrac{1}{x}}{x - z}$

c) $\dfrac{a - b}{\dfrac{1}{a} - \dfrac{1}{b}}$

d) $\dfrac{4 - x}{\dfrac{1}{x} - \dfrac{1}{4}}$

4. Simplify the expression $\dfrac{\dfrac{1}{x^2} - \dfrac{1}{x}}{1 - \dfrac{1}{x}}$ by multiplying both the numerator and denominator by x^2.

Clearly state all values of x for which your argument is not valid.

5. Simplify each of the following expressions. Use the technique presented in Example 3 in this section. Clearly state all values of x for which your argument is not valid.

a) $\dfrac{\dfrac{2}{x} - 1}{1 - \dfrac{2}{x^2}}$

b) $\dfrac{\dfrac{3}{x} - \dfrac{1}{x^2}}{9 - \dfrac{1}{x^2}}$

c) $\dfrac{1 - \dfrac{1}{x} - \dfrac{12}{x^2}}{1 - \dfrac{9}{x^2}}$

d) $\dfrac{\dfrac{4}{x} - 1}{1 - \dfrac{16}{x^2}}$

6. If $f(x) = 1 + \dfrac{1}{x}$ and $g(x) = x^2 - 1$, simplify each of the following expressions. Clearly state all values of x for which your argument is not valid.

a) $\dfrac{f(x)}{g(x)}$

b) $\dfrac{g(x)}{f(x)}$

c) $\dfrac{xf(x)}{g(x)}$

d) $\dfrac{(x-1)f(x)}{g(x)}$

7. If $f(x) = \dfrac{1}{x}$, simplify each of the following expressions. Clearly state all values of all variables for which your argument is not valid.

a) $\dfrac{f(x) - f(7)}{x - 7}$

b) $\dfrac{f(x) - f(x)}{x - a}$

8. If $f(x) = \dfrac{2x}{x+1}$, simplify the expression $\dfrac{f(x) - f(5)}{x - 5}$. Clearly state all values of x for which your argument is not valid.

9. If $f(x) = \dfrac{4}{x^2}$, simplify each of the following expressions. Clearly state all values of all variables for which your argument is not valid.

a) $\dfrac{f(x) - f(3)}{x - 3}$

b) $\dfrac{f(x) - f(a)}{x - a}$

10. The following is the graph of $f(x) = \dfrac{x}{x+1}$.

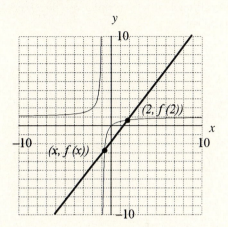

Find the slope of the line that passes through the points $(x, f(x))$ and $(2, f(2))$. Simplify your result and clearly state all values of x for which the slope of the line is undefined.

11. The following is the graph of $f(x) = \dfrac{2}{x^2}$.

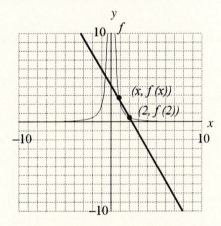

Find the slope of the line that passes through the points $(x, f(x))$ and $(2, f(2))$. Simplify your result and clearly state all values of x for which the slope of the line is undefined.

12. Simplify each of the following expressions. Clearly state all values of all variables for which your argument is not valid.

a) $\dfrac{\dfrac{a}{b} - \dfrac{b}{a}}{a - b}$

b) $\dfrac{5 + \dfrac{1}{x} - \dfrac{5}{x^2} - \dfrac{1}{x^3}}{\dfrac{1}{x^3} + \dfrac{4}{x^2} - \dfrac{5}{x}}$

c) $\dfrac{4 - \dfrac{5}{x^2} + \dfrac{1}{x^4}}{\dfrac{2}{x} + \dfrac{3}{x^2} + \dfrac{1}{x^3}}$

d) $\dfrac{\dfrac{a+1}{a-1} + 1}{1 + \dfrac{a+1}{1-a}}$

13. If $f(x) = \dfrac{x}{x-3}$ and $g(x) = \dfrac{x-3}{x+1}$, simplify each of the following expressions. *Hint:* You will need to call upon all of the skills you developed in Sections 8.3, 8.4, and 8.5 to perform well on these problems.

a) $f(x)g(x)$

b) $\dfrac{f(x)}{g(x)}$

c) $f(x) + g(x)$

d) $g(x) - f(x)$

e) $\dfrac{f(x) - f(5)}{x - 5}$

f) $\dfrac{g(x) - g(2)}{x - 2}$

g) $\dfrac{1 + f(x)}{g(x) - 1}$

h) $\dfrac{1 - f(x)}{1 - \dfrac{1}{g(x)}}$

i) $\left[1 + \dfrac{1}{f(x)}\right] \cdot \left[\dfrac{x}{2 + (x+1)g(x)}\right]$

j) $\left[f(x) + \dfrac{1}{g(x)}\right] \div (2x^2 + x)$

8.6 Equations Involving Rational Expressions

This section illustrates how to solve equations that contain rational expressions. When fractions in equations were encountered in earlier work, both sides of the equation were multiplied by the least common denominator. This technique cleared the fractions from the equation. This same technique will work equally well on the equations of this section.

Example 1   Use an analytical method to solve the following equation for x: $x - \dfrac{2x - 1}{5} = 1$.

Solution

Multiply both sides of the equation by the least common denominator.	Use a TI-82 calculator to check this solution in the following manner:
$$5\left[x - \frac{2x-1}{5}\right] = 5[1]$$ $$5x - 5\left[\frac{2x-1}{5}\right] = 5$$ $$5x - [2x - 1] = 5$$ $$5x - 2x + 1 = 5$$ $$3x = 4$$ $$x = \frac{4}{3}$$	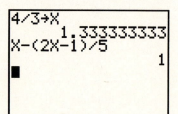

A Fundamental Principle.   Clearly, if $a = b$ and c is any real number, then $ac = bc$. However, if $ac = bc$, it is not necessarily true that $a = b$. For example, $7 \cdot 0 = 5 \cdot 0$ but $7 \neq 5$.

If $ac = bc$ and $c \neq 0$, then $\dfrac{1}{c}$ is meaningful and the following can be done:

$$ac = bc$$

$$[ac] \cdot \frac{1}{c} = [bc] \cdot \frac{1}{c}$$

$$a = b$$

The following fundamental principle has been established.

If a, b, and c are real numbers and c ≠ 0, then ac = bc if and only if a = b.

If $c \neq 0$, then the equations $a = b$ and $ac = bc$ are *equivalent*. The equations $a = b$ and $ac = bc$ will have the *same* solution set. However, if both sides of an equation are multiplied by zero, you must be on the lookout for potential trouble.

Example 2   The solution set of the equation $x = 1$ is $\{1\}$. If both sides of the equation $x = 1$ are multiplied by x, the result is the equation $x^2 = x$. Find the solution set of the equation $x^2 = x$.

Solution   Because the equation $x^2 = x$ is nonlinear, make one side of the equation zero, factor, and use the factor theorem to complete the solution (the zero product property would work as well).

$$x^2 = x$$

$$x^2 - x = 0$$

$$x(x-1) = 0$$

$$x = 0, 1$$

The solution set of $x^2 = x$ is $\{0,1\}$. An extra answer that does not satisfy the original equation $x = 1$ has been introduced.

Extra answers that do not satisfy the original equation are called extraneous answers.

What Happened?   Both sides of the equation $x = 1$ were multiplied by x. Note that x could equal zero and the resulting equation may not be equivalent to the original. The solution set of an equation can change if you multiply by zero.

Example 3   Solve the following equation for x: $\dfrac{x+1}{x-3} = 3$.

Graphical Solution   Draw the graphs of $f(x) = \dfrac{x+1}{x-3}$ and $g(x) = 3$ and note where the graphs of f and g intersect. Load the equations f and g into the Y= menu: Y1 = (X+1)/(X−3) and Y2 = 3.

Push the ZOOM button and select 6:ZStandard to capture the following image:	Use the intersect utility in the CALC menu to find the point of intersection.
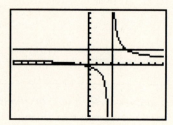	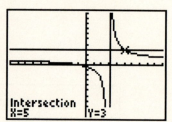
	Hint: When requested for a guess, move the cursor near the point of intersection.

It appears that the solution of the equation $\dfrac{x+1}{x-3} = 3$ is $x = 5$.

Analytical Solution

Multiply both sides of the equation by the least common denominator. $(x-3)\left[\dfrac{x+1}{x-3}\right] = [3](x-3)$ $x+1 = 3x-9$ $x-3x = -9-1$ $-2x = -10$ $x = 5$	Check your solution in the original equation. 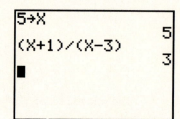

Therefore $x = 5$ is a solution of the equation $\dfrac{x+1}{x-3} = 3$.

Note: Because both sides of this equation were multiplied by $x-3$, which could equal zero, extraneous answers might be introduced. You must always check your solutions in the original equation.

It's a good idea to place your graphical solution and analytical solution side-by-side on your homework paper so that you can easily compare the results of the two methods. Note that the student whose work is shown below used the techniques of Section 8.2 to find the x-intercept of the function $f(x) = \dfrac{x+1}{x-3}$ as well as the vertical and horizontal asymptotes.

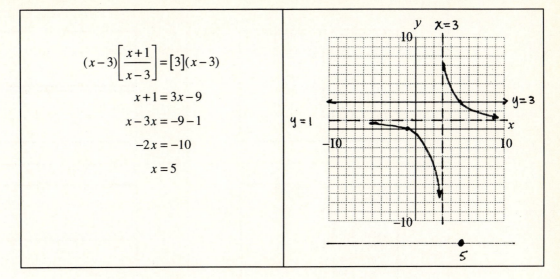

Example 4   Solve the following equation for x: $\dfrac{x}{x+1} = 2 - \dfrac{1}{x}$.

Graphical Solution   Rather than finding the points of intersection of two complicated graphs, make one side of the equation zero and find where one graph crosses the x-axis.

$$\frac{x}{x+1} + \frac{1}{x} - 2 = 0$$

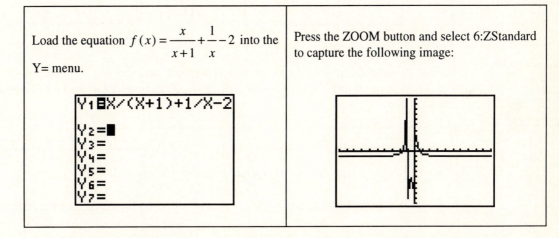

Load the equation $f(x) = \dfrac{x}{x+1} + \dfrac{1}{x} - 2$ into the Y= menu.

Press the ZOOM button and select 6:ZStandard to capture the following image:

This image is unlike any that has been drawn to date. It appears that there may be vertical asymptotes at $x = -1$ and $x = 0$. If so, they would divide the plane into three regions: to the left of $x = -1$, between $x = -1$ and $x = 0$, and to the right of $x = 0$. Investigate each of these regions separately.

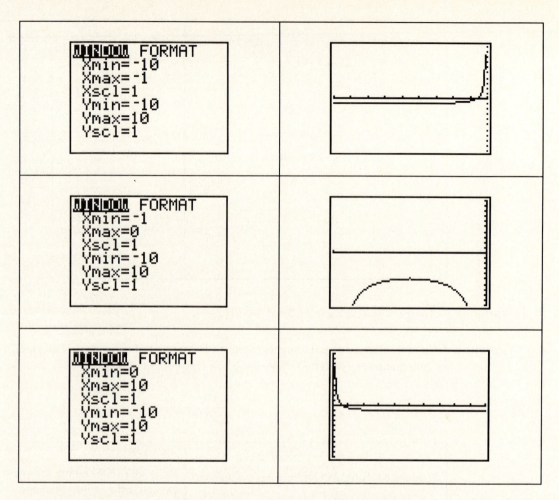

Note that this graph crosses the *x*-axis two times, once between $x = -10$ and $x = -1$ and a second time between $x = 0$ and $x = 10$. You can use the root-finding utility in the CALC menu to find these zeros.

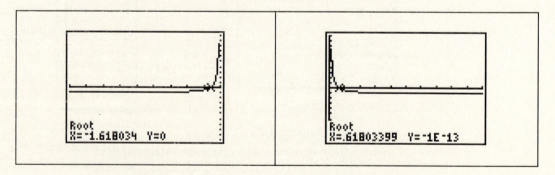

Analytical Solution Begin by multiplying both sides of the equation by the least common denominator, $x(x+1)$.

$$x(x+1)\left[\frac{x}{x+1}\right]=\left[2-\frac{1}{x}\right]x(x+1)$$

$$x(x+1)\left[\frac{x}{x+1}\right]=[2]x(x+1)-\left[\frac{1}{x}\right]x(x+1)$$

$$x^2 = 2x(x+1)-(x+1)$$

Note that multiplying by the least common denominator clears the fractions from this equation. However, $x(x+1)$ could be equal to zero, so extraneous solutions might be introduced. Because the last equation is nonlinear, make one side zero.

$$x^2 = 2x^2+2x-x-1$$

$$0 = x^2+x-1$$

| You can use the quadratic formula to find the solutions of this equation. $$x=\frac{-b\pm\sqrt{b^2-4ac}}{2a}$$ $$=\frac{-1\pm\sqrt{(1)^2-4(1)(-1)}}{2(1)}$$ $$=\frac{-1\pm\sqrt{5}}{2}$$ | Use a TI-82 calculator to find decimal approximations of these solutions.

 ```(-1-√5)/2   -1.618033989 (-1+√5)/2   .6180339887 ■``` |

Note that these approximations agree quite nicely with the approximations found by the graphing method. You can be confident that you do not have any extraneous solutions, but it is always best to check the solutions in the original equation.

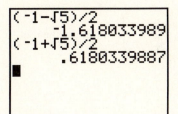

```
(-1-√5)/2→X
      -1.618033989
X/(X+1)
       2.618033989
2-1/X
       2.618033989
■
```

```
(-1+√5)/2→X
        .6180339887
X/(X+1)
        .3819660113
2-1/X
        .3819660113
```

Note that each value of x makes the left side of the original equation $\dfrac{x}{x+1}=2-\dfrac{1}{x}$ equal the right side.

Example 5   Solve the following equation for x: $\dfrac{x}{x-3}+\dfrac{1}{x-2}=\dfrac{3}{x^2-5x+6}$.

Graphical Solution Rather than finding where two rather complicated graphs intersect, make one side of the equation zero and find where one graph intersects the *x*-axis.

$$\frac{x}{x-3} + \frac{1}{x-2} - \frac{3}{x^2-5x+6} = 0$$

Load the equation $y = \dfrac{x}{x-3} + \dfrac{1}{x-2} - \dfrac{3}{x^2-5x+6}$ into the Y= menu.

Push the ZOOM button and select 6:ZStandard to capture the following image in your viewing window.

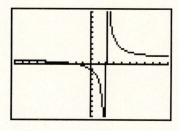

You can use the root-finding utility in the CALC menu to find where the graph crosses the *x*-axis.

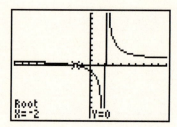

Analytical Solution Begin by factoring the denominators of the equation.

$$\frac{x}{x-3} + \frac{1}{x-2} = \frac{3}{x^2-5x+6}$$

$$\frac{x}{x-3} + \frac{1}{x-2} = \frac{3}{(x-3)(x-2)}$$

Multiply both sides of the equation by the least common denominator, $(x-3)(x-2)$. Note that if *x* were 3 or 2, you would be multiplying by zero. Keep in mind that extraneous solutions might be introduced.

$$(x-3)(x-2)\left[\frac{x}{x-3} + \frac{1}{x-2}\right] = \left[\frac{3}{(x-3)(x-2)}\right](x-3)(x-2)$$

$$(x-3)(x-2)\left[\frac{x}{x-3}\right] + (x-3)(x-2)\left[\frac{1}{x-2}\right] = \left[\frac{3}{(x-3)(x-2)}\right](x-3)(x-2)$$

$$x(x-2) + 1(x-3) = 3$$

Note that the fractions have been cleared from the equation. This last result is nonlinear. Make one side of the equation zero.

$$x^2 - 2x + x - 3 = 3$$

$$x^2 - x - 6 = 0$$

You can use the *ac* test to factor or you can use the quadratic formula. The former method is illustrated below.

$$x^2 - 3x + 2x - 6 = 0$$

$$x(x - 3) + 2(x - 3) = 0$$

$$(x + 2)(x - 3) = 0$$

$$x = -2, 3$$

Because of the graphical solution, -2 will probably be a solution. However, 3 is probably an extraneous solution.

| If you let $x = 3$ in the original equation, you can see an immediate problem with this solution. $$\frac{x}{x-3} + \frac{1}{x-2} = \frac{3}{x^2 - 5x + 6}$$ $$\frac{3}{3-3} + \frac{1}{3-2} = \frac{3}{(3)^2 - 5(3) + 6}$$ $$\frac{3}{0} + \frac{1}{1} = \frac{3}{0}$$ Division by zero is meaningless. | However, $x = -2$ works just fine. $$\frac{x}{x-3} + \frac{1}{x-2} = \frac{3}{x^2 - 5x + 6}$$ $$\frac{-2}{-2-3} + \frac{1}{-2-2} = \frac{3}{(-2)^2 - 5(-2) + 6}$$ $$\frac{2}{5} - \frac{1}{4} = \frac{3}{20}$$ $$\frac{8}{20} - \frac{5}{20} = \frac{3}{20}$$ |

The calculator does not like the solution $x = 3$ very much either.

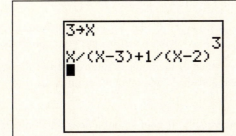

Exercises for Section 8.6

1. Use the technique of Example 1 in this section to solve each of the following equations for x.

   a) $x - \dfrac{x+2}{8} = 1$

   b) $3x - \dfrac{3-4x}{5} = 2$

   c) $\dfrac{x-1}{2} - \dfrac{x-1}{3} - \dfrac{x-1}{4} = 1$

   d) $x - \dfrac{2x+1}{5} = \dfrac{3-2x}{7}$

2. The solution set of the equation $x = 3$ is $\{3\}$. If both sides of the equation $x = 3$ are multiplied by x, the result is the equation $x^2 = 3x$. Find the solution set of the equation $x^2 = 3x$.

3. The solution set of the equation $x = -2$ is $\{-2\}$. If both sides of the equation $x = -2$ are multiplied by $x-4$, the result is the equation $x(x-4) = -2(x-4)$. Find the solution set of the equation $x(x-4) = -2(x-4)$.

   Hint: The equation is nonlinear, so make one side of the equation zero.

4. The solution set of the equation $x = 5$ is $\{5\}$. If both sides of the equation $x = 5$ are multiplied by $2x+3$, the result is the equation $x(2x+3) = 5(2x+3)$. Find the solution set of the equation $x(2x+3) = 5(2x+3)$.

   Hint: The equation is nonlinear, so make one side of the equation zero.

5. The solution set of the equation $x^2 = 4$ is $\{-2,2\}$. If both sides of the equation $x^2 = 4$ are multiplied by $4x-3$, the result is the equation $x^2(4x-3) = 4(4x-3)$. Find the solution set of the equation $x^2(4x-3) = 4(4x-3)$.

   Hint: The equation is nonlinear, so make one side of the equation zero.

6. (TI-82) Consider the equations $f(x) = \dfrac{x-3}{x+2}$ and $g(x) = 2$. Set up a coordinate system on a sheet of graph paper and clearly indicate the scale on each axis.

   a) Plot and label the x-intercept of the graph of f. Draw dotted lines representing the vertical and horizontal asymptotes of the graph of f and label each with its equation.

   b) Load the equations of f and g into Y1 and Y2 of the Y= menu. Push the ZOOM button and select 6:ZStandard. Copy the image in the viewing window onto your coordinate system.

   c) Use the intersect utility to find the x-value of the point of intersection of the graphs of f and g. Draw a number line below the graph and shade and label the solution of $f(x) = g(x)$ on your number line in the usual manner.

   d) Solve the equation $f(x) = g(x)$ analytically. If your solution does not compare favorably with that found in part (i), check your work for error.

7. Use an analytical method to find the solution of each of the following equations.

   a) $\dfrac{x-3}{x+2} = 2$

   b) $\dfrac{x+1}{4-x} = -2$

   c) $\dfrac{2x-1}{x+2} = -3$

   d) $\dfrac{1-x}{5x+2} = 1$

   e) $\dfrac{x-200}{x+50} = 10$

f) $\dfrac{500-2x}{x+100}=50$　g) $\dfrac{3}{2}=\dfrac{x}{x+1}$　h) $\dfrac{2x-3}{3x+4}=-\dfrac{5}{2}$　i) $\dfrac{x+1}{x-3}=\dfrac{x+2}{x-4}$　j) $\dfrac{3x-2}{6x+1}=\dfrac{x+5}{2x+3}$

Hint: Always begin by multiplying both sides of the equation by the least common denominator. Be sure to check for extraneous solutions.

8.　Load the equation $f(x)=1+\dfrac{3}{x}-\dfrac{10}{x^2}$ into the Y= menu. Push the ZOOM button and select 6:ZStandard. Set up a coordinate system on a sheet of graph paper and clearly indicate the scale on each axis. Copy the image in the viewing window onto your coordinate system.

a) Use the root-finding utility in the CALC menu to find where the graph of f crosses the x-axis. Draw a number line below the graph and shade and label these solutions of $f(x)=0$ on your number line in the usual manner.

b) Use an analytical method to find the solutions of the equation $f(x)=0$. If these answers do not agree favorably with the estimates found in part (a), check your work for error.

9.　Use an analytical method to find the solution of each of the following equations.

a) $4=\dfrac{5}{x}+\dfrac{6}{x^2}$　　b) $3+\dfrac{12}{x^2}=\dfrac{20}{x}$　　c) $9+\dfrac{18}{x}=\dfrac{1}{x^2}+\dfrac{2}{x^3}$　　d) $36+\dfrac{1}{x^4}=\dfrac{13}{x^2}$

Hint: Always begin by multiplying both sides of the equation by the least common denominator. Be sure to check for extraneous solutions.

10.　Load the equation $f(x)=\dfrac{x}{x-1}-\dfrac{3}{x+2}-\dfrac{15}{x^2+x-2}$ into the Y= menu. Push the ZOOM button and select 6:ZStandard. Set up a coordinate system on a sheet of graph paper and clearly indicate the scale on each axis. Copy the image in the viewing window onto your coordinate system.

a) Use the root-finding utility in the CALC menu to find where the graph of f crosses the x-axis. Draw a number line below the graph and shade and label these solutions of $f(x)=0$ on your number line in the usual manner.

b) Use an analytical method to find the solutions of the equation $f(x)=0$. If these answers do not agree favorably with the estimates found in part (a), check your work for error.

11.　Use an analytical method to find the solution of each of the following equations.

a) $\dfrac{x}{x+4}-\dfrac{1}{1-x}=\dfrac{2x+7}{x^2+3x-4}$

b) $\dfrac{x}{x+1}=\dfrac{2}{x+5}+\dfrac{8}{x^2+6x+5}$

c) $\dfrac{x}{x+3}+\dfrac{2}{x+2}=\dfrac{11}{x^2+5x+6}$

d) $\dfrac{2x}{x+1}-\dfrac{1}{5-x}=\dfrac{6}{x^2-4x-5}$

e) $\dfrac{2x}{x-3}-\dfrac{5}{x+2}=\dfrac{16}{x^2-x-6}$

f) $\dfrac{x}{x-2}=\dfrac{8}{x+2}+\dfrac{23}{x^2-4}$

Hint: Always begin by multiplying both sides of the equation by the least common denominator. Be sure to check for extraneous solutions.

12. (TI-82) Load the equation $f(x) = \dfrac{1}{x} + \dfrac{x}{x-3} - 4$ into the Y= menu. Push the ZOOM button and select 6:ZStandard. Set up a coordinate system on a sheet of graph paper and clearly indicate the scale on each axis. Copy the image in the viewing window onto your coordinate system.

   a) Use the root-finding utility in the CALC menu to find where the graph of f crosses the x-axis. Draw a number line below your graph and shade and label these solutions of $f(x) = 0$ on your number line in the usual manner.

   b) Use an analytical method to find the solutions of the equation $f(x) = 0$. Place your answers in simple radical form and use your calculator to find decimal approximations, correct to the nearest hundredth (two decimal places). If these answers do not agree favorably with the estimates found in part (a), check your work for error.

13. Use an analytical method to find the solution of each of the following equations. Place your solutions in simple radical form. Use a calculator to find decimal approximations for your solutions, correct to the nearest ten-thousandth (four decimal places). *Hint*: Always begin by multiplying both sides of the equation by the least common denominator. Be sure to check for extraneous solutions.

   a) $1 = \dfrac{4}{x} + \dfrac{2}{x^2}$

   b) $2 - \dfrac{1}{x} = \dfrac{2}{x^2}$

   c) $\dfrac{1}{x} + \dfrac{x}{x+1} = 2$

   d) $\dfrac{x}{x+2} - \dfrac{x+1}{1-x} = 4$

   e) $1 - \dfrac{4}{x+2} = \dfrac{2}{x^2+2x}$

   f) $\dfrac{x}{x+3} + \dfrac{6}{x-2} = \dfrac{-22}{6-x-x^2}$

14. (TI-82) Try to solve the equation $\dfrac{9x}{x^2+4} = 6$ analytically. Use your calculator to draw a graph that helps explain what is going on. Use complete sentences to explain what is going on.

15. Suppose that the deer population in a certain region of Idaho is increasing according to the formula $P = \dfrac{500t+500}{t+3}$, where the variable P represents the number of deer in the region and t represents the number of years that have passed since 1960.

   a) (TI-82) Use your calculator to draw the graph of the deer population from 1960 to 1980. Set up a coordinate system on a sheet of graph paper and clearly indicate the scale on each axis. Make an accurate graph of the image in the viewing window on your coordinate system.

   b) Let Y2 = 400 in the Y= menu. Draw the graph of Y2 on your coordinate system. Use the intersect utility in the CALC menu to find the approximate year when the deer population first reached 400 deer. Draw a number line below your graph and shade and label this solution of $P = 400$ in the usual manner.

   c) Use an analytical method to solve the equation $\dfrac{500t+500}{t+3} = 400$. Use your calculator to find a decimal approximation of this solution, correct to the nearest tenth of a year (one decimal place). If this solution does not compare favorably with the estimate found in part (b), check your work for error.

16. A region is hit hard by drought, and the water level in a local lake begins to drop according to the formula $h = \dfrac{15t+520}{t+8}$, where h represents the water level in feet measured from the base of a dam at one end of the lake. The variable t is the number of months that have passed since the beginning

of the drought, officially recorded on May 1, 1989. Approximate the month and the year when the water level first fell to 18 feet.

17. Max has found a new job in a machine shop. His salary is based on the number of pieces that he can cut in a day, so there is incentive for Max to produce as many pieces as possible. At first Max has miserable production levels, but soon his production begins to rise according to the formula $N = \dfrac{40t + 20}{t + 8}$, where N is the number of pieces that Max cuts in a day and t is the number of days that Max has been on the job. Approximately how many days had Max been on the job when his production level first reached 25 pieces?

18. If $a \neq 0$, then the *reciprocal* of a is the number $\frac{1}{a}$.

    a) Suppose that the sum of a number and its reciprocal is $\frac{34}{15}$. Find the number.

    b) Suppose that the sum of the reciprocals of two consecutive integers is $\frac{61}{930}$. Find these consecutive integers.

    c) Suppose that the sum of the reciprocals of two consecutive even integers is $\frac{51}{1300}$. Find these consecutive even integers.

19. A number is added to both the numerator and the denominator of $\frac{2}{3}$. The resulting fraction equals $\frac{11}{15}$. Find the number.

20. Jaime has a small business where he cuts leather pieces for ladies' handbags. Suppose that Jaime's *fixed costs*—money that goes for rent, utilities, tools, etc.—are $400. In addition, for each piece that he cuts, his costs increase by $2.50. Let x represent the number of pieces that he cuts and let C represent his total costs.

    a) Express C as a function of x.

    b) If Jaime's costs C are divided by the total pieces cut x, then the *average cost* for cutting x pieces is obtained. Let A represent Jaime's average cost for cutting x pieces. Express A as a function of x.

    c) Find Jaime's average cost if he cuts 300 pieces. Attach units to your answer and use complete sentences to explain the meaning of your answer.

    d) How many pieces does Jaime need to cut to lower his average cost to $3.55?

8.7 Summary and Review

Important Visualization Tools

The graph of $y = \dfrac{1}{x}$.

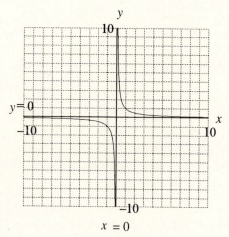

$$x = 0$$

The x-axis is a horizontal asymptote and the y-axis is a vertical asymptote.

Slide the graph of $y = \dfrac{1}{x}$ to the right 3 units and up 2 units to arrive at the graph of $y = \dfrac{1}{x-3} + 2$.

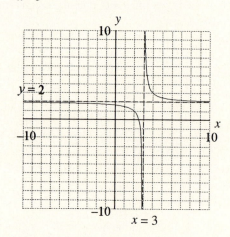

$$x = 3$$

The horizontal asymptote is the line $y = 2$ and the vertical asymptote is the line $x = 3$.

Because $x = 2$ makes the numerator zero and the denominator nonzero, the graph of $y = \dfrac{x-2}{x+3}$ will cross the x-axis at $(2,0)$.

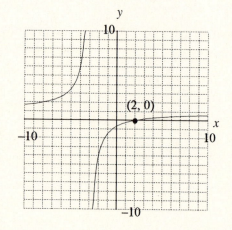

Because $x = -3$ makes the denominator zero and the numerator nonzero, the graph of $y = \dfrac{x-2}{x+3}$ will have a vertical asymptote at $x = -3$.

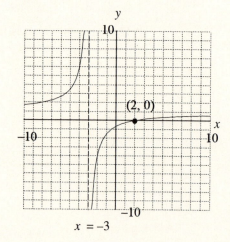

$$x = -3$$

A Review of Key Words and Phrases

☛ If p and q are polynomials, then the function f defined by $f(x) = \dfrac{p(x)}{q(x)}$ is called a rational function. (page 702)

☛ Suppose that N and D are real numbers: (page 716)

    1. If $D = 0$, then $\dfrac{N}{D}$ is undefined. Division by zero is meaningless.

    2. If $N = 0$ and $D \neq 0$, then $\dfrac{N}{D} = 0$.

☛ Suppose that we have a rational function f with equation $f(x) = \dfrac{p(x)}{q(x)}$. (page 71)

    1. If $p(x) = 0$ and $q(x) \neq 0$, then $f(x) = 0$ and x is a zero of the function f. The graph of the function f will cross the x-axis at $(x, 0)$.

    2. If $p(x) \neq 0$ and $q(x) = 0$, then the graph of the function f will have a vertical asymptote at x.

☛ To find the horizontal asymptote and the end-behavior of a rational function, investigate the behavior of the function for values of x such as 10, 100, 1,000, 10,000, ... and values of x such as $-10, -100, -1,000, -10,000, \dots$ (page 718)

☛ To investigate the behavior of the graph of a rational function near a vertical asymptote, investigate the behavior of the function for values of x that approach the asymptote from one side then the other. (page 718)

☛ The definition of multiplication: If $b \neq 0$ and $d \neq 0$, then $\dfrac{a}{b} \cdot \dfrac{c}{d} = \dfrac{ac}{bd}$. (page 728)

☛ The sign change rule: If any two parts of a fraction are negated, the fraction will remain the same. (page 732)

☛ The definition of division: If $b \neq 0$, $c \neq 0$, and $d \neq 0$, then $\dfrac{a}{b} \div \dfrac{c}{d} = \dfrac{a}{b} \cdot \dfrac{d}{c}$ or $\dfrac{\frac{a}{b}}{\frac{c}{d}} = \dfrac{a}{b} \cdot \dfrac{d}{c}$.

(page 734)

☛ When adding or subtracting rational expressions: (page 740)

    1. Find a least common denominator (LCD).

    2. Make equivalent fractions that have the LCD as their denominator.

    3. Add or subtract the numerators and place this result over the LCD.

    4. Reduce.

☛ If a, b, and c are real numbers and $c \neq 0$, then $ac = bc$ if and only if $a = b$. Equations are equivalent if and only if they have the same solution set. (page 758)

☛ Extra answers that do not satisfy the original equation are called extraneous answers. (page 759)

TI-82 Keywords and Menus

☛ Using the Dot mode versus the Connected mode when drawing graphs. (page 708)

☛ Using the Table feature to check the simplification of a rational expression. (Exercise 16, page 738)

☛ Checking the solutions of equations with your calculator. (pages 758, 760, 763, and 765)

Review Exercises

1. For the equations below, perform each of the following tasks:

   i) Set up a coordinate system on a sheet of graph paper. Clearly indicate the scale on each axis. Find the x-intercept of the graph of the equation and plot and label it on your coordinate system.

   ii) Find the horizontal and vertical asymptotes of the graph and draw them on your coordinate system as dotted lines. Label each asymptote with its equation.

   iii) Draw the graph of the equation on your coordinate system. Draw a number line below the graph and shade and label the solution of the inequality $f(x) \geq 0$ on your number line in the usual manner. Use set-builder and interval notation to describe the solution set.

   a) $f(x) = \dfrac{1}{x-4} + 2$

   b) $f(x) = -\dfrac{1}{x+3} - 4$

   c) $f(x) = \dfrac{2x-5}{x+2}$

   d) $f(x) = \dfrac{3x-1}{4-x}$

   e) $f(x) = \dfrac{180x+360}{x+3}$

   f) $f(x) = \dfrac{170-50x}{x+5}$

2. Simplify each of the following expressions. Clearly state the values of all variables for which your argument is not valid.

   a) $\dfrac{x^3 - 9x}{3x^2 - x^3}$

   b) $\dfrac{25x^3 + 50x^2 - 4x - 8}{2x - 3x^2 - 5x^3}$

   c) $\dfrac{x+2}{x^2-9} - \dfrac{x+1}{x^2+3x}$

   d) $\dfrac{x^2 - x - 6}{12 - x - x^2} \div \dfrac{x^2 - 3x - 10}{x^2 - 4x - 32}$

   e) $\dfrac{\dfrac{x}{y} - \dfrac{y}{x}}{y - x}$

   f) $\dfrac{\dfrac{5}{x} - 1}{\dfrac{1}{x} - \dfrac{25}{x^3}}$

3. If $f(x) = \dfrac{x+1}{x-5}$ and $g(x) = \dfrac{5-x}{x+3}$, simplify each of the following expressions. Clearly state all values of x for which your argument is not valid.

   a) $f(x)g(x)$

   b) $f(x) + g(x)$

   c) $\dfrac{f(x) - f(-2)}{x - (-2)}$

   d) $\dfrac{g(x) - g(-1)}{x - (-1)}$

   e) $\dfrac{1 + \dfrac{1}{f(x)}}{1 - \dfrac{1}{g(x)}}$

   f) $\dfrac{\dfrac{1}{f(x)} - \dfrac{1}{g(x)}}{\dfrac{1}{f(x)} + \dfrac{1}{g(x)}}$

4. The following is the graph of $f(x) = x^2 + 2x - 8$.

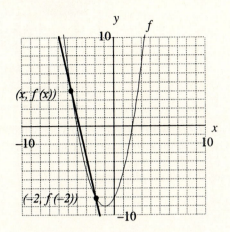

Find the slope of the line that passes through the points $(x, f(x))$ and $(-2, f(-2))$. Simplify the resulting expression and clearly state all values of x for which the slope is undefined.

5. It is not difficult to simplify the expression $\dfrac{1}{x} + \dfrac{x}{x+1}$.

$$\frac{1}{x} + \frac{x}{x+1} = \frac{x+1}{x(x+1)} + \frac{x^2}{x(x+1)}$$

$$= \frac{x^2 + x + 1}{x(x+1)}$$

Of course, this argument is valid only if x does not equal 0 or -1. Use your calculator to check this result in the following manner:

Load the original problem and your solution into the Y= menu.	Press 2nd TblSet and use the arrow and ENTER keys to make the following adjustments:
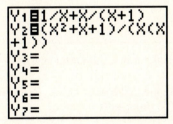	

Press 2nd TABLE to open the following window:

X	Y₁	Y₂
-3	1.1667	1.1667
-2	1.5	1.5
-1	ERROR	ERROR
0	ERROR	ERROR
1	1.5	1.5
2	1.1667	1.1667
3	1.0833	1.0833

X=-3

Note that the original expression and your solution are identical for all values except $x = -1$ and $x = 0$. Use the arrow keys to scroll up and down to view more of this table.

6. Solve each of the following equations for x.

a) $1 = \dfrac{13}{x^2} + \dfrac{9}{x^4}$

b) $4 = \dfrac{18}{x^3} + \dfrac{9}{x^2} - \dfrac{8}{x}$

c) $\dfrac{x+1}{x+3} + \dfrac{x-3}{x+1} = 1$

d) $\dfrac{2x}{x-2} - \dfrac{9}{x+1} = \dfrac{33}{x^2 - x - 2}$

7. It is not difficult to solve the equation $\dfrac{1}{x} = 2 - \dfrac{x}{x+5}$.

$$x(x+5)\left[\dfrac{1}{x}\right] = \left[2 - \dfrac{x}{x+5}\right]x(x+5)$$

$$(x+5) = 2x(x+5) - x^2$$

$$x+5 = 2x^2 + 10x - x^2$$

$$0 = x^2 + 9x - 5$$

$$x = \dfrac{-9 \pm \sqrt{81+20}}{2}$$

$$x = \dfrac{-9 \pm \sqrt{101}}{2}$$

Check the solution $\dfrac{-9 - \sqrt{101}}{2}$ on your calculator in the following manner:

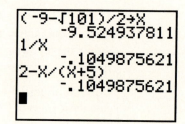

```
(-9-√101)/2→X
          -9.524937811
1/X
          -.1049875621
2-X/(X+5)
          -.1049875621
■
```

Check the second solution on your calculator in a similar manner.

8. A measles epidemic breaks out among the children of Orick, California. The number of children infected is given by the equation $N = \dfrac{1800t + 40}{5t + 14}$, where N represents the number of children infected and t represents the number of days that have passed since the outbreak of the epidemic.

a) (TI-82) Use your calculator to sketch the graph of this equation over the first 30 days of the epidemic. Set up a coordinate system on a sheet of graph paper, clearly indicate the scale on each axis, and copy the image in the viewing window onto your coordinate system.

b) How many children were infected by the tenth day of the epidemic?

c) How many days passed before the number of children infected reached 250?

d) How many children will be infected eventually?

9. If $a \neq 0$, then the reciprocal of a is the number $\frac{1}{a}$.

   a) Can the sum of a real number and its reciprocal equal 1? If so, find such a number. If not, use complete sentences and calculations to explain why it is not possible.

   b) The sum of a real number and three times its reciprocal is $\frac{79}{10}$. Find the number.

   c) The sum of a real number and the square of its reciprocal is $\frac{35}{12}$. Find the number.

10. Two smokestacks are separated by a distance of 500 yards. The total soot, in parts per million, falling on the little man shown in the diagram is given by the equation $T = \dfrac{50,000}{x} + \dfrac{2,500}{500 - x}$, where T represents the total soot and x represents the distance that the man positions himself from the smaller smokestack.

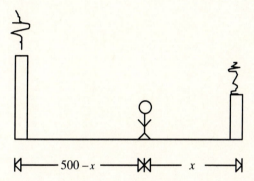

Where must the man stand so that the total soot raining down on his head is 200 parts per million?

Chapter 9

Exponential and Logarithmic Functions

This chapter introduces exponential and logarithmic functions. We'll begin with some interesting applications, then introduce the graph of $y = a^x$. The number e and the role it plays in growth and decay will be discussed. The concept of the inverse of a function will be developed and used to introduce the definition of the logarithm and the graph of $y = \log_a x$. The properties of the logarithm that are useful when solving equations and inequalities involving exponentials and logarithms will be presented. The chapter ends with some important real-world applications involving exponential and logarithmic functions.

9.1 Growth and Decay

In this section we will look at applications involving growth and decay.

Growth

Suppose that the population of Silvertown is increasing—in other words, suppose that it is *growing*.

Example 1 The population of Silvertown is growing at a rate of 2.4 percent per year. If its population is currently 5218 hearty souls, what will be the population 10 years from now?

Solution Let $P(t)$ represent the population of Silvertown after t years have passed. $P(0)$ would represent the population of Silvertown after zero years have passed. In other words, $P(0)$ represents the current population, which may be written as follows:

$$P(0) = 5218$$

The population at the end of one year will be 102.4 percent of the initial population, which may be written as follows:

$$P(1) = 1.024 P(0)$$
$$= 1.024(5218)$$

The population at the end of the second year will be 102.4 percent of the population at the end of the first year, which may be written as follows:

$$P(2) = 1.024 P(1)$$
$$= 1.024[1.024(5218)]$$
$$= (1.024)^2 (5218)$$

The population at the end of the third year will be 102.4 percent of the population at the end of the second year, which may be written as follows:

$$P(3) = 1.024\,P(2)$$

$$= 1.024\left[(1.024)^2(5218)\right]$$

$$= (1.024)^3(5218)$$

A definite pattern is emerging. The population of Silvertown after t years is given by the following equation:

$$P(t) = 5218(1.024)^t$$

To find the population 10 years from now, let $t = 10$ in the equation $P(t) = 5218(1.024)^t$. Use your calculator to approximate the answer.

$P(t) = 5218(1.024)^t$ $P(10) = 5218(1.024)^{10}$ $P(10) \approx 6614$	``` 5218*1.024^10 6614.600832 ```

Note: The original population of Silvertown was 5218. Note the position of this number in the equation $P(t) = 5218(1.024)^t$. At the end of each year, the population was 102.4 percent of the population at the end of the previous year. Note the position of the number 1.024 in the equation $P(t) = 5218(1.024)^t$.

Example 2   Use your calculator to sketch the graph of the equation $P(t) = 5218(1.024)^t$.

Solution   Load the equation $P(t) = 5218(1.024)^t$ into the Y= menu.

Examine the growth of the town over the first 10 years.

Make the following adjustments to the WINDOW parameters:	Push the GRAPH button to capture the following image in your viewing window:
WINDOW FORMAT Xmin=0 Xmax=10 Xscl=1 Ymin=0 Ymax=10000 Yscl=1000	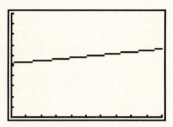

Because the population of Silvertown is growing, the graph is *increasing*.

Decay

Suppose that the population of Silvertown is decreasing—in other words, suppose that it is *decaying*.

Example 3   The population of Silvertown is decaying at a rate of 2.4 percent per year. If the population of Silvertown is currently 5218, what will be the population 10 years from now?

Solution   Let $P(t)$ represent the number of people in Silvertown after t years have passed. $P(0)$ would represent the population after zero years have passed. In other words, $P(0)$ represents the current population, which may be written as follows:

$$P(0) = 5218$$

If Silvertown loses 2.4 percent of its population each year, then it keeps 97.6 percent of its population each year. The population at the end of the first year will be 97.6 percent of the initial population.

$$P(1) = 0.976P(0)$$
$$= 0.976(5218)$$

The population at the end of the second year will be 97.6 percent of the population at the end of the first year.

$$P(2) = 0.976P(1)$$
$$= 0.976[0.976(5218)]$$
$$= (0.976)^2(5218)$$

The population at the end of the third year will be 97.6 percent of the population at the end of the second year.

$$P(3) = 0.976P(2)$$
$$= 0.976\left[(0.976)^2(5218)\right]$$
$$= (0.976)^3(5218)$$

A pattern emerges. The population of Silvertown at the end of t years is given by the following equation:

$$P(t) = 5218(0.976)^t$$

To find the population of Silvertown 10 years from now, let $t = 10$ in the equation $P(t) = 5218(0.976)^t$. Use your calculator to approximate the answer.

$P(t) = 5{,}218(0.976)^t$
 	
 $P(10) = 5{,}218(0.976)^{10}$
 	
 $P(10) \approx 4{,}092$
	```5218*0.976^10```
	``` 4092.627814```
	■

Note: The original population of Silvertown was 5218. Note the position of this number in the equation $P(t) = 5218(0.976)^t$. At the end of each year, the population of Silvertown was 97.6 percent of the population at the end of the previous year. Note the position of the number 0.976 in the equation $P(t) = 5218(0.976)^t$.

Example 4 Use your calculator to sketch the graph of the equation $P(t) = 5218(0.976)^t$.

Solution Load the equation $P(t) = 5218(0.976)^t$ into the Y= menu.

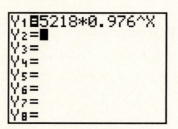

Examine the decay of the town over the first 10 years.

Make the following adjustments to the WINDOW parameters:	Push the GRAPH button to capture the following image in your viewing window:
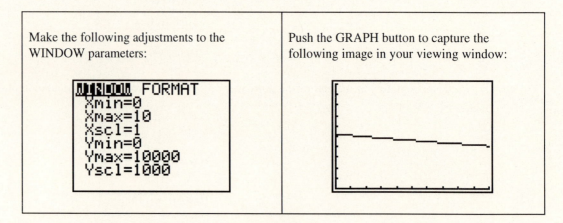	

Because the population of Silvertown is decaying, the graph is *decreasing*.

Exponential Functions

The following definition can be made:

> *If a and b are positive real numbers, the function defined by the equation $f(x) = a \cdot b^x$ is called an exponential function. We refer to b as the base of the exponential function.*

In Example 1 and Example 2, the equation $P(t) = 5218(1.024)^t$ gave the population of Silvertown as a function of time. The base of this exponential function is 1.024. Because this base is *larger* than 1, the population of Silvertown increases or grows. In Example 3 and Example 4, the equation $P(t) = 5218(0.976)^t$ gave the population of Silvertown as a function of time. The base of this exponential function is 0.976. Because this base is *smaller* than 1, the population of Silvertown decreases or decays.

> *Suppose that a and b are positive real numbers.*
> 1. *If $0 < b < 1$, then the graph of $f(x) = a \cdot b^x$ will decrease or decay.*
> 2. *If $b > 1$, then the graph of $f(x) = a \cdot b^x$ will increase or grow.*

Example 5   Jimmy deposited a sum of money in an account in 1980. Suppose that this amount is given by the equation $A(t) = 1510(1.045)^t$, where A represents the number of dollars in the account and t represents the number of years that have passed since 1980. Find the amount of Jimmy's original deposit and sketch the graph of the equation $A(t) = 1510(1.045)^t$.

Solution   To find the amount of the original deposit, you must find the amount in the account in the year 1980. Let $t = 0$ in the equation $A(t) = 1510(1.045)^t$. Recall that $a^0 = 1$, $a \neq 0$.

$$A(t) = 1510(1.045)^t$$

$$A(0) = 1510(1.045)^0$$

$$= 1510(1)$$

$$= 1510$$

Jimmy's original deposit was \$1510. Load the equation $A(t) = 1510(1.045)^t$ into the Y= menu in the following manner:

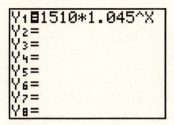

Sketch the graph of the amount in the account over the last 14 years. Because the base of the equation $A(t) = 1510(1.045)^t$ is 1.045, which is larger than 1, you can expect the graph to begin at 1510 and increase.

Make the following adjustments to the WINDOW parameters:	Push the GRAPH button to capture the following image in your viewing window:
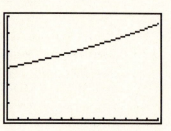	

Example 6   Jenny's doctor gives her an injection of an antibiotic. The amount of antibiotic in Jenny's bloodstream will begin to decrease as her body begins to rid itself of a certain percentage of antibiotic each hour. Suppose that the amount of antibiotic in Jenny's bloodstream is given by $A(t) = 500(0.88)^t$, where A represents the number of milligrams of antibiotic in her bloodstream and t represents the number of hours that have passed since she received the injection. Find the initial amount of antibiotic injected into Jenny's bloodstream and sketch the graph of the equation $A(t) = 500(0.88)^t$.

Solution   To find the initial amount of antibiotic injected into Jenny's bloodstream, let $t = 0$ in the equation $A(t) = 500(0.88)^t$. Recall that $a^0 = 1$, $a \neq 0$.

$$A(t) = 500(0.88)^t$$

$$A(0) = 500(0.88)^0$$

$$= 500(1)$$

$$= 500$$

The initial amount of antibiotic was 500 milligrams. Load the equation $A(t) = 500(0.88)^t$ into the Y= menu in the following manner:

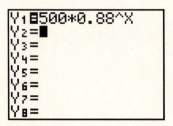

Sketch the graph of the antibiotic in Jenny's bloodstream over the first 24 hours. Because the base of the equation $A(t) = 500(0.88)^t$ is 0.88, which is less than 1, you can expect the graph to begin at 500 and decrease.

Make the following adjustments to the WINDOW parameters:

```
WINDOW FORMAT
 Xmin=0
 Xmax=24
 Xscl=6
 Ymin=0
 Ymax=500
 Yscl=100
```

Push the GRAPH button to capture the following image in your viewing window:

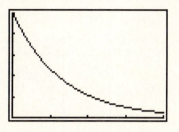

Exercises for Section 9.1

1. The population of Coppertown in 1970 was 2518. Since that time, the population has been growing at a rate of 3.2 percent per year. Therefore, at the end of any one year, the population will be 103.2 percent of the population at the end of the previous year. Let $P(t)$ represent the population of Coppertown t years after 1970. The following table charts the population for the first five years after 1970. The entries in column 4 have been rounded to the nearest person. Make a copy of the table on your homework paper and complete the missing entries.

Year	t	$P(t)$	$P(t)$
1970	0	2518	2518
1971	1	2518(1.032)	2598
1972	2	$2518(1.032)^2$	2681
1973	3		
1974	4		
1975	5		

a) Express the population of Coppertown as a function of t, where t is the number of years that have passed since 1970.

b) Use the equation developed in part (a) to find the population in 1994.

c) (TI-82) Use your calculator to sketch the graph of the population from 1970 through 1994. Set up a coordinate system on a sheet of graph paper and clearly indicate the scale on each coordinate axis. Copy the image in your viewing window onto the coordinate system.

2. (TI-82) The result of the last computation made on your calculator is stored in a memory location named Ans (short for Answer). This feature makes the TI-82 an excellent tool for carrying out the calculations made in Exercise 1.

Enter the number 2518 and press the ENTER key. Press the multiplication key, enter the number 1.032, and press the ENTER key to produce the following window:	Press the ENTER key four consecutive times to produce the following window. *Note:* Each time you press the ENTER key, the calculator replays and executes the last instruction.

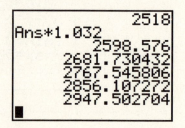

Compare these results with the last column of the table in Exercise 1.

3. The population of Coppertown in 1970 was 2518. Since that time, the population has been decaying at a rate of 3.2 percent per year. Therefore, at the end of any one year, the population will be 96.8 percent of the population at the end of the previous year. Let $P(t)$ represent the population t years after 1970. The following table charts the population for the first five years after 1970. The entries in column 4 have been rounded to the nearest person. Make a copy of the table on your homework paper and complete the missing entries.

Year	t	$P(t)$	$P(t)$
1970	0	2518	2518
1971	1	2518(0.968)	2437
1972	2	$2518(0.968)^2$	2359
1973	3		
1974	4		
1975	5		

a) Express the population of Coppertown as a function of t, where t is the number of years that have passed since 1970.

b) Use the equation developed in part (a) to find the population of Coppertown in 1994.

c) (TI-82) Use your calculator to sketch the graph of the population from 1970 through 1994. Set up a coordinate system on a sheet of graph paper and clearly indicate the scale on each coordinate axis. Copy the image in your viewing window onto your coordinate system.

4. (TI-82) The result of the last computation made on your calculator is stored in the memory location named Ans. This is the feature that makes the TI-82 an excellent tool for carrying out the calculations made in Exercise 3.

Enter the number 2518 and press the ENTER key. Press the multiplication key, enter the number 0.968, and press the ENTER key to produce the following window:	Press the ENTER key four consecutive times to produce the following window. *Note:* Each time you press the ENTER key, the calculator replays and executes the last instruction.
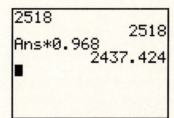	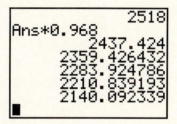

Compare these results with the last column of the table in Exercise 1.

5. In 1960 Billy Joe deposited $1250 in a savings account. Since that time, this amount has been increasing at a rate of 5 percent each year. Let A represent the number of dollars in the account and let t represent the number of years that have passed since 1960.

   a) Express A as a function of t.

   b) Use the equation developed in part (a) to find the amount in the account in 1994.

   c) (TI-82) Sketch the graph of the amount in the account over the years 1960 through 1994. Set up a coordinate system on a sheet of graph paper and clearly indicate the scale on each coordinate axis. Copy the image in your viewing window onto the coordinate system.

6. Jermaine came home with a case of strep throat after a camping trip with his boy scout troop. His doctor gives him a 500-milligram injection of antibiotic. Suppose that Jermaine's body rids itself of 4 percent of the antibiotic every hour. In other words, at the end of any hour, the amount of antibiotic remaining in Jermaine's bloodstream is 96 percent of the amount in his bloodstream at the end of the previous hour. Let A represent the amount of antibiotic in Jermaine's bloodstream and let t represent the number of hours that have passed since his initial injection.

   a) Express A as a function of t.

   b) Use the equation developed in part (a) to find the amount of antibiotic remaining in Jermaine's bloodstream after 48 hours.

   c) (TI-82) Sketch the graph of the amount of antibiotic in his bloodstream over the first 48 hours. Set up a coordinate system on a sheet of graph paper and clearly indicate the scale on each coordinate axis. Copy the image in your viewing window onto the coordinate system.

7. Initially, 100 cases of measles are reported to the health center in Boise. Thereafter, the number of cases increases at the rate of 12 percent each day. In other words, at the end of any day, the number of measles cases is 112 percent of the number of measles cases at the end of the previous day. Let N represent the number of measles cases and let t represent the number of days since the epidemic first began.

   a) Express N as a function of t.

   b) Find the number of measles cases at the end of thirtieth day of the epidemic.

   c) (TI-82) Sketch the graph of the number of measles cases over the first 30 days of the epidemic. Set up a coordinate system on a sheet of graph paper and clearly indicate the scale on each coordinate axis. Copy the image in your viewing window onto the coordinate system.

8. A hospital initially has 5000 milligrams of a certain radioactive substance on hand for certain laboratory procedures. Suppose that the substance begins to decay at rate of 6 percent. In other words, the amount remaining at the end of any day is 94 percent of the amount remaining at the end of the previous day. Let A represent the amount of the substance remaining and let t represent the number of days that have passed since the hospital received its initial amount.

   a) Express A as a function of t.

   b) Use the equation developed in part (a) to find the amount of the substance remaining at the end of 30 days.

   c) (TI-82) Sketch the graph of the amount of substance remaining over the first 30 days. Set up a coordinate system on a sheet of graph paper and clearly indicate the scale on each coordinate axis. Copy the image in your viewing window onto the coordinate system.

9. The population of gophers in a Minnesota town is given by the equation $P = 1200(1.046)^t$, where P represents the number of gophers and t represents the number of months that have passed since January 1, 1993.

   a) How many gophers were present on January 1, 1993?

   b) Use the equation to predict the gopher population on January 1, 1995.

   c) (TI-82) Sketch the graph of the gopher population over the period of time from January 1, 1993 through January 1, 1995. Set up a coordinate system on a sheet of graph paper and clearly indicate the scale on each coordinate axis. Copy the image in your viewing window onto your coordinate system.

10. The ferret population in a certain Idaho forest is given by the equation $P = 250(0.87)^t$, where P represents the number of ferrets in the region and t represents the number of months that have passed since January 1, 1994.

   a) How many ferrets were in the region on January 1, 1994?

   b) Use your equation to predict the ferret population on January 1, 1995.

   c) (TI-82) Sketch the graph of the ferret population over the period of time from January 1, 1994 through January 1, 1995. Set up a coordinate system on a sheet of graph paper and clearly indicate the scale on each coordinate axis. Copy the image in your viewing window onto the coordinate system.

11. A rumor can spread like wildfire. Suppose that the number of people that have heard a certain rumor is given by the equation $N = 3(1.75)^t$, where N is the number of people that have heard the rumor and t is the number of hours that have passed since 12 noon on May 5, 1994.

   a) How many people first heard and started the rumor?

   b) How many people had heard the rumor by 12 noon on May 6, 1994?

   c) Sketch the graph of the number of people who have heard the rumor over the first 24 hours. Set up a coordinate system on a sheet of graph paper and clearly indicate the scale on each coordinate axis. Copy the image in your viewing window onto the coordinate system.

12. Suspicious Sam buries his retirement money in a can in his backyard on the date of his retirement. The amount of money in the can is given by the equation $A = 125,500(0.85)^t$, where A is the number of dollars in the can and t is the number of years that have passed since Suspicious Sam's retirement date.

a) How much was in the can when Suspicious Sam first buried the can on the date of his retirement?

b) How much money was in the can 6 years after the date of Sam's retirement?

c) (TI-82) Sketch a graph of the amount of money in Suspicious Sam's can over the first 6 years of Sam's retirement. Set up a coordinate system on a sheet of graph paper and clearly indicate the scale on each coordinate axis. Copy the image in your viewing window onto your coordinate system.

13. Adjust your WINDOW parameters as follows:

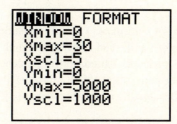

For the equations below, perform each of the following tasks:

i)   Explain how you can tell from your equation whether the graph will increase or decrease.

ii) (TI-82) Use your calculator to sketch the graph of the equation.

iii) Set up a coordinate system on a sheet of graph paper and clearly indicate the scale on each coordinate axis. Copy the image in your viewing window onto the coordinate system.

a) $y = 1213(1.08)^x$ b) $y = 4329(0.88)^x$ c) $y = 3217(0.95)^x$ d) $y = 875(1.05)^x$

14. Because of the practical nature of the problems in this section, the graph of the exponential function has been restricted to the first quadrant, where both x and y are nonnegative. However, the actual domain of the function $f(x) = a \cdot b^x$ is the set of all real numbers. Make the following adjustments to your WINDOW parameters.

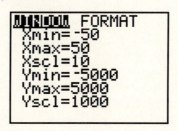

For the equations below, perform each of the following tasks:

i)   Find the y-intercept of the graph.

Hint: Let $x = 0$.

ii) Explain how you can tell from your equation whether the graph will increase or decrease.

iii) (TI-82) Use your calculator to sketch the graph of the equation.

iv) Set up a coordinate system on a sheet of graph paper and clearly indicate the scale on each coordinate axis. Copy the image in your viewing window onto the coordinate system.

v)   Discuss any asymptotic behavior you see.

a) $y = 1200(1.046)^x$ b) $y = 2312(0.89)^x$ c) $y = 1345(1.15)^x$ d) $y = 1188(0.925)^x$

9.2 Compound Interest

The exponential function lends itself quite nicely to problems involving money and interest earned on that money. In this section we develop an equation that is used in computing interest. We will also introduce the number e, which is used extensively when working with problems involving growth and decay.

Example 1   Suppose that Sundin invests $350 in an account that pays 4 percent per year, compounded semiannually. How much money will be in the account at the end of 5 years?

Solution   The interest rate is 4 percent *per year*, compounded *semiannually*. You must divide 4 percent by 2 to find the rate received at the end of every 6-month period. The amount in Sundin's account at the end of any 6-month period is therefore 102 percent of the amount in the account at the end of the previous 6-month period. Let $A(k)$ represent the amount in Sundin's account at the end of k 6-month periods. Because the initial investment is $350, the following can be written:

$$A(0) = 350$$

The amount in the account at the end of one 6-month period will be 102 percent of the initial amount in the account.

$$A(1) = \left[1 + \frac{0.04}{2}\right] A(0)$$

$$= \left[1 + \frac{0.04}{2}\right](350)$$

The expression $1 + \dfrac{0.04}{2}$ equals 1.02, which equals 102 percent. The amount in the account at the end of two 6-month periods will be 102 percent of the amount in the account at the end of one 6-month period.

$$A(2) = \left[1 + \frac{0.04}{2}\right] A(1)$$

$$= \left[1 + \frac{0.04}{2}\right]\left\{\left[1 + \frac{0.04}{2}\right](350)\right\}$$

$$= \left[1 + \frac{0.04}{2}\right]^2 (350)$$

The amount in the account at the end of three 6-month periods will be 102 percent of the amount in the account at the end of two 6-month periods.

$$A(3) = \left[1 + \frac{0.04}{2}\right]A(2)$$

$$= \left[1 + \frac{0.04}{2}\right]\left\{\left[1 + \frac{0.04}{2}\right]^2 (350)\right\}$$

$$= \left[1 + \frac{0.04}{2}\right]^3 (350)$$

A pattern emerges. The amount in the account at the end of k 6-month compounding periods is given by the following equation:

$$A(k) = 350\left[1 + \frac{0.04}{2}\right]^k$$

Because compounding occurs twice a year for 5 years, there are $k = 2(5)$, or $k = 10$, compounding periods. Therefore the amount in Sundin's account at the end of 5 years, or 10 compounding periods, is $A(10)$.

$$A(10) = 350\left[1 + \frac{0.04}{2}\right]^{10}$$

$$\approx 426.64$$

Does your bank round up or round down when awarding interest? Do you get the extra penny or do they keep the extra penny?

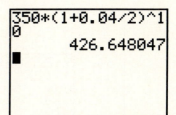

Example 2 Express the amount in Sundin's account in Example 1 as a function of t, the number of years that have passed since the initial investment was made. Use this equation to find the amount in the account at the end of 5 years.

Solution From Example 1, the amount in the account after k 6-month compounding periods was given by the following equation:

$$A(k) = 350\left[1 + \frac{0.04}{2}\right]^k$$

If you are compounding twice a year for t years, then the number of compounding periods in t years is $k = 2t$. Therefore the amount in the account at the end of t years is given by the following equation:

$$A(t) = 350\left[1 + \frac{0.04}{2}\right]^{2t}$$

This formula can be used to find the interest in Sundin's account at the end of 5 years.

$$A(5) = 350\left[1+\frac{0.04}{2}\right]^{2(5)}$$

$$= 350\left[1+\frac{0.04}{2}\right]^{10}$$

$$\approx 426.64$$

Compare this answer with the answer found in Example 1.

Example 3 In Example 2, the amount in Sundin's account as a function of the number of years that have passed is given by the equation $A(t) = 350\left[1+\frac{0.04}{2}\right]^{2t}$. Sketch the graph of the amount in Sundin's account over the first 5 years.

Solution

Load the equation $A(t) = 350\left[1+\frac{0.04}{2}\right]^{2t}$ into the Y= menu.	Draw the graph over the first 5 years of the account. You know that the initial amount in the account is \$350 and this amount will grow to \$426.64.
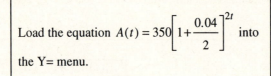	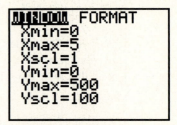

Push the GRAPH button to capture the following image in your viewing window:

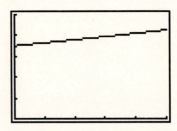

Note the growth of the money in the account.

A Formula for Compound Interest

Suppose that you invest P dollars at a yearly rate r in an account that is compounded n times a year for t years. Let $A(k)$ represent the amount in the account at the end of k compounding periods. The initial amount invested in the account is P.

$$A(0) = P$$

The yearly rate is r. Because you are compounding n times per year, $\dfrac{r}{n}$ is the rate you receive each compounding period. Therefore the amount in the account at the end of any compounding period will be $1+\dfrac{r}{n}$ times the amount in the account at the end of the previous compounding period.

The amount in the account at the end of the first compounding period is $1+\dfrac{r}{n}$ times the initial amount in the account.

$$A(1) = \left[1+\frac{r}{n}\right]A(0)$$

$$= \left[1+\frac{r}{n}\right]P$$

The amount in the account at the end of the second compounding period is $1+\dfrac{r}{n}$ times the amount in the account at the end of the first compounding period.

$$A(2) = \left[1+\frac{r}{n}\right]A(1)$$

$$= \left[1+\frac{r}{n}\right]\left\{\left[1+\frac{r}{n}\right]P\right\}$$

$$= \left[1+\frac{r}{n}\right]^2 P$$

The amount in the account at the end of the third compounding period is $1+\dfrac{r}{n}$ times the amount in the account at the end of the second compounding period.

$$A(3) = \left[1+\frac{r}{n}\right]A(2)$$

$$= \left[1+\frac{r}{n}\right]\left\{\left[1+\frac{r}{n}\right]^2 P\right\}$$

$$= \left[1+\frac{r}{n}\right]^3 P$$

Once again, a familiar pattern emerges. The amount in the account after k compounding periods is given by the following equation:

$$A(k) = P\left[1+\frac{r}{n}\right]^k$$

If you wish to find out how much is in the account after t years, you must first compute how many compounding periods there are in t years. If you compound n times per year for t years, then you will have $k = nt$ compounding periods in all. Therefore the amount in the account at the end of t years is given by the following equation:

$$A(t) = P\left[1 + \frac{r}{n}\right]^{nt}$$

The amount in an account can be computed by the formula $A = P\left[1 + \dfrac{r}{n}\right]^{nt}$.

- *P is the initial investment. It is often called the principal or the present value of the account.*
- *A is the amount in the account. It is called the amount or the future value of the account.*
- *r is the yearly interest rate.*
- *n is the number of compounding periods in 1 year.*
- *t is the number of years.*

Example 4 Suppose that Sundin invests his $350 in an account that pays 4.75 percent per year, compounded monthly. How much will be in the account at the end of 5 years?

Solution Use the formula $A = P\left[1 + \dfrac{r}{n}\right]^{nt}$. The principal P equals 350, the yearly rate r is 0.0475, the number of compounding periods per year n is 12, and the number of years t is 5.

$$A = P\left[1 + \frac{r}{n}\right]^{nt}$$

$$= 350\left[1 + \frac{0.0475}{12}\right]^{(12)(5)}$$

$$\approx 443.61$$

```
350*(1+0.0475/12
)^(12*5)
        443.6182251
■
```

Example 5 Dave and Mary just had a baby girl named Elizabeth. They wish to establish an account that will mature on their daughter's eighteenth birthday. Elizabeth could then use the money in the account for tuition and books when she goes to college. If Dave and Mary can find an account that pays 5 percent per year, compounded quarterly, how much should they put in the account now so that there will be $25,000 in the account on Elizabeth's eighteenth birthday?

Solution The future value of the money A equals $25,000, the yearly interest rate r is 0.05, the number of compounding periods n is 4, and the number of years t is 18.

$$A = P\left[1+\frac{r}{n}\right]^{nt}$$

$$25,000 = P\left[1+\frac{0.05}{4}\right]^{(4)(18)}$$

$$\frac{25,000}{\left[1+\dfrac{0.05}{4}\right]^{(4)(18)}} = P$$

$$P \approx 10,221.10$$

```
25000/(1+0.05/4)
^(4*18)
        10221.10178
■
```

If Dave and Mary invest $10,221.10 in the account now, Elizabeth will have $25,000 available on her eighteenth birthday.

Example 6   Suppose that $1,000 is invested at 5 percent. How much will be in the account at the end of 10 years if the interest is compounded the following number of times?

a) 1 time a year (annually)
b) 2 times a year (semiannually)
c) 4 times a year (quarterly)
d) 12 times a year (monthly)
e) 365 times a year (daily)
f) 8,760 times a year (hourly)
g) 31,536,000 times a year (every second)

Solution

a) $A = 1,000\left[1+\dfrac{0.05}{1}\right]^{(1)(10)} \approx 1,628.89$

b) $A = 1,000\left[1+\dfrac{0.05}{2}\right]^{(2)(10)} \approx 1,638.61$

```
1000*(1+0.05/1)^
(1*10)
        1628.894627
1000*(1+0.05/2)^
(2*10)
        1638.61644
```

c) $A = 1,000\left[1+\dfrac{0.05}{4}\right]^{(4)(10)} \approx 1,643.61$

d) $A = 1,000\left[1+\dfrac{0.05}{12}\right]^{(12)(10)} \approx 1,647.00$

```
1000*(1+0.05/4)^
(4*10)
        1643.619463
1000*(1+0.05/12)
^(12*10)
        1647.009498
```

e) $A = 1,000 \left[1 + \dfrac{0.05}{365} \right]^{(365)(10)} \approx 1,648.66$

f) $A = 1,000 \left[1 + \dfrac{0.05}{8,760} \right]^{(8,760)(10)} \approx 1,648.71$

```
1000*(1+0.05/365
)^(365*10)
       1648.664814
1000*(1+0.05/876
0)^(8760*10)
       1648.718924
■
```

g) $A = 1,000 \left[1 + \dfrac{0.05}{31,536,000} \right]^{(31,536,000)(10)}$

$A \approx 1,648.72$

```
1000*(1+0.05/315
36000)^(31536000
*10)
       1648.726678
■
```

It is important to note that as the number of compounding periods per year increases, the amount in the account reaches a limiting value.

Compounding Continuously

What happens if we compound more than a million times a year? More than a billion times a year? More than a trillion times a year? What happens if we compound *continuously*? Let's make the substitution $n = kr$ in the compound interest formula.

$$A = P \left[1 + \frac{r}{n} \right]^{nt}$$

$$= P \left[1 + \frac{r}{kr} \right]^{krt}$$

$$= P \left[1 + \frac{1}{k} \right]^{krt}$$

$$= P \left\{ \left[1 + \frac{1}{k} \right]^{k} \right\}^{rt}$$

Recall that $n = kr$ and the interest rate r is fixed. If the compounding periods n are allowed to grow without bound, then the number k must grow without bound as well. Use your calculator to investigate what happens to the expression $\left[1 + \dfrac{1}{k} \right]^{k}$ as k is allowed to grow without bound.

k	$\left[1+\dfrac{1}{k}\right]^{k}$
10	2.59374246
100	2.704813829
1,000	2.716923932
10,000	2.718145927
100,000	2.718268237
1,000,000	2.718280469
10,000,000	2.718281693
100,000,000	2.718281815
1,000,000,000	2.718281827
10,000,000,000	2.718281828

As k grows without bound, the expression $\left[1+\dfrac{1}{k}\right]^{k}$ approaches the number 2.718281828....

Now, use your calculator to compute e^{1}. Push the 2nd key, then the e^{x} key (locate above the LN key on your keyboard), enter a 1, then press the ENTER key.

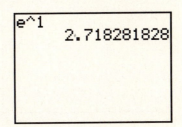

The number e is approximately 2.718281828. The actual value of e is an infinite, nonrepeating decimal, much like the decimal expansion of the number π. It should now be clear that the expression $\left[1+\dfrac{1}{k}\right]^{k}$ approaches the number e as k is allowed to increase without bound. Therefore, if you are compounding continuously, you can replace $\left[1+\dfrac{1}{k}\right]^{k}$ in the equation $A = P\left\{\left[1+\dfrac{1}{k}\right]^{k}\right\}^{rt}$ with the number e, producing the following formula for interest compounded continuously:

$$A = Pe^{rt}$$

> *If the interest is being compounded continuously, then the amount in the account is given by the formula $A = Pe^{rt}$.*
>
> - *P is the initial investment. It is often called the principal or the present value of the account.*
> - *A is the amount in the account. It is called the amount or the future value of the account.*
> - *r is the yearly interest rate.*
> - *t is the number of years.*

Example 7   Suppose that Ling-chi invests $1000 in an account at 5 percent compounded continuously. How much will be in the account at the end of 10 years?

Solution   Because the interest is being compounded continuously, use the formula $A = Pe^{rt}$. The principal P equals $1000, the rate r is 0.05, and the time t is 10 years.

<table>
<tr>
<td>

$$A = Pe^{rt}$$

$$= 1000e^{(0.05)(10)}$$

$$\approx 1648.72$$

</td>
<td>

```
1000*e^(0.05*10)
          1648.721271
```

</td>
</tr>
</table>

Compare this result with part (g) of Example 6.

Note: There are no banks that compound interest continuously, just as there are no banks that are in the habit of giving money away. However, this is a nice way to introduce the number e, because people seem to be fairly comfortable with the notions of money and interest. In later sections, we will look at some more practical applications of continuous growth or decay.

Example 8   In Example 7, Ling-chi invested $1000 in an account paying 5 percent compounded continuously. Express the amount in the account as a function of time and sketch the graph of the amount in the account over the first 10 years.

Solution   Because the interest is being compounded continuously, use the formula $A = Pe^{rt}$. The principal P equals $1000 and the rate r is 0.05. Therefore the amount in the account as a function of time is given by the equation $A = 1000e^{0.05t}$.

Load the equation $A = 1000e^{0.05t}$ into the Y= menu.

Draw the graph over the first 10 years of the account. You know that the initial amount in the account is $1000 and this amount will grow to $1648.72.

Push the GRAPH button to capture the following image in your viewing window:

Note the growth of the money in the account.

Example 9   Nemane would like to have money available for a new roof for his house. How much should he invest in an account that pays 4.25 percent compounded continuously so that $5000 will be in the account 5 years from now?

Solution   Because the interest is being compounded continuously, use the formula $A = Pe^{rt}$. The amount or future value A is $5000, the rate r is 0.0425, and the time t is 5 years.

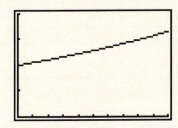

$$A = Pe^{rt}$$

$$5000 = Pe^{(0.0425)(5)}$$

$$\frac{5000}{e^{(0.0425)(5)}} = P$$

$$A \approx 4042.80$$

Nemane will have to invest $4042.80 now if he wishes to have $5000 available for his new roof 5 years from now.

Exercises for Section 9.2

1. On January 1, 1994, Shanti invests $450 in an account that pays 4.25 percent compounded quarterly. Therefore, the amount in the account at the end of any compounding period will be $1 + \dfrac{0.0425}{4}$ times the amount in the account at the end of the previous compounding period. Let $A(k)$ represent the amount in the account after k compounding periods. The following table charts the investment over the first two years. Make a copy of the table on your homework paper and complete the missing entries.

Date	Compounding Period	Amount $A(k)$	Amount $A(k)$
1/1/94	0	450	450.00
4/1/94	1	$450\left[1 + \dfrac{0.0425}{4}\right]$	454.78
7/1/94	2	$450\left[1 + \dfrac{0.0425}{4}\right]^2$	459.61
10/1/94	3		
1/1/95	4		
4/1/95	5		
7/1/95	6		
10/1/95	7		
1/1/96	8		

   a) Express the amount in the account as a function of k, the number of compounding periods that have passed since January 1, 1994.

   b) Use the equation developed in part (a) to find the amount in the account on January 1, 2004.

2. (TI-82) The result of the last computation made on your calculator is stored in a memory location named Ans. This feature makes the TI-82 an excellent tool for carrying out the calculations made in Exercise 1.

   Enter the number 450 and press the ENTER key. Press the multiplication key, enter the expression (1+0.0425/4), and press the ENTER key to produce the following window:

   Press the ENTER key four consecutive times to produce the following window.

   Note: Each time you press the ENTER key, the calculator replays and executes the last instruction.

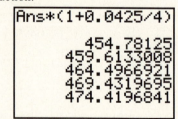

   Compare these results with the last column of your table in Exercise 1. You may need to produce a few more values by pressing the ENTER key a few more times.

3. Express the amount in Shanti's account in Exercise 1 as a function of t, the number of years that have passed since January 1, 1994.

a) Use this equation to find the amount in Shanti's account on January 1, 2004. Compare this result with the result found in Exercise 1, part (b).

b) (TI-82) Use your calculator to sketch the graph of the equation developed in part (a). Sketch the graph for $t = 0$ through $t = 10$ years, which represents the years from January 1, 1994 through January 1, 2004. Set up a coordinate system on a sheet of graph paper and clearly indicate the scale on each coordinate axis. Copy the image in your viewing window onto the coordinate system.

4. On January 1, 1994, Klee invests \$2000 dollars in an account that pays 3.75 percent compounded monthly. Therefore the amount in the account at the end of any compounding period will be $1 + \dfrac{0.0375}{12}$ times the amount in the account at the end of the previous compounding period. Let $A(k)$ represent the amount in the account after k compounding periods. The following table charts the investment over the first 6 months. Make a copy of the table on your homework paper and complete the missing entries.

Date	Compounding Period	Amount $A(k)$	Amount $A(k)$
1/1/94	0	2000	2000.00
2/1/94	1	$2000\left[1 + \dfrac{0.0375}{12}\right]$	2006.25
3/1/94	2	$2000\left[1 + \dfrac{0.0375}{12}\right]^2$	2012.51
4/1/94	3		
5/1/94	4		
6/1/94	5		
7/1/94	6		

a) Express the amount in the account as a function of k, the number of compounding periods that have passed since January 1, 1994.

b) Use the equation developed in part (a) to find the amount in the account on January 1, 2004.

5. (TI-82) The result of the last computation made on your calculator is stored in a memory location named Ans. This feature makes the TI-82 an excellent tool for carrying out the calculations made in Exercise 4.

Enter the number 2000 and press the ENTER key. Press the multiplication key, enter the expression $(1 + 0.0375/12)$, and press the ENTER key to produce the following window:

Press the ENTER key four consecutive times to produce the following window. *Note:* Each time you press the ENTER key, the calculator replays and executes the last instruction.

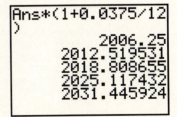

Compare these results with the last column of your table in Exercise 4. You may need to produce a few more values by pressing the ENTER key a few more times.

6. Express the amount in Klee's account in Exercise 4 as a function of t, the number of years that have passed since January 1, 1994.

   a) Use this equation to find the amount in Klee's account on January 1, 2004. Compare this result with the result found in Exercise 1(b).

   b) (TI-82) Use your calculator to sketch the graph of the equation developed in part (a). Sketch the graph for $t = 0$ through $t = 10$ years, which represents the years from January 1, 1994 through January 1, 2004. Set up a coordinate system on a sheet of graph paper and clearly indicate the scale on each coordinate axis. Copy the image in your viewing window onto your coordinate system.

7. Suppose that Yuka invests $1500 in an account on January 1, 1994, that pays 4.25 percent per year, compounded semiannually.

   a) Let A represent the number of dollars in the account and let k represent the number of compounding periods that have passed since January 1, 1994. Express A as a function of k.

   b) (TI-82) Sketch the graph of the equation developed in part (a) over the first 10 compounding periods. Set up a coordinate system on a sheet of graph paper and clearly indicate the scale on each coordinate axis. Copy the image in your viewing window onto your coordinate system.

   c) Let A represent the number of dollars in the account and let t represent the number of years that have passed since January 1, 1994. Express A as a function of t.

   d) (TI-82) Sketch the graph of the equation developed in part (a) over the first 5 years. Set up a coordinate system on a sheet of graph paper and clearly indicate the scale on each coordinate axis. Copy the image in your viewing window onto your coordinate system.

8. Suppose that Melvin invests $2500 in an account that pays 5.2 percent compounded quarterly. Find the amount in the account at the end of 5 years.

9. Suppose that Wallace invests $800 in an account that pays 4.5 percent compounded monthly. Find the amount in the account at the end of 6 years.

10. How much should Andres invest in an account that pays 3.8 percent compounded semiannually so that there will be $1800 in the account 2 years from now?

11. How much should Carmen invest in an account that pays 5.6 percent compounded quarterly so that there will be $2300 in the account three years from now?

12. Keenan and Charlene are trying to decide between two accounts. They intend to invest $1000 for 5 years. They have two choices. One account will pay 5 percent interest, compounded quarterly, while a second account will pay 5.2 percent interest, compounded annually.

    a) How much will be in the first account at the end of 5 years?

    b) How much will be in the second account at the end of 5 years?

    c) Which is the better investment?

13. Make a copy of the following table on your homework paper and use your calculator to help complete the missing entries.

n	$\left[1+\dfrac{1}{n}\right]^{n}$	$\left[1+\dfrac{1}{n}\right]^{n}$
10	$\left[1+\dfrac{1}{10}\right]^{10}$	2.59374246
100	$\left[1+\dfrac{1}{100}\right]^{100}$	2.704813829
1000		
10,000		
100,000		
1,000,000		
10,000,000		
100,000,000		
1,000,000,000		

a) The numbers in the third column of this table appear to be converging to what number?

b) Use your calculator to compute e^{1} or e. How does this number compare to the response you gave in part (a)?

c) As n increases without bound, the expression $\left[1+\dfrac{1}{n}\right]^{n}$ approaches what number?

d) As n increase without bound, the expression $\left[1+\dfrac{2}{n}\right]^{n}$ approaches what number?

14. Use your calculator to find decimal approximations for each of the following expressions, correct to the nearest hundredth (two decimal places).

   a) $100e^{(0.04)(15)}$

   b) $57e^{(0.0325)(12)}$

   c) $\dfrac{4,500}{e^{(0.06)(6)}}$

   d) $\dfrac{12,000}{e^{(0.0225)(7)}}$

   e) $1000e^{(-0.0225)(10)}$

   f) $2550e^{(-0.085)(15)}$

   g) $e^{-\frac{(2.2)^{2}}{2}}$

   h) $\dfrac{e}{1-3e}$

15. Suppose that \$2400 is invested in an account that pays 5.6 percent per year. How much will be in the account at the end of 5 years if the interest is compounded

   a) 1 time a year (annually)?

   b) 2 times a year (semiannually)?

   c) 4 times a year (quarterly)?

   d) 12 times a year (monthly)?

   e) 365 times a year (daily)?

   f) 8760 times a year (hourly)?

   g) 31,536,000 times a year (every second)?

   h) Compute $2,400e^{(0.056)(5)}$.

16. Suppose that you invest $5000 in an account that pays 7 percent per year. How much will be in the account at the end of 10 years if the interest is compounded

    a) annually?

    b) semiannually?

    c) quarterly?

    d) monthly?

    e) daily?

    f) continuously?

17. If Daryl invests $510 in an account that pays 4 percent compounded continuously, how much will be in the account 4 years from now?

18. If Ibrahim invests $425 in an account that pays 5.1 percent compounded continuously, how much will be in the account 5 years from now?

19. How much should Josie invest now in an account that pays 4.25 percent compounded continuously so that the account will contain $1200 three years from now?

20. How much should Mohammed invest in an account that pays 5.8 percent compounded continuously so that the account will contain $2600 four and one half years from now?

21. Jeremiah and Rachel would like to have $50,000 in an account when they retire, 20 years from now.

    a) They find an account that pays 6 percent interest, compounded semiannually. How much should they invest now to reach their retirement dream?

    b) Suppose that they find another account that pays 6 percent interest, compounded continuously. How much would they have to invest in this account to reach their retirement dream?

22. Suppose that Bill invests $400 in an account on January 1, 1994, that pays 4.7 percent compounded continuously.

    a) Let A represent the number of dollars in the account and let t represent the number of years that have passed since January 1, 1994. Express A as a function of t.

    b) (TI-82) Draw the graph of the equation developed in part (a) over the first 10 years of the investment. Set up a coordinate system on a sheet of graph paper and clearly indicate the scale on each coordinate axis. Copy the image in your viewing window onto the coordinate system.

23. Suppose that Manny invests $3500 in an account on January 1, 1994, that pays 5.23 percent compounded continuously.

    a) Let A represent the number of dollars in the account and let t represent the number of years that have passed since January 1, 1994. Express A as a function of t.

    b) (TI-82) Draw the graph of the equation developed in part (a) over the first 10 years of the investment. Set up a coordinate system on a sheet of graph paper and clearly indicate the scale on each coordinate axis. Copy the image in your viewing window onto the coordinate system.

24. Because of the practical nature of the problems in this section, the graph of the exponential equation $A = Pe^{rt}$ has been restricted to the first quadrant where both t and A are nonnegative. However, the actual domain of the function whose equation is $A = Pe^{rt}$ is the set of all real numbers. Make the following adjustments to your WINDOW parameters:

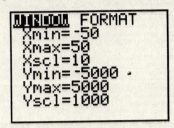

For the equations below, perform each of the following tasks.

i)   Find the A-intercept of the graph.

   Hint: Let $t = 0$.

ii) (TI-82) Use your calculator to sketch the graph of the equation.

iii) Set up a coordinate system on a sheet of graph paper and clearly indicate the scale on each coordinate axis. Copy the image in your viewing window onto the coordinate system.

iv) Discuss any asymptotic behavior you see.

a) $A = 1000e^{0.08t}$ b) $A = 800e^{0.14t}$ c) $A = 3200e^{-0.09t}$ d) $A = 500e^{-0.06t}$

Some of the graphs in this problem indicate growth and some indicate decay. Explain how you can tell from the equation whether you will have growth or decay.

9.3 The Exponential Function: $f(x) = a^x$, $a > 0$

We have already discussed the graphs of $f(x) = a \cdot b^x$ and $A = Pe^{rt}$. However, these graphs were coupled with applications that restricted our attention to the first quadrant. This section discusses the graph of $f(x) = a^x$, $a > 0$, but this is done in an abstract sense and will focus on the entire plane, not just the first quadrant. Before our discussion begins, we must define what is meant by a negative exponent.

Negative Exponents

The following discussion is intended to motivate the definition of the expression a^{-n}. Recall the following law of exponents: $a^m \cdot a^n = a^{m+n}$. When multiplying like bases together, repeat the base and add the exponents. Therefore the following is true:

$$a^n \cdot a^{-n} = a^0$$

Of course, $a^0 = 1$, which means that this last statement can be simplified.

$$a^n \cdot a^{-n} = 1$$

But isn't the following true?

$$a^n \cdot \frac{1}{a^n} = 1$$

If you compare these last two statements, you sense that the expressions a^{-n} and $\frac{1}{a^n}$ must be the same. This argument provides a pretty fair motivation for the following definition:

$$\textit{If } a \neq 0, \textit{ then } a^{-n} = \frac{1}{a^n}.$$

Example 1   Simplify the expression 4^{-3}.

Solution   Using the definition, you can write the following:

$$4^{-3} = \frac{1}{4^3} = \frac{1}{64}$$

Example 2   Simplify the expression $\left(-\frac{2}{3}\right)^{-2}$.

Solution   Using the definition, you can write the following:

$$\left(-\frac{2}{3}\right)^{-2} = \frac{1}{\left(-\frac{2}{3}\right)^2} = \frac{1}{\frac{4}{9}} = 1 \cdot \frac{9}{4} = \frac{9}{4}$$

Example 3   Simplify the expression $\left(\frac{3}{5}\right)^{-1}$.

Solution   Using the definition, you can write the following:

$$\left(\frac{3}{5}\right)^{-1} = \frac{1}{\left(\frac{3}{5}\right)^1} = \frac{1}{\frac{3}{5}} = 1 \cdot \frac{5}{3} = \frac{5}{3}$$

Example 4   Simplify the expression a^{-1}, $a \neq 0$.

Solution   Using the definition, you can write the following:

$$a^{-1} = \frac{1}{a^1} = \frac{1}{a}$$

The Essence of a Negative Exponent.   If $a \neq 0$, then $a^{-1} = \frac{1}{a}$. Perhaps you will find the following picture both helpful and amusing.

In a nutshell, raising something to the negative 1 power simply inverts the object.

If a number is raised to a negative exponent, the negative sign in the exponent simply inverts the number. For example, consider 5^{-2}. Simply invert the 5 and square the result: $5^{-2} = \left(\frac{1}{5}\right)^2 = \frac{1}{25}$. You could also square the 5 and invert the result: $5^{-2} = 25^{-1} = \frac{1}{25}$. Both of these methods are easy to execute mentally. For example, to simplify the expression $\left(\frac{3}{4}\right)^{-2}$, simply invert $\frac{3}{4}$ and square the result: $\left(\frac{3}{4}\right)^{-2} = \frac{16}{9}$.

The Laws of Exponents

The laws of exponents hold when working with negative exponents.

Example 5   When raising a product to a power, raise each factor to that power: $(ab)^n = a^n b^n$.

$$(xy)^{-2} = x^{-2} y^{-2}$$

Example 6   When raising a quotient to a power, raise both numerator and denominator to that power: $\left(\frac{a}{b}\right)^n = \frac{a^n}{b^n}$.

$$\left(\frac{x}{y}\right)^{-3} = \frac{x^{-3}}{y^{-3}}$$

Example 7   When multiplying like bases, repeat the base and add the exponents: $a^m \cdot a^n = a^{m+n}$.

$$x^{-2} x^{-4} = x^{-2+(-4)} = x^{-6}$$
$$y^{-7} y^{10} = y^{-7+10} = y^3$$

Example 8   When dividing like bases, repeat the base and subtract the exponents: $\frac{a^m}{a^n} = a^{m-n}$, $a \neq 0$.

$$\frac{x^{-2}}{x^{-7}} = x^{-2-(-7)} = x^5$$
$$\frac{y^4}{y^9} = y^{4-9} = y^{-5}$$

Example 9   When raising a power of a number to another power, multiply the exponents: $\left(a^m\right)^n = a^{mn}$.

$$\left(x^{-2}\right)^5 = x^{-10}$$

$$\left(y^{-3}\right)^{-7} = y^{21}$$

The Graph of $f(x) = a^x$, $a > 1$.   The intent of the following discussion is to thoroughly analyze the graph of $f(x) = 2^x$. Much of what you learn about the graph of $f(x) = 2^x$ will enable you to draw the graph of $f(x) = a^x$ for other values of a greater than 1.

Example 10   Sketch the graph of $f(x) = 2^x$.

Solution   The base of $f(x) = 2^x$ is 2, which is larger than 1. Therefore you can expect that the graph will indicate growth and increase. Also, $f(0) = 2^0 = 1$, so the graph must pass through the point $(0,1)$. Armed with this information and your experience with exponential functions from Sections 9.1 and 9.2, you could probably draw a pretty fair representation of the graph of $f(x) = 2^x$. However, take your time and begin with a table of points that satisfy the equation $f(x) = 2^x$.

x	-3	-2	-1	0	1	2	3
$f(x)$	$\dfrac{1}{8}$	$\dfrac{1}{4}$	$\dfrac{1}{2}$	1	2	4	8

Plot each of the points from this table on a coordinate system.

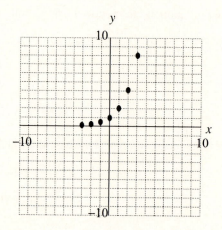

As expected, the graph is increasing. Because $2^4 = 16$, $2^5 = 32$, $2^6 = 64$, $2^7 = 128$, etc., the graph will grow without bound. As you move farther and farther to the right, the graph of $f(x) = 2^x$ rises to positive infinity.

Now investigate the behavior of the graph as you move to the left: $2^{-4} = \frac{1}{16}$, $2^{-5} = \frac{1}{32}$, $2^{-6} = \frac{1}{64}$, $2^{-7} = \frac{1}{128}$, etc. As you move to the left, the function values are approaching zero. This means that the x-axis is acting as an asymptote for the graph. What follows is a complete graph of $f(x) = 2^x$.

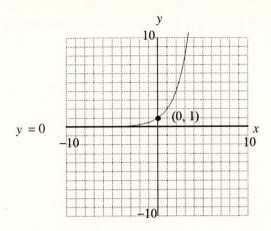

The x-axis, whose equation is $y = 0$, acts as an asymptote. The y-intercept of the graph is the point $(0,1)$.

Example 11   Use your calculator to sketch the graph of $f(x) = 3^x$.

Solution   Because the base is 3, which is larger than 1, you can expect your graph to indicate growth and increase. Because $3^0 = 1$, the y-intercept is $(0,1)$.

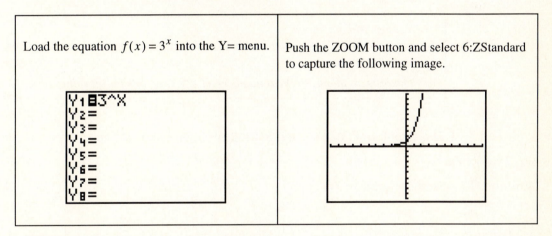

Again, the x-axis, whose equation is $y = 0$, acts as an asymptote. As you move to the left, the graph approaches the x-axis. As you move to the right, the graph rises to positive infinity. The y-intercept of the graph is $(0,1)$.

The Graph of $f(x) = a^x$, $a > 1$. When $a > 1$, the graph of $f(x) = a^x$ has the following general shape:

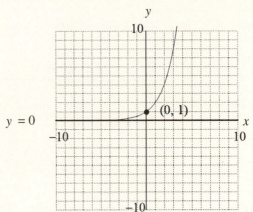

---

1. **When $a > 1$, the graph of $f(x) = a^x$ is increasing.**
2. **The x-axis acts as an asymptote and the y-intercept is $(0, 1)$.**
3. **As you move to the left, the graph approaches the x-axis and the function values approach zero.**
4. **As you move to the right, the graph rises to positive infinity and the function values increase without bound.**
5. **The domain of the function is all real numbers, which means any real number can be used as an exponent.**
6. **The range is all positive real numbers. If $a > 1$, $a^x > 0$ for all values of x.**

The Graph of $f(x) = a^x$, $0 < a < 1$. Let's begin with an example.

Example 12   Sketch the graph of $f(x) = \left(\dfrac{1}{2}\right)^x$.

Solution   The base of $f(x) = \left(\dfrac{1}{2}\right)^x$ is $\dfrac{1}{2}$, which is less than 1, so you can expect the graph of $f(x) = \left(\dfrac{1}{2}\right)^x$ to indicate decay and decrease. The easiest way to sketch the graph of $f(x) = \left(\dfrac{1}{2}\right)^x$ is to make the following observation.

$$\left(\dfrac{1}{2}\right)^x = \left(2^{-1}\right)^x = 2^{-x}$$

Therefore $f(x) = 2^{-x}$.

The graph of $y = 2^x$

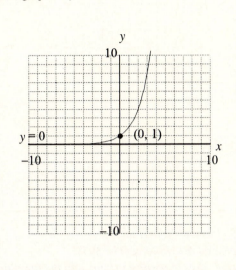

The graph of $f(x) = 2^{-x}$ is a reflection of the graph of $y = 2^x$ across the y-axis.

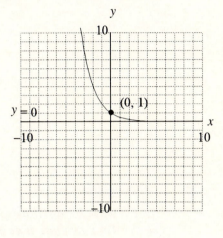

Example 13   Use your calculator to sketch the graph of $f(x) = (0.7)^x$.

Solution   The base of $f(x) = (0.7)^x$ is 0.7, which is less than 1, so you can expect the graph to decrease.

Load the equation $f(x) = (0.7)^x$ into the Y= menu.

```
Y1■0.7^X
Y2=■
Y3=
Y4=
Y5=
Y6=
Y7=
Y8=
```

Push the ZOOM button and select 6:ZStandard to capture the following image:

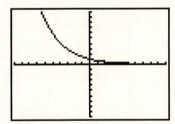

The Graph of $f(x) = a^x$, $0 < a < 1$. When $0 < a < 1$, the graph of $f(x) = a^x$ has the following general shape:

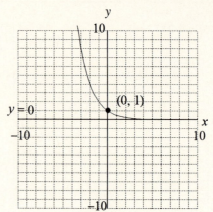

1. **When $0 < a < 1$, the graph of $f(x) = a^x$ is decreasing.**
2. **The x-axis acts as an asymptote and the y-intercept is $(0, 1)$.**
3. **As you move to the left, the graph rises to positive infinity and the function values increase without bound.**
4. **As you move to the right, the graph approaches the x-axis and the function values approach zero.**
5. **The domain of f is all real numbers, which means any real number may be used as an exponent.**
6. **The range of f is all positive real numbers. If $0 < a < 1$, then $a^x > 0$ for all values of x.**

Example 14   Sketch the graph of $y = -e^x + 4$.

Solution   A student solution is shown below.

The base of $y = e^x$ is e, which is larger than 1. The graph of $y = e^x$ has the following shape:

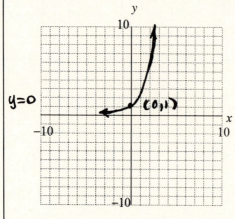

The graph of $y = -e^x$ is a reflection of the graph of $y = e^x$ across the x-axis.

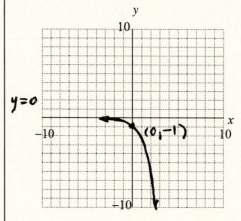

Slide the graph of $y = -e^x$ upward 4 units to arrive at the graph of $y = -e^x + 4$.

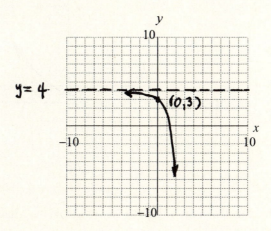

Note that the equation of the asymptote is $y = 4$ and the y-intercept is now $(0,3)$.

Example 15   Sketch the graph of $f(x) = (0.6)^{-x} - 2$.

Solution   A student solution is shown below.

The base of $y = (0.6)^x$ is 0.6, which is between 0 and 1. The graph of $y = (0.6)^x$ has the following shape: 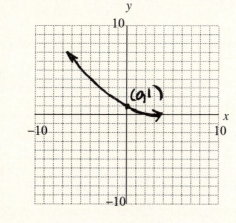	The graph of $y = (0.6)^{-x}$ is a reflection of the graph of $y = (0.6)^x$ across the y-axis. 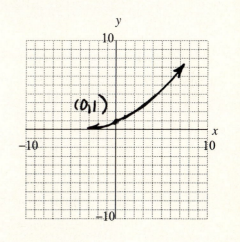

Slide the graph of $y = (0.6)^{-x}$ downward 2 units to arrive at the graph of $y = (0.6)^{-x} - 2$.

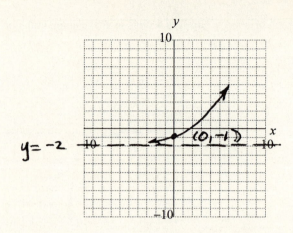

Note that the equation of the asymptote is $y = -2$ and the y-intercept is now $(0, -1)$.

Note: Although the student solutions in the last two examples are not precise, the students have certainly captured the essential behavior of each graph. Often, this will be enough when solving problems involving the exponential function. If the need should arise for greater accuracy, you can use your calculator to draw the graphs for you.

Exercises for Section 9.3

1. Simplify each of the following expressions.

   a) 3^{-2} b) 9^{-1} c) $(-5)^{-2}$ d) $(-4)^{-3}$ e) $\left(\dfrac{4}{5}\right)^{-3}$ f) $\left(-\dfrac{3}{7}\right)^{-2}$

2. Use the laws of exponents to write each of the following as a single power of x.

   a) $x^{-3}x^{-4}$ b) $x^2 x^{-7}$ c) $\dfrac{x^{-5}}{x^{-10}}$ d) $\dfrac{x^9}{x^{11}}$ e) $\left(x^{-3}\right)^{-4}$

   f) $\left(x^5\right)^{-11}$ g) $\left(x^{-2}\right)^{-3}\left(x^3\right)^{-1}$ h) $\left(x^{-2}\right)^{-3}\left(x^4\right)^{-3}$ i) $\left(\dfrac{x^2}{x^{-3}}\right)^{-3}$ j) $\left(\dfrac{x^{-1}}{x^{-3}}\right)^{-3}$

3. (TI-82) For the equations below, perform each of the following tasks:

   i)   Load the equation into the Y= menu, push the ZOOM button and select 6:ZStandard.

   ii) Set up a coordinate system on a sheet of graph paper and clearly indicate the scale on each coordinate axis. Copy the image in your viewing window onto the coordinate system.

   iii) Label the asymptote with its equation. Label the y-intercept with its coordinates.

   a) $y = 2^x$ b) $y = 3^x$ c) $y = e^x$ d) $y = 10^x$

4. (TI-82) For the equations below, perform each of the following tasks:

   i) Load the equation into the Y= menu, push the ZOOM button and select 6:ZStandard.

   ii) Set up a coordinate system on a sheet of graph paper and clearly indicate the scale on each coordinate axis. Copy the image in your viewing window onto the coordinate system.

iii)Label the asymptote with its equation. Label the y-intercept with its coordinates.

a) $y = (0.9)^x$ b) $y = (0.6)^x$ c) $y = (0.3)^x$ d) $y = (0.1)^x$

5. Consider the equation $f(x) = 3^x$. Make a copy of the following table on a sheet of graph paper and complete the missing entries. Do not use your calculator. Express you function values in fractional form. No decimals, please.

x	−3	−2	−1	0	1	2	3
$f(x)$							

a) Set up a coordinate system on your graph paper and clearly indicate the scale on each coordinate axis. Plot each point from the table on your coordinate system.

b) Use your plotted points to help draw a complete graph of the function f. Label your asymptote with its equation. Label the y-intercept with its coordinates.

6. Consider the equation $f(x) = \left(\frac{1}{2}\right)^x$. Make a copy of the following table on a sheet of graph paper and complete the missing entries. Do not use your calculator. Express your function values in fractional form. No decimals, please.

x	−3	−2	−1	0	1	2	3
$f(x)$							

a) Set up a coordinate system on your graph paper and clearly indicate the scale on each coordinate axis. Plot each point from the table on your coordinate system.

b) Use your plotted points to help draw a complete graph of the function f. Label your asymptote with its equation. Label the y-intercept with its coordinates.

7. Consider the equation $f(x) = (1.5)^x$. Make a copy of the following table on a sheet of graph paper and use your calculator to help find the missing entries. Round your function values to the nearest tenth (one decimal place).

x	−3	−2	−1	0	1	2	3
$f(x)$							

a) Set up a coordinate system on your graph paper and clearly indicate the scale on each coordinate axis. Plot each point from the table on your coordinate system.

b) Use your plotted points to help draw a complete graph of the function f. Label your asymptote with its equation. Label the y-intercept with its coordinates.

8. Consider the equation $f(x) = (0.75)^x$. Make a copy of the following table on a sheet of graph paper and use your calculator to help find the missing entries. Round your function values to the nearest tenth (one decimal place).

x	−3	−2	−1	0	1	2	3
$f(x)$							

a) Set up a coordinate system on your graph paper and clearly indicate the scale on each coordinate axis. Plot each point from the table on your coordinate system.
b) Use your plotted points to help draw a complete graph of the function f. Label your asymptote with its equation. Label the y-intercept with its coordinates.

9. Sketch the graph of $f(x) = a^x$, $a > 1$. Label the asymptote with its equation. Label the y-intercept with its coordinates.

10. Sketch the graph of $f(x) = a^x$, $0 < a < 1$. Label the asymptote with its equation. Label the y-intercept with its coordinates.

11. You will be guided through a series of steps designed to draw the graph of the function $f(x) = 2^{-x} + 3$.

    a) Set up a coordinate system on a sheet of graph paper and sketch the graph of $y = 2^x$ on your coordinate system.

       Hint: The base is larger than 1.

    b) Set up second coordinate system and sketch the graph of $y = 2^{-x}$ on this system.

       Hint: The graph of $y = 2^{-x}$ is the reflection of the graph of $y = 2^x$ across the y-axis.

    c) Set up a third coordinate system and sketch the graph of $y = 2^{-x} + 3$ on this system.

       Hint: The graph of $y = 2^{-x} + 3$ is translated 3 units upward from the graph of $y = 2^{-x}$.

       Draw a dotted line representing the asymptote and label this asymptote with its equation. Label the y-intercept with its coordinates.

12. You will be guided through a series of steps designed to draw the graph of the function $f(x) = -2^x - 1$.

    a) Set up a coordinate system on a sheet of graph paper and sketch the graph of $y = 2^x$ on your coordinate system.

       Hint: The base is larger than 1.

    b) Set up second coordinate system and sketch the graph of $y = -2^x$ on this system.

       Hint: The graph of $y = -2^x$ is the reflection of the graph of $y = 2^x$ across the x-axis.

    c) Set up a third coordinate system and sketch the graph of $y = -2^x - 1$ on this system.

       Hint: The graph of $y = -2^x - 1$ is translated 1 unit downward from the graph of $y = -2^x$.

       Draw a dotted line representing the asymptote and label this asymptote with its equation. Label the y-intercept with its coordinates.

13. For each of the following equations, perform each of the following tasks:

    i) Set up a sequence of graphs as in Exercise 11 that lead to the graph of the given equation. Draw the asymptote in your final sketch as a dotted line and label it with its equation. Label the y-intercept in your final sketch with its coordinates.

    ii) Check your solution with your calculator.

    a) $f(x) = 2^x + 3$ b) $f(x) = 3^x - 4$ c) $f(x) = -2^{-x}$ d) $f(x) = -10^x$

    e) $f(x) = -3^x + 2$ f) $f(x) = 4^{-x} - 4$ g) $f(x) = 5 - 2^x$ h) $f(x) = 12 - 3^{-x}$

i) $f(x) = (0.65)^{-x} + 2$ j) $f(x) = -(1.2)^x - 5$

Note: Use your calculator to check your solution, not to perform the problem's calculations.

14. (TI-82) Because $e = 2.718281828...$, it seems reasonable that the graph of $y = e^x$ would more closely resemble the graph of $y = 3^x$ than the graph of $y = 2^x$. Load the equations $y = 2^x$, $y = e^x$, and $y = 3^x$ into the Y= menu.

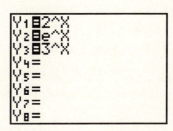

Push the ZOOM button and select 6:ZStandard. Set up a coordinate system on a sheet of graph paper and clearly indicate the scale on each coordinate axis. Copy the image in your viewing window onto your coordinate system. Label each graph with its equation. Label the common asymptote of each graph with its equation. Label the common y-intercept of each graph with its coordinates.

15. For the equations below, perform each of the following tasks:

i) Set up a sequence of graphs as in Exercise 11 that lead to the graph of the given equation. Draw the asymptote in your final sketch as a dotted line and label it with its equation. Label the y-intercept in your final sketch with its coordinates.

ii) Check your solution with your calculator.

a) $f(x) = e^{-x}$ b) $f(x) = -e^x$ c) $f(x) = e^x - 2$ d) $f(x) = e^x + 3$

e) $f(x) = -e^x + 5$ f) $f(x) = e^{-x} - 3$ g) $f(x) = e^{-x} + 5$ h) $f(x) = -e^x - 6$

i) $f(x) = 10 - e^x$ j) $f(x) = 9 - e^{-x}$

Note: Use your calculator to check your solution, not to solve the problem for you.

16. Use interval notation to describe the domain and range of the functions defined by each of the following equations.

Hint: Sketch the graph of the equation and examine your graph to find the domain and range of the function.

a) $f(x) = e^{-x} - 4$ b) $f(x) = e^x + 2$ c) $f(x) = 12 - e^x$ d) $f(x) = 6 - e^{-x}$

17. (TI-82) Set up a coordinate system on a sheet of graph paper and clearly indicate the scale on each coordinate axis. Use your calculator to sketch the graph of the equation $f(x) = e^{kx}$ for each of the following k-values: $k = 0.05$, $k = 0.15$, and $k = 0.30$. Copy the image in your viewing window onto your coordinate system. Label each graph on the coordinate system with its equation. The value k is sometimes called the *growth rate* of the function. Can you explain why?

18. (TI-82) Set up a coordinate system on a sheet of graph paper and clearly indicate the scale on each coordinate axis. Use your calculator to sketch the graph of the equation $f(x) = e^{kx}$ for each of the following k-values: $k = -0.05$, $k = -0.15$, and $k = -0.30$. Copy the image in your viewing window onto your coordinate system. Label each graph on the coordinate system with its equation. The value k is sometimes called the *decay rate* of the function. Can you explain why?

19. (TI-81) Use your graphing calculator to draw the graph of each of the following functions. Set up a coordinate system on a sheet of graph paper and copy the image in your viewing window onto your coordinate system. Label the asymptote with its equation. Label the y-intercept with its coordinates.

a) $f(x) = 100e^{0.12x}$ b) $g(x) = 100e^{-0.80x}$ c) $h(x) = 100e^{0.07x}$ d) $k(x) = 100e^{-1.44x}$

Some of the graphs are increasing and some are decreasing. In some instances we have growth, while in others we have decay. In your opinion, what determines whether the function will grow or decay?

20 A potato is removed from an oven and it immediately begins to cool. Suppose the temperature of the potato is given by the equation $T = 60 + 510e^{-0.82t}$, where T is measured in degrees Fahrenheit and t is the number of minutes that have passed since the potato was removed from the oven.

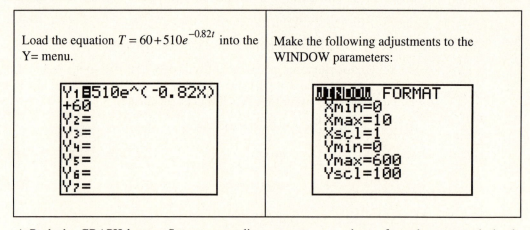

Load the equation $T = 60 + 510e^{-0.82t}$ into the Y= menu.

Make the following adjustments to the WINDOW parameters:

a) Push the GRAPH button. Set up a coordinate system on a sheet of graph paper and clearly indicate the scale on each coordinate axis. Copy the image in your viewing window onto your coordinate system.

b) What is the temperature of the potato 5 minutes after it is removed from the oven?

c) What is the temperature of the room, the eventual temperature to which the potato must cool?

21. Monte is attending secretary school and is learning to type. Suppose that the number of words per minute that he can type is given by the equation $W = 100 - 100e^{-0.12t}$, where W is measured in words per minute and t represents the number of weeks that have passed since Monte first started at the school.

a) Load the equation into the Y= menu and adjust the WINDOW parameters so that the graph of the equation is visible over the first 30 weeks that Monte attended school. Set up a coordinate system on a sheet of graph paper and clearly indicate the scale on each coordinate axis. Copy the image in your viewing window onto the coordinate system.

b) How many words per minute was Monte typing after 15 weeks in school?

c) How many words per minute will he eventually learn to type?

9.4 Composition of Functions

This section introduces a method of combining functions called the *composition of functions*.

Function Machines. Let's begin this section with a little construct that might help you better understand function notation.

Example 1 If $f(x) = 2x$, find $f(5)$ and $f(a)$.

Solution Consider the function represented by the equation $f(x) = 2x$. It is helpful to think of this function as a machine that doubles anything that is dropped into it.

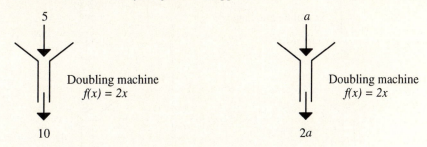

Hence $f(5) = 10$ and $f(a) = 2a$.

Example 2 If $g(x) = x + 3$, find $g(10)$ and $g(a)$.

Solution Consider the function represented by the equation $g(x) = x + 3$. It is helpful to think of this function as a machine that adds 3 to anything that is dropped into it.

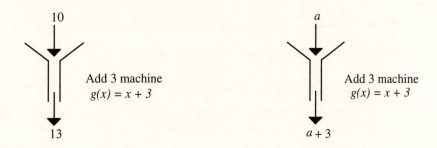

Hence, $g(10) = 13$ and $g(a) = a + 3$.

Now suppose you drop a number into one of the function machines, then immediately drop the output of the first function machine into a second function machine. This process is called the *composition of functions*.

Example 3 If $f(x) = 2x$ and $g(x) = x + 3$, find $g(f(2))$ and $f(g(2))$.

Solution

To find $g(f(2))$, you must first drop 2 into the f-machine to find $f(2)$, which is 4. Then you must drop 4 into the g-machine to find $g(4)$, which is 7.

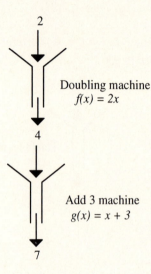

In the language of function notation, you would write the following:

$$g(f(2)) = g(4)$$
$$= 7$$

To find $f(g(2))$, you must first drop 2 into the g-machine to find $g(2)$, which is 5. Then you must drop 5 into the f-machine to find $f(5)$, which is 10.

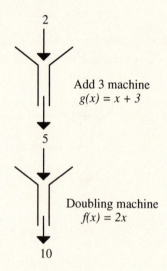

In the language of function notation, you would write the following:

$$f(g(2)) = f(5)$$
$$= 10$$

In this example, note that $f(g(2))$ and $g(f(2))$ are not equal.

Example 4 If $f(x) = 3x + 2$ and $g(x) = 6x - 2$, find $f(g(1))$ and $g(f(1))$.

Solution To find $f(g(1))$, you must first drop 1 into the g-machine to find $g(1)$, which is 4. Then you must drop 4 into the f-machine to find $f(4)$, which is 14.

$$f(g(1)) = f(4)$$
$$= 14$$

To find $g(f(1))$, you must first drop 1 into the f-machine to find $f(1)$, which is 5. Then you must drop 5 into the g-machine to find $g(5)$, which is 28.

$$g(f(1)) = g(5)$$
$$= 28$$

In this example, note that $f(g(1))$ and $g(f(1))$ are not equal.

Example 5 If $f(x) = 10x + 7$, find $f(8 - x)$.

Solution   Think of *f* as a machine. If you drop an object into this machine, the machine will first multiply the object by 10, then add 7.

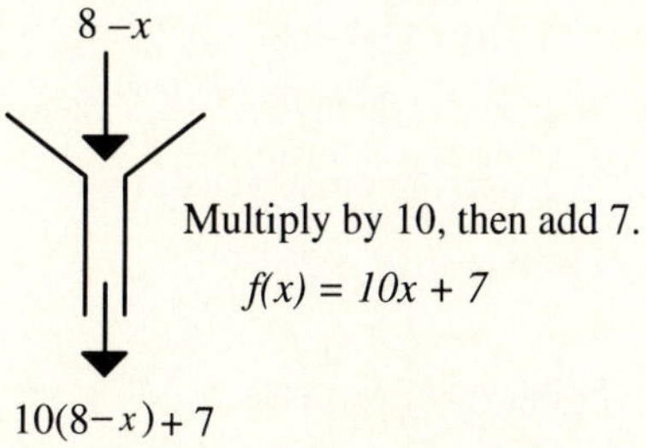

$$8 - x$$

Multiply by 10, then add 7.
f(x) = 10x + 7

$$10(8-x)+7$$

In the language of function notation, you may write the following.

$$f(8-x) = 10(8-x)+7$$
$$= 80-10x+7$$
$$= 87-10x$$

Note that the result has been simplified.

Example 6   If $f(x) = 3-2x$ and $g(x) = 4x+1$, find $f(g(x))$ and $g(f(x))$.

Solution

To find $f(g(x))$, you must first drop *x* into the *g*-machine to find $g(x)$, which is $4x+1$. Then you must drop $4x+1$ into the *f*-machine to find $f(4x+1)$, which is $3-2(4x+1)$.	To find $g(f(x))$, you must first drop *x* into the *f*-machine to find $f(x)$, which is $3-2x$. Then you must drop $3-2x$ into the *g*-machine to find $g(3-2x)$, which is $4(3-2x)+1$.
$$f(g(x)) = f(4x+1)$$ $$= 3-2(4x+1)$$ $$= 3-8x-2$$ $$= 1-8x$$	$$g(f(x)) = g(3-2x)$$ $$= 4(3-2x)+1$$ $$= 12-8x+1$$ $$= 13-8x$$

In this example, note that $f(g(x))$ and $g(f(x))$ are not equal.

Example 7   If $f(x) = \sqrt[3]{x-2}$ and $g(x) = x^3+2$, find $f(g(x))$ and $g(f(x))$.

Solution

To find $f(g(x))$, you first drop x into the g-machine to find $g(x)$, which is x^3+2. Then you drop x^3+2 into the f-machine to find $f(x^3+2)$, which is $\sqrt[3]{(x^3+2)-2}$.	To find $g(f(x))$, you first drop x into the f-machine to find $f(x)$, which is $\sqrt[3]{x-2}$. Then you drop $\sqrt[3]{x-2}$ into the g-machine to find $g(\sqrt[3]{x-2})$, which is $(\sqrt[3]{x-2})^3+2$.
$$\begin{aligned} f(g(x)) &= f(x^3+2) \\ &= \sqrt[3]{(x^3+2)-2} \\ &= \sqrt[3]{x^3} \\ &= x \end{aligned}$$	$$\begin{aligned} g(f(x)) &= g(\sqrt[3]{x-2}) \\ &= (\sqrt[3]{x-2})^3+2 \\ &= (x-2)+2 \\ &= x \end{aligned}$$

In this example, note that $f(g(x))$ and $g(f(x))$ both equal x.

Example 8 If $f(x)=\dfrac{x}{x-3}$ and $g(x)=\dfrac{x+1}{2-3x}$, find $g(f(x))$.

Solution To find $g(f(x))$, you first drop x into the f-machine to find $f(x)$, which is $\dfrac{x}{x-3}$. Then you drop $\dfrac{x}{x-3}$ into the g-machine to find $g\left(\dfrac{x}{x-3}\right)$.

$$g(f(x)) = g\left(\frac{x}{x-3}\right)$$

$$= \frac{\dfrac{x}{x-3}+1}{2-3\left(\dfrac{x}{x-3}\right)}$$

To simplify this complex fraction, multiply both numerator and denominator by $x-3$.

$$g(f(x)) = \frac{(x-3)\left[\dfrac{x}{x-3}+1\right]}{(x-3)\left[2-3\left(\dfrac{x}{x-3}\right)\right]}$$

$$= \frac{(x-3)\left[\dfrac{x}{x-3}\right]+(x-3)[1]}{(x-3)[2]-3(x-3)\left[\dfrac{x}{x-3}\right]}$$

$$= \frac{x+(x-3)}{2(x-3)-3x}$$

Finally, simplify both the numerator and denominator.

$$g(f(x)) = \frac{2x-3}{2x-6-3x}$$

$$= \frac{2x-3}{-x-6}$$

$$= -\frac{2x-3}{x+6}$$

Example 9   Suppose that the area of an oil spill is given by the equation $A(r) = \pi r^2$, where r is the radius of the oil spill in meters and A is the area of the oil spill in square meters. Suppose that the radius of the oil spill is increasing with time according to the equation $r(t) = 30t + 10$, where t is the number of hours that have passed since the spill began.

a) Express the area of the oil spill as a function of time.

b) Use the result of part (a) to find the area of the oil spill 10 hours after the spill began.

Solution   First insert the time into the r-machine to find the radius at time t. Then drop this radius into the A-machine to find the area of the spill. This is a classic application of the composition of functions.

$$A(r(t)) = A(30t + 10)$$

$$= \pi(30t + 10)^2$$

Therefore the area of the spill as a function of time is given by the following equation:

$$A = \pi(30t + 10)^2$$

To find the area of the spill at 10 hours, insert $t = 10$ into this last equation.

$$A = \pi(30(10) + 10)^2$$
$$A = \pi(50)^2$$
$$A = 2500\pi$$
$$A \approx 7854 \text{ square meters}$$

Example 10   Consider the following graph of the functions *f* and *g*. Use the graph to help find $g(f(2))$.

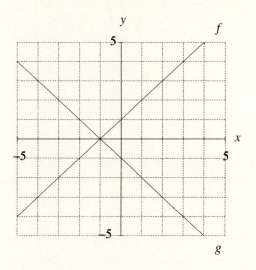

Solution   To compute $g(f(2))$, first drop 2 into the *f*-machine to find $f(2)$.

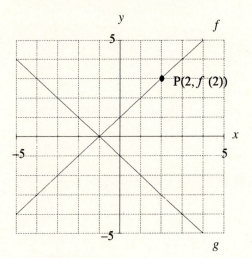

The *y*-value of the point P is $f(2)$. Therefore $f(2) = 3$. You must now drop 3 into the *g*-machine to find $g(3)$.

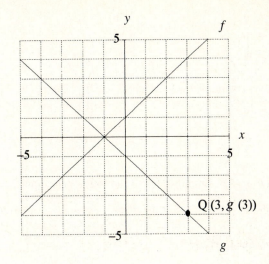

The y-value of the point Q is $g(3)$. Therefore $g(3) = -4$, which means that $g(f(2)) = g(3) = -4$.

Exercises for Section 9.4

1. If $f(x) = x + 5$, it is helpful to think of the function f as a machine that adds 5 to any number or object that is dropped into it. Copy each of the following function machines onto your homework paper and include the output from each machine in your sketch.

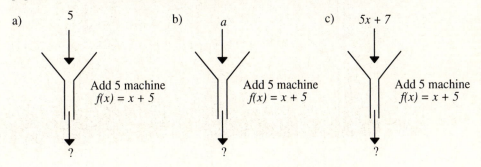

2. If $f(x) = x + 5$, simplify each of the following as much as possible.

   a) $f(5)$ b) $f(a)$ c) $f(5x+7)$ d) $f(11-6x)$

   Hint: See Exercise 1.

3. If $g(x) = 3x$, it is helpful to think of the function g as a machine that multiplies any number or object that is dropped into it by 3. Copy each of the following function machines onto your homework paper and include the output from each machine in your sketch.

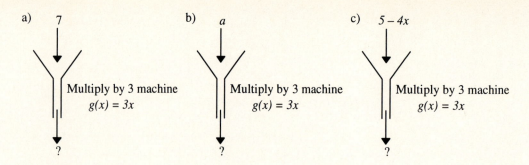

4. If $g(x) = 3x$, simplify each of the following as much as possible.

   a) $g(7)$ b) $g(a)$ c) $g(5-4x)$ d) $g(8x-7)$

Hint: See Exercise 3.

5. The composition of functions can be determined by letting the output of one function machine fall into a second function machine. Make a copy of each of the following composition machines on your homework and clearly indicate the final output in your sketch.

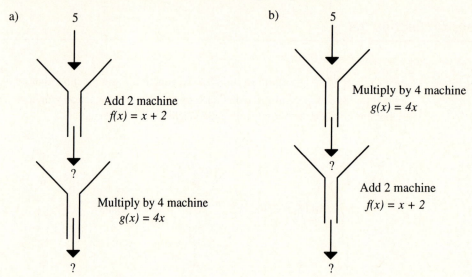

6. If $f(x) = x+2$ and $g(x) = 4x$, simplify each of the following as much as possible.

   a) $f(g(5))$ b) $g(f(5))$ c) $f(g(7))$ d) $g(f(7))$

Hint: See Exercise 5.

7. Suppose that $f(x) = x^2$ and $g(x) = x+1$.

   a) Create a function composition machine that depicts the calculation of $f(g(-4))$.
   b) Create a function composition machine that depicts the calculation of $g(f(-4))$.

8. Suppose that $f(x) = x+5$ and $g(x) = \dfrac{1}{x}$.

   a) Create a function composition machine that depicts the calculation of $f(g(12))$.
   b) Create a function composition machine that depicts the calculation of $g(f(12))$.

9. If $f(x) = x^3$ and $g(x) = x - 3$, simplify each of the following as much as possible.

   a) $f(g(2))$ b) $g(f(2))$ c) $f(g(-5))$ d) $g(f(-5))$

10. If $f(x) = 3x + 8$, it is helpful to think of f as a machine. If an object falls into the f-machine, it is first multiplied by 3, then 8 is added to the result. Copy each of the following function machines onto your homework paper and include the output from each machine in your sketch. Simplify your result as much as possible.

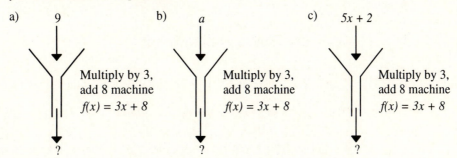

a)   9   b) a c) $5x + 2$

Multiply by 3, add 8 machine $f(x) = 3x + 8$ Multiply by 3, add 8 machine $f(x) = 3x + 8$ Multiply by 3, add 8 machine $f(x) = 3x + 8$

      ?   ?   ?

11. If $f(x) = 3x + 8$, simplify each of the following as much as possible.

    a) $f(9)$ b) $f(a)$ c) $f(5x+2)$ d) $f(11-7x)$

    Hint: See Exercise 10.

12. $f(x) = 9 - 4x$, simplify each of the following as much as possible.

    a) $f(10)$ b) $f(a)$ c) $f(2x+11)$ d) $f(x^2 - 3)$

13. If $f(x) = 2x^2 + 5$, simplify each of the following as much as possible.

    a) $f(11)$ b) $f(a)$ c) $f(2x+1)$ d) $f(1-x^2)$

14. If $f(x) = 2x + 3$ and $g(x) = \dfrac{x-3}{2}$, simplify each of the following as much as possible.

    a) $f(g(4))$ b) $g(f(4))$ c) $f(g(x))$ d) $g(f(x))$

15. If $f(x) = 4x + 11$ and $g(x) = 3 - x$, simplify each of the following as much as possible.

    a) $f(g(10))$ b) $g(f(10))$ c) $f(g(x))$ d) $g(f(x))$

16. If $f(x) = 2x^3$ and $g(x) = \sqrt[3]{\dfrac{x}{2}}$, simplify each of the following as much as possible.

    a) $f(g(2))$ b) $g(f(2))$ c) $f(g(x))$ d) $g(f(x))$

17. If $f(x) = \dfrac{1}{x}$ and $g(x) = \dfrac{x}{x+1}$, simplify each of the following as much as possible.

    a) $f(g(2))$ b) $g(f(2))$ c) $f(g(x))$ d) $g(f(x))$

    e) $f(2)g(2)$ f) $f(x)g(x)$ g) $f(g(-1))$ h) $g(f(-1))$

18. If $f(x) = \dfrac{x+2}{x-3}$ and $g(x) = \dfrac{2}{x+2}$, simplify each of the following as much as possible.

    a) $f(g(1))$ b) $g(f(1))$ c) $f(g(x))$ d) $g(f(x))$

    e) $f(5)g(5)$ f) $f(x)g(x)$ g) $f(g(3))$ h) $g(f(3))$

19. If $y = x^2 + 2x$ and $x = 3 - 2t$, express y as a function of t. Simplify your result as much as possible.

20. If $w = e^u$ and $u = -\dfrac{1}{2}x^2$, express w as a function of x.

21. If $f(x) = x^2$ and $g(x) = x + 3$ and $h(x) = 3x$, simplify each of the following as much as possible.

    a) $f(g(h(x)))$ b) $g(f(h(x)))$ c) $h(f(g(x)))$ d) $h(g(f(x)))$

22. If $z = x^2 + 4x$ and $x = t^2 + 1$ and $t = u - 3$, express z as a function of u. Simplify your result as much as possible.

23. If $z = \sqrt{u}$ and $u = e^w$ and $w = 2t - 1$, express z as a function of t.

24. As Julie begins to blow up a balloon, it inflates in the shape of a sphere. Suppose that the volume of the balloon is given by the equation $V = \dfrac{4}{3}\pi r^3$, where V is the volume in cubic inches and r is the radius of the balloon in inches. Suppose that the radius of the balloon is increasing with time according to the equation $r = 0.25t + 0.125$, where t is the number of seconds that have passed since Julie began inflating her balloon.

    a) Express the volume of the balloon as a function of time.

    b) Use the result of part (a) to find the volume of the balloon 5 seconds after Julie began blowing up her balloon.

25. An airplane is flying at an elevation of 6 miles on a flight path that will take it directly over a radar installation.

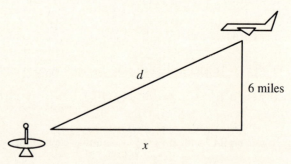

    a) Let d represent the number of miles between the plane and the radar installation and let x be the horizontal distance in miles between the plane and the radar installation. Express d as a function of x.

    b) Suppose that the horizontal distance between the plane and the radar installation is decreasing according to the equation $x = 200 - 6.7t$, where t is measured in minutes. Express d as a function of t.

c) Use the equation developed in part (a) to find the distance d between the plane and the radar installation at $t = 10$ minutes.

26. Willie, who is 6 feet tall, is walking away from a lamppost.

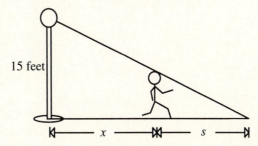

The length of his shadow is given by the equation $s = \frac{2}{3}x$, where s represents the length of the shadow in feet and x represents Willie's distance from the lamppost in feet. Suppose that Willie's distance from the lamppost is given by the equation $x = 2.7t + 10$, where t represents time in seconds.

a) Express the length of Willie's shadow as a function of time.

b) Use the equation developed in part (a) to find the length of Willie's shadow at $t = 7$ seconds.

27. Set up a coordinate system on a sheet of graph paper and clearly indicate the scale on each coordinate axis. Duplicate the following graphs of f and g on the coordinate system.

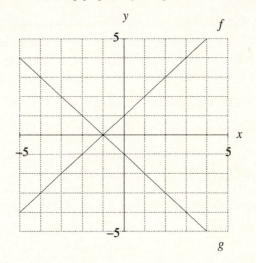

Use your graph to help find each of the following.

a) $f(g(2))$ b) $g(f(2))$ c) $f(g(0))$ d) $g(f(0))$

28. Set up a coordinate system on a sheet of graph paper and clearly indicate the scale on each coordinate axis. Duplicate the following graphs of f and g on your coordinate system.

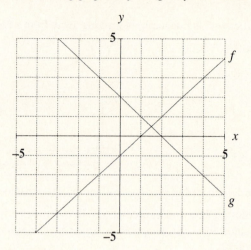

Make a copy of the following table on your graph paper. Use your graph to help complete the missing entries in the table.

x	-3	-2	-1	0	1	2	3	4	5
$f(g(x))$	4				0				
Points to plot	$(-3,4)$				$(1,0)$				

a) Use a red pencil to plot each of the points in the table on your coordinate system. Use these points to draw a complete graph of $y = f(g(x))$. What is the equation of the graph of $y = f(g(x))$?

b) What is the equation of the function f? $f(x) = ?$

c) What is the equation of the function g? $g(x) = ?$

d) Use the functions from parts (b) and (c) to find $f(g(x))$. Does this result match the result in part (a)? If not, check your work for error.

29. Set up a coordinate system on a sheet of graph paper and clearly indicate the scale on each axis.

a) Sketch the graph of $f(x) = 2x + 3$ on your coordinate system.

b) Sketch the graph of $g(x) = 2 - \frac{1}{2}x$ on your coordinate system.

c) Find $f(g(x))$ and use a red pencil to sketch the graph of this equation on your coordinate system.

30. Suppose that $f(x) = -x$ and $g(x) = \sqrt{x}$.

a) Sketch the graph of $y = f(g(x))$.
b) Sketch the graph of $y = g(f(x))$.

9.5 Reflecting Across the Line $y = x$

A new and very important graphing shortcut will be introduced in this section. You'll find it extremely useful in upcoming sections, particularly when sketching the inverse of a function.

Interchanging Coordinates

What happens when we interchange the coordinates of an ordered pair? Set up a coordinate system on a sheet of graph paper and draw the line $y = x$ on your coordinate system. Plot the points P(2,1), Q(5,3), and R(7,3). Now plot the points P'(1,2), Q'(3,5), and R'(3,7) and draw the line segments $\overline{PP'}$, $\overline{QQ'}$, and $\overline{RR'}$.

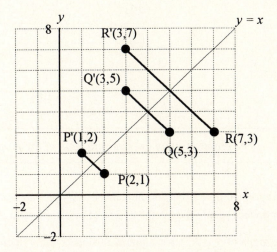

The line $y = x$ is the perpendicular bisector of the segments $\overline{PP'}$, $\overline{QQ'}$, and $\overline{RR'}$. The points P', Q', and R' are the *reflections* of the points P, Q, and R across the line $y = x$.

The point (y,x) is a reflection of the point (x,y) across the line $y = x$.

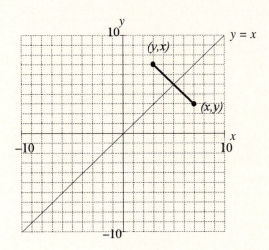

The Graph of $x = f(y)$.   If you interchange the x- and y-value of each point on the graph $y = f(x)$, the resulting graph of $x = f(y)$ will be a reflection of the graph of $y = f(x)$ across the line $y = x$.

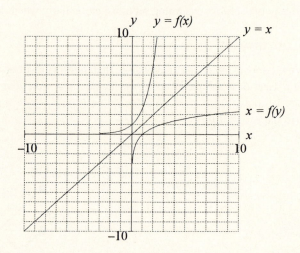

The graph of $x = f(y)$ is a reflection of the graph of $y = f(x)$ across the line $y = x$.

Example 1   Set up a coordinate system on a sheet of graph paper and draw the graph of the line $y = x$ on your coordinate system. Sketch the graphs of $y = 2x + 3$ and $x = 2y + 3$ on the same coordinate system.

Solution   The graph of $y = 2x + 3$ is a line with slope 2 and y-intercept 3.

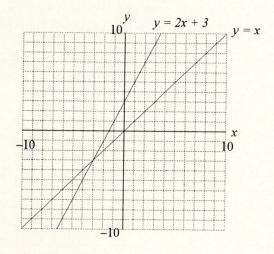

Next, solve the equation $x = 2y + 3$ for y.

$$x = 2y + 3$$
$$x - 3 = 2y$$
$$y = \frac{x - 3}{2}$$
$$y = \frac{1}{2}x - \frac{3}{2}$$

The graph of the equation $y = \frac{1}{2}x - \frac{3}{2}$, or $x = 2y + 3$, is a line with slope $\frac{1}{2}$ and y-intercept $-\frac{3}{2}$.

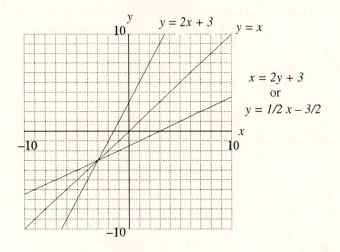

It is important to note that the graph of $x = 2y + 3$, or $y = \frac{1}{2}x - \frac{3}{2}$, is a reflection of the graph of $y = 2x + 3$ across the line $y = x$. It is also important to note that the graph of $x = 2y + 3$ passes the vertical line test and *does* represent a function.

Whenever you interchange x and y in an equation, the graph of the resulting equation is a reflection of the graph of the original equation across the line $y = x$.

Example 2   Sketch the graph of $x = y^2$.

Solution   If you interchange x and y, the equation $y = x^2$ becomes $x = y^2$. Therefore the graph of $x = y^2$ is a reflection of the graph of $y = x^2$ across the line $y = x$.

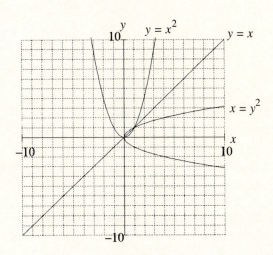

Note that the graph of $x = y^2$ fails the vertical line test and does *not* represent a function.

Seeing the Mirror Image Across Line $y = x$: A Great Visual Trick. Sometimes it is difficult to visualize the reflection of a graph across the line $y = x$. When this happens, use a darker pen or pencil than usual to draw the graph. Hold your paper up to a strong source of light and reflect your paper across the line $y = x$. Look through the back of your paper to see the reflection of the graph across the line $y = x$. Give this technique a try with the graph of $y = x^2$.

Example 3   The graph of $y = f(x)$ has been sketched below. Sketch the graph of $x = f(y)$.

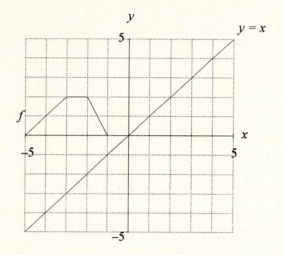

Solution   If you interchange x and y, the equation $y = f(x)$ becomes $x = f(y)$. Therefore the graph of $x = f(y)$ is a reflection of the graph of $y = f(x)$ across the line $y = x$.

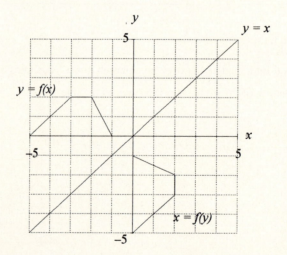

It is important to note that the graph of $x = f(y)$ fails the vertical line test and does *not* represent a function.

One-to-One Functions and the Horizontal Line Test. You already know that if any vertical line cuts the graph more than once, then the graph does *not* represent a function. This is known as the vertical line test. A similar test, called the *horizontal line test*, determines whether a function is one-to-one.

> *If no horizontal line cuts the graph of the function more than once, then the function is called a one-to-one function.*

Example 4

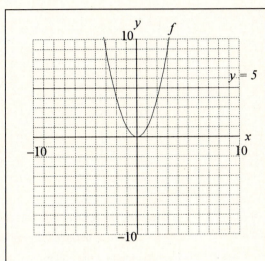

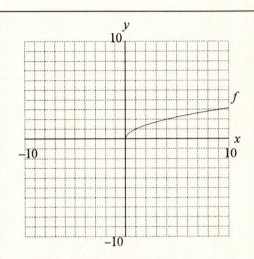

The horizontal line $y = 5$ cuts the graph of $f(x) = x^2$ twice. Therefore f is *not* a one-to-one function.

No horizontal line cuts the graph of $f(x) = \sqrt{x}$ more than once. Therefore f *is* a one-to-one function.

What's so special about one-to-one functions? When a horizontal line is reflected across the line $y = x$, it becomes a vertical line. Therefore, if a graph fails the horizontal line test, then its reflection across the line $y = x$ will fail the vertical line test and will not represent a function.

Example 5

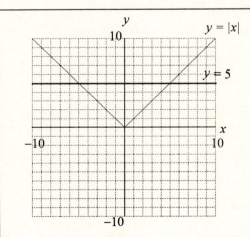

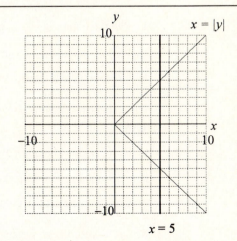

The graph of $y = |x|$ fails the horizontal line test and does *not* represent a one-to-one function.

Therefore the graph of $x = |y|$ fails the vertical line test and does *not* represent a function.

Example 6

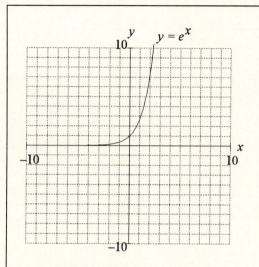
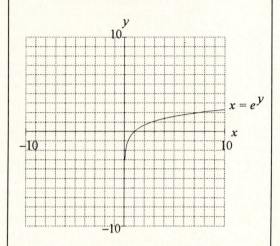

No horizontal line cuts the graph of $y = e^x$ more than once. Therefore the function represented by the equation $y = e^x$ *is* a one-to-one function.

Therefore no vertical line cuts the graph of $x = e^y$ more than once. The relation represented by the equation $x = e^y$ *is* a function.

Exercises for Section 9.5

1. Set up a coordinate system on a sheet of graph paper and clearly indicate the scale on each of the coordinate axes. Draw the line $y = x$ on your coordinate system.

   a) Plot the points P(5,1) and P'(1,5). Draw the line segment $\overline{PP'}$.

   b) Plot the points R(9,3) and R'(3,9). Draw the line segment $\overline{RR'}$.

   c) Plot the points Q(−5,−2) and Q'(−2,−5). Draw the line segment $\overline{QQ'}$.

   d) Plot the points S(−7,1) and S'(1,−7). Draw the line segment $\overline{SS'}$.

   e) Use complete sentences to explain what you learned in this exercise.

2. On a sheet of graph paper, set up a coordinate system. Draw the line $y = x$ on your coordinate system.

   a) Draw the line that goes through the points P(−3,−2) and Q(7,3).

   b) Draw the line that goes through the points P'(−2,−3) and Q'(3,7).

   c) Use complete sentences to explain what you learned in this exercise.

3. Set up a coordinate system on a sheet of graph paper and clearly indicate the scale on each coordinate axis. Draw the graph of the line $y = x$ on this same coordinate system.

   a) Make a copy of the following table on your graph paper and complete the missing entries.

x	0	1	4	9	16	25
$y = \sqrt{x}$						5
Points to plot						(25,5)

b) Plot each of the points in the table on your coordinate system. Use these points to draw a complete graph of the equation $y = \sqrt{x}$.

c) Make a copy of the following table on your graph paper. Complete the missing entries in this second table by reversing all of the points from the table in part (a).

$x = \sqrt{y}$						5
y	0	1	4	9	16	25
Points to plot						(5,25)

d) Use a red pencil to plot all of the points in this second table on your coordinate system. Use these points to help draw a complete graph of the equation $x = \sqrt{y}$.

e) Use complete sentences to explain what you learned in this exercise.

4. Set up a coordinate system on a sheet of graph paper and clearly indicate the scale on each coordinate axis. Draw the graph of the line $y = x$ on this same coordinate system.

a) Make a copy of the following table on your graph paper and complete the missing entries.

x	−3	−2	−1	0	1	2	3
$y = x^2$							
Points to plot							

b) Plot each of the points in the table on your coordinate system. With a dark marking pen, darken the x-axis and y-axis, the line $y = x$, and the graph of the equation $y = x^2$.

c) Hold your paper up to a strong light source. Grasp your paper at the ends of the line $y = x$ and flip your paper across the line $y = x$. Stare through the back of your paper to see the graph of the equation $x = y^2$.

5. Some graphs are symmetric with respect to the line $y = x$. If you reflect the graph across the line $y = x$, there is absolutely no change. Set up a coordinate system on a sheet of graph paper and clearly indicate the scale on each axis.

a) Sketch the graph of the equation $y = \dfrac{1}{x}$ on your coordinate system.

b) Using a dark pen, darken the x-axis and y-axis, the line $y = x$, and the graph of $y = \dfrac{1}{x}$.

c) Hold your paper up to a strong light source. Grasp your paper at the ends of the line $y = x$ and flip your paper across the line $y = x$. Look through the back of your paper to see the graph of $x = \dfrac{1}{y}$.

d) Use complete sentences to explain what you learned in this exercise.

6. Set up a coordinate system on a sheet of graph paper and clearly indicate the scale on each coordinate axis. Duplicate the following graph of $y = f(x)$ on your coordinate system. Draw the graph of the line $y = x$ on this same coordinate system.

a) Make a copy of the following table on your graph paper. Use the graph of f to help complete the missing entries in the table.

x	−4	−3	−2	−1	0	1	2
$y = f(x)$	5						
Points to plot	(−4,5)						

b) Make a copy of the following table on your graph paper. Complete the missing entries in this second table by reversing all of the points from the table in part (a).

$x = f(y)$	5						
y	−4	−3	−2	−1	0	1	2
Points to plot	(5,−4)						

c) Use a red pencil to plot all of the points in this second table on your coordinate system. Use these points to help draw the graph of $x = f(y)$.

d) Use complete sentences to explain what you learned in this exercise.

7. On a sheet of graph paper, sketch the graph of each of the following equations on its own coordinate system.

a) $x = -|y|$ b) $x = y^3$ c) $x = -y^2$ d) $x = -\sqrt{y}$ e) $x = 2^y$ f) $x = 10^y$

8. Set up a coordinate system on a sheet of graph paper and clearly indicate the scale on each coordinate axis. Draw the graph of the line $y = x$ on this same coordinate system.

a) Sketch the graph of $y = 3x + 6$ on your coordinate system.

b) If you interchange x and y, the equation $y = 3x + 6$ becomes $x = 3y + 6$. Solve the equation $x = 3y + 6$ for y and use your result to sketch the graph of $x = 3y + 6$ on your coordinate system.

c) Are the graphs of $y = 3x + 6$ and $x = 3y + 6$ symmetric with respect to the line $y = x$? If not, check your work for error.

9. Set up a coordinate system on a sheet of graph paper and clearly indicate the scale on each coordinate axis. Draw the graph of the line $y = x$ on this same coordinate system.

a) Sketch the graph of $y = \dfrac{1}{2}x + 4$ on your coordinate system.

b) If you interchange x and y, the equation $y = \frac{1}{2}x + 4$ becomes $x = \frac{1}{2}y + 4$. Solve the equation $x = \frac{1}{2}y + 4$ for y and use your result to sketch the graph of $x = \frac{1}{2}y + 4$ on your coordinate system.

c) Are the graphs of $y = \frac{1}{2}x + 4$ and $x = \frac{1}{2}y + 4$ symmetric with respect to the line $y = x$? If not, check your work for error.

10. Set up a coordinate system on a sheet of graph paper and clearly indicate the scale on each coordinate axis. Duplicate the following graph of $y = f(x)$ on your coordinate system. Draw the graph of the line $y = x$ on this same coordinate system.

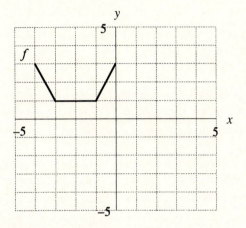

a) Is f a one-to-one function? Justify your answer.
b) Sketch the graph of $x = f(y)$ on your coordinate system.
c) Is the graph of $x = f(y)$ a function? Justify your answer.

11. Set up a coordinate system on a sheet of graph paper and clearly indicate the scale on each coordinate axis. Duplicate the following graph of $y = f(x)$ on your coordinate system. Draw the graph of the line $y = x$ on this same coordinate system.

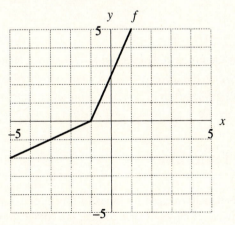

a) Is f a one-to-one function? Justify your answer.
b) Sketch the graph of $x = f(y)$ on your coordinate system.
c) Is the graph of $x = f(y)$ a function? Justify your answer.

12. Set up a coordinate system on a sheet of graph paper and clearly indicate the scale on each coordinate axis. Duplicate the following graph of $y = f(x)$ on your coordinate system. Draw the graph of the line $y = x$ on this same coordinate system.

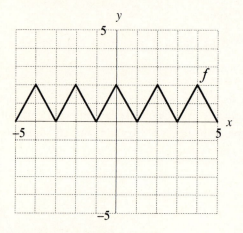

a) Is f a one-to-one function? Justify your answer.
b) Sketch the graph of $x = f(y)$ on your coordinate system.
c) Is the graph of $x = f(y)$ a function? Justify your answer.

13. Set up a coordinate system on a sheet of graph paper and clearly indicate the scale on each coordinate axis. Duplicate the following graph of $y = \sin x$ on your coordinate system. Draw the graph of the line $y = x$ on this same coordinate system.

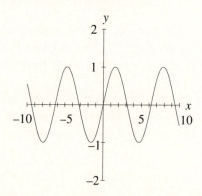

a) Is the function represented by $y = \sin x$ a one-to-one function? Justify your answer.
b) Sketch the graph of $x = \sin y$ on your coordinate system.
c) Does the graph of $x = \sin y$ represent a function? Justify your answer.

9.6 The Inverse Function

This section introduces the inverse function, a topic for which we have been steadily laying the groundwork. You know how to take the composition of functions and you also know that if x and y are interchanged in an equation, the graph of the resulting equation will be a reflection of the graph of the original equation across the line $y = x$. These concepts will be put to good use in this section.

Example 1   Consider the function WRAP, which is defined by the following sequence of operations:

   1. Put on the paper.
   2. Put on the tape.
   3. Put on the ribbon.

Use your intuition to define the function UNWRAP.

Solution   If you are to proceed with proper decorum, you must unwrap in *inverse* order. Therefore the function UNWRAP is defined by the following sequence of operations.

   1. Take the ribbon off.
   2. Take the tape off.
   3. Take the paper off.

Do not take this example lightly! The thought processes used in this example are the heart and soul of the inverse function.

Example 2   Consider the function f represented by the equation $f(x) = x - 3$. It is helpful to think of this function as a machine that subtracts 3 from any input. For example, if you input the number 9, the f-machine outputs the number 6.

Find a machine that will undo the effects of this subtract 3 machine.

Solution   What you need is a machine that adds 3 to any input. Consequently, choose the function g defined by $g(x) = x + 3$.

When 9 is input, the *f*-machine outputs the number 6. When 6 is put into the *g*-machine, the number 9 is returned.

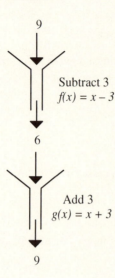

In function notation, $g(f(9)) = 9$.

Note that the *f*-machine will also undo the effects of the *g*-machine.

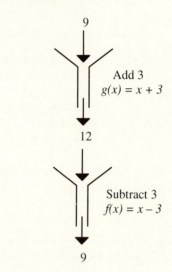

In function notation, $f(g(9)) = 9$.

Example 3   Consider the function *f* represented by the equation $f(x) = 2x$. It is helpful to think of this function as a machine that multiplies any input by 2. For example, if you input the number 8, the *f*-machine outputs the number 16.

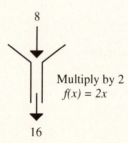

Find a machine that will undo the effects of this doubling machine.

Solution   What you need is a machine that divides its input by 2. Consequently, choose the function *g* defined by $g(x) = \dfrac{x}{2}$.

When 8 is input, the *f*-machine outputs the number 16. When 16 is put into the *g*-machine, the number 8 is returned.

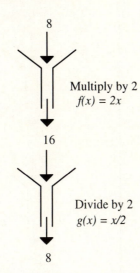

In function notation, $g(f(8)) = 8$.

Note that the *f*-machine will also undo the effects of the *g*-machine.

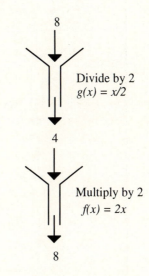

In function notation, $f(g(8)) = 8$.

Example 4   Consider the function *f* represented by the equation $f(x) = 2x + 3$. It is helpful to think of the function *f* as a machine. If a number is input, the *f*-machine first multiplies by 2, then adds 3. For example, if you input the number 7, the *f*-machine outputs the number 17.

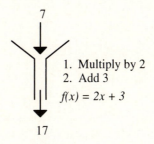

Find a machine that will undo the effects of the *f*-machine.

Solution   Recall how you unwrapped in Example 1. You must undo the effects of the *f*-machine in *inverse order*. Therefore, you must perform the following sequence of operations to undo the effects of the *f*-machine.

1. Subtract 3
2. Divide by 2

You must first subtract 3, then divide by 2. Therefore define the function *g* by $g(x) = \dfrac{x-3}{2}$.

When 7 is input, the *f*-machine outputs the number 17. When 17 is put into the *g*-machine, the number 7 is returned.

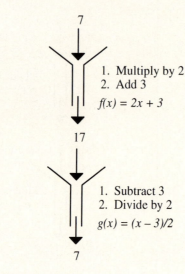

In function notation, $g(f(7)) = 7$.

Note that the *f*-machine will also undo the effects of the *g*-machine.

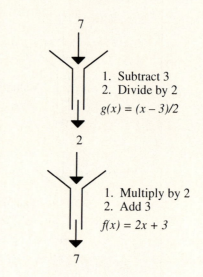

In function notation, $f(g(7)) = 7$.

Example 5   In Example 4, you saw that the functions *f* and *g*, defined by $f(x) = 2x + 3$ and $g(x) = \dfrac{x-3}{2}$, undid the effects of each other. Show that $f(g(x)) = x$ and $g(f(x)) = x$ for all values of *x*.

Solution

$$f(g(x)) = f\left(\frac{x-3}{2}\right)$$

$$= 2\left(\frac{x-3}{2}\right) + 3$$

$$= (x-3) + 3$$

$$= x$$

$$g(f(x)) = g(2x + 3)$$

$$= \frac{(2x+3) - 3}{2}$$

$$= \frac{2x}{2}$$

$$= x$$

The Definition of the Inverse Function.   In Example 1, we saw that the function UNWRAP undid the effects of the WRAP function. In Examples 2, 3 and 4, the function *g* undid the effects of the function *f*. In general, if a function *g* undoes the effects of a function *f*, and vice versa, the functions *f* and *g* are called *inverses* of one another.

> *Let f and g be functions that satisfy the following conditions.*
>
> 1. $f(g(x)) = x$ *for all x in the domain of g.*
> 2. $g(f(x)) = x$ *for all x in the domain of f.*
>
> *The function g is called the inverse of the function f. By symmetry, the function f must be the inverse of the function g. Therefore the functions f and g are called inverses of each other.*

Example 6   Find the inverse of the function f defined by $f(x) = x^3 + 5$ and use the definition to verify that your answer is the inverse of the function f.

Solution   It is helpful to think of the function f as a machine that performs the following sequence of operations.

1. Cube
2. Add 5

To undo the effects of this function, you must undo this sequence of operations in *inverse order*. Therefore you must perform the following sequence of operations.

1. Subtract 5
2. Take the cube root

Let $g(x) = \sqrt[3]{x-5}$. You must use the definition to verify that f and g are inverses of each other.

$$f(g(x)) = f\left(\sqrt[3]{x-5}\right) \qquad\qquad g(f(x)) = g(x^3 + 5)$$

$$= \left(\sqrt[3]{x-5}\right)^3 + 5 \qquad\qquad = \sqrt[3]{(x^3 + 5) - 5}$$

$$= (x-5) + 5 \qquad\qquad\qquad = \sqrt[3]{x^3}$$

$$= x \qquad\qquad\qquad\qquad\qquad = x$$

Because $f(g(x)) = x$ for all x in the domain of g and $g(f(x)) = x$ for all x in the domain of f, the functions f and g are inverses of each other.

> *If the inverse of the function f exists, mathematicians sometimes use the notation f^{-1} to represent the inverse function*

Example 7   If $f(x) = x^3 + 5$, then you may write $f^{-1}(x) = \sqrt[3]{x-5}$.

The Graph of the Inverse Function

Suppose that the functions f and f^{-1} are inverses of each other. Therefore, whatever the f-machine does, the f^{-1}-machine must undo. For example, suppose that the number 7 is put into the f-machine and the f-machine outputs the number 11. If the number 11 is placed in the f^{-1}-machine, then the f^{-1}-machine must output the number 7. This observation can be simplified with function notation: If

$f(7) = 11$, then $f^{-1}(11) = 7$. The inverse function f^{-1} simply reverses the ordered pairs of the function f. Consequently, the graph of f^{-1} must be a reflection of the graph of f across the line $y = x$.

Example 8   Suppose that the function f is defined by the equation $f(x) = 2x + 4$.

     a) Find $f^{-1}(x)$.

     b) Sketch the graph of f and f^{-1} on the same coordinate system.

Solution   a) It is helpful to think of the function f as a machine that first multiplies by 2, then adds 4. Therefore the inverse must first subtract 4, then divide by 2. Let $f^{-1}(x) = \dfrac{x-4}{2}$.

     b) The graph of $f(x) = 2x + 4$ is a line with slope 2 and y-intercept 4. The graph of $f^{-1}(x) = \dfrac{x-4}{2}$, or $f^{-1}(x) = \dfrac{1}{2}x - 2$, is a line with slope $\dfrac{1}{2}$ and y-intercept -2.

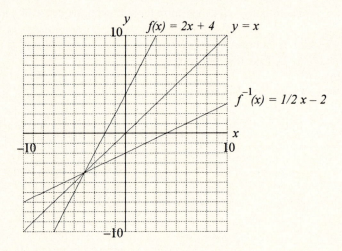

Note that the graph of f^{-1} is a reflection of the graph of f across the line $y = x$.

If the function f^{-1} is the inverse of the function f, then the graph of f^{-1} will be a reflection of the graph of f across the line $y = x$.

An Algebraic Method for Finding f^{-1}

If $f(8) = 12$, then $f^{-1}(12) = 8$. The inverse function f^{-1} simply switches the ordered pairs of the function f. This fact can be used to help find the inverse function.

Example 9   If $f(x) = 3x - 8$, find $f^{-1}(x)$.

Solution   Because y and $f(x)$ are interchangeable, you may write the following.

$$f(x) = 3x - 8$$
$$y = 3x - 8$$

Because f^{-1} simply reverses the ordered pairs of f, you can find the inverse by switching x and y and solving for y.

$$x = 3y - 8$$
$$x + 8 = 3y$$
$$\frac{x + 8}{3} = y$$

Therefore, $f^{-1}(x) = \dfrac{x + 8}{3}$.

Example 10   Use your calculator to observe the symmetry of the graphs of $f(x) = 3x - 8$ and $f^{-1}(x) = \dfrac{x + 8}{3}$ with respect to the line $y = x$.

Load the equations of f, f^{-1}, and the equation $y = x$ into the Y= menu. 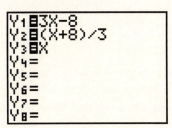	Push the ZOOM button and select 5:ZSquare to capture the following image: 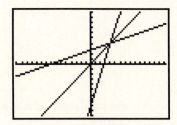 Note the symmetry with respect to the line $y = x$.

Example 11   Find the inverse of the function whose equation is $f(x) = x^2$.

Solution   First, sketch the graph of $f(x) = x^2$.

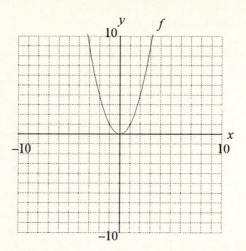

The graph of f fails the horizontal line test, so the reflection of the graph of f across the line $y = x$ will fail the vertical line test and cannot be a function. Therefore the inverse of the function f does not exist.

Example 12   Suppose that you use only the left half of the parabola whose equation is $f(x) = x^2$, $x \le 0$.

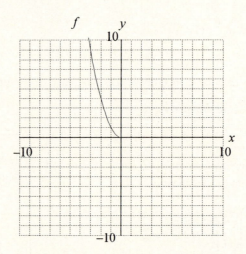

The domain of f has been restricted so that its graph now satisfies the horizontal line test and the function f is a one-to-one function. Consequently, the reflection of this graph across the line $y = x$ will satisfy the vertical line test and will represent a function. Therefore the inverse of the function f exists.

Note that the domain of f is $\{x : x \le 0\}$. It is helpful to think of the function f as a machine that is fed *nonpositive* real numbers that it squares. Therefore, to undo the effects of this machine, the function f^{-1} must return a *nonpositive* square root. Therefore let $f^{-1}(x) = -\sqrt{x}$.

To verify that the functions f and f^{-1} are inverses of each other, you must show that $f(f^{-1}(x)) = x$ for all x in the domain of f^{-1}, and you must show that $f^{-1}(f(x)) = x$ for all x in the domain of f.

The domain of $f^{-1}(x) = -\sqrt{x}$ is $\{x : x \geq 0\}$. If x is in the domain of f^{-1}, then $x \geq 0$.	The domain of $f(x) = x^2$, $x \leq 0$ is $\{x : x \leq 0\}$. If x is in the domain of f, then $x \leq 0$.
$$f(f^{-1}(x)) = f(-\sqrt{x})$$ $$= (-\sqrt{x})^2$$ $$= x$$	$$f^{-1}(f(x)) = f^{-1}(x^2)$$ $$= -\sqrt{x^2}$$ $$= -|x|$$ $$= -(-x)$$ $$= x$$ *Note:* Because $x \leq 0$, $|x| = -x$.

If the graphs of f, f^{-1}, and of the line $y = x$ are placed on the same coordinate system, you can see the symmetry with respect to the line $y = x$.

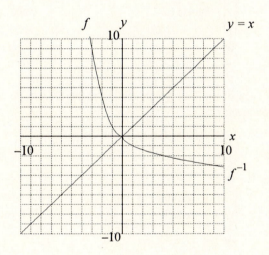

Exercises for Section 9.6

1. Before going out into a cold and windy Chicago evening, Matthew must dress warmly to protect himself from the cold. Define the function BUNDLE by the following sequence of operations.

   1. Put on suit coat.
   2. Put on overcoat.
   3. Put on scarf.

   Use your intuition to define the function UNBUNDLE.

2. Marcie rents a safe deposit box at the bank where she keeps valuable papers. Define the function LOCK by the following sequence of operations.

   1. Close the door of the box.
   2. Turn the key to the right to lock the door.

3. Remove the key from the lock.

Use your intuition to define the function UNLOCK.

3. For each of the following function machines, create a machine that will undo the effects of the given machine.

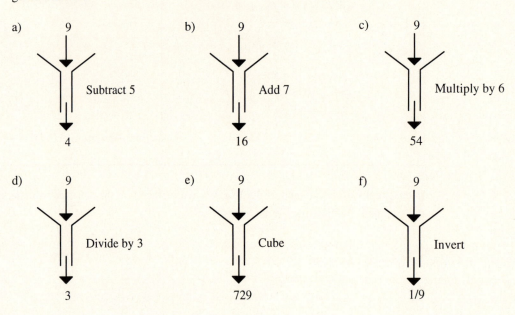

a) 9 → Subtract 5 → 4

b) 9 → Add 7 → 16

c) 9 → Multiply by 6 → 54

d) 9 → Divide by 3 → 3

e) 9 → Cube → 729

f) 9 → Invert → 1/9

4. For each of the following function machines, create a machine that will undo the effects of the given machine.

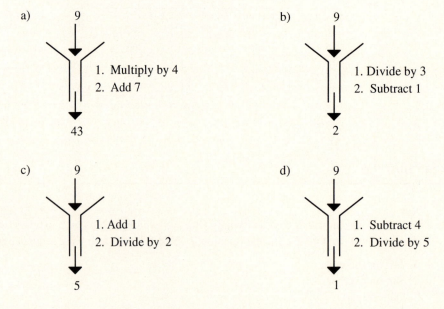

a) 9 → 1. Multiply by 4 2. Add 7 → 43

b) 9 → 1. Divide by 3 2. Subtract 1 → 2

c) 9 → 1. Add 1 2. Divide by 2 → 5

d) 9 → 1. Subtract 4 2. Divide by 5 → 1

5. Verify that each of the following pairs of functions are inverses of each other by showing that $f(g(x)) = x$ for all x in the domain of g and that $g(f(x)) = x$ for all x in the domain of f.

a) $f(x) = 7x$, $g(x) = \dfrac{x}{7}$

b) $f(x) = x + 5$, $g(x) = x - 5$

c) $f(x) = 4x - 9$, $g(x) = \dfrac{x+9}{4}$

d) $f(x) = \dfrac{x-7}{2}$, $g(x) = 2x + 7$

e) $f(x) = x^3 + 9$, $g(x) = \sqrt[3]{x-9}$

f) $f(x) = 3x^5 - 7$, $g(x) = \sqrt[5]{\dfrac{x+7}{3}}$

6. For the equations below, perform each of the following tasks:

 i)   Use your intuition to find a function g that will undo the effects of the function f.

 ii) Verify that $f(g(x)) = x$ for all x in the domain of g and verify that $g(f(x)) = x$ for all x in the domain of f.

 a) $f(x) = 5x$

 b) $f(x) = x + 7$

 c) $f(x) = 5x + 1$

 d) $f(x) = 3x - 2$

 e) $f(x) = \dfrac{x+10}{7}$

 f) $f(x) = \dfrac{x-8}{11}$

 g) $f(x) = x^3 - 5$

 h) $f(x) = \sqrt[3]{x+7}$

7. The functions f and g, defined by the equations $f(x) = 2x$ and $g(x) = \dfrac{x}{2}$, are inverses of each other.

 a) On a sheet of graph paper, make a copy of the following table and complete the missing entries.

x	-3	-2	-1	0	1	2	3
$f(x) = 2x$							

 b) Make a copy of the following table and complete the missing entries.

x	-6	-4	-2	0	2	4	6
$g(x) = \dfrac{x}{2}$							

 c) Set up a coordinate system on your graph paper and clearly indicate the scale on each coordinate axis. Draw the line $y = x$ on your coordinate system.

 d) Plot each of the points from tables in parts (a) and (b) on your coordinate system. Use the plotted points to draw complete graphs of the functions f and g.

Note: The point of this problem is to recognize that the function g reverses the ordered pairs of the function f. Therefore the graph of the function g must be a reflection of the graph of the function f across the line $y = x$.

8. (TI-82) The functions $f(x) = x^3 + 4$ and $g(x) = \sqrt[3]{x-4}$ are inverses of each other.

Load the equations of the functions f and g into the Y= menu along with the equation $y = x$. 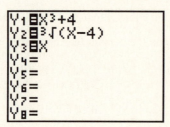	Push the ZOOM button and select 6:ZStandard. Push the ZOOM button and select 5:ZSquare to capture the following image. 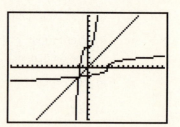

Note that the functions f and g are symmetric with respect to the line $y = x$. Each of the following pairs of functions are inverses of each other. Load the equations of f and g into the Y= menu along with the equation of the line $y = x$. Push the ZOOM button and select 6:ZStandard. Push the ZOOM button and select 5:ZSquare. Set up a coordinate system on a sheet of graph paper and clearly indicate the scale on each coordinate axis. Make an accurate copy of the image in your viewing window on your coordinate system.

a) $f(x) = 3x$, $g(x) = \dfrac{x}{3}$

b) $f(x) = x + 4$, $g(x) = x - 4$

c) $f(x) = 2x + 1$, $g(x) = \dfrac{x-1}{2}$

d) $f(x) = 4x - 1$, $g(x) = \dfrac{x+1}{4}$

e) $f(x) = x^3$, $g(x) = \sqrt[3]{x}$

f) $f(x) = x^3 - 3$, $g(x) = \sqrt[3]{x+3}$

9. For the equations below, perform each of the following tasks:

i)   Set up a separate coordinate system on a sheet of graph paper and clearly indicate the scale on each coordinate axis. Draw the line $y = x$ on your coordinate system.

ii) Use the slope and y-intercept to help sketch the graph of f on your coordinate system.

iii) Use your intuition to find $f^{-1}(x)$. Use the slope and y-intercept to help sketch the graph of f^{-1}.

iv) Verify that $f(f^{-1}(x)) = x$ for all x in the domain of f^{-1} and verify that $f^{-1}(f(x)) = x$ for all x in the domain of f.

a) $f(x) = 2x + 1$ b) $f(x) = 3x - 4$ c) $f(x) = \dfrac{x+3}{2}$ d) $f(x) = \dfrac{x-4}{3}$

10. Solve each of the following equations for y.

a) $x = 2y + 7$ b) $x = 3 - 2y$ c) $x = \dfrac{1}{2}y + 5$ d) $x = 5 - \dfrac{2}{3}y$ e) $x = 2y^3 + 1$

f) $x = 1 - 3y^3$ g) $x = \sqrt[3]{2y + 3}$ h) $x = \sqrt[3]{3 - 4y}$ i) $x = \dfrac{y}{y+2}$ j) $x = \dfrac{2y+5}{4-3y}$

11. Use the method of switching x and y and solving for y to find the inverse of each of the following functions.

a) $f(x) = \dfrac{2}{5}x - 3$

b) $f(x) = 5 - \dfrac{1}{4}x$

c) $f(x) = \dfrac{3}{2}x^3 + 1$

d) $f(x) = 3 - \dfrac{3}{4}x^5$

e) $f(x) = \sqrt[3]{3x+4}$

f) $f(x) = \sqrt[5]{2x-7}$

g) $f(x) = \dfrac{x}{x+1}$

h) $f(x) = \dfrac{3-2x}{5x+1}$

12. Let $f(x) = x^2$. It is helpful to think of the function f as a machine that squares its input. Let $g(x) = \sqrt{x}$. It is helpful to think of the function g as a machine that takes the *nonnegative* square root of its input. Make a copy of each of the following composition machines on your homework paper. Include the intermediate and final output with your sketch.

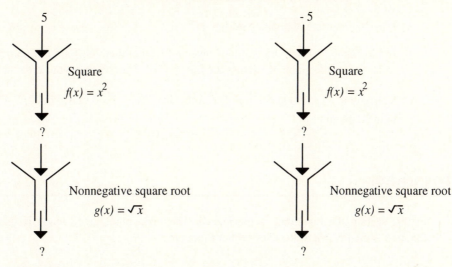

a) What is $g(f(5))$? What is $g(f(-5))$?

b) Does $g(f(x)) = x$ for all x in the domain of f? Are f and g inverses of each other?

13. Set up a coordinate system on a sheet of graph paper and clearly indicate the scale on each coordinate axis. Draw the line $y = x$ on your coordinate system. In red pencil, draw the graph of $f(x) = x^2$ on your coordinate system. In blue pencil, draw the graph of $g(x) = \sqrt{x}$ on your coordinate system.

a) Use your graphs to explain why f and g are not inverses of each other.

b) Show that $g(f(x)) \ne x$ for all x in the domain of f.

14. Set up a coordinate system on a sheet of graph paper and clearly indicate the scale on each coordinate axis. Draw the line $y = x$ on your coordinate system. In red pencil, draw the graph of $f(x) = x^2$, $x \ge 0$, on your coordinate system. In blue pencil, draw the graph of $g(x) = \sqrt{x}$ on your coordinate system.

a) Use your graphs to explain why f and g are inverses of each other.

b) Verify that $g(f(x)) = x$ for all x in the domain of f and verify that $f(g(x)) = x$ for all x in the domain of g.

15. Set up a coordinate system on a sheet of graph paper and clearly indicate the scale on each coordinate axis. Draw the line $y = x$ on your coordinate system. In red pencil, draw the graph of $f(x) = x^2 + 4$, $x \geq 0$, on your coordinate system. In blue pencil, draw the graph of $g(x) = \sqrt{x - 4}$ on your coordinate system.

   a) Use your graphs to explain why f and g are inverses of each other.

   b) Verify that $g(f(x)) = x$ for all x in the domain of f and verify that $f(g(x)) = x$ for all x in the domain of g.

16. Set up a coordinate system on a sheet of graph paper and clearly indicate the scale on each coordinate axis. Draw the line $y = x$ on your coordinate system. In red pencil, draw the graph of $f(x) = x^2 + 4$, $x \leq 0$, on your coordinate system.

   a) Use a method of your choice to find $f^{-1}(x)$.

   b) In blue pencil, sketch the graph of inverse function that you found in part (a). Is your graph a reflection of the graph of f across the line $y = x$? If not, correct your error before moving on.

   c) Verify that $f(f^{-1}(x)) = x$ for all x in the domain of f^{-1}. Verify that $f^{-1}(f(x)) = x$ for all x in the domain of f.

9.7 The Logarithm

In this section the logarithm is introduced. The main point to be made is this: The logarithmic function is the inverse of the exponential function.

Example 1 (TI-82) Locate the LOG key on your calculator. Note that the function 10^x is located on the calculator case directly above this key. You will have to use the 2nd key to access the 10^x function. Let $f(x) = 10^x$ and let $g(x) = \log x$.

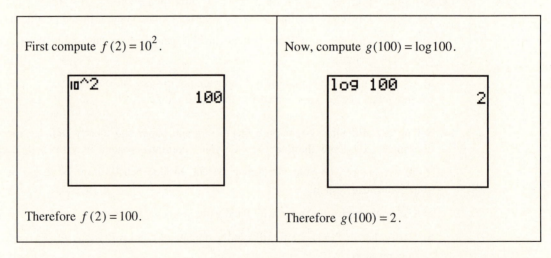

It appears that the logarithmic function is undoing the effects of the exponential function. Now try to experiment a bit further.

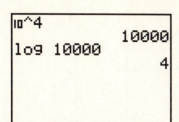

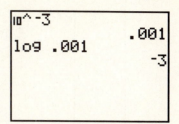

Therefore $f(4) = 10,000$ and $g(10,000) = 4$. | Therefore $f(-3) = 0.001$ and $g(0.001) = -3$.

The logarithmic function is undoing the effects of the exponential function. Now, look at the graphs of $f(x) = 10^x$ and $g(x) = \log x$.

Load the equations $f(x) = 10^x$ and $g(x) = \log x$ into the Y= menu. Also load the equation $y = x$.

Push the ZOOM button and select 6:ZStandard. Push the ZOOM button and select 5:ZSquare to capture the following image. Note the symmetry with respect to the line $y = x$.

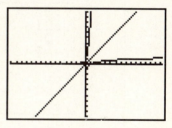

Some pretty convincing evidence has been compiled showing that the functions f and g are inverses of each other. The logarithmic function $g(x) = \log x$ is the inverse of the exponential function $f(x) = 10^x$.

Example 2   *(TI-82)* Locate the LN key on your calculator. Note that the function e^x is located on the calculator case directly above this key. You will have to use the 2nd key to access the e^x function. Let $f(x) = e^x$ and let $g(x) = \ln x$. Locate ANS above the (-) key, near the ENTER key. You have to use the 2nd key to access this variable. The variable ANS always contains the answer to the last computation made by the calculator.

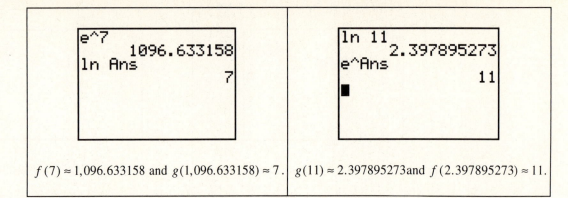

$f(7) \approx 1,096.633158$ and $g(1,096.633158) \approx 7$. $g(11) \approx 2.397895273$ and $f(2.397895273) \approx 11$.

The logarithmic function g is again undoing the effects of the exponential function f. Now look at the graphs of f and g.

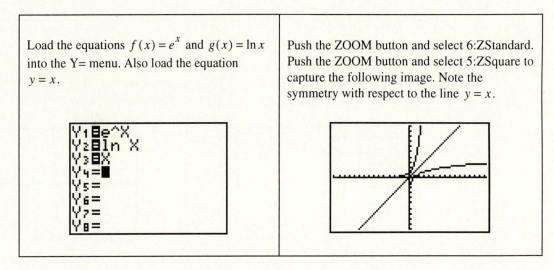

Load the equations $f(x) = e^x$ and $g(x) = \ln x$ into the Y= menu. Also load the equation $y = x$.

Push the ZOOM button and select 6:ZStandard. Push the ZOOM button and select 5:ZSquare to capture the following image. Note the symmetry with respect to the line $y = x$.

Some pretty convincing evidence has been compiled showing that the functions f and g are inverses of each other. The logarithmic function $g(x) = \ln x$ is the inverse of the exponential function $f(x) = e^x$.

Definitions

Let $f(x) = 10^x$. *The function* $g(x) = \log x$ *is the inverse of the function f.*

The graph of $f(x) = 10^x$. Note that the x-axis acts as an asymptote and the y-intercept is $(0,1)$.

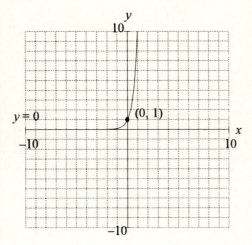

The domain of f is $(-\infty, +\infty)$. The range of f is $(0, +\infty)$.

The graph of $g(x) = \log x$. Note that the y-axis acts as an asymptote and the x-intercept is $(1,0)$.

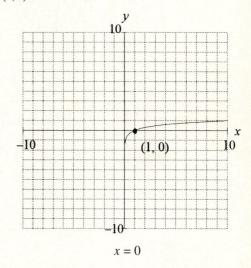

The domain of g is $(0, +\infty)$. The range of g is $(-\infty, +\infty)$.

Because f and g are inverses of each other, $f(g(x)) = x$ for all values of x in the domain of g. Thus, if $x > 0$, we can make the following argument:

$$f(g(x)) = x$$
$$f(\log x) = x$$
$$10^{\log x} = x$$

Note that this property says that $\log x$ is an exponent, the exponent to which you raise 10 to get x.

Because f and g are inverses of each other, $g(f(x)) = x$ for all values of x in the domain of f. Thus, if x is any real number, you can make the following argument:

$$g(f(x)) = x$$
$$g(10^x) = x$$
$$\log 10^x = x$$

We now make a second definition, similar to our first.

Let $f(x) = e^x$. *The function* $g(x) = \ln x$ *is the inverse of the function f.*

The graph of $f(x) = e^x$. Note that the x-axis acts as an asymptote and the y-intercept is $(0,1)$.

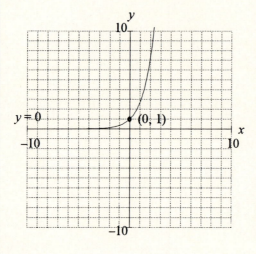

The domain of f is $(-\infty, +\infty)$. The range of f is $(0, +\infty)$.

The graph of $g(x) = \ln x$. Note that the y-axis acts as an asymptote and the x-intercept is $(1,0)$.

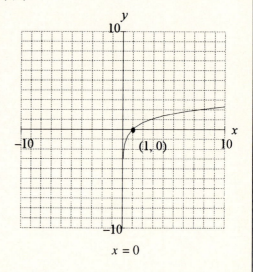

The domain of g is $(0, +\infty)$. The range of g is $(-\infty, +\infty)$.

Because f and g are inverses of each other, $f(g(x)) = x$ for all values of x in the domain of g. Thus, if $x > 0$, you can make the following argument.

$$f(g(x)) = x$$
$$f(\ln x) = x$$
$$e^{\ln x} = x$$

Note that this property says that $\ln x$ is an exponent, the exponent to which you raise e to get x.

Because f and g are inverses of each other, $g(f(x)) = x$ for all values of x in the domain of f. Thus, if x is any real number, you can make the following argument.

$$g(f(x)) = x$$
$$g(e^x) = x$$
$$\ln e^x = x$$

Which Goes with Which? The logarithmic function $\ln x$ is sometimes called the *natural logarithm* by mathematicians. It is important to remember that this function is paired with e^x. The logarithmic function $\log x$ is sometimes called the *common logarithm* by mathematicians. It is important to remember that this function is paired with 10^x.

> **If $f(x) = e^x$, then $f^{-1}(x) = \ln x$.**
> **If $f(x) = 10^x$, then $f^{-1}(x) = \log x$.**

Example 3   Sketch the graph of the function $f(x) = -\log(x+3)$. State the domain and range of this function.

Solution

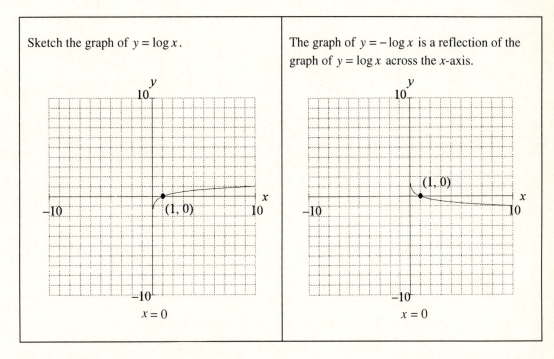

To draw the graph of $y = -\log(x+3)$, slide the graph of $y = -\log x$ to the left 3 units.

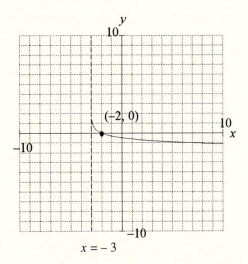

Note that the equation of the vertical asymptote is $x = -3$ and the x-intercept is now at $(-2, 0)$. The domain of the function f is $\{x : x > -3\}$. The range of the function f is all real numbers.

Example 4   Sketch the graph of $f(x) = \ln(-x)$. State the domain and range of this function.

Solution   In Example 3, the impression may have been left that you can draw extremely precise graphs of the logarithm. This is not the case. However, you can draw the basic shape, identify the asymptote and the x-intercept, and find the domain and range from the graph. Perhaps a student solution would be helpful at this point.

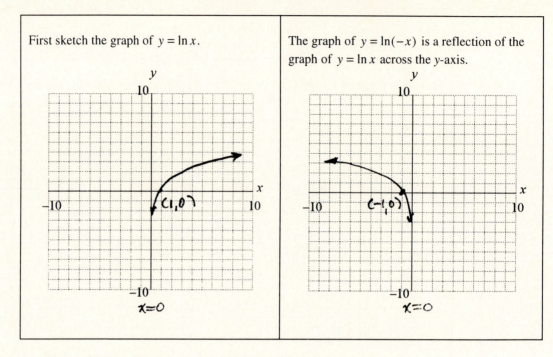

First sketch the graph of $y = \ln x$.

The graph of $y = \ln(-x)$ is a reflection of the graph of $y = \ln x$ across the y-axis.

The domain of $f(x) = \ln(-x)$ is $(-\infty, 0)$. The range of f is $(-\infty, +\infty)$.

Other Bases

Although most of your work will require the use of the natural logarithm and the common logarithm, the exponential function $y = a^x$, $a > 0$, also has an inverse. For example, suppose $a > 1$.

Let $f(x) = a^x$, $a > 1$. The function $g(x) = \log_a x$ is the inverse of the function f.

The graph of $f(x) = a^x$, $a > 1$. Note that the x-axis acts as an asymptote and the y-intercept is $(0,1)$.

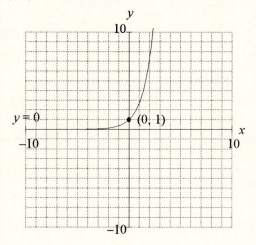

The domain of f is $(-\infty, +\infty)$. The range of f is $(0, +\infty)$.

The graph of $g(x) = \log_a x$, $a > 1$. Note that the y-axis acts as an asymptote and the x-intercept is $(1,0)$.

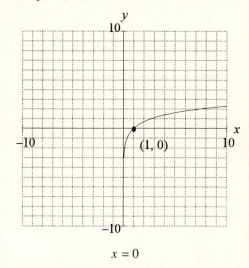

$x = 0$

The domain of g is $(0, +\infty)$. The range of g is $(-\infty, +\infty)$.

Because f and g are inverses of each other, $f(g(x)) = x$ for all values of x in the domain of g. Thus, if $x > 0$, the following argument can be made:

$$f(g(x)) = x$$
$$f(\log_a x) = x$$
$$a^{\log_a x} = x$$

Note that this property says that $\log_a x$ is an exponent, the exponent to which you raise a to get x.

Because f and g are inverses of each other, $g(f(x)) = x$ for all values of x in the domain of f. Thus, if x is any real number, the following argument can be made:

$$g(f(x)) = x$$
$$g(a^x) = x$$
$$\log_a a^x = x$$

The function $f(x) = a^x$, $0 < a < 1$, also has an inverse. However, logarithms with bases between 0 and 1 will not be considered in this book.

> **If $f(x) = a^x$, $a > 1$, then $f^{-1}(x) = \log_a x$.**

The Logarithm as an Exponent

The statement $a^{\log_a x} = x$ declares that $\log_a x$ is an exponent, the exponent to which you raise a to get x. If you let $y = \log_a x$, then $a^y = x$.

$$y = \log_a x \text{ if and only if } a^y = x, \text{ provided } a, x > 0$$

Example 5   Rewrite each of the following exponential equations in exponential form.

     a) $2^3 = 8$ b) $3^0 = 1$ c) $4^{-1} = \dfrac{1}{4}$ d) $9^{\frac{1}{2}} = 3$

Solution   Note which number is the base and use the fact that the logarithm is the exponent.

     a) $\log_2 8 = 3$ b) $\log_3 1 = 0$ c) $\log_4 \dfrac{1}{4} = -1$ d) $\log_9 3 = \dfrac{1}{2}$

Example 6   Rewrite each of the following logarithmic equations in exponential form.

     a) $\log_2 16 = 4$ b) $\log_3 81 = 4$ c) $\log_5 \dfrac{1}{25} = -2$ d) $\log_8 2 = \dfrac{1}{3}$

Solution   Note which number is the base and use the fact that the logarithm is the exponent.

     a) $2^4 = 16$ b) $3^4 = 81$ c) $5^{-2} = \dfrac{1}{25}$ d) $8^{\frac{1}{3}} = 2$

$$\log x \text{ means } \log_{10} x$$
$$\ln x \text{ means } \log_e x$$

Example 7   Rewrite each of the following exponential equations in logarithmic form.

     a) $10^5 = 100,000$ b) $10^{-2} = 0.01$ c) $e^0 = 1$ d) $e^1 = e$

Solution   Note which number is the base and use the fact that the logarithm is the exponent.

     a) $\log 100,000 = 5$ b) $\log 0.01 = -2$ c) $\ln 1 = 0$ d) $\ln e = 1$

Example 8   Rewrite each of the following logarithmic equations in exponential form.

     a) $\log 100 = 2$ b) $\log 0.00001 = -5$ c) $\log 1 = 0$ d) $\ln 8 \approx 2.079441542$

Solution   Note which number is the base and use the fact that the logarithm is the exponent.

     a) $10^2 = 100$ b) $10^{-5} = 0.00001$ c) $10^0 = 1$ d) $e^{2.079441542} \approx 8$

When evaluating $\log_a x$, you are being asked to find the exponent to which you raise a to get x.

Example 9 Evaluate each of the following logarithms.

a) $\log_2 32$ b) $\log_4 \dfrac{1}{16}$ c) $\log_{16} 2$ d) $\log_8 4$

Solution

a) Because $2^5 = 32$, $\log_2 32 = 5$.

b) Because $4^{-2} = \dfrac{1}{16}$, $\log_4 \dfrac{1}{16} = -2$.

c) Because $16^{\frac{1}{4}} = \sqrt[4]{16} = 2$, $\log_{16} 2 = \dfrac{1}{4}$.

d) Because $8^{\frac{2}{3}} = \left(\sqrt[3]{8}\right)^2 = 4$, $\log_8 4 = \dfrac{2}{3}$

Exercises for Section 9.7

1. (TI-82) Use your calculator to evaluate $10^{\log 9}$ and $\log 10^9$. Enter the following in your calculator and verify the results.

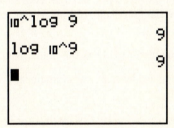

a) Use your calculator to evaluate each of the following pairs of expressions. Write the results on your homework paper.

   i) $10^{\log 7}$, $\log 10^7$ ii) $10^{\log 11}$, $\log 10^{11}$ iii) $10^{\log 3}$, $\log 10^3$ iv) $10^{\log 15}$, $\log 10^{15}$

b) Use complete sentences to explain the relationship that exists between the functions f and g defined by $f(x) = 10^x$ and $g(x) = \log x$.

c) Simplify each of the following expressions.

   i) $10^{\log x}$ ii) $\log 10^x$

2. (TI-82) Load the following equations $y = 10^x$, $y = \log x$, and $y = x$ into the Y= menu.

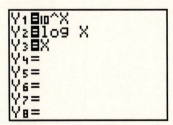

Push the ZOOM button and select 6:ZStandard. Push the ZOOM button and select 5:ZSquare. Set up a coordinate system on a sheet of graph paper and clearly indicate the scale on each coordinate axis. Copy the image in your viewing window onto your coordinate system.

a) Use interval notation to describe the domain and range of the function defined by the equation $y = 10^x$.

b) Use interval notation to describe the domain and range of the function defined by the equation $y = \log x$

3. (TI-82) Enter each of the following expressions into your calculator and press the ENTER key. Use complete sentences to explain the result of each calculation.

a)

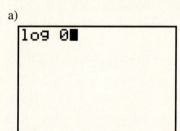

b)

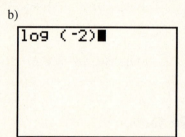

4. For the equations below, perform each of the following tasks:

i) (No calculator allowed) Set up a separate coordinate system on a sheet of graph paper and clearly indicate the scale on each coordinate axis. Use graphing shortcuts to sketch the graph of the equation.

ii) (TI-82) Use your calculator to check the graph from part (i).

iii) Use interval notation to describe the domain and range of the function represented by the equation.

a) $f(x) = \log(x-5)$ b) $f(x) = \log(x+2)$ c) $f(x) = -\log x$

d) $f(x) = \log(-x)$ e) $f(x) = -\log(x+6)$ f) $f(x) = \log[-(x-8)]$

Note: It is important that you use your calculator to *check* your graph, not to *draw* it.

5. Simplify each of the following expressions. Clearly state the values of x for which your solution is valid.

a) $10^{\log x^2}$ b) $\log 10^{x-3}$ c) $\log 10^{\sqrt{x}}$ d) $10^{\log(4-x)}$

6. (TI-82) Use your calculator to evaluate $e^{\ln 8}$ and $\ln e^8$. Enter the following in your calculator and verify the results.

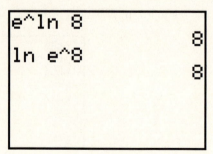

a) Use your calculator to evaluate each of the following pairs of expressions. Write the results on your homework paper.

i) $e^{\ln 14}$, $\ln e^{14}$ ii) $e^{\ln 9}$, $\ln e^9$ iii) $e^{\ln 23}$, $\ln e^{23}$ iv) $e^{\ln 18}$, $\ln e^{18}$

b) Use complete sentences to explain the relationship between the functions f and g defined by $f(x) = e^x$ and $g(x) = \ln x$.

c) Simplify each of the following expressions.

   i) $e^{\ln x}$ ii) $\ln e^x$

7. (TI-82) Load the following equations $y = e^x$, $y = \ln x$, and $y = x$ into the Y= menu.

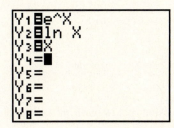

Push the ZOOM button and select 6:ZStandard. Push the ZOOM button and select 5:ZSquare. Set up a coordinate system on a sheet of graph paper and clearly indicate the scale on each coordinate axis. Copy the image in your viewing window onto the coordinate system.

a) Use interval notation to describe the domain and range of the function defined by the equation $y = e^x$.

b) Use interval notation to describe the domain and range of the function defined by the equation $y = \ln x$

8. (TI-82) Enter the following sequence into your calculator and evaluate the last expression. The command $4 \to X$ stores the number 4 in the variable X. Use the STO▷ key to produce the $\to$. Finally, use complete sentences to explain the result of each calculation.

a) 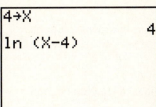   b)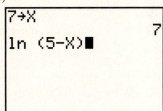

9. For the equations below, perform each of the following tasks:

  i) (No calculator allowed) Set up a separate coordinate system on a sheet of graph paper and clearly indicate the scale on each coordinate axis. Use your graphing shortcuts to sketch the graph of the equation.

  ii) (TI-82) Use your calculator to check your graph from part (i).

  iii) Use interval notation to describe the domain and range of the function represented by the equation.

      a) $f(x) = \ln(x+9)$ b) $f(x) = \ln(x-1)$ c) $f(x) = \ln(-x)$

      d) $f(x) = -\ln x$ e) $f(x) = -\ln(x-7)$ f) $f(x) = \ln\left[-(x+5)\right]$

Note: It is important that you use your calculator to *check* your graph, not to *draw* it.

10. Simplify each of the following expressions. Clearly state the values of x for which your solution is valid.

   a) $e^{\ln(x+11)}$

   b) $\ln e^{\frac{1}{x}}$

   c) $e^{\ln(7-x)}$

   d) $\ln e^{\sqrt{x-6}}$

11. Recall the definition of the logarithm: $y = \log_a x$ if and only if $a^y = x$. Use the definition to write each of the following exponential equations in logarithmic form.

   a) $2^4 = 16$

   b) $3^2 = 9$

   c) $5^{-1} = \frac{1}{5}$

   d) $6^{-2} = \frac{1}{36}$

   e) $4^{\frac{1}{2}} = 2$

   f) $8^{-\frac{1}{3}} = \frac{1}{2}$

   g) $4^{\frac{3}{2}} = 8$

   h) $27^{-\frac{2}{3}} = \frac{1}{9}$

12. For the expressions below, perform each of the following tasks:

   i)   Simplify the expression.
   ii) Write the resulting exponential equation in logarithmic form.

   a) 7^2

   b) 2^7

   c) 4^{-1}

   d) 6^{-3}

   e) $49^{\frac{1}{2}}$

   f) $64^{\frac{2}{3}}$

   g) $125^{-\frac{1}{3}}$

   h) $16^{-\frac{3}{4}}$

13. For the expressions below, perform each of the following tasks:

   i)   Simplify the expression.
   ii) Write the resulting exponential equation in logarithmic form.

   a) 10^{-4}

   b) 10^{-3}

   c) 10^{-2}

   d) 10^{-1}

   e) 10^0

   f) 10^1

   g) 10^2

   h) 10^3

14. Write each of the following logarithmic equations in exponential form.

   a) $\log(x-8) = 2$

   b) $\log 7 = 3 - x$

   c) $\log x^2 = 2$

   d) $\log 9 = 3 - 2t$

15. Write each of the following exponential equations in logarithmic form.

   a) $10^x = 12$

   b) $10^{-2} = x$

   c) $10^{2-x} = 7$

   d) $10^{-1} = 5x + 1$

16. Write each of the following logarithmic equations in exponential form.

   a) $\ln x = 4$

   b) $\ln 3 = 6t$

   c) $\ln(3 - x) = -2$

   d) $\ln 4 = x - 5$

17. Write each of the following exponential equations in logarithmic form.

   a) $e^x = 4$

   b) $e^3 = x$

   c) $e^{t-3} = 5$

   d) $e^{-2} = x - 5$

18. Evaluate each of the following logarithms.

   a) $\log_2 32$

   b) $\log_4 16$

   c) $\ln 1$

   d) $\log 1{,}000{,}000{,}000$

   e) $\log_7 \frac{1}{7}$

   f) $\log_{11} \frac{1}{121}$

   g) $\log_{25} 5$

   h) $\log_{125} 5$

i) $\log 0.0000001$ j) $\ln e$ k) $\log_8 16$ l) $\log_4 \dfrac{1}{8}$

19. Set up a coordinate system on a sheet of graph paper and clearly indicate the scale on each coordinate axis. Draw the line $y = x$ on your coordinate system.

   a) Make a copy of the following table on your graph paper and complete the missing entries.

x	-3	-2	-1	0	1	2	3
$f(x) = 2^x$							

   b) Plot each point in the table on your coordinate system and use these plotted points to draw a complete graph of the equation $f(x) = 2^x$. Use interval notation to describe the domain and range of the function f.

   c) Make a copy of the following table on your graph paper and complete the missing entries.

x	$\dfrac{1}{8}$	$\dfrac{1}{4}$	$\dfrac{1}{2}$	1	2	4	8
$g(x) = \log_2 x$							

   d) Plot each point in the table on the same coordinate system used in part (b) and use these plotted points to draw a complete graph of the equation $g(x) = \log_2 x$. Use interval notation to describe the domain and range of the function g.

20. Use your calculator to find decimal approximations for each of the following expressions, correct to the nearest ten-thousandth (four decimal places).

   a) $\dfrac{\log 7}{\log 5}$ b) $\log \dfrac{7}{5}$ c) $\dfrac{\log 7}{5}$ d) $\dfrac{2 + \log 3}{3 - 2\log 3}$

   e) $\dfrac{\ln 2}{0.08}$ f) $\dfrac{1}{5}\ln\dfrac{4}{5}$ g) $\dfrac{\ln 10}{2\ln\left[1 + \dfrac{0.05}{2}\right]}$ h) $-\dfrac{1}{0.05}\ln\dfrac{4}{97}$

21. If money in an account is compounded continuously, then the time that it takes this money to double is given by the equation $T = \dfrac{\ln 2}{r}$, where T is the time in years and r is the yearly interest rate. If the interest rate is 4 percent per year, find the time required for the account to double in value.

22. If money in an account is compounded yearly, then the time that it takes this money to double is given by the equation $T = \dfrac{\ln 2}{\ln(1+r)}$, where T is the time in years and r is the yearly interest rate. If the interest rate is 5.25 percent per year, find the time required for the account to double in value.

23. Sandy is attending a professional school where he is learning how to type. The time that it will take him to learn how to type P words per minute is given by the equation $T = -40\ln\left[1 - \dfrac{P}{100}\right]$, where T is the time in days. How long will it take Sandy to learn how to type 60 words per minute?

9.8 The Laws of Logarithms

Recall the following laws of exponents:

$$a^x \cdot a^y = a^{x+y}$$

$$\frac{a^x}{a^y} = a^{x-y}$$

$$\left(a^x\right)^y = a^{xy}$$

Remember that these properties are not valid if certain conditions arise. For example, division by zero is meaningless, so the second property is not valid if $a = 0$. Zero raised to the zero power is also meaningless and must be avoided. Taking the even root of a negative number is meaningless, so avoid expressions such as $(-2)^{\frac{1}{2}}$. We know that the logarithmic function is the inverse of the exponential function. Therefore, it is reasonable to expect that there must also be laws of logarithms.

Example 1   *(TI-82)* Use your calculator to help introduce the first law of logarithms.

Enter the following expressions and verify the results:	Enter the following expressions and verify the results:

`ln 4+ln 7`
`        3.33220451`
`ln (4*7)`
`        3.33220451`
	`ln 5+ln 11`
	`        4.007333185`
	`ln (5*11)`
	`        4.007333185`

$\ln 4 + \ln 7 = \ln(4 \cdot 7)$	$\ln 5 + \ln 11 = \ln(5 \cdot 11)$

Do you see the pattern? It appears that the sum of the natural logarithms is the natural logarithm of the product; that is, $\ln x + \ln y = \ln xy$. However, two examples on a calculator do not constitute a proof.

Prove: If $x > 0$ and $y > 0$, then $\ln x + \ln y = \ln xy$.

Proof: Assume that $x > 0$ and $y > 0$. Because $x > 0$, $e^{\ln x} = x$; because $y > 0$, $e^{\ln y} = y$. If $x > 0$ and $y > 0$, then $xy > 0$ and you can write the following:

$$e^{\ln xy} = xy$$

In this last expression, you can replace x and y on the right side of the equation with $e^{\ln x}$ and $e^{\ln y}$, respectively.

$$e^{\ln xy} = e^{\ln x} \cdot e^{\ln y}$$

Repeat the base and add the exponents on the right-hand side of this last equation.

$$e^{\ln xy} = e^{\ln x + \ln y}$$

If you equate the exponents in this last equation, you arrive at the following law of logarithms:

If $x > 0$ and $y > 0$, then $\ln xy = \ln x + \ln y$.
The natural logarithm of a product is the sum of the natural logarithms.

Example 2　Express $\ln x(x+1)$ as a sum of natural logarithms.

Solution　The natural logarithm of a product is the sum of the natural logarithms.

$$\ln x(x+1) = \ln x + \ln(x+1)$$

Example 3　Express $\ln x + \ln(x-3) + \ln(x+3)$ as the natural logarithm of a single expression.

Solution　The sum of the natural logarithms is the natural logarithm of the product.

$$\ln x + \ln(x-3) + \ln(x+3) = \ln x(x-3)(x+3)$$

Example 4　*(TI-82)*　Use your calculator to help introduce the second law of logarithms.

Enter the following expressions and verify the results:	Enter the following expressions and verify the results:
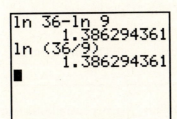	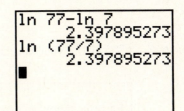
$\ln 36 - \ln 9 = \ln \dfrac{36}{9}$	$\ln 77 - \ln 7 = \ln \dfrac{77}{7}$

Do you see the pattern? It appears that the difference of the natural logarithms is the natural logarithm of the quotient; that is, $\ln x - \ln y = \ln \dfrac{x}{y}$. However, two examples on a calculator do not constitute a proof.

Prove: If $x > 0$ and $y > 0$, then $\ln x - \ln y = \ln \dfrac{x}{y}$.

Proof: If $x > 0$ and $y > 0$, then $\dfrac{x}{y} > 0$ and you can argue in the following manner.

$$e^{\ln\frac{x}{y}} = \frac{x}{y}$$

$$= \frac{e^{\ln x}}{e^{\ln y}}$$

$$= e^{\ln x - \ln y}$$

If you equate the exponents in this last expression, you will obtain the second law of logarithms.

If $x > 0$ and $y > 0$, then $\ln\dfrac{x}{y} = \ln x - \ln y$

The natural logarithm of a quotient is the difference of the natural logarithms.

Example 5 Express $\ln\dfrac{x-3}{x+5}$ as a difference of natural logarithms.

Solution The natural logarithm of a quotient is the difference of the natural logarithms.

$$\ln\frac{x-3}{x+5} = \ln(x-3) - \ln(x+5)$$

Example 6 Express $\ln x - \ln(x+5)$ as the natural logarithm of a quotient.

Solution The difference of the natural logarithms is the natural logarithm of the quotient.

$$\ln x - \ln(x+5) = \ln\frac{x}{x+5}$$

Example 7 (TI-82) Use your calculator to help introduce the third law of logarithms.

Enter the following expressions and verify the results:	Enter the following expressions and verify the results:
| ```
2ln 5
        3.218875825
ln 5²
        3.218875825
``` | ```
3ln 4
        4.158883083
ln 4³
        4.158883083
■
``` |
| $2\ln 5 = \ln 5^2$ | $3\ln 4 = \ln 4^3$ |

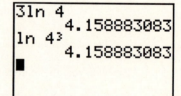

Do you see the pattern? It appears that $c\ln x = \ln x^c$. However, two examples on a calculator do not constitute a proof.

Prove: If $x > 0$ and c is any real number, then $c\ln x = \ln x^c$.

Proof: If $x > 0$ and c is any real number, you can argue in the following manner:

$$e^{\ln x^c} = x^c$$
$$= \left(e^{\ln x}\right)^c$$
$$= e^{c\ln x}$$

If you equate the exponents in this last equation, you will obtain the third law of logarithms.

If $x > 0$ and c is any real number, then $\ln x^c = c\ln x$.

Example 8   Here are few example of this last law of logarithms in action.

$$\ln x^3 = 3\ln x$$
$$\ln x^{-1} = -\ln x$$
$$\ln x^{\frac{1}{2}} = \frac{1}{2}\ln x$$

Summarizing the Laws of Logarithms.   Although the laws of logarithms were proved only for the natural logarithm, each of the laws of logarithms is valid for any base that is larger than zero.

The Laws of Logarithms. Let $a > 0$. If $x > 0$ and $y > 0$ and c is any real number, then the following laws hold.

- $\log_a xy = \log_a x + \log_a y$

- $\log_a \dfrac{x}{y} = \log_a x - \log_a y$

- $\log_a x^c = c\log_a x$

Example 9   Write the expression $\log_3 \dfrac{x^2}{(x+4)(x-7)}$ as sums, differences, and multiples of logarithms.

Solution

$$\log_3 \frac{x^2}{(x+4)(x-7)} = \log_3 x^2 - \log_3 (x+4)(x-7)$$

$$= 2\log_3 x - \left[\log_3 (x+4) + \log_3 (x-7)\right]$$

$$= 2\log_3 x - \log_3 (x+4) - \log_3 (x-7)$$

Example 10   Write the expression $3\log_2 x - 4\log_2 y + 2\log_2 w - 7\log_2 z$ as the logarithm of a single expression.

Solution   Terms in the form $c\log_2 x$ are first placed in the form $\log_2 x^c$. Then the terms are arranged in two distinct groups: (1) terms with plus signs in one group, and (2) terms with minus signs in a second group.

$$3\log_2 x - 4\log_2 y + 2\log_2 w - 7\log_2 z = \log_2 x^3 - \log_2 y^4 + \log_2 w^2 - \log_2 z^7$$

$$= \left[\log_2 x^3 + \log_2 w^2\right] - \left[\log_2 y^4 + \log_2 z^7\right]$$

$$= \log_2 x^3 w^2 - \log_2 y^4 z^7$$

$$= \log_2 \frac{x^3 w^2}{y^4 z^7}$$

Example 11   Simplify the expression $e^{2\ln x + \ln y}$.

Solution

$$e^{2\ln x + \ln y} = e^{\ln x^2 + \ln y}$$

$$= e^{\ln x^2 y}$$

$$= x^2 y$$

Note that $e^{\ln x^2 y} = x^2 y$ because the exponential function and the natural logarithmic function are inverses of each other.

Example 12   *(TI-82)* Simplify the expression $10^{\log 24 - \log 3 - \log 2}$ and check the result on your calculator.

Solution

$$10^{\log 24 - \log 3 - \log 2} = 10^{\log 24 - [\log 3 + \log 2]}$$

$$= 10^{\log 24 - \log 6}$$

$$= 10^{\log \frac{24}{6}}$$

$$= 10^{\log 4}$$

$$= 4$$

```
10^(log 24-log 3-
log 2)
                   4
```

Example 13   Express $\ln\sqrt[4]{x^2(x+1)^3}$ as sums and multiples of natural logarithms.

Solution   First, recall that taking the fourth root of a number is the same as raising that number to the $\frac{1}{4}$ power. The laws of logarithms can be used to complete the expansion.

$$\ln\sqrt[4]{x^2(x+1)^3} = \ln\left[x^2(x+1)^3\right]^{\frac{1}{4}}$$

$$= \frac{1}{4}\ln x^2(x+1)^3$$

$$= \frac{1}{4}\left[\ln x^2 + \ln(x+1)^3\right]$$

$$= \frac{1}{4}\left[2\ln x + 3\ln(x+1)\right]$$

$$= \frac{1}{2}\ln x + \frac{3}{4}\ln(x+1)$$

Example 14   Write the expression $\frac{1}{2}\ln x - \frac{3}{2}\ln(x+3)$ as the natural logarithm of a single expression.

Solution   Begin by factoring a $\frac{1}{2}$ from each term.

$$\frac{1}{2}\ln x - \frac{3}{2}\ln(x+3) = \frac{1}{2}\left[\ln x - 3\ln(x+3)\right]$$

$$= \frac{1}{2}\left[\ln x - \ln(x+3)^3\right]$$

$$= \frac{1}{2}\ln\frac{x}{(x+3)^3}$$

$$= \ln\left[\frac{x}{(x+3)^3}\right]^{\frac{1}{2}}$$

$$= \ln\sqrt{\frac{x}{(x+3)^3}}$$

Paying Attention to the Domain. You cannot take the logarithm of zero, nor can you take the logarithm of any negative number. When applying the laws of logarithms, you must always keep this in mind.

Example 15   Consider the functions f and g defined by $f(x) = \ln x^2$ and $g(x) = 2\ln x$. Are the functions f and g the same function?

Solution   If the third law of logarithms is used, it appears that the following argument can be made:

$$f(x) = \ln x^2 = 2\ln x = g(x)$$

At first glance, it appears that the functions f and g are equal. Use your calculator to take a second glance.

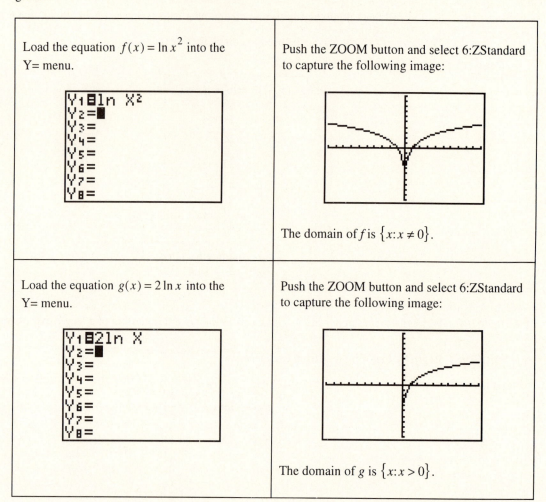

Load the equation $f(x) = \ln x^2$ into the Y= menu.	Push the ZOOM button and select 6:ZStandard to capture the following image: The domain of f is $\{x : x \neq 0\}$.
Load the equation $g(x) = 2 \ln x$ into the Y= menu.	Push the ZOOM button and select 6:ZStandard to capture the following image: The domain of g is $\{x : x > 0\}$.

Since the graph of the function f is *not* the same as the graph of the function g, the functions f and g are *not* equal. In the first examination of the functions f and g, the argument was made that $\ln x^2 = 2 \ln x$. That argument is valid only if $x > 0$. It is interesting to note that the graphs of f and g coincide for $x > 0$.

Exercises for Section 9.8

1. (TI-82) Enter the following expressions on your calculator and verify the results:

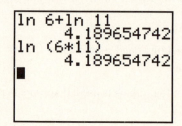

Enter each of the following pairs of expressions in your calculator and compare the results. Record each expression and its decimal approximation on your homework paper.

a) $\ln 4 + \ln 9$, $\ln(4 \cdot 9)$

b) $\log 8 + \log 11$, $\log(8 \cdot 11)$

c) $\log 13 + \log 3$, $\log(3 \cdot 13)$

d) $\ln 3 + \ln 17$, $\ln(3 \cdot 17)$

What can you conclude from this exercise?

2. Express each of the following as a sum of logarithms.

a) $\ln 4x$

b) $\log 3x$

c) $\log x(x+10)$

d) $\ln(x-3)(x-5)$

3. Express each of the following as a logarithm of a single expression.

a) $\ln x + \ln 10$

b) $\log 5 + \log x$

c) $\ln x + \ln(x+6)$

d) $\log(x+5) + \log(x-5)$

4. (TI-82) Enter the following expressions on your calculator and verify the results.

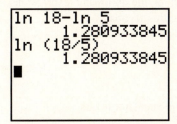

Enter each of the following pairs of expressions in your calculator and compare the results. Record each expression and its decimal approximation on your homework paper.

a) $\ln 27 - \ln 3$, $\ln \dfrac{27}{3}$

b) $\log 100 - \log 2$, $\log \dfrac{100}{2}$

c) $\log 23 - \log 5$, $\log \dfrac{23}{5}$

d) $\ln 53 - \ln 7$, $\ln \dfrac{53}{7}$

What can you conclude from this exercise?

5. Express each of the following as a difference of logarithms.

a) $\ln \dfrac{5}{x}$

b) $\log \dfrac{x}{10}$

c) $\log \dfrac{x+1}{x-2}$

d) $\ln \dfrac{x-2}{x}$

6. Express each of the following as a logarithm of a single expression.

a) $\ln x - \ln 10$

b) $\log 11 - \log 2x$

c) $\ln 2x - \ln(3-x)$

d) $\log(x-5) - \log(3-2x)$

7. (TI-82) Enter the following expressions on your calculator and verify the results.

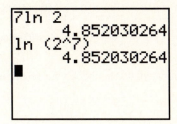

Enter each of the following pairs of expressions in your calculator and compare the results. Record each expression and its decimal approximation on your homework paper.

a) $3\ln 6$, $\ln 6^3$

b) $2\log 9$, $\log 9^2$

c) $-2\ln 7$, $\ln 7^{-2}$

d) $-\log 5$, $\log 5^{-1}$

e) $5\log 2$, $\log 2^5$

f) $-6\ln 3$, $\ln 3^{-6}$

What can you conclude from this exercise?

8. Express each of the following as a multiple of a logarithm.

a) $\ln x^5$

b) $\log x^{-1}$

c) $\ln x^7$

d) $\log x^{-3}$

9. Express each of the following as a logarithm of a single expression.

a) $2\ln x$

b) $-\log x$

c) $9\ln x$

d) $\log x$

10. Simplify each of the following expressions, then verify the results on your calculator as in Example 12 of this section.

a) $10^{\log 5 + \log 7}$

b) $e^{\ln 15 - \ln 3}$

c) $10^{2\log 5}$

d) $e^{\ln 4 + \ln 6}$

e) $10^{\log 27 - \log 9}$

f) $e^{3\ln 4}$

Note: Use your calculator to check your result, not to solve the problem.

11. Expand each of the following as sums, differences, and multiples of logarithms.

a) $\ln x(x+2)(x-2)$

b) $\log x^2(x+3)$

c) $\ln x^2 y^3$

d) $\log_2 (x+1)^2 (x-3)^3$

e) $\log \dfrac{x^3}{(x+1)}$

f) $\ln \dfrac{x^2}{y^3 z^5}$

g) $\log \dfrac{x^2}{(x+1)(x-3)^3}$

h) $\log_3 \dfrac{2x-3}{x^3(x+1)^5}$

12. Write each of the following expressions as a logarithm of a single expression.

a) $\ln x + \ln(x+2) + 3\ln(x-3)$

b) $\log x - \log(x+1) + \log(x-5)$

c) $\ln x - \ln y + \ln z - \ln w$

d) $3\log_3 x - 2\log_3 (x+1) - 4\log_3 (2x-5)$

e) $2\log x - 3\log y - 5\log w - \log z$

f) $3\log_2 x - \log_2 (x+1) - 2\log_2 (x-5) + \log_2 (x-7)$

13. Simplify each of the following expressions, then verify your result on your calculator as in Example 12 of this section.

a) $e^{2\ln 3 + \ln 4}$

b) $10^{\log 40 - 3\log 2}$

c) $10^{\log 6 - \log 2 + \log 4}$

d) $e^{\ln 18 - \ln 3 - \ln 2}$

e) $10^{2\log 5 + 3\log 2}$

f) $e^{-\ln 8}$

Note: Use your calculator to check your result, not to perform the calculations.

14. Expand each of the following as sums, differences, and multiples of logarithms.

    a) $\log \sqrt[3]{xy^2}$
    b) $\log_2 \sqrt{\dfrac{x+5}{x-5}}$
    c) $\ln \sqrt{x(x-4)}$
    d) $\log_4 \sqrt{\dfrac{x^2}{y^3}}$

15. Express each of the following as a logarithm of a single expression.

    a) $\dfrac{1}{2}\ln x + \dfrac{1}{2}\ln(x+10)$
    b) $\dfrac{1}{3}\ln(x-1) - \dfrac{1}{3}\ln(x+1)$

    c) $\dfrac{1}{2}\log_3 x + \dfrac{3}{4}\log_3 y$
    d) $\dfrac{1}{2}\log(x+1) - \dfrac{1}{3}\log(x-5)$

16. Simplify each of the following expressions.

    a) $e^{2\ln x}$
    b) $10^{\log x + \log(x+3)}$
    c) $10^{\frac{1}{2}\log x}$
    d) $e^{\ln 5 - 2\ln x}$

17. (TI-82) Consider the functions f and g defined by $f(x) = e^{\ln x}$ and $g(x) = x$.

    a) Set up a coordinate system on a sheet of graph paper and clearly indicate the scale on each coordinate axis. Use your calculator to sketch the graph of f and copy the image in your viewing window onto the coordinate system. Use interval notation to describe the domain of the function f.

    b) Set up a second coordinate system on your graph paper and clearly indicate the scale on each coordinate axis. Use your calculator to sketch the graph of g and copy the image in your viewing window onto the coordinate system. Use interval notation to describe the domain of the function g.

    c) Clearly state the conditions under which f and g are identical.

18. (TI-82) Consider the functions f and g defined by $f(x) = \ln(x+2)(x-2)$ and $g(x) = \ln(x+2) + \ln(x-2)$.

    a) Set up a coordinate system on a sheet of graph paper and clearly indicate the scale on each coordinate axis. Use your calculator to sketch the graph of f and copy the image in your viewing window onto the coordinate system. Use interval notation to describe the domain of the function f.

    b) Set up a second coordinate system on your graph paper and clearly indicate the scale on each coordinate axis. Use your calculator to sketch the graph of g and copy the image in your viewing window onto the coordinate system. Use interval notation to describe the domain of the function g.

    c) Clearly state the conditions under which f and g are identical.

19. (TI-82) Consider the functions f and g defined by $f(x) = \ln\dfrac{x}{x-3}$ and $g(x) = \ln x - \ln(x-3)$.

    a) Set up a coordinate system on a sheet of graph paper and clearly indicate the scale on each coordinate axis. Use your calculator to sketch the graph of f and copy the image in your viewing window onto your coordinate system. Use interval notation to describe the domain of the function f.

b) Set up a second coordinate system on your graph paper and clearly indicate the scale on each coordinate axis. Use your calculator to sketch the graph of g and copy the image in your viewing window onto your coordinate system. Use interval notation to describe the domain of the function g.

c) Clearly state the conditions under which f and g are identical.

9.9 Solving Equations and Inequalities

In this section equations and inequalities that involve exponential and logarithmic functions will be solved using both graphical and analytic techniques. Let's start with an example.

Example 1 Solve the following equation for x: $e^{2x-3} = 7$.

Solution

First, rewrite the equation in logarithmic form.	Use your calculator to find a decimal approximation and check your solution.
$$e^{2x-3} = 7$$ $$2x - 3 = \ln 7$$ $$2x = 3 + \ln 7$$ $$x = \frac{3 + \ln 7}{2}$$	``` (3+ln 7)/2 2.472955075 Ans→X 2.472955075 e^(2X-3) 7 ```

Example 2 Solve the following equation for x: $\ln(4x-5) = 3$.

Solution

First, rewrite the equation in exponential form.	Use your calculator to find a decimal approximation and check your solution.
$$\ln(4x-5) = 3$$ $$4x - 5 = e^3$$ $$4x = 5 + e^3$$ $$x = \frac{5 + e^3}{4}$$	``` (5+e^3)/4 6.271384231 Ans→X 6.271384231 ln (4X-5) 3 ```

Example 3 Solve the following equation for x: $2^x = 5$.

Guess, and Check Solution. Try to guess. Since $2^2 = 4$ and $2^3 = 8$, your solution must lie between 2 and 3. The solution probably lies closer to 2.

You are looking for a solution of $2^x = 5$; try 2.3.

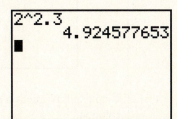

Because $2^{2.3} \approx 4.924577653$, 2.3 is too low.

You are looking for a solution of $2^x = 5$; try 2.4.

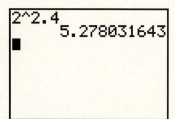

Because $2^{2.4} \approx 5.278031643$, 2.4 is too high.

The solution of $2^x = 5$ lies between 2.3 and 2.4. You could continue in this manner to add more significant digits to your answer.

Graphical Solution

Load the equations $y = 2^x$ and $y = 5$ into the Y= menu.

Push the ZOOM button and select 6:ZStandard. Use the intersect utility in the CALC menu to find the point of intersection.

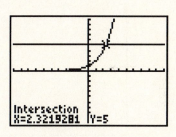

If you *immediately* press the $\boxed{\text{X,T,}\theta}$ button, you can add a few significant digits to your solution. If you enter the expression 2^ X, you can check your solution.

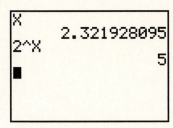

Analytical Solution Begin by taking the natural logarithm of both sides of the equation, then apply the third law of logarithms, $\ln x^c = c \ln x$.

$$2^x = 5$$

$$\ln 2^x = \ln 5$$

$$x \ln 2 = \ln 5$$

$$\frac{x \ln 2}{\ln 2} = \frac{\ln 5}{\ln 2}$$

$$x = \frac{\ln 5}{\ln 2}$$

```
ln 5/ln 2
          2.321928095
```

Note that this solution matches the graphical solution.

Place your analytical solution and the graphical solution side-by-side for comparison.

$$2^x = 5$$

$$\ln 2^x = \ln 5$$

$$x \ln 2 = \ln 5$$

$$\frac{x \ln 2}{\ln 2} = \frac{\ln 5}{\ln 2}$$

$$x = \frac{\ln 5}{\ln 2}$$

$$x \approx 2.321928095$$

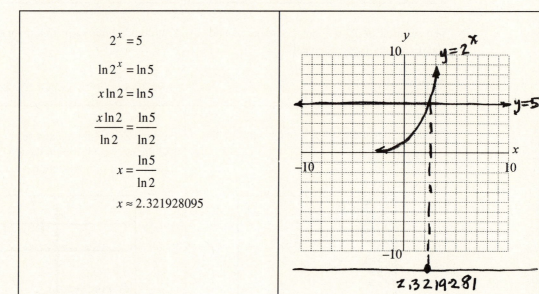

The Change of Base Rule

In the analytical solution in Example 1, the natural logarithm of both sides of the equation $2^x = 5$ was taken.

You could easily have taken the common logarithm of both sides of the equation.

$$2^x = 5$$

$$\log 2^x = \log 5$$

$$x \log 2 = \log 5$$

$$\frac{x \log 2}{\log 2} = \frac{\log 5}{\log 2}$$

$$x = \frac{\log 5}{\log 2}$$

Note the similar answer. Only the base of the logarithm is different.

```
log 5/log 2
        2.321928095
```

Note that the decimal approximation of $\dfrac{\log 5}{\log 2}$ is identical to that of $\dfrac{\ln 5}{\ln 2}$.

In general, a logarithm in one base can always be transformed into a logarithm with base of your choice. Suppose that you wish to express $\log_a x$ in terms of base b logarithms. You can argue as follows. Let $\log_a x = y$ and apply the definition of the logarithm.

$$\log_a x = y$$

$$a^y = x$$

Take the logarithm of both sides of this equation, using a base b of your choice.

$$\log_b a^y = \log_b x$$

$$y \log_b a = \log_b x$$

$$y = \frac{\log_b x}{\log_b a}$$

Because $\log_a x = y$ and $y = \dfrac{\log_b x}{\log_b a}$, you can come to the following conclusion.

> **The change of base rule for logarithms:** $\log_a x = \dfrac{\log_b x}{\log_b a}$.

Example 4　The following expressions are identical:

$$\log_2 5 = \frac{\ln 5}{\ln 2} = \frac{\log 5}{\log 2} = \frac{\log_3 5}{\log_3 2} = \frac{\log_7 5}{\log_7 2} = \frac{\log_9 5}{\log_9 2} = \frac{\log_{11} 5}{\log_{11} 2}$$

Note the pattern. You take logarithm of 5 and divide by the logarithm of the original base 2, and you can do this in the base of your choice.

Example 5 Find a decimal approximation of $\log_3 30$.

Solution Because $3^3 = 27$ and $3^4 = 81$, $\log_3 30$ must lie between 3 and 4. You can use the change of base rule and your calculator to find an even better approximation.

Because your calculator has two logarithm buttons—LN and LOG—you have two options: $$\log_3 30 = \frac{\ln 30}{\ln 3}$$ or $$\log_3 30 = \frac{\log 30}{\log 3}$$	Use your calculator to find decimal approximations for each of these expressions. 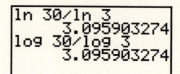

Example 6 Solve the following equation for t: $500(0.96)^{4t} = 200$.

Graphical Solution If $t = 0$, then $y = 500(0.96)^0 = 500$ and the y-intercept is 500. The base of the exponential equation $y = 500(0.96)^{4t}$ is 0.96. Because the base is less than 1, the graph will exhibit exponential decay. This sort of analysis will help you to make appropriate WINDOW settings on your calculator.

Load the equations $y = 500(0.96)^{4t}$ and $y = 200$. 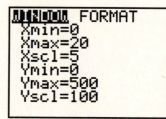	Adjust the WINDOW parameters. 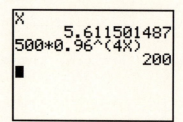
Use the intersect utility in the CALC menu. 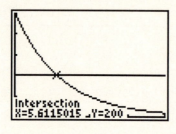	Press X *immediately* and check your solution. 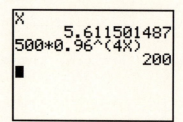

Analytical Solution Begin by dividing both sides of the equation by 500, then take the natural logarithm of both sides of the equation.

$$500(0.96)^{4t} = 200$$

$$(0.96)^{4t} = \frac{200}{500}$$

$$(0.96)^{4t} = \frac{2}{5}$$

$$\ln(0.96)^{4t} = \ln\frac{2}{5}$$

$$4t\ln 0.96 = \ln\frac{2}{5}$$

$$t = \frac{\ln\frac{2}{5}}{4\ln 0.96}$$

Use your calculator to find a decimal approximation of the solution.

```
ln (2/5)/(4ln 0.
96)
        5.611501487
```

Note that this matches the graphical solution.

Example 7 Solve the following equation for x: $\log x + \log(x-3) = 1$.

Graphical Solution The domain of $y = \log + \log(x-3)$ is $\{x : x > 3\}$. The graph of $y = \log + \log(x-3)$ must lie entirely to the right of $x = 3$.

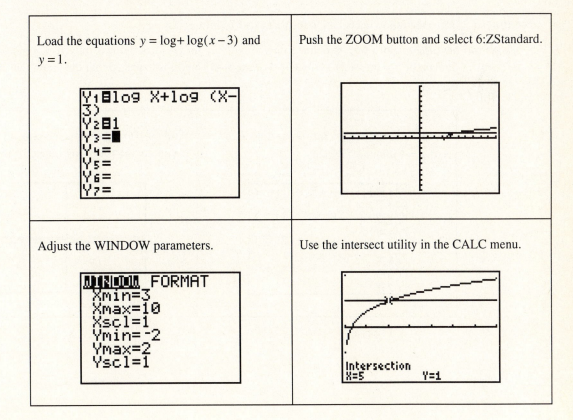

Load the equations $y = \log + \log(x-3)$ and $y = 1$.

```
Y₁Blog X+log (X-
3)
Y₂B1
Y₃=■
Y₄=
Y₅=
Y₆=
Y₇=
```

Push the ZOOM button and select 6:ZStandard.

Adjust the WINDOW parameters.

```
WINDOW FORMAT
Xmin=3
Xmax=10
Xscl=1
Ymin=-2
Ymax=2
Yscl=1
```

Use the intersect utility in the CALC menu.

```
Intersection
X=5          Y=1
```

Analytical solution First, write the sum of the logarithms as the log of a product.

$$\log x + \log(x-3) = 1$$
$$\log x(x-3) = 1$$
$$\log(x^2 - 3x) = 1$$

Change this equation to exponential form.

$$10^1 = x^2 - 3x$$
$$10 = x^2 - 3x$$

This equation is nonlinear. Make one side of the equation zero and use the *ac* test to factor.

$$0 = x^2 - 3x - 10$$
$$0 = x^2 - 5x + 2x - 10$$
$$0 = x(x-5) + 2(x-5)$$
$$0 = (x+2)(x-5)$$
$$x = -2, 5$$

Because the solution -2 is not in the domain of $y = \log x + \log(x-3)$, it must be discarded. Note that the remaining solution, $x = 5$, matches the solution found by the graphing method.

Example 8 Solve the following equation for x: $\ln(x+5) - \ln(x+7) = -1$.

Graphical Solution $\ln(x+5)$ is defined only if $x > -5$; $\ln(x+7)$ is defined only if $x > -7$. Therefore the domain of $y = \ln(x+5) - \ln(x+7)$ is $\{x : x > -5\}$ and it is expected that the graph will lie entirely to the right of $x = -5$.

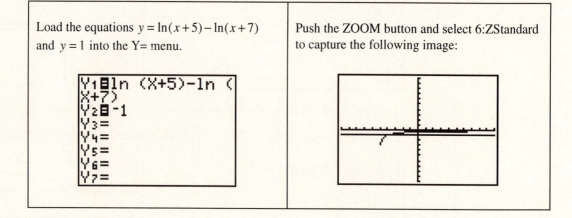

Load the equations $y = \ln(x+5) - \ln(x+7)$ and $y = 1$ into the Y= menu.

Push the ZOOM button and select 6:ZStandard to capture the following image:

Adjust the WINDOW parameters.	Use the intersect utility in the CALC menu.
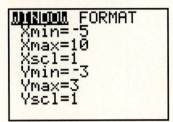	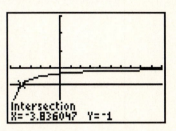

If you press the ⌊X,T,θ⌋ *immediately,* you can obtain a few more significant digits. You can also check your solution.

```
X
            -3.836046586
ln (X+5)-ln (X+7
)
                      -1
```

Analytical Solution   Begin by expressing the difference of the logarithms as the logarithm of a quotient.

$$\ln(x+5) - \ln(x+7) = -1$$

$$\ln\frac{x+5}{x+7} = -1$$

Change this equation to exponential form.

$$e^{-1} = \frac{x+5}{x+7}$$

$$\frac{1}{e} = \frac{x+5}{x+7}$$

Multiply both sides by the least common denominator.

$$e(x+7)\left[\frac{1}{e}\right] = \left[\frac{x+5}{x+7}\right]e(x+7)$$

$$x+7 = e(x+5)$$

$$x+7 = ex+5e$$

This last equation is linear. Isolate the terms containing x on one side of the equation.

$$x - ex = 5e - 7$$

$$(1-e)x = 5e - 7$$

$$x = \frac{5e - 7}{1 - e}$$

Use your calculator to find a decimal approximation for this solution.

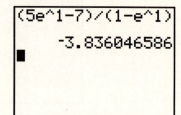

Note that this solution matches the solution found by the graphing method.

Example 9   Solve the following inequality for x: $\ln(x-2) \le 0$.

Solution   A student solution is shown below.

First draw the graph of $y = \ln x$.

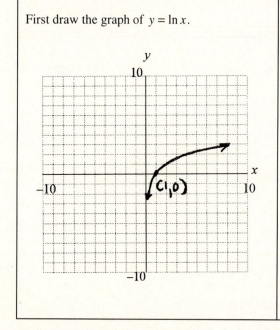

To draw the graph of $y = \ln(x-2)$, shift the graph of $y = \ln x$ 2 units to the right.

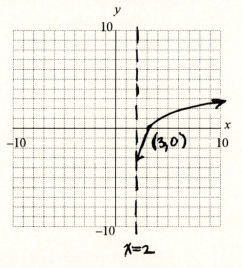

To solve the inequality $\ln(x-2) \le 0$, you must find where the graph of $y = \ln(x-2)$ lies below or on the x-axis. A number line is drawn below the graph and the solution is shaded on the number line in the usual manner.

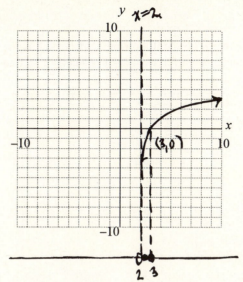

In interval notation, the solution set is $(2,3]$. In set-builder notation, the solution set is $\{x : 2 < x \le 3\}$

Exercises for Section 9.9

1. Change each of the following logarithmic equations into exponential equations. Solve the resulting equation for x. Use your calculator to find a decimal approximation and to check your solution.

Example:	
$$\ln(x+3) = 2$$ $$x+3 = e^2$$ $$x = -3 + e^2$$	``` -3+e^2 4.389056099 Ans→X 4.389056099 ln (X+3) 2 ■ ```

a) $\ln x = 3$ b) $\ln 4x = 1$ c) $\ln(x-3) = 2$

d) $\ln(x+5) = 4$ e) $\ln(3x+2) = -1$ f) $\ln(4-5x) = -2$

2. Change each of the following exponential equations into logarithmic equations. Solve the resulting equation for x. Use your calculator to find a decimal approximation and to check your solution.

Example:	
$$e^{2x-1} = 3$$ $$2x-1 = \ln 3$$ $$2x = 1 + \ln 3$$ $$x = \frac{1+\ln 3}{2}$$	``` (1+ln 3)/2 1.049306144 Ans→X 1.049306144 e^(2X-1) 3 ```

a) $e^x = 7$ b) $e^{2x} = 11$ c) $e^{x+1} = 10$ d) $e^{x-3} = 30$ e) $e^{3x+5} = 100$ f) $e^{2-x} = 3$

3. Change each of the following exponential equations into logarithmic equations. Solve the resulting equation for x. Use your calculator to find a decimal approximation and to check your solution.

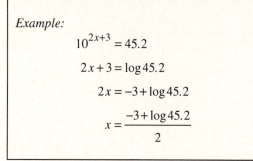

Example:

$$10^{2x+3} = 45.2$$

$$2x+3 = \log 45.2$$

$$2x = -3 + \log 45.2$$

$$x = \frac{-3 + \log 45.2}{2}$$

a) $10^x = 23$ b) $10^{5x} = 11$ c) $10^{x+5} = 200$ d) $10^{x-4} = 150$ e) $10^{2x+5} = 50$ f) $10^{5-x} = 15.8$

4. For the logarithms below, perform each of the following tasks:

   i)   Without the aid of a calculator, state two consecutive integers between which the given logarithm lies.

   ii) Use the change of base rule to write the logarithm in terms of the natural logarithm.

   iii) Use your calculator to find a decimal approximation for the logarithm.

   a) $\log_3 20$ b) $\log_5 10$ c) $\log_4 3$ d) $\log_2 35$

5. For the logarithms below, perform each of the following tasks:

   i)   Without the aid of a calculator, state two consecutive integers between which the given logarithm lies.

   ii) Use the change of base rule to write the logarithm in terms of the common logarithm.

   iii) Use your calculator to find a decimal approximation for the logarithm.

   a) $\log_3 90$ b) $\log_4 50$ c) $\log_5 200$ d) $\log_2 100$

6. For the pairs of equations below, perform each of the following tasks:

   i)   (TI-82) Load each equation into the Y= menu. Push the ZOOM button and select 6:ZStandard. Use the intersect utility in the CALC menu to find the coordinates of the point of intersection.

   ii) Set up a coordinate system on a sheet of graph paper and clearly indicate the scale on each coordinate axis. Copy the image in your viewing window onto your coordinate system. Draw a number line below your graph and shade and label the solution of $f(x) = g(x)$ on the number line in the usual manner.

   iii) Use an analytical method to find an *exact* solution of $f(x) = g(x)$. Use your calculator to find a decimal approximation for your solution. If this approximation does not compare favorably with the estimate from part (ii), check your work for error.

   a) $f(x) = 3^x$, $g(x) = 7$ b) $f(x) = 2^{3x}$, $g(x) = 5$

   c) $f(x) = 4^{-x}$, $g(x) = 7$ d) $f(x) = 5^{-2x}$, $g(x) = 8$

7. For the equations below, perform each of the following tasks.

   i) Use an analytical method to find an *exact* solution of the equation.
   ii) Use your calculator to find a decimal approximation and to check your solution.

   a) $2^{4x} = 30$ b) $3^{-2x} = 7$ c) $5^{x+3} = 7$

   d) $7^{3-x} = 4$ e) $2^{x+3} = 3^{x-5}$ f) $3^{x-4} = 5^{2x}$

8. For the pairs of equations below, perform each of the following tasks:

   i) (TI-82) Load each equation into the Y= menu. Adjust the WINDOW parameters so that the point of intersection of the two graphs is visible in the viewing window. Use the intersect utility in the CALC menu to find the coordinates of the point of intersection.

   ii) Set up a coordinate system on a sheet of graph paper and clearly indicate the scale on each coordinate axis. Copy the image in your viewing window onto your coordinate system. Draw a number line below your graph and shade and label the solution of $f(x) = g(x)$ on your number line in the usual manner.

   iii) Use an analytic method to find an *exact* solution of $f(x) = g(x)$. Use your calculator to find a decimal approximation for your solution. If this approximation does not compare favorably with the estimate from part (ii), check your work for error.

   a) $f(x) = 100(1.02)^x$, $g(x) = 500$ b) $f(x) = 1000(0.85)^x$, $g(x) = 500$

   c) $f(x) = 1200(1.06)^{2x}$, $g(x) = 1500$ d) $f(x) = 5000(0.92)^{4x}$, $g(x) = 1000$

   e) $f(x) = 1500e^{0.09x}$, $g(x) = 2000$ f) $f(x) = 3200e^{-0.11x}$, $g(x) = 1800$

   g) $f(x) = 100(1.02)^x$, $g(x) = 50(1.04)^x$ h) $f(x) = 1000(0.76)^x$, $g(x) = 700(0.87)^x$

   i) $f(x) = 3500(0.85)^{2x}$, $g(x) = 2000(0.95)^{3x}$ j) $f(x) = 500(1.03)^{2x}$, $g(x) = 200(1.04)^{3x}$

   k) $f(x) = 1000e^{0.08x}$, $g(x) = 1500e^{0.05x}$ l) $f(x) = 10000e^{-0.10x}$, $g(x) = 6000e^{-0.06x}$

9. For the pairs of equations below, perform each of the following tasks:

   i) (TI-82) Load each equation into the Y= menu. Adjust the WINDOW parameters so that the point of intersection of the two graphs is visible in the viewing window. Use the intersect utility in the CALC menu to find the coordinates of the point of intersection.

   ii) Set up a coordinate system on a sheet of graph paper and clearly indicate the scale on each coordinate axis. Copy the image in your viewing window onto the coordinate system. Draw a number line below your graph and shade and label the solution of $f(x) = g(x)$ on your number line in the usual manner.

   iii) Use an analytical method to find an *exact* solution of $f(x) = g(x)$. Use your calculator to find a decimal approximation for your solution. If this approximation does not compare favorably with the estimate from part (ii), check your work for error.

   a) $f(x) = \log x + \log(x+21)$, $g(x) = 2$ b) $f(x) = \ln(x+6) - \ln(x+4)$, $g(x) = 1$,

   c) $f(x) = \log x + \log(x+48)$, $g(x) = 2$ d) $f(x) = \ln x - \ln(x-3)$, $g(x) = 1$

10. (TI-82) Suppose that you are asked to solve the equation $\log_6 x + \log_6(x-5) = 2$. You will have to use the change of base rule if you wish to enter this logarithmic function into the Y= menu. Begin by changing to the natural logarithm.

$$\log_6 x + \log_6(x-5) = 2$$

$$\frac{\ln x}{\ln 6} + \frac{\ln(x-5)}{\ln 6} = 2$$

Enter each side of this last equation into the Y= menu in the following manner:	Because the domain of this logarithmic function is all *x*-values greater than 5, make the following adjustments to the WINDOW parameters:
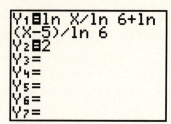	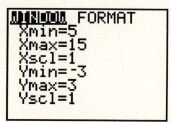

a) Push the GRAPH button. Set up a coordinate system on a sheet of graph paper and clearly indicate the scale on each coordinate axis. Copy the image in your viewing window onto the coordinate system.

b) Use the intersect utility in the CALC menu to find the coordinates of the point of intersection. Draw a number line below your graph and shade and label the solution of the equation $\log_6 x + \log_6(x-5) = 2$ on your number line in the usual manner.

c) Use an analytical method to find an *exact* solution of the equation $\log_6 x + \log_6(x-5) = 2$. If this solution does not compare favorably with the solution found in part (b), check your work for error.

11. For the equations below, perform each of the following tasks:

 i) Use an analytical method to find an *exact* solution of the equation.
 ii) Use your calculator to find a decimal approximation and to check your solution.

 a) $e^{2\ln x} = 7$ b) $10^{\log x + \log(x-2)} = 2$ c) $\log x + \log(x-2) = 2$

 d) $e^{\ln x - \ln(x-5)} = 4$ e) $\ln(x+4) - \ln(x+5) = -1$ f) $\log(x+2) + \log(x-2) = 2$

12. For each of the following equations, perform each of the following tasks:

 i) Set up a coordinate system on a sheet of graph paper and clearly indicate the scale on each coordinate axis. Use graphing shortcuts to help sketch the graph of the function. Label the asymptote with its equation and the *x*-intercept with its coordinates.

 ii) Draw a number line below your graph and shade and label the solution of $f(x) \le 0$ on your number line in the usual manner. Use both set-builder and interval notation to describe your solution.

 a) $f(x) = \ln x$ b) $f(x) = -\log x$ c) $f(x) = \ln(-x)$

 d) $f(x) = \ln(x-2)$ e) $f(x) = -\log(x+3)$ f) $f(x) = \ln(3-x)$

13. (TI-82) It is not possible to solve all equations analytically. There many equations that defy an algebraic solution. Use the intersect utility in your calculator to find approximate solutions of each of the following equations. Try to solve the problems algebraically and see what happens.

a) $e^{-0.25x} = x$ b) $3 - x = \ln x$

9.10 Applications of the Logarithm

In this section, the techniques learned in this chapter will be applied to some real-world models that involve exponential and logarithmic functions.

Example 1   Melissa invests $1,000 in an account that pays 5 percent interest, compounded semiannually. She would like to know how long it will take her money to double in value.

Graphical Solution   Begin with the formula for interest compounded a finite number of times per year and substitute the given values. The principal P is $1,000, the yearly rate r is 0.05, and the number of compounding periods per year n is 2.

$$A = P\left[1 + \frac{r}{n}\right]^{nt}$$

$$= 1000\left[1 + \frac{0.05}{2}\right]^{2t}$$

$$= 1000[1.025]^{2t}$$

Find out how long it will take Melissa's money to double—that is, to reach $2000.

Load the equations $A = 1000[1.025]^{2t}$ and $A = 2000$ into the Y= menu.	Make the following adjustments to the WINDOW parameters:
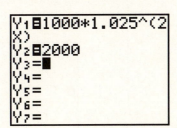	

Use the intersect utility in the CALC menu to find the point of intersection.	Press X *immediately* to capture more significant digits, then check your solution.
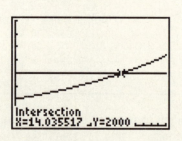	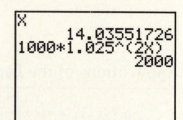

It will take about 14 years for the account to double.

Analytical Solution   First let $A = 2000$ in the equation.

$$A = 1000[1.025]^{2t}$$

$$2000 = 1000[1.025]^{2t}$$

Divide both sides of this last equation by 1000.

$$2 = [1.025]^{2t}$$

Take the natural logarithm of both sides of the equation and use the property $\ln x^c = c \ln x$.

$$\ln 2 = \ln[1.025]^{2t}$$

$$= 2t \ln 1.025$$

Divide both sides of the last equation by $2\ln 1.025$.	Use your calculator to find a decimal approximation of this solution.
$$\frac{\ln 2}{2\ln 1.025} = \frac{2t \ln 1.025}{2\ln 1.025}$$ $$\frac{\ln 2}{2\ln 1.025} = t$$	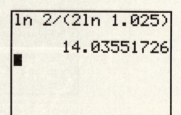

This last solution compares favorably with the estimate found with the graphing method.

Example 2   The population of Allentown is presently 7000 people and is estimated to be decreasing at a rate of 5 percent per year. The population of Davistown is presently 10,000 people and is estimated to be decreasing at a rate of 11 percent per year. If these trends continue, when will the population of Allentown equal the population of Davistown?

Graphical Solution   Since Allentown is losing 5 percent of its population each year, 95 percent of its population returns at the end of each year. The population of Allentown is given by the equation $P = 7000(0.95)^t$, where P represents the population t years from now. The population of Davistown is given by the equation $P = 10,000(0.89)^t$, where P represents the population of Davistown t years from now.

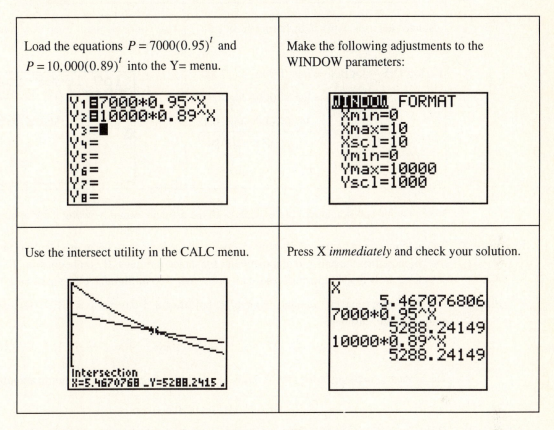

Load the equations $P = 7000(0.95)^t$ and $P = 10,000(0.89)^t$ into the Y= menu.

Make the following adjustments to the WINDOW parameters:

Use the intersect utility in the CALC menu.

Press X *immediately* and check your solution.

In approximately 5.47 years the population of Allentown will equal the population of Davistown. Note that the population of each town will be approximately 5288 people.

Analytical Solution   Begin by setting the equations equal to each other.

$$\text{Population of Allentown} = \text{Population of Davistown}$$

$$7000(0.95)^t = 10,000(0.89)^t$$

The natural logarithm was used in the last example. Just to be different, take the common logarithm of both sides of the equation in this example.

$$\log 7000(0.95)^t = \log 10,000(0.89)^t$$

The common logarithm of a product is the sum of the common logarithms:

$$\log 7000 + \log 0.95^t = \log 10,000 + \log 0.89^t$$

Next, use the property $\log x^c = c \log x$.

$$\log 7000 + t \log 0.95 = \log 10,000 + t \log 0.89$$

Because no power of t is higher than 1, this last equation is linear.

Isolate the terms with t in them on one side of the equation. $$t \log 0.95 - t \log 0.89 = \log 10,000 - \log 7000$$ $$(\log 0.95 - \log 0.89)t = \log 10,000 - \log 7000$$ $$t = \frac{\log 10,000 - \log 7000}{\log 0.95 - \log 0.89}$$	Use your calculator to approximate this solution. 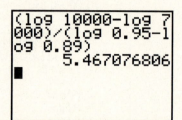

Note that this solution agrees quite nicely with the estimate found by using the graphing method.

Example 3 The amount of time required for half of a particular substance to decay is called the substance's *half-life*. Suppose that the half-life of a particular substance is 5 days. If you begin with 1000 grams of the substance, how much time must pass until fewer than 100 milligrams of the substance remains?

Solution Let $A(n)$ represent the amount of material left after n 5-day periods. $A(0)$ represents the initial amount of the substance.

$$A(0) = 1000$$

The amount remaining at the end of one 5-day period is 50 percent of the initial amount.

$$A(1) = 0.50A(0)$$
$$= 0.50(1000)$$

The amount remaining at the end of two 5-day periods is 50 percent of the amount remaining at the end of one 5-day period.

$$A(2) = 0.50A(1)$$
$$= 0.50[0.50(1000)]$$
$$= (0.50)^2(1000)$$

If you continue in this manner, the amount remaining at the end of n 5-day periods is given by the following equation:

$$A = 1000(0.50)^n$$

Let t represent the number of days that have passed. Then the number of 5-day periods can be found by dividing the number of days by 5. Therefore $n = \dfrac{t}{5}$ and the amount remaining after t days have passed is given by the following equation:

$$A = 1000(0.50)^{\frac{t}{5}}$$

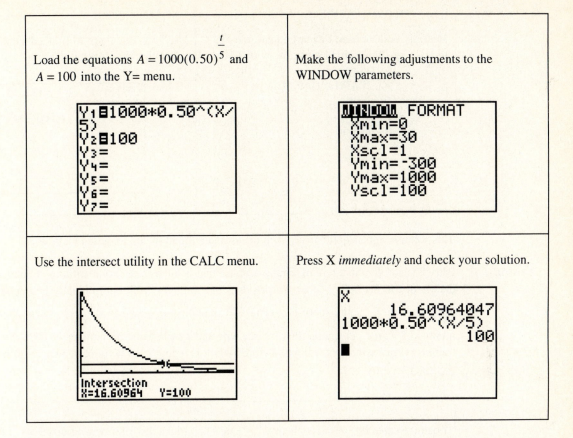

Load the equations $A = 1000(0.50)^{\frac{t}{5}}$ and $A = 100$ into the Y= menu.

Make the following adjustments to the WINDOW parameters.

Use the intersect utility in the CALC menu.

Press X *immediately* and check your solution.

Approximately 16.6 days must pass, at which point approximately 100 grams of the substance remains.

Analytical Solution   Let $A = 100$ in the equation and divide both sides of the resulting equation by 1000.

$$A = 1000\,(0.50)^{\frac{t}{5}}$$

$$100 = 1000\,(0.50)^{\frac{t}{5}}$$

$$\frac{100}{1000} = (0.50)^{\frac{t}{5}}$$

$$0.1 = (0.50)^{\frac{t}{5}}$$

Take the natural logarithm of both sides of the equation and use the property $\ln x^c = c \ln x$.

$$\ln 0.1 = \ln(0.50)^{\frac{t}{5}}$$

$$\ln 0.1 = \frac{t}{5}\ln 0.50$$

Multiply both sides of this last equation by 5, then divide the resulting equation by $\ln 0.50$.	Use your calculator to find a decimal approximation for this solution.
$$5\ln 0.1 = t\ln 0.50$$ $$\frac{5\ln 0.1}{\ln 0.50} = t$$	``` (5ln 0.1)/ln 0.5 0 16.60964047 ```

This solution agrees quite nicely with the estimate found by the graphing method.

Example 4   Patrick is enrolled in a typing school. The training is intense, the competition is fierce, and Patrick is determined to do well. At first, Patrick does not test very well, but gradually he begins to improve. Suppose that his typing speed is given by the equation $W = 100\left(1 - e^{-0.10t}\right)$, where W represents the number of words per minute that he can type and t represents the number of days that he has been training. How many days will pass before Patrick is typing at a rate of 65 words per minute?

Graphical Solution

Load the equations $W = 100\left(1 - e^{-0.10t}\right)$ and $W = 65$ into the Y= menu.	Make the following adjustments to the WINDOW parameters:
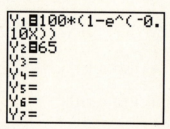	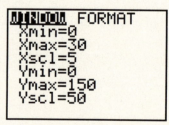
Use the intersect utility in the CALC menu.	Press X *immediately* and check your solution.
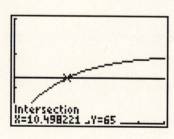	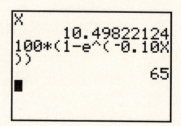

It appears that it will take approximately 10.5 days before Patrick is typing at a rate of 65 words per minute.

Analytical Solution   Let $W = 65$ in the equation.

$$W = 100\left(1 - e^{-0.10t}\right)$$

$$65 = 100\left(1 - e^{-0.10t}\right)$$

There are a variety of ways that you could proceed. You can distribute the 100, then subtract 100 from both sides of the resulting equation.

$$65 = 100 - 100e^{-0.10t}$$

$$-35 = -100e^{-0.10t}$$

Divide both sides of this last equation by -100, then express the resulting equation in logarithmic form.

$$\frac{35}{100} = e^{-0.10t}$$

$$0.35 = e^{-0.10t}$$

$$-0.10t = \ln 0.35$$

Finally, multiply both sides of this last equation by -10.	Use your calculator to find a decimal approximation for this solution.
$$t = -10\ln 0.35$$	``` -10ln 0.35 10.49822124 ```

This solution agrees quite nicely with the estimate found using the graphing method.

Example 5   Imagine a deer population in a forest is allowed to grow without bound. Soon there would be deer everywhere. This is not realistic as the forest will only be able to sustain a certain number of deer. Let's look at a more realistic model, one that takes into account the capacity of the forest to care for the needs of the deer population. Suppose that a certain section of forest was closed to deer hunters in 1970. Since that time, the deer population in the forest is given by the equation $N = \dfrac{500}{1 + 9e^{-0.29t}}$, where N is the number of deer and t is the number of years that have passed since 1970.

a) What was the deer population in 1970?
b) What will be the eventual deer population?
c) When did the deer population reach 350?

Solution, Part (a) To find the deer population in 1970, insert $t = 0$ into the equation.

$$N = \frac{500}{1+9e^{-0.29t}}$$

$$= \frac{500}{1+9e^{-0.29(0)}}$$

$$= \frac{500}{1+9e^{0}}$$

$$= \frac{500}{1+9}$$

$$= 50$$

The deer population in 1970 was 50 deer.

Graphical Solution of Part (b)

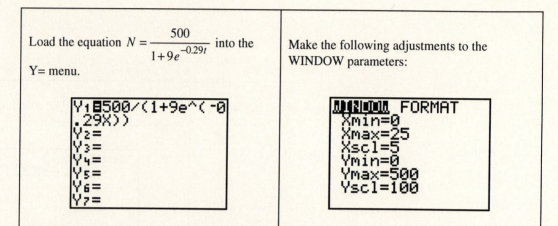

Load the equation $N = \dfrac{500}{1+9e^{-0.29t}}$ into the Y= menu.

Make the following adjustments to the WINDOW parameters:

Push the GRAPH button to capture the following image in your viewing window:

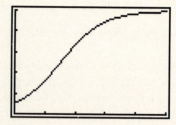

This growth of the deer population is quite realistic. At first, because plenty of food and space are available, the deer population grows rapidly. As the years pass, space and food are less available, and the growth begin to slow. The deer population begins to level off. According to this graph, it appears that the deer population will level off at 500 deer.

Graphical Solution of Part (c)

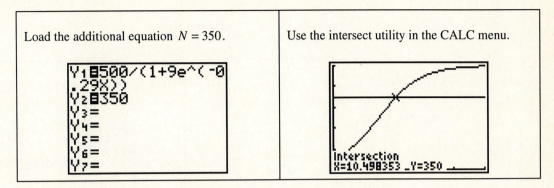

| Load the additional equation $N = 350$. | Use the intersect utility in the CALC menu. |

Press X *immediately* and check your solution.

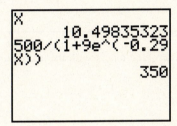

It appears that the deer population reaches 350 approximately 10.5 years after 1970. Therefore somewhere near the middle of 1980, the deer population was 350.

Analytical Solution for Part (c) Let $N = 350$ in the equation.

$$N = \frac{500}{1+9e^{-0.29t}}$$

$$350 = \frac{500}{1+9e^{-0.29t}}$$

Multiply both sides of this equation by the least common denominator.

$$\left(1+9e^{-0.29t}\right)[350] = \left[\frac{500}{1+9e^{-0.29t}}\right]\left(1+9e^{-0.29t}\right)$$

$$350 + 3150e^{-0.29t} = 500$$

Subtract 350 from both sides of the equation, then divide both sides of the resulting equation by 3150.

$$3150e^{-0.29t} = 150$$

$$e^{-0.29t} = \frac{150}{3150}$$

Write this last equation in logarithmic form.

$$-0.29t = \ln \frac{150}{3150}$$

Multiply both sides of this last equation by -100, then divide both sides of the resulting equation by 29. $$29t = -100 \ln \frac{150}{3150}$$ $$t = \frac{-100 \ln \dfrac{150}{3150}}{29}$$	Use your calculator to find a decimal approximation for this solution. 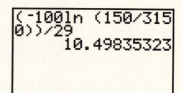

This solution agrees quite nicely with the estimate found using the graphing method.

Exercises for Section 9.10

1. (TI-82) Hans has won a $1000 prize for his pig at the state fair. He finds a savings account that pays 4 percent per year, compounded quarterly. Hans wants to know how long it will take the money in his account to double. Let A represent the amount in the account after t years.

   a) Express A as a function of t.

   b) Load the equation developed in part (a) in Y1 and the equation $A = 2000$ in Y2 of the Y= menu. Adjust the WINDOW parameters so that the point of intersection of the two graphs is visible in your viewing window. Set up a coordinate system on a sheet of graph paper and clearly indicate the scale on each coordinate axis. Copy the image in your viewing window onto the coordinate system.

   c) Use the intersect utility in the CALC menu to find the coordinates of the point of intersection. Clearly state the amount of time required for the account to double.

   d) Let $A = 2000$ in the equation developed in part (a). Use an analytical method to solve the resulting equation for t. Use a calculator to find a decimal approximation for this solution. If this solution does not compare favorably with your estimate from part (c), check your work for error.

2. Amani has won a $500 prize at his school science fair. He finds a savings account that pays 5 percent interest compounded semiannually and invests his prize money. Use both a graphical method and analytical method to find the amount of time required for the money in Amani's account to double.

3. Amanda invests $4000 in an account that pays 4.5 percent interest, compounded semiannually. Use both a graphical method and an analytical method to find the amount of time required for the money in Amanda's account to grow to $6000.

4. Shaquille invests $2000 in an account that pays 4.2 percent interest, compounded quarterly. Use both a graphical method and an analytical method to find the amount of time required for the money in Shaquille's account to grow to $3000.

5. (TI-82) Each year, the population of Fortuna increases by 2 percent. The present population is 9000 hearty souls, but the town council would like to investigate how long it will take the population of the town to grow to 10,000. Let P represent the population of Fortuna after t years.

   a) Express P as a function of t.

   b) Load the equation developed in part (a) in Y1 and the equation $A = 10,000$ in Y2 of the Y= menu. Adjust the WINDOW parameters so that the point of intersection of the two graphs is visible in your viewing window. Set up a coordinate system on a sheet of graph paper and clearly indicate the scale on each coordinate axis. Copy the image in your viewing window onto your coordinate system.

   c) Use the intersect utility in the CALC menu to find the coordinates of the point of intersection. Clearly state the amount of time required for the population in Fortuna to grow to 10,000 people.

   d) Let $A = 10,000$ in the equation developed in part (a). Use an analytical method to solve the resulting equation for t. Use a calculator to find a decimal approximation for this solution. If this solution does not compare favorably with your estimate from part (c), check your work for error.

6. The silver mine in Silver Springs has played out, and the town of Silver Springs is losing people at the rate of 5 percent per year. The present population is 10,000 people. Use both a graphical method and an analytical method to find the number of years that must pass before the population reaches a level of 8000.

7. Population of the town of Bangor is decreasing at the rate of 3.2 percent per year. The present population is 12,000 people. Use both a graphical method and an analytical method to find the number of years that must pass before the population reaches a level of 10,000.

8. Simeon Springs is booming due to the influx of several new computer chip manufacturing plants that have relocated in the area. If the present population is 15,000 and the town population is increasing at the rate of 5 percent per year, use both a graphical method and an analytical method to find the number of years that must pass before the population reaches a level of 19,000.

9. The population of Diggstown is presently 4000 and is increasing at a rate of 6 percent per year. The population of River Bend is presently 8000 and decreasing at a rate of 3 percent per year. Use both a graphical method and an analytical method to find the number of years that must pass before the population of Diggstown equals the population of River Bend.

10. The population of Dead End is presently 7500 and is increasing at a rate of 2 percent per year. The number of automobiles in Dead End is presently 5000 and is increasing at a rate of 9 percent per year. Use both a graphical method and an analytical method to find the number of years that must pass before there is an average of one automobile for each person in Dead End.

11. (TI-82) A hospital purchases a certain radioactive substance for its radiology lab on January 1, 1995. Suppose that the amount of the radioactive substance is given by the equation $A = 5000e^{-0.02t}$, where A is the remaining number of milligrams of the substance and t is the number of days that have passed since January 1, 1995. This substance is essential to diagnostic work done at the hospital and the hospital will buy a fresh supply of the substance when the amount remaining falls below 1000 milligrams.

a) Load the equations $A = 5000e^{-0.02t}$ and $A = 1000$ into the Y= menu. Adjust the WINDOW parameters so that the point of intersection of the two graphs is visible in your viewing window. Set up a coordinate system on a sheet of graph paper and clearly indicate the scale on each coordinate axis. Copy the image in your viewing window onto your coordinate system.

b) Use the intersect utility in the CALC menu to find the coordinates of the point of intersection. Clearly state the number of days that must pass for the amount of substance remaining to fall below 1000 milligrams.

c) Let $A = 1000$ in the equation $A = 5000e^{-0.02t}$. Use an analytical method to solve the resulting equation for t. Use a calculator to find a decimal approximation for this solution. If this solution does not compare favorably with your estimate from part (b), check your work for error.

12. The population in Springtown is given by the equation $P = 1176e^{0.07t}$, where P represents the number of people in the Springtown t years after 1980.

a) What was the population in Springtown in 1980?

b) What was the population in Springtown in 1985?

c) Use both a graphical method and an analytical method to determine the year when the population of Springtown first reaches 2500 people.

13. The size of tumor in a laboratory rat is given by $d = 7.2e^{rt}$, where d represents the diameter of the tumor in millimeters and t represents the number of days that have passed since the tumor was first noticed. The variable r is called the growth rate. If the tumor grows to a diameter of 10.3 millimeters in 10 days, use both a graphical method and an analytical method to determine the growth rate r.

14. (TI-82) The half-life of a particular radioactive substance is 12 hours. Initially, there are 200 grams of this substance present. Find out how much time must pass until only 40 grams of the substance remains. Let A represent the number of grams of this substance remaining and let n represent the number of 12-hour periods that have passed. Let t represent the number of hours that have passed.

a) Express A as a function of n.

b) Express n as a function of t.

c) Use the results from parts (a) and (b) to help express A as a function of t.

d) Load the equation developed in part (c) in Y1 and the equation $A = 40$ in Y2 of the Y= menu. Adjust the WINDOW parameters so that the point of intersection of the two graphs is visible in your viewing window. Set up a coordinate system on a sheet of graph paper and clearly indicate the scale on each coordinate axis. Copy the image in your viewing window onto your coordinate system.

e) Use the intersect utility in the CALC menu to find the coordinates of the point of intersection. Clearly state the amount of time that must pass before there is only 40 milligrams of the substance remaining.

f) Let $A = 40$ in the equation developed in part (a). Use an analytical method to solve the resulting equation for t. Use a calculator to find a decimal approximation for this solution. If this solution does not compare favorably with your estimate from part (e), check your work for error.

15. The half-life of a particular radioactive substance is 10 days. Initially, there are 2,000 milligrams of the substance present. Let A represent the number of milligrams of the substance remaining and let t represent the number of days that have passed.

   a) Express A as a function of t.

   b) Use both a graphical and an analytic method to find the number of days that must pass until only 210 milligrams of the radioactive substance remain.

16. Suppose that the half life of a particular radioactive substance is 25,000 years. Initially, suppose that there are 2,000 kilograms of the substance present. Let A represent the number of kilograms remaining and let t represent the number of years that have passed.

   a) Express A as a function of t.

   b) Use both a graphical and an analytic method to find the number of years that must pass until only 200 kilograms of the radioactive substance remain.

17. A potato is placed in an oven and its temperature is given by the equation $T = 450\left(1 - 0.87e^{-0.15t}\right)$, where T is measured in degrees Fahrenheit and t is the number of minutes that the potato has been in the oven.

   a) What is the initial temperature of the potato?

   b) (TI-82) Load the equation $T = 450\left(1 - 0.87e^{-0.15t}\right)$ into the Y= menu. Adjust the WINDOW parameters so that you can view the temperature of the potato over its first 30 minutes in the oven. Clearly state what you feel is the temperature of the oven.

   c) Use both a graphical method and an analytical method to find the amount of time required for the potato to warm to 350 degrees Fahrenheit.

18. Francisco has a new job working at a sheet metal factory cutting parts for elevators. His pay is based on the number of parts that he completes each day, so he is highly motivated to improve. At first, the job is difficult to learn, so the number of parts that he manages to cut is very low. However, as the days go by, he improves rapidly. Suppose that the number of parts he can cut in a day is given by the equation $N = 100 - 70e^{-0.08t}$, where N represents the number of parts and t represents the number of days that he has been on the job.

   a) How many parts did Francisco cut the first day on the job?

   b) (TI-82) Load the equation $N = 100 - 70e^{-0.08t}$ into the Y= menu. Adjust the WINDOW parameters so that you can view the number of parts that he cuts over his first 60 days on the job. Clearly state the level at which he will eventually work.

   c) Use both a graphical and an analytical method to find the number of days that Francisco was on the job before he was cutting parts at a rate of 80 per day.

19. The population of golden mantled ground squirrels in a certain section of Sequoia National Park is given by the equation $P = \dfrac{2000}{1 + 9e^{-0.07t}}$, where P represents the number of squirrels and t is the number of months that have passed since January 1, 1980.

   a) How many squirrels were present on January 1, 1980?

   b) How many squirrels were present on January 1, 1985?

c) Load the equation $P = \dfrac{2000}{1+9e^{-0.07t}}$ into the Y= menu. Adjust the WINDOW parameters so that the squirrel population is visible in your viewing window over the time interval from January 1, 1980 to January 1, 1990. Clearly state what appears to be the maximum number of squirrels that this particular section of Sequoia National Park can sustain.

d) Use both a graphical method and an analytical method to find the year and the month when the squirrel population first reached a level of 1500 squirrels.

20. A measles epidemic breaks out among the children of Bangor, Maine. Suppose that the number of children infected is given by the equation $N = \dfrac{800}{1+99e^{-0.25t}}$, where N represents the number of children infected and t represents the number of days that have passed since the outbreak of the epidemic.

a) What was the original number of children infected at the outbreak of the epidemic?

b) How many children were infected by the twentieth day of the epidemic?

c) (TI-82) Load the equation $N = \dfrac{800}{1+99e^{-0.25t}}$ into the Y= menu. Adjust the WINDOW parameters so that the number of children infected is visible in your viewing window over the first 50 days of the epidemic. Clearly state what appears to be the eventual number of children that will be infected.

d) Use both a graphical and analytical method to find the day of the epidemic when the number of children infected first reached a level of 700 children.

9.11 Summary and Review

Important Visualization Tools

The graph of $f(x) = a^x$, $a > 1$. Note that the x-axis acts as an asymptote and the y-intercept is $(0,1)$.

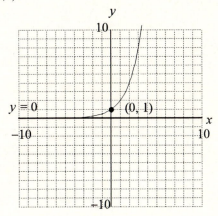

The domain of f is $(-\infty, +\infty)$. The range of f is $(0, +\infty)$.

The graph of $f(x) = a^x$, $0 < a < 1$. Note that the x-axis acts as an asymptote and the y-intercept is $(0,1)$.

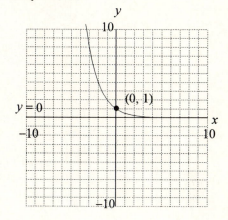

The domain of f is $(-\infty, +\infty)$. The range of f is $(0, +\infty)$.

The point (y,x) is a reflection of the point (x,y) across the line $y = x$.

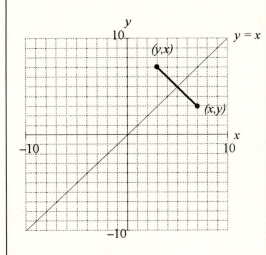

The graph of $x = f(y)$ is a reflection of the graph of $y = f(x)$ across the line $y = x$.

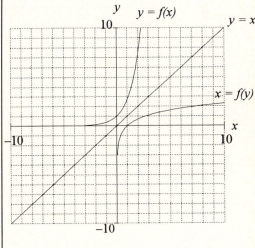

To find $g(f(2))$, you must first drop 2 into the f-machine to find $f(2)$, which is 4. Then you must drop 4 into the g-machine to find $g(4)$, which is 7.

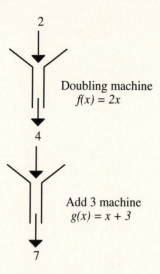

Doubling machine
$f(x) = 2x$

Add 3 machine
$g(x) = x + 3$

In the language of function notation, you would write the following:

$$g(f(2)) = g(4)$$
$$= 7$$

To find $f(g(2))$, you must first drop 2 into the g-machine to find $g(2)$, which is 5. Then you must drop 5 into the f-machine to find $f(5)$, which is 10.

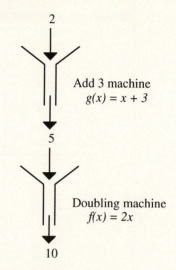

Add 3 machine
$g(x) = x + 3$

Doubling machine
$f(x) = 2x$

In the language of function notation, you would write the following:

$$f(g(2)) = f(5)$$
$$= 10$$

The graph of $g(x) = \log_a x$, $a > 1$. Note that the y-axis acts as an asymptote and the x-intercept is $(1, 0)$.

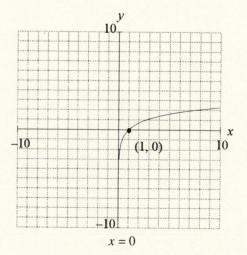

$x = 0$

The domain of g is $(0, +\infty)$. The range of g is $(-\infty, +\infty)$.

A Review of Key Words and Phrases

☛ Suppose that a and b are positive real numbers. The function defined by the equation $f(x) = a \cdot b^x$ is called an exponential function. We refer to b as the base of the function. (page 713)

☛ Suppose that a and b are positive real numbers. (page 713)

1. If $0 < b < 1$, then the graph of $f(x) = a \cdot b^x$ will decrease or decay.

2. If $b > 1$, then the graph of $f(x) = a \cdot b^x$ will increase or grow.

☛ The amount in an account can be computed by the formula $A = P\left[1 + \dfrac{r}{n}\right]^{nt}$. (page 724)

1. P is the initial investment; it is often called the principal or the present value of the account.
2. A is the amount in the account; it is called the amount or the future value of the account.
3. r is the yearly interest rate.
4. n is the number of compounding periods in 1 year.
5. t is the number of years.

☛ The number e is the limiting value of the expression $\left[1 + \dfrac{1}{k}\right]^{k}$ as k grows without bound. (page 727)

☛ The number e is approximately equal to 2.7818281828 . . . . (page 727)

☛ If interest is being compounded continuously, then the amount in an account is given by the formula $A = Pe^{rt}$. (page 728)

1. P is the initial investment; it is often called the principal or the present value of the account.
2. A is the amount in the account; it is called the amount or the future value of the account.
3. r is the yearly interest rate.
4. t is the number of years.

☛ If $a \neq 0$, then $a^{-n} = \dfrac{1}{a^n}$. (page 736)

☛ In a nutshell, raising something to the negative 1 power simply inverts the object. (page 737)

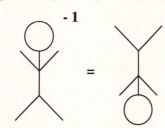

☛ When $a > 1$, the graph of $f(x) = a^x$ is increasing. (page 740)

1. The x-axis acts as an asymptote and the y-intercept is $(0,1)$.
2. As you move to the left, the graph approaches the x-axis and the function values approach zero.
3. As you move to the right, the graph rises to positive infinity and the function values increase without bound.
4. The domain of the function is all real numbers, which means any real number can be used as an exponent.
5. The range is all positive real numbers. If $a > 1$, then $a^x > 0$ for all values of x.

☞ When $0 < a < 1$, the graph of $f(x) = a^x$ is decreasing. (page 742)

1. The x-axis acts as an asymptote and the y-intercept is $(0,1)$.
2. As you move to the left, the graph rises to positive infinity and the function values increase without bound.
3. As you move to the right, the graph approaches the x-axis and the function values approach zero.
4. The domain of the function is all real numbers, which means any real number can be used as an exponent.
5. The range is all positive real numbers. If $0 < a < 1$, $a^x > 0$ for all values of x.

☞ Whenever you interchange x and y in an equation, the graph of the resulting equation is a reflection of the graph of the original equation across the line $y = x$. (page 763)

☞ If no horizontal line cuts the graph of the function more than once, then the function is called a one-to-one function. (page 765)

☞ Let f and g be functions that satisfy the following conditions. (page 775)

1. $f(g(x)) = x$ for all x in the domain of g.
2. $g(f(x)) = x$ for all x in the domain of f.

The function g is called the inverse of the function f. By symmetry, the function f must also be the inverse of the function g. Therefore the functions f and g are often called inverses of each other.

☞ If the inverse of the function f exists, mathematicians sometimes use the notation f^{-1} to represent the inverse function (page 775)

☞ If the function f^{-1} is the inverse of the function f, then the graph of f^{-1} will be a reflection of the graph of f across the line $y = x$. (page 776)

☞ To find the inverse of a function, switch x and y in the equation and solve for y. (Example 9, page 776)

☞ Let $f(x) = 10^x$. The function g defined by $g(x) = \log x$ is the inverse of the function f. (page 786)

1. Because f and g are inverses, $f(g(x)) = x$ or $10^{\log x} = x$.
2. Because f and g are inverses, $g(f(x)) = x$ or $\log 10^x = x$.

☞ Let $f(x) = e^x$. The function g defined by $g(x) = \ln x$ is the inverse of the function f. (page 787)

1. Because f and g are inverses, $f(g(x)) = x$ or $e^{\ln x} = x$.
2. Because f and g are inverses, $g(f(x)) = x$ or $\ln e^x$.

☞ Let $f(x) = a^x$, $a > 1$. The function g defined by $g(x) = \log_a x$ is the inverse of the function f. (page 790)

1. Because f and g are inverses, $f(g(x)) = x$ or $a^{\log_a x} = x$.
2. Because f and g are inverses, $g(f(x)) = x$ or $\log_a a^x = x$.

☞ The statement $a^{\log_a x} = x$ implies that $\log_a x$ is an exponent, the exponent to which one raises the base a to get x. (page 791)

☞ $y = \log_a x$ if and only if $a^y = x$, provided $a, x > 0$. (page 792)

☞ $\log x$ means $\log_{10} x$ (page 792)

☞ $\ln x$ means $\log_e x$. (page 792)

☞ When evaluating $\log_a x$, you are being asked to find the exponent to which you raise a to get x. (page 792)

☛ The Laws of Logarithms: Let $a > 0$. If $x > 0$ and $y > 0$ and c is any real number, then the following laws hold. (page 801)

1. $\log_a xy = \log_a x + \log_a y$

2. $\log_a \dfrac{x}{y} = \log_a x - \log_a y$

3. $\log_a x^c = c \log_a x$

☛ The change of base rule: $\log_a x = \dfrac{\log_b x}{\log_b a}$. (page 811)

Chapter Review Exercises

1. Suppose that \$500 is invested in an account that pays 4.25 percent per year compounded quarterly.

   a) How much will be in the account at the end of 2 years?

   b) Use both a graphing method and an analytic method to find the number of years required for the amount in the account to double.

2. Suppose that \$1,000 is invested in an account that pays 4 percent per year compounded continuously.

   a) How much will be in the account at the end of 2 years?

   b) Use both a graphing method and an analytical method to find the number of years required for the amount in the account to double.

3. Ginny is given an injection of 500 milligrams of antibiotic. Thereafter, her body rids itself of 15 percent of the antibiotic every 6 hours.

   a) How much antibiotic remains in Ginny's system after 2 days?

   b) Use both a graphing method and an analytical method to find the amount of time required for Ginny's system to rid itself of all but 50 milligrams of the antibiotic.

4. The town of Rockridge has 1,576 people living within its borders. A new ski resort has just been built in the area and the population of the town is increasing at a rate of 12 percent per year.

   a) If this trend continues, how many people will be in the town at the end of 5 years?

   b) Use both a graphing method and an analytical method to find the amount of time required for the population of Rockridge to reach 5,000 people.

5. Simplify each of the following expressions.

   a) $(-5)^{-4}$

   b) $\left(-\dfrac{2}{5}\right)^{-2}$

   c) $\left(3x^2\right)\left(-2x^{-5}\right)$

   d) $\left(\dfrac{x^{-2}}{x^{-5}}\right)^{-2}$

   e) $\left(3x^{-2}\right)^{-2}$

   f) $\left(3x^{-1}\right)^2\left(4x^{-2}\right)^{-3}$

6. Find $f(g(x))$ and $g(f(x))$ for each of the following pairs.

   a) $f(x) = 2x + 11$, $g(x) = \dfrac{x-11}{2}$

   b) $f(x) = 3x + 5$, $g(x) = \dfrac{x+2}{7}$

   c) $f(x) = \dfrac{3}{x}$, $g(x) = \dfrac{x+2}{3-2x}$

   d) $f(x) = e^{-x} + 5$, $g(x) = -\ln(x-5)$

7. For the equations below, perform each of the following tasks:

   i)   Set up a coordinate system on a sheet of graph paper and clearly indicate the scale on each coordinate axis. Use graphing shortcuts to sketch the graph of the equation. Label the asymptote with its equation.

   ii) Use interval notation to describe the domain and range of the function represented by the equation.

   a) $f(x) = -e^x + 8$

   b) $f(x) = 10^{-x} - 9$

   c) $f(x) = -\ln(x+9)$

   d) $f(x) = \log[-(x-7)]$

8. For the equations below, perform each of the following tasks:

   i)   Use your intuition to find $f^{-1}(x)$.

   ii) Set up a coordinate system on a sheet of graph paper and clearly indicate the scale on each coordinate axis. Draw the line $y = x$ on your coordinate system. Draw the graphs of f and f^{-1} on your coordinate system.

   a) $f(x) = 3x + 6$

   b) $f(x) = \dfrac{x-4}{2}$

9. For the equations below, perform each of the following tasks:

   i)   Use the method of switching x and y and solving for y to find $f^{-1}(x)$.

   ii) Set up a coordinate system on a sheet of graph paper and clearly indicate the scale on each coordinate axis. Draw the line $y = x$ on your coordinate system. Draw the graphs of f and f^{-1} on your coordinate system.

   a) $f(x) = e^x + 3$

   b) $f(x) = -\ln(x-2)$

10. Consider the following graph of the function f.

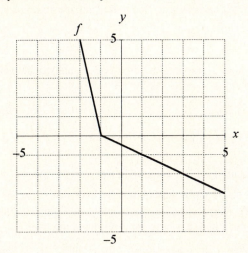

Perform each of the following tasks.

a) Find $f(3)$.   b) Find $f(f(1))$.   c) Find $f^{-1}(-1)$.   d) Sketch the graph of f^{-1}.

11. Write each of the following expressions as a logarithm of a single expression. Clearly state all values of x for which your result is valid.

a) $2\log(x+1) - \log x - 3\log(x-1)$

b) $\dfrac{1}{2}\ln x - \dfrac{3}{4}\ln(x+3)$

12. Simplify each of the following expressions.

a) $e^{2\ln 17}$ b) $10^{\log 11 + \log 17}$ c) $10^{2\log 8 + 3\log 2 - \log 4}$ d) $e^{\frac{1}{2}\ln 4}$

13. Use your calculator to sketch the graphs of $f(x) = e^{\ln(x-3)}$ and $g(x) = x - 3$.

a) Set up a coordinate system on a sheet of graph paper and clearly indicate the scale on each coordinate axis. Sketch the graph of f on this coordinate system. Use interval notation to describe the domain of this function.

b) Set up a second coordinate system on a sheet of graph paper and clearly indicate the scale on each coordinate axis. Sketch the graph of g on this coordinate system. Use interval notation to describe the domain of this function.

c) For what values of x do the graphs of the functions coincide? In other words, for what values of x is $e^{\ln(x-3)}$ equal to $x - 3$?

14. Use the change of base rule and your calculator to find decimal approximations for each of the following logarithms.

a) $\log_2 56$ b) $\log_3 100$

15. Change each of the following exponential equations into logarithmic equations, then solve for x. Use your calculator to find a decimal approximation and to check your solution.

a) $e^{7x-11} = 123$ b) $10^{5-8x} = 1.23$

16. Change each of the following logarithmic equations into exponential equations, then solve for x. Use your calculator to find a decimal approximation and to check your solution.

    a) $\ln(2x - 11) = 5$

    b) $\ln(x^2 - 3) = 1$

17. Use both a graphical method and an analytical method to solve each of the following equations for x.

    a) $2^{5x} = 100$

    b) $100(1.05)^x = 200(1.01)^x$

    c) $\log x + \log(x + 4) = 1$

    d) $\ln(x + 2) - \ln(x + 4) = -2$

18. A potato has been cooking in an oven for over an hour and is removed from the oven and begins to cool. Suppose that the temperature of the potato is given by the equation $T = 40 + 410e^{-0.22t}$, where T represents the temperature in degrees Fahrenheit and t represents the number of minutes that the potato has been removed from the oven.

    a) What is the temperature of the oven?

    b) What is the temperature of the potato 5 minutes after its removal from the oven?

    c) What is the temperature of the room?

    d) Use both a graphical method and an analytical method to find the amount of time required for the potato to cool to 100 degrees Fahrenheit.

19. An epidemic of chicken pox breaks out at the Fortuna school. Suppose the number of children infected is given by the equation $N = \dfrac{100}{1 + 49e^{-0.20t}}$, where N is the number of children that have been infected and t is the number of days that have passed since the epidemic began.

    a) Use your calculator to sketch a graph of the equation. Adjust your WINDOW parameters so that the number of children infected is visible over the first 60 days of the epidemic. Set up a coordinate system on a sheet of graph paper and clearly indicate the scale on each coordinate axis. Copy the image in your viewing window onto your coordinate system.

    b) How many children will eventually be infected?

    c) Use both a graphing method and an analytical method to find the number of days that must pass before 50 children are infected.

Appendix A
TI-81 Programs

This appendix contains a number of programs for the TI-81 calculator. These programs will enable the TI-81 to emulate some of the better features of the TI-82. Let's begin by introducing some programming fundamentals on the TI-81. (These fundamentals are similar on the TI-82.)

To enter a new program, first push the PRGM key. A menu will open. Across the top of the menu are the words EXEC (for execute), EDIT, and ERASE—on the TI-82 you'll see EXEC, EDIT, NEW. There may also be a listing of programs in your calculator. This list will be blank if you do not have any programs in your calculator. To enter a new program, use your arrow keys to highlight the word EDIT, then use your down arrow key to select a blank position on your list of programs. Press the ENTER key. You are now ready to enter your program. (On the TI-82, highlight NEW, press ENTER, type the program name, and press ENTER.)

The EVAL Program

This program will enable you to fill in the numerous tables in this book with ease. This program will run equally well on a TI-82.

Prgm1: EVAL	The assumption here is that you are entering this program in Prgm1. This is not required. You may enter it in another position. Type in the program name EVAL using the letters on your keyboard. Press ALPHA, then E, ALPHA, then V, etc.
:ClrHome	Clears the home screen. Push the PRGM key, then use the arrow keys to highlight I/O. Select 5:ClrHome.
:Lbl 0	A label signifying the beginning of a loop. Push the PRGM key, then use the arrow keys to highlight CTL. Select 1:Lbl. Enter the zero from the keyboard.
:Disp "X"	Displays a message. Push the PRGM key, then use the arrow keys to highlight I/O. Select 1:Disp. Press ALPHA + to capture a quote. Press ALPHA X to capture the X.
:Input X	Push the PRGM key, then use the arrow keys to highlight I/O. Select 2:Input. Press ALPHA X to capture the X.
:Disp "Y"	Displays a message. Press ALPHA + to capture a quote. Press ALPHA Y to capture the Y.
:Disp Y1	Displays the value of Y1. The variable Y1 can be found in the Y-VARS menu. Press 2nd Y-VARS and select 1:Y1.
:Goto 0	Go to label zero (Lbl 0). Press the PRGM key, then use the arrow keys to highlight CTL. Select 2:Goto. Enter the zero from the keyboard.

To finish your editing session press 2nd QUIT. The QUIT command is above the CLEAR key on your keyboard (above the MODE key on the TI-82).

Running the Program. Enter the equation $y = x + 3$ in the Y= menu as follows: Y1 = X + 3. Press the PRGM key and use the arrow keys to highlight EXEC (for execute). Use the down-arrow key to highlight 1:Prgm1 EVAL. Press ENTER to select the program. Press ENTER a second time to start the program. You are prompted for a value of X. Enter a 5 and press the ENTER key. If the program is running properly, the viewing window should look as follows.

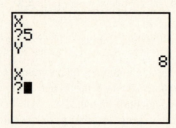

Note that you are prompted for the next value of X. The calculator is in what programmers call a loop. You may enter the next value of X and the calculator will respond with the associated value of Y. You terminate the program by pressing the ON key on your keyboard and selecting 2:Quit.

The Root Program

This program finds zeros of functions.

Prgm2:ROOT	Type in the program name ROOT using the appropriate keys on your keyboard.
:(Xmax – Xmin)/100 → D	Initialize delta for NDeriv. You will find Xmax and Xmin in the RNG submenu of the VARS menu. The / is captured by pressing the division key. The → is captured by pressing the STOΠ key on your calculator. Press ALPHA D to capture the D.
:Lbl 1	Main loop begins here. The Lbl command is in PRGM CTL menu. Enter a one after Lbl.
:X – Y1/ NDeriv(Y1, D) → R	Y1 is in the Y-VARS menu. The NDeriv command is in the MATH menu. Scroll down with the arrow keys and slect 8:NDeriv(. On the TI-82, replace this command with X – Y1/nDeriv(Y1, X, X) → R.
:If abs (X – R) 7 abs(X/1E10)	Test for accuracy. The absolute value symbol, ABS, is on your calculator keyboard. You will find the 7 symbol in the TEST menu. 1E10 means 1×10^{10}. Enter this from your keyboard by pressing a 1, followed by pressing the EE key on your keyboard, then enter the number 10.
:Goto 2	If test is true, go to label two (Lbl 2). The Goto command is in the PRGM CTL menu. Enter a two after Goto.
:R → X	Otherwise, store R in X.
:Goto 1	Go to label one.
:Lbl 2	
:ClrHome	The ClrHome command is located in the PRGM I/O menu.
:Disp "ROOT"	A message. the Disp command IS in the PRGM I/O menu. Press ALPHA + to capture a quote.
:Disp R	Display the value of the root.

Running the Program. Let's find the zero of $y = 2x + 3$, which is $-\frac{3}{2}$ or -1.5. Load the equation $y = 2x + 3$ in the Y= menu as follows: Y₁ = 2X + 3. Press the ZOOM button and select 6:Standard. Press the TRACE button and use the arrow keys to move the cursor reasonably close to the x-intercept of the graph. Press the PRGM key, highlight EXEC, and highlight 2:Prgm2 ROOT. Press ENTER to select the program. Press ENTER again to run the program. If the program is running properly, the viewing window should appear as follows.

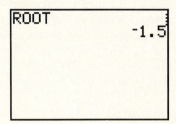

The INTERSCT Program

This program will find the coordinates of the point of intersection of two graphs.

Prgm3:INTERSCT	Type in the program name INTERSCT. You are only allowed eight letters for a program name.
:"Y₁ – Y₂" → Y₃	Store the difference of the functions in Y₁ and Y₂ in Y₃. The variables Y₁, Y₂, and Y₃ can be found in the Y-VARS menu. Press ALPHA + to capture a quote. Press the STO▷ key to produce the →.
:(Xmax – Xmin)/100 → D	Store a hundredth of a screen width in the variable D. You can find the variable Xmax and Xmin in the VARS RNG menu. You can press the STO▷button to produce the → .
:Lbl 1	The beginning of a loop. You can find the Lbl command in the PRGM CTL menu. Enter the one from the keyboard.
:X – Y₃/NDeriv(Y₃, D) → R	Newton's method for finding roots. You can find the NDeriv command in the MATH menu. It's number 8 in the menu. Press ALPHA then the decimal point to generate the comma. On the TI-82, replace this command with X – Y₃/nDeriv(Y₃, X, X) → R.
:If abs (X – R) ≤ abs (X/1E10)	If the error is small enough, then we'll stop the iterative process. You can find the If command in the PRGM CTL menu. The abs (absolute value) is located on the keyboard (2nd x^{-1}). You can produce the E by pressing the EE key on the keyboard. The / is made by pressing the ÷ symbol on your keyboard. The ≤ is located in the TEST menu.
:Goto 2	Go to label two (Lbl 2). The Goto command can be found in the PRGM CTL menu. Enter the 2 from the keyboard.
:R → X	Store the contents of the variable R in the variable X.
:Goto 1	Go to label one (Lbl 1).
:Lbl 2	A label.
:ClrHome	The ClrHome command is found in the PRGM I/O menu.
:Disp "X,Y"	A message. The Disp command is located in the PRGM I/O menu.
:Disp X	Display the contents of the variable X.
:Y₁ → Y	Store the function value in Y₁ in the variable Y.
:Disp Y	Display the contents of the variable Y.

:Y3 – Off	Turn the function in Y3 off so that it will not graph. The Y3 – Off command is located in the Y-VARS menu. On the TI-82 replace this command with FnOff 3.

Running the Program. Enter the following equations in the Y= menu: Y1 = $X^2 - X$ and Y2 = $X + 3$. Press the ZOOM button and select 6:Standard. Push the TRACE button and use the arrow keys to center the cursor over the first point of intersection. Push the PRGM key and use the arrow keys to highlight EXEC. Use the arrow keys to highlight 3:Prgm3 INTERSCT. Push the ENTER key to select the program. Push the ENTER key again to run the program. If the program is running properly, it should respond with the following image in your viewing window.

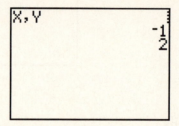

Press the TRACE button and use the arrow keys to center the cursor over the second point of intersection. Run the program. The program should respond with the following image in the viewing window.

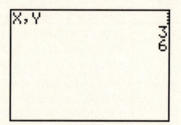

Appendix B

Answers to Selected Exercises

Chapter 1

Section 1.1

1 abc) (Answers may vary)

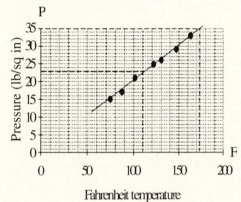

Fahrenheit temperature

1d)　i) 23 lb/in^2　ii) 173 $^{\circ}$F

6ab) (Answers may vary)

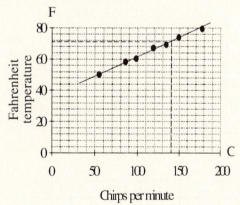

Chirps per minute

6c) 71 $^{\circ}$F

8ab) (Answers may vary)

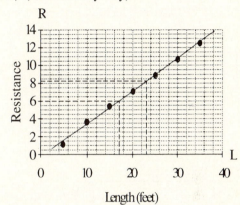

Length (feet)

8c)　i) 8.2×10^{-2} ohms　ii) 17.5 feet

Section 1.2

3)

x	−1	0	1	2	3	4	5	6
y	4	3	2	1	0	1	2	3

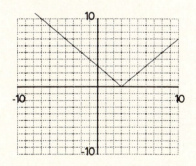

5)

x	−4	−3	−2	−1	0	1	2	3	4
y	−16	9	16	11	0	−11	−16	−9	16

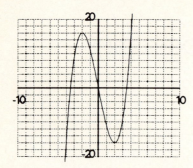

9)

x	–4	–3	–2	–1	0	1	2	3	4
y	1	2	3	4	5	4	3	2	1

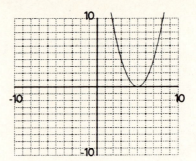

14e)

x	–4	–3	–2	–1	0	1	2	3	4
y	–3	–1	1	3	5	3	1	–1	–3

11)

x	–3	–2	–1	0	1	2	3	4	5	6	7
y	–16	–7	0	5	8	9	8	5	0	–7	–16

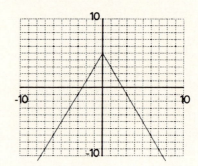

14g)

x	3	4	7	12	19
y	0	–1	–2	–3	–4

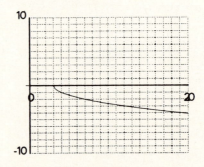

11)

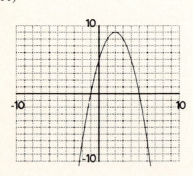

14b)

x	–4	–3	–2	–1	0	1	2	3	4
y	81	64	49	36	25	16	9	4	1

16)

x	–5	–4	–3	–2	–1	0	1	2	3	4	5
y	6	5	4	3	2	1	0	–1	–2	–3	–4

20)

x	–5	–4	–3	–2	–1	0	1	2	3	4	5
y	–13	–11	–9	–7	–5	–3	–1	1	3	5	7

22)

x	–7	–6	–5	–4	–3	–2	–1	0	1	2	3	4	5	6	7
y	7	6	5	4	3	2	1	0	1	2	3	4	5	6	7

22a)

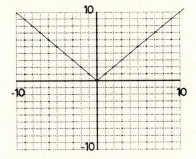

22b)

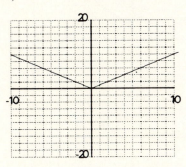

22c)

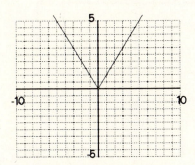

22d) The different scales stretch or shrink the graph in the vertical direction.

Section 1.3

1c) 2   1d) 175 1e) 7

4a) −18 4b) 18

6b) −2 6d) 7   6f) −41

8a) 3   8b) 1

10a) 5   10b) 2

13a) 1   13b) 3

14c) 2.419   14e) −8.4346894

16)

x	−6	−5	−4	−3	−2	−1	0
y	3	2	1	0	1	2	3

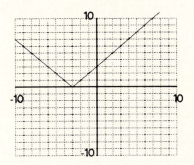

19)

x	−3	−2	−1	0	1	2	3	4	5	6	7
y	−16	−7	0	5	8	9	8	5	0	−7	−16

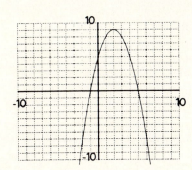

20a)

x	−5	−4	−3	−2	−1	0	1	2	3	4	5
y	15	13	11	9	7	5	3	1	−1	−3	−5

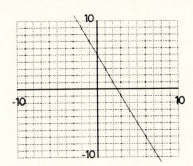

20e)

x	−2	−1	0	1	2	3	4
y	10	−1	−10	−17	−22	−25	−26
x	5	6	7	8	9	10	
y	−25	−22	−17	−10	−1	10	

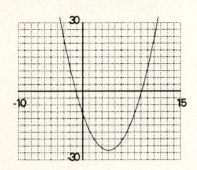

22)

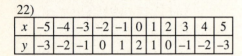

x	−5	−4	−3	−2	−1	0	1	2	3	4	5
y	−3	−2	−1	0	1	2	1	0	−1	−2	−3

24)

x	−5	−4	−3	−2	−1	0	1	2	3	4	5
y	1	0	−1	−2	−3	−4	−3	−2	−1	0	1

27) $L = 100 - w$

29) $A = 16\pi - \pi r^2$

33) $P = 42x - 200$

Section 1.4

3a)

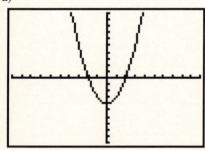

3f)

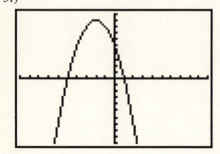

4)

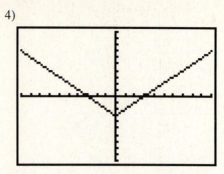

5)

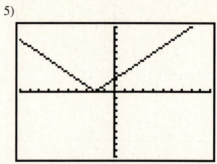

9a)

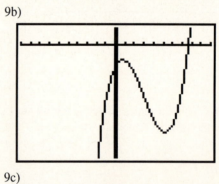

9b)

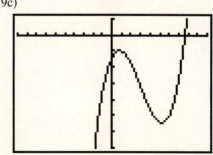

9c)

11a)

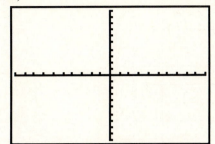

11b)

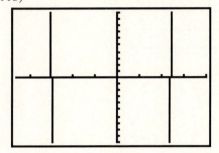

11c)

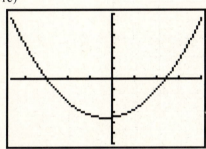

15) (Answers may vary)

```
WINDOW FORMAT
 Xmin=-15
 Xmax=20
 Xscl=5
 Ymin=-7000
 Ymax=3000
 Yscl=1000
```

16b) (Answers may vary)

```
WINDOW FORMAT
 Xmin=-10
 Xmax=10
 Xscl=1
 Ymin=-30
 Ymax=30
 Yscl=5
```

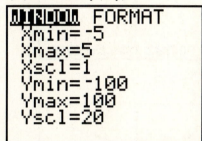

16f) (Answers may vary)

```
WINDOW FORMAT
 Xmin=-5
 Xmax=5
 Xscl=1
 Ymin=-100
 Ymax=100
 Yscl=20
```

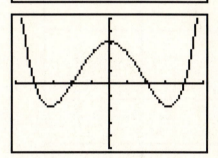

Section 1.5

2)

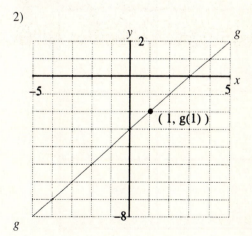

$(1, g(1))$

5) $g(2) = -1$

8) $h(-2) = 4$

11) $x = 1$

14a) About 305 lines.
14b) About 395 lines.
14c) A little over 500 lines.
14d) About 8 days.

15) (Answers may vary)

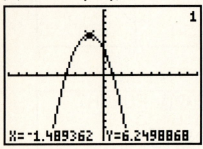

17)

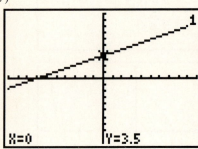

19ab) (Answers may vary)

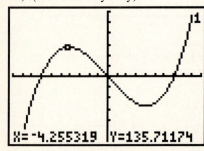

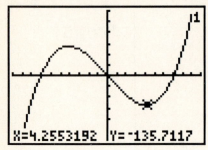

22) 6

24) $A = 10x - 2x^2$

Section 1.6

1)

x	f(x)	g(x)
−1	1	13
0	3	12
1	5	11
2	7	10
3	9	9
4	11	8
5	13	7
6	15	6
7	17	5

1a) $x = 3$ 1b) 4, 5, 6, 7
1c) −1, 0, 1, 2

3)

x	f(x)	g(x)
−4	18	−3
−3	10	−2
−2	4	−1
−1	0	0
0	−2	1
1	−2	2
2	0	3
3	4	4
4	10	5

3a) −1, 3   3b) −4, −3, −2, 4
3c) 0, 1, 2

6)

x	g(x)	h(x)
−5	3	−3
−4	2	−2
−3	1	−1
−2	0	0
−1	−1	1
0	−2	2
1	−3	3
2	−4	4
3	−5	5

6a) $x = -2$ 6b) −5, −4, −3
6c) −1, 0, 1, 2, 3

8)

x	f(x)	h(x)
–5	2	–4
–4	1	–3
–3	0	–2
–2	–1	–1
–1	–2	0
0	–3	1
1	–2	0
2	–1	–1
3	0	–2
4	1	–3
5	2	–4

8a) $x = -2, 2$ 8b) $-5, -4, -3, 3, 4, 5$

8c) $-1, 0, 1$ 8d) $-2, -1, 0, 1, 2$

10)

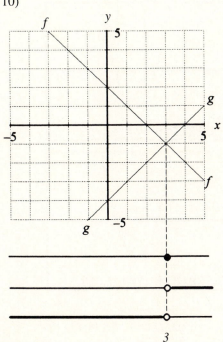

10a) $\{x : x = 3\}$ 10b) $\{x : x > 3\}$

10c) $\{x : x < 3\}$

12)

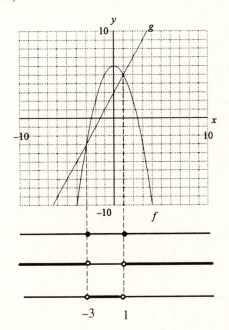

12a) $\{x : x = -3, 1\}$

12b) $\{x : x < -3 \text{ or } x > 1\}$

12c) $\{x : x > -3 \text{ and } x < 1\}$

16)

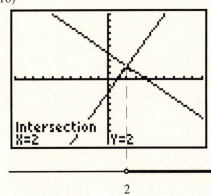

16c) $\{x : x > 2\}$

Section 1.7

1) 2

5)

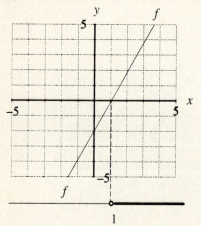

5b) $\{x : x > 1\}$

8)

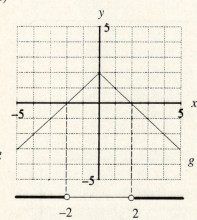

8b) $\{x : x < -2 \text{ or } x > 2\}$

10)

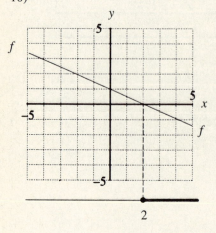

10b) $\{x : x \geq 2\}$

12)

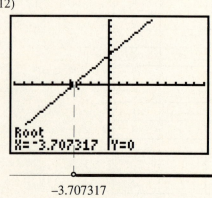

−3.707317

12c) $\{x : x > -3.707317\}$

14)

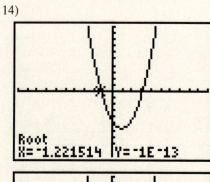

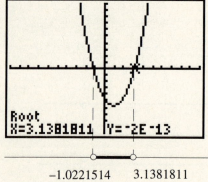

−1.0221514   3.1381811

14c) $\{x : x > -1.0221514 \text{ and } x < 3.1381811\}$

Chapter 1 Review Exercises

1) (Answers may vary)

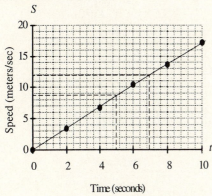

b)   i) 8.7 meters/second
        ii) 7 seconds

2f)

x	2	3	6	11	18
y	0	1	2	3	4

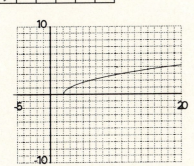

2g)

x	0	1	4	9	16
y	-2	-1	0	1	2

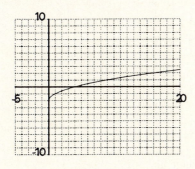

3b) 3   3c) 3   3f) −72

5) $P = 10x - x^2$

6a) $L = 2w + 4$
6b) $P = 6w + 8$
6c) $A = 2w^2 + 4w$

8b) −4.750506

9b) 3.9375

10a) 2.569046516   10d) −1.159292035

11a) $f(-3) = 5$ 11d) $f(3) = -1$

14) $A = 32x - 2x^3$

17) (Answers may vary)

17b) It takes about 8 seconds to return to the ground.

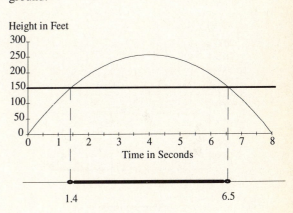

17c) From about 1.4 seconds after release until about 6.5 seconds after release.

Chapter 2

Section 2.1

1)

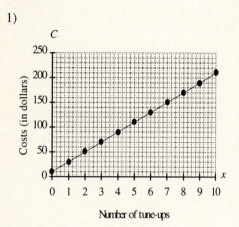

Number of tune-ups

1b) (1, 30)
1e) $170
1f) $C = 10 + 20x$, $170

7)

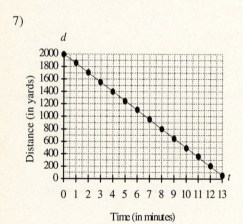

Time (in minutes)

7b) Between 300 and 400 yards.
7c) $d = 2000 - 150t$, 350 yards

8)

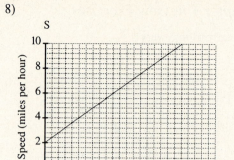

Time (seconds)

8b) 10 miles per hour
8c) $S = 2 + 0.4t$, 10 miles per hour

11a) $40
11b) $5 per basket
11c) $C = 40 + 5x$

14a) $10,000
14b) $2000 per year
14c) $V = 10,000 - 2,000t$

Section 2.2

2a) Slope $= \dfrac{\text{Change in } y}{\text{Change in } x} = \dfrac{-5}{3} = -\dfrac{5}{3}$

2b) Slope $= \dfrac{\text{Change in } y}{\text{Change in } x} = \dfrac{-10}{6} = -\dfrac{5}{3}$

3a)

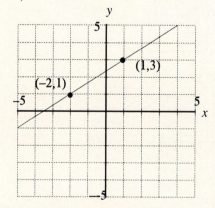

Slope $= \dfrac{\text{Change in } y}{\text{Change in } x} = \dfrac{2}{3}$

5a) 5   5e) 0   5f) Undefined

6b)

P(−12, 250)

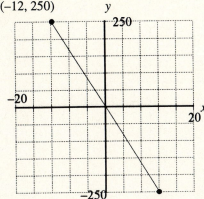

ii) Slope $= \dfrac{\text{Change in } y}{\text{Change in } x} = \dfrac{-500}{24} = -\dfrac{125}{6}$

iii) Slope $= \dfrac{-250 - 250}{12 - (-12)} = \dfrac{-500}{24} = -\dfrac{125}{6}$

Q(12, −250)

7b) −50

13a)

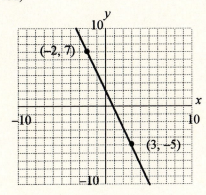

13b) Slope $= \dfrac{7 - (-5)}{-2 - 3} = -\dfrac{12}{5}$

14)

x	-3	-2	-1	0	1	2	3	4	5
y	12	5	0	-3	-4	-3	0	5	12

14ab)

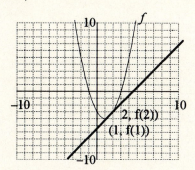

Slope $= \dfrac{f(2) - f(1)}{2 - 1} = \dfrac{-3 - (-4)}{2 - 1} = 1$

14c) $\dfrac{f(3) - f(1)}{3 - 1} = \dfrac{0 - (-4)}{3 - 1} = 2$ and represents

the slope of the line through the points $(1, f(1))$
and $(3, f(3))$.

16a) −3   16d) 2.2

18a)

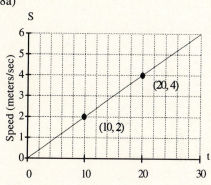

18b) Slope = 0.2 meters per second per second.
This is the acceleration of the wagon, meaning
that the wagon's speed increase 0.2 meters per
second every second.

Section 2.3

1a) $y = -\dfrac{1}{3}x + 1$

1d) $y = x + 1$

7ab)

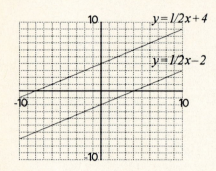

8)

x	−12	−9	−6	−3	0	3	6	9	12
y	−7	−5	−3	−1	1	3	5	7	9

8a)

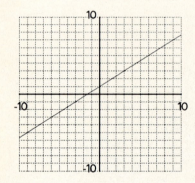

8b) Slope $= \dfrac{\text{Change in } y}{\text{Change in } x} = \dfrac{2}{3}$

8c) If equal increments in the x-values produce equal increments in the y-values in the table, then the function must be linear.

8d) i and ii are linear, iii is not linear.

10a)

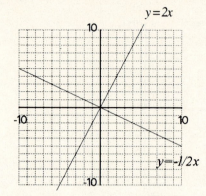

10b)

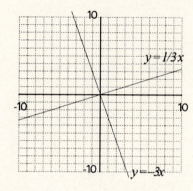

10c)

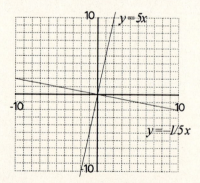

10d) The lines are perpendicular.

10e) They are negative reciprocals.

14ab)

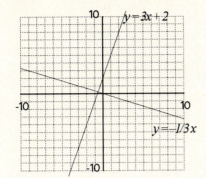

15ab)

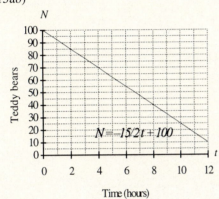

18ab)

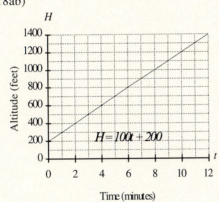

19abc)

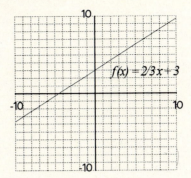

Section 2.4

1c)

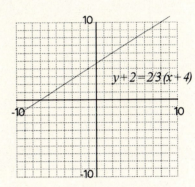

1d)

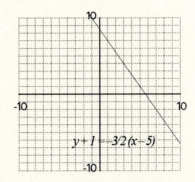

1f)

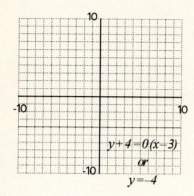

$$y+4=0(x-3)$$
or
$$y=-4$$

2)

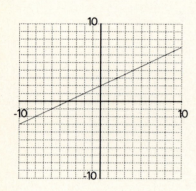

2a) 2

2b) $y-3=\dfrac{1}{2}(x-2)$

2c) $y=\dfrac{1}{2}x+2$

5) $y=-\dfrac{5}{4}x+5$

8d)

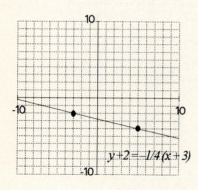

$$y+2=-1/4(x+3)$$

9d)

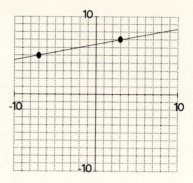

ii) 6.5

iii) $y=\dfrac{1}{5}x+\dfrac{32}{5},\ b=\dfrac{32}{5}=6.4$

10a) $\dfrac{17}{3}$

12)

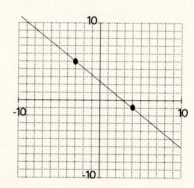

12a) 0

12b) $m=-\dfrac{6}{7}$

12c) $f(x)=-\dfrac{6}{7}x+\dfrac{17}{7}$

12d) $f(3)=-\dfrac{1}{7}$

14)

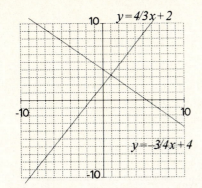

$y = 4/3x + 2$

$y = -3/4x + 4$

16ab)

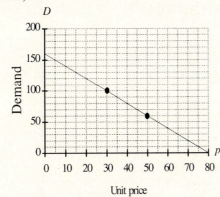

16c) $m = -2$ kettles / dollar. For every increase of \$1 in the unit price, the public will buy 2 fewer kettles.

16d) $D - 100 = -2(p - 30)$

16e) $D = -2p + 160$

18a)

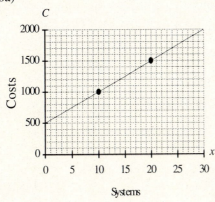

18b) $m = 50$ dollars / system. For each system she installs, her costs increase by \$50.

18c) $C - 1000 = 50(x - 10)$

18d) $C = 50x + 500$, \$1250

18e) \$500

19a)

i)

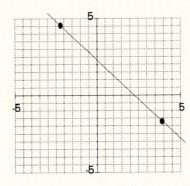

ii) $m = -1.02$

iii) $y = -1.02x + 2.29$

Section 2.5

2)

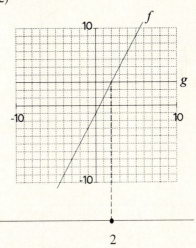

4a) 5   4b) $-\dfrac{13}{3}$

6)

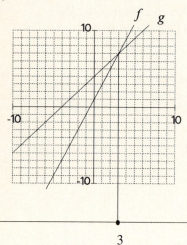

3

8b) $-\dfrac{1}{2}$ 8c) $\dfrac{11}{8}$

8f) 2   8g) $\dfrac{17}{3}$

10)

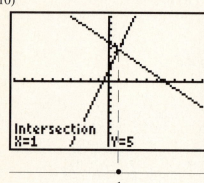

1

12b) $-\dfrac{25}{8}$ 12c) $\dfrac{24}{17}$

12f) $-\dfrac{15}{22}$

15)

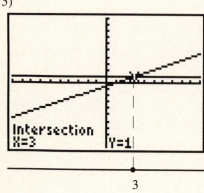

3

16b) $\dfrac{10}{3}$ 16d) $\dfrac{5}{7}$

18a) 2.20   18d) 1.19
18e) 0.03

20a) $R = 50 + 0.10x$
20b) $H = 40 + 0.29x$
20c) 52.6 miles

22a) $D = 150 - 20.3t$
22bcd)

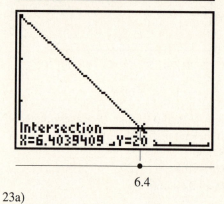

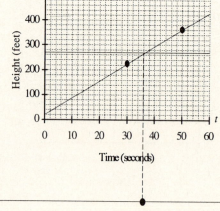

6.4

23a)

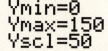

36

23b) 6.8 feet/second
23c) 36 seconds
23d) $H = 6.8t + 20$
23e) $t = 36.8$ seconds

Section 2.6

4a) $\left\{x : x < 8\right\}$ 4d) $\left\{x : x \geq -3\right\}$

6)

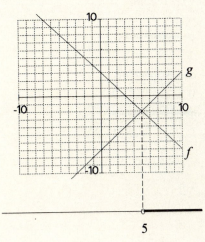

8c) $\left\{x : x < -\dfrac{1}{2}\right\}$ 8f) $\left\{x : x \geq \dfrac{5}{3}\right\}$

8g) $\left\{x : x \geq \dfrac{5}{2}\right\}$ 8h) $\left\{x : x \leq 1\right\}$

12a) $\left\{x : x > -\dfrac{48}{17}\right\}$ 12d) $\left\{x : x \geq -\dfrac{26}{15}\right\}$

12e) $\left\{x : x \geq \dfrac{1}{24}\right\}$

14a) $\left\{x : x > \dfrac{5}{2}\right\}$ 14c) $\left\{x : x \leq \dfrac{67}{5}\right\}$

14e) $\left\{x : x < 1\right\}$

15a)

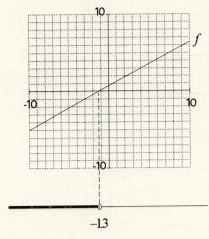

15b) $f(x) = \dfrac{4}{7}x + \dfrac{5}{7}$

15c) $\left\{x : x < -\dfrac{5}{4}\right\}$

18) $\left\{x : x > 0 \text{ and } x < 20\right\}$

19ab)

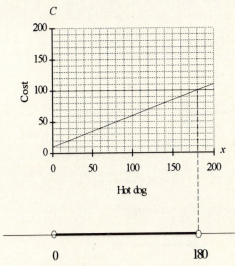

19c) $C = 10 + 0.50x$
19d) $\left\{x : x > 0 \text{ and } x < 180\right\}$

21) $\left\{t : t < \dfrac{50}{3}\right\}$. She must keep her playing time under 16 hours, 40 minutes.

Section 2.7

3a) ———————○━━━━━━━━━
                4

3d) ━━━━━━●———————————
                4

3g) ━━━━━━━━━○———————
                   −7

3h) ———————●━━━━━━━━━━
                −5

4a) $(-\infty,4) = \{x: x < 4\}$

4b) $[-3,+\infty) = \{x: x \geq -3\}$

5a) ━━━━━━━━○———————
                4

5b) ━━━━━━━━○———————
                8

5e) ———————●━━━━━━━━━
                4

5f) ———————●━━━━━━━━━
                2

6c) $x > 5$ and $x < 1$

7a) $(2,4) = \{x: 2 < x < 4\}$

7b) $(-\infty,2) \cup (4,+\infty) = \{x: x < 2 \text{ or } x > 4\}$

7e) $(-5,8] = \{x: -5 < x \leq 8\}$

8a) ━━━━━○————————○━━━━━
          4   7

$(-\infty,4) \cup (7,+\infty)$

8b) ———————————————————

The solution set is empty.

8e) ———————●━━━━━━━━○———
          −1   8

$(-1,8)$

8f) ━━━━━━━━━━━━━━━━━━━

The entire line is shaded. $(-\infty,+\infty)$

9d) ———————●━━━━━━━━○———
          −3   8

10c) $[-8,-5) = \{x: -8 \leq x < -5\}$

11c) $(-\infty,-1) \cup [9,+\infty)$

12b) $(-\infty,-8) \cup [-5,9)$
$= \{x: x < -8 \text{ or } -5 \leq x < 9\}$

13a) ━━━━━○———————○━━━━●———
          0   5   7

13b) ———————————————————

The solution set is empty.

13d) ———————○━━━━━━━━○———
          2   3

14a) ———————○━━━━━━━━○———
          0   9

14d) ———————●━━━━━━━━○———
          5   11

Section 2.8

3)

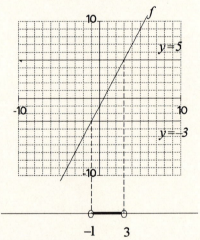

$(-1,3) = \{x: -1 < x < 3\}$

5a) $(0,8] = \{x: 0 < x \leq 8\}$

5d) $(-3,2] = \{x: -3 < x \leq 2\}$

5e) $(-4,-1) = \{x: -4 < x < -1\}$

5h) $\left[-\dfrac{6}{5}, \dfrac{8}{5}\right] = \left\{x: -\dfrac{6}{5} \leq x \leq \dfrac{8}{5}\right\}$

7)

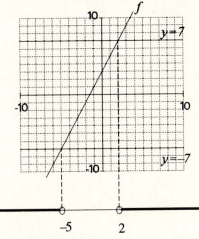

$$\left(-\infty,-3\right]\cup\left[\frac{7}{3},+\infty\right)$$

9a)
$$=\left\{x:x\le-3 \ \text{ or } \ x\ge\frac{7}{3}\right\}$$

11a) $\left(-\dfrac{2}{3},\dfrac{11}{3}\right]=\left\{x:-\dfrac{2}{3}<x\le\dfrac{11}{3}\right\}$

11d)
$$\left(-\infty,-3\right)\cup(5,+\infty)$$
$$=\left\{x:x<-3 \ \text{ or } \ x>5\right\}$$

11e) $\left[\dfrac{9}{5},3\right]=\left\{x:\dfrac{9}{5}\le x\le 3\right\}$

13b) $\left[\dfrac{7}{4},\dfrac{23}{4}\right]=\left\{x:\dfrac{7}{4}\le x\le\dfrac{23}{4}\right\}$

13c) $\left(\dfrac{5}{6},\dfrac{25}{6}\right)=\left\{x:\dfrac{5}{6}<x<\dfrac{25}{6}\right\}$

14a) $(-0.24,1.39)=\left\{x:-0.24<x<1.39\right\}$

14d) $(9.42,13.25)=\left\{x:9.42<x<13.25\right\}$

14f)
$$\left(-\infty,-1.29\right)\cup(3.55,+\infty)$$
$$=\left\{x:x<-1.29 \ \text{ or } \ x>3.55\right\}$$

16)

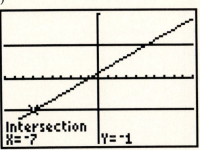

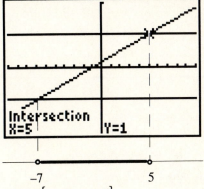

$(-7,5)=\left\{x:-7<x<5\right\}$

17b) $\left[-\dfrac{13}{11},\dfrac{3}{11}\right]=\left\{x:-\dfrac{13}{11}\le x\le\dfrac{3}{11}\right\}$

17c) $(-43,-1)=\left\{x:-43<x<-1\right\}$

20)

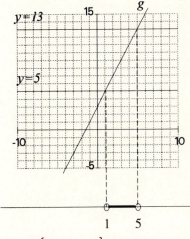

$(1,5)=\left\{x:1<x<5\right\}$

24)

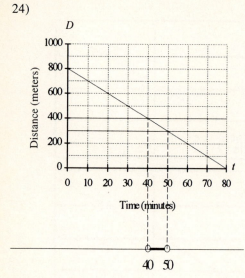

$(40,50) = \{t : 40 < t < 50\}$

Section 2.9

1a) $|-4| = 4$ 1b) $|-1| = 1$
1c) $|5| = 5$

3a) If $x > 0$, then $|x| = x$.
3b) if $x < 0$, then $|x| = -x$.

5)

x	-12	#	-8	-6	#	#	0	2	4	6	8
y	5	4	3	2	1	0	1	2	3	4	5

5a)

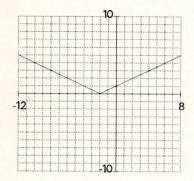

5b) The slope of the right side of the V is $\dfrac{1}{2}$.

5b) The slope of the left side of the V is $-\dfrac{1}{2}$.

7c)

x	-2	-1	0	1	2	3	4	5	6	7	8
y	15	12	9	6	3	0	3	6	9	12	15

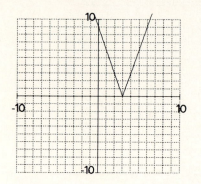

The slope of the right side of the V is 3.
The slope of the left side of the V is –3.

11a) $(-9,9) = \{x : -9 < x < 9\}$
11d) $[-4,4] = \{x : -4 \leq x \leq 4\}$
11f) $[-15,7] = \{x : -15 \leq x \leq 7\}$

12b) $\begin{aligned}&(-\infty,-1] \cup [1,+\infty) \\ &= \{x : x \leq -1 \ \text{or} \ x \geq 1\}\end{aligned}$

12e) $\begin{aligned}&(-\infty,-3) \cup (19,+\infty) \\ &= \{x : x < -3 \ \text{or} \ x > 19\}\end{aligned}$

14)

x	-6	-5	-4	-3	-2	-1	0	1	2	3	4
y	5	4	3	2	1	0	1	2	3	4	5

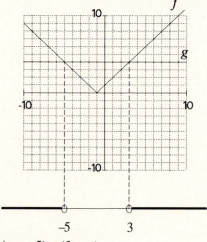

$(-\infty,-5) \cup (3,+\infty)$
$= \{x : x < -5 \ \text{or} \ x > 3\}$

16a) $(-2,3) = \{x: -2 < x < 3\}$

16c) $(-\infty, -6] \cup [16, +\infty)$
$= \{x: x \le -6 \text{ or } x \ge 16\}$

16e) $\left[-\dfrac{18}{5}, -\dfrac{4}{5}\right] = \left\{x: -\dfrac{18}{5} \le x \le -\dfrac{4}{5}\right\}$

19a) $(-\infty, -6) \cup (10, +\infty)$
$= \{x: x < -6 \text{ or } x > 10\}$

19c) $\left(-\dfrac{4}{9}, \dfrac{8}{9}\right) = \left\{x: -\dfrac{4}{9} < x < \dfrac{8}{9}\right\}$

19e) $\left[\dfrac{28}{45}, \dfrac{52}{45}\right] = \left\{x: \dfrac{28}{45} \le x \le \dfrac{52}{45}\right\}$

20a) $(-1.04, 2.26) = \{x: -1.04 < x < 2.26\}$

20d) $(-\infty, -0.66] \cup [1.11, +\infty)$
$= \{x: x \le -0.66 \text{ or } x \ge 1.11\}$

21a)
$(3.9999, 4.0001) = \{x: 3.9999 < x < 4.0001\}$

21c)
$(3.9995, 4.0005) = \{x: 3.9995 < x < 4.0005\}$

24b) $\left[-\dfrac{1}{4}, \dfrac{3}{2}\right] = \left\{x: -\dfrac{1}{4} \le x \le \dfrac{3}{2}\right\}$

24c) $(-\infty, -7] \cup [13, +\infty)$
$= \{x: x \le -7 \text{ or } x \ge 13\}$

25f)

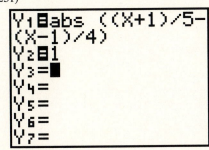

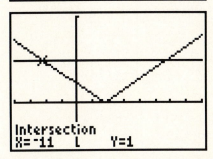

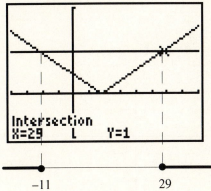

$(-\infty, -11] \cup [29, +\infty)$
$= \{x: x \le -11 \text{ or } x \ge 29\}$

Chapter 2 Review Exercises

2a) 200 feet

2b) 10 feet/minute

2c) $D = 200 - 10t$, 115 feet

5)

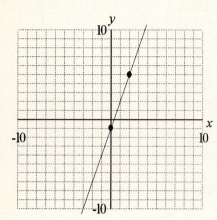

5a) $f(x) = 3x - 1$

5b) $f(10) = 29$

8)

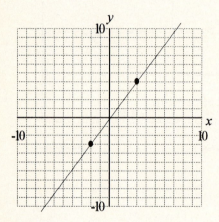

8a) $m = \dfrac{7}{5}$

8b) $f(x) = \dfrac{7}{5}x - \dfrac{1}{5}$

8c) $f(10) = 13.8$

9)

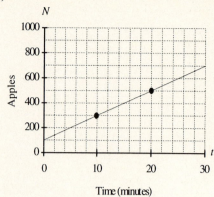

9b) 20 apples/minute

9c) $N - 300 = 20(t - 10)$

9d) $N = 20t + 100$, 100 apples

10f) $-\dfrac{17}{10}$ 10g) $\dfrac{31}{20}$

10i) 2   10j) $-\dfrac{253}{7}$

11d)

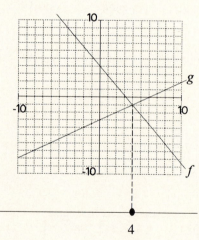

13d) $\left(\dfrac{1}{7}, +\infty\right) = \left\{x : x > \dfrac{1}{7}\right\}$

13e) $\left(-\infty, -\dfrac{31}{2}\right] = \left\{x : x \le -\dfrac{31}{2}\right\}$

13g) $\left(\dfrac{17}{44}, +\infty\right) = \left\{x : x > \dfrac{17}{44}\right\}$

16ab)

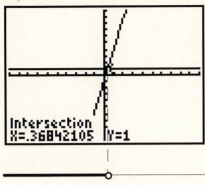

0.36842105

16c) $\left(-\infty,\dfrac{7}{19}\right)=\left\{x:x<\dfrac{7}{19}\right\}$

20a) $(43,82)=\left\{x:43<x<82\right\}$

20b) $\left(\dfrac{65}{3},\dfrac{115}{3}\right)=\left\{x:\dfrac{65}{3}<x<\dfrac{115}{3}\right\}$

22d) $\left(-\infty,-\dfrac{2}{5}\right]\cup[4,+\infty)$
$=\left\{x:x\le-\dfrac{2}{5}\ \text{ or }\ x\ge4\right\}$

22g) $\left(-\dfrac{7}{2},1\right)=\left\{x:-\dfrac{7}{2}<x<1\right\}$

23)

x	1	2	3	4	5	6	7	8	9	10	11	9	10	11
y	15	12	9	6	3	0	3	6	9	12	15	9	12	15

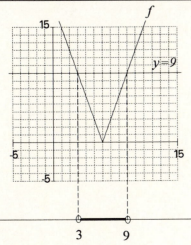

$(3,9)=\left\{x:3<x<9\right\}$

Chapter 3

Section 3.1

1a)

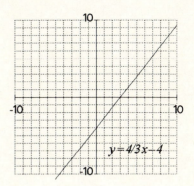

1b)

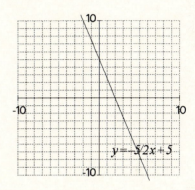

1f)

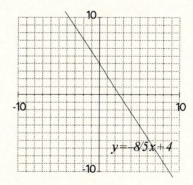

2b)

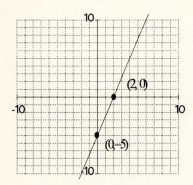

2c)

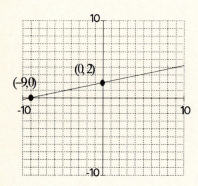

2f)

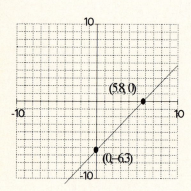

5a)

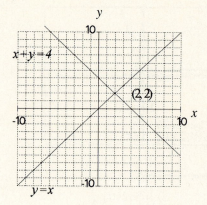

5d)

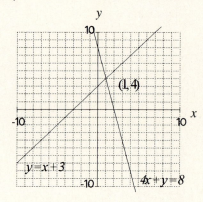

5e)

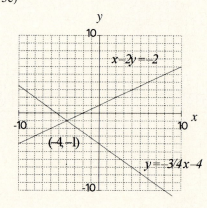

7a)
$$L + S = 7$$
$$L = 1 + 2S$$

7b)

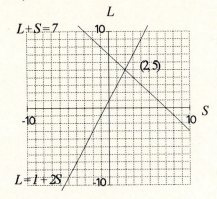

9a)
$$N + Q = 14$$
$$5N + 25Q = 170$$
9b)

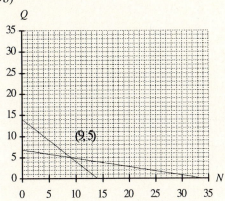

12a) $y = -\dfrac{2}{3}x + 2$

12bc)

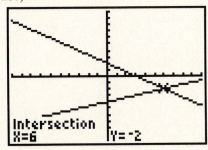

14)
$$L - S = 9$$
$$L = 2S$$

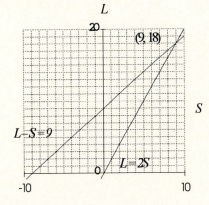

16)
$$2L + 2W = 26$$
$$L = 1 + 2W$$

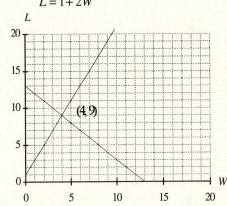

19)

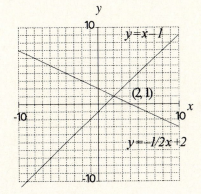

Section 3.2

1a)

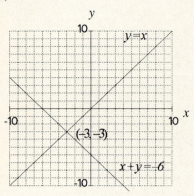

2a)

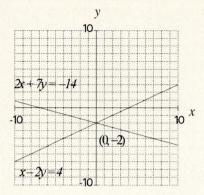

1d)

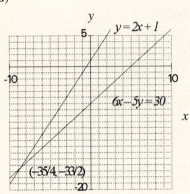

2d)

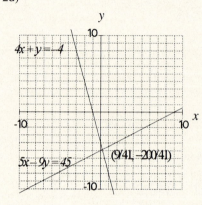

1e)

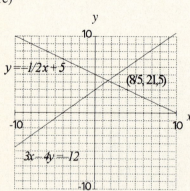

2e)

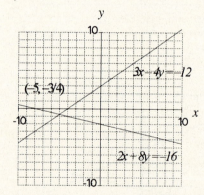

4abc)

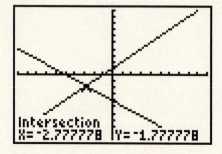

4d) $\left(-\dfrac{25}{9},-\dfrac{16}{9}\right)$

7)
$$N+Q=50$$
$$5N+25Q=650$$

7ab)

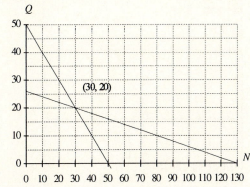

10a)
$\left(\dfrac{91}{86},-\dfrac{36}{43}\right)$

10d)
$\left(\dfrac{95}{44},-\dfrac{20}{11}\right)$

11a)
$(3.03,1.15)$

11d)
$(27.66,-8.94)$

14a) $L=1+2W$
14b) $2L+2W=20$
14c) $L=7,W=3$

16)

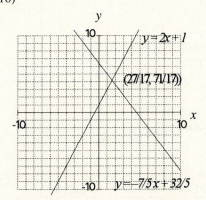

19)
$$Q=1+2D$$
$$25Q+10D=925$$
$$D=15,Q=31$$

Section 3.3

2a)

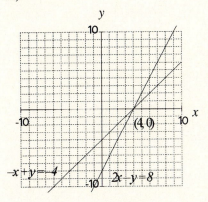

2d)

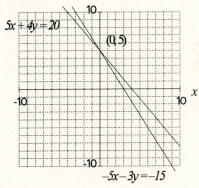

3a)

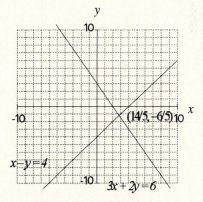

3d)

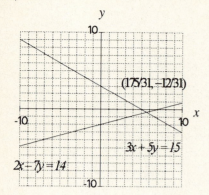

3e)

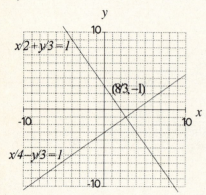

4a)

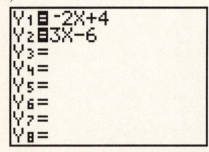

4d)

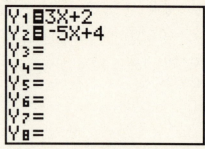

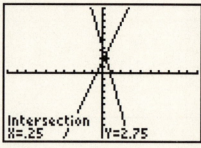

$$\left(\frac{1}{4}, \frac{11}{4}\right)$$

8a) $\left(\dfrac{130}{61}, -\dfrac{6}{61}\right)$ 8d) $\left(\dfrac{184}{79}, \dfrac{34}{79}\right)$

10a)

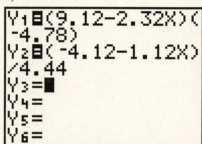

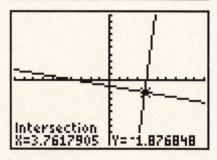

16)
$$C + R = 200$$
$$15C + 12R = 2760$$

$$C = 120, R = 80$$

19)
$$t_1 + t_2 = 110$$
$$10t_1 + 7t_2 = 800$$

$$t_1 = 10, t_2 = 100$$

20a) $2x + y = 1$

20b) $x - y = 2$

Chapter 3 Review Exercises

3a)

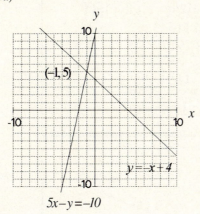

3d)

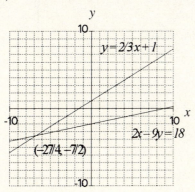

3e)

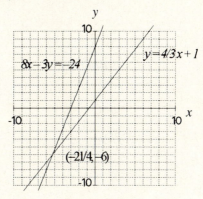

4a)

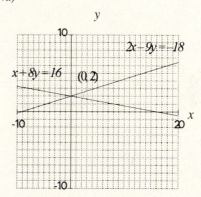

4d)

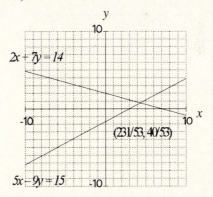

4f)

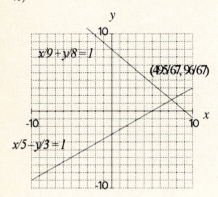

7a) $\left(\dfrac{155}{167}, -\dfrac{245}{167} \right)$

9) $\begin{aligned} 2S + 3L &= 24 \\ L &= 2S \end{aligned}$

$L = 6,\ S = 3$

12) $\begin{aligned} P + T &= 54 \\ P &= T - 3 \end{aligned}$

$P = 35,\ T = 19$

13) $\begin{aligned} S + B &= 210{,}000 \\ B &= 10{,}000 + 4S \end{aligned}$

```
Y₁■210000-X
Y₂■10000+4X
Y₃=
Y₄=
Y₅=
Y₆=
Y₇=
Y₈=
```

```
WINDOW FORMAT
Xmin=0
Xmax=210000
Xscl=50000
Ymin=0
Ymax=210000
Yscl=50000
```

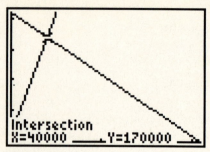

```
Intersection
X=40000      Y=170000
```

15a) $C = 2.5x + 500$

15b) $R = 4.75x$

15c)

```
WINDOW FORMAT
Xmin=0
Xmax=500
Xscl=100
Ymin=0
Ymax=2000
Yscl=500
```

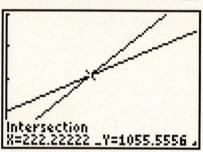

```
Intersection
X=222.22222  Y=1055.5556
```

17) $\begin{aligned} x + y &= 200 \\ 0.10x + 0.20y &= 0.125(200) \end{aligned}$

$x = 150,\ y = 50$

18) $\begin{aligned} t_1 + t_2 &= 2.5 \\ 2t_1 + 3t_2 &= 6 \end{aligned}$

$t_1 = 1.5,\ t_2 = 1$

22) $\begin{aligned} R &= E - 4 \\ R + E &= 158 \end{aligned}$

$E = 81,\ R = 77$

23) $\begin{aligned} C + G &= 1000 \\ 0.10C + 0.08G &= 0.09(1000) \end{aligned}$

$C = 500,\ G = 500$

Chapter 4

Section 4.1

1a) 4,096   1c) 65,536
1d) −11,390,625

2a) 1   2d) Undefined

4a) 256   4b) 16
4c) 256   4d) 16

6b) 5,764,801   6d) 5,038,848

7b) y^{10} 7d) $6x^7 y^{11}$
7f) $-10x^{8n}$

8b) y^{40} 8d) x^{6a}

9b) a^7 9d) $\dfrac{a^2 b^3}{c^2}$

9f) $\dfrac{1}{x^{5a}}$

10b) $a^4 b^4$ 10d) $81y^4$
10f) $125a^3 b^3 c^3$

11b) $\dfrac{x^4}{81}$ 11d) $\dfrac{y^{3b}}{z^{3b}}$

12b) $81y^{12}$ 12d) $\dfrac{x^4 y^{10}}{16}$
12f) $-72a^{23}b^{16}$ 12h) $27x^{3n}y^{6n}$
12j) $9x^{4n}$ 12l) y^{5n^3}

14b) $-20y$ 14d) Already simplified
14f) $-5x^{3n}$ 14h) $-22x^{2a}$

15b) $x^3 + 8x^2 + 4x - 21$
15d) $-5x^{2n} + 5x^n - 4$

16b) $-3 - 10y^2 - 8y^3 + 2y^4$
16d) $-16 - 4y^a - y^{2a} - 3y^{3a}$

17b) $6x^3 + 5x^2 y + 2xy^2 - 9y^3$

18b) $2x^5$ 18d) Already simplified

18f) $2x^n$ 18h) $-x^{3n}$

19b) $-54x^{7n+1}$ 19d) x^{3n^2+6n+4}

20bd)

```
250*0.98^20
         166.9019929
900*(1-0.05/6)^1
2
         814.0125367
```

21) 5.2 feet

24) $14,166.54

25) $65,966.49

Section 4.2

2b) 1   2d) 3

3b) $p(x) = 3 - 4x - 2x^2 + x^3$

4b) $p(x) = x^3 - 3x^2 + 4x + 2$

5b) $x^2 + 7x - 15$
5d) $3x^3 + x^2 + 4x - 17$

6b)
$p(x) + q(x) = -x^2 - 2x + 2$
$p(x) - q(x) = x^2 + 6x - 8$
6d)
$p(x) = q(x) = -10x^2 + 4x + 6$
$p(x) - q(x) = 2x^3 - 6x^2 - 6x - 8$

7a) $6x^3 - 12x^2 + 4x - 7$
7c) $-x^3 - 23x^2 + 5x + 12$

8b) $-2x^2 - 8xy - 13y^2$
8d) $-x^4 - 6x^3 y - 7x^2 y^2 + 7xy^3 - 7y^4$

9b) $\left(-\infty, -\dfrac{20}{13}\right] = \left\{x : x \le -\dfrac{20}{13}\right\}$

10b) $(-\infty,-2)\cup(0,+\infty)$
$=\{x:x<-2 \text{ or } x>0\}$

11b) $2x^2+5x$
11d) $x^4-2x^3-5x^2$
11f) $6x^6-9x^4-27x^3$
11h) $x^{4n}-3x^{3n}-7x^{2n}$

12b) $6x^2-x-35$
12d) $x^2-6x-16$
12f) $6x^3-17x^2+14x-3$
12h) $2x^4+7x^3+9x^2-11x-7$
12j) $2x^3-17x^2+41x-30$
12l) $x^6-2x^4-2x^3+2x+1$

14b) $-x^2+9x+9$
14d) x^2+2x+1
14f) $x^3-2x^2+24x+24$
14h) $x^4-16x^3+48x^2+128x+64$

15b) $x^{4n}-x^{3n}-6x^{2n}+11x^n+5$
15d) $x^{4n}-y^{4n}$

17) 217.5 seconds

19a) $A=x^2-(x-6)^2$
19b) $A=12x-36$
12c) 8 inches

22a) 41, 43, 47, 53, 61, 71
22b) No, try $p(41)$.

Section 4.3

2b) $5(2x+3)$
2d) $8(y^2-2y+3)$
2f) $4x(3x+4)$
2h) $4z(2z^2+z-2)$
2j) $3x^n(x^{2n}+2x^n-3)$

4b) $(8y-9)(2y-3)$
4d) $(3z-1)(5z-1)$

6b) $(y+6)(3y+4)$
6d) $(z-4)(3z+2)$
6f) $(2x+7)(8x+3)$
6h) $(3x+5)(4x+1)$

8b) $x=11,12$
8d) $x=-\dfrac{7}{4},-\dfrac{11}{7}$

9)

x	-2	-1	0	1	2	3	4	5	6
y	12	5	0	-3	-4	-3	0	5	12

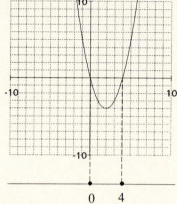

11b) $x=0,14$
11d) $x=0,-\dfrac{8}{5}$
11f) $x=0,-\dfrac{3}{2}$

14b) $x=-11,7$
14d) $x=\dfrac{7}{9},-\dfrac{8}{5}$

15)

x	-2	-1	0	1	2	3	4	5	6	7
y	21	6	-5	-12	-15	-14	-9	0	13	30

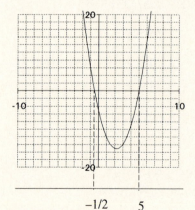

17b) $x=-4,\dfrac{7}{2}$
17d) $x=\dfrac{2}{3},-\dfrac{11}{5}$

19b) $6z(2z-3)$

19d) $3a^2b^2(2a+3b)$

19f) $10y(2y^2-4y+3)$

19h) $y^{2b}(9y^b-2)$

19j) $2x^{2n}(3x^{2n}-x^n+7)$

19l) $(21x-4)(3x+11)$

19n) $(4x-3)(5x+2)$

19p) $(6z^a+5)(3z^a-1)$

Section 4.4

1b) $(x-8)(x+3)$

1d) $(x+2)(3x+8)$

1f) $(x+4)(x+12)$

1h) $(x-4)(2x-9)$

1j) $(2x^n-3)(2x^n+5)$

3b) $(x-12)(x+3)$

3d) $(2x+1)(5x-2)$

3f) Does not factor.

3h) $(2x+3)(8x-1)$

3j) $(3x^a-2)(5x^a-1)$

5)

x	-5	-4	-3	-2	-1	0	1	2	3	4
y	35	18	5	-4	-9	-10	-7	0	11	26

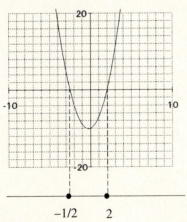

6b) $x=-\dfrac{5}{2},9$

6d) $x=-2,10$

6f) $x=\dfrac{2}{5},-\dfrac{2}{3}$

7b)

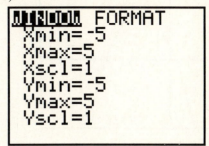

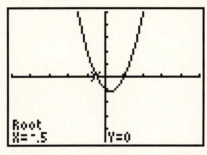

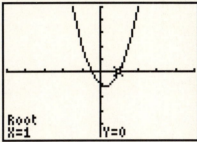

$x=-\dfrac{1}{2},1$

7f)

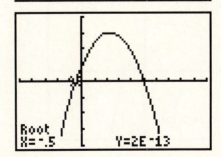

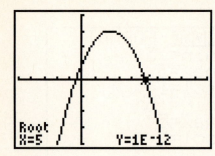

$x = -\dfrac{1}{2}, 5$

8b) $x = -1, \dfrac{9}{2}$

8d) $x = -\dfrac{1}{2}, 7$

8f) $x = 8, -10$

9b) $x = 1, 3$

9d) $x = 3, 6$

9f) $x = -5, 12$

9h) $x = -\dfrac{1}{2}, -\dfrac{4}{3}$

11a) $b = 4 + h$

11b) $A = \dfrac{1}{2} h (4 + h)$

11c) $h = 2$

11d) base = 6 inches, height = 2 inches

14b)

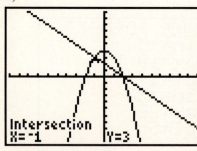

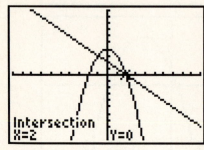

$x = -1, 2$

14d)

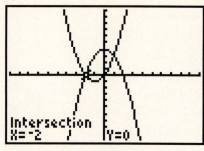

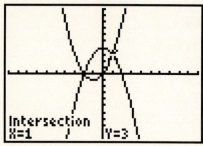

$x = -2, 1$

15b) $x = -2, 5$

15c) $x = -\dfrac{1}{4}, \dfrac{2}{3}$

17a) $A = 20 - (5 - 2x)(4 - 2x)$

17b) $x = \dfrac{1}{2}, 4$

17c) The maximum possible width of the border is 2, so the solution 4 must be discarded.

19a) $R = xp$

19b) $R = (54 - p)p$

19c) $p = 4, 50$

19d) Both prices work equally well.

Section 4.5

1b) $81 - x^2$

1d) $225a^2 - 289b^2$

1f) $x^6 - 1$

1h) $x^{10} - 4$

1j) $x^{2a} - y^{2a}$

1l) $49 - 4x^8$

2b) $(2x + 1)(2x - 1)$

2d) $(6x + 11y)(6x - 11y)$

2f) $(y^b + z^b)(y^b - z^b)$

2h) $(10xy + 7z^2)(10xy - 7z^2)$

2j) $(z^2 + 9)(z + 3)(z - 3)$

2l) $(x^{4n}+1)(x^{2n}+1)(x^n+1)(x^n-1)$

4b) $x=-\dfrac{3}{2},\dfrac{3}{2}$

4d) $x=-\dfrac{6}{5},\dfrac{6}{5}$

6)

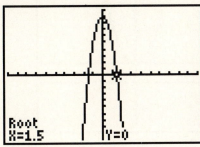

$x=-\dfrac{3}{2},\dfrac{3}{2}$

7b) $x^2-12xy+36y^2$

7d) $16x^4-72x^2+81$

7f) $x^{2n}-22x^n+121$

7h) $x^4y^4-6x^2y^2z^2+9z^4$

8b) $4x^2-12x+9$

8d) $x^{2n}+4x^n+4$

10b) $(x-6)^2$

10d) $(3x-4)(3x-1)$

10f) $(x^n-4)^2$

10h) $(2x^b-y^b)^2$

12d)

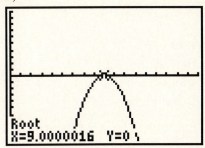

$x=9$ is a double root and the graph appears to be tangent to the x-axis at this double root.

13b) $x=-\dfrac{5}{4}$

13d) $x=\dfrac{1}{5},\dfrac{4}{5}$

13f) $x=\dfrac{1}{4},9$

13h) $x=-\dfrac{5}{12}$

14a) $A=xy$

14b)
$xy=12$
$3x+y=12$

14c) $x=2,\ y=6$

17a) $A=\pi(r+7)^2$

17b) $r=3$

Chapter 4 Review Exercises

1b) $9x^{6a}$ 1d) $-72x^{8n}$

1f) $\dfrac{9}{x^{4a}}$

2b) $2x^{3n}+3x^{2n}-14x^n+12$

3b) x^3-64

3d) $2x^3+5x^2-28x-15$

3f) $x^{2a}+4x^ay^a+4y^{2b}$

4b) $x^4-x^3-12x^2-7x+5$

4d) $x^3-14x-15$

4f) x^3-x^2+2x+2

4h) $x^4-4x^3-8x^2-x+1$

5b) $\left(-\infty,\dfrac{9}{2}\right)=\left\{x:x<\dfrac{9}{2}\right\}$

5d) $\left(-\infty, \dfrac{1}{5}\right) \cup \left(\dfrac{7}{5}, +\infty\right) = \left\{x : x < \dfrac{1}{5} \text{ or } x > \dfrac{7}{5}\right\}$

6b) $(2x-1)(x-8)$

6d) $(x-9)(x-3)$

6f) $(2x^a + 3)(3x^a - 8)$

6h) $(10x^a - 7y^b)^2$

7b) $x = \dfrac{1}{2}, \dfrac{11}{3}$

7d) $x = \dfrac{3}{5}, 12$

7f) $x = \dfrac{3}{10}$

10a) \$40   10b) $x = 3, 10$

10c) If Jerome makes either 3 handbags or 10 handbags, his costs will be \$50.

12a) $P = 2W + 2L$

12b) $28 = 2W + 2L$

12c) $A = LW$

12d) $45 = LW$

12e) $L = 5, W = 9$ or $L = 9, W = 5$

12f) The dimensions of the rectangle are 5 feet by 9 feet.

13) \$1,923.05

Chapter 5

Section 5.1

1c)

x	-3	-2	-1	0	1	2	3
y	81	16	1	0	1	16	81

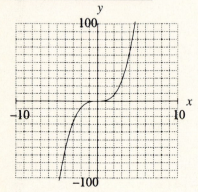

1d)

x	-3	-2	-1	0	1	2	3
y	-243	-32	-1	0	1	32	243

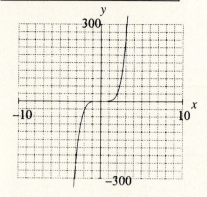

4b)

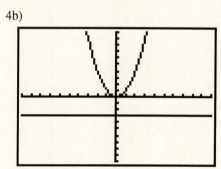

There are no real solutions.

4d)

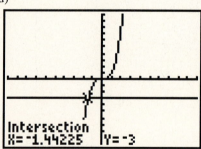

$x = \sqrt[3]{-3}$

6b) $x = \sqrt[3]{13}$

6d) $x = \sqrt[5]{-9}$

6e) No real solutions.

6g) $x = \pm\sqrt[8]{21}$

7bdf)

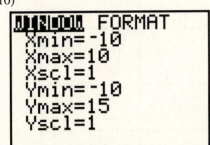

```
5×√20
          1.820564203
7×√-50
          -1.748678622
-8×√40
          -1.585833175
■
```

12bdf)

```
(-√17)²
                    17
(³√11)³
                    11
(³√-11)³
                   -11
```

10)

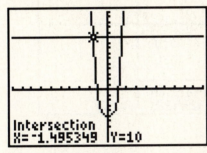

```
ZOOM  FORMAT
 Xmin=-10
 Xmax=10
 Xscl=1
 Ymin=-10
 Ymax=15
 Yscl=1
```

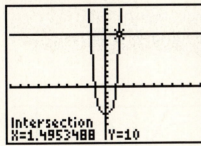

```
Intersection
X=-1.495349  Y=10
```

```
Intersection
X=1.4953488  Y=10
```

$x = \pm\sqrt[4]{15}$

11a) $x = \pm\sqrt{7}$

11b) $x = \sqrt[3]{2}$

11e) $x = \pm\sqrt[4]{9}$

17)

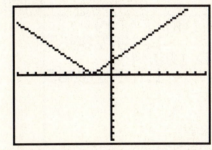

```
Y₁=6×√((X+2)^6)
Y₂=X+2
Y₃■abs (X+2)
Y₄=
Y₅=
Y₆=
Y₇=
Y₈=
```

The graph of $y = \sqrt[6]{(x+2)^6}$.

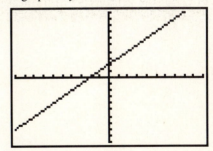

The graph of $y = x+2$.

The graph of $y = |x+2|$.

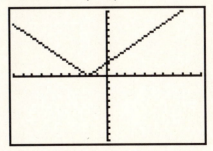

Therefore $\sqrt[6]{(x+2)^6} = |x+2|$.

18b)
$(-\infty, -10) \cup (10, +\infty) = \{x : x < -10 \text{ or } x > 10\}$

18c) $(-3, 7) = \{x : -3 < x < 7\}$

20a) 160 feet   20b) 96 feet

20c) $t = \sqrt{10} \approx 3.2$ seconds

Section 5.2

2b)

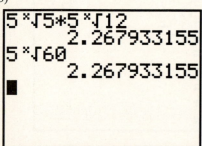

4d)

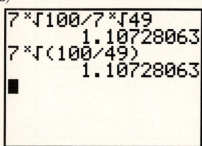

5b) $5\sqrt{3}$ 5d) $6\sqrt{2}$

5f) $3\sqrt[3]{3}$ 5h) $2\sqrt[5]{2}$

6b) $\dfrac{\sqrt{14}}{8}$ 6d) $\dfrac{\sqrt[3]{20}}{2}$

6f) $\dfrac{\sqrt[5]{2}}{2}$

7b) $\dfrac{5\sqrt{2}}{6}$ 7d) $3\sqrt[3]{4}$

7f) $5\sqrt[5]{8}$

8b)

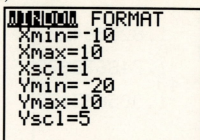

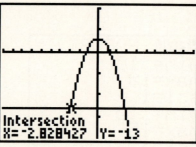

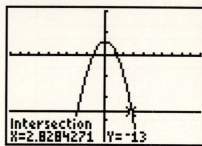

$x = \pm 2\sqrt{2}$

12a) If $x \geq 0$, then $\sqrt{25x^2} = 5x$.

12b) If $x < 0$, then $\sqrt{25x^2} = -5x$.

13)

The graph of $f(x) = \sqrt{9x^2}$.

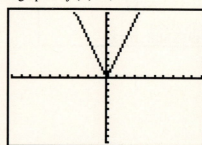

The graph of $g(x) = 3x$.

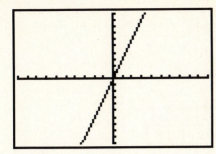

The graph of $h(x) = |3x|$.

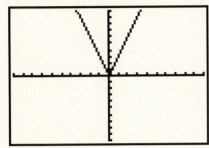

Therefore $\sqrt{9x^2} = |3x|$.

15b) $\{x : x \geq 0\}$ 15d) $\{x : x \neq 0\}$

16b) $3|x|\sqrt{3x}$, $\{x : x \geq 0\}$

16d) $6x^2\sqrt{2x}$, $\{x : x \geq 0\}$

16f) $\dfrac{2\sqrt{3x}}{x^2}$, $\{x : x > 0\}$

16h) $\dfrac{\sqrt{6x}}{2x^2}$, $\{x : x > 0\}$

16j) $|x - 5|\sqrt{x - 5}$, $\{x : x \geq 5\}$

19)
The graph of $f(x) = \sqrt[5]{x^5}$.

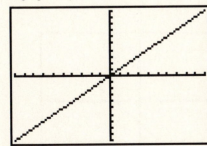

The graph of $g(x) = x$.

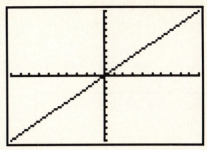

The graphs of f and g coincide for all values of x.
Therefore $\sqrt[5]{x^5} = x$.

Section 5.3

1b) $12\sqrt{7}$

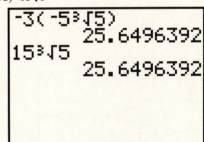

1d) $15\sqrt[3]{5}$

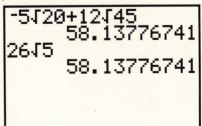

1f) 180   1h) -12

4b) 64

5b) $18\sqrt[3]{2}$ 5d) $8\sqrt[4]{4}$

6b) $-4\sqrt[3]{5}$ 6d) $3\sqrt[3]{2}$

6f) $26\sqrt{5}$

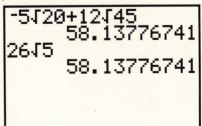

6h) $3\sqrt{6}-3\sqrt[3]{6}$ 6j) $\dfrac{\sqrt[3]{2}}{2}$

7b) $11\sqrt{6}-18$ 7d) $-\sqrt{10}-2\sqrt{6}+3\sqrt{15}+18$

7e) 3

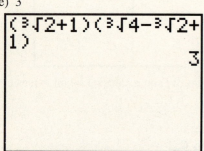

8b) $8-2\sqrt{15}$

8d) $25-30\sqrt[3]{2}+9\sqrt[3]{4}$

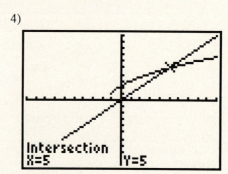

9b) 10

9d) $\sqrt[3]{25}-\sqrt[3]{4}$

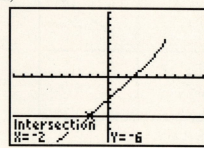

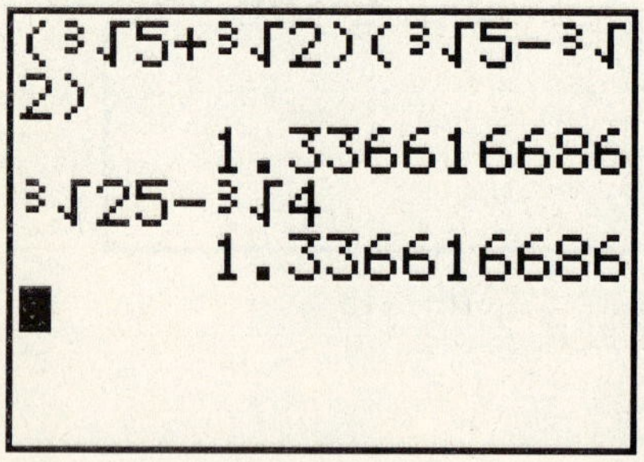

11b) $74+40\sqrt{3}\approx143.3$ square inches

11d) $37+16\sqrt{6}\approx76.2$ square inches

11f) $\pi(13+4\sqrt{10})\approx80.6$ square inches

13) $-2-\sqrt{2}\approx3.4142$

15b) $y-\sqrt{2}=\dfrac{\sqrt{2}}{4}(x-2)$

15c) $y=\dfrac{\sqrt{2}}{4}x+\dfrac{\sqrt{2}}{2}$

17) $\dfrac{2\pi}{g}\sqrt{Lg}$

Section 5.4

2)

x	-3	-2	1	6	13
y	0	1	2	3	4

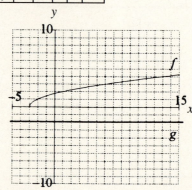

There are no solutions.

4)

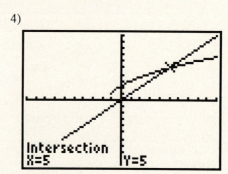

7b) $x=-12$

7d) No solutions.

7f) $x=11$

7h)

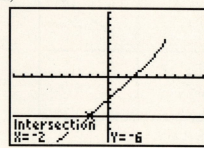

$x=-2$

7j)

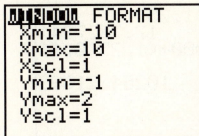

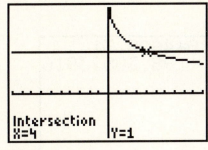

$x = 4$

7l)

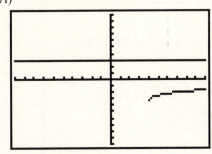

No solutions. *Author's note: Will the curve ever rise to meet the line $y = 3$? This is note easy to see graphically, although the analysis says that the graph will not rise to the line $y = 3$. Experiment to find how high the curve will rise.*

9a) $y = \dfrac{gt^2}{2}$

9b) 44,1 meters

11b) $m = \dfrac{k}{\omega^2}$

11d) $r = \dfrac{d^2 - h^2}{2h}$

Section 5.5

2b)

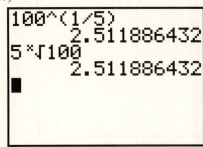

4)

The graph of $y = x^{\frac{1}{5}}$.

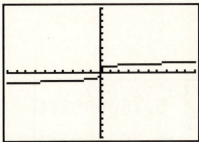

The graph of $y = \sqrt[5]{x}$.

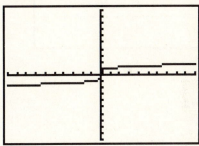

Therefore $x^{\frac{1}{5}} = \sqrt[5]{x}$.

5b) 6   5d) Undefined

5f) 2

8b)

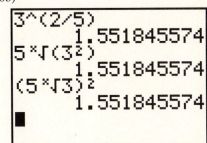

9b) 64   9d) 25

9f) −8

Author's note: The TI–82, like many other calcualtors, does not like raising a negative number to the 3/5 power. This will not be a problem as most of our work will deal with bases that are positive.

10b) $5\sqrt[3]{5}$

10d) $4\sqrt[3]{3}$

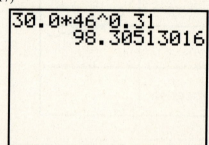

11b) $x\sqrt[3]{x}$

11d) $\dfrac{12\sqrt{x}}{x^2}$, $x > 0$

12b) $\sqrt[12]{177147}$ 12d) $\sqrt[12]{2}$

12f) x^2, $x \geq 0$

12h) $\dfrac{x^3}{y^4}$, $x \geq 0$, $y \neq 0$

13b) $(5-x)^{\frac{1}{3}}$

13d) $(1-x)^{\frac{4}{3}}$

14bd)

17)

![calculator screen showing: 30.0*46^0.31 98.30513016]

Chapter 5 Review Exercises

2d)

![calculator screen showing: ERR:DOMAIN 1:Goto 2:Quit]

5b) $3\sqrt{11}$

5d) $2\sqrt[4]{5}$

5f) $\dfrac{\sqrt{30}}{6}$

![calculator screen showing: √(5/6) .9128709292 √30/6 .9128709292]

5h) $\dfrac{\sqrt[4]{10}}{2}$

5j) $\sqrt{2}$

5l) $3\sqrt[4]{2}$

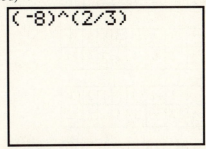

6b) $4|x|\sqrt{3}, \{x:x \in R\}$

6d) $2x^2\sqrt{11}, \{x:x \in R\}$

6f) $2x\sqrt[3]{11x}, \{x:x \in R\}$

6h) $\dfrac{\sqrt{3x}}{6x^2}, \{x:x > 0\}$

6j) $\dfrac{\sqrt[3]{9x^2}}{3x}, \{x:x \neq 0\}$

7a) $7 + 2\sqrt{10}$

7d) $5\sqrt{6} - 5\sqrt{5}$

7f) 1

7h) $-\dfrac{14 + 5\sqrt{10}}{3}$

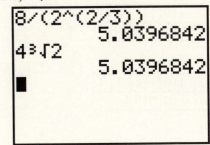

8) The equation of the line is $y - 3 = \sqrt{2}(x - 2)$. The y–intercept is $3 - 2\sqrt{2}$. The x–intercept is $\dfrac{4 - 3\sqrt{2}}{2}$.

12b) $d = \dfrac{I_0^2}{I^2}$

13b) 4

13d) 4

```
( -8)^(2/3)
```

```
ERR:DOMAIN
1:Goto
2:Quit
```

Author's note: The TI–82, like many calculators, does not like raising a negative base to the 2/3 power. This will not be a problem as most of our work will deal with positive bases.

14b) $2\sqrt[3]{9}$

14d) $4\sqrt[3]{2}$

```
8/(2^(2/3))
         5.0396842
4³√2
         5.0396842
```

14f) $2\sqrt{2}$

15b) $(x + 6)^{\frac{2}{3}}$

Chapter 6

Section 6.1

2)

x	-3	-2	-1	0	1	2	3
$y = x^2$	9	4	1	0	1	4	9

x	-3	-2	-1	0	1	2	3
$y = x^2 + 1$	10	5	2	1	2	5	10

x	-3	-2	-1	0	1	2	3
$y = x^2 + 2$	11	6	3	2	3	6	11

x	-3	-2	-1	0	1	2	3
$y = x^2 + 3$	12	7	4	3	4	7	12

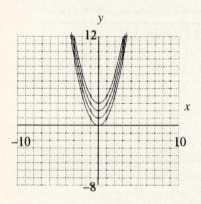

To draw the graph of $y = x^2 + c$ shift the graph of $y = x^2$ upward c units if c is positive.

7a)

x	-3	-2	-1	0	1	2	3	4
y	0	1	2	2	2	2	1	0

7b)

x	-3	-2	-1	0	1	2	3	4
y	-2	-1	0	0	0	0	-1	-2

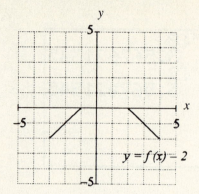

7c)

x	-1	0	1	2	3	4	5	6
y	0	1	2	2	2	2	1	0

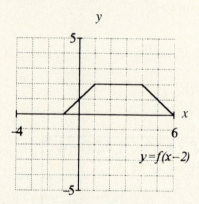

7d) To draw the graph of $y = f(x) - 2$ shift the graph of $y = f(x)$ downward 2 units. To draw the graph of $y = f(x - 2)$ shift the graph of $y = f(x)$ to the right 2 units.

9a)

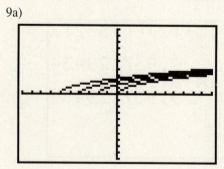

9b) To draw the graph of $y = \sqrt{x + 2}$ shift the graph of $y = \sqrt{x}$ to the left 2 units. A similar statement can be made about the remaining graphs.

11b) $y = |x| - 4$

11d) $y = x^3 + 3$

13a)

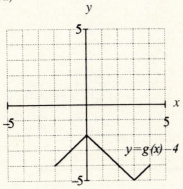

$y = g(x) - 4$

13b)

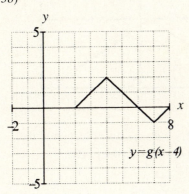

$y = g(x - 4)$

15a) $y = \sqrt{x+2} - 3$

15c) $y = (x-3)^2 + 1$

17a)

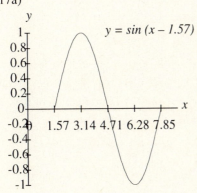

$y = \sin(x - 1.57)$

17c)

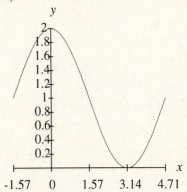

Section 6.2

3a)
The graph of $y = \sqrt{x}$.

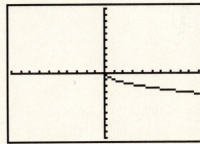

The graph of $y = -\sqrt{x}$.

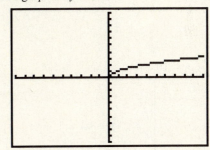

3b) The graph of $y = -\sqrt{x}$ is a reflection of the graph of $y = \sqrt{x}$ across the x–axis.

6a)

x	−2	−1	0	1	2	3	4
$f(x)$	0	2	0	2	4	2	0

6b)

x	−4	−3	−2	−1	0	1	2
$y = f(-x)$	0	2	4	2	0	2	0

6c)

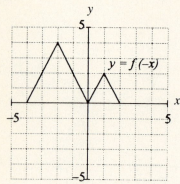

6d) The graph of $y = f(-x)$ is a reflection of the graph of $y = f(x)$ across the y–axis.

9b) $y = -\sqrt{x}$

9d) $y = \sqrt{x+4}$

11a)

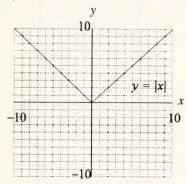

11b)

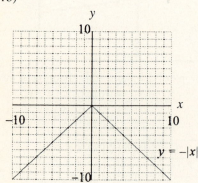

11c)

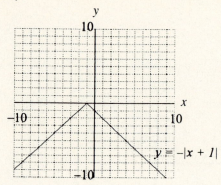

11d)

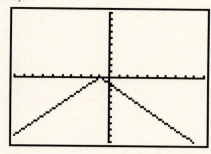

15b)
The graph of $y = h(-x)$ is a reflection of the graph of $y = h(x)$ across the y–axis.

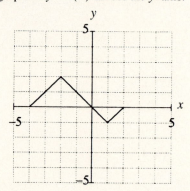

The graph of $y = -h(-x)$ is a reflection of the graph of $y = h(-x)$ across the x–axis.

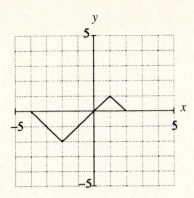

15d)
The graph of $y = -h(x)$ is a reflection of the
graph of $y = h(x)$ across the x–axis.

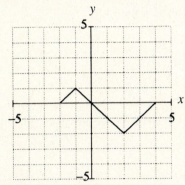

To draw the graph of $y = -h(x+2)-1$ shift the
graph of $y = -h(x)$ to the left 2 units and down
1 unit.

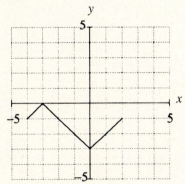

16b) $y = -(x+4)^2 + 5$

16d) $y = -(x-4)^3 - 3$

17b) The graph of $y = \sin(-x)$ is a reflection of
the graph of $y = \sin x$ across the y–axis.

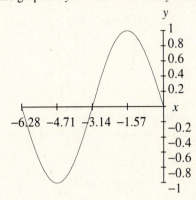

17d)
The graph of $y = -\sin x$ is a reflection of the
graph of $y = \sin x$ across the x–axis.

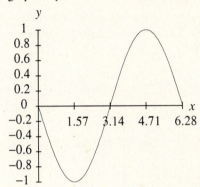

To draw the graph of $y = -\sin(x+1.57)$ shift
the graph of $y = -\sin x$ to the left 1.57 units.

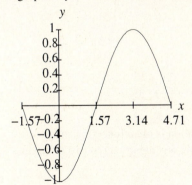

Section 6.3

1b)
i) Domain = {1, 2, 3}, Range = {0}
ii)

iii) This relation is a function because each domain object is sent to exactly one range object.

1d)
i) Domain = {1, 4}, Range = {0, 5}
ii)

iii) This relation is *not* a function because the domain object 1 is sent to *two* different range objects.

2a)

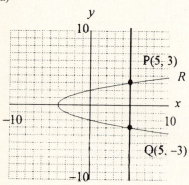

$$x = 5$$

2b)

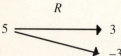

2c) The relation is *not* a function because the domain object 5 is being sent to two different range values.

4b) Not a function.
4d) Function.

5) Domain = $[0, +\infty) = \{x : x \geq 0\}$
Range = $[0, +\infty) = \{y : y \geq 0\}$

7) Domain = $(-\infty, +\infty) = \{x : x \in R\}$

Range = $[0, +\infty) = \{y : y \geq 0\}$

10a) The graph of $y = \sqrt{x}$.

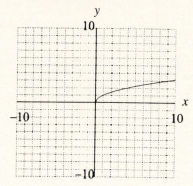

10b) The graph of $y = \sqrt{-x}$.

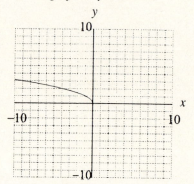

10c) The graph of $y = \sqrt{-(x-4)}$.

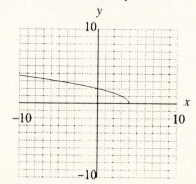

10d) g is a function because no vertical line cuts its graph more than once.

10e) Domain = $(-\infty, 4] = \{x : x \leq 4\}$
Range = $[0, +\infty) = \{y : y \geq 0\}$

15b) Domain = $(-\infty, 5] = \{x : x \leq 5\}$
15d) Domain = $(-\infty, +\infty) = \{x : x \in R\}$

15f) Domain $= (-\infty, 6] = \{x : x \le 6\}$

Section 6.4

1b)

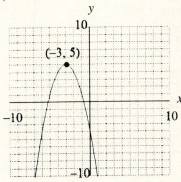

2c) $y = (x+5)^2 - 4$

2d) $y = -(x-3)^2 + 6$

3b) $x^2 - 6x + 9$ 3d) $x^2 + 7x + \dfrac{49}{4}$

5b) $y = (x+3)^2 - 27$

5d) $y = \left(x + \dfrac{5}{2}\right)^2 - \dfrac{49}{4}$

7b) $y = -(x+5)^2 + 36$

7d) $y = -\left(x + \dfrac{7}{2}\right)^2 + \dfrac{81}{4}$

8b) $y = (x+4)^2 - 25$

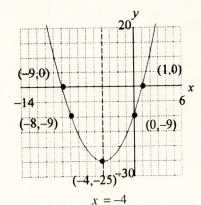

$x = -4$

8f) $y = -(x-5)^2 + 36$

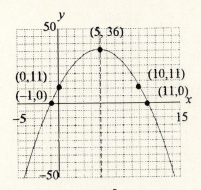

$x = 5$

8h) $y = -\left(x + \dfrac{9}{2}\right)^2 + \dfrac{49}{4}$

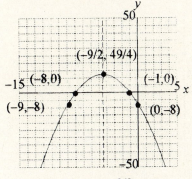

$x = -9/2$

11a)

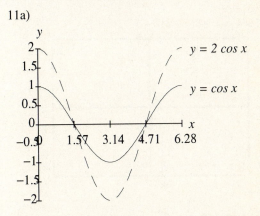

14b) $y = -3(x+2)^2 + 27$, Vertex $= (-2, 27)$

14d) $y = -\dfrac{1}{3}(x+3)^2 + 1$, Vertex $= (-3, 1)$

15b) $y = -3(x-1)^2 + 48$

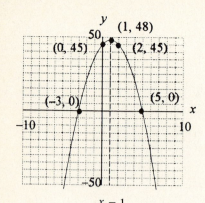

$$x = 1$$

15d) $y = -\dfrac{1}{3}(x-3)^2 + 12$

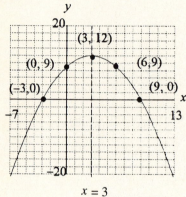

$$x = 3$$

16b) $y = -3(x+3)^2 + 5$

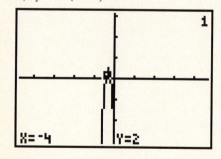

Section 6.5

1b) $y = -(x+4)^2 + 26$, Vertex $= (-4, 26)$

1c) $y = 2(x+3)^2 - 42$, Vertex $= (-3, -42)$

1d) $y = -\dfrac{1}{2}(x+3)^2 + \dfrac{19}{2}$, Vertex $= \left(-3, \dfrac{19}{2}\right)$

3b) $1, -6$ 3d) $-\dfrac{1}{3}, 1$

4b)

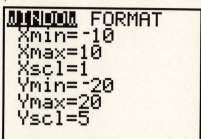

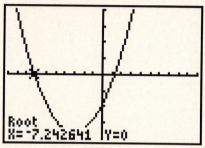

$$x = -3 \pm 3\sqrt{2}$$

4d)

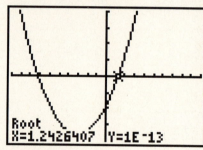

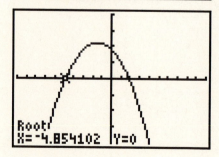

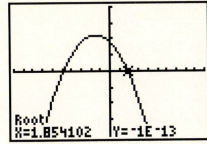

$$x = \frac{-3 \pm 3\sqrt{5}}{2}$$

4f)

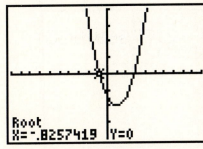

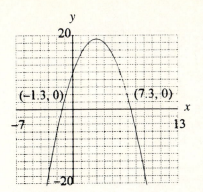

$$x = 3 \pm \sqrt{19}$$

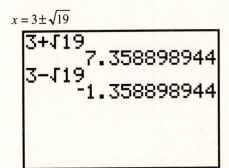

9a)

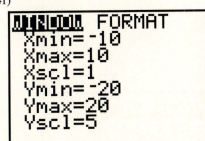

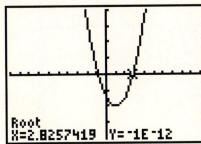

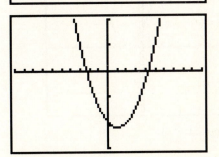

$$x = \frac{3 \pm \sqrt{30}}{3}$$

5b) $x = 1 \pm \sqrt{5}$ 5d) $x = 1 \pm \sqrt{3}$

Discriminant = 44

7)

x	−2	−1	0	1	2	3	4	5	6	7	8
y	−6	3	10	15	18	19	18	15	10	3	−6

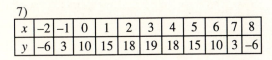

9b)

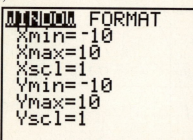

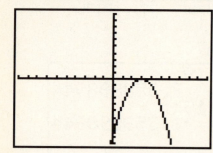

Discriminant = 0

9c)

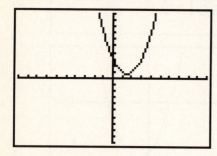

Discriminant = −3

12b) $\left\{k:k>-\dfrac{1}{2}\right\}$

13b) $\{k:k<-1\}$

15b) $x=\dfrac{4\pm\sqrt{10}}{2}$ 15d) $x=\dfrac{5\pm\sqrt{57}}{4}$

15f) $x=-2\pm\sqrt{6}$

17b) $x=\dfrac{3\pm\sqrt{105}}{8}$

17d) No real solutions.

18b) $x\approx-7.57,1.49$

Section 6.6

1a) 12 feet   1b) 36 feet

1cd)

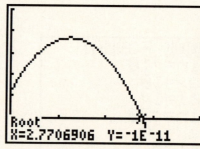

1e) $t=\dfrac{5\pm\sqrt{37}}{4}$

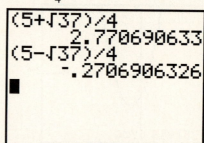

1f)

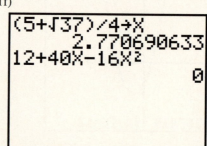

At approximately 2.77 seconds the height of the ball is zero.

2b) $x=6\sqrt{3}$ 2d) $x=8\sqrt{7}$

3a) $L=\sqrt{100-x^2}$

4c) $(50t)^2+(80t-80)^2=500^2$

$t=6$

$$W = \frac{-3 + \sqrt{809}}{4} \approx 6.4 \text{ feet}$$

8)
$$L = 2\left(\frac{-3 + \sqrt{809}}{4}\right) + 3 \approx 15.7 \text{ feet}$$

11a) $(x+4)^2 + x^2 = 40^2$

11b) $x = -2 + 2\sqrt{199} \approx 26.2$ inches

11c)

```
-2+2√199→X
          26.21347196
X²+(X+4)²
                  1600
40²
                  1600
```

14a) $R = (50 - p)p$

14b) $p = 20, 30$

14c) At $p = \$20$, the public will buy 30 calculators. At $p = \$30$, the public will buy 20 calculators.

Section 6.7

2)

x	−2	−1	0	1	2	3	4
y	−5	0	3	4	3	0	−5

2ab)

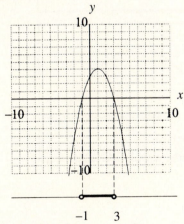

2c) $(-1, 3) = \{x : -1 < x < 3\}$

5b)

i) $y = (x-1)^2 - 9$

ii) y–intercept $= -8$

iii) x–intercepts $= -2, 4$

iv,v)

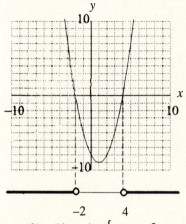

$(-\infty, -2) \cup (4, +\infty) = \{x : x < -2 \text{ or } x > 4\}$

5d)

i) $y = \frac{1}{2}(x-2)^2 - 18$

ii) y–intercept $= -16$

iii) x–intercepts $= -4, 8$

iv,v)

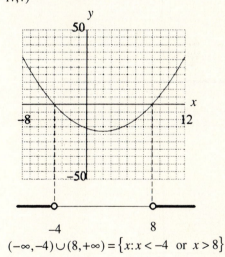

$(-\infty, -4) \cup (8, +\infty) = \{x : x < -4 \text{ or } x > 8\}$

7b)

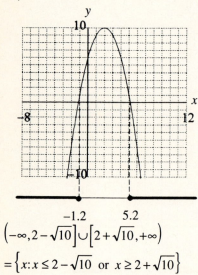

$$\left(-\infty, 2-\sqrt{10}\right] \cup \left[2+\sqrt{10}, +\infty\right)$$
$$=\left\{x : x \le 2-\sqrt{10} \ \text{ or } \ x \ge 2+\sqrt{10}\right\}$$

7d)

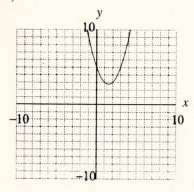

$$\left(-\infty, +\infty\right) = \left\{x : x \in R\right\}$$

7e)

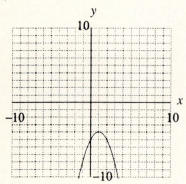

There are no solutions.

9b) $[-6,1] = \left\{x : -6 \le x \le 1\right\}$

9d)
$$\left[\frac{2-\sqrt{19}}{3}, \frac{2+\sqrt{19}}{3}\right] = \left\{x : \frac{2-\sqrt{19}}{3} \le x \le \frac{2+\sqrt{19}}{3}\right\}$$

10b)
$$\left(-\infty, -4\right] \cup \left[4, +\infty\right) = \left\{x : x \le -4 \ \text{ or } \ x \ge 4\right\}$$

10d)
$$\left(-\infty, -\frac{1}{2}\right] \cup \left[1, +\infty\right) = \left\{x : x \le -\frac{1}{2} \ \text{ or } \ x \ge 1\right\}$$

15)

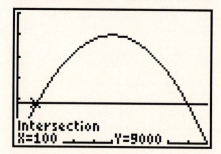

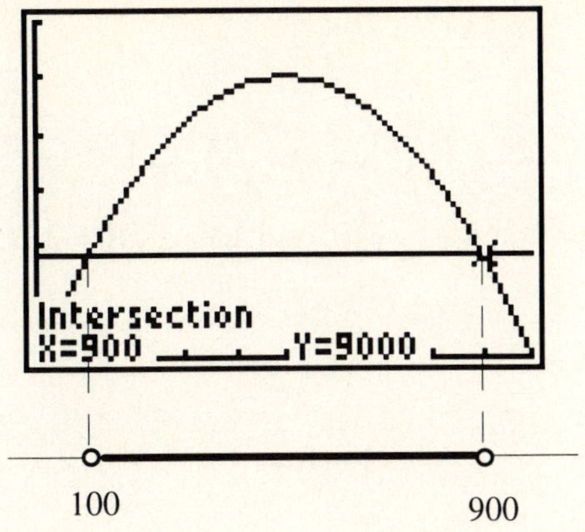

100   900

Set the unit price between \$100 and \$900.

Section 6.8

1b) Maximum value is 4, which occurs at $x = -1$.

2b) Minimum value is 1, which occurs at $x = -2$.

4)

x	-5	-4	-3	-2	-1	0	1
y	1	6	9	10	9	6	1

4a) The maximum function value in the table is 10, which occurs when x equals -2.

5b) The maximum function value is 8, which occurs when $x = 2$.

5d) The minimum function value is $-\dfrac{19}{8}$, which occurs when $x = \dfrac{3}{4}$.

5f) The maximum function value is 20,250,000, which occurs when $x = 45,000$.

8a)

x	y	P
0	12	0
1	11	11
2	10	20
3	9	27
4	8	32
5	7	35
6	6	36
7	5	35
8	4	32
9	3	27
10	2	20
11	1	11
12	0	0

The maximum product in the table is 36, which is attained when $x = 6$ and $y = 6$.

8b) $12 = x + y$

8c) $P = -x^2 + 12x$

8d) The maximum P–value is 36, which occurs when $x = 6$.

8e) The maximum product is 36, which occurs when $x = 6$ and $y = 6$.

12a) $S = x^2 + (100 - x)^2$

12b) The minimum S–value is 5000 which occurs when $x = 50$.

13a) $y = 30 + 256t - 16t^2$

13b) (8, 1054)

13c) The maximum height is 1054 feet which occurs at $t = 8$ seconds.

Chapter 6 Review Exercises

4b)

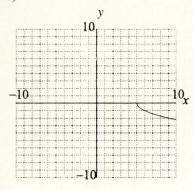

Domain $= [5, +\infty) = \{x : x \geq 5\}$
Range $= (-\infty, 0] = \{y : y \leq 0\}$

5b) Domain $= \left[-\dfrac{11}{2}, +\infty\right) = \left\{x : x \geq -\dfrac{11}{2}\right\}$

6b) $y = -(x+3)^2 + 49$

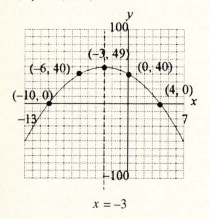

$$x = -3$$

7) $y = -3(x+2)^2 + 8$

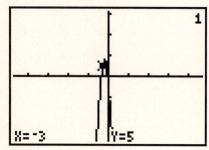

9) $k = \pm 4$

11b) $x = \dfrac{1 \pm \sqrt{17}}{2}$

13b) $\left(-\infty, -2-\sqrt{10}\,\right] \cup \left[-2+\sqrt{10}, +\infty\right)$
$= \left\{x : x \leq -2-\sqrt{10}\ \text{ or }\ x \geq -2+\sqrt{10}\right\}$

14) $w = \dfrac{-5+5\sqrt{31}}{2} \approx 11.4$ feet
$l = \dfrac{-5+5\sqrt{31}}{2} + 5 \approx 16.4$ feet

16b) Domain $= [-5, 10] = \{x : -5 \leq x \leq 10\}$

Chapter 7

Section 7.1

3b)

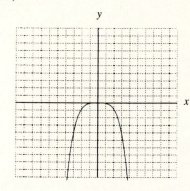

3d)

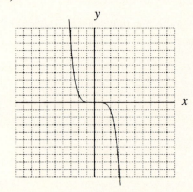

6b) The graph of $y = x^3 - 4x^2 + 2x + 2$.

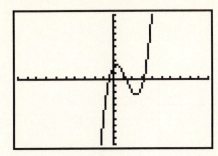

The graph of $y = x^3$.

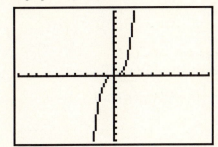

Moving from left to right, both graphs rise from negative infinity and rise to positive infinity.

7b) The graph of $y = x^4 - x - 5$.

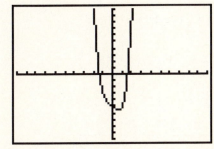

The graph of $y = x^4$.

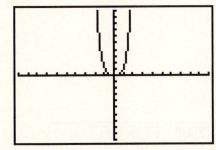

Moving from left to right, both graphs fall from positive infinity then rise to positive infinity.

9b) The graph of
$y = -x^4 + 6x^3 - 7x^2 + 6x + 11$.

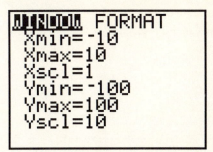

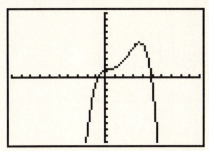

9d) The graph of $y = -2x^3 + 2x^2 + 16x - 24$.

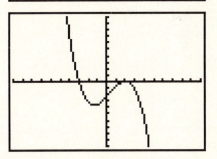

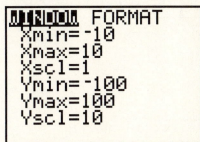

10b) odd   10d) odd
10e) even

Section 7.2

1b) $p(x) = -x^3 + 4x^2 + 7x - 10$, degree = 3

1d) $p(x) = -2x^4 + 8x^3 + 28x^2 - 72x - 90$,
degree = 4

2b) Zeros are $-4, -1, 6$

2d) Zeros are $-3, -\dfrac{1}{2}, \dfrac{4}{3}, 5$

3b) Zeros are -2, -1, 3, and 6.

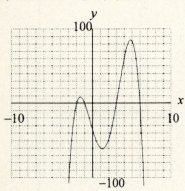

3d) Zeros are 0, -4, -2, 3, and 6.

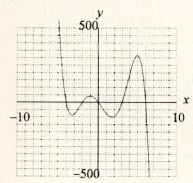

5b) $p(x) = (x+4)(2x-1)(2x+1)$

Zeros are -4, $\dfrac{1}{2}$, and $-\dfrac{1}{2}$.

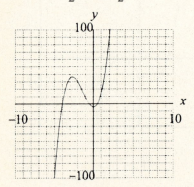

5d) $p(x) = -(x-2)(x+2)(x+5)$

Zeros are 2, -2, and -5.

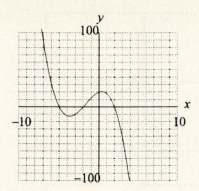

8b)
$p(x) = (x+4)(x-2)(x-5)$
$p(x) = 2(x+4)(x-2)(x-5)$

$p(x) = -\dfrac{1}{3}(x+4)(x-2)(x-5)$

9b)
$p(x) = (x+7)(x+2)(x-3)(x-8)$
$p(x) = -5(x+7)(x+2)(x-3)(x-8)$

$p(x) = -\dfrac{2}{5}(x+7)(x+2)(x-3)(x-8)$

12) $p(x) = \dfrac{1}{7}(x+5)(x+1)(x-3)$

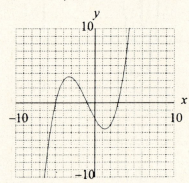

15a) $p(x) = x(x-8)(2x-5)$

Zeros are 0, 8, and $\dfrac{5}{2}$.

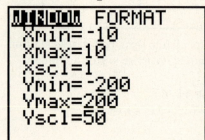

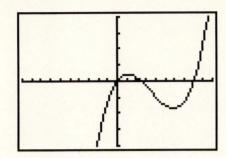

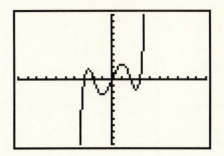

15b) $p(x) = -(x-6)(2x-3)(2x+3)$

Zeros are 6, $\dfrac{3}{2}$, and $-\dfrac{3}{2}$.

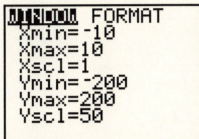

15c) $p(x) = x(x-3)(x+3)(x-2)(x+2)$

Zeros are 0, 3, -3, 2, and -2.

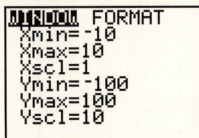

Section 7.3

3b)
quotient = $x+3$
remainder = -15

3d)
quotient = $3x^2 - 17x + 44$
remainder = -122

3f)
quotient = $x^3 + 2x^2$
remainder = -9

5b)
quotient = $-x+2$
remainder = $-10x - 5$

5d)
quotient = $x^2 + x + 5$
remainder = $16x - 6$

5f)
quotient = x
remainder = $-9x - 9$

7a) $\dfrac{x^3 - 1}{x - 1} = x^2 + x + 1$

7b) $\dfrac{x^4 - 1}{x - 1} = x^3 + x^2 + x + 1$

7c) $\dfrac{x^5 - 1}{x - 1} = x^4 + x^3 + x^2 + x + 1$

7d) $\dfrac{x^n - 1}{x - 1} = x^{n-1} + x^{n-2} + \cdots + x + 1$

9b)
i,ii) $\begin{aligned} p(x) &= (x-2)(x^2 - 3x - 40) \\ &= (x-2)(x-8)(x+5) \end{aligned}$

iii) Zeros are 2, 8, and -5.

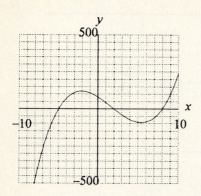

9d)

i,ii) $p(x) = (x-1)(x^2 - 6x - 16)$
$= (x-1)(x-8)(x+2)$

iii) Zeros are 1, 8, and −2.

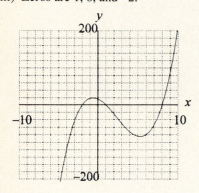

9f)

i,ii) $p(x) = (x+2)(x^2 - 12x + 27)$
$= (x+2)(x-9)(x-3)$

iii) Zeros are −2, 9, and 3.

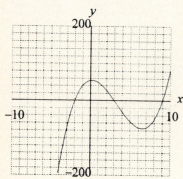

11b) $p(x) = (x-5)(2x+1)(3x-4)$

Zeros are 5, $-\dfrac{1}{2}$, $\dfrac{4}{3}$

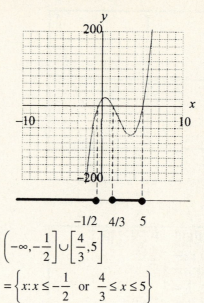

$$\left(-\infty, -\frac{1}{2}\right] \cup \left[\frac{4}{3}, 5\right]$$

$$= \left\{ x : x \le -\frac{1}{2} \ \text{ or } \ \frac{4}{3} \le x \le 5 \right\}$$

Section 7.4

1b)
i) −4
ii,iii)

$p(x) = (x+4)(-6x^2 + 23x - 10)$
$= -(x+4)(2x-1)(3x-10)$

iv) Zeros are −4, $\dfrac{1}{2}$, and $\dfrac{10}{3}$

2b)
i)

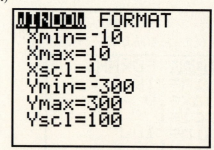

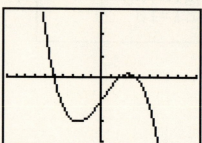

ii,ii,iv)

$$p(x) = (x+5)(-3x^2 + 17x - 22)$$
$$= -(x+5)(x-2)(3x-11)$$

Zeros are $-5, 2, \dfrac{11}{3}$

2d)

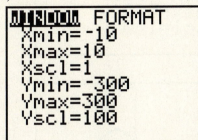

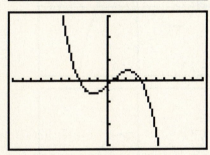

ii,iii,iv)

$$p(x) = (x+3)(-4x^2 + 16x - 7)$$
$$= -(x+3)(2x-7)(2x-1)$$

Zeros are $-3, \dfrac{7}{2}$, and $\dfrac{1}{2}$.

3b)

i) 2

ii) $p(x) = (x-2)(-x^2 - 4x + 6)$

iii) Zeros are $2, -2 \pm \sqrt{10}$

5b)

i) Zeros are $3, -1, -\dfrac{26}{5}$

ii)

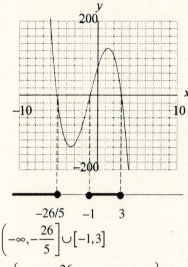

$$\left(-\infty, -\dfrac{26}{5}\right] \cup [-1, 3]$$
$$= \left\{ x : x \le -\dfrac{26}{5} \ \text{ or } \ -1 \le x \le 3 \right\}$$

6b)

i) Zeros are $-1, 2$, and -5.

ii)

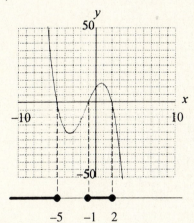

iii)

$$(-\infty, -5] \cup [-1, 2]$$
$$= \left\{ x : x \le -5 \ \text{ or } \ -1 \le x \le 2 \right\}$$

6d)

i) Zeros are $-1, \dfrac{15}{4}$, and $-\dfrac{17}{4}$.

ii)

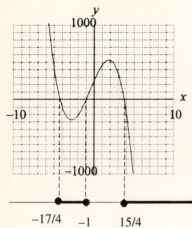

$$\left[-\frac{17}{4},-1\right]\cup\left[\frac{15}{4},+\infty\right)$$

$$=\left\{x:-\frac{17}{4}\le x\le -1 \text{ or } x\ge\frac{15}{4}\right\}$$

7b)

i) Zeros are $-6,\ 3\pm\sqrt{3}$

ii)

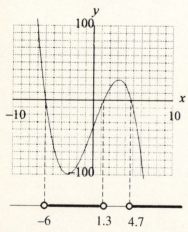

iii)

$$\left(-6, 3-\sqrt{3}\right)\cup\left(3+\sqrt{3},+\infty\right)$$

$$=\left\{x:-6<x<3-\sqrt{3} \text{ or } x>3+\sqrt{3}\right\}$$

9b)

i)

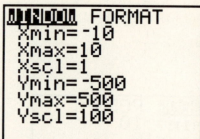

ii)

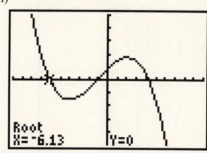

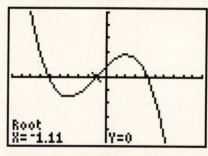

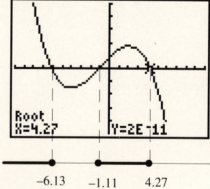

iii)

$$(-\infty,-6.13]\cup[-1.11,4.27]$$

$$=\left\{x:x\le-6.13 \text{ or } -1.11\le x\le 4.27\right\}$$

12a) $V = x(16-2x)(12-2x)$

Zeros are 0, 8, and 6.

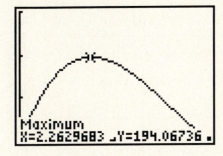

12b) x must be between 0 and 6 inches.

12cd)

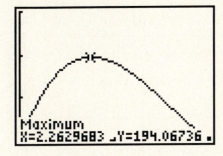

12e) If you cut squares of about 2.26 inches in length, then the box will have a maximum volume of about 194.1 cubic inches.

Chapter 7 Review Exercises

4) $p(x) = -\dfrac{1}{72}(x+5)(x+3)(x-2)(x-6)$

5b) Zeros are -4, $2 \pm 2\sqrt{2}$.

6b)
$$(-\infty, -4] \cup \left[2 - 2\sqrt{2}, 2 + 2\sqrt{2}\right]$$
$$= \left\{x : x \le -4 \ \text{ or } \ 2 - 2\sqrt{2} \le x \le 2 + 2\sqrt{2}\right\}$$

7a)
$$(-8, -5) \cup (1, +\infty)$$
$$= \left\{x : -8 < x < -5 \ \text{ or } \ x > 1\right\}$$

7d)
$$[-9, -7] \cup [1, +\infty)$$
$$= \left\{x : -9 \le x \le -7 \ \text{ or } \ x \ge 1\right\}$$

9) If you were to look at the graph of the polynomial over all real numbers, you would see the following.

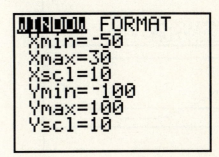

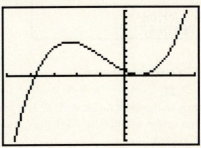

However, t represents the months of the year, so t must be a number between 1 and 12.

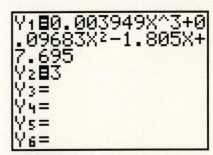

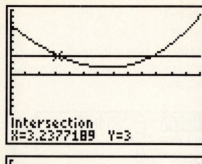

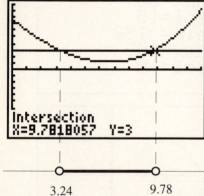

3.24   9.78

The t-values in this interval are 4, 5, 6, 7, 8, and 9, which represent the months April, May, June, July, and August.

Chapter 8

Section 8.1

3b)

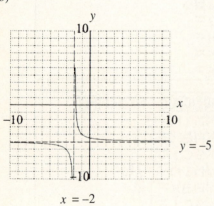

$x = -2$

3d)

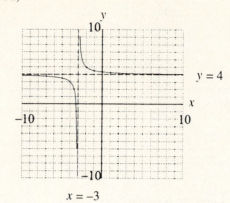

$x = -3$

5b)

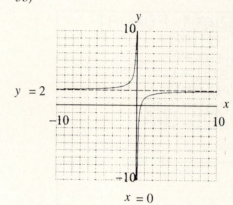

$x = 0$

5d)

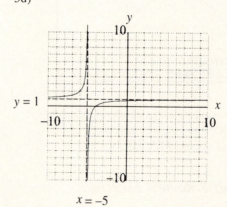

$x = -5$

6a) $y = \dfrac{1}{x+3} - 4$

6b) $y = -\dfrac{1}{x+2} + 1$

7b)

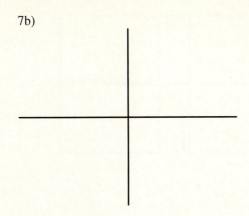

Author's note: It is difficult to find a viewing window on your calculator which will show this image.

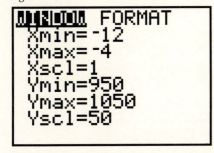

7d)

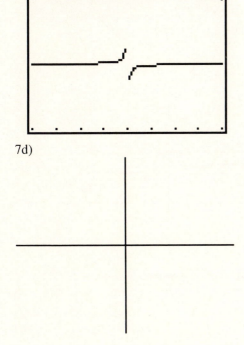

Author's note: It is difficult to find a viewing window that will show this image.

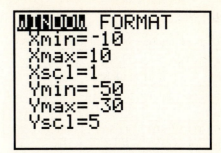

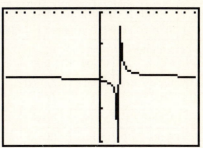

10a)

10b) 50 degrees Fahrenheit

10c)

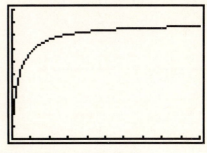

$T = 450$

Section 8.2

1b) Numbers in tables rounded to nearest tenth.

x	–1	0	1	2	3	4	5
y	–0.7	–1.5	–4.0	Undefined	6.0	3.5	2.7

1c) Numbers rounded to nearest ten–thousandth.

x	10	100	1000	10000	100000
y	1.6250	1.0510	1.0050	1.0005	1.0001

x	–10	–100	–1000	–10000	–100000
y	0.5833	0.9510	0.9950	0.9995	1.0000

1d) Numbers rounded to nearest tenth.

x	1.7	1.8	1.9	1.99	1.999
y	–15.7	–24.0	–49.0	–499.0	–4999.0

x	2.3	2.2	2.1·	2.01	2.001
y	17.7	26.0	51.0	501.0	5001.0

1e)

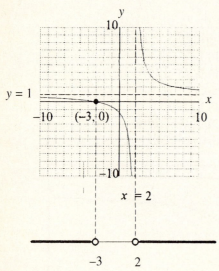

$(-\infty, -3) \cup (2, +\infty)$
$= \{x : x < -3 \text{ or } x > 2\}$

3b) $y = \dfrac{x-1}{x+3}$

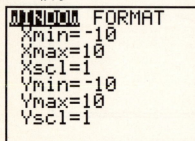

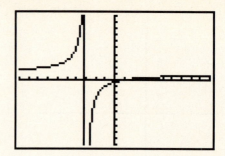

4b)

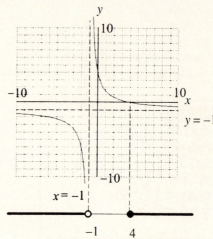

$(-\infty, -1) \cup [4, +\infty)$
$= \{x : x < -1 \text{ or } x \geq 4\}$

4d)

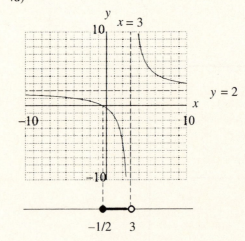

$\left[-\dfrac{1}{2}, 3\right) = \left\{x : -\dfrac{1}{2} \leq x < 3\right\}$

4f)

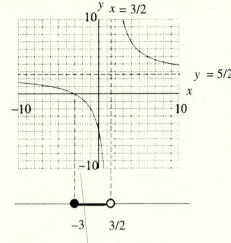

$$\left[-3,\frac{3}{2}\right) = \left\{x: -3 \leq x < \frac{3}{2}\right\}$$

5b) $$\begin{array}{l}(-\infty, -4) \cup (2, +\infty) \\ = \left\{x: x < -4 \ \text{ or } \ x > 2\right\}\end{array}$$

5d) $$\begin{array}{l}\left(-\infty, -\frac{1}{2}\right] \cup \left(\frac{3}{2}, +\infty\right) \\ = \left\{x: x \leq -\frac{1}{2} \ \text{ or } \ x > \frac{3}{2}\right\}\end{array}$$

6b)

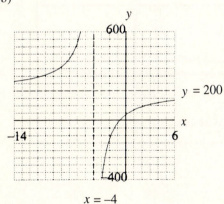

Section 8.3

1b) $\dfrac{a^3}{b^7}$

1d) $\dfrac{(x^2+4)^3}{x^3(x^2+9)^4}$

3b) $\dfrac{5}{6}$

3d) $\dfrac{2}{3}$

6) $x-1, \ x \neq 3$

7b) $-\dfrac{1}{x^2}, \ x \neq 0, -5$

7d) $-\dfrac{x+1}{x}, \ x \neq 0, 3$

7f) $-\dfrac{(x+1)(2x-1)}{x}, \ x \neq 0, 1, -\dfrac{1}{2}$

9) $-2x-9, \ x \neq 3$

11b) $-\dfrac{4x}{x-1}, \ x \neq 0, 1, -1$

11d) $\dfrac{(x-4)^2}{(x+4)(x+2)}, \ x \neq -2, -3, -4, 4$

11f) $1, \ x \neq -2, 2, -3, -4$

11h) $-\dfrac{x-1}{x^2}, \ x \neq 0, 1, -1$

13b) $\dfrac{1}{x-1}, \ x \neq -1, 1$

13d) $-\dfrac{1}{x}, \ x \neq -3, 0, 3$

13f) $\dfrac{1}{x}, \ x \neq -5, -3, 0, 5$

15) $2x+2a+1, \ x \neq a$

Section 8.4

1b) LCD $= 2^2 \cdot 3^3 = 108$
1d) LCD $= 2 \cdot 3^2 \cdot 11 = 198$
1f) LCD $= 2^2 \cdot 3^4 = 324$

2b) LCD $= x(x-6)(x+1)$
2d) LCD $= (x-8)(x+2)(2x+1)$
2f) LCD $= x(x-2)(x+2)(x-1)(x+1)$

3b) $\dfrac{2 \cdot 7}{2 \cdot 2 \cdot 3 \cdot 3 \cdot 5}$

3d) $\dfrac{3 \cdot 7 \cdot 11}{2 \cdot 3 \cdot 3 \cdot 3 \cdot 11 \cdot 11}$

3f) $\dfrac{(x+2)(x+1)}{x(x+2)(x-5)}$

3h) $\dfrac{(x^2-4)(2x-3)}{x(x+5)(2x-3)}$

5b) $\dfrac{5x+1}{x(x+1)}, \ x \neq 0, -1$

5d) $\dfrac{t^2-t+1}{t^2}, \ t \neq 0$

5f) $\dfrac{-t^3+t^2+2t+1}{t^2(t+1)^2}$, $t \neq 0,-1$

5h) $\dfrac{6x-3}{(x-2)(x-3)(x+1)}$, $x \neq 2,3,-1$

5j) $\dfrac{6x+9}{x(x+9)(x+3)}$, $x \neq 0,-9,-3$

7b) $\dfrac{-h}{x(x+h)}$, $x \neq 0,-h$

8b) $-\dfrac{10x+5}{(x-4)(x-2)(x+3)}$, $x \neq 4,2,-3$

9b) 3, $x \neq 3$

9d) $\dfrac{2}{x(x+2)(x-2)}$, $x \neq 0,-2,2$

11b) $\dfrac{3}{x+1}$, $x \neq 2,-1,\dfrac{1}{2}$

11d) $-\dfrac{1}{x(x+h)}$, $x \neq 0,-h$, $h \neq 0$

11f) $\dfrac{x^3-x^2-x-3}{(x+1)(x-3)(x-1)}$, $x \neq -1,3,1$

12b) $-\dfrac{5}{2(h+2)}$, $h \neq 0,-2$

13a) $-\dfrac{2(x+a)}{a^2 x^2}$, $x \neq 0,a$, $a \neq 0$

Section 8.5

1b) $\dfrac{1}{25}$ 1d) $\dfrac{2}{5}$

1f) $\dfrac{44}{3}$

3b) $\dfrac{1}{xz}$, $x \neq 0,z$, $z \neq 0$

3d) $4x$, $x \neq 0,4$

5b) $\dfrac{1}{3x+1}$, $x \neq 0,-\dfrac{1}{3},\dfrac{1}{3}$

5d) $\dfrac{1}{x}$, $x \neq 0,-1,1$

7b) $-\dfrac{1}{ax}$, $x \neq 0,a$

9b) $-\dfrac{4x+4a}{a^2 x^2}$, $x \neq 0,a$

10) $\dfrac{1}{3(x+1)}$, $x \neq -1,2$

12b) $-x-1$, $x \neq 0,1,-\dfrac{1}{5}$

12d) $-a$, $a \neq 1$

13b) $\dfrac{x(x+1)}{(x-3)^2}$, $x \neq 3,-1$

13d) $-\dfrac{7x-9}{(x+1)(x-3)}$, $x \neq -1,3$

13f) $\dfrac{4}{3(x+1)}$, $x \neq 2,-1$

13h) $\dfrac{3}{4}$, $x \neq 3,-1,\dfrac{1}{8}$

13j) $\dfrac{1}{x(x-3)}$, $x \neq 0,3,-1,-\dfrac{1}{2}$

Section 8.6

1b) $\dfrac{13}{19}$ 1d) $\dfrac{22}{31}$

3) $x = -2,4$

5) $x = -2,2,\dfrac{3}{4}$

6)

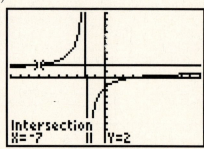

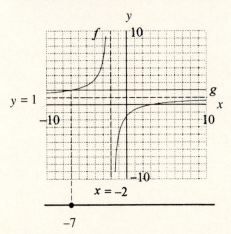

7b) $x = 9$

7d) $x = -\dfrac{1}{6}$

7f) $x = -\dfrac{1125}{13}$

7h) $x = -\dfrac{14}{19}$

7j) $x = -\dfrac{11}{26}$

9b) $x = \dfrac{2}{3}, 6$

9d) $x = \pm\dfrac{1}{3}, \pm\dfrac{1}{2}$

11b) $x = 2$

11d) $x = -\dfrac{1}{2}$

11f) $x = -1, 7$

13b) $x = \dfrac{1 \pm \sqrt{17}}{4}$
      $x \approx -0.7808, 1.2808$

13d) $x = \dfrac{-1 \pm \sqrt{21}}{2}$
      $x \approx -2.7913, 1.7913$

13f) $x = -2 \pm 2\sqrt{2}$
      $x \approx -4.8284, 0.8284$

15a)

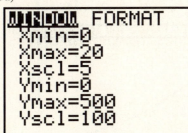

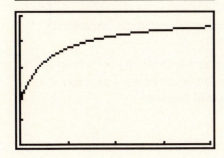

15b)

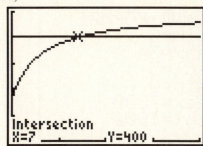

15c) $t = 7$ years, which would be about 1967.

17) $t = 12$ days

19) $\dfrac{3}{4}$

Chapter 9

Section 9.1

1a) $P = 2{,}518(1.032)^t$

1b) $P = 5{,}363$

1c)

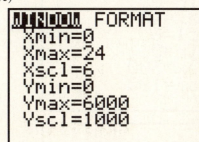

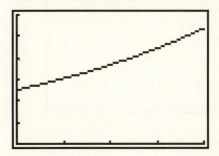

5a) $A = 1250(1.05)^t$

5b) $\$6566.68$

5c)

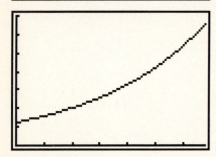

9c)

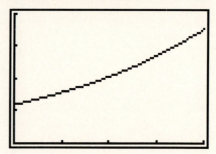

6a) $A = 500(0.96)^t$

6b)   70.5 milligrams

6c)

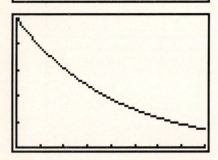

9a)   1200 gophers

9b)   3531 gophers

10a)   250 ferrets

10b)   47 ferrets

10c)

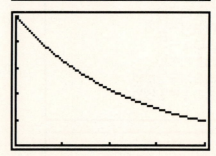

13b)   Because the base of $y = 4329(0.88)^x$ is 0.88—which is less than 1—the graph of the equation will decrease.

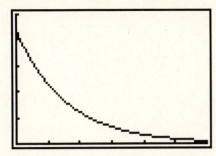

13d) Because the base of $y = 875(1.05)^x$ is 1.05—which is larger than 1—the graph of the equation will increase.

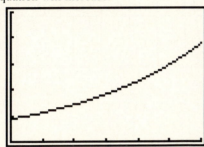

Section 9.2

1a) $A = 450\left[1 + \dfrac{0.0425}{4}\right]^k$

1b)

```
450*(1+0.0425/4)
^40
        686.7742508
```

$A \approx \$686.77$

8)

```
2500(1+0.052/4)^
(4*5)
        3236.897267
■
```

10)

```
1800/((1+0.038/2
)^(2*2))
        1669.459044
■
```

12ab)

```
1000*(1+0.05/4)^
(4*5)
        1282.037232
1000*(1+0.052)^5

        1288.483018
```

12c) 5.2% compounded annually is the better investment.

14bd)

```
57e^(0.0325*12)
        84.18790525
12000/e^(-0.085*
15)
        42944.41692
■
```

14fh)

```
2550e^(-0.085*15
)
        712.548969
e^1/(1-3e^1)
        -.3799218074
```

16ab)

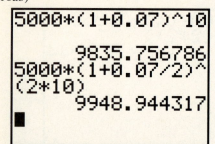

16cd)

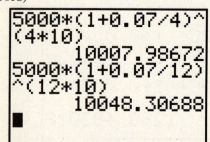

16ef)

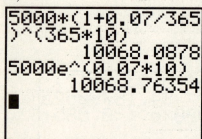

18)

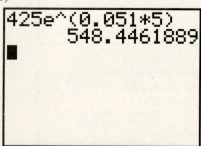

20)

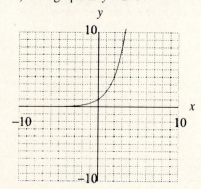

Section 9.3

1b) $\dfrac{1}{9}$ 1d) $-\dfrac{1}{64}$

1f) $\dfrac{49}{9}$

2b) x^{-5} 2d) x^{-2}

2f) x^{-55} 2h) x^{-6}

2j) x^{-6}

5)

x	-3	-2	-1	0	1	2	3
$f(x)$	$\dfrac{1}{27}$	$\dfrac{1}{9}$	$\dfrac{1}{3}$	1	3	9	27

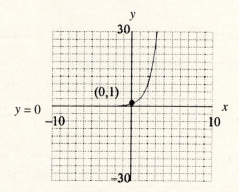

11a) The graph of $y = 2^{x}$.

The graph of $y = 2^{-x}$.

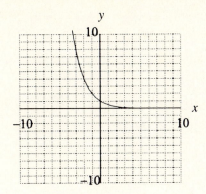

The graph of $y = 2^{-x} + 3$.

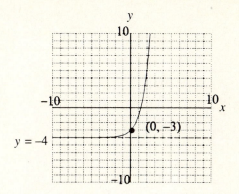

13f) The graph of $y = 4^x$.

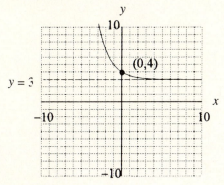

13b) The graph of $y = 3^x$.

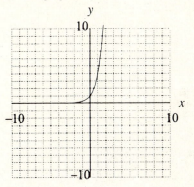

The graph of $y = 4^{-x}$ is a reflection of the graph
of $y = 4^x$ across the y-axis.

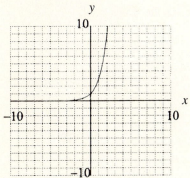

To draw the graph of $y = 3^x - 4$, shift the graph
of $y = 3^x$ downward 4 units.

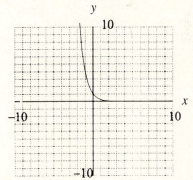

To draw the graph of $y = 4^{-x} - 4$, shift the
graph of $y = 4^{-x}$ downward 4 units.

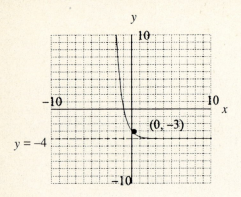

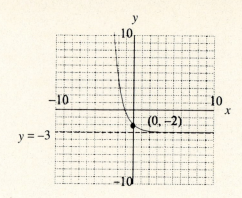

15f) The graph of $y = e^x$.

15h) The graph of $y = e^x$.

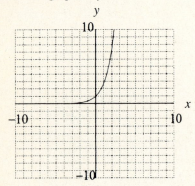

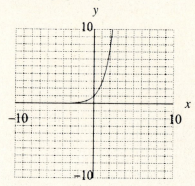

The graph of $y = e^{-x}$ is a reflection of the graph of $y = e^x$ across the y-axis.

The graph of $y = -e^x$ is a reflection of the graph of $y = e^x$ across the x-axis.

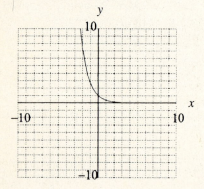

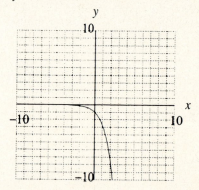

To draw the graph of $y = e^{-x} - 3$, shift the graph of $y = e^{-x}$ downward 3 units.

To draw the graph of $y = -e^x - 6$, shift the graph of $y = -e^x$ downward 6 units.

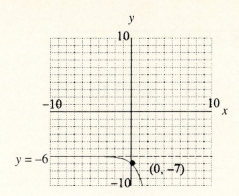

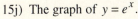

15j) The graph of $y = e^x$.

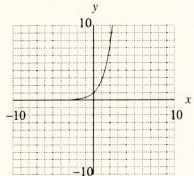

The graph of $y = e^{-x}$ is a reflection of the graph of $y = e^x$ across the y-axis.

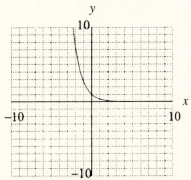

The graph of $y = -e^{-x}$ is a reflection of the graph of $y = e^{-x}$ across the x-axis.

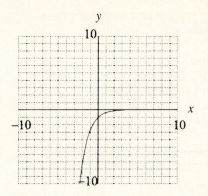

To draw the graph of $y = 9 - e^{-x}$, shift the graph of $y = -e^{-x}$ upward 9 units.

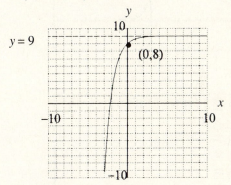

19c) Because the growth rate 0.07 is positive, the graph of $h(x) = 100e^{0.07x}$ increases.

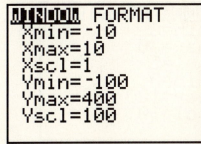

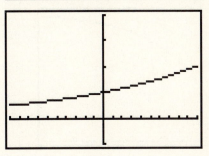

19d) Because the growth rate -1.44 is negative, the graph of $k(x) = 100^{-1.44x}$ will decrease.

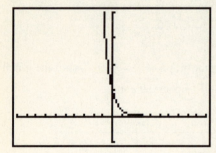

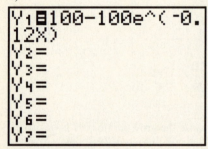

21a)

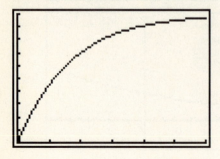

21b)

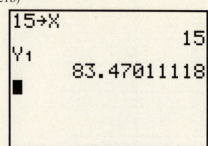

21c)

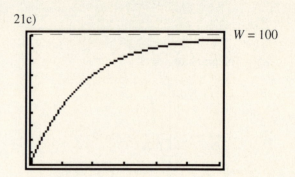

$W = 100$

Section 9.4

2b) $a + 5$ 2d) $16 - 6x$

4b) $3a$ 4d) $24x - 21$

6b) 28 6d) 36

8a)

12

↓

Invert machine
$g(x) = 1/x$

1/12

Add 5 machine
$f(x) = x + 5$

61/12

8b)

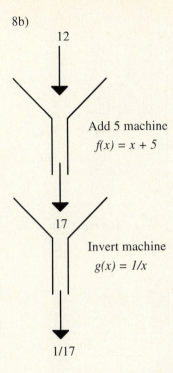

1/17

9c) −512

9d) −128

11b) $3a+8$

11d) $41-21x$

13b) $2a^2+5$

13d) $2x^4-4x^2+7$

15c) $23-4x$

15d) $-4x-8$

17b) $\dfrac{1}{3}$

17d) $\dfrac{1}{x+1}$

17f) $\dfrac{1}{x+1}$

17h) Undefined

18b) 4

18d) $\dfrac{2x-6}{3x-4}$

18f) $\dfrac{2}{x-3}$

18h) 0

20) $w=e^{-\frac{1}{2}x^2}$

21b) $9x^2+3$

21d) $3x^2+9$

23) $x=\sqrt{e^{2t-1}}$

25a) $d=\sqrt{x^2+36}$

25b) $d=\sqrt{44.89t^2-2680t+40036}$

25c) $d=133.1$ miles

27b) −4   27c) 0

x	−3	−2	−1	0	1	2	3
y	4	3	2	1	0	−1	−2

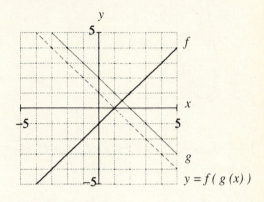

$y=f(g(x))$

28a) $f(g(x))=1-x$

28b) $f(x)=x-1$

28c) $g(x)=-x+2$

28d) $f(g(x))=1-x$

Section 9.5

3a)

x	0	1	4	9	16	25
$y=\sqrt{x}$	0	1	2	3	4	5

3b)

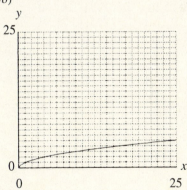

3c)

$x=\sqrt{y}$	0	1	2	3	4	5
y	0	1	4	9	16	25

3d)

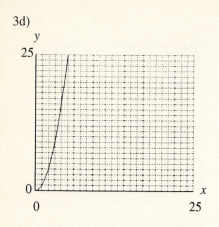

3e) The graph of $x = \sqrt{y}$ is a reflection of the graph of $y = \sqrt{x}$ across the line $y = x$.

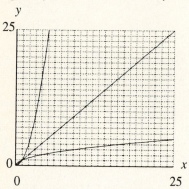

7b) The graph of $y = x^3$.

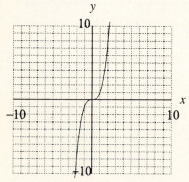

To draw the graph of $x = y^3$, reflect the graph of $y = x^3$ across the line $y = x$.

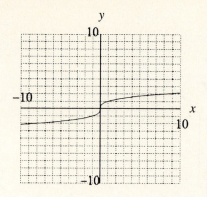

7d) The graph of $y = -\sqrt{x}$ is a reflection of the graph of $y = \sqrt{x}$ across the x-axis.

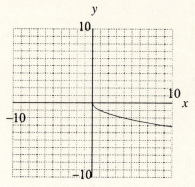

The graph of $x = -\sqrt{y}$ is a reflection of the graph of $y = -\sqrt{x}$ across the line $y = x$.

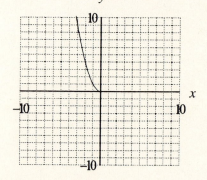

7f) The graph of $y = 10^x$.

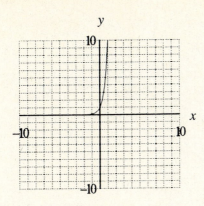

The graph of $x = 10^y$ is a reflection of the graph of $y = 10^x$ across the line $y = x$.

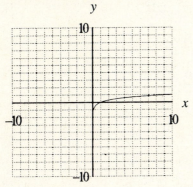

9a) $y = \dfrac{1}{2}x + 4$

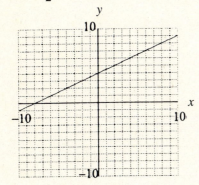

9b) $y = 2x - 8$

9c) They appear to be symmetric with respect to the line $y = x$.

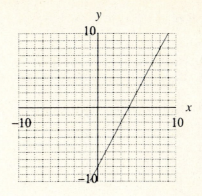

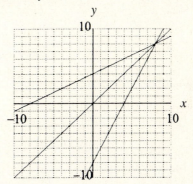

10a) The graph of f fails the horizontal line test. It is *not* a one-to-one function.

11a) No horizontal line cuts the graph of $y = f(x)$ more than once. The function f *is* a one-to-one function.

11b)

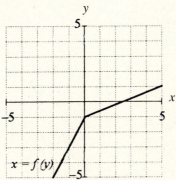

11c) No vertical line cuts the graph of $x = f(y)$ more than once. The graph of $x = f(y)$ represents a function.

Section 9.6

2)

      1. Put the key in the lock.
      2. Turn the key to the left to unlock the door.
      3. Open the door of the box.

3b)

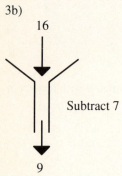

16

Subtract 7

9

3d)

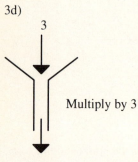

3

Multiply by 3

9

3f)

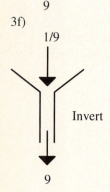

1/9

Invert

9

9b) $f(x) = 3x - 4$, $f^{-1}(x) = \dfrac{x+4}{3} = \dfrac{1}{3}x + \dfrac{4}{3}$

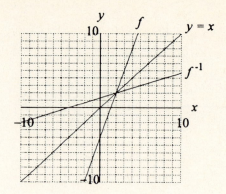

10b) $y = -\dfrac{1}{2}x + \dfrac{3}{2}$

10d) $y = -\dfrac{3}{2}x + \dfrac{15}{2}$

10f) $y = \sqrt[3]{\dfrac{1-x}{3}}$

10h) $y = \dfrac{3 - x^3}{4}$

10j) $y = \dfrac{4x - 5}{3x + 2}$

11b) $f^{-1}(x) = 20 - 4x$

11d) $f^{-1}(x) = \sqrt[5]{\dfrac{12 - 4x}{3}}$

11f) $f^{-1}(x) = \dfrac{x^5 + 7}{2}$

11h) $f^{-1}(x) = \dfrac{3 - x}{5x + 2}$

13a)

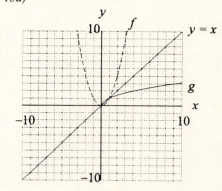

Because g is *not* a reflection of the graph of f across the line $y = x$, f and g are *not* inverses of each other.

13b) Note that $g(f(-5)) = g(25) = 5$.

Therefore $g(f(x)) \neq x$ for all x in the domain of f.

15a)

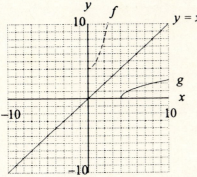

Note that g is a reflection of f across the line $y = x$. Therefore f and g are inverses of each other.

15b)

$$f(g(x)) = f(\sqrt{x-4})$$
$$= \left(\sqrt{x-4}\right)^2 + 4$$
$$= x - 4 + 4$$
$$= x$$

This argument is valid for all x in the domain of g, which is $\{x : x \geq 4\}$.

$$g(f(x)) = g(x^2 + 4)$$
$$= \sqrt{(x^2 + 4) - 4}$$
$$= \sqrt{x^2}$$
$$= |x|$$
$$= x$$

This argument is valid for all x in the domain of f, which is $\{x : x \geq 0\}$.

Section 9.7

2)

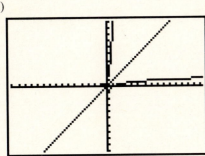

2a) Domain of $y = 10^x$ is $(-\infty, +\infty)$.

Range of $y = 10^x$ is $(0, +\infty)$

2b) Domain of $y = \log x$ is $(0, +\infty)$.

Range of $y = \log x$ is $(-\infty, +\infty)$

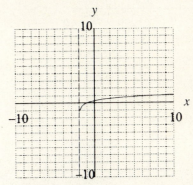

$$x = -2$$

Domain $= (-2, +\infty)$

Range $= (-\infty, +\infty)$

5b) $x - 3, \{x : x \in R\}$

5d) $4 - x, \{x : x < 4\}$

9b)

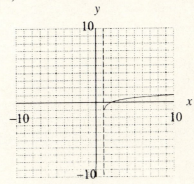

$$x = 1$$

Domain $= (1, +\infty)$

Range $= (-\infty, +\infty)$

9d)

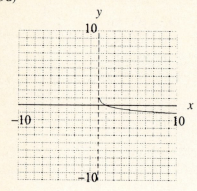

$x = 0$

Domain $= (0, +\infty)$

Range $= (-\infty, +\infty)$

9f)

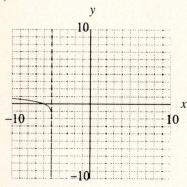

$x = -5$

Domain $= (-\infty, -5)$

Range $= (-\infty, +\infty)$

10b) $\dfrac{1}{x}$, $\{x: x \neq 0\}$

10c) $\sqrt{x-6}$, $\{x: x \geq 6\}$

11a) $\log_2 16 = 4$

11c) $\log_5 \dfrac{1}{5} = -1$

11f) $\log_8 \dfrac{1}{2} = -\dfrac{1}{3}$

11h) $\log_{27} \dfrac{1}{9} = -\dfrac{2}{3}$

12a) $7^2 = 49$, $\log_7 49 = 2$

12c) $4^{-1} = \dfrac{1}{4}$, $\log_4 \dfrac{1}{4} = -1$

12f) $64^{\frac{2}{3}} = 16$, $\log_{64} 16 = \dfrac{2}{3}$

12h) $16^{-\frac{3}{4}} = \dfrac{1}{8}$, $\log_{64} \dfrac{1}{8} = -\dfrac{3}{4}$

13b) $10^{-3} = 0.001$, $\log 0.001 = -3$

13e) $10^0 = 1$, $\log 1 = 0$

13f) $10^3 = 1000$, $\log 1000 = 3$

14a) $10^2 = x - 8$

14d) $10^{3-2t} = 9$

15a) $\log 12 = x$

15d) $\log(5x + 1) = -1$

16a) $e^4 = x$

16f) $e^{x-5} = 4$

17b) $\ln x = 3$

17d) $\ln(x - 5) = -2$

18a) 4   18c) 0

18e) -1 18g) $\dfrac{1}{2}$

18i) -7 18k) $\dfrac{4}{3}$

Section 9.8

2a) $\ln 4 + \ln x$ 2d) $\ln(x-3) + \ln(x-5)$

3a) $\ln 10x$ 3d) $\log(x+5)(x-5)$

5a) $\ln 5 - \ln x$ 5c) $\log(x+1) - \log(x-2)$

6a) $\ln \dfrac{x}{10}$ 6d) $\log \dfrac{x-5}{3-2x}$

8a) $5\ln x$ 5d) $-3\log x$

9a) $\ln x^2$ 9d) $\log x^{-4}$

11b) $2\log x + \log(x+3)$

11d) $2\log_2(x+1) + 3\log_2(x-3)$

11f) $2\ln x - 3\ln y - 5\ln z$

11h) $\log_3(2x-3) - 3\log_3 x - 5\log_3(x+1)$

12b) $\log \dfrac{x(x-5)}{x+1}$

12d) $\log_3 \dfrac{x^3}{(x+1)^2(2x-5)^4}$

12f) $\log_2 \dfrac{x^3(x-7)}{(x+1)(x-5)^2}$

13b) 5

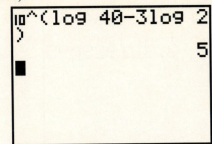

13d) 3

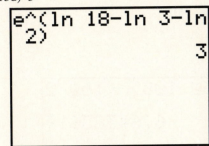

13f) $\dfrac{1}{8}$

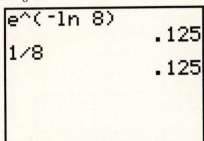

14b) $\dfrac{1}{2}\log_2(x+5)-\dfrac{1}{2}\log_2(x-5)$

14d) $\dfrac{1}{2}\log x-\dfrac{3}{4}\log y$

15b) $\ln\sqrt[3]{\dfrac{x-1}{x+1}}$

15c) $\log_3\sqrt[4]{x^2y^3}$

16b) $x(x+3),\ \{x:x>0\}$

16d) $\dfrac{5}{x^2},\ \{x:x>0\}$

17a) The graph of $f(x)=e^{\ln x}$.

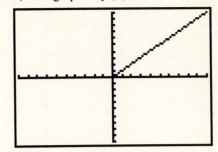

Domain $=(0,+\infty)$

17b) The graph of $g(x)=x$.

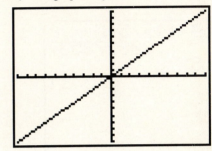

Domain $=(-\infty,+\infty)$

17c) If $x>0$, then $f(x)=g(x)$.

Section 9.9

1b) $\dfrac{e}{4}$

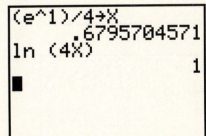

1c) e^4-5

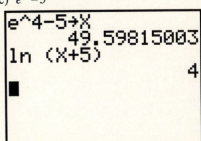

1f) $\dfrac{4 - e^{-2}}{5}$

```
(4-e^-2)/5→X
          .7729329434
ln (4-5X)
                   -2
■
```

iii)

```
(log 50)/(log 4)
       2.821928095
```

5d)

i) $6 < \log_2 100 < 7$

ii) $\log_2 100 = \dfrac{\log 100}{\log 2}$

iii)

```
(log 100)/(log 2
)
       6.64385619
■
```

3b) $\dfrac{\log 11}{5}$

```
(log 11)/5→X
          .208278537
10^(5X)
                   11
```

3d) $4 + \log 150$

```
4+log 150→X
       6.176091259
10^(X-4)
                  150
```

7d) $x = 3 - \dfrac{2\ln 2}{\ln 7}$

```
3-(2ln 2)/(ln 7)
→X
       2.287585626
7^(3-X)
                    4
```

3f) $5 - \log 15.8$

```
5-log 15.8→X
       3.801342913
10^(5-X)
                 15.8
■
```

7f) $x = \dfrac{4\ln 3}{\ln 3 - 2\ln 5}$

```
(4ln 3)/(ln 3-2l
n 5)→X
          -2.072595731
3^(X-4)
          .0012665884
5^(2X)
          .0012665884
```

5b)

i) $2 < \log_4 50 < 3$

ii) $\log_4 50 = \dfrac{\log 50}{\log 4}$

9a)

i)

```
Y₁▪log X+log (X+
21)
Y₂▪2
Y₃=
Y₄=
Y₅=
Y₆=
Y₇=
```

```
WINDOW FORMAT
 Xmin=-10
 Xmax=10
 Xscl=1
 Ymin=-10
 Ymax=10
 Yscl=1
```

```
WINDOW FORMAT
 Xmin=-10
 Xmax=10
 Xscl=1
 Ymin=-10
 Ymax=10
 Yscl=1
```

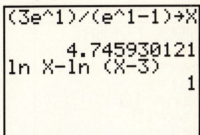

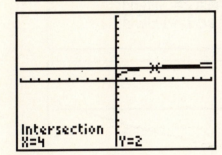

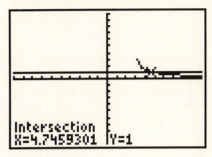

iii) $x = \dfrac{3e}{e-1}$

```
(3e^1)/(e^1-1)→X
      4.745930121
ln X-ln (X-3)
              1
```

iii) $x = 4$

```
4→X
            4
log X+log (X+21)
            2
■
```

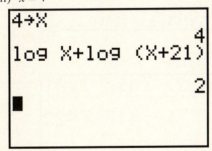

9d)

i)

```
Y₁▪ln X-ln (X-3)

Y₂▪1
Y₃=
Y₄=
Y₅=
Y₆=
Y₇=
```

11b) $x = 1 + \sqrt{3}$

```
1+√3→X
     2.732050808
10^(log X+log (X-
2)
              2
■
```

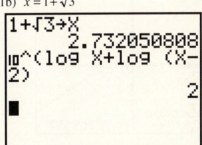

11d) $x = \dfrac{20}{3}$

```
20/3→X
      6.666666667
e^(ln X-ln (X-5)
)
              4
```

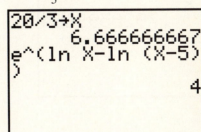

11f) $x = 2\sqrt{26}$

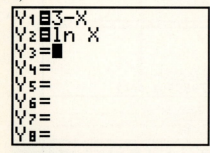

13b)

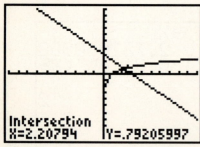

Section 9.10

1a) $A = 1000\left[1 + \dfrac{0.04}{4}\right]^{4t}$

1bc)

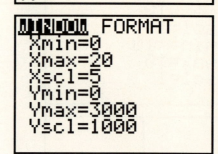

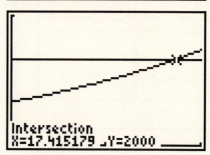

1d) $t = \dfrac{\ln 2}{4\ln 1.01}$

3)

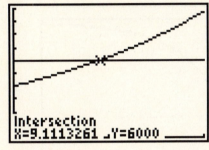

$$t = \frac{\ln\dfrac{3}{2}}{2\ln\left[1+\dfrac{0.045}{2}\right]}$$

```
(ln (3/2))/(2ln
(1+0.045/2))
        9.111326106
```

9)

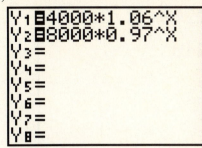

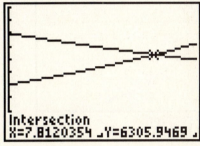

$$t = \frac{\ln 2}{\ln 1.06 - \ln 0.97}$$

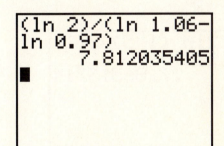

14a) $A = 200\left(\dfrac{1}{2}\right)^{n}$

14b) $n = \dfrac{t}{12}$

14c) $A = 200\left(\dfrac{1}{2}\right)^{\frac{t}{12}}$

14d)

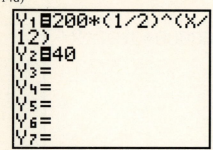

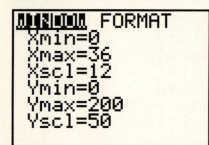

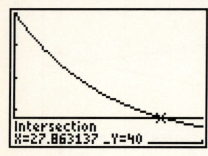

14f) $t = \dfrac{12\ln\dfrac{1}{5}}{\ln\dfrac{1}{2}}$

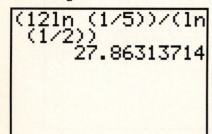

17a)

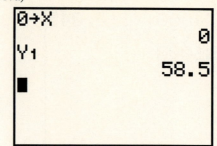

17b)

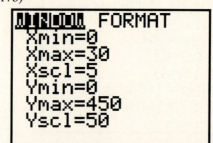

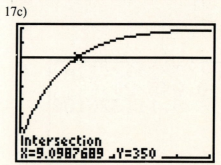

$T = 450$

17c)

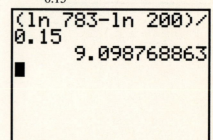

$t = \dfrac{\ln 783 - \ln 200}{0.15}$

Chapter 9 Review Exercises

3a)

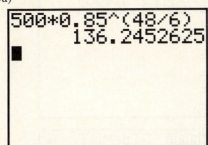

3b)

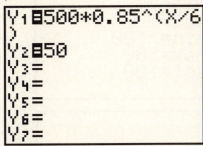

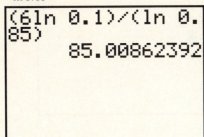

$$t = \frac{6\ln 0.1}{\ln 0.85}$$

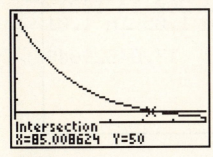

5b) $\dfrac{25}{4}$ 5d) x^{-6}

5f) $\dfrac{9}{64}x^4$

7a)

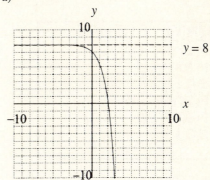

$y = 8$

Domain $= (-\infty, +\infty)$
Range $= (-\infty, 8)$

7c)

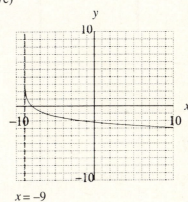

$x = -9$

Domain $= (-9, +\infty)$
Range $= (-\infty, +\infty)$

9b) $f(x) = -\ln(x-2), f^{-1}(x) = e^{-x} + 2$

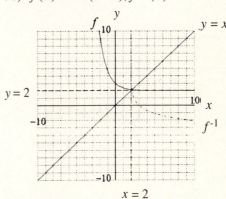

$y = x$
f
$y = 2$
f^{-1}
$x = 2$

11b) $\ln \sqrt[4]{\dfrac{x^2}{(x+3)^3}}$

12b) 187   12d) 2

13)

13a) The graph of $f(x) = e^{\ln(x-3)}$.

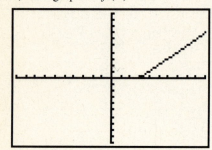

Domain = $(3, +\infty)$

13b) The graph of $g(x) = x - 3$.

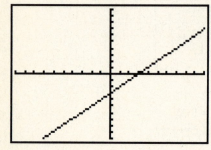

Domain = $(-\infty, +\infty)$

13c) If $x > 3$, then $f(x) = g(x)$.

15b) $x = \dfrac{5 - \log 1.23}{8}$

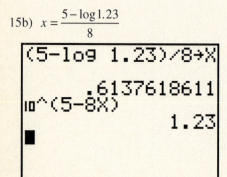

17b)

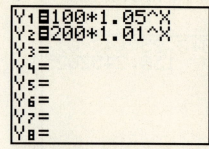

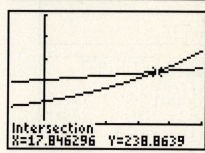

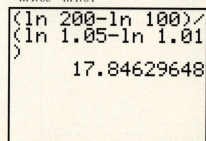

$x = \dfrac{\ln 200 - \ln 100}{\ln 1.05 - \ln 1.01}$

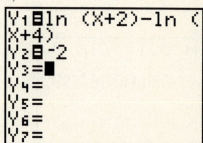

17d)

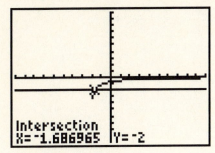

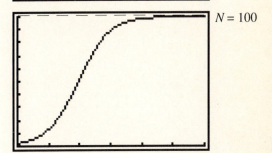

$N = 100$

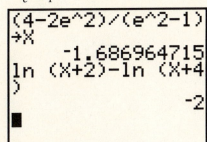

$x = \dfrac{4 - 2e^2}{e^2 - 1}$

19c)

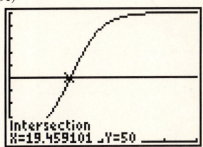

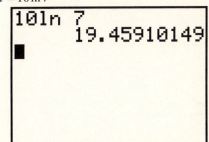

$t = 10\ln 7$

```
10ln 7
        19.45910149
```

19ab)

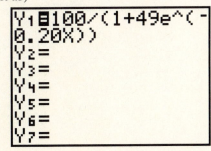